Handbook of Climate Change Resilience

Walter Leal Filho
Editor

Handbook of Climate Change Resilience

Volume 4

With 563 Figures and 494 Tables

Springer

Editor
Walter Leal Filho
International Climate Change Information
and Research Programme (ICCIRP)
Faculty of Life Sciences
Hamburg University of Applied Sciences
Hamburg, Germany

Research and Transfer Centre Sustainable Development and
Climate Change Management
Hamburg University of Applied Sciences
Hamburg, Germany

ISBN 978-3-319-93335-1 ISBN 978-3-319-93336-8 (eBook)
ISBN 978-3-319-93337-5 (print and electronic bundle)
https://doi.org/10.1007/978-3-319-93336-8

This Springer imprint is published by the registered company Springer Nature Switzerland AG.
The registered company address is: Gewerbestrasse 11, 6330 Cham, Switzerland

Preface

As the Fifth Assessment Report (AR5) produced by the Intergovernmental Panel on Climate Change (IPCC) has shown, climate change is expected to have widespread impacts on societies.

The fact that the impacts of climate change are already being felt, not only in terms of increases in temperature but also in respect of agriculture (with lower crop yields) and the availability of water resources, among others, suggests that action is urgently needed, in order to increase the resilience of both the physical environment and communities.

In particular, there is a perceived need to identify, test, and deploy appropriate processes, methods, and tools that may help societies and nations to better adapt to the many impacts of a changing climate and become more resilient. There is also a perceived need to showcase successful examples of how this can be achieved and to illustrate how the quest to pursue climate resilience may face the many social, economic, and political problems associated with it.

It is against this background that the *Handbook of Climate Change Resilience* has been prepared by numerous scientists from across all geographical regions, under the auspices of the International Climate Change Information and Research Programme (ICCRIP). This handbook is a truly interdisciplinary publication, mobilizing scholars from over 30 countries undertaking research and/or executing climate change projects around the world.

This book focuses on "fostering resilience and capacity to adapt," meaning that it serves the purpose of showcasing experiences from research, field projects, and best practice in climate change adaptation across many countries, which may be useful and suitable for replication of perhaps even upscaling elsewhere.

Consistent with the need for more cross-sectoral interactions among the various stakeholders working in the field of climate change adaptation, this book aims to

(i) Provide research institutions, universities, NGOs, and enterprises with an opportunity to display and present their works in the field of climate change adaptation and resilience
(ii) Disseminate information, ideas, and experiences acquired in the execution of climate change adaptation projects, especially successful initiatives and good practice

(iii) Introduce methodological approaches and experiences deriving from case studies and projects, which aim to show how climate change adaptation may be implemented in practice

Last but not least, a further aim of this book is to document and disseminate the wealth of experiences on climate change resilience available today.

Enjoy your reading!

August 2019
Germany

Walter Leal Filho

Acknowledgments

The editor would like to thank the valuable support of the Deputy Editor, Professor Desalegn Ayal, as well as the many members of the Editorial Board. Thanks to your hard work, expertise, and constructive criticisms, we were able to produce a publication with the highest standards.

I would also thank the authors for their willingness to share their knowledge, know-how, and experiences, as well as the many peer reviewers, who have helped us to ensure the quality of the manuscripts.

Finally, thanks are due to the International Climate Change Information and Research Programme (ICCIRP) at the Hamburg University of Applied Sciences, Germany, for the coordination of the project and mobilization of a wide range of stakeholders and authors around the world.

Contents

Volume 1

Part I Climate Change Resilience in Transportation, Energy, Forestry, and Water/Coastal Infrastructure 1

1 **Sequestrated Carbon Content in Tree Species and Diurnal Temperature Influence for Adaptive Climate Change Resilience in Nigeria** 3
Mustapha Adeojo Ibrahim, Bashir Yusuf Abubakar, and Mohammed Lawal Balarabe

2 **Climate Resilience in African Coastal Areas: Scaling Up Institutional Capabilities in the Niger Delta Region** 29
Chika Ogbonna, Eike Albrecht, Collins Ugochukwu, Chinedum Nwajiuba, and Robert Onyeneke

3 **Livelihood Resilience of the Indigenous Munda Community in the Bangladesh Sundarbans Forest** 51
Sajal Roy

4 **Building Community Capacity in Fragile Environments: Case Study of the Mara Serengeti Ecosystem** 73
Rebekah Karimi, Albanus Mutiso, and Lippa Wood

5 **Building Resilience of Urban Ecosystems and Communities to Sea-Level Rise: Jamaica Bay, New York City** 95
A. Saleem Khan, Kytt MacManus, Jane Mills, Malgosia Madajewicz, and Laxmi Ramasubramanian

6 **Risk Perception and Action to Reduce the Impact of Floods in the Czech Republic** 117
Mohan Kumar Bera and Petr Daněk

7 **Water Management as a Means for Climate Change Adaptation and Sustainable Development** 133
Ghrmawit Haile

8 **Urbanization, Climate Linked Water Vulnerability as Impediments to Gender Equality: A Case Study of Delhi, India** 149
Jagriti Kher, Savita Aggarwal, Geeta Punhani, and Sakshi Saini

9 **Vulnerability of Uganda's Electricity Sector to Climate Change: An Integrated Systems Analysis** 177
Vignesh Sridharan, Eunice Pereira Ramos, Constantinos Taliotis, Mark Howells, Paul Basudde, and Isaac V. Kinhonhi

10 **Managing the Increasing Heat Stress in Rural Areas** 207
Adithya Pradyumna, Ramkumar Bendapudi, Dipak Zade, Marcella D'Souza, and Premsagar Tasgaonkar

11 **Resettlement and Relocation Options for Coastal Communities** 229
Peni Hausia Havea, Sarah L. Hemstock, and Helene Jacot Des Combes

12 **Building Resilience to Climate Change: Water Stewardship in Rainfed Agrarian Villages in Maharashtra, India** 245
Eshwer Kale and Marcella D'Souza

13 **Resilience in Climate Stressed Environment Through Water Grabbing** 269
Makarius C. S. Lalika

14 **Impacts of Some Climatic Factors on Soil Quality of Tropical Acid-Sand Soils** 295
Nkem J. Nwosu and Paul B. Okon

15 **Bioremediation Effects of Nitrogen Fixing Trees on Nutrients and Heavy Metals in Spent Engine Oil Polluted Soil** 319
Bamidele Adanikin and Joshua Kayode

16 **Building Vulnerable Islander Resilience to Natural Hazard: A Participatory Approach** 329
Riffat Mahmood and A. Q. M. Mahbub

17 **Climate Variability and Water Availability in Riparian Rural Communities of the Ebrié Lagoon in Côte d'Ivoire** 359
Isimemen Osemwegie and Adjoua Nadège Boko-Koiadia

18 **Microbial Inoculants for Improving Carbon Sequestration in Agroecosystems to Mitigate Climate Change** 381
Abeer Ahmed Qaed Ahmed, Kehinde Abraham Odelade, and Olubukola Oluranti Babalola

19 Area Exclosure as a Strategy for Climate Change Mitigation: Case Study from Tigray Region, Northern Ethiopia 403
Samson Shimelse, Tamrat Bekele, and Sileshi Nemomissa

20 Balancing Adaptation and Mitigation in the Building Sector of New York State 429
Yasmein Okour, Nicholas B. Rajkovich, and Martha Bohm

21 Biodiversity, Ecosystem Degradation, and Climate Change Effects on Livelihoods in the Bitumen Area of Nigeria 447
Mosunmola Lydia Adeleke, O. J. Oluwatosin, O. A. Fagbenro, T. T. Amos, and I. A. Ajibefun

22 Climate Change and Coastal Resilience in Nigeria 459
Mosunmola Lydia Adeleke, Bosede Olufunmilayo Akinwalere, Kehinde Oluwatosin Olajubu, and Gbenga Emmanuel Onibi

23 Drivers of Deforestation and Land-Use Change in Southwest Nigeria 475
Mayowa Fasona, Peter Adegbenga Adeonipekun, Oludare Agboola, Akinlabi Akintuyi, Adedoyin Bello, Oluwatoyin Ogundipe, Alabi Soneye, and Ademola Omojola

24 Tradition Versus Climate Change: Cultural Importance of Indigenous Fruit Tree and Adaptation in Benin, West Africa . . . 499
Laurent Gnonlonfin, Christine N. A. Ouinsavi, Gérard Gouwakinnou, Belarmain A. Fandohan, and Towanou Olivier Houetchegnon

25 Forest Cover Change and Its Impacts on Ecosystem Services in Katimok Forest Reserve, Baringo County, Kenya 515
A. Jebiwott, G. M. Ogendi, S. M. Makindi, and M. O. Esilaba

26 Ecotourism as an Adaptation Strategy for Mitigating Climate Change Impacts on Local Communities Around Protected Areas in Ghana 537
Yaw Boakye Agyeman

27 Influence of REDD+ Pilot Project on Livelihoods of Mangrove Forest Adjacent Communities in Zanzibar, Tanzania 557
Rukia A. Kitula

28 Impacts of Climate Change on Health: Evidences from Multi-stakeholders in the Western Region of Cameroon 571
Hubert Fudjumdjum

29 Ecological Limits: Implications for Sustainable Human Capital Development 585
Oluwabunmi Opeyemi Adejumo

30 Relationship Between Climate Change, Natural Disasters, and Resilience in Rural and Urban Societies 607
Safieh Javadinejad, Saeid Eslamian, Kaveh Ostad-Ali-Askari, Mohsen Nekooei, Neda Azam, Hosein Talebmorad, Ali Hasantabar-Amiri, and Mohammad Mousavi

Volume 2

Part II Climate Resilience and Public Safety, Health, Agriculture, Food Security, and Emergency Management **633**

31 Resilience of Greek Communities Hosting Climate Migrants: The Perceptions of Environmental Educators 635
Aristea Kounani and Constantina Skanavis

32 Effect of Climate Variability on Crop Income in Central Highlands of Ethiopia 661
Arega Shumetie, Kassa Belay, Degye Goshu, and Majaliwa Mwanjalolo

33 Sustainability of Climate Change Adaptation Measures in Rivers State, South-South, Nigeria 675
A. Henri-Ukoha and O. M. Adesope

34 Climate and Climate Change Scenarios in the Indian Thar Region 685
Surendra Poonia and A. S. Rao

35 Ecosystem-Based Adaptation and Gender Perspectives from a Participatory Vulnerability Assessment in Mountainous Rural Vietnam 699
Tran Thi Kim Lien and Kathryn Brown

36 Climate Change: Depression in Egg Production in Chickens During the Hot Season with Long-Term Honey Administration 717
M. O. Abioja and M. O. Adekunle

37 Climate Variability and Change in Guinea Savannah Ecological Zone, Nigeria: Assessment of Cattle Herders' Responses 729
Ayansina Ayanlade and Stephen M. Ojebisi

38 Assessment of the Degree of Households' Vulnerability to Climate Variability in Nigeria 749
Abiodun Emmanuel Awoyemi and O. Adeola Olajide

39 Impacts of Climate Change on Agriculture, Livelihoods, and Women in Nile Delta, Egypt 765
Marwa R. Hafez

40 **Impacts of Climate Change on Food Security: An Appraisal** . . . 785
Damilola Grace Ogunrotimi and Joshua Kayode

41 **Climate Change, Vulnerability, and Adaption Under the Small Farming Households of Konso Community, Southern Ethiopia** . . . 797
Yohannes GebreMichael

42 **Role of Climate Justice in Strengthening Adaptive Capacities in Developing Countries** . . . 823
Ruchi Sachan

43 **Toward Climate-Resilient African Indigenous Vegetable Production in Kenya** . . . 845
Winifred Chepkoech, Nancy W. Mungai, Hillary K. Bett, Silke Stöber, and Hermann Lotze-Campen

44 **Agricultural Impacts of the 2015/2016 Drought in Ethiopia Using High-Resolution Data Fusion Methodologies** . . . 869
James M. Warner and Michael L. Mann

45 **Intensifying Maize Production Under Climate Change Scenarios in Central West Burkina Faso** . . . 895
Omonlola Nadine Worou, Jérôme Ebagnerin Tondoh, Josias Sanou, Thomas Gaiser, Pinghouinde Michel Nikiema, Jules Bayala, Paulin Bazié, Catherine Ky-Dembele, and Antoine Kalinganiré

46 **Fertilization Strategies Based on Climate Information to Enhance Food Security Through Improved Dryland Cereals Production** . . . 917
Komla Kyky Ganyo, Bertrand Muller, Aliou Guissé, and Myriam Adam

47 **Drip Watering Regimes on Growth Performance, Yield, and Water Use Efficiency of Sorghum in Semi-arid Environment of Tanzania: Effects** . . . 935
Athuman Juma Mahinda, Method Kilasara, and Charles K. K. Gachene

48 **Managing Vulnerability to Drought and Enhancing Smallholder Farmers Resilience to Climate Change Risks in Zimbabwe** . . . 961
Obert Jiri and Paramu L. Mafongoya

49 **Building Pastoral and Agro-pastoral Community Resilience Against Drought in the Context of the Paris Agreement: The Case of Isiolo County, Kenya** . . . 979
David Walunya Ong'are and Anne Nyatichi Omambia

50 Climate Change Adaptation Among Female-Led Micro, Small, and Medium Enterprises in Semiarid Areas: A Case Study from Kenya 1001
Joanes Atela, Kate Elizabeth Gannon, and Florence Crick

51 Assessment of Barriers to Food Security in North-Eastern Nigeria 1019
Hubert Fudjumdjum, Walter Leal Filho, and Desalegn Yayeh Ayal

52 Rural Farmers' Adaptation Decision to Climate Change in Niger Delta Region, Nigeria 1035
Nsikak-Abasi A. Etim and NseAbasi N. Etim

53 Economic Viability of Alternative Small-Scale Irrigation Systems Used in Vegetables Production in Koulikoro and Mopti Regions, Mali 1051
A. M. Kane, J. K. Lagat, T. Fane, J. K. Langat, and B. Teme

54 Cultivar Selection and Management Strategies for *Lablab purpureus* (L.) Sweet in Africa 1083
Neil R. Miller, Wilfred Mariki, Alison Nord, and Sieglinde Snapp

55 Introduction of a Biodiversity Farming Concept in Amuru District, Uganda 1097
Nyeko Pen-Mogi and Martine Nyeko

56 Scaling Up Climate-Smart Agricultural (CSA) Solutions for Smallholder Cereals and Livestock Farmers in Zambia 1115
Idowu Kolawole Odubote and Oluyede Clifford Ajayi

57 Multidimensional Framework for Achieving Sustainable and Resilient Food Systems in Nigeria 1137
Kyle Frankel Davis and Olawale Emmanuel Olayide

58 Potential of Smart Aquaculture Technologies on Improving Fish Production in Malawi 1161
M. Mulumpwa

59 Climate Change and Variability in the Mixed Crop/Livestock Production Systems of Central Ethiopian Highland 1169
Nigatu Alemayehu, Mary Masafu, Abule Ebro, Azage Tegegne, and Getachew Gebru

60 Impact of Smart Crop-Livestock Diversification as Climate Change Adaptation Strategies on Farmers' Living Conditions, Tahoua State, Niger Republic 1193
Zakou Amadou and Zakari Saley Bana

61 Climate Change Implication on Airborne Infections: Roadmap for Nigeria's Health Sector Development 1215
Tosin K. W. Gbadegesin

62 Herdsmen on the Move: The Burdens of Climate Change and Environmental Migration in Nigeria 1225
Toyib Aremu and Praise Abraham

63 Assessing Socioeconomic Factors Influencing Production and Commercialization of Bambara Groundnut as an Indigenous Climate-Resilient Crop in Nigeria 1237
Olawale Emmanuel Olayide, Samuel A. Donkoh, Isaac Gershon Kodwo Ansah, William Adzawla, Patrick J. O'Reilly, Sean Mayes, Aryo Feldman, Razlin Azman Halimi, George Nyarko, Christopher O. Ilori, and Tunrayo Alabi

64 Perceptions of Challenges Facing Pastoral Small Ruminant Production in a Changing Climate in Kenya 1257
Rosemary N. Ngotho-Esilaba, J. O. Onono, J. N. Ombui, J. F. Lindahl, and H. O. Wesonga

65 Assessment of Climatic Variability Effect on Millet Production and Yield 1269
Olumuyiwa Idowu Ojo, Abdulafeez Olalekan Olaniyan, Adeniyi Suleiman Gbadegesin, and Masengo Francois Ilunga

66 Climate Change Vulnerability Among Pastoralists and Farmers in Ethiopia 1291
Desalegn Yayeh Ayal, Walter Leal Filho, Muluneh Woldetisadik, Solomon Desta, and Chunlan Li

Volume 3

Part III Climate Resilience Measures Integrated with Policies and Governance at City, Regional, and National Level 1315

67 Domestic Water and Disaster Management Legislation: A Key Tool for the Implementation of the Paris Agreement in LDCs . . . 1317
Tales Carvalho Resende

68 Enhancing Resilience of Livelihoods and Production Systems to Climate Variability and Other Related Risks in Africa 1343
Herve Alain Napi Wouapi and Maruf Sanni

69 Financing Projects for Improving Climate Change Resilience: The Cases of Djibouti and Yemen 1357
Imad Antoine Ibrahim

70 Fostering Climate-Resilient Biodiversity Framework in West Africa and Implications for Regional Development 1377
Samuel Aderemi Igbatayo

71 Causes and Impacts of Climate Change in Asia Pacific: Integrated Responses and Need for Change in Decision Making in Australia 1395
Maureen Papas

72 Community-Based Climate Change Adaptation Action Plans to Support Climate-Resilient Development in the Eastern African Highlands 1417
Tino Johansson, Emmah Owidi, Sarah Ndonye, Sarah Achola, Weyessa Garedew, and Claudia Capitani

73 Working Together to Build Climate Resilience in Hudson Riverfront Communities 1443
Libby Zemaitis, Nava Tabak, Kristin Marcell, Bennett Brooks, and Julie Noble

74 Planning and Climate Change in Serbia: Integrated Responses to the Causes and Impacts 1463
Tijana Crnčević and Jelena Živanović Miljković

75 Urban Climate Change Vulnerability, Responses, and Policies in Qatar: An Assessment 1477
Mohammad Al-Saidi, Sara Abdelhakim Mohammad, and Amina Nihad Awartani

76 From Vulnerability Assessments to Low/No Regret Resilience Planning in Rural Contexts 1501
Anuradha Phadtare, Swapnil Vyas, Marcella D'Souza, Dipak Zade, and Yogesh Shinde

77 Quality Management System for Climate Change Adaptation for the Tri-City Area Bergisches Städtedreieck 1531
Kathrin Prenger-Berninghoff, Alice Neht, and Andreas Witte

78 Resilience to Extreme Events in the City of La Paz, Mexico 1557
Antonina Ivanova

79 Institutional Barriers in Local Adaptation Policies 1577
Stéphane La Branche and P. Bosbœuf

80 Environmental Sustainability and Systems Thinking: A Foundation for More Effective Climate Policy 1597
G. D. Bothun

81 Policy and Institutional Dimensions in Climate-Smart Agriculture Adoption: Case of Rural Communities in Zimbabwe 1611
Angeline Mujeyi

82 Climate Change and Population Growth in Pastoral Communities of Ngorongoro District, Tanzania 1627
Sef Slootweg

83 Determinants of Adoption of Climate-Smart Agriculture Technologies at Farm Plot Level: An Assessment from Southern Tanzania 1647
Chris M. Mwungu, Caroline Mwongera, Kelvin M. Shikuku, Mariola Acosta, and Peter Läderach

84 Understanding Kenyan Farmers' Perceptions of and Responses to Climatic Variability to Build their Resilience 1661
Fiona Mwaniki and Humphrey M. Ngibuini

85 Mitigating and Adapting to "Arid" Climate Extremes: Impacts of Locally Prioritized Ecosystem-Based Adaptations at the Water-Energy Nexus 1681
Caroline King-Okumu, Jarso Harou, Victor A. Orindi, and Molu Tepo

86 Household Head-Related Social Capital: The Trump Card for Facilitating Actual Uptake of Innovation in Rural Smallholder Systems 1711
Felix Kwabena Donkor and Kevin Mearns

87 Climate Change Adaptation Through Science-Farmer-Policy Dialogue in Mali 1727
Lin Bautze, Gian L. Nicolay, Matthias Meier, Andreas Gattinger, and Adrian Muller

88 Livestock Market Improvement with Anthropological Approach in Drought Resilience Project in Northern Kenya 1743
Fumiaki Murakami and Monicah Kinuthia

89 Determinants of Household Food Security Status and Challenges of Building Resilience to Climate Variability and Change Posed by Drought in Tharaka Nithi, Kenya 1757
Victoria Gioto, Shem Wandiga, and Christopher Oludhe

90 Implications for Mitigation and Adaptation Measures: Rice Farmers' Response and Constraints to Climate Change in Ivo Local Government Area of Ebonyi State 1787
K. N. N. Ezike

91 Building Eco-social Resilience in Rural Communities: Benefits of Permaculture Pedagogy and Praxis 1801
David Yisrael Epstein HaLevi, Greg William Misiaszek, Hugh Kelly, Sheena Shah, Charles Mugarura, and Liam James Walsh

92 Climate Finance as a Catalyst for Initiating an African Green Revolution to Enhance Drought Resilience: Policy Prospects and Challenges 1843
Dumisani Chirambo

93 Green Bonds: A Catalyst for Sustainable Development in Nigeria 1859
Nelson Ifeanyi Obine

94 Isolation as a Barrier for the Climate Change Actions of Insurers 1883
Lara Johannsdottir and James R. Wallace

95 Early Warning System and Ecosystem-Based Adaptation to Prevent Flooding in Ibadan Metropolis, Nigeria 1909
Olusegun Michael Ogundele and Rosemary Egodi Ubaekwe

96 Effects of Mentoring and Field Study Instructional Strategies on Students' Climate Change Reduction Practices in Social Studies in Lagos State 1935
Olawale Oyemade Oyekanmi, Peter Adewale Amosun, and Ibidun O. Adelekan

97 Strengthening Climate Change Adaptation in the Cities of West Africa: Policy Implications for Urban Resilience 1957
Maruf Sanni, Abdulai Jalloh, Aliou Diouf, Musiliyu K. Atoyebi, Oluwatosin E. Ilevbare, and Sunday J. Olotu

98 Climate Change Awareness and Adaptation Measures Among Farmers in Cross River and Akwa Ibom States of Nigeria 1983
A. U. Ogogo, M. U. Ekong, and N. M. Ifebueme

99 Sustainability and the Effectiveness of BNRCC Community-Based Adaptation (CBA) to Address Climate Change Impact in Nigeria 2003
Emmanuel C. Nzegbule, Chinedum Nwajiuba, Gloria Ujor, Robert Onyeneke, Samson Samuel Ogallah, and David Okali

100 Protecting Climate Change-Induced Internally Displaced Persons in Africa: Relevance of the Kampala Convention 2023
Romola Adeola

101 Leveraging Best Practices for Climate Adaptation in the West African Sahel: The Emergence of Global Alliance for Resilience 2035
Samuel Aderemi Igbatayo and Oladapo Opeyemi Babalola

102 Governance Dimensions of Climate Change Adaptation: The Case of Didahara, Borana, Southern Ethiopia 2059
Desalegn Yayeh Ayal, Solomon Desta, and Lance Robinson

103 Urban Water Reclamation with Resource Recovery as a Cornerstone of Urban Climate Change Resilience 2075
Daphne Gondhalekar, Mohammed Al-Azzawi, and Jörg E. Drewes

104 Gender Analysis Approach to Analyzing Gender Differentiated Impacts of Coping Strategies to Climate Change 2097
Jagriti Kher and Savita Aggarwal

105 Addressing Slow Onset Disasters: Lessons from the 2015–2016 El Niño in the Philippines 2125
Erwin A. Alampay and Dennis dela Torre

Volume 4

Part IV Technological and Science-Based Approaches to Climate Resilience **2143**

106 Climate Change Resilience: Role of Insurers in Bridging the Gap Between Climate Change Science and Heterogeneous Stakeholders 2145
Lara Johannsdottir and James R. Wallace

107 Climate-Proof Retrofitting of Urban Areas for the Same Cost . . . 2171
L. Kleerekoper, R. Loeve, and J. Kluck

108 Persuading Pacific Maritime Supply Chain Stakeholders to Prioritize Climate Change Resilience 2197
Jack Alban Dyer

109 Fostering Climate Smartness in Smallholder Farming Systems: Business Promotional Approaches for Improved Maize Varieties in Eastern and Southern Africa 2225
Kingstone Mujeyi and Angeline Mujeyi

110 Innovation Systems to Adapt to Climate Change: Lessons from the Kenyan Coffee and Dairy Sectors 2249
Kinfe Asayehegn, Ludovic Temple, Philippe Vaast, and Ana Iglesias

111 **Optimizing Outputs of Nigerian Research Institutes to Achieve National Food Security in the Face of Climate Change** 2273
Samuel Adelowo Olakojo

112 **Saline Agriculture: A Climate Smart Integrated Approach for Climate Change Resilience in Degraded Land Areas** 2287
Muhammad Saqib, Javaid Akhtar, Ghulam Abbas, and Hafiz Abdul Wahab

113 **Discussion with Farmers in Southwestern Nigeria on Climate Change and Agriculture** 2307
Obasanjo Joseph Oyedele

114 **Cross-Disciplinary Drivers: Benefit to Smallholder Farmers and to Achieve SDGs by Various Means** 2325
Ijaz Rasool Noorka and J. S. (Pat) Heslop-Harrison

115 **Climate-Resilient Agricultural Practices in Different Agro-ecological Zones of Bangladesh** 2337
Muhammad Abdur Rahaman, Mohammad Mahbubur Rahman, and Md. Shahadat Hossain

116 **Agro-met Services and Farmer Responsiveness to Advisories: Implications for Climate Smart Agriculture** 2365
Ramkumar Bendapudi, Nitin Kumbhar, Prithviraj Gaikwad, and Crispino Lobo

117 **Vulnerability and Adaptation Levels of the Construction Industry in Kenya to Climate Change** 2383
Onkangi N. Ruth, Mwangi Peter Njiiri, Erick Maklago, and Ondari Lilian

118 **Drought-Tolerant Crops in Kirinyaga County, Kenya: Climate-Smart Agriculture Adaptation Strategies** 2401
Peterson N. M. Njeru, Jayne Mugwe, Monicah Mucheru-Muna, Immaculate Maina, Stephen K. Kimani, and David K. Lelgut

119 **Effect of Mulching on Soil Temperature and Moisture for Potato Production in Agro-ecological Zones of Central Highlands of Kenya** 2415
Eva N. Gacheru, Charles K. K. Gachene, Patrick T. Gicheru, and Lieven Claessens

120 **Ecosystem-Based Adaptation in Tigray, Northern Ethiopia: A Systematic Review of Interventions, Impacts, and Challenges** 2439
Mulubrhan Balehegn, Mitiku Haile, Chao Fu, and Wu Liang

121 Seasonal Climate Forecasts Among Farmers in Nigeria: Assessment of the Usability 2485
Glory Ikponmwosa Edwards and Dominic Kniveton

122 Asphalt Making Potential of Pyrolytic Bitumen from Waste Rubber Tyres: An Adaptive Measure to Climate Change 2515
J. G. Akinbomi, S. O. Asifat, A. Ajao, and O. Oladeji

123 Local Knowledge of Climate Change Among Arable Farmers in Selected Locations in Southwestern Nigeria 2531
David Olabanjo Sanni

124 Adaptation of Sorghum (*Sorghum bicolor* L. Moench) Crop Yield to Climate Change in Eastern Dryland of Sudan 2549
Haitham R. Elramlawi, Hassan I. Mohammed, Ali W. Elamin, Omer A. Abdallah, and Abdel Aziz A. M. Taha

125 Agriculture in the Face of Climate-Mediated Flooding in Tropical Africa: Technical Innovations of Fish Farmers in Southwestern Nigeria 2575
Oyediran O. Oyebola, Jackson Efitre, Augustine E. Falaye, Taiwo M. Dada, and Funmilayo C. Idowu

126 Investigation of Rainfall Characteristics in Sudano-Sahelian Region of Nigeria (1971–2006) 2595
Aliyu Tambuwal Umar, Sani Abubakar Mashi, and Mansur Matazu Bako

127 Climate Change Adaptation: Adoption of GMOs in Africa 2623
Emmanuel Ejim-Eze

128 Climate Smart Adaptations in the African Tropics: Scaling Weather Information for Decision Support Outcomes in Nigeria Savannahs 2651
David Olufemi Awolala

129 Climate Change Education for Sustainable Development: Lessons for Nigeria 2685
Daniel Mailumo, Simeon Igbe, and Pius Mailumo

130 Enhancing the Capacity of Vulnerable Community to Climate Change: Role of Quality Declared Seed Production Model in Semi-arid Areas of Central Tanzania 2699
E. Y. Swai, F. Njau, and M. Farrely

131 Assessment of Information on Successful Climate-Smart Agricultural Practices/Innovations in Tanzania 2721
Paschal Arsein Mugabe

132 Impact of Climate Change on Growing Season in Nigeria: Seasonal Rainfall Prediction (SRP) as Assessment and Adaptation Tool 2743
Paul Akeh Ugbah, Olumide Olaniyan, Sabastine Dekaa Francis, and Adamu James

133 Educating Farmers in Rural Areas on Climate Change Adaptation for Sustainability in Nigeria 2771
Benjamin Anabaraonye, Joachim Chukwuma Okafor, and James Hope

134 Water Harvesting Technology for Enhancing Food Security Livelihood: The Case of Northern Katsina State, Nigeria 2791
S. Jari, M. Muntaka, and R. T. Nabinta

Index 2805

About the Editor

Prof. Walter Leal Filho (B.Sc., Ph.D., D.Sc., D.Phil., D.L., D.Litt., D.Ed.) is Professor and Head of the Research and Transfer Centre "Sustainable Development and Climate Change Management" at the Hamburg University of Applied Sciences in Germany and holds the Chair of Environment and Technology at Manchester Metropolitan University, UK. He is a Lead Author at AR6's Working Group II (Climate Change Adaptation) at the Intergovernmental Panel on Climate Change (IPCC), is Founding Editor of the *International Journal of Climate Change Strategies and Management*, and heads the International Climate Change Information and Research Programme (ICCIRP). He is also Editor-in-Chief of the Climate Change Management series with Springer. Professor Leal serves on the editorial board of various journals. He has in excess of 400 publications to his credit, among which are ground-breaking books such as *Universities and Climate Change*; *The Economic, Social and Political Aspects of Climate Change*; the *Handbook of Climate Change Adaptation*; and the *Handbook of Climate Change Communication*. He has nearly 30 years of field experience on project management and has a particular interest in the connections between sustainability, climate change adaptation, and human behavior.

About the Deputy Editor

Desalegn Yayeh Ayal is Assistant Professor of Geography and Environmental Studies at the Center for Food Security Studies, College of Development Studies, Addis Ababa University. Desalegn holds a Ph.D. in Geography. Desalegn serves as the Deputy Editor of *International Journal of Climate Change Strategies and Management*. He has published more than 30 publications including books, book chapters, and refereed journal articles and is East Africa Vice President for Interconnections for Making Africa Great Empowered and Sustainable Initiative. Desalegn has also presented papers on climate adaptation and related issues at many international and national conferences. His principal areas of research interest include climate change and adaptation, climate resilience, climate change mitigation and related issues, indigenous weather forecasting, integrated natural resources rehabilitation and management, livelihoods, and food security nexus, among others. He thoroughly understands the link between natural and human-induced hazards with sustainable development and works hard to familiarize with current tools of climate change impact assessment on livelihood and the wider environment. Desalegn has been actively involved in climate resilience and integrated natural resources rehabilitation and management research, and development interventions to improve food security.

Editorial Board

Editorial Advisory Board

Contributors

Ghulam Abbas Department of Environmental Science, COMSATS University Islamabad, Vehari Campus, Vehari, Pakistan

Omer A. Abdallah Department of Agricultural Engineering, Faculty of Agriculture, University of Khartoum, Khartoum, Sudan

M. O. Abioja Department of Animal Physiology, College of Animal Science and Livestock Production, Federal University of Agriculture, Abeokuta, Ogun, Nigeria

Praise Abraham Department of Social Sustainability, Centre for Sustainable Development, University of Ibadan, Ibadan, Oyo, Nigeria

Bashir Yusuf Abubakar Department of Botany, Ahmadu Bello University, Zaria, Nigeria

Sarah Achola International Centre of Insect Physiology and Ecology, Nairobi, Kenya

Mariola Acosta International Institute of Tropical Agriculture (IITA), Kampala, Uganda

Myriam Adam Centre de Coopération Internationale en Recherche Agronomique pour le Développement (CIRAD), UMR AGAP/PAM, Montpellier, France

Institut de l'Environnement et de la Recherche Agricole (INERA), Bobo Dioulasso, Burkina Faso

International Crops Research Institute for Semi-Arid Tropics (ICRISAT), Bamako, Mali

Bamidele Adanikin Department of Plant Science and Biotechnology, Ekiti State University, Ado Ekiti, Ekiti State, Nigeria

Oluwabunmi Opeyemi Adejumo Institute for Entrepreneurship and Development Studies, Obafemi Awolowo University, Ile-Ife, Nigeria

M. O. Adekunle Department of Animal Physiology, College of Animal Science and Livestock Production, Federal University of Agriculture, Abeokuta, Ogun, Nigeria

Ibidun O. Adelekan Department of Geography, Faculty of Social Science, University of Ibadan, Ibadan, Oyo State, Nigeria

Mosunmola Lydia Adeleke Department of Fisheries and Aquaculture Technology, FUTA, Akure, Nigeria

Romola Adeola Centre for Human Rights, Faculty of Law, University of Pretoria, Pretoria, South Africa

Peter Adegbenga Adeonipekun Department of Botany, University of Lagos, Akoka-Yaba, Lagos, Nigeria

O. M. Adesope Department of Agricultural Economics and Extension, University of Port Harcourt, Choba/Port Harcourt, Nigeria

William Adzawla Faculty of Agribusiness and Communication Sciences, University for Development Studies, Tamale, Ghana

Oludare Agboola Department of Botany, University of Lagos, Akoka-Yaba, Lagos, Nigeria

Savita Aggarwal Department of Development Communication and Extension, Institute of Home Economics, University of Delhi, New Delhi, India

Yaw Boakye Agyeman Department of Ecotourism, Recreation and Hospitality, University of Energy and Natural Resources, UENR, Sunyani, Brong Ahafo Region, Ghana

Abeer Ahmed Qaed Ahmed Pharmacy Department, Al-Saeed University, Osaifrah, Taiz, Yemen

A. Ajao Amsol Bio Company, Lagos, Nigeria

Oluyede Clifford Ajayi Technical Centre for Agriculture and Rural Co-operation, Wageningen, The Netherlands

I. A. Ajibefun Department of Agriculture and Resource Economics, FUTA, Akure, Nigeria

The Federal University of Technology, (FUTA), Akure, Nigeria

Javaid Akhtar Institute of Soil and Environmental Sciences, University of Agriculture, Faisalabad, Pakistan

J. G. Akinbomi Department of Chemical and Polymer Engineering, Faculty of Engineering, Lagos State University, Lagos, Nigeria

Akinlabi Akintuyi Department of Geography, University of Lagos, Akoka-Yaba, Lagos, Nigeria

Bosede Olufunmilayo Akinwalere Department of Agricultural Extension and Communication, FUTA, The Federal University of Technology, Akure, Nigeria

Tunrayo Alabi Geographic Information Systems Unit, International Institute of Tropical Agriculture, Ibadan, Nigeria

Erwin A. Alampay National College of Public Administration and Governance, University of the Philippines, Diliman, Quezon City, Philippines

Mohammed Al-Azzawi Department of Civil, Geo and Environmental Engineering, Technical University of Munich, Garching, Germany

Eike Albrecht Brandenburg University of Technology Cottbus – Senftenberg, Senftenberg, Germany

Nigatu Alemayehu Consultant, Addis Ababa, Ethiopia

Mohammad Al-Saidi Center for Sustainable Development, College of Arts and Sciences, Qatar University, Doha, Qatar

Zakou Amadou Faculty of Agricultural Sciences, Department of Rural Economics and Sociology, Tahoua University, Tahoua, Niger

T. T. Amos Department of Agriculture and Resource Economics, FUTA, Akure, Nigeria

Peter Adewale Amosun Department of Teacher Education, Faculty of Education, University of Ibadan, Ibadan, Oyo State, Nigeria

Benjamin Anabaraonye Benjy Poetry and Music Global Concepts, Awka, Anambra State, Nigeria

Isaac Gershon Kodwo Ansah Faculty of Agribusiness and Communication Sciences, University for Development Studies, Tamale, Ghana

Toyib Aremu Department of Environmental Sustainability, Centre for Sustainable Development, University of Ibadan, Ibadan, Oyo, Nigeria

Kinfe Asayehegn School of environment, Gender and development Studies, Hawassa University, Hawassa, Ethiopia

Center for International Cooperation in Agricultural Research and Development (CIRAD), Montpellier, France

Department of Agricultural Economics, Universidad Politécnica de Madrid, Madrid, Spain

S. O. Asifat Lopek Engineering and Construction Ltd, Lagos, Nigeria

Joanes Atela African Centre for Technology Studies, Nairobi, Kenya

Musiliyu K. Atoyebi National Centre for Technology Management, Federal Ministry of Science and Technology, Obafemi Awolowo University, Ile-Ife, Nigeria

Amina Nihad Awartani College of Arts and Sciences, Qatar University, Doha, Qatar

David Olufemi Awolala Department of Agricultural and Resource Economics, Federal University of Technology Akure, Akure, Nigeria

Research Fellow, Adaptation Finance Fellowship Programme of the Thailand Research Development Institute Bangkok, Thailand and Frankfurt School of Finance and Management, Frankfurt, Germany

Abiodun Emmanuel Awoyemi Department of Agricultural Economics, Faculty of Agriculture and Forestry, University of Ibadan, Ibadan, Oyo State, Nigeria

Desalegn Yayeh Ayal Center for Food Security Studies, College of Development Studies, Addis Ababa University, Addis Ababa, Ethiopia

Ayansina Ayanlade Department of Geography, Obafemi Awolowo University, Ile-Ife, Nigeria

Neda Azam Department of Water Engineering, College of Agriculture, Isfahan University of Technology, Isfahan, Iran

Olubukola Oluranti Babalola Food Security and Safety Niche Area, Faculty of Natural and Agricultural Sciences, North-West University, Mmabatho, Mafikeng, South Africa

Oladapo Opeyemi Babalola Afe Babalola University, Ado-Ekiti, Nigeria

Mansur Matazu Bako Nigerian Meteorological Agency (NiMet), Abuja, Nigeria

Mohammed Lawal Balarabe Department of Biology, Ahmadu Bello University, Zaria, Nigeria

Mulubrhan Balehegn Department of Animal, Rangeland and Wildlife Sciences, Mekelle University, Mekelle, Ethiopia

Paul Basudde Sector Planning and Policy Analysis Department, Ministry of Energy and Mineral Development (MEMD), Kampala, Uganda

Lin Bautze Research Institute of Organic Agriculture (FiBL), Frankfurt am Main, Germany

Jules Bayala World Agroforestry Centre, West and Central Africa Regional Office – Sahel Node, Bamako, Mali

Paulin Bazié Institut de l'Environnement et de Recherche Agricole (INERA), Ouagadougou, Burkina Faso

Tamrat Bekele Department of Plant Biology and Biodiversity Management, College of Natural and Computational Sciences, Addis Ababa University, Addis Ababa, Ethiopia

Kassa Belay Pan African University Africa Union, Addis Ababa, Ethiopia

Adedoyin Bello Department of Urban and Regional Planning, Crescent University, Abeokuta, Ogun State, Nigeria

Ramkumar Bendapudi Watershed Organisation Trust (WOTR), Pune, India

Mohan Kumar Bera Institute of Economic Growth, New Delhi, India

Hillary K. Bett Department of Agricultural Economics and Agribusiness Management, Egerton University, Njoro, Kenya

Martha Bohm School of Architecture and Planning, State University of New York at Buffalo, Buffalo, NY, USA

Adjoua Nadège Boko-Koiadia Department of Sociology, University of Félix Houphouët-Boigny, Abidjan, Côte d'Ivoire

P. Bosbœuf Lab'Urba – Paris-Est University, Marne-la-Vallée, France

G. D. Bothun Department of Physics, University of Oregon, Eugene, OR, USA

Bennett Brooks Consensus Building Institute, New York, NY, USA

Kathryn Brown Marine Research Centre, School of Environment, Science and Engineering, Southern Cross University, Lismore, NSW, Australia

Claudia Capitani Environment Department, University of York, York, UK

Tales Carvalho Resende University of Strathclyde, Glasgow, UK

Winifred Chepkoech Center for Rural Development (SLE), Humboldt Universität zu Berlin, Berlin, Germany

Dumisani Chirambo Department of Civil and Public Law with references to Law of Europe and the Environment, Brandenburg University of Technology Cottbus-Senftenberg, Cottbus, Germany

Lieven Claessens International Institute of Tropical Agriculture (IITA), Arusha, Tanzania, Kenya

Helene Jacot Des Combes PaCE-SD, Lower Campus, Laucala, USP, Suva, Fiji

Florence Crick Grantham Research Institute on Climate Change and the Environment, London School of Economics and Political Science, London, UK

Tijana Crnčević Institute of Architecture and Urban and Spatial Planning of Serbia, Belgrade, Serbia

Marcella D'Souza The WOTR-Centre for Resilience Studies (W-CReS), Pune, India

Taiwo M. Dada Department of Aquaculture and Fisheries Management, University of Ibadan, Ibadan, Nigeria

Petr Daněk Department of Geography, Masaryk University, Brno, Czech Republic

Kyle Frankel Davis The Earth Institute, Columbia University, New York, NY, USA

The Nature Conservancy, New York, NY, USA

Data Science Institute, Columbia University, New York, NY, USA

Dennis dela Torre Center for Local and Regional Governance, NCPAG, University of the Philippines, Diliman, Quezon City, Philippines

Solomon Desta MARIL (Managing Risk for Improved Livelihood) Consulting Firm, Addis Ababa, Ethiopia

Aliou Diouf Climate Change, Natural Resource Management, Rural and Urban Development Enda Energie-Environnement-Développement, Dakar, Senegal

Samuel A. Donkoh Faculty of Agribusiness and Communication Sciences, University for Development Studies, Tamale, Ghana

Felix Kwabena Donkor College of Agriculture and Environmental Sciences, University of South Africa (UNISA), UNISA Science Campus. Corner of Christiaan de Wet Road and Pioneer Avenue, Florida, South Africa

School of Animal, Plant and Environmental Sciences, University of the Witwatersrand WITS, Johannesburg, South Africa

Jörg E. Drewes Department of Civil, Geo and Environmental Engineering, Technical University of Munich, Garching, Germany

Jack Alban Dyer Australia Maritime College, University of Tasmania, Launceston, TAS, Australia

Abule Ebro International Livestock Research Institute, Addis Ababa, Ethiopia

Glory Ikponmwosa Edwards Climate Change, Development and Policy, School of Global Studies, University of Sussex, Falmer, UK

Jackson Efitre Department of Zoology, Entomology and Fisheries Sciences, Makerere University, Kampala, Uganda

Emmanuel Ejim-Eze National Centre for Technology Management, Obafemi Awolowo University, Ile-Ife, Ile-Ife, Nigeria

M. U. Ekong Department of Forestry and Wildlife Resources Management, University of Calabar, Calabar, Cross River State, Nigeria

Ali W. Elamin Department of Agricultural Engineering, Faculty of Agriculture, University of Khartoum, Khartoum, Sudan

Haitham R. Elramlawi Center of Dryland Farming Research and Studies (CDFRS), Faculty of Agricultural and Environmental Sciences, University of Gadarif, Gadarif, Sudan

David Yisrael Epstein HaLevi Department of Educational Policy and Leadership, University at Albany, Albany, NY, USA

M. O. Esilaba Department of Environmental Science, Egerton University, Egerton, Kenya

Saeid Eslamian Department of Water Engineering, College of Agriculture, Isfahan University of Technology, Isfahan, Iran

Nsikak-Abasi A. Etim Department of Agricultural Economics and Extension, University of Uyo, Uyo, Akwa Ibom State, Nigeria

NseAbasi N. Etim Department of Animal Science, Akwa Ibom State University, Obio Akpa Campus, Abak, Akwa Ibom State, Nigeria

K. N. N. Ezike Department of Agricultural Economics, Management and Extension, Ebonyi State University, Abakaliki, Ebonyi State, Nigeria

O. A. Fagbenro Department of Fisheries and Aquaculture Technology, FUTA, Akure, Nigeria

Augustine E. Falaye Department of Aquaculture and Fisheries Management, University of Ibadan, Ibadan, Nigeria

Belarmain A. Fandohan Université nationale d'agriculture École des sciences et techniques de conservation et de transformation des produits agricoles BP 114, Sakété, Bénin

T. Fane Faculty of Economic Sciences and Management, University of Bamako, Bamako, Mali

M. Farrely Tanzania Organic Movement for Agriculture, Dar es Salaam, Tanzania

Mayowa Fasona Department of Geography, University of Lagos, Akoka-Yaba, Lagos, Nigeria

Aryo Feldman Crops for the Future (CFF), Jalan Broga, Selangor Darul Ehsan, Malaysia

Sabastine Dekaa Francis National Weather Forecasting and Climate Research Centre, Nigerian Meteorological Agency, Abuja, Nigeria

Chao Fu Key Laboratory of Ecosystem Network Observation and Modeling, Institute of Geographic Sciences and Natural Resources Research, Chinese Academy of Sciences, Beijing, China

Hubert Fudjumdjum Research and Transfer Centre Sustainable Development and Climate Change Management, Hamburg University of Applied Sciences, Hamburg, Germany

Charles K. K. Gachene Department of Land Resource Management and Agricultural Technology (LARMAT), University of Nairobi, Nairobi, Kenya

Kenya Agriculture and Livestock Research Organisation, Nairobi, Kenya

Eva N. Gacheru Kenya Agriculture and Livestock Research Organisation, Nairobi, Kenya

Prithviraj Gaikwad Watershed Organisation Trust (WOTR), Pune, India

WOTR Centre for Resilience Studies (W-CReS), Pune, India

Thomas Gaiser Institute of Crop Science and Resource Conservation, University of Bonn, Bonn, Germany

Kate Elizabeth Gannon Grantham Research Institute on Climate Change and the Environment, London School of Economics and Political Science, London, UK

Komla Kyky Ganyo Centre d'Etude Régional pour l'Amélioration de l'Adaptation à la Sécheresse (CERAAS), BP, Thiès, Sénégal

Institut Togolais de Recherche Agronomique (ITRA), Lomé, Togo

Weyessa Garedew Jimma University College of Agriculture and Veterinary Medicine, Jimma, Ethiopia

Andreas Gattinger Faculty 9 – Agricultural Sciences, Nutritional Sciences, and Environmental Management, Justus Liebig University (JLU), Institute of Agronomy and Plant Breeding II, Gießen, Germany

Tosin K. W. Gbadegesin Centre for Sustainable Development (CESDEV), University of Ibadan, Ibadan, Nigeria

Yohannes GebreMichael Department of Geography and Environmental Studies, Addis Ababa University, Addis Ababa, Ethiopia

Getachew Gebru International Livestock Research Institute, Addis Ababa, Ethiopia

Patrick T. Gicheru Kenya Agriculture and Livestock Research Organisation, Embu, Kenya

Victoria Gioto Institute of Climate Change and Adaptation (ICCA), University of Nairobi, Nairobi, Kenya

Laurent Gnonlonfin Doctoral School of Agronomic and Water Sciences/Department of Sustainable Management of Natural Resources, Laboratory of Ecology and Forest Research (LERF), University of Parakou, Parakou, Borgou, Republic of Benin

Daphne Gondhalekar Department of Civil, Geo and Environmental Engineering, Technical University of Munich, Garching, Germany

Degye Goshu Department of Economics, Faculty of Business and Economics, Kotebe Metropolitan University (Asso.), Addis Ababa, Ethiopia

Gérard Gouwakinnou Laboratory of Ecology, Botanic and Plant Biology (LEB), University of Parakou, Parakou, Borgou, Republic of Benin

Aliou Guissé Département de Biologie Végétale, Université Cheikh Anta Diop de Dakar (UCAD), Dakar, Sénégal

Marwa R. Hafez Institute of Graduate Studies and Research (IGSR), Alexandria University, Alexandria, Egypt

Ghrmawit Haile Ministry of Environment Forest and Climate Change and Addis Ababa University, Addis Ababa, Ethiopia

Mitiku Haile Department of Land Resource Management and Environmental Protection, Mekelle University, Mekelle, Ethiopia

Razlin Azman Halimi Crops for the Future (CFF), Jalan Broga, Selangor Darul Ehsan, Malaysia

Southern Cross Plant Science, Southern Cross University, East Lismore, NSW, Australia

Jarso Harou Ministry of Water, Energy and Climate Change, Isiolo County Government, Isiolo, Kenya

Ali Hasantabar-Amiri Department of Civil Engineering, Lenjan Branch, Islamic Azad University, Lenjan, Iran

Peni Hausia Havea PaCE-SD, Lower Campus, Laucala, USP, Suva, Fiji

Sarah L. Hemstock Bishop Grosseteste University, Lincoln, UK

A. Henri-Ukoha Department of Agricultural Economics and Extension, University of Port Harcourt, Choba/Port Harcourt, Nigeria

J. S. (Pat) Heslop-Harrison Department of Genetics and Genome Biology, University of Leicester, Leicester, UK

James Hope Ahmadu Bello University, Zaria, Kaduna State, Nigeria

Md. Shahadat Hossain Institute of Water Modeling, Dhaka, Bangladesh

Towanou Olivier Houetchegnon Doctoral School of Agronomic and Water Sciences/Department of Sustainable Management of Natural Resources, Laboratory of Ecology and Forest Research (LERF), University of Parakou, Parakou, Borgou, Republic of Benin

Mark Howells Unit of Energy Systems Analysis, KTH – Royal Institute of Technology, Stockholm, Sweden

Imad Antoine Ibrahim Institute of Law, Politics, and Development (Dirpolis), Sant'Anna School of Advanced Studies, Pisa, Italy and gLAWcal – Global Law Initiative for Sustainable Development, Essex, UK

Mustapha Adeojo Ibrahim Department of Biology, Ahmadu Bello University, Zaria, Nigeria

Funmilayo C. Idowu National Biotechnology Development Agency, Ogbomoso, Nigeria

N. M. Ifebueme Department of Forestry and Wildlife Resources Management, University of Calabar, Calabar, Cross River State, Nigeria

Samuel Aderemi Igbatayo Department of Economics and Management Studies, Afe Babalola University, Ado-Ekiti, Nigeria

Simeon Igbe Association of Entrepreneurs and Technology Managers of Nigeria (AETMAN), Eco-system Based Adaptation For Food Security Assembly (EBAFOSA), Makurdi, Nigeria

Ana Iglesias Department of Agricultural Economics, Universidad Politécnica de Madrid, Madrid, Spain

Oluwatosin E. Ilevbare National Centre for Technology Management, Federal Ministry of Science and Technology, Obafemi Awolowo University, Ile-Ife, Nigeria

Christopher O. Ilori Crop Protection and Environmental Biology, University of Ibadan, Ibadan, Nigeria

Masengo Francois Ilunga Department of Civil Engineering, University of South Africa, Pretoria, South Africa

Antonina Ivanova Universidad Autonoma de Baja California Sur, La Paz, Mexico

Abdulai Jalloh West and Central African Council for Agricultural Research and Development (CORAF), Dakar, Senegal

Adamu James National Weather Forecasting and Climate Research Centre, Nigerian Meteorological Agency, Abuja, Nigeria

S. Jari Department of Crop Production and Protection, Federal University Dutsin-Ma Katsina State, Katsina, Nigeria

Safieh Javadinejad Water Resource Management, University of Birmingham, Edgbaston, UK

A. Jebiwott Department of Environmental Science, Egerton University, Egerton, Kenya

Obert Jiri Agricultural Practice, Faculty of Agriculture, University of Zimbabwe, Harare, Zimbabwe

School of Agricultural, Earth and Environmental Sciences, University of KwaZulu-Natal, Pietermaritzburg, South Africa

Lara Johannsdottir Environment and Natural Resources, School of Business, University of Iceland, Reykjavík, Iceland

Tino Johansson Department of Geosciences and Geography, University of Helsinki, Helsinki, Finland

Eshwer Kale The WOTR-Centre for Resilience Studies (W-CReS), Pune, India

Antoine Kalinganiré World Agroforestry Centre, West and Central Africa Regional Office – Sahel Node, Bamako, Mali

A. M. Kane Egerton University, Njoro, Kenya

Institut d'Economie Rurale (IER), Bamako, Mali

Rebekah Karimi Enonkishu Conservancy, Narok County, Kenya

Joshua Kayode Department of Plant Science and Biotechnology, Ekiti State University, Ado Ekiti, Ekiti State, Nigeria

Hugh Kelly Permezone, Santa Barbara, CA, USA

Jagriti Kher Department of Development Communication and Extension, Institute of Home Economics, University of Delhi, New Delhi, India

Method Kilasara Department of Soil and Geological Sciences, Sokoine University of Agriculture, Morogoro, Tanzania

Stephen K. Kimani KALRO- Muguga South, Nairobi, Kenya

Caroline King-Okumu The Borders Institute (TBI), Nairobi, Kenya

GeoData Institute, Southampton, UK

Geography and Environment, University of Southampton, Southampton, UK

Isaac V. Kinhonhi Electricity Regulatory Authority (ERA), Kampala, Uganda

Monicah Kinuthia Ministry of Devolution and Planning, Nairobi, Kenya

Rukia A. Kitula Institute of Marine Sciences, University of Dar es Salaam, Zanzibar, Tanzania

L. Kleerekoper Amsterdam University of Applied Sciences, Amsterdam, Netherlands

J. Kluck Amsterdam University of Applied Sciences, Amsterdam, Netherlands

Dominic Kniveton Climate Science and Society (Geography), University of Sussex, Falmer, UK

Aristea Kounani Research Centre of Environmental Education and Communication, Department of Environment, University of the Aegean, Mytılene, Greece

Nitin Kumbhar Watershed Organisation Trust (WOTR), Pune, India

WOTR Centre for Resilience Studies (W-CReS), Pune, India

Catherine Ky-Dembele World Agroforestry Centre, West and Central Africa Regional Office – Sahel Node, Bamako, Mali

Stéphane La Branche PACTE – Grenoble Institute of Political Studies, Grenoble, France

Peter Läderach International Centre for Tropical Agriculture (CIAT), International Center for Tropical Agriculture (CIAT) – Asia Regional Office c/o Agricultural Genetics Institute, Hanoi, Vietnam

J. K. Lagat Egerton University, Njoro, Kenya

Makarius C. S. Lalika Department of Geography and Environmental Studies, Solomon Mahlangu College of Science and Education, Sokoine University of Agriculture, Morogoro, Tanzania

J. K. Langat Egerton University, Njoro, Kenya

Walter Leal Filho International Climate Change Information and Research Programme (ICCIRP), Faculty of Life Sciences, Hamburg University of Applied Sciences, Hamburg, Germany

Research and Transfer Centre Sustainable Development and Climate Change Management, Hamburg University of Applied Sciences, Hamburg, Germany

David K. Lelgut KALRO- Muguga South, Nairobi, Kenya

Chunlan Li School of Geographic Sciences, East China Normal University, Shanghai, China

Wu Liang Research Center for World Geography and Resources, Institute of Geographic Sciences and Natural Resources Research, Chinese Academy of Sciences, Beijing, China

Tran Thi Kim Lien Deutsche Gesellschaft für Internationale Zusammenarbeit (GIZ) GmbH, Ha Noi, Vietnam

Ondari Lilian University of Toledo, Toledo, OH, USA

Clayco Construction, Inc., Chicago, IL, USA

J. F. Lindahl International Livestock Research Institute, Uppsala University and Swedish University of Agricultural Sciences, Nairobi, Kenya

Crispino Lobo Watershed Organisation Trust (WOTR), Pune, India

R. Loeve Amsterdam University of Applied Sciences, Amsterdam, Netherlands

Hermann Lotze-Campen Potsdam Institute for Climate Impact Research (PIK), Potsdam, Germany

Department of Sustainable Land Use and Climate Change, Humboldt-Universität zu Berlin, Berlin, Germany

Kytt MacManus Center for International Earth Science Information Network (CIESIN), The Earth Institute, Columbia University, New York, NY, USA

Malgosia Madajewicz Center for Climate Systems Research, The Earth Institute, Columbia University, New York, NY, USA

Paramu L. Mafongoya School of Agricultural, Earth and Environmental Sciences, University of KwaZulu-Natal, Pietermaritzburg, South Africa

A. Q. M. Mahbub Department of Geography and Environment, Faculty of Earth and Environmental Sciences, University of Dhaka, Dhaka, Bangladesh

Athuman Juma Mahinda Department of Agricultural Engineering, University of Dar es Salaam, Dar es Salaam, Tanzania

Department of Land Resource Management and Agricultural Technology (LARMAT), University of Nairobi, Nairobi, Kenya

Environmental Science and Technology, Kyoto University, Kyoto, Japan

Tanzania Agricultural Research Institute (TARI)-Makutupora Centre, Dodoma, Tanzania

Riffat Mahmood Department of Geography and Environment, Faculty of Life and Earth Sciences, Jagannath University, Dhaka, Bangladesh

Daniel Mailumo Department of Government and Public Administration, College of Advanced and Professional Studies, Makurdi, Nigeria

Association of Entrepreneurs and Technology Managers of Nigeria (AETMAN), Eco-system Based Adaptation For Food Security Assembly (EBAFOSA), Makurdi, Nigeria

Pius Mailumo Association of Entrepreneurs and Technology Managers of Nigeria (AETMAN), Eco-system Based Adaptation For Food Security Assembly (EBAFOSA), Makurdi, Nigeria

University of Calabar, College of Education, Katsina-Ala, Nigeria

Immaculate Maina KALRO Headquarters, Nairobi, Kenya

S. M. Makindi Department of Environmental Science, Egerton University, Egerton, Kenya

Erick Maklago Research and Business Development, National Construction Authority, Nairobi, Kenya

Michael L. Mann Department of Geography, George Washington University, Washington, DC, USA

Kristin Marcell NYS Department of Environmental Conservation and Cornell University, New Paltz, NY, USA

Wilfred Mariki Selian Agriculture Research Institute, Arusha, Tanzania

Mary Masafu University of South Africa, Pretoria, South Africa

Sani Abubakar Mashi Nigerian Meteorological Agency (NiMet), Abuja, Nigeria

Sean Mayes Crops for the Future (CFF), Jalan Broga, Selangor Darul Ehsan, Malaysia

Kevin Mearns College of Agriculture and Environmental Sciences, University of South Africa (UNISA), UNISA Science Campus, Florida, South Africa

Matthias Meier Department of Socio-Economic Sciences, Research Institute of Organic Agriculture (FiBL), Frick, Switzerland

Neil R. Miller Canadian Foodgrains Bank, Arusha, Tanzania

Jane Mills Center for International Earth Science Information Network (CIESIN), The Earth Institute, Columbia University, New York, NY, USA

Greg William Misiaszek Institute of Education Theories, Beijing Normal University, Beijing, People's Republic of China

Sara Abdelhakim Mohammad College of Engineering, Qatar University, Doha, Qatar

Hassan I. Mohammed Department of Agricultural Engineering, College of Agricultural Studies, Sudan University of Science and Technology, Khartoum, Sudan

Mohammad Mousavi Department of Civil Engineering, Najafabad Branch, Islamic Azad University, Najafabad, Iran

Monicah Mucheru-Muna Kenyatta University, Nairobi, Kenya

Paschal Arsein Mugabe College of Engineering and Technology (COET), University of Dar es Salaam, Dar es Salaam, Tanzania

Charles Mugarura Broadfield Enterprises Uganda – Permaculture Group (Resilient African Network- MUK), Kampala, Uganda

Jayne Mugwe Kenyatta University, Nairobi, Kenya

Kingstone Mujeyi International Maize and Wheat Improvement Center (CIMMYT), Southern Africa Regional Office (SARO), Harare, Zimbabwe

Angeline Mujeyi College of Agricultural Engineering and Science, Discipline of Agricultural Economics, University of KwaZulu-Natal, Pietermaritzburg Campus, Scottville, South Africa

Adrian Muller Department of Socio-Economic Sciences, Research Institute of Organic Agriculture (FiBL), Frick, Switzerland

Bertrand Muller Centre d'Etude Régional pour l'Amélioration de l'Adaptation à la Sécheresse (CERAAS), BP, Thiès, Sénégal

Centre de Coopération Internationale en Recherche Agronomique pour le Développement (CIRAD), UMR AGAP/PAM, Montpellier, France

M. Mulumpwa Sengabay Fisheries Research Centre, Salima, Malawi

Nancy W. Mungai Department of Crops, Horticulture and Soils, Njoro, Kenya

M. Muntaka Department of Agricultural Extension and Rural Sociology, Federal University Dutsin-Ma, Katsina State, Nigeria

Fumiaki Murakami Rural and Agriculture Department, Nippon Koei Co., Ltd, Tokyo, Japan

Albanus Mutiso Enonkishu Conservancy, Narok County, Kenya

Fiona Mwaniki Climate Change Communication, Kilimo Media International, Nairobi, Kenya

Majaliwa Mwanjalolo Department of Geography, Geo Informatics and Climatic Sciences, College of Agriculture and Environmental Science, School of Forestry, Environmental and Geographical Science, Makerere University (Asso.), Kampala, Uganda

Caroline Mwongera International Centre for Tropical Agriculture (CIAT), Africa Regional Office, Nairobi, Kenya

Chris M. Mwungu International Centre for Tropical Agriculture (CIAT), Africa Regional Office, Nairobi, Kenya

R. T. Nabinta Department of Agricultural Extension and Rural Sociology, Federal University Dutsin-Ma, Katsina State, Nigeria

Herve Alain Napi Wouapi Department of Agricultural Extension and Rural Sociology, Dschang School of Agronomy and Environmental Sciences, University of Dschang (UDs), Dschang, Cameroon

Sarah Ndonye International Centre of Insect Physiology and Ecology, Nairobi, Kenya

Alice Neht Institute of Urban and Transport Planning, RWTH Aachen University, Aachen, Germany

Mohsen Nekooei Department of Water Engineering, College of Agriculture, Isfahan University of Technology, Isfahan, Iran

Sileshi Nemomissa Department of Plant Biology and Biodiversity Management, College of Natural and Computational Sciences, Addis Ababa University, Addis Ababa, Ethiopia

Humphrey M. Ngibuini Forestry Development Trust, Iringa, Tanzania

Rosemary N. Ngotho-Esilaba Veterinary Sciences Research Institute, Veterinary Research Centre, Kenya Agricultural and Livestock Research Organisation, Nairobi, Kenya

Gian L. Nicolay Department of International Cooperation, Research Institute of Organic Agriculture (FiBL), Frick, Switzerland

Pinghouinde Michel Nikiema Agence Nationale de la Météorologie du Burkina Faso, Ouagadougou, Burkina Faso

F. Njau Institute of Rural Development Planning (IRDP), Dodoma, Tanzania

Peterson N. M. Njeru KALRO- Muguga South, Nairobi, Kenya

Mwangi Peter Njiiri Kenya Wildlife Service, Nairobi, Kenya

Julie Noble City of Kingston, Kingston, NY, USA

Ijaz Rasool Noorka Department of Plant Breeding and Genetics College of Agriculture, University of Sargodha, Sargodha, Pakistan

Alison Nord Department of Plant, Soil and Microbial Sciences, Michigan State University, East Lansing, MI, USA

Chinedum Nwajiuba Alex Ekwueme Federal University Ndufu-Alike Ikwo (AE-FUNAI), Ikwo, Nigeria

Nkem J. Nwosu Department of Agronomy, University of Ibadan, Ibadan, Oyo State, Nigeria

George Nyarko Faculty of Agriculture, University for Development Studies, Tamale, Northern Region, Ghana

Martine Nyeko Faculty of Agriculture and Environment, Gulu University, Gulu, Uganda

Emmanuel C. Nzegbule Nigerian Environmental Study Action Team (NEST), Ibadan, Nigeria

Michael Okpara University of Agriculture Umudike, Umudike, Nigeria

Patrick J. O'Reilly Crops for the Future (CFF), Jalan Broga, Selangor Darul Ehsan, Malaysia

Nelson Ifeanyi Obine Centre for Sustainable Development, University of Ibadan, Ibadan, Nigeria

Kehinde Abraham Odelade Food Security and Safety Niche Area, Faculty of Natural and Agricultural Sciences, North-West University, Mmabatho, Mafikeng, South Africa

Idowu Kolawole Odubote School of Agricultural Sciences, Zambian Open University, Lusaka, Zambia

Samson Samuel Ogallah Solidaridad/Pan African Climate Justice Alliance (PACJA), Nairobi, Kenya

Chika Ogbonna Brandenburg University of Technology Cottbus – Senftenberg, Senftenberg, Germany

G. M. Ogendi Department of Environmental Science, Egerton University, Egerton, Kenya

Dryland Research Training and Ecotourism Centre, Chemeron, Egerton, Kenya

A. U. Ogogo Department of Forestry and Wildlife Resources Management, University of Calabar, Calabar, Cross River State, Nigeria

Olusegun Michael Ogundele Research Group in Health and Environment- Africa, (RGHE-AFRICA), Ibadan, Nigeria

International University Network on Cultural and Biological Diversity, Nigeria International Students' Coordination Unit, (IUNCBD N-ISCU), University of Ibadan, Ibadan, Nigeria

Oluwatoyin Ogundipe Department of Botany, University of Lagos, Akoka-Yaba, Lagos, Nigeria

Damilola Grace Ogunrotimi Department of Plant Science and Biotechnology, Faculty of Science, Ekiti State University, Ado Ekiti, Nigeria

Stephen M. Ojebisi Department of Geography, Obafemi Awolowo University, Ile-Ife, Nigeria

Olumuyiwa Idowu Ojo Department of Civil Engineering, University of South Africa, Pretoria, South Africa

Ladoke Akintola University of Technology, Ogbomoso, Nigeria

Joachim Chukwuma Okafor Department of Political Science, University of Nigeria, Nsukka, Enugu State, Nigeria

David Okali Nigerian Environmental Study Action Team (NEST), Ibadan, Nigeria

Paul B. Okon Department of Soil Science, University of Calabar, Calabar, Nigeria

Yasmein Okour Department of City Planning and Design, Jordan University of Science and Technology, Irbid, Jordan

O. Oladeji Amsol Bio Company, Lagos, Nigeria

O. Adeola Olajide Department of Agricultural Economics, Faculty of Agriculture and Forestry, University of Ibadan, Ibadan, Oyo State, Nigeria

Kehinde Oluwatosin Olajubu Department of Fisheries and Aquaculture Technology, FUTA, Akure, Nigeria

Samuel Adelowo Olakojo Institute of Agricultural Research and Training, Obafemi Awolowo University, Ibadan, Nigeria

Abdulafeez Olalekan Olaniyan Ladoke Akintola University of Technology, Ogbomoso, Nigeria

Olumide Olaniyan National Weather Forecasting and Climate Research Centre, Nigerian Meteorological Agency, Abuja, Nigeria

Olawale Emmanuel Olayide Centre for Sustainable Development, University of Ibadan, Ibadan, Nigeria

Sunday J. Olotu National Centre for Technology Management, Federal Ministry of Science and Technology, Obafemi Awolowo University, Ile-Ife, Nigeria

Christopher Oludhe Department of Meteorology and Institute of Climate Change and Adaptation (ICCA), University of Nairobi, Nairobi, Kenya

O. J. Oluwatosin Department of Fisheries and Aquaculture Technology, FUTA, Akure, Nigeria

Anne Nyatichi Omambia National Environment Management Authority, Nairobi, Kenya

J. N. Ombui Department of Public Health, Pharmacology and Toxicology, College of Agriculture and Veterinary Sciences, University of Nairobi, Nairobi, Kenya

Ademola Omojola Department of Geography, University of Lagos, Akoka-Yaba, Lagos, Nigeria

Gbenga Emmanuel Onibi Department of Animal Production and Health, FUTA, The Federal University of Technology, Akure, Nigeria

J. O. Onono Department of Public Health, Pharmacology and Toxicology, College of Agriculture and Veterinary Sciences, University of Nairobi, Nairobi, Kenya

Robert Onyeneke Alex Ekwueme Federal University Ndufu-Alike Ikwo (AE-FUNAI), Agriculture (Agricultural Economics and Extension Programme), Ikwo, Nigeria

Victor A. Orindi ADA, National Drought Management Authority (NDMA), Nairobi, Kenya

Isimemen Osemwegie International University of Grand Bassam, Grand-Bassam, Côte d'Ivoire

Kaveh Ostad-Ali-Askari Department of Civil Engineering, Isfahan (Khorasgan) Branch, Islamic Azad University, Isfahan, Iran

Christine N. A. Ouinsavi Doctoral School of Agronomic and Water Sciences/ Department of Sustainable Management of Natural Resources, Laboratory of Ecology and Forest Research (LERF), University of Parakou, Parakou, Borgou, Republic of Benin

Emmah Owidi International Centre of Insect Physiology and Ecology, Nairobi, Kenya

Oyediran O. Oyebola Department of Aquaculture and Fisheries Management, University of Ibadan, Ibadan, Nigeria

Obasanjo Joseph Oyedele Department of Mass Communication, Faculty of Social and Management Sciences, Bowen University Iwo, Osun State, Nigeria

Olawale Oyemade Oyekanmi Department of Teacher Education, University of Ibadan, Ibadan, Oyo State, Nigeria

Maureen Papas Faculty of Law, The University of Western Australia, Perth, WA, Australia

Nyeko Pen-Mogi Gulu University, Gulu, Uganda

Anuradha Phadtare The WOTR-Centre for Resilience Studies (W-CReS), Pune, India

Surendra Poonia ICAR-Central Arid Zone Research Institute, Jodhpur, Rajasthan, India

Adithya Pradyumna Watershed Organisation Trust (WOTR), Pune, India

Society for Community Health Awareness Research and Action (SOCHARA), Bangalore, India

Kathrin Prenger-Berninghoff Institute of Urban and Transport Planning, RWTH Aachen University, Aachen, Germany

Geeta Punhani Department of Development Communication and Extension, Institute of Home Economics, University of Delhi, New Delhi, India

Muhammad Abdur Rahaman Climate Change Adaptation, Mitigation, Experiment and Training (CAMET) Park, Noakhali, Bangladesh

Mohammad Mahbubur Rahman Network on Climate Change, Bangladesh (NCC,B) Trust, Dhaka, Bangladesh

Nicholas B. Rajkovich School of Architecture and Planning, State University of New York at Buffalo, Buffalo, NY, USA

Laxmi Ramasubramanian Department of Urban Policy and Planning and The Institute for Sustainable Cities, Hunter College, CUNY, New York, NY, USA

Eunice Pereira Ramos Unit of Energy Systems Analysis, KTH – Royal Institute of Technology, Stockholm, Sweden

A. S. Rao ICAR-Central Research Institute for Dryland Agriculture, Hyderabad, India

Lance Robinson International Livestock Research Institute, Nairobi, Kenya

Sajal Roy Institute for Culture and Society, Western Sydney University, Sydney, NSW, Australia

Department of Women and Gender Studies, Begum Rokeya University, Rangpur (BRUR), Bangladesh

Onkangi N. Ruth Research and Business Development, National Construction Authority, Nairobi, Kenya

Ruchi Sachan School of International Studies, Centre for African Studies, Jawaharlal Nehru University, New Delhi, India

Sakshi Saini Department of Development Communication and Extension, Institute of Home Economics, University of Delhi, New Delhi, India

A. Saleem Khan Center for International Earth Science Information Network (CIESIN), The Earth Institute, Columbia University, New York, NY, USA

Zakari Saley Bana Disaster Risks Reduction Program Officer (DRR/O), Catholic Agency For Overseas Development (CAFOD), Niamey, Niger

David Olabanjo Sanni Department of Geography, Obafemi Awolowo University, Ile Ife, Nigeria

Maruf Sanni National Centre for Technology Management, Federal Ministry of Science and Technology, Obafemi Awolowo University, Ile-Ife, Nigeria

Josias Sanou Institut de l'Environnement et de Recherche Agricole (INERA), Ouagadougou, Burkina Faso

Muhammad Saqib Institute of Soil and Environmental Sciences, University of Agriculture, Faisalabad, Pakistan

Sheena Shah Permaculture Research Institute Kenya, Nairobi, Kenya

Kelvin M. Shikuku International Centre for Tropical Agriculture (CIAT), Africa Regional Office, Nairobi, Kenya

Samson Shimelse Department of Plant Biology and Biodiversity Management, College of Natural and Computational Sciences, Addis Ababa University, Addis Ababa, Ethiopia

College of Dryland Agriculture and Natural Resources, Mekelle University, Mekelle, Ethiopia

Yogesh Shinde The WOTR-Centre for Resilience Studies (W-CReS), Pune, India

Arega Shumetie Department of Economics, College of Business and Economics, Haramaya University, Haramaya, Ethiopia

Constantina Skanavis Research Centre of Environmental Education and Communication, Department of Environment, University of the Aegean, Mytilene, Greece

Sef Slootweg Natural Resources Management Project – Ngorongoro District Council, Deutsche Gesellschaft für Internationale Zusammenarbeit (GIZ) GmbH, Loliondo, Tanzania

Sieglinde Snapp Department of Plant, Soil and Microbial Sciences, Michigan State University, East Lansing, MI, USA

Alabi Soneye Department of Geography, University of Lagos, Akoka-Yaba, Lagos, Nigeria

Vignesh Sridharan Unit of Energy Systems Analysis, KTH – Royal Institute of Technology, Stockholm, Sweden

Silke Stöber Center for Rural Development (SLE), Humboldt Universität zu Berlin, Berlin, Germany

Adeniyi Suleiman Gbadegesin Ladoke Akintola University of Technology, Ogbomoso, Nigeria

E. Y. Swai Agricultural Research Institute (ARI) Hombolo, Dodoma, Tanzania

Nava Tabak Scenic Hudson, Poughkeepsie, NY, USA

Abdel Aziz A. M. Taha Department of Agricultural Engineering, University of Gadarif, Gadarif, Sudan

Hosein Talebmorad Department of Water Engineering, College of Agriculture, Isfahan University of Technology, Isfahan, Iran

Constantinos Taliotis Unit of Energy Systems Analysis, KTH – Royal Institute of Technology, Stockholm, Sweden

Premsagar Tasgaonkar Watershed Organisation Trust (WOTR), Pune, India

Azage Tegegne International Livestock Research Institute, Addis Ababa, Ethiopia

B. Teme Institut d'Economie Rurale (IER), Bamako, Mali

Ludovic Temple Center for International Cooperation in Agricultural Research and Development (CIRAD), Montpellier, France

Molu Tepo Merti Integrated Development Programme (Mid-P), Nairobi, Kenya

Jérôme Ebagnerin Tondoh WASCAL and UFR des Sciences de la Nature, Nandjui Abrogoua University, Abidjan, Côte d'Ivoire

Rosemary Egodi Ubaekwe Research Group in Health and Environment- Africa, (RGHE-AFRICA), Ibadan, Nigeria

International University Network on Cultural and Biological Diversity, Nigeria International Students' Coordination Unit, (IUNCBD N-ISCU), University of Ibadan, Ibadan, Nigeria

Paul Akeh Ugbah National Weather Forecasting and Climate Research Centre, Nigerian Meteorological Agency, Abuja, Nigeria

Collins Ugochukwu Alberta Environmental and Parks, Government of Alberta, Edmonton, AB, Canada

Gloria Ujor Nigerian Environmental Study Action Team (NEST), Ibadan, Nigeria

Aliyu Tambuwal Umar Department of Geography, Usmanu Danfodiyo University, Sokoto, Nigeria

Philippe Vaast Center for International Cooperation in Agricultural Research and Development (CIRAD), Montpellier, France

Swapnil Vyas The WOTR-Centre for Resilience Studies (W-CReS), Pune, India

Hafiz Abdul Wahab Institute of Soil and Environmental Sciences, University of Agriculture, Faisalabad, Pakistan

James R. Wallace Group Communications and Corporate Responsibility, Allianz, Munich, Germany

Liam James Walsh Sydney, NSW, Australia

David Walunya Ong'are National Environment Management Authority, Nairobi, Kenya

Shem Wandiga Department of Chemistry, University of Nairobi, College of Biological and Physical Sciences, Nairobi, Kenya

James M. Warner International Food Policy Research Institute, Addis Ababa, Ethiopia

H. O. Wesonga Veterinary Sciences Research Institute, Veterinary Research Centre, Kenya Agricultural and Livestock Research Organisation, Nairobi, Kenya

Andreas Witte Institute of Urban and Transport Planning, RWTH Aachen University, Aachen, Germany

Muluneh Woldetisadik Addis Ababa University, Addis Ababa, Ethiopia

Lippa Wood Mara Training Centre, Narok County, Kenya

Omonlola Nadine Worou The International Crops Research Institute for the Semi-Arid Tropics (ICRISAT), Bamako, Mali

Dipak Zade The WOTR-Centre for Resilience Studies (W-CReS), Pune, India

Libby Zemaitis NYS Department of Environmental Conservation and Cornell University, New Paltz, NY, USA

Jelena Živanović Miljković Institute of Architecture and Urban and Spatial Planning of Serbia, Belgrade, Serbia

Part IV

Technological and Science-Based Approaches to Climate Resilience

Climate Change Resilience: Role of Insurers in Bridging the Gap Between Climate Change Science and Heterogeneous Stakeholders

106

Lara Johannsdottir and James R. Wallace

Contents

Introduction: Some Fact Gaps Between Climate Change Knowledge and Actions 2146
Climate Change and the Insurance Sector 2148
Identification of Knowledge-Action Gaps 2150
Scientific Knowledge and Policy Making 2151
The North and the South 2152
The Affluent and the Less Affluent 2153
Global and Local Knowledge 2154
Scientific Knowledge, Public Awareness, and Actions 2154
Findings: The Role of Insurers in Bridging the Knowledge-Action Gap 2156
Insurers and Policy Making 2157
Insurers and the North and the South 2158
Insurers and the Affluent and the Less Affluent 2159
Insurers and Global and Local Knowledge 2160
Insurers and Public Awareness 2161
Discussion and Conclusion 2163
Cross-References 2164
References 2164

Abstract

Climate change science, including the analysis of weather-related trends, is of importance to reduce risks threatening sustainable development. For a long time this has been acknowledged by the insurance sector. Through their expertise in

L. Johannsdottir (✉)
Environment and Natural Resources, School of Business, University of Iceland, Reykjavík, Iceland
e-mail: laraj@hi.is

J. R. Wallace
Group Communications and Corporate Responsibility, Allianz, Munich, Germany
e-mail: james.wallace@allianz.com

W. Leal Filho (ed.), *Handbook of Climate Change Resilience*,
https://doi.org/10.1007/978-3-319-93336-8_16

policgecy design, risk assessment, management, loss prevention, claims handling, and responsible investment, the insurance sector has the potential to bridge the gap between climate change science and stakeholder expectations. Therefore, their role is critical with regard to knowledge transfer and climate change resilience. This chapter investigates the role of insurers in bridging the gap between climate change science, manifested in international climate change targets and goals, and actors on different institutional levels in different locations around the world that need to respond to the climate targets and goals. Five different types of knowledge-action gaps which insurers may bridge are explored: (1) the gap between scientific knowledge and policy making, (2) between "North" and the "South," (3) between the rich and the poor, (4) between global, regional, and local knowledge, and (5) between scientific knowledge and public awareness. The role of insurers in bridging the gaps is explored in the context of climate change impacts on the insurance sector and the consequent actions. The paper draws on literature from multiple disciplines and by using practical examples from the insurance sector.

Keywords
Climate change · Insurer · Insurability · Knowledge transfer · Resilience · Stakeholder

Introduction: Some Fact Gaps Between Climate Change Knowledge and Actions

Nature and society are interwoven. Nature and ecosystems allow life to thrive, being the foundation for human well-being and economic development. However, mounting evidence from the Intergovernmental Panel on Climate Change (IPCC) suggest that nature is deteriorating, the climate is changing, and the globe is warming. It is "evident from observations of increases in global average air and ocean temperatures, widespread melting of snow and ice and rising global average sea level" (IPCC 2007b, p. 30), changed wind patterns, ocean currents, distribution of plants and animals, rainfall, etc. The main cause, now stated with 95% certainty, is human influence on the system, mainly caused by population and economic growth. Climate change influences are clear and growing, and impacts are observed across the globe (IPCC 2014). It is furthermore pointed out by IPCC (2014, p. v) that "that the more human activities disrupt the climate, the greater the risks of severe, pervasive and irreversible impacts for people and ecosystems, and long-lasting changes in all components of the climate system." It is furthermore stated in the report that there are ways to limit the risks of climate change, but this means that temperature increase needs to be stabilized below 2 °C, compared with preindustrial levels. This means that business as usual is not acceptable, and there is a need for departing from this pathway. The IPCC report (2014, p. 7) also emphasizes that "many extreme weather and climate events have been observed since about 1950." Complementary strategies of mitigation (reducing emissions) and adaptation (reducing exposure and vulnerability) can limit risks of climate change. What is also noted by the IPCC (2014) is that existing risks will be amplified, and new risks for human and natural systems will be created. Risks will not be distributed evenly, there are greater risks for communities and people who are disadvantaged, in countries at all development levels. There are, however, gaps between the knowledge of the scientific community and different actors on different institutional levels in different locations. While the knowledge of scientists is highly regarded, knowledge and expectations of the public are viewed unfavorably (PewResearchCenter 2009). Moreover, according to the United Nations Population Fund, there is also a gender imbalance, where "women and girls are disproportionately disadvantaged and less prepared or empowered to survive or recover" (UNFPA 2015, p. 5).

The economics of climate change were discussed in the Stern review. In the review, strong global actions were urged, as they will help avoid the most harmful impact of climate change (Stern 2006). The review (2006), furthermore, states that inaction will be more costly, with an annual loss of at least 5–20% of global GDP. Actions where the loss can be limited to 1% of global GDP annually still suggest serious impacts on development and growth. Extreme weather-related costs have been rising due to droughts, floods, and storms (Stern 2006). This is evident in upward trends in overall and insured losses registered for the period of 1980–2016 (Munich Re NatCatSERVICE 2016). The Stern review (2006) claims that the poorest populations and countries are most vulnerable, although all countries will be affected. They "will suffer earliest and most, even though they have contributed least to the causes of climate change" (Stern 2006, p. vii). The percentage of uninsured losses in the year 2016 shows a great variety between regions, as the percentage of uninsured losses was 99% in Africa, 89% in Asia, 82% in Mid- and South America, 70% in Europe, 51% in Australia, and 49% in North America (Munich Re NatCatSERVICE 2016). This suggests a great need for adaptation by reducing exposure and vulnerability in all continents.

However, there is a vast difference in GHG emissions in differing countries and regions. According to the US Environmental Protection Agency (EPA), the top emitters of carbon dioxide (CO_2) from fossil fuel combustion, gas flaring, and cement manufacturing were China (30%), followed by the United States (15%), the European Union (EU-28), (9%), India (7%), the Russian Federation (5%), Japan (4%), and other countries collectively (30%) (EPA 2017). The accumulative numbers from 1850 to 2002, however, tell a different story, but according to these numbers the United States is responsible for most of the emissions worldwide (29.3%), followed by the European Union, (EU-25), (26.5%), Russia (8.1%), China (7.6%), Germany (7.3%), and Japan (4.1%) (Baumert et al. 2005; Engelman 2009).

On a global scale climate change is seen as a key risk factor, which in addition to weather extremes can set off political volatility as a result of water and food scarcity (Dlugolecki 2008), with the potential for climate refugees. To facilitate action, scientific climate knowledge has to be transferred to political and practical levels, and different actors need to be mobilized. The need for action is evident in the Kyoto Protocol binding reduction emission targets. The first commitment period started in 2008 and ended in 2012, and the second commitment period started in 2013 ending in 2020 (UNFCCC n.d.-b). This was followed by another global effort to deal with climate emissions, the Paris Agreement, where the key aim is to keep "a global temperature rise this century well below 2 degrees Celsius above pre-industrial levels and to pursue efforts to limit the temperature increase even further to 1.5 degrees Celsius." Of 197 parties to the convention, 172 have ratified the agreement (UNFCCC n.d.-c). The urgency of actions to ensure environmental sustainability (goal 7) is also manifested in the United Nations Millennium Sustainable Development Goals (MDGs), targets to be reached by 2015 (United Nations n.d.-a), followed by the United Nations' Sustainable Development Goals (SDGs), targets to be reached by 2030, which specifically address climate actions (goal 13) (United Nations n.d.-b).

The aim of this chapter is to identify potential gaps between scientific knowledge about climate change to fulfil international climate change targets and goals. Gaps are also assessed and the role of insurers in bridging this context highlighted. Literature from various sources is reviewed, including academic, governmental, IGOs, and insurance. The following section outlines how climate change affects the insurance sector. Then five different gaps are identified, followed by a section explaining how the insurance sector can help bridge each of the gap the paper has identified. Finally there is a conclusion section.

Climate Change and the Insurance Sector

The insurance sector has been identified as one of the sectors highly exposed to climate change because of changing weather variability, with more frequent and severe weather events (Allianz Group and WWF 2006; Arent et al. 2014; Botzen et al. 2009; CEA 2009; Dlugolecki 2009; Lloyd's 2006; Mills 2009a; Munich Re Group 2008b; United Nations Environment Programme Finance Initiative (UNEP FI) 2007) causing more claims. Climate change jeopardizes different lines of insurance business, including life and health, crop, property and casualty, business interruption, liability, and investments (Epstein and Mills 2005; Mills 2009a; Mills et al. 2005). Furthermore, different global surveys (Ernst and Young and Oxford Analytica 2009; PriceWaterhouseCoopers 2001) identify climate change as a major strategic threat for the insurance sector, but the strategic risks that insurers need to address include the risk of physical damage to properties they cover, liability, litigation, regulatory and reputational risks to the business, as well as exposure to their assets and loan portfolios (Arent et al. 2014). To address increased risk exposure, insurers need a variety of climate prediction models (Dlugolecki 2009) and adequate procedures for managing climate risk and setting premiums accordingly (Mills et al. 2005). Insurers will be challenged to offer coverage for premiums that are still affordable while at the same time requiring more risk-based capital. The recent developments in the Taskforce on Climate-related Financial Disclosures (TCFD) highlight this point and the need for models to be applied for climate-related exposure to different types of insurers (TCFD n.d.). Adequate insurance coverage will be needed in low- and middle-income countries (Arent et al. 2014).

The response-options insurance companies have are to adjust premiums to the risk exposure and/or to change the policy conditions. This transforms the risk to those insured, which need to be encouraged to take adaptive measures to reduce their vulnerability. Factors in risk models may have to be amplified, risks diversified in different markets, and by basing prices on risk, it may foster adaptation in index-based or other novel solutions (Arent et al. 2014). Other possible factors relate to influencing other actors, including authorities, with regard to disaster risk management, land-use planning, and the quality of building codes (IPCC 2014).

It has been emphasized that various stakeholders expect businesses to address the climate change issue (Hoffman 2000), and for those failing to adapt to a changing business landscape there is additional risk and opportunities for those willing to act as leaders (Hoffman 2000, 2006). Climate change offers new opportunities to the insurance sector (Bacani 2008; Dlugolecki 2009; LeBlanc and Linkin 2010; Mills 2007a, 2009a; Munich Re Group 2008b; The Geneva Association 2009). Climate-related opportunities include crafting of new insurance products, building awareness and participating in public policy, disclosing carbon risk, promoting loss prevention, aligning terms and conditions with risk reducing behavior (Mills 2009a), and offering advisory services, risk management, and more. Investment is clean technology that is also considered to be an opportunity, as well as benchmarking of best practice between insurers (Bosse et al. 2009; Johannsdottir et al. 2014; Jóhannsdóttir et al. 2012) and across industries.

To emphasize the importance of the insurance sector in tackling climate issues, the IPCC TAR report included a separate chapter (▶ Sustainability of Climate Change Adaptation Measures in Rivers State, South-South, Nigeria) on insurance and financial services (Vellinga et al. 2001). For short the reports are called FAR (First Assessment Report), SAR (Second Assessment Report), TAR (Third Assessment Report), AR4 (Fourth Assessment Report), and AR5 (Fifth Assessment Report). In the IPCC report, the importance of the insurance sector is based on its potential to serve as an indicator for socioeconomic impacts of climate change. Furthermore, it is the role of the insurance sector to spread risk of extreme weather events between different stakeholders, throughout societies and across borders and time, thereby influencing macroeconomic outcome of climate change (Epstein and Mills 2005). Insurance risk has for a long time been based on historical trends, but currently future scenarios must be used to assess the risk (Meadows 2009). If higher premiums are to be justified, insurers must educate their clients about the scale of risk which will be a great deal larger than previously observed (Meadows 2009). Potential losses might include property (Munich Re Group 2008a), lives, health, business interruptions, restrictions on trade, travel and tourism, and potential market instability, and small-scale events are rarely accounted for (Epstein and Mills 2005). However, in this context insurers must bear in mind not to price themselves out of the market (Jóhannsdóttir 2012).

For decades reinsurers such as Swiss Re and Munich Re have been active in climate change research and debate, but more recently the primary insurance sector has acknowledged its role in this respect (Jóhannsdóttir et al. 2012; PriceWaterhouseCoopers 2001). Based on a literature review the insurance sector is regarded as a leading sector with regard to climate change adaptation (Nitkin et al. 2009). Additionally the insurance industry is considered to be a "key stakeholder in climate change issues and that it can and should play an active role in helping to ensure that climate change is managed decisively" (Dlugolecki 2009, Chap. 2, p. 14; The Geneva Association 2009). Risk sharing and transferring mechanisms the insurance sector offer can enhance resilience to climate extremes at local, national, regional, and local levels (IPCC 2012). Action taken by insurance companies to integrate mitigation and adaption to climate change into core businesses is expected to play an important role in reducing insurance losses (Mills 2005), thereby mitigating the risk of insurers, clients, and communities.

The insurance sector is one of the largest industries in the world. Insurance spending, defined as the ratio of direct gross premiums to GDP, between 2000 and 2015 ranged between 8.6% and 8.6%, with peaks in 2007 and 2009 (9.3%) (Epstein and Mills 2005; OECD n.d.). Various environmental and climate initiatives focusing on the importance of the insurance sector witness an increasing understanding and interest in leveraging the size and connectivity of the insurance sector in mobilizing necessary changes on global, regional, national, and local levels. Among the initiatives are the United Nations Environmental Programme Financial Initiative, the Chartered Insurance Institute, Geneva Association, Munich Climate Insurance Initiative (MCII), Swiss Re Climate Change Initiatives, Lloyd's of London 360 Risk Project, ClimateWise, and others. Insurers have also been identified, on regional and local levels, as a key sector for addressing climate change, by taking part in the climate debatem, lobbying for more effective climate policies and by collaborating with scientists, suppliers, NGOs, and other critical actors. The public-private debate is essential because if governments fail to act on the climate issues, sizable markets become uninsurable in the near future (Dlugolecki 2009; Mbogo 2009). Passiveness threatens the insurance industry as well as the wider economy (Mills 2007b). The association of Nordic insurers (Danish Insurance Association, Federation of Finnish Financial Services, Norwegian Financial Services Association, and Swedish Insurance Association), for example, is taking an active role in attaining political and industry-wide attention by sponsoring conference among industry leaders, specialists, and politicians. Nonetheless, individual insurers (private and federal) are responding differently, but the scale of actions suggested range across inactive, reactive, proactive, developed, and integrated (Johannsdottir 2014; The Geneva Association 2009). According to the Geneva Association (2009), the majority of insurers are still considered to be reactive, and in the Nordic region the smaller insurance companies are seen to be inactive/reactive (Johannsdottir 2014, 2015).

Identification of Knowledge-Action Gaps

In this chapter the knowledge-action gap (or the gap) is defined as the difference between the accumulative scientific knowledge about climate change, published in the IPCC reports and manifested in international climate change targets and goals, and the knowledge of those that need to be mobilized in order to prevent the negative climate change consequences from happening. The actors are policy makers, governments, organizations, individuals, IGOs, NGOs, organizations, and others in the position to act on the scientific knowledge. The different levels discussed in the paper are the global, regional, national, and local levels. The locations are the regions and countries around the world. The knowledge-action gaps discussed in the following sections are (1) the gap between scientific knowledge and policy making, (2) the gap between the North and the South, (3) the gap between the more affluent and the less affluent, (4) the gap between the global and the local knowledge, and (5) the gap between scientific knowledge and public awareness. This section derives from the literature, bringing forth some critical questions. These questions will be explored in the following section using practical examples.

Scientific Knowledge and Policy Making

The scientific community and policy makers, on different institutional levels, have until recently mainly been focusing on technical and financial aspects of climate change and the mitigation efforts of moving toward low-carbon communities (Hoffmann and Busch 2008). This is evident by the establishment of the United Nations Framework Convention on Climate Change (UNFCCC), a treaty with near-worldwide membership of 194 nations. The main objective is to "stabilise greenhouse gas concentrations in the atmosphere at a level that will prevent dangerous human interference with the climate system" (UNFCCC n.d.-a). Linked to the UNFCCC is the Kyoto Protocol and the Paris Agreement discussed in the introduction. The MDGs and SDGs also have put forth environmental sustainability and climate change goals. Although binding agreements have been reached and targets and goals set, fulfillment of the Kyoto Protocol and the MDGs has not been satisfactory with regard to tackling the issue of climate change. The SDGs are increasingly visible (more so than the MDGs) and being taken into account by more corporations. Already companies are attempting to highlight how their sustainability activities relate to the SDGs and their core business. This is being partly driven by responsible investors, like the Dow Jones Sustainability Index, whose questionnaires are looking for how companies are approaching the SDGs and even how much of the revenue is aligned. Policy makers involved in insurance are often disconnected from the climate discourse, but weather-/catastrophe-related issues remain a high priority. The indirect nature of climate versus the direct exposure of weather means that weather is easier concept to evaluate in the context of an annual insurance policy or by the regulators being able to check how exposed an insurer is. Insurance regulators are often not involved in these discourses which is why the Sustainable Insurance Forum is a unique collaboration between regulators from different countries on different environmental and social issues often cutting out central government regulators who are not always successful in translating these international or national policy issues into relevance for industry sectors (Sustainable Insurance Forum n.d.). This knowledge-action gap will be addressed in the paper by exploring the question, "How can insurers help bridge the gap between scientific knowledge and climate change policy making?"

The North and the South

Wealth enhances the capacity of societies and individuals to adapt to climate change. Climate resilience is defined as "[t]he ability of a social or ecological system to absorb disturbances while retaining the same basic structure and ways of functioning, the capacity for self-organisation, and the capacity to adapt to stress and change" (UNFCCC n.d.-a). The UNFCCC makes a distinction between "Annex I" (developed countries and economies in transition) and "non-Annex I" countries (developing countries). This corresponds to the international political literature of the North (high-income countries) and the South (low-income countries) (Harris 2001). Furthermore, the Kyoto Protocol makes a distinction between GHG reduction of the industrialized countries and developing countries (United Nations 1998). The IPCC Fourth Assessment Report (AR4) predicts that regional-scale impacts will be more compelling than the previous report (TAR) expected, thereby reinforcing the North-South debate. The developing regions furthermore state that there are insufficient financial resources for implementing adaptation measures (Hewitson et al. 2014). The poorest regions will suffer the most from a wide range of climate-linked issues, e.g., from water stress, floods, droughts, mortality by reason of diarrhea diseases, and loss of biodiversity and natural resources (IPCC 2007a). This has been an obstacle in climate change negotiations, because the South wants the North to take actions before they start sharing the burden which the North is responsible for (Harris 2001). This can be classified as environmental justice (Athanasiou and Baer 2002; Norgaard 2006) or justice dilemma where the developing world needs support in order to adapt to consequences of historical emissions of the North. Additionally, it has been stated that "[t]he climate change regime firmly establishes the duty of developed countries to assist developing countries by financing, technology transfer and insurance" (Paavola and Adger 2006, p. 599).

The concept of North and South is also one which applies to other geographical variations and therefore perhaps linked to the extent to which a place is developed in economic terms. This would then highlight regions within the North that are less developed or even within the same country or community. Solutions are available which help to tackle climate change at different geographic scales. There are, however, challenges in expanding different types of insurance business around the world, but these challenges are more complicated in the South (low-mid-income countries) than in the North. These include regulatory constraints, lack of re(insurance) expertise and capacity, high cost of distribution and claims settlement, limited or lack of technical expertise and financial infrastructure, data, risk modeling tools and risk management culture, and more (The Geneva Association n.d.). The question to explore in the context of this knowledge-action gap is therefore "How can insurers help bridge the gap between the North and the South in the context of climate change?"

The Affluent and the Less Affluent

In the debate between the North and the South, climate change can be seen as a social issue in the developed countries (Wilbanks et al. 2007) negatively affecting some more than others, e.g., people in poverty. Inequality has an impact on local adaptive capacity and the ability to cope with the situation, but this shifts challenges regarding adaptation and risk management from the local or private sector to national or international levels (IPCC 2012). Regional differences within Europe will increasingly see inland or coastal flooding, increased erosion, melting glaciers, less snow, and consequently less winter tourism. Parts of Europe will also suffer from loss of species, water availability, and crop yield (Wilbanks et al. 2007). Furthermore, heat waves and wildfires will threaten lives. According to the IPCC variability among regions in North America is also evident, where the coastal areas will suffer from climate change impacts and interrelations of development and coincident pollution. Other impacts expected are heat waves, decreased snowpack, winter flooding, and aggregated crop yields in some areas (Wilbanks et al. 2007).

Insurance premiums are rising as a result of increased overall risk (Epstein and Mills 2005; Mills et al. 2005; Wilbanks et al. 2007) due to weather extremes. People living in high-risk areas prone to flood risk and hurricanes (Botzen and van den Bergh 2008) are already affected. Premiums for businesses operating in those areas are also rising. In many cases low-income homes and smaller firms will not be able to afford to buy insurance coverage due to rising of premiums and will have to rely on their social network if possible (Wilbanks et al. 2007). Therefore adapting by the means of buying insurance coverage for weather impacts will be disproportional, and "[p]oor communities can be especially vulnerable, in particular those concentrated in high-risk areas" (IPCC 2007a, p. 12). Insurers have already started to cancel insurance policies in risk-prone areas or where their risk appetite has changed, such as the US Gulf Coast area and in Florida (Epstein and Mills 2005; Mills 2007b). Other consequences are new exclusions in insurance terms, reduced limits on risks covered, and increased deductibles (Mills 2007b; Mills et al. 2005). These issues mean that those with the least chance of coping financially will have to bear more of the risk themselves or rely on state support. As a consequence, local and regional economies will be adversely affected, including homes and businesses, unless governments offer new ways of coping with the risk (Wilbanks et al. 2007). In the case of reduced precipitation, water stress might negatively affect individuals and companies who have few alternatives ways of getting new water supplies (Wilbanks et al. 2007). Future losses are expected to be immense (Mills 2007b), which will then widen the gap between the rich and the poor (Wilbanks et al. 2007) within Western societies. Multi-stakeholder approaches involving regulators are needed to help address some of these wealth disparity issues, and insurance can play a role. This leads to the question, "How can insurers help bridge the gap between the affluent and the less affluent by means of offering climate insurance solutions?"

Global and Local Knowledge

Climate change impacts on a global scale have been studied more thoroughly, than on regional, national, and local levels. Predictions are being adjusted (Harris 2001), and climate extremes can now be more accurately be simulated (Hayhoe et al. 2008; Racherla et al. 2012). Results from climate models predicting global impacts are not adequate for adaptation efforts on national or local levels. An example of data that is lacking or inadequate relates to climate and crop models (Whitfield 2013) where global, regional, or national scales are often used (Thornton et al. 2010), as well as data on disasters and how disaster risks are reduced, but this can "constrain improvements in local vulnerability reduction" (IPCC 2012, p. 10). Even though the resolution of climate models is improving, as well as the confidence in the results, the ambiguity about smaller-scale impacts still exists, and it makes it difficult for policy makers to act accordingly (Harris 2001), and scientific uncertainty may also be used as an justification for inaction (Maslin 2013). In addition to scientific climate modeling, a balanced approach between top-down decisions of policy makers and bottom-up approaches, based on local know-how and experience, is needed (Blanco 2006; Engelman 2009). Furthermore, the vast amount of global scientific knowledge available has to be presented in the way that locals in vulnerable areas understand in order to come up with appropriate solutions for local communities (Jasanoff 1997). Additionally, the science community has to respond to local experience, practices, and know-how. But, scientists have a propensity to discount such knowledge, and the inputs perish when developing policies (Engelman 2009; Jasanoff 1997) on different levels, for example, the knowledge of indigenous peoples has been under-represented in IPCC assessment reports (Ford et al. 2016).

This gap between global and local levels is also evident in the context of sustainability, where Agenda 21 had to be brought down to local levels through Local Agenda 21 (UNEP n.d.), as the Agenda 21 did not provide sufficient guidance on smaller scales. Moreover, until rather recently, the main research focus has been on a mitigation, while relatively small body, but a growing body, of research focuses on climate change adaptation (National Academy of Sciences 2010) which often is more relevant on smaller scales, particularly in vulnerable communities. While many insurers are aware of the medium- and longer-term risk of climate change and the potential impacts on customers, it remains difficult to communicate or price this in appropriately for customers. The short-term nature of insurance is a barrier to this, and the relevance of the priced risk to the local consumer is also a barrier. An easier strategic step is perhaps with the role of insurers as asset managers where the steering of policies away from fossil fuels or industries which in turn impact customers is more controllable in some respects and already occurring. In this context the next research question is "How can insurers help bridge the gap between global and local climate knowledge?"

Scientific Knowledge, Public Awareness, and Actions

Research has shown that scientific knowledge about climate change is not enough for the public to act; there are also psychological and sociological factors which inhibit the public to accept the threat and act according to their national climate policies (Sandvik 2008). Therefore the importance of social science literature (social, ethical, cultural, behavioral, organizational, etc.) on climate change is growing in importance, as well as importance of psychological studies on the same topic. Additionally measurement of GDP wealth correlates to a skepticism about global warming in countries contributing the most to climate change (Sandvik 2008). In many cases the reason is believed to be *lack of information* and *lack of concern* (Norgaard 2006, p. 355), but in other cases the willingness relates to the "inconvenient truth" (Gore 2006) or different types of denials surface. Those denials are *socially organized denial*, where the message is ignored; *literal denial*, where there is a denial due to counterclaims of stakeholders such as oil producers; and *interpretive denial,* where facts are not denied but given dissimilar meaning (Norgaard 2006). Adding to this could be a *political skepticism and denial*, as evident in current US administrations and President Trump decision regarding the withdrawal of the Unite States from the Paris Agreement on climate change mitigation (Jameton 2014). This means that the climate debate is a part of the political debate in the United States where democrats are more concern about climate change than republicans (Oppenheimer and Todorov 2006).

Research have shown that people experiencing in person climate-related events such as extreme weathers (Krosnick et al. 2006) and higher average temperatures (Lorenzoni and Pidgeon in Oppenheimer and Todorov 2006) are more concerned about global warming, thus according to Hersh and Viscusi and Leiserowitz (in Oppenheimer and Todorov 2006) more willing to support climate policies such as energy taxes. Other scholars point out that as the consequences of climate change become more evident and the social pressure on the corporate world intensifies (Hill 2010; Hoffman 2000), "stakeholders increasingly require companies to disclose their strategies for addressing climate change" (Hoffmann and Busch 2008, p. 506). This is now becoming evident, for instance, with regulatory requirements on nonfinancial reporting requiring large companies to publish regular reports where the social and environmental impacts of their activities are disclosed in line with TCFD (European Commission n.d.) and in the recent report by the European Commission from the High Level Expert Group on Sustainable Finance noting that insurers look at their long-term climate risk exposure and not just annual pricing (European Commission 2018).

Climate risk ranks high as a business risk, for the insurance sector in various management consultant and insurance surveys. On the impact scale it ranks as high as the credit crunch, and above risks such as deepening of the recession and regulation and compliance (Ernst and Young and Oxford Analytica 2009), but in some cases it is listed after global financial instability, emerging markets, growth of corporate liability, and political risk (Dlugolecki 2009). In some cases responding to social pressure due to climate change, it is seen as moral obligation, (Hill 2010) while in other cases it is seen more strategically as a way of exploiting new opportunities (Hoffman 2000). The latest annual risk report from the World Economic Forum shows the top risks in terms of impact and likelihood to be extreme weather, natural disasters, and failure of climate mitigation and adaptation. Water and biodiversity loss are also in the top-ranked risks (World Economic Forum 2018).

With respect to climate change and public awareness, the last research question is "How can insurers help bridge the gap between climate scientific knowledge and public awareness?"

Findings: The Role of Insurers in Bridging the Knowledge-Action Gap

Owing to climate risks and opportunities, the insurance sector needs to take an active role in solving the climate issues (Dlugolecki 2009), by means of absorbing, pricing, and spreading climate risk across policy-holders, other insurers, and reinsurers. The industry is in a position to make recommendations, thus helping societies to mitigate and adapt to the new risks (Epstein and Mills 2005). In addition to the expert knowledge, the size and connectivity of the insurance sector are of vital importance. Insurers are realizing that there is a connection between their financial risk and operations of their corporate clients which may be regarded environmentally unsafe, which then translates as risks if not addressed in underwriting (Hoffman 2000). The insurance sector is however only one of the stakeholders in tackling climate risk, so solutions will have to be developed in "conjunction with governments, individuals and civil society organizations" (United Nations Environment Programme Finance Initiative (UNEP FI) n.d.). Other potential partners in tackling weather risks are national and local governments, foreign governments and various UN treaties and mechanism, NGOs and private donors, and individuals and organizations that are "self-insured" (Epstein and Mills 2005), which emphasize the importance of collaboration and sharing of best practices. In general insurers act as bridge between the insured and claimants (see Fig. 1). Insurance contracts usually collect periodical payments of premium in exchange for protection against sudden unexpected events with the potential of drastic losses. Claimants receive payments based on "pay-out mechanism for everybody contributing to the [insurance] scheme" (The Geneva Association 2009, p. 10). In this respect it is the insurance clients that insure each other. This method reduces the likelihood of economic losses and potential bankruptcies. Furthermore, it allows the customers to take on higher risks, with higher rates of return than they would otherwise have done, thus reinforcing the economic development.

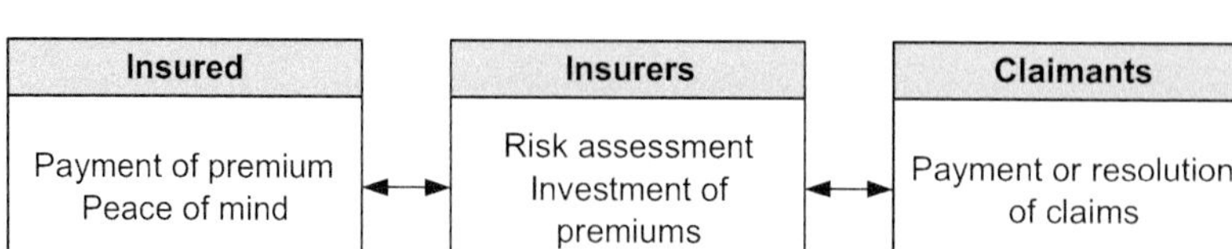

Fig. 1 The bridge between insured and claimants

Insurers and Policy Making

As pointed out previously, policy makers have been able to reach climate change agreements and develop the SDGs. This has not yet resulted in a satisfactory outcome for tackling climate change. Figure 2 explains how insurers can bridge the gap between the science community and policy makers by means of collaborating with, or putting pressure on, policy makers and/or local governments. The Geneva Association, an international insurance think tank, states that it is important to use public-private partnerships to raise "awareness for socio-economic benefits of preventative and risk transfer measures, broadening the use of modern risk modeling tools in the public sector, and building resilience of critical infrastructure and urban-systems (including mega cities)" (The Geneva Association n.d.). Furthermore, the Geneva Association (n.d.) suggests that in case of managing risks of climate and extreme events, the role of government would be to (1) provide enabling environments, including policies and regulatory frameworks, (2) establish institutional foundations, (3) ensure systematic collection of relevant socioeconomic and hazard data, (4) seek risk reduction and risk transfer opportunities, including preparedness and preventive measures, and (5) invest in risk awareness and education. The role of in insurance sector, according to the Geneva Association (n.d.), would be (1) to share their expertise in risk knowledge and risk pricing; (2) share their own knowledge about loss prevention and research; (3) innovation with regard to building programs and products enhancing financial protections for various stakeholders, including individuals, business, communities, and governments, e.g., by using incentives to change behavior; (4) settle claims faster and more efficiently; and (5) take part in public-private partnership projects where the aim is to develop risk transfer programs. An integrated approach, including government and insurance, would include risk assessment, risk reduction, financial protection, and building of resilience by means of recovery and reconstruction (The Geneva Association n.d.). One example of a public-private partnership project is a joint publication of the Geneva Association and the Insurance Development Forum, which is between the United Nations, the World Bank, and the insurance sector. This publication is titled Guidelines

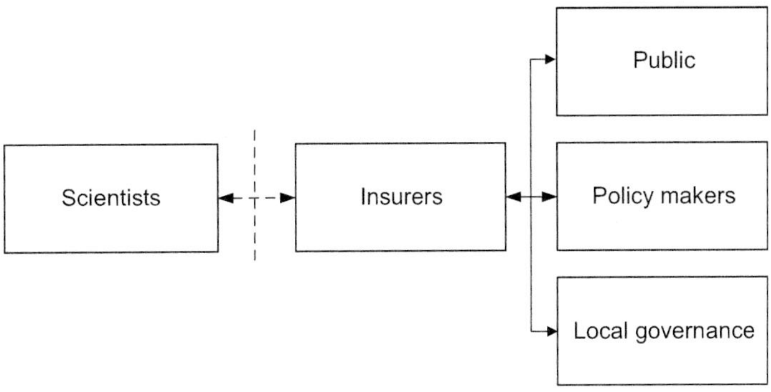

Fig. 2 The bridge between science and policy making

for Risk Assessment to Support Sovereign Risk Financing and Risk Transfer (Golnaraghi et al. 2017). The purpose of the publication was to provide a risk assessment framework for governments to support developing programs on sovereign risk finance and risk transfer and to offer checklist for designing such programs. Scientific knowledge about climate change does not only flow directly through insurers (see dotted line in Fig. 2) to their stakeholders, as scientists talk directly too, but outcome of climate change models and research has an input into various insurers' initiatives and actions, and this understanding is transformed into a message communicated to various groups of insurers-stakeholders.

In cases where insurers start withdrawing from markets, raising premiums and deductibles, the burden shifts automatically over to the government, which "becomes insurers of last resort" (Mills et al. 2005, p. 2). In many cases, government responses establish a national catastrophic insurance fund, to prepare for disasters and convene recovery mechanism (Mills et al. 2005). Insurers, particularly in Europe, taking a strategic view on climate change as an opportunity, proactively try to collaborate or engage governments to initiate necessary adaptation measures (Mills et al. 2001). Claims statistics, forecasts of climate scenarios, and expertise in risk management make insurance knowledge valuable in fields such as loss prevention, land-use planning, building regulations, energy, transport, resilience of infrastructure, and health-care provision (Krebs et al. 2010). Due to size, bargaining power of insurers also gives them weight in negotiating governmental spending on adaptation measures such as in flood defenses. In cases where insurers are not willing to collaborate on solutions and abandon markets, their bargaining power fades out (The Geneva Association 2009). Investment in climate friendly technologies for low-carbon economies (Johannsdottir et al. 2014) also helps mitigate future vulnerabilities. Therefore, reinforcement of public-private partnership in assessing vulnerability and adapt to the situation is critical for insurers as well as for policy makers and local authorities as they have common interest in maintaining sustainable development.

Insurers and the North and the South

In the South, the focus has to be on the immediate need to prevent deaths when developing insurance products. In the North, the focus is more on environmental issues that can put economical (Swiss Re 2010) and ecological stability at risk. In the North, the sophisticated insurance market covers 40% of total costs, while this is 4% in the South (Epstein and Mills 2005). The justice dilemma demands that governments and other stakeholder worldwide help the South to cope with the situation (ClimateWise et al. 2010). This is evident by the Cancun COP 16 establishment of a special climate change funds to administer assistance for the least-developed countries (Banda, December 11).

The insurance sector is starting to accept this role and is already offering capacity for understanding, managing, and spreading risks in developing countries (Mills 2007c), although coverage for natural hazards is still quite limited (Arnold 2008).

However, changes are gradually happening, and in September 2010, the MCII, ClimateWise, the Geneva Association, and UNEP FI, whose collective membership includes more than hundreds of leading insurers across six continents, launched a Global Insurance Industry Statement on “Adaptation to Climate Change in Developing Countries” (ClimateWise et al. 2010). The statement suggests that the insurance industry can support adaptation efforts in developing countries through expertise in risk management, prioritizing adaptation measures, by incentivizing loss reduction, by developing new insurance products, and by raising awareness among the many stakeholders of the insurance industry. While governments and international climate negotiators look at insurers as a part of the solution in solving the climate issue, business environment which enables insurers to operate effectively must also be established (ClimateWise et al. 2010).

Insurance is already providing risk management solutions in developing countries, such as the Caribbean Catastrophe Risk Insurance Facility (CCRIF), the Horn of Africa Risk Transfer for Adaptation (HARITA), micro-insurance coverage in Mongolia, weather insurance in Malawi, and index-based insurance in Bolivia (ClimateWise et al. 2010). The mechanism of microfinance has been offered in the developing world for around 20 years. Very small amount of capital is lend to people for investments in local projects, which strengthens their livelihood and creates more job opportunities. More recently this mechanism has expanded to insurance products called micro-insurance, in countries where the insurance markets are underdeveloped. So far this development is reinforced by European insurers (Mills 2009b), supported by the United Nations’ Millennium Development Goals (The Geneva Association 2009). Index-based insurance is another type of mechanism which uses measurable events such as average crop yields or revenue, cumulative rainfall, or temperature or flood levels to proxy losses. Still those insurance solutions have limitations; therefore sharing of best practice on how to develop them further is of vital importance (Fig. 3).

Insurers and the Affluent and the Less Affluent

Inequality of income is not only an issue in the South or developing countries. Even within developed countries, there are people in poverty who do not have the benefit of being insured. This can be an issue for individuals, small businesses, or even government where they would benefit from innovative disaster risk transfer schemes.

Focusing on developed countries, the wealth inequality is increasing. This is therefore creating a growing group of potential customers who cannot be insured by conventional products. Therefore, there are increasing opportunities to look for larger groups of these customers which may generate less premium, but offering the product via a third party or in bulk can make the costs neutral or to generate a profit.

In the United Kingdom, the lack of home content insurance coverage received a lot of attention after the 2007 floods and consequent damages (Palmer n.d.). While

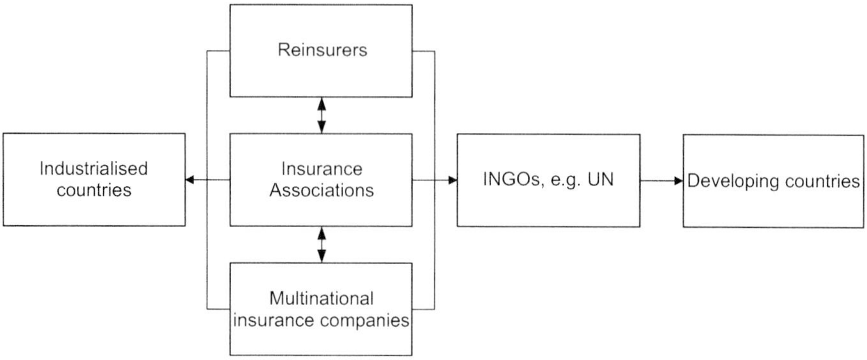

Fig. 3 The bridge between the North and the South

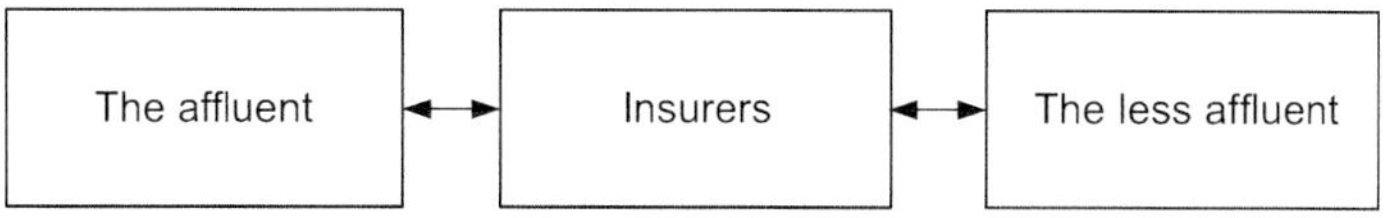

Fig. 4 The bridge between the affluent and the less affluent

less than 10% of the richest fifth (the affluent) of the population were without home content insurance, it was around 50% for the poorest fifth (less affluent). However, financial losses will affect those households most as they have less financial means to cope with losses (Palmer n.d.). For less affluent households, insurance is comparatively expensive as a proportion of their incomes compared to the risk of not being insured (Brainard n.d.). Climate hazards will also widen the gap between the affluent and the less affluent and hinder economic development (Leary et al. 2007). Studies have shown that there is a social value in offering of micro-insurers, which combines philanthropic and commercial elements, to low-income homes and entrepreneurs (Brainard n.d.). For those reasons, insurers should focus on solutions that suit different types of customers and the local situation (see Fig. 4), which helps to support economic development in the North and the South.

Social housing insurance is an example of solution offered by the RSA to households without home content insurance. Weekly payments of premiums are as low as £1.50 per week paid via rent insurance schemes in cash so there is no need for a bank account. Further development of the product is taking place, e.g., by offering the product through a wider distribution net such as retailers and credit unions (RSA 2010).

Insurers and Global and Local Knowledge

The most accurate climate predictions are on global and continental scales. When scaling down and simulating on small geographical scales, the models become less accurate. This also brings about implications for insurers as impacts may vary from region to region and country to country (Epstein and Mills 2005), therefore influencing climate change risks of insurers and other stakeholders along their value chains.

The relevance of these models also varies by type of insurance. At the overall exposure to natural catastrophe, these large-scale models are arguably more relevant. When considering the individual customer, such as insurance for their home, the assessment of the annual risk of climate change is different from likelihood of exposure of extreme weather impact for the duration of their policy. Likewise, some other types of insurance may have no or little relevance to climate change, e.g., pet insurance, cyber insurance, etc. Certain types of insurance are more exposed like property or energy business as highlighted by the Lloyd's Stranded Asset report (2017).

As indicated by the High Level Expert Group (HLEG 2018) report to the European Commission, insurers are encouraged to look beyond the annual pricing of risk, to longer-term consideration of investments and strategy. This seems to suggest that there is limited materiality or ability to pass costs onto the consumer within the annual policy cycle and implicitly stating that climate change needs to be considered more amid to long-term strategic issue for insurers due to the nature of the way insurance contracts work.

By nature the insurance sector is global as reinsurers spread their risk across large geographical areas; however majority of insurance companies operate on regional and local levels (see Fig. 5). Where data gaps exist, insurers on all these levels offer a way to fine-tune scientific models and quality of the data with claims statistics via collaboration with climate modelers (Epstein and Mills 2005). This type of collaboration is also vital for the insurers as it will enhance their ability to respond to extreme weather events and rising number of smaller-scale events (Mills et al. 2005).

Insurers and Public Awareness

It is essential to use scientific knowledge to raise public awareness about climate change and the need for mitigation and adaptation. Insurers are among the actors that can help in bridging this gap (see Fig. 6). Insurance companies can, for instance, play an instrumental role in the shift toward the circular economy and enhancing the resilience of societies. This is vital for the insurance sector, because if individuals and societies are vulnerable, so are insurers and reinsurers which insure them (Epstein and Mills 2005).

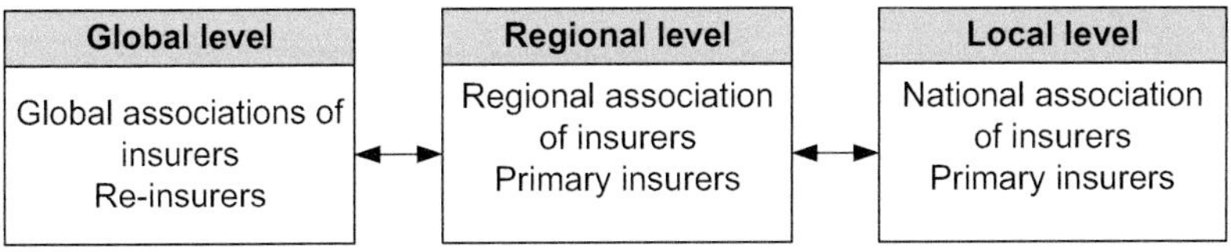

Fig. 5 The bridge between the global and the local climate knowledge

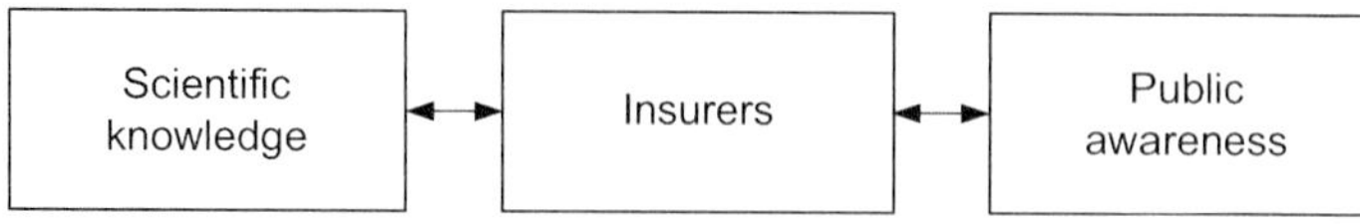

Fig. 6 The bridge between scientific knowledge and public awareness

Proactive insurance companies have accepted the role of raising climate awareness and to offer new solutions based on scientific knowledge about climate change, therefore bridging the gap by influencing their clients' responsible behavior and preventing losses from happening. Insurers which still are inactive or reactive are lagging behind their clients and mainstream science and need to catch up fast (The Geneva Association 2009) because outside pressure from shareholders, customers, and regulators is escalating (Ernst and Young and Oxford Analytica 2009; Mills 2009b; PriceWaterhouseCoopers 2001).

The Unfriend Coal campaign is an example of NGO awareness increasing pressure on the insurance industry, if not necessarily public consumer driven. The campaign which initially focused on encouraging insurers to divest assets in coal energy generating and mining companies has shifted in focus toward insurers. This is generating internal discussions and momentum internally in the largest companies, but the vast majority of the industry and brokers are largely unaffected. As they are generally not dealing with the public (B2C) and only with other corporations (B2B), there is little reputation exposure or downside to them initially. Part of the challenge is that often coal generation is part of the portfolio of companies active in many different energy sources, e.g., renewables, nuclear, etc. The dynamic of working through brokers complicates the ability to minimize or exclude the coal part of the business. Arguments are also increasingly raised on how can the insurance industry support the transition to a low-carbon economy rather than just excluding an energy source which is likely to play a major part of our economies until 2050 (Unfriend Coal 2017).

As explained in section "Scientific Knowledge, Public Awareness, and Actions," there are different types of barriers that must be overcome. Insurers can signal climate risk through premiums, insurance terms, and conditions (Mills 2009b). This does not necessarily mean raising prices; understanding the risk profile and with the customer taking steps to lower their risk can lower premiums, e.g., responding to an environmental questionnaire (RSA 2010) or motor policy based on distance traveled (The Brookings Institution 2008). Offering higher premiums in concentrated high-risk areas under development might discourage development and thus the risks insurers undertake. Furthermore, guidance for behavior is also valuable, for example, on how to build or rebuild energy-saving or climate resilience buildings. Insurers can lead by example (Bosse et al. 2009; The Geneva Association 2009), campaign for stronger emission limits, use less energy or green energy themselves, develop early warning systems, exchange best practices, and so forth. This expert knowledge can then be shared with their suppliers and collaborators, as well as with their corporate clients.

Discussion and Conclusion

The aim of this chapter was to identify potential gaps between scientific knowledge about climate change, urgently expressed in the need for fulfilling international climate change targets and goals, and those that need to act upon the knowledge, targets, and goals and to point out the role of insurers in bridging this context.

To help identify this gap and the role of insurers in bridging the gap, five specific research questions were proposed:

1. How can insurers help bridge the gap between *scientific knowledge and climate change policy making*?
2. How can insurers help bridge the gap between the *North and the South* in the context of climate change?
3. How can insurers help bridge the gap between the *affluent and the less affluent* by means of offering climate insurance solutions?
4. How can insurers help bridge the gap between *global and local climate knowledge*?
5. How can insurers help bridge the gap between climate *scientific knowledge and public awareness*?

This chapter demonstrates that the insurance sector has a role to play on bridging the gap between climate change science and heterogeneous actors, both when it comes to mitigating GHG emissions and just as importantly in adapting to climate change consequences because even though GHG emissions were reduced substantially, the climate will continue to change for some time to come (National Academy of Sciences 2010).

So far, on a scientific and political level, mitigation has received far more attention than adaptation (Dlugolecki 2009), where the insurers' expert knowledge on risk management is of vital importance. Additionally climate science has been focusing more on the technical side of climate change, while the human dimension must increasingly be taken into account. In this respect the size and connectivity of insurers are an advantage.

Not all risk can be insured by private insurers, which means that public insurance systems must be available in such cases (Epstein and Mills 2005) or some sort of public-private solutions as societies become more vulnerable and the gap between economic and insured losses increases.

Barriers on regulatory, institutional, and behavioral levels must vanish, and collaboration and dialogue between different actors on different levels are fundamental as well as sharing of information. The limitation of this chapter is that it only provides an overview of how insurers can help bridge several different climate gaps. However, each gap must be explored more thoroughly, and there might be other gaps which need to be explored as well.

The paper helps to define the gap between what the insurance industry is actually able to do and the high-level policy aspirations. Initial indications are given toward the materiality of climate change to different lines of business, sector exposure, and product types. This underlines the fact that for some stakeholders in the insurance industry, climate change is very relevant. For other and certain types of insurance, there is no material impact. At present, the industry is viewed as a single consistent provider of the same products and services. Further work would be beneficial in identification of which insurance industry stakeholders, lines of underwriting, and product types are most relevant to climate change. This would then help to identify partners in future multi-stakeholder initiatives to tackle climate adaptation to a greater extent.

Cross-References

▶ Balancing Adaptation and Mitigation in the Building Sector of New York State
▶ Building Resilience of Urban Ecosystems and Communities to Sea-Level Rise: Jamaica Bay, New York City
▶ Climate Change Resilience: Role of Insurers in Bridging the Gap Between Climate Change Science and Heterogeneous Stakeholders
▶ Climate Resilience in African Coastal Areas: Scaling up Institutional Capabilities in the Niger Delta Region
▶ Climate Finance as a Catalyst for Initiating an African Green Revolution to Enhance Drought Resilience: Policy Prospects and Challenges
▶ Domestic Water and Disaster Management Legislation: A Key Tool for the Implementation of the Paris Agreement in LDCs
▶ Financing Projects for Improving Climate Change Resilience: The Cases of Djibouti and Yemen
▶ Isolation as a Barrier for the Climate Change Actions of Insurers
▶ Planning and Climate Change in Serbia: Integrated Responses to the Causes and Impacts
▶ Resettlement and Relocation Options for Coastal Communities
▶ Risk Perception and Action to Reduce the Impact of Floods in the Czech Republic
▶ Vulnerability and Adaptation Levels of the Construction Industry in Kenya to Climate Change
▶ Working Together to Build Climate Resilience in Hudson Riverfront Communities

References

Allianz Group and WWF (2006) Climate change and insurance: an agenda for action in the United States. Retrieved from http://www.knowledge.firemansfund.com/pdf/studies_climatechange_insurance.pdf

Arent DJ, Tol RSJ, Faust E, Hella JP, Kumar S, Strzepek KM, … Yan D (2014) Key economic sectors and services – supplementary material. In: Climate Change 2014: Impacts, Adaptation, and Vulnerability. Part A: Global and Sectoral Aspects. Contribution of Working Group II to the Fifth Assessment Report of the Intergovernmental Panel on Climate Change [Field, C.B.,V.R. Barros, D.J. Dokken, K.J. Mach, M.D. Mastrandrea, T.E. Bilir, M. Chatterjee, K.L. Ebi, Y.O.

Estrada, R.C. Genova, B. Girma, E.S. Kissel, A.N. Levy, S. MacCracken, P.R. Mastrandrea, and L.L.White (eds.)]. Retrived from Cambridge and New York. Available from www.ipcc-wg2.gov/AR5 and www.ipcc.ch

Arnold M (2008) The role of risk transfer and insurance in disaster risk reduction and climate change adaptation. In: Policy brief for the commission on climate change and development. Commission on Climate Change and Development, Stockholm

Athanasiou T, Baer P (2002) Dead heat: global justice and global warming. Seven Stories Press, New York

Bacani B (2008) Insuring climate change and sustainability. Unleasing innovation and intelligent partnerships. UNEP FI. Retrieved from http://www.unepfi.org/fileadmin/documents/bacaniarticle.pdf

Banda S (December 11) (2010) Cancun agrees on 'climate deal', funds for developing countries. Retrieved from http://www.ibtimes.com/articles/91217/20101211/cancun-mexico-climate-change-deal-developing-countries-united-nations-global-warming-copenhagen-mexi.htm

Baumert KA, Herzog T, Pershing J (2005) Navigating the numbers: greenhouse gas data and international climate policy. World Resources Institute, Washington, DC

Blanco AVR (2006) Local initiatives and adaptation to climate change. Disasters 30(1):140–147

Bosse S, Rasmussen PB, Lindenius C, Bjerke R, Skauge A, Kauppi P-N (2009) Nordic insurance companies meet the challenge of climate change: Statement from the Nordic Insurance Industry adopted at the Nordic Insurers' Climate Conference in Copenhagen 17th September 2009. Retrieved from http://www.nicc.dk/news/Documents/Pan-Nordic%20Statement%20on%20climate%20change.pdf

Botzen WJW, van den Bergh JCJM (2008) Insurance against climate change and flooding in the Netherlands: present, future, and comparison with other countries. Risk Anal 28 (2):413–426

Botzen WJW, van den Bergh JCJM, Bouwer LM (2009) Climate change and increased risk for the insurance sector: a global perspective and an assessment for the Netherlands. Nat Hazards 2010(52):577–598

Brainard L (n.d.) What is the role of insurance in economic development? Retrieved from Zurich. http://zdownload.zurich.com/main/reports/What_is_the_role_of_economic_developement.pdf

CEA (2009) Tackling climate change: the vital contribution of insurers. Retrieved from Brussels. www.cea.eu

ClimateWise, The Geneva Association, Munich Climate Insurance Initiative (MCII), & United Nations Environment Programme Finance Initiative (UNEP FI) (2010) Global insurance industry statement on adapting to climate change in developing countries. Retrieved from http://www.unepfi.org/fileadmin/documents/insurance_climatechange_statement.pdf

Dlugolecki A (2008) Climate change and the insurance sector. Geneva Pap Risk Insur-Issues Pract 33(1):71–90. https://doi.org/10.1057/palgrave.gpp.2510152

Dlugolecki A (ed) (2009) Coping with climate change: risks and opportunities for insurers. The Chartered Insurance Institute (CII), London

Engelman R (2009) Facing a changing world: women, population and climate. Retrieved from New York. http://www.unfpa.org/swp/2009/en/pdf/EN_SOWP09.pdf

EPA (2017) Global greenhouse gas emissions data. Retrieved from https://www.epa.gov/ghgemissions/global-greenhouse-gas-emissions-data

Epstein PR, Mills E (eds) (2005) Climate change futures: health, ecological and economic dimensions. Harvard Medical School, Boston

Ernst & Young, & Oxford Analytica (2009) The 2009 Ernst & Young business risk report: the top 10 risks for global business. Retrieved from London: http://www.ey.com/IS/en/SearchResults?query=the+2009+ernst+%26+young+business+risk+report&search_options=country_name

European Commission (2018) Financing a sustainable European economy. Retrieved from Brussels

European Commission (n.d.) Non-financial reporting: EU rules require large companies to publish regular reports on the social and environmental impacts of their activities. Retrieved from https://ec.europa.eu/info/business-economy-euro/company-reporting-and-auditing/company-reporting/non-financial-reporting_en

Ford JD, Cameron L, Rubis J, Maillet M, Nakashima D, Willox AC, Pearce T (2016) Including indigenous knowledge and experience in IPCC assessment reports. Nat Clim Change 6:349–353. https://doi.org/10.1038/nclimate2954

Golnaraghi M, Branagan I, Fraser S, Gascoigne J, Gordon AM (2017) Guidelines for risk assessment to support sovereign risk financing and risk transfer: a joint publication of the Geneva Association and the Insurance Development Forum (IDF). Retrieved from Zurich. https://www.genevaassociation.org/sites/default/files/research-topics-document-type/pdf_public//ga2017_eecr_guidelines_for_risk_assessment.pdf

Gore A (Writer) (2006) An inconvenient truth: a global warning. Paramount Classics and Participant Production, Beverly Hills, Lawrence Bender Productions Inc is located in Los Angeles, CA

Harris PG (2001) Assessing climate change: international co-operation and predictions of environmental change. Politics 21(1):11–22

Hayhoe K, Wake C, Anderson B, Liang XZ, Maurer E, Zhu J, ... Wuebbles D (2008) Regional climate change projections for the Northeast USA. Mitig Adapt Strateg Glob Chang 13(5–6):425–436. https://doi.org/10.1007/s11027-007-9133-2

Hewitson B, Janetos AC, Carter TR, Giorgi F, Jones RG, Kwon W-T, ... van Aalst M (2014) Regional context. In: Barros VR, Field CB, Dokken DJ, Mastrandrea MD, Mach KJ, Bilir TE, Chatterjee M, Ebi KL, Estrada YO, Genova RC, Girma B, Kissel ES, Levy AN, MacCracken S, Mastrandrea PR, White LL (ed) Climate change 2014: impacts, adaptation, and vulnerability. Part B: regional aspects. Contribution of Working Group II to the Fifth Assessment Report of the Intergovernmental Panel on Climate Change. Cambridge University Press, Cambridge/New York, pp 1133–1197

Hill S (2010) Climate change: primer. Retrieved from http://www.nbs.net/wp-content/uploads/Climate-Change-Primer_noline.pdf

High-Level Expert Group on Sustainable Finance (HLEG) (2018) Financing a sustainable European economy: Final report 2018. https://ec.europa.eu/info/sites/info/files/180131-sustainable-finance-final-report_en.pdf

Hoffman AJ (2000) Competitive environmental strategy: a guide to the changing business landscape. Island Press, Washington, DC

Hoffman AJ (2006) Getting ahead of the curve: corporate strategies that address climate change. Retrieved from Arlington. http://www.pewclimate.org/docUploads/PEW_CorpStrategies.pdf

Hoffmann VH, Busch T (2008) Corporate carbon performance indicators: carbon intensity, dependency, exposure, and risk. J Ind Ecol 12(4):505–520

http://www.nap.edu/catalog/12783.html

IPCC (2007a) Summary for policymakers. In: Parry ML, Canziani OF, Palutikof JP, van der Linden PJ, Hanson CE (eds) Climate change 2007: impacts, adaptation and vulnerability. Contribution of Working Group II to the Fourth Assessment Report of the Intergovernmental Panel on Climate Change. Cambridge University Press, Cambridge

IPCC (2007b) Synthesis report. Contribution of Working Groups I, II and III to the Fourth Assessment Report of the Intergovernmental Panel on Climate Change. In: Core Writing Team, Pachauri RK, Reisinger A (eds) IPCC Fourth Assessment Report (AR4). Participant Media (or Participant Production) is locaed in Beverly Hills, CA p 104

IPCC (2012) Managing the risks of extreme events and disasters to advance climate change adaptation. In: Field CB, Barros V, Stocker TF, Dahe Q, Dokken DJ, Ebi KL, Mastrandrea MD, Mach KJ, Plattner G-K, Allen SK, Tignor M, Midgley PM (eds) A special report of Working Groups I and II of the Intergovernmental Panel on Climate Change. Cambridge University Press, Cambridge/New York

IPCC (2014) Climate change 2014: synthesis report. Contribution of Working Groups I, II and III to the Fifth Assessment Report of the Intergovernmental Panel on Climate Change. Retrieved from Geneva

Jameton A (2014) The Importance of physician climate advocacy in the face of political denial. AMA J Ethics 19(12):1222–1237. https://doi.org/10.1001/journalofethics.2017.19.12.sect1-1712

Jasanoff S (1997) NGOs and the environment: from knowledge to action. Third World Q 18(3):579–594

Jóhannsdóttir L (2012) Nordic non-life insurer's interest in, and response to, environmental issues. PhD thesis. Faculty of Business, Reykjavík, pp 562

Johannsdottir L (2014) The Geneva Association framework for climate change actions of insurers: a case study of Nordic insurers. J Clean Prod 75:20–30. https://doi.org/10.1016/j.jclepro.2014.03.074

Johannsdottir L (2015) Drives of proactive environmental actions of small, medium and large Nordic non-life insurance companies–and insurers as a driving force of actions. J Clean Prod. Available online 29 June 2015. https://doi.org/10.1016/j.jclepro.2015.06.055

Jóhannsdóttir L, Wallace J, Jones A (2012) The primary insurance industry's role in managing climate change risks and opportunities. In: Stoner JAF, Wankel C (eds) Managing climate change business risks and consequenses: leadership for global sustainability. Palgrave Macmillan, New York

Johannsdottir L, Davidsdottir B, Goodsite ME, Olafsson S (2014) Insurers' role in enhancing development and utilization of environmentally sound technologies: a case study of Nordic insurers. J Clean Prod 65:526–538. https://doi.org/10.1016/j.jclepro.2013.09.043

Krebs J, Dlugolecki A, Fankhauser S, Hall J, Johnson A, Palmer T, ... Young B (2010) How well prepared is the UK for climate change? Retrieved from London. http://downloads.theccc.org.uk.s3.amazonaws.com/ASC/CCC_ASC_Report_web_1.pdf

Krosnick JA, Holbrook AL, Lowe L, Visser PS (2006) The origins and consequences of democratic citizens' policy agendas: a study of popular concern about global warming. Clim Chang 77:7–43

Leary N, Adejuwon J, Barros V, Batimaa P, Biagini B, Burton I, ... Wehbe M (2007) A stitch in time: lessons for Climate Change Adaptation from the AIACC Project. AIACC working paper no. 48. Assessments of Impacts and Adaptations to Climate Change (AIACC). Washington, DC. Retrieved from http://www.aiaccproject.org/working_papers/Working%20Papers/AIACC_WP48_Leary_etal.pdf

LeBlanc A, Linkin M (2010) Chapter 6: Insurance industry. Ann N Y Acad Sci 1196(1):113–126

Lloyd's (2006) Climate change adapt or bust. Retrieved from London. http://www.lloyds.com/NR/rdonlyres/38782611-5ED3-4FDC-85A4-5DEAA88A2DA0/0/FINAL360climatechangereport.pdf

Maslin M (2013) Cascading uncertainty in climate change models and its implications for policy. Geogr J 179(3):264–271. https://doi.org/10.1111/j.1475-4959.2012.00494.x

Mbogo S (2009) Natural disaster damage may be uninsurable due to climate. Business Day. Retrieved from http://www.businessday.co.za/

Meadows D (2009) Insurance in a world without growth. Paper presented at the NORIS™, Helsinki

Mills E (2005) Insurance in a climate of change. Science 309:1040–1044

Mills E (2007a) From risk to opportunity: 2007 insurer responses to climate change. Retrieved from Boston. http://www.knowledge.firemansfund.com/pdf/risk-to-opportunity-2007.pdf

Mills E (2007b) Responding to climate change – the insurance industry perspective. In: C Action (ed) Climate action. Sustainable Development International in partnership with United Nations Environmental Programme, London, UK pp 100–103

Mills E (2007c) Synergisms between climate change mitigation and adaptation: an insurance perspective. Mitig Adapt Strateg Glob Chang 12(5):809–842. https://doi.org/10.1007/s11027-007-9101-x

Mills E (2009a) From risk to opportunity: 2008 – insurer responses to climate change. Retrieved from Boston: http://insurance.lbl.gov/opportunities/risk-to-opportunity-2008.pdf

Mills E (2009b) A global review of insurance industry responses to climate change. Geneva Pap Risk Insur 34(3):323–359

Mills E, Lecomte E, Peara A (2001) U.S. insurance industry perspectives on global climate change. University of California, Lawrence Berkeley National Laboratory. Retrieved from Berkeley.

Mills E, Roth RJ, Lecomte E (2005) Availability and affordability of insurance under climate change: a growing challenge for the U.S. Retrieved from Boston: http://www.pewclimate.org/docUploads/Ceres%20%2D%2D%20Insurance%20&%20Climate%20Change%202005.pdf

Munich Re Group (2008a) Catastrophe figures for 2008 confirm that climate agreement is urgently needed. Retrieved from http://www.munichre.com/en/press/press_releases/2008/2008_12_29_press_release.aspx

Munich Re Group (2008b) Opportunities from climate change: innovative insurance solutions for renewable energies. Retrieved from Munich: http://www.munichre.com/en/service/search.aspx

Munich Re NatCatSERVICE (2016) Natural disasters: the year in figures. Retrieved from https://www.munichre.com/topics-online/en/2017/topics-geo/overview-natural-catastrophe-2016

National Academy of Sciences (2010) Adapting to the impacts of climate change: report in brief. Retrieved from Washington, DC: http://dels.nas.edu/resources/static-assets/materials-based-on-reports/reports-in-brief/Adapting_Report_Brief_final.pdf

Nitkin D, Foster R, Medalye J (2009) Business adaptation to climate change. Retrieved from London, Ontario. http://www.nbs.net/News.aspx?iid=0914450b-3f1a-4d38-bbd8-171a82a07065&aid=433f22bf-3275-4326-9d57-7436e5d94521

Norgaard KM (2006) “We don’t really want to know”: environmental justice and socially organized denial of global warming in Norway. Organ Environ 19(347):347–370. https://doi.org/10.1177/1086026606292571

OECD (n.d.) Insurance spending: total, % of GDP, 2000–2015. Retrieved from https://data.oecd.org/insurance/insurance-spending.htm

Oppenheimer M, Todorov A (2006) Global warming: the psychology of long term risk. Clim Chang 77(1–2):1–6

Paavola J, Adger WN (2006) Fair adaptation to climate change. Ecol Econ 56(4):594–609

Palmer G (n.d.) Without home contents insurance. Retrieved from http://www.poverty.org.uk/74/index.shtml

PewResearchCenter (2009) Public praises science; scientists fault public, media: scientific achievements less prominent than a decade ago. Retrieved from http://pewresearch.org/pubs/1276/science-survey

PriceWaterhouseCoopers (2001) The role of Australia’s financial sector in sustainability: a report prepared for Environment Australia. Retrieved from http://www.environment.gov.au/settlements/industry/publications/pubs/role-fin-sector.pdf

Racherla PN, Shindell DT, Faluvegi GS (2012) The added value to global model projections of climate change by dynamical downscaling: a case study over the continental U.S. using the GISS-ModelE2 and WRF models. J Geophys Res-Atmos 117(D20). https://doi.org/10.1029/2012JD018091

RSA (2010) Corporate responsible report 2009. Retrieved from London: http://corporateresponsibility.rsagroup.com/rsacr/dlibrary/documents/RSACRReport.pdf

Sandvik H (2008) Public concern over global warming correlates negatively with national wealth. Clim Chang 90(3):333–341

Stern N (2006) The economics of climate change – the Stern review. Cambridge University Press, Cambridge

Sustainable Insurance Forum (n.d.) Sustainable Insurance Forum: members. Retrieved from http://unepinquiry.org/sif/members/

Swiss Re (2010) Natural catastrophes and man-made disasters in 2009: catastrophes claim fewer victims, insured losses fall. Sigma 1:1–37

TCFD (n.d.) Task force on climate-related financial disclosures. Retrieved from https://www.fsb-tcfd.org/

The Brookings Institution (2008) Pay-as-you-drive auto insurance: a simple way to reduce driving-related harms and increase equity, Policy brief no. 2008-06. The Brookings Institution, Washington, DC

The Geneva Association (2009) The insurance industry and climate change – contribution to the global debate. Retrieved from Geneva. http://www.genevaassociation.org/PDF/Geneva_Reports/Geneva_report[2].pdf

The Geneva Association (n.d.) Building socio-economic resilience and insurance protection with a concerted public-private approach. Retrieved from https://www.genevaassociation.org/sites/default/files/research-topics-document-type/pdf_public/20160909_ecoben4_final_1.pdf

Thornton PK, Jones PG, Alagarswamy G, Andresen J, Herreroa M (2010) Adapting to climate change: agricultural system and household impacts in East Africa. Agric Syst J 103(2):73–82. https://doi.org/10.1016/j.agsy.2009.09.003

UNEP (n.d.) Local authorities' initiatives in support of Agenda 21. Retrieved from http://www.unep.org/documents.multilingual/default.asp?DocumentID=52&ArticleID=76&l=en

UNFCCC (n.d.-a) First steps to a safer future: introducing the United Nations Framework Convention on Climate Change. Retrieved from http://unfccc.int/essential_background/convention/items/6036.php

UNFCCC (n.d.-b) Kyoto protocol. Retrieved from http://unfccc.int/kyoto_protocol/items/2830.php

UNFCCC (n.d.-c) The Paris agreement. Retrieved from http://unfccc.int/paris_agreement/items/9485.php

UNFPA (2015) Shelter from the storm: A transformative agenda for women and girls in a crisis-prone world. State of world population 2015. Retrieved from New York. https://www.unfpa.org/swop-2015

Unfriend Coal (2017) Climate campaigners found a new way to stop coal. Retrieved from https://unfriendcoal.com/

United Nations (1998) Kyoto protocol to the United Nations framework convention on climate change. Retrieved from http://unfccc.int/resource/docs/convkp/kpeng.pdf

United Nations (n.d.-a) Millennium development goals. Retrieved from http://www.un.org/millenniumgoals/

United Nations (n.d.-b) Sustainable development goals. Retrieved from https://sustainabledevelopment.un.org/sdgs

United Nations Environment Programme Finance Initiative (UNEP FI) (2007) Insuring for sustainability. Why and how the leaders are doing it. Retrieved from Paris. http://www.unepfi.org/fileadmin/documents/insuring_for_sustainability.pdf

United Nations Environment Programme Finance Initiative (UNEP FI) (n.d.) Statement of environmental commitment by the insurance industry. Retrieved from http://www.unepfi.org/statements/ii/index.html

Vellinga P, Mills E, Berz G, Bowers L, Huq S, Kozak LA, ... Soler G (2001) Insurance and other financial services. Climate Change 2001: Impacts, Vulnerability, and Adaptation. Working Group II, chapter 8. Retrieved from Cambridge

Whitfield S (2013) Uncertainty, ignorance and ambiguity in crop modelling for African agricultural adaptation. Clim Chang 120:325–340. https://doi.org/10.1007/s10584-013-0795-3

Wilbanks TJ, Lankao PR, Bao M, Berkhout F, Cairncross S, Ceron J-P, ... Zapata-Marti R (2007) Industry, settlement and society. In: Parry ML, Canziani OF, Palutikof JP, van der Linden PJ, Hanson CE (eds) Climate change 2007: impacts, adaptation and vulnerability. Contribution of Working Group II to the Fourth Assessment Report of the Intergovernmental Panel on Climate Change. Cambridge University Press, Cambridge, pp 357–390

World Economic Forum (2018) The global risks report 2018: 13th edition. Retrieved from Geneva, World Economic Forum. http://www3.weforum.org/docs/WEF_GRR18_Report.pdf.

Climate-Proof Retrofitting of Urban Areas for the Same Cost

107

L. Kleerekoper, R. Loeve, and J. Kluck

Contents

Introduction 2172
Methodology 2174
  Choosing an Extreme Rainfall Event 2174
  Cost 2175
  Benefits 2176
  Neighborhood Typologies 2176
Climate-Resilient Retrofitting 2179
  High-Rise Post-war Garden City (Flat) 2179
  Urban City Block (Flat) 2184
  More Examples of Case Studies in Flat Areas 2189
  Examples of Case Studies in Sloping Areas 2189
Discussion and Conclusion 2191
Summary 2194
References 2194

Abstract

Due to climate change the frequency of extreme precipitation increases. To reduce the risk of damage by flooding, municipalities will need to retrofit urban areas in a climate-resilient way. To justify this investment, they need insight in possibilities and costs of climate-resilient urban street designs. This chapter focused on how to retrofit characteristic (Dutch) typologies of urban residential areas. For ten cases alternative street layouts were designed with a determination of the life cycle costs and benefits. All designs are resilient to extreme rain events. The results show that most flat urban typologies can easily be retrofitted in a climate-resilient way without additional costs compared to the standard way of retrofitting. Climate proofing sloping areas are highly dependent on the situation downstream.

L. Kleerekoper (✉) · R. Loeve · J. Kluck
Amsterdam University of Applied Sciences, Amsterdam, Netherlands
e-mail: l.kleerekoper@hva.nl; r.loeve@hva.nl; j.kluck@hva.nl

W. Leal Filho (ed.), *Handbook of Climate Change Resilience*,
https://doi.org/10.1007/978-3-319-93336-8_20

When there is no space downstream to divert the water into waterways or parks, costs to provide storage easily rise above traditional levels for retrofitting. In addition to reducing flood risk, for each case one variant includes resilience to extreme heat events making use of green. The life cycle costs and benefits of the green variants showed that especially green designs in high-density urban areas result in a better value for money.

Keywords
Extreme rainfall events · Climate adaptation · Urban retrofitting · Urban typologies · Cost and benefits

Introduction

Our climate is changing. The world is gradually warming up, and more extreme weather events, from rainfall to drought and heat, are expected. The impact on the world as we know it will be tremendous. Sea level rise can have devastating effects on our coastal habitats, where over half a billion urban residents live and population growth is expected to be high (Merkens et al. 2016; Neumann et al. 2015) affecting even more people in the future. Extremes of drought, resulting in desertification or crop loss, and flooding will become more common, causing displacement and conflict. Heat and heat waves will impact human health, biodiversity, and ecosystems. No matter what you are passionate about, something you value will be affected by climate change.

Our urban areas will be affected by climate change as well. Today, 54% of the world's population lives in urban areas, a proportion that is expected to increase to 66% by 2050 (UN 2014). "Managing urban areas has become one of the most important development challenges of the 21st century," according to John Wilmoth, Director of UN DESA's Population Division. Urban areas will mainly be affected in three ways by climate change: (i) change in rainfall patterns, (ii) increase in (occurrence of) extreme heat, and (iii) sea level rise leading to inundations or problems with water discharge. This chapter will not discuss the issue and problems associated with sea level rise nor will it analyze comprehensively the effects of heat waves. This chapter focusses on the change in rainfall patterns and its impact on urban areas. Due to climate change, the force and frequency of extreme rainfall events are increasing in large parts of Europe (and other parts of the world), and more water will fall in shorter spells of time. People have been building sewer systems for wastewater and stormwater in cities for more than centuries, and there have always been extreme rainfall events that the drainage system could not cope with, causing flooding. So, nothing new? The force and occurrence of extreme rainfall events are increasing so fast that adaptation to climate change is required. In this case adaptation means higher water storage capacity of excess water on the surface and water conservation to cover dry periods and to reduce excessive groundwater level changes. The urban design focus is shifting from direct discharge to storing and retaining water.

Even today, more and more cities are experiencing extreme weather situations such as cloudbursts and the associated damage and repairs. Many cities around the world have started investigating the local impact of climate change and particularly of the hazards of extreme weather. Nevertheless, there seem to be structural obstacles to integrating climate change solutions in all refurbishment and maintenance operations and to moving from "knowing" to "doing." Changing the infrastructure requires a new perspective on urban planning. One of the problems is the lack of a standard definition of what climate resilience actually is – at least in the Netherlands this is an obstacle.

It is a political decision to tell if and how often disruptive effects of extreme rainfall are considered acceptable. The essence is that urban planning must take heed of the increasing frequency of extreme weather (cloudbursts, drought, and heat waves) and its consequences. Urban design must adapt to the changes in extreme weather events. In this chapter climate-resilient urban design means taking initiatives to encourage soft surfacing, greening, and creating space for water and buffers to store it for dry spells.

In the Netherlands, insight in consequences of climate change has resulted in a new policy aiming at climate proofing all built-up areas. Every construction, reconstruction, or refurbishment of urban areas should be done in a climate-resilient way from the year 2020 onward (Deltaprogramma 2014). Assuming a reconstruction of the roads in urban areas at least once in 30 years, this should result in a climate-proof urban environment by 2050.

In order to investigate how this climate proofing should and could be done in residential streets, a research has been carried out in 2015 and 2016. This research aimed at showing the possibilities and showing the cost of the different options, as this often is of main importance.

Institutions in the field, such as municipalities and consultancies, express the need to move forward and tackle the structural obstacles to integrating climate change solutions in all refurbishment and maintenance operations. The professionals indicate a great need for inspiring practical examples with reliable technical underpinnings and an estimate of the cost and benefits. This chapter therefore aims to inform designers, technicians, and governors alike. This chapter provides examples of illustrative designs and financial substantiation. The conducted research is not only meant for professionals, but thanks to these same professionals it was made possible. The Dutch municipalities of Almere, Almelo, Amsterdam, Arnhem, Beverwijk, Deventer, Eindhoven, Enschede, Groningen, Haaksbergen, Hoogeveen, and Houten and the (regional) water authorities Waternet, Waterschap Drents Overijsselse Delta, RWS, and the Dutch Ministry of Infrastructure and Environment, STOWA, SBRCurnet, and Tauw supported the research by providing the case studies or values comments on methodology and research outcomes.

Ideally, every case of urban retrofitting or maintenance should consider options for climate resilience when refurbishing streets and when renovating or building new residential areas and business districts. This chapter presents examples and offers knowledge and practical information to inspire and convince urban planners to take measures.

Methodology

This paragraph describes the different methods used in the research. After a discussion on the choice of an extreme rainfall event follows an explanation of the calculations of cost and benefits of climate-proof retrofitting of urban areas. Finally, the common neighborhood typologies that are typical and representative of the Dutch infrastructure (e.g., the urban city block, post-war garden cities, and community neighborhoods) are given. It is important to remember that all the examples we describe in this chapter are case studies, suggested by professionals working for municipalities, where climate-proof retrofitting is planned or will be executed in the near future.

Choosing an Extreme Rainfall Event

Currently, there is a lively debate in the Netherlands about a possible standard for an "extreme rainfall event norm." With regard to retrofitting public space in built-up areas, the researchers have the opinion that the extreme rainfall norm is less consequential than the urgency to consider extreme events in the street design. In the coming years, knowledge of extreme weather events will develop further and therefore, urban design should maintain its adaptability. There will be plenty of opportunities to implement adjustments by piggybacking on, for example, replacement of sewer systems, cables, pipelines, hard surfaces, and with the refurbishment of public spaces. Adaptation to climate change is therefore a continuous process, rather than a one-off operation. Nevertheless, in order to indicate effects of climate-resilient design variants and in order to be able to compare them, a reference value (in mm) for the concept "extreme rainfall event" in the Netherlands has been chosen. The reference value is not intended to set the norm for "extremity" as that is considered a policy issue. Reference values for the concept "extreme rainfall event" in other areas of the world will be different from the value for the Netherlands.

An extreme rainfall event is defined by three aspects: duration, frequency of occurrence, and amount of rainfall during the event.

Duration

The shorter the rainfall peak, the higher the rainfall intensity. Nevertheless, in a short period, the amount of rainfall is less than during a longer period. For surface water systems (e.g., canals), rainfall periods of several days are most critical. For stormwater discharge in a city by sewer system and above ground, shorter rainfall periods of some 15 min to several hours are most critical. Kluck et al. (2013) chose a rainfall duration of 1 h, a time period in which the rainwater can flow a considerable distance on the (street) surface. They indicated a flow distance of approximately 360 m with an average flow velocity of 0.1 m/s. Within this distance the surface streamflow might reach the surface water where storage or discharge is possible.

Frequency of Occurrence or Return Period

The second aspect, the return period, is in fact a policy decision. The question is what frequency of flooding or damage due to inundation is believed to be acceptable. In general, when asked, most people consider water inside their homes as totally unacceptable. However it is impossible, or only at very high cost, to eliminate all risks. A return period of once in a 100 years seems to be a good choice. It gives insight into a "once in a lifetime" event.

Amount of Rainfall (Volume) During the Event

As stated earlier, climate change has an impact on rainfall patterns and extreme events will increase in time with climate change. For this research the year 2050 has been chosen as the relevant time horizon to quantify the amount of future rainfall, because most measures in spatial urban adaptation, like the refurbishments of streets and pavements, have a life span of 30 years.

Based on (Kluck et al. 2013), a rainfall volume of 60 mm in 1 h has been accepted as the best estimate of an extreme rainfall event in one place expected to occur once per 100 years for the year 2050. Kluck et al. (2013) calculated this volume of 60 mm in 1 h based on 10 years of rainfall radar data (hourly readings) (Overeem and Buishand 2012), expected temperature rise (KNMI 2006), and impact on rainfall intensities (Lenderink et al. 2011). A more recent study of the extreme events for durations of 2 h and longer (Beersma et al. 2015) came to rainfall volumes which are in line with the above hourly value in 1 h. More information is available in Kluck et al. (2017a, in Dutch).

Cost

An often-heard argument against climate-proof retrofitting of public spaces is the belief that climate-resilient measures are more costly. However also many urban professionals think this is not true. To investigate this, in this research a methodology has been developed to estimate the costs and benefits of different variants. This methodology includes construction and maintenance costs of all relevant physical parts of the streets (e.g., sewer systems and permeable paving). Furthermore, it includes variation in the life span of the particular refurbishments. Calculations are based on the cost ratios and prizes per element from the Dutch Sewer Guidelines (Leidraad Riolering D1100, Stichting Rioned 2015) as well as empirical evidence provided by individual professionals from municipalities.

All investments have been based on a 100-year period in order to provide a realistic comparison of variants with differences in maintenance costs. This should provide a deeper insight into the financial consequences of various options to allow policy makers, designers, administrators, and other experts to make well-informed decisions.

The annual cost for each variant is calculated based on investments, periodical reinvestments, maintenance, and expected benefits, over a period of 100 years, assuming that sewer systems have a life span of 60 years and streets require major reconstruction every 30 years.

Costs due to flooding were also expressed in cost per annum based on estimated frequency and the magnitude of the disruption (amount of houses flooded and flood depth).

Benefits

Climate-resilient refurbishment of public space has many benefits. This research only takes into account those benefits, which can be quantified, such as reduced cost for flood damage and a saving due to a reduction of water discharge to the wastewater treatment system. Other benefits, which cannot easily be quantified, such as reduced or delayed drainage to surface water, groundwater recharge, heat stress reduction, and increased water availability for urban green, have not been taken into account into the cost comparison. The reason for this is that the research was set up in collaboration with professionals from municipalities and the whish was to create a line of argument for climate proofing residential streets without doubtful arguments. Those benefits can however be important. In Kluck et al. (2017b) the benefits of greening public space on public comfort and health, water quality, reduction of energy consumption, and increased biodiversity were studied and (partly) quantified. We have limited ourselves to a rough description of the benefits of greening because an exact description of the background factors and uncertainties would not fit within the scope of this chapter.

Neighborhood Typologies

Street design in the Netherlands and other parts of Europe is often based on a particular philosophy of its time. Ideas and technologies that were available at the time of constructing are captured in the authentic details of these streets, such as the size of the houses, gardens, public space for greens and playgrounds, the width of the streets, and the architecture of the buildings.

Characteristic features that were found across the Netherlands were used to distinguish neighborhood typologies. The set of neighborhood typologies that we have applied is listed in Table 1 (based on Kleerekoper 2016). The neighborhood typologies give direction to the approach to combat more extreme climate effects. For instance, the abundance of public space in post-war neighborhoods can easily be employed for climate adaptation, whereas in the dense urban housing blocks and pre-war blocks, underground solutions are more important. The structure of garden cities offers space for swales to absorb heavy rainfall locally. The opportunities for climate-proof measures will be more or less the same for cities (around the world) in

Table 1 Neighborhood classification based on Kleerekoper (2016)

Dutch neighborhood typology[a]	Period	Features
Urban city block	Before 1930	No front garden nor green skirting, 4–5 layers
Pre-war city block	1900–1940	Occasional front garden, 3–4 layers, wider streets than urban blocks, and occasional green skirting
Garden village	1910–1930	Spacious front and back garden, 2–3 layers, ample parking space,1930s architecture, limited public green, and rarely street trees
Working-class neighborhood	1930–1940	No front garden, little public green, 2–3 layers, single-family units
Low-rise post-war garden city	1945–1955	Open building block with ample green, 2–3 layers, single-family units
High-rise post-war garden city	1950–1960	Open building blocks with ample green, 4–6 layers, apartments, storage on the ground level
Post-war neighborhood	1940–1990	Front and back garden, 2–3 layers, single-family terraced houses, semidetached or detached
Community neighborhood	1975–1980	Single-family unit with front- and back garden, meandering street pattern, courtyards, wide green skirting around the neighborhood
High-rise city center	1960-present	More than 10 layers in grid formation
Suburbanization – Vinex	1990–2005	Single-family unit, terraced, semidetached or detached apartments

[a]In this chapter we discuss the case studies of urban city block and the high-rise post-war garden city. More case studies and practical examples of climate-proof retrofitting of urban areas can be found on www.hva.nl/klimaatbestendigestad

streets of the same typology. Nevertheless, specific characteristics, such as the slope, type of soil, and groundwater level, may affect local solutions.

Knowledge of the neighborhood typology, gradient (flat or sloping), type of soil, and the groundwater level enables us to give a reliable projection of the possibilities and effectivity of local climate adaptation. They apply to many of the streets and neighborhoods across the Netherlands. In every country common typologies can be determined to present climate adaptations that generally fit in. Throughout Europe, typologies will vary strongly, especially from North to South due to difference in climate. Northern countries tend to have more spacious streets to allow sunlight entering the houses during winter.

Case Studies

This chapter aims to show the possibilities for climate-resilient street design. Through a systematic research by design approach, we developed realistic solutions that provide insight in (i) how ordinary residential streets can be made climate resilient in practice, (ii) the costs involved, and (iii) the benefits. The type of research by design can be labeled *typological research* according de Jong and Voordt (2002). In the case studies the context is variable (different neighborhoods) and the object (climate-resilient street) is determined. Each separate case study is, however, a "design study" in itself, with one specific neighborhood and thus determined context and variable climate adaptation solutions (variable objects). In Table 2 the different types of research methods are presented.

An important selection criterion for the cases was the commonality of the typology of the neighborhood. This ensures the cases are examples for many municipalities. For each case we present a traditional design and three more climate-resilient alternatives. Each variant is presented with a cost and benefit balance.

The research described in this chapter includes ten cases, presented in Table 3, all with three climate-resilient variants. All case studies are based on planned or completed refurbishment projects in public spaces where the sewer system needed to be replaced or adapted. In each case the focus is on anticipated extreme rainfall. The examples may include more than one street to represent the scale at which projects, such as road construction and sewer replacement, are generally planned.

Cases include flat and sloping surfaces and differences in soil permeability and groundwater levels. The Netherlands has a predominantly flat surface with some sloping areas in the eastern and southern regions. The soil is predominantly sandy with clay or peat, and the groundwater tables are usually high. These features can

Table 2 Types of research by design by de Jong and Voordt (2002)

		Object	
		Determined	Variable
Context	Determined	Design research	Design study
	Variable	Typological research	Study by design

Table 3 The ten case studies used to present climate-resilient retrofitting for areas with flat and sloping surfaces (Kluck et al. 2018)

Case study	Neighborhood typology	Incline
1	Pre-war city block	Flat
2	Urban city block	Flat
3	Post-war neighborhood	Sloping
4	Low post-war garden city	Flat
5	High-rise post-war garden city	Sloping
6	Suburbanization – Vinex	Flat
7	Urban city block	Sloping
8	High-rise post-war garden city	Flat
9	Community neighborhood	Flat
10	Garden village	Sloping

vary significantly depending on the location. Furthermore, some of the proposed solutions may not be replicable one-on-one in terms of implementation, operationalization, and costing. The case studies do not aim to cover all situations but provide examples for the most common cases.

In this chapter only two case studies of the elaborate research by design study are presented: high-rise post-war garden city (flat) (case study 8) and the urban city block (flat) (case study 2). The two examples provide a clear image of the approach and climate-resilient variants applied in this research. A complete overview of all cases is presented in Kluck et al. (2018).

The methodology and examples presented in this chapter are based on real cases in the Netherlands. These examples are designed for extreme rainfall intensities for the Dutch situation, which are illustrative for Europe and other temperate climates. In countries with a more humid climate the rainfall intensities could be much larger which requires different dimensions and solutions.

Climate-Resilient Retrofitting

For ten case studies of residential streets which needed retrofitting, alternative climate-resilient street layouts have been designed and compared to the traditional redesign. This paragraph presents two of the ten cases and the results of the cost and benefits comparison for all of the cases. The two selected cases are both in flat areas and are in paragraphs 3.1 and 3.2. Paragraph 3.3 gives the cost comparison of all cases in flat areas. Paragraph 3.4 explains the results for the four cases in sloping areas.

High-Rise Post-war Garden City (Flat)

The flat post-war garden city with high-rise buildings is characterized by a relatively large public (green) space between the buildings. The neighborhoods often have a limited variation in type of housing stock, often with small and cheap houses. These areas are mostly seen as disadvantaged neighborhoods where municipalities see the need to upgrade the urban space, buildings, and green. More variety in housing stock with especially larger, comfortable, and energy-efficient dwellings would increase attractiveness. In addition, the organization around buildings needs to change from one entrance and storage on the ground floor toward buildings with functions and dwellings on the ground floor with many entrance points. This change increases safety and liveliness outside. Green is an essential part of the structure of these neighborhoods. Between building blocks there is alternately a green area or a parking space, sometimes the two are combined. Although the amount of green is abundant, it has a sole visual function with only a meadow. Here people do not experience the green nor is it functional for water buffering or heat resilience.

Neighborhood typology characteristics	
Building form	Open building block
Average building height	30 m
Street	
H/W ratio	0.3
Ventilation	Very good
Shadowing	Moderate
Presence of street trees	Moderate
Inside building block	
H/W ratio	0.4
Ventilation	Good
Shadowing	Moderate
Presence of trees	Moderate

The high buildings with large open spaces allow a good ventilation. Wind is directed downward by tall buildings. This can lead to uncomfortable and even dangerous situations during stormy weather, while during warm periods extra ventilation is very welcome.

The surface water system in the post-war garden cities was designed to drain and discharge water. However, the spacious layout and many green spaces offer an opportunity to store, retain, and infiltrate water locally.

Case Study

The study location is on flat terrain and consists of an average permeable soil. The street foundation (cunette) has good permeability, and there is a separate sewer system (sanitary sewer and stormwater sewer). The study location is located on flat terrain, which has the benefit of relatively easy water retention. With extreme rainfall the water will not flow freely but it will be collected at the lowest points.

Case properties:
217 m street length
0 ground floor homes
Approximately 50% of public space is paved
Paving consists of bricks and concrete paving stones
The street level has no slope
16 trees in the street
Approximately 30 m distance between facades

This high-rise post-war garden city case in a flat area shows that water storage in a public area is uncomplicated and ensures that retrofitting can be climate resilient at the same price as traditional refurbishment. Public space provides sufficient storage space for large quantities of water.

In the following variants, first the traditional way of retrofitting is presented, followed by three climate-resilient variants. In Fig. 1 the variants are illustrated.

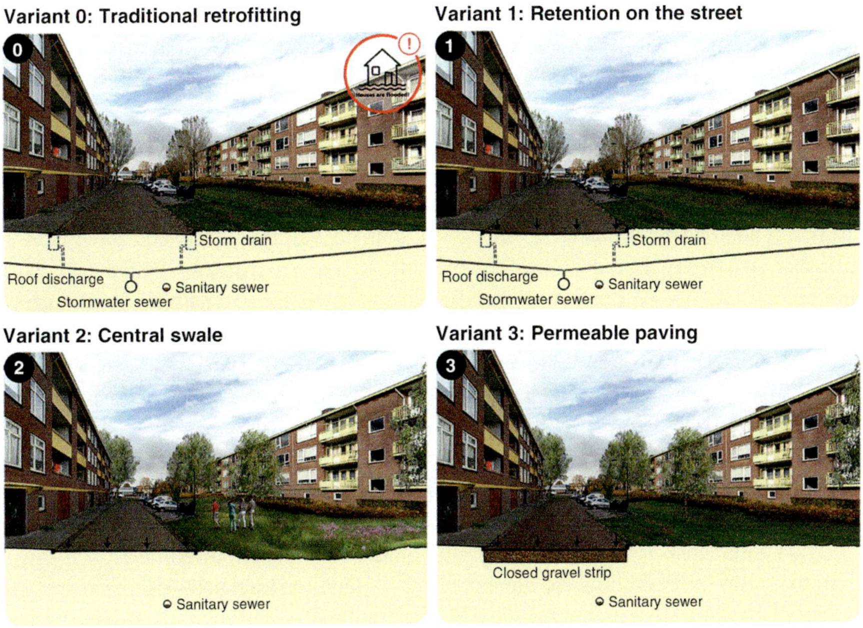

Fig. 1 Climate-resilient designs for a high-rise post-war garden city with a flat surface (Kluck et al. 2018)

Variant 0: Traditional Refurbishment

The municipality raises the subsided public space to its original construction height. The existing separate sewer system (sanitary sewer and stormwater sewer) and paving are renewed. The sewer system can cope with heavy rainfall once per 1 or 2 years on average. There is some space for stormwater storage in the street, but it has not been designed for that purpose. With extreme rainfall (40 mm or more in 1 h), stormwater can flood the buildings. The green area is situated at a higher level than the road.

Variant 1: Retention in the Street

The municipality lowers the road level to 9 cm lower than the level of variant 0. The pavements are then adjusted accordingly, providing storage space in the streets. When a cloudburst exceeds 60 mm in 1 h, water will flood buildings. The existing sewer system and paving are replaced. The sewer system can cope with a rainstorm once per 1 or 2 years on average. The green area is situated at a higher level than the road.

Variant 2: Retention in the Swale

In this variant there is no stormwater sewer system. The municipality builds a swale in the green area. This swale can deal with 20 mm of rain in 1 h. The road slopes toward the swale and is 4 cm lower than variant 0, which means that in an area with soil subsidence, there is no need to raise the ground as much as in variant 0. Finally, because the swale is positioned at such a low level, flooding of the houses will only occur in case of a cloudburst larger than 60 mm in 1 h.

Variant 3: Permeable Paving

In this variant there is no stormwater sewer system. Instead, the municipality refurbishes the road with permeable paving that can cope with 20 mm of rainfall in 1 h. The road is situated 12 cm lower than in variant 0, and the pavement is adjusted accordingly and slopes toward the street. This creates space for water storage during heavy rainfall. Water will only enter houses with a cloudburst of more than 60 mm in 1 h. Consequently, the ground does not need to be raised much in areas with soil subsidence.

Cost-Benefits of Rainwater

Figure 2 shows the annual costs for each variant, including construction, maintenance, and flooding costs. The annual costs are based on estimates over a period of 100 years.

The annual costs for variant 1 (retention on the street) are approximately 3% lower than the costs for traditional refurbishing. For variant 2 annual costs decrease approximately 6% when the existing meadow is equipped with infiltration in a swale. Variant 3 with infiltration through permeable pavement has higher annual costs. All three variants have the advantage of the capacity to store stormwater in the ground when it is permeable. Or, if the ground is poorly permeable, stormwater drainage will be delayed. Drainage in the street foundation can also work for higher natural groundwater levels as they delay stormwater drainage (Kluck et al. 2018). When variants with infiltration

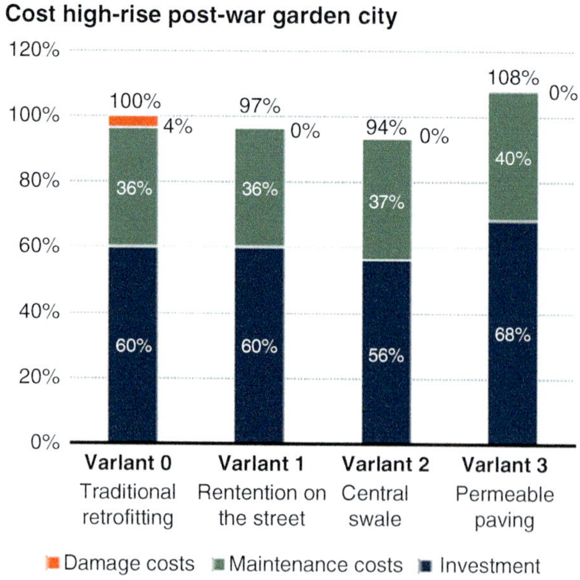

Fig. 2 Cost of high-rise post-war garden city (Kluck et al. 2018)

(2 and 3) are disconnected from the stormwater system, they will delay discharge onto the surface water and alleviate stress on the system. The possible benefits of this have not been included in the cost and benefit analyses.

Climate Adaptation in High-Rise Post-war Garden Cities

Extreme rainfall events can result in a lot of water in the streets in these neighborhoods because the green space is not designed to contribute to infiltration, storage, and drainage. Damage on buildings or dwellings will generally be minimal because dwellings are situated above ground level. In some cases the storage rooms (ground floor level) have been laid out half-deep; these are vulnerable to flooding with heavy precipitation.

Not only the vulnerability of the neighborhood itself but also the passing of water to other areas that may be (more) vulnerable is important. The green space in this neighborhood can easily be used as part of the water system: green with water infiltration and buffer zones. Here, water and heat resistance can come together, for example, a combination of a water playground and a swale or a fruit orchard with benches that may flood. Figure 3 shows an example of such a combination.

The current green is mostly sole grass meadow to keep maintenance costs low. While with certain perennials or flower mixtures, management costs can be reduced (Hop 2011) transforming the image into a diverse green area that varies per season. The same goes for the planting of swales. For this purpose, specific plant mixtures are available that can also support the operation of a swale (Blänsdorf 2015). In addition, perennials provide more cooling and more space and food for bees,

Fig. 3 Combination of a swale and a playground. (Source: Ronald Wentink)

butterflies, and birds. The right choice of vegetation in a swale contributes to perception, pollution removal, and long-term functioning (Boogaard et al. 2003).

The sensitivity to heat stress in flat post-war garden city neighborhoods is not very large, because of the good ventilation due to the structure of the neighborhood and the abundance of green space. However, the temperature can rise considerably on the large parking spaces and on the grassy areas with few trees. Especially on windless days, there are insufficient places in these neighborhoods that provide cooling.

An important aspect of heat stress in these neighborhoods is the dwellings where sun protection is lacking and indoor spaces become extremely hot during a heat wave. An important step toward heat resistance here is the adaptation of the housing. Outdoor sun protection is the best way to prevent heating indoors. Cool roofs by reflective roofing, green roofs, solar panels, or water-storing roofs mean a big difference for the top floor. Facade greening on the east, south, and west facades contribute to both the indoor and outdoor climate. From a green-blue layout of the ground floor, the first four residential floors benefit most.

The large green areas in these neighborhoods can increase in value when they offer multiple functions. In addition, more functions in green contribute to heat resistance. For example, design the green space as an outdoor space with shady and sunny places. Other functions of the large green areas between the building blocks can serve the entire neighborhood: a water purification park with reed plants, a petting zoo, urban agriculture with a neighborhood restaurant, or a sloping lawn where in the summer there are picnic tables and barbecues that can (partly) flood under heavy precipitation. The green areas can be better connected with activity routes, for example, a bicycle and skate route that connects to a route outside the neighborhood, or walking paths for the daily 30-min exercise.

Conclusions

In the flat high-rise post-war garden city neighborhood typology, streets can be made climate resilient at no extra costs by lowering the surface level of the street or transforming part of the green into a swale. To achieve this, it is important to piggyback on planned operations, such as replacing the sewer system and periodical redesign operations.

The climate-resilient designs cause less inconvenience and flood damage and can increase the attractiveness of the neighborhood with more functions in green spaces.

Urban City Block (Flat)

The traditional building blocks of the Dutch cities have a relatively high density and a very high percentage of paved surface. On the other hand, the spaces within the building block are often very green with large mature trees. The green inside these building blocks is under pressure due to the trend of tiled terraces, extension of indoor space, and parking spaces. In addition to the green inside building blocks, historic part of cities include monumental canals, avenues, and parks with large trees.

Neighborhood typology characteristics	
Building form	Closed building block
Average building height	12 m
Street	
H/W ratio	2.4
Ventilation	Little
Shadowing	Much
Presence of street trees	Moderate
Inside building block	
H/W ratio	0.4–0.3
Ventilation	Little
Shadowing	Moderate
Presence of trees	Little – moderate

Urban city blocks are characterized by multilevel stories and an organic street pattern. The paved streets leave little space for public green, although there are a few large trees. The closed building blocks prevent water discharge via the back gardens. Solutions to this problem are left to private initiative.

Case Study

The study location is on flat terrain and consists of poorly permeable soil. The street foundation (cunette) has good permeability, and there is a separate sewer system (sanitary sewer and stormwater sewer). Flat terrain has the benefit of relatively easy water retention. With extreme rainfall, the water will not flow freely, but it will be collected at the lowest points.

Case properties:
299 m street length
40 ground floor homes
Approximately 90% of public space is paved
Paving consists of bricks and concrete paving stones
The street level has no slope
43 trees in the street
Approximately 15 m distance between facades

In the following variants, first the traditional way of retrofitting is presented, followed by three climate-resilient variants (Fig. 4).

Traditional Refurbishment

With a traditional refurbishment, we expect flooding of houses and buildings at rainfall intensity rates of approximately 40 mm in 1 h. This image illustrates the traditional refurbishment.

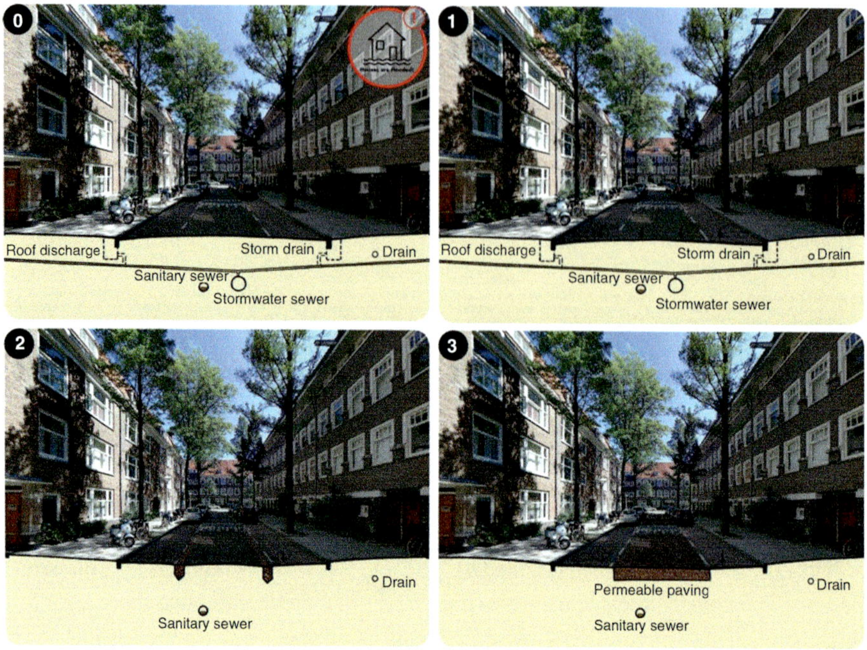

Fig. 4 Climate-resilient designs for an urban city block with a flat surface (Adjusted from Kluck et al. 2018)

Variant 0: Traditional Refurbishment

The municipality raises the level of the public space to its original construction height to correct for the soil subsidence. The existing separate sewer system (sanitary sewer and stormwater sewer) and the pavement are renewed. The sewer system can cope with heavy rainfall once per 1 or 2 years on average. The street is not specifically designed to retain water, but it can hold a small amount. Water will enter buildings with extreme rainfall of 40 mm or more in 1 h.

Variant 1: Retention in the Street

The municipality rebuilds the road level 10 cm lower than the level of variant 0. Therefore, in case of soil subsidence, the costs for raising the street are lower. The pavement is adjusted accordingly and slopes toward the street. Consequently, it provides water storage space in the streets. A cloudburst of more than 60 mm in 1 h may cause water to enter the buildings. The existing sewer system and paving are replaced. The sewer system can cope with rainfall once 1–2 years on average.

Variant 2: Retention in the Gully

In this case, there will be no separate stormwater sewer system. Instead, infiltration gutters are added to the road paving, allowing stormwater to flow into the street foundation. These infiltration gutters have a capacity of up to 20 mm of rainfall in

1 h. The municipality builds the road 10 cm lower than in variant 0. Therefore, in case of soil subsidence, the costs for raising the street are lower. The pavement is adjusted accordingly and slopes toward the street. The lower street level creates space for heavy rainfall to bc stored in the street. Water will enter the houses only with a cloudburst causing more rainfall than 60 mm in 1 h.

Variant 3: Storage in the Street Foundation (Cunettes) via Permeable Paving

There is no stormwater sewage system. Instead, the road has permeable paving that can cope with 20 mm of rainfall in 1 h. The municipality builds the road 10 cm lower than in variant 0. Therefore, in case of soil subsidence, the costs for raising the street are lower. The pavement is adjusted accordingly and slopes toward the street. The lower street level creates space for heavy rainfall to be stored in the street. Water will enter the houses only with a cloudburst causing more rainfall than 60 mm in 1 h.

Cost-Benefits of Rainwater

Figure 5 shows the annual costs for each variant, including construction, maintenance, and flooding costs. The annual costs are based on estimates over a period of 100 years.

The annual costs for variant 1 (retention in the street) are approximately 9% lower than the costs for traditional refurbishing. Variants with infiltration are more expensive. However, the advantage is that these variants have a capacity to store stormwater in the ground when it is permeable. Or, if the ground is poorly permeable, stormwater drainage will be delayed. Drainage in the street foundation can also

Fig. 5 Cost of urban city block (Kluck et al. 2018)

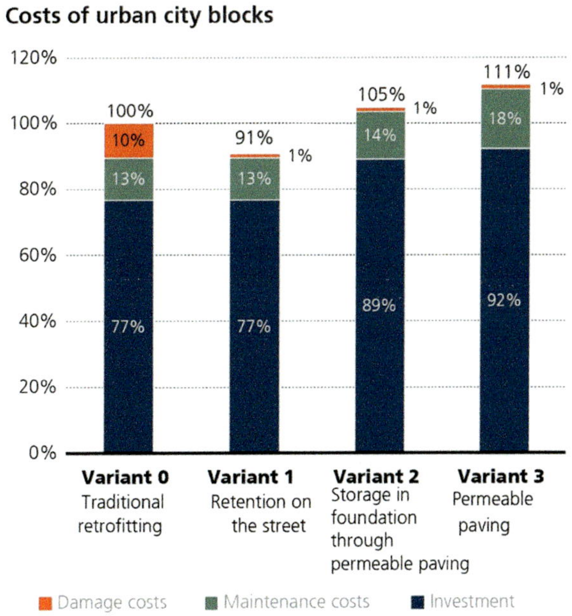

work for higher natural groundwater levels as they delay stormwater drainage (see Kluck et al. 2017a [in Dutch]). When variants with infiltration (2 and 3) are disconnected from the stormwater system, they will delay discharge onto the surface water and alleviate stress on the system. The possible benefits of this have not been taken into account (Kluck et al. 2018).

As a consequence of infiltrating or storing rainwater, the groundwater level might be influenced. Attention is needed for the resistance of the existing foundations and trees. Building foundations made of wooden piles are sensitive to degradation. In this regard it is important to prevent low groundwater levels to avoid water pressure differences in the piles (Klaassen 2008). For trees the resistance to high groundwater levels due to infiltration or water storage varies per specie. The drainage system can be adjusted according to the species resistance. The same approach can be used to preserve archaeological remains (Boogaard et al. 2016).

Climate Adaptation in Urban City Blocks

The structure and organization of historic city centers cause a lot of heat storage in stony materials, and there is little cooling from green and little ventilation. This leads to high temperatures in streets and on squares.

In this type of neighborhoods, the vulnerability to extreme rainfall events is highly dependent on the presence, location, and size of surface water such as canals. There is a lot of paved surface, which means that there is also a lot of run-off rainwater. If sufficient surface water is present within a reasonable distance from the drained paved surface and this run-off water is properly guided, then there will be few problems with extreme rainfall. However, if there are no large water elements present, problems can arise quickly.

Historic city centers are prone to flooding for the following reasons:

- The houses are regularly equipped with basements where considerable damage can occur when water flows in.
- The older parts of the city have grown organically with often varying height differences between street level and threshold height, some buildings being equal or lower than the street level.
- Shopping streets often do not have curbs and ground level differences, many shops do not have thresholds at the entrance, and there is hardly any difference in height between street and indoor floor level of the ground floor.

Because of the large percentage of pavement in relation to green and water, inner cities are sensitive to heat stress. The extent to which streets warm up or retain heat depends on the aspect ratio (height and width) and orientation of the streets and buildings. Due to the heavy pressure on public space in city centers, measures that occupy public space are not easy to realize in these neighborhoods. It is desirable to search for solutions that use surface that is not fully utilized or that can fulfill a dual function. It is logical to start combining measures that contribute to water storage and limit heat stress, for example, with the design of a "tiny park," which creates a cool spot where water is also infiltrated. Another example is a green roof on an historic building. A substrate layer of 8 cm thickness holds 40 mm or 40 l of water per m^2. Here the vegetation on the roof ensures cooling by evaporation.

In fact, trees and plants always fulfill multiple functions and therefore offer several advantages, such as a positive effect on air quality, health and well-being, habitat for flora and fauna, higher real estate values, and so on. Green can often be combined with facilities in the public space, such as shelters, bicycle racks, seating areas, or lighting.

Conclusions

In the flat urban city block typology, residential streets can be made climate resilient at no extra cost by lowering the surface level. To achieve this, it is important to piggyback on planned operations, such as replacing the sewer system and periodical redesign operations. The climate-resilient designs cause less inconvenience and flood damage and can be combined with more green in the street.

More Examples of Case Studies in Flat Areas

In addition to the two case studies described above, four more variants for residential neighborhood typologies in flat areas have been developed and compared. They have been described in Kluck et al. (2018). The four case studies are of the following typologies:

- A pre-war city block (1900–1920)
- A low-rise post-war garden city (1945–1955)
- A community neighborhood (1975–1980)
- A suburbanization or Vinex expansion (1990–2005)

Figure 6 shows that also in those case studies, a climate-resilient design can be achieved at no extra cost by lowering the surface level.

Green Benefits

More green in the city contributes to the reduction of heat stress and the prevention of drought. According to Kluck et al. (2017b), the benefits to health, comfort, economic value, and energy use are many times higher than the annual costs of refurbishing the entire street. Moreover, the benefits are considerably higher than the additional cost (investment, management, and maintenance) of the green areas. Besides green benefits, additional green areas provide opportunities for stormwater drainage.

Examples of Case Studies in Sloping Areas

Next to the residential streets in flat areas, streets in sloping areas have also been examined. Four more neighborhoods case study have been investigated:

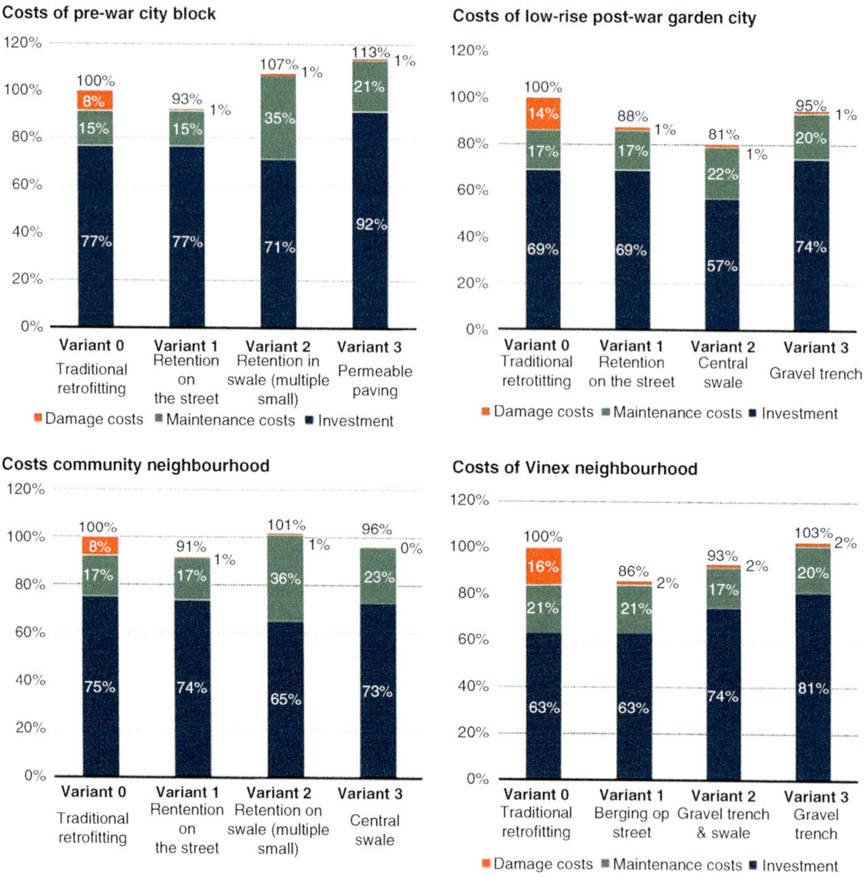

Fig. 6 Cost and benefits of six cases with a flat surface (Kluck et al. 2018)

- Urban city block (before 1900)
- Garden village (1910–1930)
- High-rise post-war garden city (1945–1955)
- Post-war neighborhood (1940–1990)

The definition of more climate-proof variants for the sloping areas appeared more complicated than for the flat areas. The slope makes it on the one hand harder to retain water on the slope itself than it is in flat terrain. On the other hand, it makes it easier to discharge the water. If there are no flood risks downstream of the case study street, then it is a climate-proof solution to discharge the water, e.g., to a pond or a park. If, however, downstream of the case study area there is a flood risk (e.g., houses or shops), then discharging is not a climate-proof solution. The design principle was not to increase flood risk downstream of the case study area. In the same way, the assumption was that no water from upstream would flow in (increasing flood risk in the case study area).

This resulted in the conclusion that the local water retention in the sloping areas is essential if downstream flood risks are present. As flood risk in and downstream of sloping areas varies significantly from one situation to another, it was decided in the variants how to retain three different amounts of the assumed extreme storm event, namely, 33% of the event and 66% and 100% of the once in a lifetime storm event (60 mm in 1 h). In fact a general standard Dutch drainage system can retain and discharge the 33% of this extreme event (20 mm of rainfall in 1 h) as this is more or less equal to the standard design once per 2 years storm for drainage systems. The 66% variant (40 mm in 1 h) is in between the traditional redesign and the storage of the complete extreme event (60 mm in 1 h).

This research therefor only illustrates the possibilities of water retention in these sloping areas and shows the cost and maintenance in comparison to traditional refurbishment. The comparison of the variants showed that storing water on the slopes in underground infiltration facilities) appeared in most cases quite costly (resulting in a doubling of investment and maintenance cost as compared to the traditional variant for 20 mm of extra underground water storage), as is shown in Fig. 7. The costs increase as the storage increases.

One of the case study resulted in a remarkable different result. The case study for the sloping residential street in the typology high-rise post-war garden city is not more expensive as the storage increases. The reason is that in this case, the ample green public spaces are situated along the elevation contours, allowing above water retention in the green at low cost. The climate-proof variants appear to be possible for the same price as the traditional design.

Discussion and Conclusion

In this research ten case studies for residential streets (from eight neighborhood typologies) have been investigated. For each case three climate-resilient variants have been compared to a traditional design for retrofitting the street and the accompanying costs and benefits.

The focus in climate proofing the streets was on preventing stormwater flooding. Consequently, the design principles for the more climate-resilient variants were increasing the sponge capacity of the street (retaining and storing more water), guiding surplus water above ground – but without increasing flood risks downstream – and greening the street. The climate-resilient variants can cope with a once in 100 years extreme storm event without flooding. More extreme rain events still might cause flooding. The cost assessments for the variants have been carried out for the situation that a street and possibly also the sewer system need to be reconstructed. Combining the climate proofing with the moment of retrofitting, refurbishment or reconstruction is important to save on investment cost.

The ten case studies show that climate-resilient retrofitting can be quite easy and is not necessarily more costly than traditional retrofitting. This appeared true for the flat residential areas in which lowering the street level is a particularly easy and cheap solution with roughly the same maintenance and management costs, while at

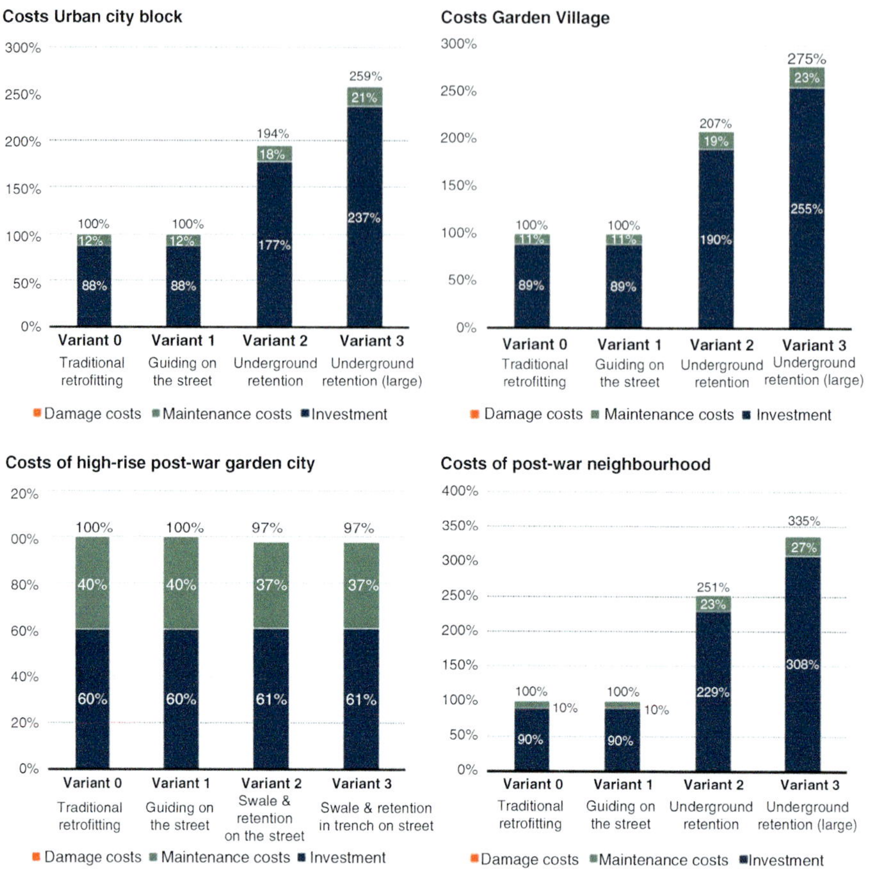

Fig. 7 Cost and benefits of four cases with a sloping surface (Kluck et al. 2018)

the same time flood damage costs are reduced significantly. The variants without stormwater system appear not much more expensive than the traditional variant. The extra costs for the stormwater system appear about equal to the savings on flood damage.

For residential streets in sloping areas the situation is more complicated. On the one hand, it is harder to retain water on the slope itself than it is in flat terrain. On the other hand, it is easier to discharge the water. The flood risk in sloping areas should be carefully considered as it varies significantly from one situation to another and predominantly occurs downstream. The need to retain water on slopes depends very much on the downstream situation. The design principle not to cause flood problems downstream is important. If there are vulnerable buildings downstream, water retention in the sloping areas is essential. Because the situation varies very much per case, this research only illustrates the possibilities of water retention in these sloping areas and shows the cost and maintenance in comparison to traditional refurbishment. In assessing the flood risk, the assumption was that no water enters from upstream areas. The comparison of the variants showed that storing water on the slopes in underground infiltration facilities) appeared in most cases quite costly (resulting in a doubling of investment and maintenance cost as compared to the traditional variant for 20 mm of extra underground water storage). However, if the area has a lot of public space, as, for example, in the high-rise post-war garden city, with green areas along the elevation contours, it is very well possible to retain water for the same price as the traditional design.

Next to the conclusion that retrofitting in a climate-proof way is not so complicated and often not (much) more expensive than the traditional reconstruction of a residential street, the research reveals that neighborhood typologies help to order the possibilities for climate-resilient refurbishment. The case studies show that more spacious typologies with more public space, e.g., in post-war garden cities and community neighborhoods, provide ample space to store water in the green areas. Even in such neighborhoods in sloping areas, it is easy to retain the water in the green areas. The more densely built-up typologies such as historic and pre-war city blocks have little space in the streets and hardly any green. However, in those typologies, lowering the street level creates a lot of water storage space in the streets and reduces the flood risk of houses, resulting in a climate-proof alternative for the same cost as the traditional reconstruction. Even in those densely built-up areas, it is possible to come up with a more climate-resilient variant (than the traditional variant) with permeable pavement and other infiltration measures. The use of neighborhood typologies helps to quickly identify measures for climate proofing the residential streets.

The measures in the climate-proof variants mainly aim at preventing stormwater flooding, but some also have some effect in reducing drought problems as more infiltration results in more groundwater. This might be important in areas where drought might result in damage to the constructions and buildings or extra maintenance cost for watering the urban green. The reduction of damage due to drought could not be estimated within the research, and therefore those benefits have not been taken into account in the comparison.

Climate change also results in more hot days. In the Netherlands, there is no clear goal for urban heat management yet; therefore it was decided not to take this into account in the comparison of variants.

A methodology to estimate the costs and benefits of the variants was developed. This methodology has also been used to perform a sensitivity analysis: what factors are uncertain and have a large impact on the annual cost. This revealed that permeable pavements and small swales have relatively high maintenance costs and that there are large differences for those costs between municipalities. It is recommended to further research the possibilities for cost-effective maintenance.

The research of the ten case studies has been published in a book of examples (Kluck et al. 2018), which is meant to inspire municipalities and contractors to choose for climate-resilient variants and to help those who already know that climate proofing is often easy and cost-effective in convincing others. Each reconstruction of refurbishment of a road should be taken as an opportunity for climate proofing. The stormwater discharge of a real extreme event should be evaluated and the possibilities to prevent flooding of houses investigated. Often above ground retention and redirection to local green or surface water are possible. This is how the municipalities can become more climate resilient, one at a time.

Summary

The main conclusions of the research are the following:

- In flat areas climate-resilient refurbishment is perfectly possible.
- Lowering the street level is the cheapest option – even cheaper than a traditional refurbishment.
- Greening is multifunctional (it benefits the infiltration capacity and reduces heat stress and wastewater treatment costs).
- The variant without stormwater sewer system, but local storage of rainwater are somewhat more expensive than the variant with a lower street level. However, such solutions are attractive from a holistic point of view as, in the case of flooding, prevented flood costs will equal the extra cost for investment and maintenance. Water retention also relieves the water system and sewage treatment plant.
- Retrofitting by piggybacking on planned refurbishment and maintenance operations is an essential to keep costs low.
- The case study examples with cost calculations provide designers and policy makers with a strong argument to choose for climate-resilient street design.

References

Beersma J, Bessembinder J, Brandsma T, Versteeg R, Hakvoort H (2015) Actualisatie meteogegevens voor waterbeheer 2015. STOWA rapport 2015-10, november 2015. p 78

Blänsdorf E (2015) Waterberging in de stad, natuurlijke wadi's. www.molinia.net

Boogaard FC, Jeurink N, Gels JHB (2003) Vooronderzoek natuurvriendelijke wadi's: inrichting, functioneren en beheer. STOWA/Rioned Publicatie, Utrecht. 2003–04, 36p

Boogaard F, Wentink R, Vorenhout M, de Beer J (2016) Implementation of sustainable urban drainage systems to preserve cultural heritage – pilot Motte Montferland. Conserv Manag Archaeol Sites 18(1–3):328–341

d. Jong TM, v.d. Voordt D (2002) Ways to study and research: urban, architectural and technical design. DUP Science, Delft

Deltaprogramma Ruimtelijke adaptatie (2014) Deltabeslissing Ruimtelijke adaptatie: Het Deltaprogramma: een nieuwe aanpak. Ministerie van Infrastructuur en Milieu en Ministerie van Economische Zaken, Den Haag

Hop MECM (2011) Vaste planten in openbaar groen. Voor functionele en onderhoudsvriendelijke toepassingen. Praktijkonderzoek Plant en Omgeving (PPO), WageningenUR, Boskoop, 47p

Klaassen RKWM (2008) Water flow through wooden foundation piles: a preliminary study. Int Biodeter Biodegr 61(1):61–68

Kleerekoper L (2016) Urban climate design. Improving thermal comfort in Dutch neighbourhood typologies. Thesis, Delft University of Technology

Kluck J, van Hogezand RJP, van Dijk E, van der Meulen J, Straatman JHM (2013) Extreme neerslag: Anticiperen op extreme neerslag in de stad. Hogeschool van Amsterdam: Kenniscentrum Techniek, Amsterdam

Kluck J, Loeve R, Bakker WJ, Kleerekoper L, Rouvoet MM, Wentink R, Viscaal J, Klok EJ, Boogaard F (2017a) Het klimaat past ook in uw straatje. De waarde van klimaatbestendig inrichten. Achtergronden [Background documentation]. Amsterdam University of Applied Sciences (HvA), Amsterdam, 49 pp

Kluck J, Kleerekoper L, Klok EJ, Loeve R, Bakker WJ, Boogaard FC (2017b) De klimaatbestendige wijk: Onderzoek voor de praktijk. Urban Technology paper number 10. Amsterdam University of Applied Sciences (Hva), Amsterdam

Kluck J, Loeve R, Bakker WJ, Kleerekoper L, Rouvoet MM, Wentink R, Viscaal J, Klok EJ, Boogaard F (2018) The climate is right up your street. The value of retrofitting in residential streets, a book of examples. Amsterdam University of Applied Sciences (HvA), Amsterdam, 77 pp

KNMI (2006) Klimaat in de 21e eeuw: vier scenario's voor Nederland. mei 2006, 16p

Lenderink G, van Oldenborgh GJ, van Meijgaard E, Attema J (2011) Intensiteit van extreme neerslag in een veranderend klimaat. Meteorologica 20(2):17–20

Merkens J, Reimann L, Hinkel J, Vafeidis AT (2016) Gridded population projections for the coastal zone under the Shared Socioeconomic Pathways. Glob Planet Chang 145:57–66. https://doi.org/10.1016/j.gloplacha.2016.08.009

Neumann B, Vafeidis AT, Zimmermann J, Nicholls RJ (2015) Future coastal population growth and exposure to sea-level rise and coastal flooding – a global assessment. PLoS One 10(3): e0118571. https://doi.org/10.1371/journal.pone.0118571

Overeem A, Buishand TA (2012) Statistiek van extreme gebiedsneerslag in Nederland. KNMI, Technical report; TR-332, 4 mei 2012, 34p + bijlagen

Stichting Rioned (2015) Leidraad Riolering D1100 (Sewer guide). Rioned, Ede

United Nations, Department of Economic and Social Affairs, Population Division (2014) World urbanization prospects: The 2014 revision, highlights. (ST/ESA/SER.A/352)

108 Persuading Pacific Maritime Supply Chain Stakeholders to Prioritize Climate Change Resilience

Jack Alban Dyer

Contents

Introduction and Background 2198
Cook Islands Background and Field Research Method 2199
Climate Change Risk Perceptions Versus Actual Risks for Cook Islands Stakeholders 2206
Experiencing Impact Cost Consequences 2209
Climateproofing Adaptation Strategies: Psychology and Proactive Expectations Theory 2213
Conclusions, Limitations, and Directions for Future Research 2221
References 2223

Abstract

From religious extremism to financial crises, cybersecurity to mass migration, political instability, social tensions, technological disruptions, crimes, wars, and chaos, problems demanding our attention appear immeasurable and unresolvable. Yet from 1970–2014, Earth lost in estimated $2.8 trillion in climate change-related natural disasters and direct consequences. From Caribbean and South Pacific cyclones to a storm and tsunami ravaging Africa and the Southern hemisphere's greatest port, to drought induced, mass migration conflicts towards a fracturing Europe, climate change resilience presents our most urgent need for integrated solutions. How can stakeholders personally contribute towards saving the only life inhabited world from climate change uncertainty? How can reluctant investors convert from recovery to prosperity and opportunity? In response, this chapter proposes the single major threat is psychological! How humans react will determine the extent of risk and impact costs. While previous studies focus on mitigation and adaptation, it draws on field research experience using Cook Islands, maritime supply chain stakeholders. By mainstreaming psychology

J. A. Dyer (✉)
Australia Maritime College, University of Tasmania, Launceston, TAS, Australia
e-mail: Jack.Dyer@utas.edu.au

W. Leal Filho (ed.), *Handbook of Climate Change Resilience*,
https://doi.org/10.1007/978-3-319-93336-8_22

through managing proactive risk expectations, it proposes a theory explaining how this nation integrated climate change resilience into coordinated policies across its population. This could persuade individuals, businesses, and organizations to invest in ecological capital and psychology, prioritizing effective global resilience.

Keywords
Climate change psychology · Resilience · Pacific maritime economies · Supply chain adaptation

Introduction and Background

12th December 2015: 195 countries specifically committed to the Paris Agreement, restricting global climate change to 1.5–2 °C. From 1998–2017, the Middle East experiences its worst drought in 900 years. In 2017 were the worst Paris floods in 50 years, a record US blizzard and Australia anticipates complete Great Barrier coral reef collapse by 2033. April 14th 2018: Day Zero – Cape Town becomes the first advanced world city to run out of water. Kiribati, Maldives, and Tuvalu are projected to be the first three sovereign states, economies, and communities in Earth's history to be completely submerged by 2020. Despite increasing evidence (agreed to by 97% of global scientists and the Intergovernmental Panel on Climate Change (IPCC)), our planet and humanity face an increasingly uncertain future from climate change.

In response, this chapter considers the first challenge is to overcome a refusal to recognize the problem. It does not matter how many (non-binding) agreements are pledged; how much venture funding is available; how much is devoted to research, technology, climateproofing infrastructure, aid projects, and token efforts at mitigation, if we as human stakeholders do not accept this single most significant barrier is psychological! Beyond other factors, it influences how effective mitigation, adaptation, and actions will be. From a commercial perspective, stakeholder requirements, assets, and resources will not survive under business as usual, IPCC scenarios. Yet until this chapter, the above psychological constraint has lacked sufficient critical appraisal to consider how to convince individuals and supply chain stakeholders, to be proactive at managing risk expectations. Evidence from the Cook Islands and other Pacific adaptation strategies seeks to overcome human global inertia and stagnation, failing to prioritize effective, binding action for climate change.

Section "Cook Islands Background and Field Research Method" introduces the Cook Islands, the maritime supply chain (MSC), stakeholder demographic profile, and mixed method approach. This derives from PHD field research during 14th November–5 December 2017. Section "Climate Change Risk Perceptions Versus Actual Risks for Cook Islands Stakeholders" evaluates climate change risk perceptions for stakeholders; compared to actual risk events. This considers the extent to which stakeholders are psychologically over or underestimating risk. Section "Experiencing Impact Cost Consequences" outlines existing and projected impact costs, experienced by stakeholders. These partially highlight why Cook Islands and other stakeholders historically experiencing climate change risk events are far more likely to prioritize effective mitigation and adaptation actions, in contrast to many developed nations and emerging economies. Section “Climateproofing Adaptation Strategies: Psychology and Proactive Expectations Theory” outlines specific solution examples; other individuals and nations could effectively learn from to survive and prosper by focusing on psychology. As part of an integrated theory, this chapter utilize the Cook Islands to consider why stakeholders are so reluctant to prioritize climate change. It identifies what the constraints are. It ascertains what information stakeholders require financially and psychologically to climateproof. This aims to convince Pacific maritime and other global stakeholders to prioritize climate change.

Cook Islands Background and Field Research Method

While existing climate change impact studies are land centered, this paper emphasizes how a significant ocean presence, especially acclimatizes South Pacific, island developing states to prioritize active resilience. It forms the center for Pacific and global trade, resources, climate change risks, economies, ecosystems, supply chains, and stakeholders. The Cook Islands economy is specifically maritime orientated. It is a small 15 island archipelago with over 90% seaborne trade dependent on its ports, for exports, the highest proportions of GDP and economic activity. Figure 1 illustrates this Maritime Supply Chain example, contributing to 13.9% GDP directly and 62.3% indirectly through marine tourism. Figure 2 emphasizes 44.05% of 2016 employment is MSC linked. Marine, cruise and coastal tourism (including game/recreational fishing) provides 85% of foreign exchange earnings (83.5% of GDP). Pearls, fisheries access fees, aquaculture, and remittances are significant remaining contributors of revenue.

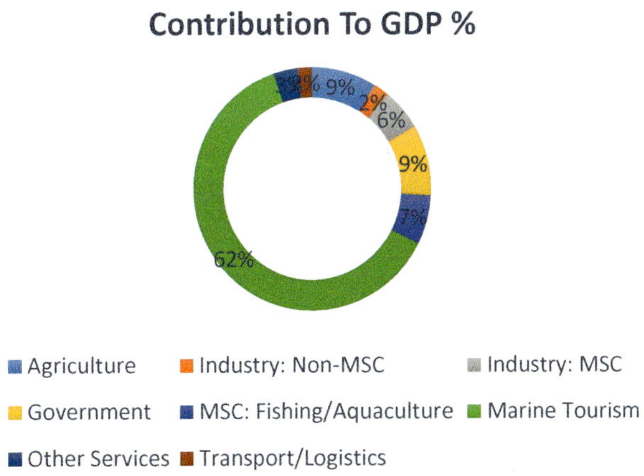

Fig. 1 Cook Islands MSC/Maritime economy contribution as a percentage of total GDP

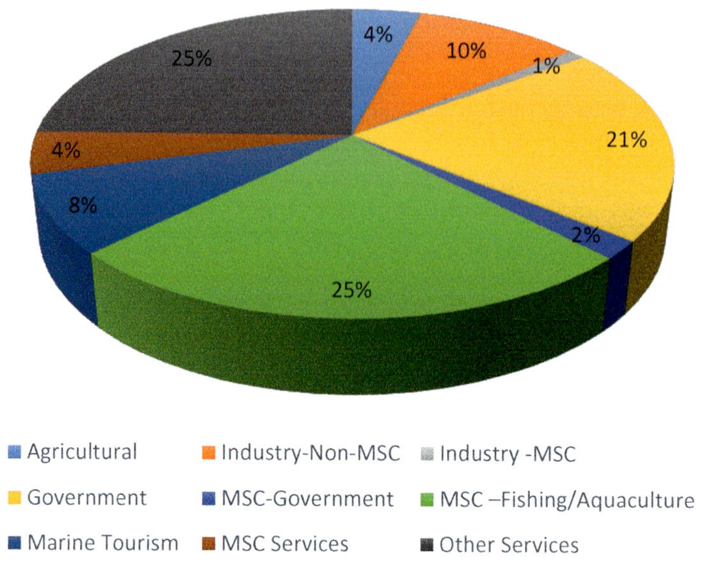

Fig. 2 Cook Islands 2016, MSC employment percentages of total employment

Agriculture is more marginal, contributing 4.2% to GDP and 4.01% of employment. This economy concentrates on noni, pawpaw, citrus fruits, coconuts, and maire but without hinterland connections is highly vulnerable to ports and shipping. Limited industry exists focused on food processing, clothing, marine resource handicrafts, jewellery, gifts, and aquaculture. Other commercial/taxation prospects are restricted without a local tertiary education/research sector, a small, decreasing population, restricted land and customary land ownership. It possesses large ocean distances and no terrestrial substantial forestry/mineral resources, except for potential, underwater cobalt/manganese mining currently being investigated (World Bank 2016). The Outer Islands are highly conditional on 2 cargo vessels every 1–3 months. They are especially vulnerable when climate conditions close ports on average 57–72 days per year. Foreign Direct Investment/Aid has primarily concentrated on developing aquaculture and maritime infrastructure including climateproofing ports and logistics (Fig. 2).

This chapter conducted a mixed method, utilizing qualitative interviews conducted by field research, contrasted with a more empirical time series method approach. Australian Survey Research (2017) defines it online as: "Qualitative research is aimed at gaining a deep understanding of a specific organization or event, rather than a surface description of a large sample of a population. It aims to provide an explicit rendering of the structure, order, and broad patterns found among a group of participants. It does not introduce treatments or manipulate variables, or impose the researcher's operational definitions of variables on the participants. Rather, it lets the meaning emerge from the participants. Concepts, data collection tools, and data collection methods can be adjusted as the research progresses."

Previous wors have identified four main qualitative methods and approaches. Methods include ethnography (anthropology and culture); field research/physical observation with testing, phenomenology (human perceptions and psychology), and grounded theory. Qualitative method characteristics depend on non-calculable data and information based on subjective, stakeholder perceptions. Qualitative methods develop from an established question/s or set of research hypotheses. Methods can consist of theoretical rather than empirical data with predefined variables. Methods derive from certain questions and subsequent grounding/observations. These include recording researcher and stakeholder insights to provide insight into a problem. It can include interactive or interpretive methods to connect theory with data. These often involve a literature review or form of stakeholder consultation through common method, data collection instruments of surveys, workshops, or semi-structured/ structured interviews. Method characteristics are summarized in Table 1.

In reviewing literature from the above search criteria and key research objectives, qualitative methods form most established, method approaches. Ethnography is overwhelmingly ignored in sources in favor of the other three methods. Existing field research method, examples include linking regional, climate change risk, and qualitative analysis with stakeholders in public forums. This presents an inclusive, interactive approach to obtain data and develop theories. Examples include IPCC (2010) and McNaught et al. (2011) for Pileni Island, Vanuatu. Dumaru et al. (2011) offers a grounded theory example for three Fiji villages: Bavu, Druadrua, and Navukailagi. It illustrates the method's comparative effectiveness for similar Pacific research (Characteristics are summarized in Table 3.1). This method is simple, repeatable, and transferable, following selective coding. 85 of 122 provided a reasonable sample size, given the population. A 69.7% response rate offered greater validity. It increases validity, reliability, and consistency with multiple case studies and follow up surveys for the re-iterative process. This approach is also particularly favored for the Pacific region, specific nations, or technical case studies. Examples of research via collaboration include Anderson and Wongbusarakum (2009) and Emaurois (2012). These involved stakeholder workshops for Micronesian communities. Stakeholders complement literature through a specific, local, context. They avoid repeating previous research limits. Field research compensates for insufficient data available. It is necessary, given developing nation constraints. This is essential where previous studies have not provided specific information/insight for MSCs.

However, no methods have been previously applied to Pacific MSCs and climate change, risk management in past studies (Hanson and Nicholls 2012; Camacho-Lim et al. 2014). This chapter proposes a mixed method, including key qualitative method characteristics, approaches, advantages, and stakeholder recruitment for data collection. Including this grounded method and direct data collection, approach presents certain advantages. Qualitative method approaches have several research advantages (Bryman 2001; Gibbs 2007). These include providing background and insight into factors affecting risks, challenges, costs, and constraints for MSCs and stakeholders. It also provides advantages of directly accessing information. This data is otherwise difficult to obtain for many remote, small, developing, Pacific island nations with limited institutional capacity. Grounded theory advantages include linking theory to field research to validate results and observations. Far fewer grounded method examples exist focusing on psychological aspects for climate change. This paper identifies the need to consult key supply chain stakeholders for grounded theory, to obtain accurate risk perceptions, impact costs and adaptation solution information. This obtains specific insight into key research objectives. This is increasingly favored by Pacific field research reports, e.g., the Secretariat of the Pacific Community (2015), with its recent Pacific Resilience Program but this chapter specifically develops a theory applicable to individuals, communities, and private sector businesses elsewhere, experiencing but not reacting to more frequent climate extreme events. This grounded theory approach improves research credibility and accuracy when accessing primary information for an integrated, mixed method.

General climate change, risk management, and impact studies focus on ecosystems, populations, communities, agriculture, and aggregate economic sectors. These frequently ignore methods for specific stages, stakeholders and products/commodities. Only a few recent supply chain examples exist. These have mainly concentrated on ports, shipping, intermodal transport, and general supply chains as separate stages. None exists for assessing climate change impacts on general or Pacific, maritime supply chains. Becker et al. (2011) proved among the first to utilize a qualitative research, perception-based survey to identify climate change, disruption risks, impacts and adaptation solutions for seaports. The method consulted 208 International Association of Ports and Harbours stakeholders with 93 responses. Nearly 53% considered climate change would affect them, and 88% agreed more research was necessary to understand its impacts. The sampling size of 1056 ports from 195 countries ensured the method was valid being of sufficient size. This article considers this crucial based on Table 1 criteria. Results were consistent across regions, countries, and question types. The method is transferable to other case studies. It is transparent and repeatable in clearly specifying questions as an example for grounded theory case studies to follow. However, with such a large sample size, inspiring follow up studies for a re-iterative process may be constrained. The source ignored a re-iterative process to consider if the results changed over time. It did not specify details over coding of follow up interviews, complicating stakeholder content analysis in determining thematic context.

Kretsch (2016) evaluated adaptation leadership in a Providence case study. Using Table 1 criteria, this phenomenology example consulted a small sample size of 30. Twenty-five responses and only 7 interviews limits the validity, transferability, and reliability of the results. Stakeholders considered pre-set data and theory relating to resilience strategies rather than influencing theory, as for grounded methods. No indication was given over coding. A once off workshop, avoided the reiterative process's value. This survey method provided advantages of analyzing stakeholder perceptions towards climate change, the extent of their awareness and to which they are prioritizing adaptation solutions. Advantages are incorporated into this thesis's integrated research design. UNCTAD (2011) similarly used a survey instrument among its 200 members to identify port vulnerability to climate change risks and associated adaptation responses. However, existing sources primarily concentrates

Table 1 Qualitative method characteristics

Theory	Data collection	Characteristics	Evaluation criteria
Ethnography	Secondary data analysis Field work, observation, Surveys/interviews, recordings, Anthropology	Individual factors not overall theory Purpose sampling Narratives Extended contact	Coding Consistency Credibility Reiterative process Reliability Repeatability Sample size Transferability Validity – internal/ external Ethics
Field research	Case studies, testing, experiments, surveys, interviews	Analytical framework	
Grounded theory	Surveys, interviews, various data	Multiple sources influence literature/data/ theory Pre-testing Results develop theory Theoretical sampling Conceptual framework	
Phenomenology	Case studies Theoretical models	Theory develops results Data is selected from single source Theoretical framework	

on port administrators ignoring entire supply chains. Being qualitative/phenomenology based, these sources do not provide specific, ground method advantages.

The following studies affirm existing, phenomenology method approaches with similar data collection instruments outlined in Table 1. These apply conventional risk management theory, rather than grounded theory. However, they fail to provide any empirical means to quantify risk probabilities, perceptions, or impact costs. McEvoy and Mullet (2013) utilize workshops as an alternative qualitative method designed to ascertain Australian port stakeholder perceptions and actions over adapting to climate change risks and impacts. It uses case studies for Gladstone, Sydney, and Port Kembla, seeking to pre-validate its proposed risk-vulnerability method. Its sample size of 41 appears reasonable to validate the model but low given the number of workshops. It provided four stakeholder consultation workshops to test the model and three follow up workshops to incorporate feedback, adhering to the re-iterative process. The model appears to be theoretically valid, based on favorable stakeholder perceptions. However, it is inaccessible for separate external validity and reliability. However, the source provides insufficient information to determine if the results are repeatable, consistent, or transferable to other ports, stakeholders, risk types, and scenarios. In failing to provide individual port specific issues and insight, the method ignored its theoretical value, which would have improved its application and value. It ignores extending a vulnerability-risk assessment method and stakeholder consultation, approach across the entire supply chain system. In contrast, industry assessments of adapting supply chains to climate change include CSR (2011) and BSR (2015). These ignore theory in favor of specific case studies based on qualitative descriptions of risks, impact costs, and adaptation solutions via surveys and interviews.

A number of climate change survey studies including UNCTAD (2011) and Kreie (2013 for global supply chains) also selectively ignore key supply chain stakeholders as economically peripheral. However, this chapter agrees with Becker and Caldwell (2015 for a Gulfport and Providence, USA, seaport case study). It argues climate change impact and adaptation strategies can no longer be constrained to just consulting direct port stakeholders involved. Ports alone are insufficient and uninformed to resolve potential event disruptions on MSCs. The studies' method weaknesses consider stakeholders involve too much time, resources, and effort to incorporate a risk event's economic impact in methods. Kreie provides a phenomenology approach using a case study and interviews based on Table 1 criteria. Its sample size includes 17 stakeholders across a coffee supply chain. This increases validity as comparable studies frequently interview only 5–15. Its external and construct validity is extended through being pre-tested for tea (with similar risks). These test a pre-conceived theory model for interorganzsational learning, using inductive criteria to validate its theory from aggregated results. Its reliability and transparency are increased through a database of results and clearly published questions. The interpretivist perspective considers the method provides greater individual insight and background context into understanding specific issues. Relying on interviews, however, reduces model consistency and validity within a re-iterative process, as results are more likely to vary. The results are less likely to be transferable to other applications. However, they can further validate an overall theory, especially for exploratory research such as climate change and MSCs.

Disadvantages include limited, external reliability and validity, creating a response bias. This bias assumes stakeholder familiarity with climate change, estimating risks and impacts, not scientific evidence or empirical data. Methods may be biased from stakeholder perceptions, leading to risk underestimation or overestimation. For example, McNamara, Hemstock, and Holland (2013) used interviews and a community survey on the effectiveness of adaptation strategies. It also points out a psychological bias affects certain survey answers. Aid agencies and communities may seek greater funding and publicity through myriad workshops rather than focusing on implemented, adaptation solutions. They may be ignorant or exaggerate costs, consequences and solutions (Trundle 2015 for Port Vila, Vanuatu). Qualitative research method characteristics also ignore results that are more inconsistent over quantitative approaches. Some consulted stakeholders may face asymmetrical information relating to climate change. This complicates calculating risks, costs, benefits and consequences (IPCC 2013; United States Accountability Office 2015). Very few data methods involving empirical, case studies exist In response, section "Climate Change Risk Perceptions Versus Actual Risks for Cook Islands Stakeholders," presents actual time series data, the first coordinated database of all risk events from 1900–2015 in the Cook Islands. An empirical approach can assess the extent to which prioritizing and experiencing climate change events, more accurately prepares stakeholders psychologically. Accurate risk perceptions therefore enable stakeholders to minimize potential disruption costs (Section "Experiencing Impact Cost Consequences") and reluctance to be proactive

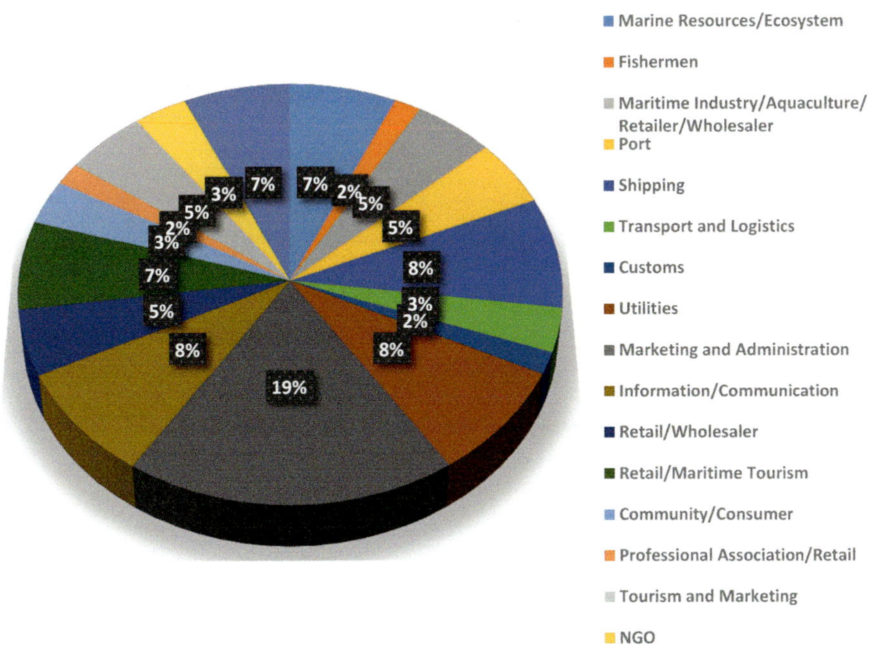

Fig. 3 Cook Island stakeholder demographic profile

at surviving uncertainty (Section "Climateproofing Adaptation Strategies: Psychology and Proactive Expectations Theory").

This research included 59 Cook Islands interviews. The demographic profile of MSC stakeholders in Fig. 3 targeted a cross-section across a maritime supply chain, to understand climate change psychology more effectively. Marketing and administration with 19% of total response rate can be attributed to the majority of marine tourism operators and government ministries more concerned with climate change. In contrast, individuals (1% consumers, 2% fishermen) and smaller private sector operators were less concerned to participate, lacking resources and frequently preoccupied, as observed globally. From Fig. 4, most stakeholders had experience of at least 2 major cyclone and 1 drought risk event in their present role (41% 5–10 years experience). However the next highest proportion (27%), were less psychologically prepared with 1 or no event and less than 5 years experience. Only 8% were really conscious of historic impacts. The average company/organization however was more resilient and familiar, with 21.16 years existence and 56% over 10 years experience. (Fig. 5). To enhance the validity and accuracy of psychological risk expectations, the sample size specifically recruited qualified professionals (29%) for technical aspects in Fig. 6. Almost 56% had at least a degree/diploma to sufficiently understand the research aspect. The few secondary education qualified candidates were experts such as pearl farmers and fisherfolk/consumers

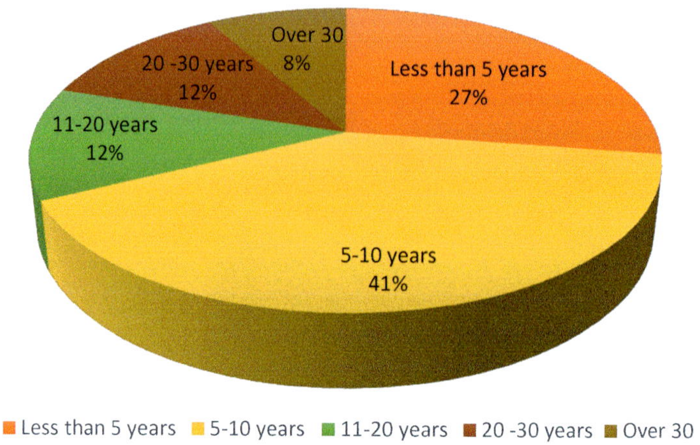

Fig. 4 No of years stakeholder experience

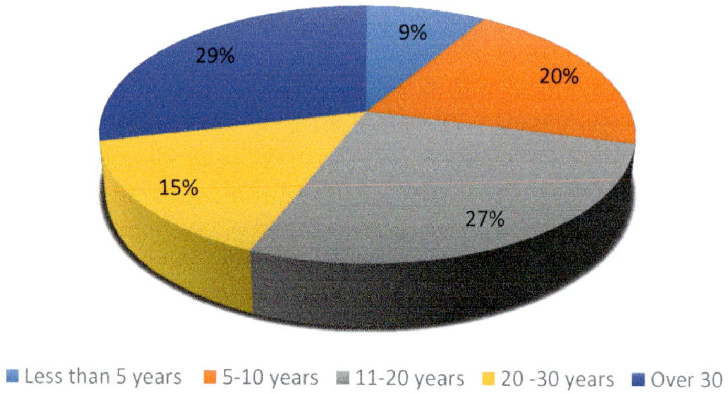

Fig. 5 No of years company is established (2018)

directly perceiving climate change risks, influencing their psychology to have proactive expectations.

Climate Change Risk Perceptions Versus Actual Risks for Cook Islands Stakeholders

This chapter's core objective is to understand why stakeholders are so reluctant to prioritize climate risks. It hypothesizes risk perceptions and experiences influence the extent to which Cook Islands and other Pacific stakeholders are so willing to climateproof and adapt. If other nations and individuals are to minimize disruption consequences, similar or more intense to those outlined in section "Experiencing Impact Cost Consequences," this approach is recommended to overcome

Fig. 6 Stakeholder education qualifications

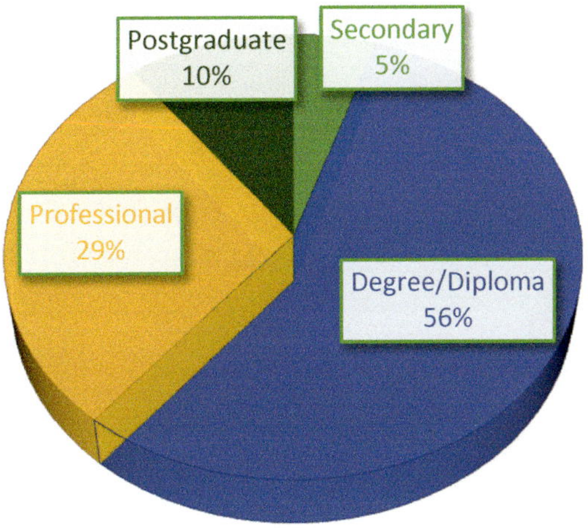

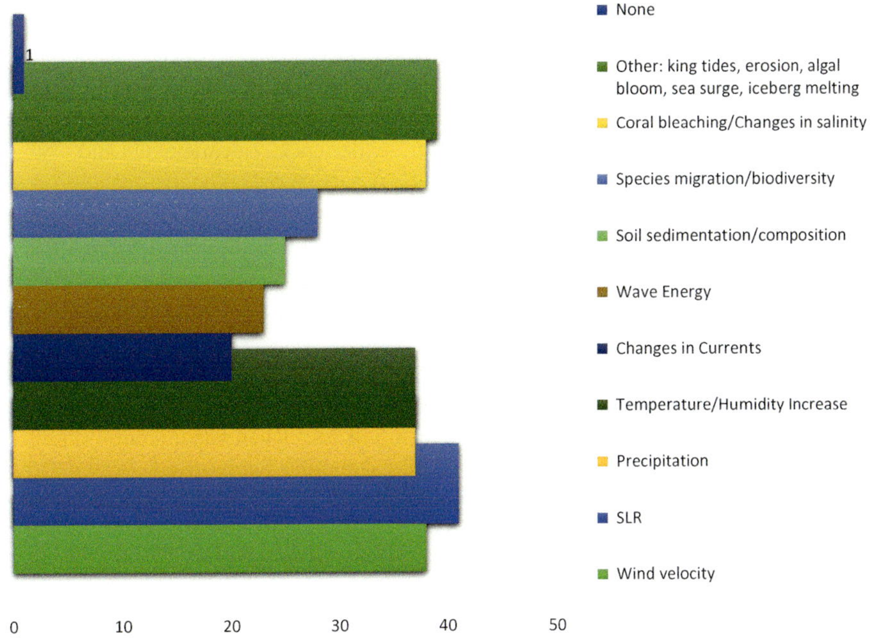

Fig. 7 MSC stakeholder interview long-term risk perceptions

psychological barriers. Figure 7 illustrates various long-term, climate change risk perceptions for Cook Islands, MSC stakeholders. Of a sample size of 59, 41 (69.5%) perceived sea level rise as most significant, followed by coral bleaching and wind velocity (64.04%). These were attributable to high recognition of the contribution of coral reefs to marine tourism, ecosystems, and aquaculture, while wind and

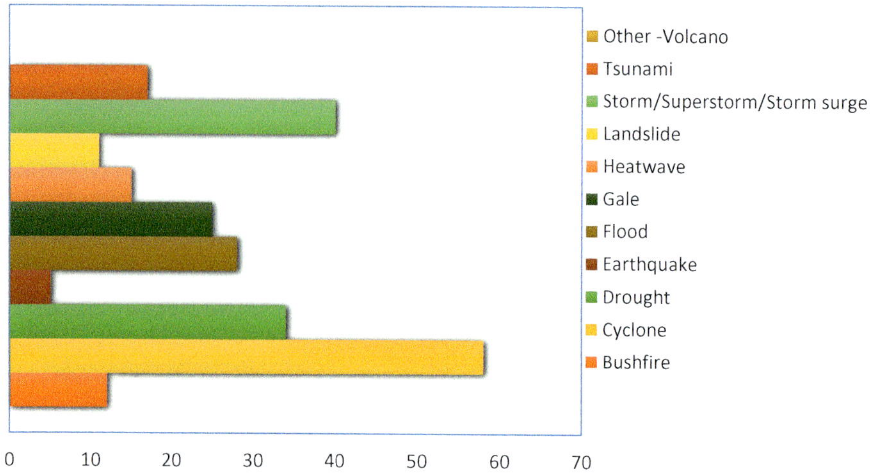

Fig. 8 MSC stakeholder interview short-term risk perceptions

precipitation were often traced to concerns for cyclones and storms. In contrast, changes in currents, waves, and species received less priority, often from a lack of interest, knowledge, or direct impact identified. Figure 8 highlighted short-term climate change risk perceptions. From experience, awareness, and other mainstreamed adaptation strategies, forming a proactive risk expectations approach, there was far more significant consensus on the most significant risks. Cyclones received 58 out of 59 (98%) as the foremost concern, with the sole other respondent not identifying any risk concerns. Storms followed for 40 stakeholders (67.79%), and 34 indicated droughts (57.62%). Earthquakes with 5 (8.47%), Landslides with 15 (25.42%), and Bushfires with 12 (20.39%) were perceived as far less probable to occur. In contrast to 66% of stakeholders aware enough to pinpoint other long-term risk concerns, none identified any short-term events. This was possibly due to uncertainty as to what constituted a long-term risk and what was a natural disaster. Risks remained consistent across the type of stakeholder interviewed.

Including measurements of risk consider the extent to which stakeholders perceived risk accurately plus their ability to learn and experience events. This validates the theory proposed, risk perception management stage, unestablished in conventional theories of climate change psychology (Dyer 2017). Field research provided the first centralized time series data of all recorded short-term risk events from 1900–2015 across 17 Pacific nations (Dyer 2017). It calculated the probability of a historic risk event occurring. Figure 9 illustrates cyclones represent the highest, existing proportion of recorded risks (33%), followed by storms (27%); unlike non-occurring bushfires, earthquakes and volcanoes with 0 events. A comparison with perceptions in Figs. 7 and 8 indicates the extent to which risk can be accurately determined with cyclones followed by storms as most frequent and no examples of volcanoes. However, although only two historic droughts had adversely been recorded, an astonishing 57.62% (34/59) of stakeholders perceived drought to be

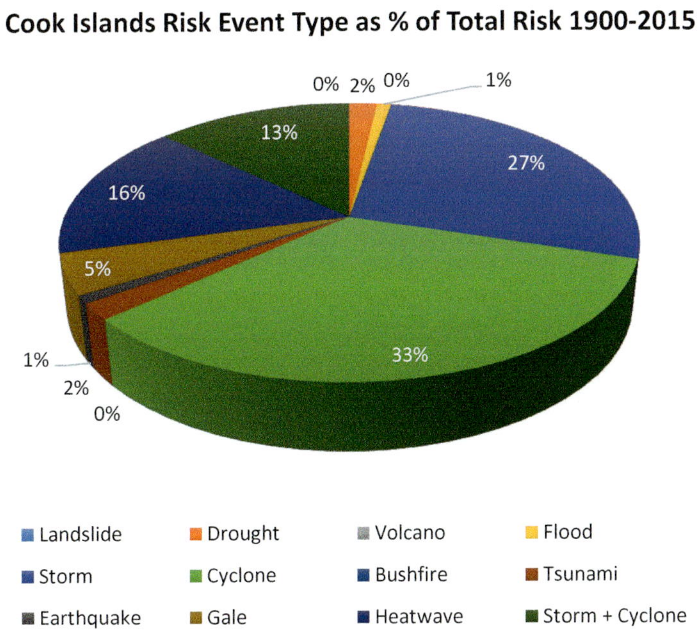

Fig. 9 Cook Islands historic actual risk events

the third most significant risk. This possibly overestimated it due to both droughts occurred recently (within the past 8 years). Several stakeholders were concerned with bushfires, tsunamis, and earthquakes, far more than their historic occurrence suggested. Fewer risks were underestimated. Providing data of actual events and probabilities therefore will produce more accurate risk perceptions and estimation to plan and allocate resources. For low probability, high impact risk events, this paper advocates initially measuring event increases from historic data, combined with climate change scenarios (adjusted for an increased frequency, duration, and intensity). The advantages of gauging historic and future risk through probabilities and this grounded theory approach utilizing psychological risk perceptions, counteracts stakeholder uncertainty, as new risk types are considered highly unlikely by this thesis to emerge. Stakeholders then consciously ensure they know how risks threaten individual operations, underlying ecosystem, and entire supply chains. As the proactive risk expectations theory advances, this ensures those projected to be most significant for MSC stakeholders, correlates to their perceptions and expectations.

Experiencing Impact Cost Consequences

Once stakeholders know how vulnerable and resilient maritime supply chains (MSCs) actually are, the probabilities of particular failures, and which risk types will affect them, they then experience uncertainty over projected consequences. This paper proposes many impact costs are unnecessarily amplified; as even when stakeholders perceive climate change as an actual or potential risk, they may not be personally familiar and psychologically resilient. This paper contends that physical resilience has received far more attention and resources, than how humans will react and why. Therefore, part of proactive risk expectations consists of simple awareness of projected impact consequences for oneself, one's family, one's community, ecosystem, supply chain, and homeland. For those facing a future of uncertain climate change, with an ever soaring frequency and intensity of low probability, high impact events (Black Swans), this section and mixed method approach proposes learning from the following experiences.

How does an economy and how do people survive? "*Storms and cyclones lead to loss of stock and business generation as well as outright destruction of property. Cyclones and coral bleaching events disturbing coral reefs are also followed by ciguatera fish poisoning meaning we have not been able to harvest reef fish here on the main island of Rarotonga since the 1990's. We have to import fish instead*" (Respondent 7). "*It interrupts business, it impacts on the tourism markets affecting numbers and the nation's ability to tend to the local community. It creates local food security issues, which forces the islands to rely more on the importation of food goods at high retail prices affecting our livelihoods.* Without recovery (further pressurised by invasive species including crown of thorn starfish) and prioritising future ecological capital, maritime ecosystem resources decline. This increases extinction rates as catches unsustainably increase, multiplying ecosystem pressure further. *The Cook Islands relies on coral reefs as a nursery and habitat for fish, increased PH levels and warmth levels. This largely impacts on fishing communities that rely on this food supply for their household and trade. It forces fishermen to travel further out into the ocean to catch fish, adding extra financial burdens on a person to maintain an outboard motor and fuel supply, which can be very expensive. Heavier precipitation causes flooding that inundates the road and other infrastructures' capacity to cope. It impacts on agricultural yields and has the ability to increase vector borne diseases, including the downward flow of sedimentation in the lagoon, SLR and the change in current and wave energy.*" This chapter proposes affected stakeholders utilize sources when checking risks to maritime ecosystems and species. They can ascertain which are rarest, most vulnerable and most crucial for MSCs. This prioritize conservation, resources, and reserves with eco agencies, local community, and government stakeholders. "*This impacts maritime industry from local fishermen to national and international shipping suppliers. It also increases the rate of coastal erosion and natural flushing cycles within lagoons. These impacts can also affect the growth of biodiversity and potentially increase the rate of invasive species*" (Respondent 37). The Cook Islands pearl farming industry's collapse re-iterates the need for MSC and other supply chain dependents to incorporate sustainable resources, when planning for business as usual scenarios. It illuminates just how costly disruption can be not just to individuals but to stages and entire economies based on direct and indirect impact costs.

Many stakeholders became psychologically acclimatized referring to five major events, incentivizing them to prioritize climate change from an actual risk event being directly or indirectly affected. These formative events included Cyclone Sally in 1987, Cyclone Martin (1997), five cyclones in 2 weeks (2004), Cyclone Pat (2010), and a recent drought, as influential in altering their risk perceptions. Cyclone Martin in particular illustrated how unaware and under-prepared psychologically and physically stakeholders were. The population dropped from 650 to 240 people. The pearl farm industry that for a decade contributed the most economic activity, collapsed production from 200 farms to below 20. Pearls collapsed from 300–40,000 to around 10,000–30,000 per year by 2017. Respondent 24 emphasizes this total loss "*You are living on narrow strips of atolls and once the waves fold over from the ocean to the lagoon, everything on that narrow strip of land will get taken out to sea. So I don't know how to specify what type of assets were affected from climate change. Anything tied down or not was vulnerable. Coconut trees were being ripped out so anything! Your capital infrastructure –pearl lines, floats, spats, trestles, dive gear, boats, and seedling houses – They were gone – blown away by the cyclone or taken away by the waves. But more importantly, the psychological and social impacts on individuals were devastating. So you have to start from scratch again and it is a very expensive business.* Stakeholders expressed doubts the market is ever likely to thrive to the level it's ecologically and commercially viable/sustainable." Respondent 55 emphasized: "*The most vulnerable cost is loss of equipment, infrastructure as well as morale and psychological strength to keep going. Infrastructure alone the foreshore are most likely to fall during cyclone events.*" Respondent 36 warned: "*Development in high-risk areas is really prone to natural disasters. Removing natural barriers or changing landforms has increased the risks as development continues to expand.*"

Cook Islands stakeholders reveal just how essential it is to prioritize climate change resilience psychologically, to climateproof effectively. As described by the voices of tourism operators, *tourism contributes around 60% – all our eggs in one basket – government takes steps to try and diversify the economy limited. The problem is we are running out of time. I suspect we are still vulnerable Nov–April cyclone season – that would undermine the future tourism business. With uncertainty, you cannot make a firm risk assessment. The tourism industry would be supportive to any research and adaptation initiatives but at the end of the day realize we will look after ourselves. However, we must be mindful of those who say climate change is a lot of rubbish.* "Tourism representatives rather than visitors are concerned about anything threatening sustainable guest experiences. They value a pristine marine environment, basing their marketing strategy on its continued existence. They remain conscious about past cyclones creating reputational, marketing, and other undetermined impact costs. *Extreme events mean they can credibly say this no longer e.g. political figures in US. Externally we do not know what will happen and the global response. All we can do is see what is happening elsewhere concerning climate change and learn from it. A lot of aid money comes from Australia, New Zealand and EEU. They tend to be guided by their use. At least we are not so much aid dependent. But this is probably a good thing because left to our own devices we probably advocate a more extreme position over climate change. Most of our tourism industry is vulnerable on the coast with extensive capital investment. For adaptation – what can be done, what should be done, what are the limits and constraints?* After 2010, it took 2–3 years for tourism to recover. One concern involved competing among Pacific nations for the same tourism pool. Any failing to prepare for climate change faces diversion costs to those more resilient. *"The presumption is that the problem is there – its not going to go away and that any countermeasure that we take to adapt will only be partially effective. It is going to be really necessary to adapt – otherwise businesses will not survive. Hopefully given support/aid funding, communities are more likely to endure. As a business, we have to be realistic by cutting down costs."*

Respondent XXXIII considers most domestic tourism operators would probably be in agreement over climate change awareness and concern, although it remains dominated by less aware expatriates, lacking historic familiarity. "*Climate change as a problems may never be solved. We may be around the tipping point already. How do we manage our way and prosper – without responding? Well, we don't have a business otherwise. We would experience a downturn in tourism numbers as among the first to directly encounter global climate change locally.* Many do not understand personal implications or costs with a naïve view of risks and coral bleaching. *In the Cook Islands, there lurks the catastrophic perception that if we've done the risk assessment – we are fine, even though it's once off. We have to overcome the dangers of over-complacency. That is, I won't be around so it's not my problem. This is not true. I don't think risk assessments are rigorous enough. There is the problem of not forecasting for long-term projections. A risk assessment is a continuous ongoing process. The risks will be different next year. There is a grave sense of complacency. Of course we know what the main problems are. That is, for Kiribati it is sea level rise, but being gradual, but we really don't see it.* For MSCs to effectively climateproof, far more marine tourism operators need to be conscious of threats to their physical environment, assets, and ecosystem resources rather than expecting government action or relying on insurance.

Stakeholders themselves recognized many psychological barriers when prioritizing climate change resilience, indicating "*The nature of climate change appears so slow"* (Respondent 6). In 2017, when relying on the rest of the world to lead through an example, the problem of "fake news" and ideologies has challenged other regions and individuals globally to recognize the need to adapt. Respondent 3 pointed out, "*Tourism and other operators are businesses – have to weigh up long term and short term priorities and long term and short term investment decisions."* If there was less pressure to provide good short-term results rather than long-term values, then they would be more likely to respond to mitigation measures. Yet others are mobilizing psychology, so stakeholders aware of potential impact costs, vow to not gamble on surviving such uncertainty, adrift. "Being resilient means being able to address it better – and having the resources to do so (Respondent 47). *"We didn't play that much of a part of it and contribute, yet we are facing it. With only 6.5 km^2 total limited land area – land is so important to survival. The oceans are going to be even more so. What are we going to do to minimize the effect on us and those around us?" (Respondent 39) "I believe everyone should be thinking about it. 'When the ocean and world are destroyed – we will be destroyed!'"* They need to consider how prosperity will be retained rather than complacently assuming it's not their long-term problem. There appears little evidence of mutual cooperation and information sharing with Pacific tourism operators viewing each other as competitors, rather than curtailing existing ecological pressures, pursuing eco-tourism and communicating to pool resources during risk events. Marketing approaches need to register their roles in promoting environmentally conscious actions upon tourism, logistics sector, producers, and retailers (Respondent 22). *"We should not ignore those or expect everyone else to do something about it but ignore them. People in the future will be able to say that we did something that matters – we secured and dreamed of our future"* (Respondent 55). *As the younger generations, you will have to pay and live with our mistakes* (Respondent 58). *"For the outer Island atolls it is extremely worrying and not sure what we can honestly do about it – except to plan? It doesn't matter so much what we are doing here – but it depends so much on what other countries do or fail to do"* (Respondent 15).

While existing research concentrates on top down methods providing aggregate costs to entire economies, this chapter's theory distinguishes itself further by a bottom up approach. This specifically consults participants to determine impact costs to individual stakeholders and stages. This aims to aid individuals to more accurately understand personal risk consequences not just to entire economies, ecosystems, and supply chains. It seeks to incentivize risk action as soon as possible, via proposed specific strategies in section "Climateproofing Adaptation Strategies: Psychology and Proactive Expectations Theory." It enables participants to consider possible partner and competitor's costs, given commercially sensitive data, limited information, and other adaptation constraints and adjust trade accordingly.

Climateproofing Adaptation Strategies: Psychology and Proactive Expectations Theory

Aside from proposing the proactive risk expectations theory to overcome psychological barriers to climate change, Cook Islands stakeholders identified a core adaptation constraint to effective climateproofing, included limited private sector capital and funding access. This section also clarifies reasons why stakeholders are reluctant to invest for climate change psychologically. It details information they specifically require to effectively finance adaptation. It proposes physical indicators measuring the extent to which an investment is effectively climateproofed. This is essential to consider how any investment is or will be influenced by climate change. This chapter proposes these measures to overcome existing research limitations, which recognize this constraint but ignore these factors, (Inoue 2013; Averchinkova et al. 2015). This aims to encourage uncertain stakeholders to become more proactive, recognizing opportunities and the need to effectively respond to risks and impact costs. Climate change is worth investing in the short, medium, and long term, being more profitable, sustainable, and preserving stakeholder requirements more reliably than the opportunity cost of inaction. These indicators can validate investment decisions over alternatives.

Factors Clarifying Stakeholder Reluctance to Prioritize Climate Change

- Access to capital/finance or awareness of options.
- Asymmetrical information/climate change uncertainty.
- Lack of concern.
- Lack of incentive/legal uncertainty.
- No investment/information criteria.
- Originality – novelty – resilience to chance and change.
- Other priorities – Finance, risks, business operations.
- Perceived relevance.
- Risk ambiguity.
- Skepticism.
- Timing.
- Uncertain short run profitability and benefits.
- Uncertainty over cost effective, sustainable adaptation responses.
- Unknown inaction, maladaptation, and opportunity costs.

Information Specifically Required for MSC Stakeholders to Invest and Adapt

- Asset resilience under IPCC/downscaled projections.
- Conditional probability of failure, timing, intensity.
- Extent of vulnerability/risk.
- Event experience.
- Impact costs.
- Implications for cashflow, gearing ratio, liquidity, solvency, profit, return on investment.
- P/E to growth ratio, P/cashflow, P/E ratio, profit margin.
- Portfolio exposure.
- Projected recovery time.
- Projected performance.
- Opportunity/inaction costs.
- Regulations, taxes, incentives and penalties.
- Targeted benefits, cost savings, revenue and opportunities.

Physical Indicators Measuring the Extent to Which an Investment is Climateproofed

- Business awareness over climateproofing resilience.
- Change in asset performance.
- Change in conditional probability of asset failure/asset resilience.
- Change in communication, financial, information, physical, psychological exposure, and leadership.
- Change in percentage of assets exposed.
- Change in risk perception/people trained.
- Competitors.
- Coordination/cooperation with other stakeholders.
- Demand/supply/market changes.
- Development of a business continuity plan.

- Extent of mitigation/adaptation investment.
- Extent of globalization, asset interdependency and supply chain exposure.
- Future earning power.
- Liability.
- Projected risk/vulnerability –long run/short run.
- Recovery time changes.
- Resources allocated/reserves.
- Resource sustainability.
- Stakeholder reactions/reputation/ requirements.

Figure 10 illustrates 12% of major constraints to climate change, identified by Cook Islands stakeholders is psychological, 14% information/communication, 11% education, and many include human elements such as communication/cooperation. Respondent 16 summarized reasons *why many are not prioritizing climate change, despite risk perceptions and spectacular impact costs.* Those identified include a lack of knowledge, lack of understanding – just disseminate. Yet communication is insufficient – we all live in a community with different languages. "We *all experience different levels of literacy – what you say to a group of adults or intellectuals is not what you will say to a bunch of kids –the message is different – we need to find communication to understand. Sometimes we have set ways – sometimes it's really hard – have to be really dynamic to understand whether it has reached that community. But sometimes they create barriers themselves because they feel their authority needs to be challenged or compromised – they take it as a threat – the language, the tone – takes a special kind of person."* To improve communications and psychological responses to barriers, they suggested*"Someone fluent in both languages – not just a scientist who research and understand it; diverse background; cultural aware of changes – who has historical knowledge as well. If you were just a researcher – outsider – but don't understand the stories – then the barrier goes up, But if you can make comparisons in the past; in the future; regionally what is happening to other islands and come with people who compliment your work in other areas."* (Respondent 30) *When you see the effect of climate change – then we don't know when to start. Climate change is not going back. The severity of climate change is increasing and increasing. We have to move forward.... We need to continuously look at options or ways in which we can coincide with changes in climate because as history stated in that case, we see what will affect the port. We can decide what we can do to improve it – so as not to do it perfectly. This is so that it leaves something for us to rebuild it. Not when something happens, nothing is left....*

This proactive risk expectations theory formalizes a psychological approach to climateproofing MSC'S and stakeholders. Recognizing the significance of human risk expectations and behavior on impact costs, it favors proactive intervention for effective adaptation. It ponders core factors as to why stakeholders do not prioritize adaptation. It advances a response strategy. This would utilize the above core strategies to motivate psychological willingness and behavior to change. It provides several techniques as to how expectations can be rationally managed for MS's. The Proactive Expectations Theory proposes to mainstream climate change risk

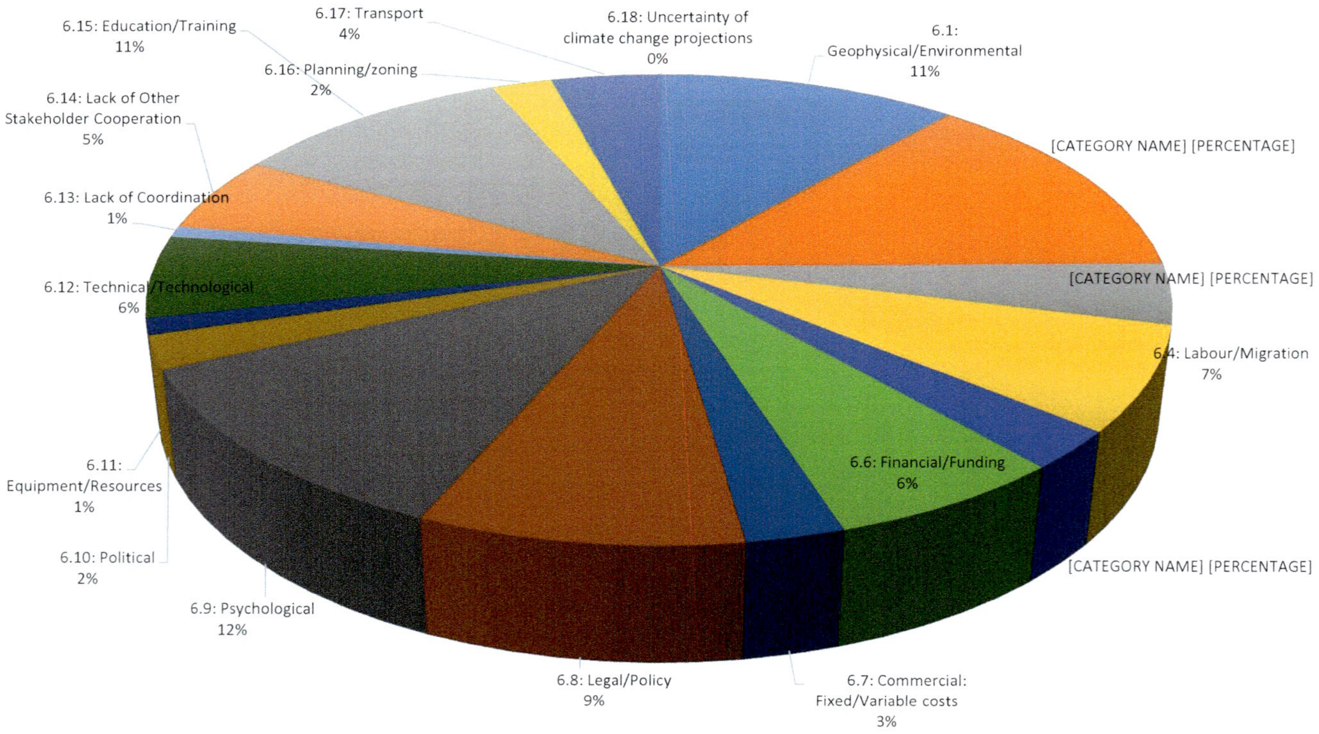

Fig. 10 Constraints to Cook Islands MSC stakeholders' climate change adaptation action

management as one risk. Stakeholders should be legally bound to consider this Pacific approach across the supply chain, when registering, operating, and physically creating a business. As Stakeholder 23 states, "*People's lives are not only for today but for tomorrow. People's lives should be nothing to compromise on in the future.*" "*People need to start planning and making investment choices for the future and need to listen to and engage with the people in the industry. What works for people is being able to see the visual impact – information as video documentaries about different topics,*" another states. Access to time series, historic risk data, the extent of public sector adaptation and future projections needs to be available. Greater enforcement capacity would ensure this with incentives to encourage greater mitigation/adaptation. Violations should be swiftly penalized. The optimal legislation would ensure stakeholders are responsible for local ecosystem defenses, to lower vulnerability, and increase resilience where practically possible. It overcomes market failure to estimate and prioritize climate change. Based on the precautionary principle and scientific evidence, this seeks to overcome denialism, inaction, and expectations, which solely rely on the public rather than commercial sector to proactively adapt. The following psychological constraints present core factors as to why adaptation is not prioritized.

MSC Psychological Constraints to Rectify

- Apathy/indifference.
- Asymmetrical information/uncertainty.
- Character assassination – Rather than evidence based critiqueDespair/haplessness.
- Faith in others/technology.
- Fear of ostracism/powerlessness.
- Lethargy/inaction.
- Moral hazard.
- Pedantry.
- Present moment orientated – Short term immediacy syndrome.
- Risk aversion – The status quo.
- Self-advantage/gain.
- Self-interest.

To convince people to modify human behavior represents a significant global challenge this paper proposes for future research, given these factors. However, uncertainty and asymmetrical information are resolvable by producing simplified, local data and criteria based on these guidelines. Increasing stakeholder participation and awareness, facilitates more accurate risk estimation for their perceptions. Offering actual methods and case studies emphasizing how risks affect stakeholders personally and across a supply chain can minimize apathy and lethargy. Thinking about risks could be conditional for investment and development. Assessing risk, vulnerability, the extent of adaptation, and systematic interdependency for individuals, partners and competitors provides empirical metrics as to whether having faith in others is a reliable climateproofing measure. The conditional probability of an asset/system failure from a specific event provides a second measure. The more stakeholders are involved, the more they will feel personally interested, motivated, and willing to act. If solutions and opportunities are pointed out, rather than obsessing over costs, they are less likely to feel despair, indifferent, or powerless. If a formal policy is actively encouraged, fear of ostracism or loss of self-gain and self-advantage is diminished. Limited research has evaluated the feasibility, rate of investment return, and consequences of technology solutions including geo-engineering and carbon capture and storage. Technological progress is more uncertain than investments in measures that not only adapt but provide extensive co-benefits. The following Cook Islands adaptation strategies illuminate efforts to mainstream and prioritize climate change (Cook Islands Government 2010; Cook Islands Government 2012b; Cook Islands Office of the Prime Minister, 2012). Previously, certain hard engineering, adaptation strategies including seawalls were attempted but failed. The Cook Islands prefers to utilize its experience and physical exposure to risks to ensure the survival of future infrastructure with 100 year time horizons, not merely seeking "business as usual."

- **1987 cyclone Sally**: Avarua breakwater was disassembled to Armour the foreshore revetment; relocate fuel depots inland, partial breakwater, selective protection – Runway/harbour. A coastal breakwater was relocated after experiencing cyclone damage.
- **1992**: Ratified and implemented Kyoto.
- **2005**: The airport protected its radar and other electronics and secured 3 months, fuel supply after 5 cyclones – Construction of Avarua Marina, Aviatiu western breakwater, triad petroleum depot offshore breakwater/sea wall; offshore breakwater to protect the runway end.
- **2010 cyclone pat recovery plan:** This established $530,416 committed to improving disaster risk reduction in foreign aid with $200,000 to improving waste management, $24,000 on a climate risk warning system, and $200,000 on improving general infrastructure resilience. It also allocated $1,356,870 to local economic recovery, aid being conditional on stakeholders considering climate change awareness to enhance future resilience, which developed countries have yet to consider. $1,000,000 of this was allocated to small entrepreneurs, $194,870 to restore livestock and $55,000 for more climate resistant taro production. It also allocated $1,356,870 to local economic recovery. Aid was conditional on stakeholders considering climate change awareness to enhance future resilience, which this chapter notes developed countries have yet to consider as a major climateproofing recruitment strategy. $200,000 was committed to improving waste management, $24,000 on a climate risk warning system, and $200,000 on improving general infrastructure resilience.
- **2012: The National Infrastructure and preventive infrastructure investment plan** requires considering climate change for any potential maintenance/upgrades to preserve and enhance performance, flexibility, and productivity.
- **2013**: $512,343 was invested in improving disaster risk event, response training to emergency management Cook Islands.

Climateproofing involved a series of various adaptation strategies to address various issues. To secure marine ecosystems' future, from 2012–2016, the "Ridge to Reef" approach extended maritime sanctuaries, imposed fishing restrictions in days and quotas, increased community-based enforcement, ecotourism, and prioritized aquaculture to ensure future maritime food and resource security in its National Aquaculture Development Plan. For aquaculture, it believed in enhancing resilience through income source diversification from wild fisheries and improving controlled environment conditions. It received funding to attend a 2012 aquaculture adaptation workshop, technical assistance, documents, and training courses. A Cook Islands National Marine Park (costing $32.91 million), with biodiversity species survey has been formed. It covers 1,100,000 km^2 of southern EEZ. Its policy framework specifically aims to prevent species extinction. In 2016, it ratified COP21 Paris and passed a new marine resources bill with stringent enforcement powers, bans, and fines from $50,000 to $1,000,000 for violations. This believes in rights based resource management, marine reserves, Muri community management policy, species harvest bans, and coastal monitoring ecosystem survey. The 2005 and 2016 marine resources acts call for resource conservation and management. The Cook Islands has established designated fishing rights, aquaculture and fisheries management plans. Its national vision Te Kaveinga Nui aims, "To enjoy the highest quality of life consistent with the aspirations of our people and in harmony with our culture and environment" with the following core goals.

- **Goal 5: Resilient and sustainable communities. "A Cook Islands where our people are resilient to climate change and disasters to create sustainable livelihoods."**
- **Goal 6: Environment for living: "A Cook Islands where we sustain our ecosystems and use our natural resources efficiently to create sustainable livelihoods."**
- **Goal 7: Good governance.**

The strategies involved climateproofing public infrastructure such as Mangaia Harbor: This extended concrete hardstand width precast concrete quay walls to minimize corrosion. It added complete channel widening to minimize wave impacts, relocating the boat ramp south with a beach spending zone north of existing ramp providing a natural ecosystem resilience zone. This provides benefits of avoided cyclone damage and extended harbor lifespan. It avoided losses to Taio Shipping; benefits from reduced injury/death; avoided losses to fisherman – safe access. The Port Authority invested $ 574,203 for wharf strengthening and durability. It allocated $19,660 to magnify safe wharf access during rough seas. Since 2012, risk expectations consider these to reduce maintenance costs. The port constructed a concrete slab with 250 mm thickness, a double reinforcing mesh, a perimeter quay platform with concrete wall, wave scour, and energy minimization, a concrete barrier wall to reduce cross currents plus a wave energy dissipation zone. The climateproofed infrastructure project provides improved monitoring and maintenance programs/training/funding for additional sustainability for the Mangaia local government plus disaster emergency funds.

Other strategies included financial preparation such as the ETRF Fund/Loan Repayment Fund/contingency budget of ($819,000), which needs a state of disaster/emergency to access. From 2011 an established disaster response trust fund increased from $NZ 200,000 to 2,100,000 in 2017, PCARFI (Pacific Catastrophe and Risk Financing Initiative provides a centralized public infrastructure insurance scheme and asset register for Pacific islands governments since 2013). Climateproofing the Cook Islands included improved communication and information awareness: curriculum, training, video, workshops, plus the Ministry of Marine Resources website providing information on environment, future of resources, central research, business opportunities, careers, ecosystem, and fisheries management. Stockpiling/reserves are only informally adopted and indicated from a few evaluated stakeholders. Other tools include Cook Islands Climate Change Biography (Online source centralizing past research), disaster risk management plan/profile, online geo-portal with data, training, and materials. It includes effective coordination through disaster risk management, climate change and stakeholders via a NDRMCCC/local island specific DRMs' linking stakeholders, responses, stockpiling, and emergency contact points. Te Terai Vaka (Coordinated Development/Aid Project System) specifically set up with $150,000 maritime.

Other efforts involve improving early warning systems to anticipate risks and information gathering. This includes Integrated Coastal Zone Management. SPC (2014) provide a marine habitat map and baseline data, hydrodynamic modeling, geospatial framework, training, research, improved environmental education, and institutional capacity building. It included ($188,131) on a Cook Islands Coastal Calculator. Coastal risk modeling for Mangaia identified coastal exposure of specific assets given climate change projections. Oneroa village agreed to prevent further building in risk exposed area and to prioritize adaptation via evacuation routes, restricting land use, and restoring natural vegetation, moving infrastructure inland, elevating structures where possible. It formed a Mangaia Coastal Policy framework. Pacific Climate Futures provided an online tool to connect Cook Islands and other Pacific climate/climate change information) plus climate early warning, monitoring, and evaluation systems. PACMAS (Pacific Media Assistance Scheme) recommended induced increasing media awareness and training for information/communication. The Cook Islands finally believed in strengthening Institutional Capacity. This extended to the formation of Ministry of Marine Resources in 1984 (policies, research, species information; projects) and creation of Climate Change Division/Emergency Management, Cook Islands. In 2011, this moved to the Office of Prime Minister. It incorporated this into legislation and policy. Regional Pacific plus mainstreaming into various laws including National Adaptation Plan of Action/2011 National Sustainable Development Plan and aligned to sustainable development/co-benefit opportunities. The Joint National Disaster Risk Management and Climate Change Adaptation Plan (2011–2015) promotes cooperation, coordination, and collaboration – governance and monitoring across stakeholders to prioritize climate change risk. Each island has prepared a Disaster Risk Management Plan for cyclones/droughts, identified key stakeholders, conducted awareness simulations and shelters.

Over decades of risk research, comparatively few solutions incorporate changing human psychology and perceptions of risk, despite this influence. Other than experiencing an event and accessing information, training through simulated single and multiple events provides an effective adaptation measure. Willingness to adapt is as significant as physical capacity. This chapter can assess how individuals will psychologically react and prepare for any scenario type (as undertaken for the Cook Islands). Climateproofing via psychology and expectations provides mechanisms to determine if business would resume normally. It is more cost-effective, given increasing average and Black Swan risk event probabilities, then to await an event as a final test. Self-interest needs to become a part of this, especially since personal behavior from a risk event not only affects the business and ecosystem but across the MSC's. It can emphasize how these minimize maladaptation, impact, and opportunity cost. Understanding actions and inaction presents definable consequences could prompt more environmentally and risk sustainable behavior. Hope, profit, empathy, ambition, responsibility, efficient resource utilization facilitate productivity and far more probable survival. Effective information, participation and media communication over systematic risks and consequences can counter skepticism, particularly among developed nations. Decisive leadership and mass participation are pivotal, plus supporting successes for pervasive risk. These can more credibly persuade those uncertain or reluctant to change.

Conclusions, Limitations, and Directions for Future Research

In conclusion, this chapter specifically identified the foremost barrier to climate change is psychological. It conducted 59 interviews for Cook Islands maritime supply chain stakeholders, assessing short- and long-term risk perceptions, impact costs experienced, and proposed solutions implemented. These were assessed against empirical time series data to determine how accurately risk was underestimated, overestimated, or determined. It highlighted personal experiences before proposing an integrated response to the causes of climate change. This aims to convince Pacific and other global stakeholders to prioritize climate change resilience through climateproofing adaptation strategies including psychology. This paper's limitations is that physical constraints did not permit a more effective examination of impact costs, other solutions and across risk types, additional locations, IPCC scenarios and stakeholders or factors other than psychological. These could provide further directions for future research.

This theory considers stakeholders need realistic expectations viewing climate change proactively, to mitigate and adapt. This avoids reactive behavior where disruption costs throughout all time phases are maximized; as psychologically, many stakeholders are unprepared for climate change. Personal relevance promotes cooperation once they know how to identify risks, which information to collect, which resources are necessary, and what they can do about it. Stakeholders need to recognize risks but value assets, assess impacts, understand, and prioritize. Conventional climate change, economics, or risk management theory often marginalizes these psychological factors. Proposing a theory fails if it does not reflect why stakeholders initially created the conditions necessary for the strategy to be proposed and why these conditions often remain. Stakeholders are frequently present moment orientated, concentrating on risk events, costs, and benefits within an immediate time horizon. Providing time series data, probabilities and methods enabled them to concentrate decision-making over greater periods. Risk aversion, apathy, and inaction can be countered by advising stakeholders further consider how requirements will be affected plus inaction, maladaptation, and opportunity costs. Ignoring psychology ensures the status quo of inaction, and limited reactive adaptation is preserved. Risk averse humans prefer certainty, despite significant impact costs to the unknown.

Targeting psychological expectations and behavior ensures any adaptation benefits become routine and permanent, not temporary. Moral hazard is marginalized with mutual cooperation, interest, and adaptation. Indicating opponents lack empirical evidence can persuade others through common sense and reason. Inquiries could focus on what they are unsure about and what would convince them to react. To persuade people, climateproofing adaptation needs to indicate personal lifestyle, interests, employment, income, and environment are far more secure and cost effective, than losing access when events occur. For those who have pursued adaptation, How many have actually regretted it? How many wish they had responded after a risk event? Theory needs to distinguish between producers, consumers, other stakeholders, and the public sector in adaptation. Stakeholders need to provide as much specific information as possible, to replicate a scenario as accurately as possible and to link adaptation and experience to training and to coordinate across supply chains.

When climateproofing, how each stakeholder is projected to react influences the strategy's effectiveness on performance, resilience, and disruption. Information, training, and awareness lessen this risk. Currently, they react with maximum disruption, focusing on other priorities. Many act post-event, – those that manage to survive. Focusing on recovery advised by conventional risk adaptation theory doesn't prepare stakeholders psychologically to effectively climateproof, invest in ecological capital, and exploit other opportunities. Climate change needs emphasis over other more familiar risks, with conventional tools for risk reduction. Yes problems exist – but hope remains if action is prioritized. Interest needs to be sustained so momentum isn't lost. Effective information needs to be obtained and utilized over risks and impact costs. This chapter and proposed theory incorporates psychological responses, improving training, processes, and facilities for future events. It provides advantages of an integrated response to the causes of climate change. Stakeholders need clear cost and benefit comparisons of various adaptation strategies to understand self-advantage and self-gain more convincingly. It needs to ensure stakeholder requirements and functions are retained. Although not all human behavior is certain to change, achieving these strategies, provides greater inducements to persuade other stakeholders, given pervasive skepticism and urgency. Ultimately, reality will impoverish skepticism, through attrition, war, failure, and bankruptcy. Choosing uncertainty over stability of resources, revenue, infrastructure, expectations, supply chain activity, and policy (as favored by the Pacific) will determine if these psychological constraints to effective climateproofing is less valid than this sections' theory favoring proactive expectations of climate change.

References

Anderson C, Wongbusarakum S (2009) Climate change and the Micronesia challenge: ways forward in collaboration and adaptation. University of Hawaii Hazards, Climate and Environment Program Report, viewed 29 March 2015, http://www.hazards-climate-environment.org

Averchinkova A, Crick F, Kocornnik-Mina K, Leck H, Surminski S (2015) Multinational corporations and climate change: are we asking the right questions? A review of current knowledge and a new research perspective. Grantham Research Institute on Climate Change and the Environment, Centre for Climate Change and Policy Working Paper 208, viewed 3 June 2015 http://www.cccep.ac.uk

Becker A, Satoshi I, Fischer M and Schwegler B (2011) Climate change impacts on international seaports: knowledge, perceptions, and planning efforts among port administrators. Springer Journal of Science and Business Media, viewed 22 April 2015, http://unctad.org/sections/wcmu/doc

Bryman A (2001) Social research methods. Oxford University Press, Oxford

Camacho-Lim L, Hobday A, Bustamante R, Farmery A, Fleming A, Frusher S, Bridget S, Green B, Norman-Lopez A, Pecl G, Plagányi E, Schrobback P, Thebaud O, Thomas L, van Putten I (2014) Facing the wave of change: stakeholder perspectives on climate adaptation for Australian seafood supply chains. Regional Environmental Change, viewed 5 Aug 2016, http://www.springer.com/gp

Cook Islands Government (2010) Cook Islands National Disaster Management Plan. Cook Islands Government Press, Rarotonga

Cook Islands Government (2012a) Joint National Action Plan for disaster risk reduction and climate change adaptation 2011–2015. Cook Islands Government Press, Rarotonga

Cook Islands Government (2012b) Joint National Action Plan for disaster risk reduction and climate change adaptation 2011–2015. Cook Islands Government Press, Rarotonga

Cook Islands Office of the Prime Minister (2012) An assessment of National Climate Change Priorities Alignment to the regional framework. Cook Islands Government Report, Avarua

Dumaru P, Jeke V, Simpson R, Ratukalou V (2011) The adaptive capacity of three Fijian Village communities: Bavu, Druadrua and Navukailagi. Australia Department of Climate Change and Energy Efficiency Report, Canberra

Dyer, 2017 Projecting the future of Pacific Ports, shipping and maritime supply chains: evolving climate change risk management into risk opportunities. IAME 2017 Conference Proceedings 27–30 June 2017, Kyoto

Emaurois C (2012) Rapid assessment report – community engagement component. USP-EU Global Climate Change Alliance Project, Pacific Centre for Environment and Sustainable Development University of the South Pacific Report, Suva

Gibbs G (2007) Analysing qualitative data. Sage Publications, London

Hanson S, Nicholls R (2012) Extreme flood events and port cities throughout the twenty first century: implications of climate change and other drivers. In: Asariotis R, Benamara H (eds) Maritime transport and the climate change challenge. Routledge Taylor and Francis Group Press, London

Inoue S (2013) Incorporating climate adaptation in Port Dynamism. European Commission Joint Research Centre, The Scoping Workshop on Seaports and Climate Change, 4–5 March 2013, Brussels, viewed 2 July 2015, https://ec.europa.eu

Intergovernmental Panel on Climate Change (2010) IPCC TGICA expert meeting report: integrating analysis of regional climate change and response options. IPCC Report, Geneva

Intergovernmental Panel on Climate Change, Prather M, Flato G, Friedlingstein P, Jones C, Lamarque J, Liao H, Rasch P (2013) Annex II: climate system scenario tables. In: Stocker T, Qin D, Plattner G, Tignor M, Allen S, Boschung J, Nauels A, Xia Y, Bex V, Midgley P (eds) Climate change 2013: the physical science basis. Contribution of working group I to the fifth assessment report of the intergovernmental panel on climate change. Cambridge University Press, Cambridge

Kretsch E (2016) Stakeholder perceptions of climate adaptation leadership Port of Providence case study. University of Rhode Island Master's Theses

McEvoy D, Mullet J (2013) Enhancing the resilience of seaports to a changing climate: research synthesis and implications for policy and practice. National Climate Change Adaptation Research Facility Report, viewed 15 May 2015, http://www.nccarf.edu.au

McNamara K, Hemstock S and Holland E 2013 Practises of climate change adaptation in the Pacific. Survey of implementing agencies (phase II) European Union, Global Climate Change Alliance (EUGCCA)/University of South Pacific Report, Suva

McNaught R, Ngatulu C, Tego G, Nicholson T (2011) The adaptive capacity of Pileni Island community, Viakau Ward, Temotu Province, Solomon Islands. Australian Department of Climate Change and Energy Efficiency Report, Canberra

Pacific Catastrophe Risk Assessment and Financing Initiative (2013) Better risk information for smarter developments catastrophe risk assessment methodology. World Bank Press, New York

Secretariat of the Pacific Regional Environmental Programme (SPREP) (2015) Decision support tools for climate-resilient coastal development: a case study from the Cook Islands. Pacific Adaptation to Climate Change (PACC) Project Report, Suva

Trundle A (2015) Draft Port Vila, Vanuatu climate vulnerability assessment abridged report. RMIT University and UN Habitat Joint Report, viewed 11 July 2015, http://www.fukuoka.unhabitat.org

United Nations Conference on Trade and Development (2011) Ad hoc expert meeting summary on climate change impacts and adaptation: a challenge for global ports. UNCTAD Report, Geneva

United States Government Accountability Office (2015) Preparing for climate risk: lessons from the private sector. United States Comptroller General, Congressional Report, viewed 3 Feb 2016, http://www.gao.gov/assets

Fostering Climate Smartness in Smallholder Farming Systems: Business Promotional Approaches for Improved Maize Varieties in Eastern and Southern Africa

109

Kingstone Mujeyi and Angeline Mujeyi

Contents

Introduction 2226
Key Concepts and Insights from Literature 2228
Scaling Up of CSA Innovations 2229
Scalability of CSA Technologies and Practices 2230
Theoretical Underpinnings 2231
Data and Methods 2232
The CIMMYT Approach to Scaling Up DTM Varieties 2233
Impact of the Seed Business Development Approach 2234
Case Studies 2238
  Case Study 1: AgriSeed in Tanzania 2238
  Case Study 2: Phoenix Seed Company in Mozambique 2239
  Case Study 3: Zimbabwe Super Seeds 2240
Drivers and Enabling Factors for DTM Scaling 2241
Conclusion and Recommendations 2243
Cross-References 2244
Appendix 1: DTMASS Project Partners 2017 2244
References 2247

Abstract

In the face of increasing risk from climate change, the availability and accessibility of climate smart agricultural technologies to the most vulnerable smallholder farming households is critical for adaptation and resilience. In this regard,

K. Mujeyi (✉)
International Maize and Wheat Improvement Center (CIMMYT), Southern Africa Regional Office (SARO), Harare, Zimbabwe
e-mail: kingmujeyi@gmail.com

A. Mujeyi
College of Agricultural Engineering and Science, Discipline of Agricultural Economics, University of KwaZulu-Natal, Pietermaritzburg Campus, Scottville, South Africa
e-mail: munogona@gmail.com

W. Leal Filho (ed.), *Handbook of Climate Change Resilience*,
https://doi.org/10.1007/978-3-319-93336-8_24

the International Maize and Wheat Improvement Center (CIMMYT) has been working with many partners in Eastern and Southern Africa (ESA) to promote climate smartness in smallholder farming systems by generating, for the farmers, maize varieties that are higher-yielding, drought-tolerant, stress-resilient, and resistant to common pests and diseases. To enhance the faster and wider spread of these improved maize varieties to reach the target beneficiaries (smallholder farm families), CIMMYT adopted a business development approach to promote the scaling up and out of promising drought-tolerant maize (DTM) varieties. This has been achieved through engaging and capacitating seed companies and agro-dealer seed stockists with knowledge on available DTM varieties and their characteristics as well as seed business management and marketing skills. This chapter analyzes trends and tendencies of the DTM variety promotional strategies and highlights key lessons from success factors driving this approach for scaling up to enhance climate smartness of the predominantly smallholder farming systems that characterize African agriculture.

Keywords
Seed business development · Agro-dealers · DTM varieties · Capacity building

Introduction

In the face of climate change, the availability, accessibility, and affordability of climate smart agricultural technologies to the most vulnerable smallholder farming households are critical for adaptation and resilience to risks. It is estimated that around 40% of maize-growing areas in Africa are riddled with occasional drought-related stress, resulting in yield losses of between 10% and 25% (Fischer et al. 2015). Around 25% of the maize crop suffers frequent drought, with losses of up to half the harvest (Fisher et al. 2015). According to MacRobert (2009), access to good quality improved seed is critical for achieving food, nutrition, and income security among the vast majority of the African population. Over the past four to five decades, there have been numerous well-intentioned interventions in rural areas (by a diverse range of players) aimed at supporting rural communities in many Sub-Saharan Africa (SSA) to adapt to the changing climatic conditions. The interventions have had mixed results, which have been used as lessons learnt for improvement in the design, targeting, and implementation of successive interventions. Thus, with changes in social, economic, political, and technological landscapes, approaches have also shifted and evolved over time. As noted by Best, Ferris and Wheatley (2015), the role of market development and linkage, particularly in the agricultural sector, has emerged as a powerful tool in fostering climate smartness for rural development and alleviation of poverty and food insecurity among smallholder farming communities.

Like many development agents, the International Maize and Wheat Improvement Center (CIMMYT) has also intervened through various programs and projects aimed at alleviating poverty and food insecurity and lessening the impact of climate change in SSA. Through its mandate of seeking to improve livelihoods through maize and wheat science, CIMMYT develops improved maize varieties that are adaptable to the farmer's conditions. It has, over the years, been working to generate maize varieties that are higher-yielding, drought-tolerant, stress-resilient, and resistant to common pests and diseases in SSA. This tremendous breeding work has also been targeted at replacing inferior traditional varieties with superior drought-tolerant maize (DTM) varieties that are input-responsive, stress-tolerant, and high-yielding (Abate et al. 2015). Evaluations conducted by CIMMYT and its partners under controlled trials over the years (see Lunduka et al. 2017; Fisher et al. 2015) have revealed that drought-tolerant maize (DTM) varieties do not only exhibit drought- and stress-tolerance but also superior yielding capacity compared to most traditionally grown commercial hybrids. For instance, Setimela et al. (2017) carried out an evaluation using on-farm trials and estimated a 4–19% yield advantage for the DTM varieties over other commercial checks under drought conditions. Thus, the DTM varieties offer some kind of insurance cover for occasional mid-season dry spells and moisture stresses that have become synonymous with smallholder agriculture in East and Southern Africa (ESA). These drought- and stress-tolerance attributes of the DTM varieties, therefore, provide some guarantee to the farmers for a harvest under mild drought conditions where the traditionally grown commercial hybrids would normally fail. The full potential and benefits from the DTM varieties have not been fully exploited and realized among smallholder farmers in ESA due to the lack of knowledge about the existence of such varieties coupled by lack of capacity and resources on the part of seed companies to effectively serve seed markets in remote areas (Fisher et al. 2015; Lunduka et al. 2017). The improved maize varieties have not been spreading fast and wide enough to reach the target beneficiaries (smallholder farm families) for their food, livelihood, and nutritional security. Results of adoption monitoring surveys conducted in 2013 for the new DTM varieties released under the Drought Tolerant Maize for Africa (DTMA) project revealed variedly lower levels of adoption rates for selected ESA countries (Mozambique 44%, Uganda 32%, Tanzania 18%, Ethiopia 17% and Kenya 6%) (Fisher et al. 2015). One major contributing factor to the slow spread of the improved maize varieties in ESA has been noted as the reluctance by established seed companies to trade their high-market performing traditional maize varieties for the drought-tolerant varieties (Kaimenyi 2017).

Various approaches and mechanisms have been put in place to enhance diffusion and adoption of the improved varieties, particularly among smallholder farmers in drought- and stress-prone regions of SSA. One such approach is the use of market-based and private-sector-driven seed business development principles to facilitate the development and functioning of agricultural input markets in rural areas. The core of CIMMYT's business development promotional approaches to scaling (i.e., enhancing increased uptake) of maize seed for drought-tolerant varieties in Eastern and Southern Africa is capacity strengthening of agro-dealers, who stock and sell agricultural inputs to farmers. Under the Drought Tolerant Maize for Africa Seed Scaling (DTMASS) project, CIMMYT adopted a business development approach as one of its strategies to promote the scaling up and out of drought-tolerant maize

(DTM) varieties. The DTM varieties are new maize cultivars with higher yield potential and stability across varying levels of rainfall and moisture availability and growing conditions (Takim 2017). Both open-pollinated varieties and hybrids have been developed, several of which combine tolerance to drought, low soil nitrogen, and resistance to common pests and diseases. Capacity strengthening of rural-based agro-dealers with business managerial and technical skills is seen as an effective way to ensure increased availability, accessibility, and affordability of high-quality climate smart agricultural technologies to the most vulnerable smallholder farming households (Mthetwa et al. 2010). By engaging and capacitating small- and medium-scale enterprise (SME) seed companies, agro-dealers, and public sector seed inspectors, the project has proved to be effective in speeding up the production, distribution, and adoption of the DTM varieties in the selected ESA countries targeted by the project.

Key Concepts and Insights from Literature

Climate Smart Agriculture (CSA) is a concept in agriculture that seeks to deliver multiple benefits, including enhancing agricultural productivity and food security, poverty alleviation, and climate change adaptation and mitigation for socioeconomic development (AGRA 2014). The concept integrates the three dimensions of sustainable development (social, economic, and environmental) to address the challenges of food insecurity and climate change (FAO 2013). While CSA is composed of three main pillars, the focus of this chapter is on two of them, which are (i) sustainable increment of agricultural productivity and incomes and (ii) climate change adaptation and resilience building through business (private sector-driven) promotional approaches for scaling up the uptake and use of improved maize varieties in ESA. It is important to note that CSA is not confined to a single specific agricultural technology or practice that can be universally applied. Instead, it is a concept that requires for site-specific assessments to be conducted to identify suitable agricultural production technologies and practices (Ibid.). CSA brings together institutions, policies, and practices that may not necessarily be new but rather can be used in the context of changing climatic conditions, which may be unfamiliar to the farmers. What may be new with CSA as a concept is the potential for the numerous challenges faced in agriculture and food systems to be addressed simultaneously and holistically. Thus, CSA is an approach that seeks to address complex interrelated climate change-induced challenges in agriculture and food systems by identifying integrated solutions through creation of synergies and benefits that reduce trade-offs. The priority in CSA is to strengthen livelihoods, particularly those of smallholders, through enhancing their access to knowledge, resources (including genetic), and services (especially financial and marketing). Improved access to these services helps to build their adaptive capacity and resilience to climate-related shocks, with mitigation coming in as a secondary benefit. Key CSA technologies and practices most commonly encountered in SSA are conservation agriculture, mixed livestock and cropping systems, agroforestry, and the use of improved crop varieties (AGRA 2014).

According to AGRA (2014), inclusive partnerships involving governments, private sector (agribusinesses), development organizations, and farmers are critical for the development, dissemination, and adoption of CSA technologies. The same sentiments are echoed by Dayo et al. (2018), who assert that for agricultural productivity to improve and poverty and hunger to decline among smallholder farming households in SSA, it is not enough to generate superior agricultural technologies and innovations, but efforts must be made to scale up the innovations for wider uptake and utilization by the target beneficiaries. Scaling of CSA technological innovations can take different forms and approaches (vertical, horizontal, functional, and combinations) but requires concerted efforts from different stakeholders including from national governments, private sector, NGOs, donor agencies, research institutions, and extension agencies among others (Ibid.).

Scaling Up of CSA Innovations

The term "scaling up" has been defined in many different but related ways in literature within the domain of agriculture. In development work, scaling up is linked to the need to reach a wider clientele base of members of a target group for the attainment of a particular desired outcome. For instance, Anandajayasekeram (2016) defines scaling up as the act of "expanding, adapting and sustaining" successful projects, programs, and policies in spatially and temporally to impact on a greater number of people. It is regarded as a deliberate effort meant to increase the impact of successfully tested innovations, in order to benefit more people on a sustainable basis. The key emphasis in this definition is on achieving greater reach with successful interventions for adaptation on a sustainable basis. The United States Government (2017) defines scaling for widespread adoption of proven CSA technologies and practices as the process of sustainably increasing the adoption of a credible technology or practice, or a package of technologies and practices, with quality to retain or improve upon the demonstrated positive impact of the technology or practice and achieve widespread use by stakeholders. Despite it being difficult to achieve, scaling up is key to attaining improvement in effectiveness of any development intervention (Cooley and Linn 2014). According to Dayo et al. (2018), scaling up ensures expansion, replication, adaptation, and sustainability of successful projects, programs, or policies to reach a greater mass of intended beneficiaries. This implies that the benefits of successful CSA and DTM technologies are widely disseminated to reach more farmers. Thus, scaling up entails wider coverage, greater impact, and enhanced sustainability of successful technological interventions (Linn 2012a).

According to Appadurai et al. (2015), diffusion of innovative technology or practice normally moves "either vertically or horizontally across local, meso, and macro levels on a defined pathway." While horizontal scaling occurs when an innovation is replicated across individuals and geographies, vertical scaling induces policy and regulatory changes at the national or local levels. There are basically four generic pathways through which scaling may occur, according to Cooley and Linn (2014):

- Centralized scaling, which is initiated and driven by a central or local government actor.
- Multi-actor scaling is driven by multiple actors working in partnership. It may be initiated by an NGO but will require engagement and involvement of other actors for scaling to be achieved.
- NGO-driven scaling starts at the local level driven by a single NGO but can achieve decentralized replication through several NGOs.
- Spontaneous scaling takes place when individual beneficiaries replicate an innovation informally without guidance from external actors like the government or NGO.

Effective scaling up of high-performing agricultural technologies is an important measure of success for the technological innovation or development intervention (Anandajayasekeram 2016). Normal practice is that before a new technology is disseminated for wider adoption, it gets trial-tested through a pilot project that normally has limited reach. Lessons learnt from monitoring and evaluation of the pilot project can then be used to scale up the model from the pilot to a larger catchment for greater developmental impacts. Scaling up of beneficial agricultural technological innovations and practices has increasingly become important due to ambitiously big national, regional, and global development goals and targets (e.g., CAADP, SDGs) that continue to be set and reviewed upward. However, design and implementation of development programs have been found to be problematic in that they involve numerous once-off efforts, which Linn (2014) refers to as "pilots to nowhere." These fragmented development interventions (too few partnerships) do not allow for adequate learning from and building on successes and mistakes of past efforts. As a result, no meaningful benefits from economies of scale are realized from learning, replication with adaptation, and partnership (Fatunbi et al. 2015; Cooley and Linn 2014).

Scalability of CSA Technologies and Practices

Anandajayasekeram (2016) contends that in order to enhance scalability of CSA technologies and practices, or potential for their scaling up, the innovations should be credible in the sense that they are based on sound evidence or else they should be promoted by reputable organizations or people. Evidence and practicability of beneficial results of a technology have to be observable to and testable by potential adopters in the real world, and these have to prove relevance for addressing their felt need. The intended users need to be convinced that the new technology being advocated for is superior to existing ones, particularly in terms of net benefits realized from implementation of the new practice. The technology should also allow for testability and evaluation of results on a small-scale basis before a potential adopter can commit to full adoption.

The Netherlands Development Organization (SNV) and the Royal Tropical Institute (KIT) experimented with three market-based scaling approaches to input supply over the years. These approaches are the rural agro-dealership capacitation, the value chain leader model, and the local traders support. The approaches have originated from experiences and lessons learnt from agricultural value chain development work that the two organizations had been collaborating on across nine African countries (Belt et al. 2015). The SNV and KIT criticize the business development models or market-based approaches used by most donors, NGOs, and development agencies in scaling input-based agricultural technologies for concentrating their focus and operations in peri-urban areas, where there is higher concentration of already existing agro-dealer shops (Belt et al. 2015). The criticism emanates from the fact that the approaches fail to effectively target the poorest of farming households in remote areas, who will then need to commute to the growth points to access the inputs. For such farmers, the expected trickle-down effect is less evident since the inputs end up becoming much more expensive due to the transaction costs involved, particularly with regard to transportation. The same applies to the agro-dealers; they end up operating their businesses closer to suppliers of input assortments and in areas with critical masses of clientele as a way of enhancing viability and profit margins. Thus, in areas where the presence of agro-dealers is limited, there is need to identify alternative channels, such as village-based agents (including vendors, lead farmers, informal traders, etc.) and provide them with support and incentives in the form of business management training, investment in infrastructure upgrading, and provision of start-up capital.

Some beneficial CSA technologies and practices may take longer periods to produce benefits resulting in them not receiving favorable responses from intended beneficiaries. Many farmers, agro-dealers, and seed companies may either be reluctant or are unable to invest substantial resources and time in trying out new crop varieties, inputs, technologies, or practices that do not seem to provide definite results in the long term (Hartmann and Linn 2008). The most preferred technologies and practices are where risks are minimized and returns from minimal investment in time and resources are maximized (Chandy et al. 2013). Thus, some CSA technologies like DTM varieties may require extra incentives to promote dissemination and uptake. Such CSA technologies need to be incorporated into already existing structures in order to overcome the challenges inherent in the conventional scaling up approaches. In such circumstances, it may become necessary to partner and collaborate with players who have already achieved scale and only invest in adding value to what they are already doing.

Theoretical Underpinnings

The thrust of this chapter is anchored on the theory of change where the expectation is that when viable in-country seed systems are created and coupled with innovative dissemination and risk reduction strategies, they have potential to create incentives for both producers and consumers (farmers) of seed to take up beneficial DTM varieties. While the key measure of success is farmer adoption, sustainability will be measured by the viability of private actors within a national seed system. Viability in this case implies having quality production systems, where actors have little or no dependence on donor funding but are able to generate sufficient profits as incentives for continued scaling up of production and achieve wider dissemination of the DTM varieties to reach a critical mass of maize growers in the target countries. Accessibility and use of the CSA technologies will be determined by the level of awareness creation, technical knowledge generation and co-creation, physical access through various marketing channels, and affordability of seed for the improved DTM varieties. The approach adopted under the DTMASS project puts emphasis on incentive-based scaling up of the improved DTM varieties. To engender sustained uptake of these beneficial technologies, the approach considers both the supply and demand sides of the maize seed value chain by engaging identified relevant stakeholders to collaborate in the promotion of the DTM varieties. On the supply side, the project works with seed companies and regulatory agencies to promote production of good quality seed, while on the demand side, the focus is on collaborating with in-country partners to identify and capacitate networks of agro-dealers, marketing personnel for seed companies and local farmer groups/cooperatives with relevant marketing and seed business management skills. The thrust is on creating enhanced awareness of the benefits of DTM seed through various promotional strategies including conducting field days and on-farm demonstration plots to showcase how maize productivity can be maximized through adoption of appropriate seed and growing techniques. This, coupled with strengthening of marketing capacity of the seed companies, has great potential for maximizing the distribution of and access to the DTM seed through market channels.

Data and Methods

Using secondary data and empirical evidence from the DTMASS project at CIMMYT, the chapter conducts a critical appraisal of the business development approach to promotion and scaling up of DTM varieties in six project countries in ESA (Ethiopia, Kenya, Mozambique, Tanzania, Uganda, and Zambia). The review analyzes the impact that the approach has had on availability and accessibility of the DTM to poor smallholder farmers in ESA. In addition to appraisal of various project reports, the chapter also adopts a case study approach (three cases) to highlighting success stories and analyzing factors that have influenced these outcomes. In all the cases, lessons have been drawn for informing effective policy and programmatic interventions aimed at enhancing climate resilience and food and nutritional security outcomes for smallholder farm families. Best practice models and approaches have also been reviewed and recommended accordingly.

The CIMMYT Approach to Scaling Up DTM Varieties

To popularize the DTM varieties as a CSA technology in ESA, CIMMYT, with support from the United States Agency for International Development (USAID), has been implementing a business development approach to seed production, processing, and marketing under the DTMASS project. This has been achieved through engagement of public and private sector partners to develop and promote improved DTM varieties in six ESA countries. The DTMASS project seeks to significantly enhance the adoption of new, climate-resilient maize seed in the targeted zones of influence in the seven ESA countries, through public-private partnerships and by addressing issues related to seed demand, accessibility, and availability. The aim is to increase both production (total availability) and productivity (yield per unit area) of maize, leading to improved food security and increased incomes for smallholder farmers. Central to the implementation of the project has been the organization's partnership with SME seed companies and agro-dealers in these countries for the production, processing, and marketing of the DTM varieties. The identified partner seed companies, which exhibit potential for growth, receive financial and technical support (sub-grants) from CIMMYT for development of their seed businesses. This seed business development work of CIMMYT in ESA has witnessed an upward trajectory in the production, dissemination, and uptake of certified DTM seed through involvement of the SME seed companies. With this support from CIMMYT, the seed companies have been able to access high-quality germplasm for free, which they then multiply and commercialize. The support goes further to building their capacity for production, processing, and marketing of the certified DTM seed to ensure that the seed becomes accessible and affordable to smallholder farmers in a sustainable manner.

According to Kandiwa and Lweya (2017), partner seed companies use a combination of promotional approaches to create awareness of improved maize seed varieties among farmers. These approaches include establishment of demonstration plots (demos), hosting of field days, small seed packs for distribution as free sample handouts to agro-dealers and farmers, branded regalia (corporate wear), national and community radios, television, advertisements in newspapers and magazines, printed materials such as brochures and leaflets, mobile road shows, agricultural shows, seed fairs, billboards and other forms of signage, lead farmers, videos, documentaries, bulk text messaging, and increasingly social media such as Twitter, Facebook, Instagram, and so forth. For accomplishment of these promotional activities, the seed companies receive support from CIMMYT as part of its business promotional strategy. To ensure sustainability of these promotional activities beyond the DTMASS project, CIMMYT also strengthens the capacity of country-level seed trade associations as well as agro-dealers and their respective associations. These capacity-building activities also include government seed inspectors to ensure sustainable improvement in the accessibility and affordability of not only good quality seed to the farmers but seed that is appropriate and adaptable to their conditions. Affordability of the DTM seed to farmers is enhanced through increased competition in the seed market by capacitating more seed companies, particularly SMEs, for participation in the production and supply of the seed. The main competitive strategy for the new entrants (SME seed companies) in the market is through low-cost leadership and pricing strategies. Otherwise, they might not succeed in wrestling market shares from the established seed companies.

Impact of the Seed Business Development Approach

Evaluation of effectiveness of seed business development approach to promotion of scaling up DTM varieties in ESA is still scarce. There is scant literature on economic impact evaluation of such projects as DTMASS to generate empirical evidence for the promotion of scaling up and out of promising technologies. In Ethiopia, the growing liberalization and investment in marketing systems has facilitated increased participation of both the private sector and farmers' cooperatives in the seed industry, and in turn the adoption of modern technologies by farmers has accelerated (Abate et al. 2015). One such technology is the new DTM variety, BH661, promoted under the auspices of the DTMASS project. This variety is of particular significance due to its tolerance to stress and drought conditions, higher-yielding potential, resistance to major diseases, and wider adaptability. Due to these superior characteristics, its market performance has enabled it to surpass the traditionally reliable varieties like BH660 and BH540. There has also been rapidly growing demand for early generation seed of this new hybrid by SME seed companies. While only 6 t of certified seed were produced in 2012 by one state enterprise, by 2014 about 2,900 mt were produced by five SME companies, mainly due to the DTMASS intervention (Abate et al. 2015).

To reduce vulnerability and improve food security, the DTMASS project and its predecessor, the Drought Tolerant Maize for Africa (DTMA) project, have witnessed the release of over 200 drought-tolerant maize varieties between 2007 and 2016 (CIMMYT 2017). These have been tested in experimental and farmers' fields and disseminated to farmers in 13 African countries through national agricultural research systems and private seed companies. Yields of the new DTM varieties have been proven to be superior to those of currently available commercial maize varieties under both stress and optimum growing conditions (Kaimenyi 2016, 2017; CIMMYT 2016). Although the benefits of DT maize for African farmers have been repeatedly predicted, realization of those benefits depends on farmer uptake, which has not yet been adequately studied. Under the DTMASS project, CIMMYT provides capacity building through training and technical backstopping for private (seed companies and agro-dealers, including their respective associations) and public sector staff to enhance their skills in seed production, processing, and marketing. Additionally, targeted sub-grants are provided to these seed companies and other national partners to support expansion activities such as procurement of special seed processing and packing equipment, seed storage infrastructure, and other facilities as well as assisting them with establishment of demonstration sites, holding of field days, and production of promotional materials for their seed.

Since its launch in 2007 by CIMMYT, the DTMA project has fostered partnerships for production and dissemination of drought- and other stress-tolerant maize varieties to smallholder farmers across Africa. The frequency of releases and commercialization of adaptable maize varieties has been enhanced tremendously through increased involvement of private sector players in the business of seed production, processing, and marketing. This business promotional approach has substantially enhanced the availability and accessibility of improved maize seed. Under the DTMASS project, CIMMYT has consolidated the tradition of fostering partnerships with public and private players in the seed industry, including nongovernmental organizations (NGOs) and community-based organizations (CBOs). To achieve greater scale in terms of diffusion and uptake of the DTM varieties, the DTMASS project is advocating for the use of various media and promotional strategies for wider dissemination of information and products. While on-farm demonstrations have been central to awareness-building activities of previous CIMMYT projects, under the DTMASS project, these have been combined with other innovative demand creation activities like the use of ICT tools, electronic and print media, and free distribution of small seed packs and branded promotional materials to enable seed companies to create market demand for the improved seed. Partner seed companies and agro-dealers are also capacitated to implement these promotional activities as a way of increasing awareness and stimulating demand among the farmers.

Scaling of DTM varieties has substantially been aided by product promotion and delivery efforts implemented through public-private partnerships (PPPs). Establishment and strengthening of partnerships with an array of public and private actors engaged in production and marketing of improved maize varieties has become a key scaling strategy for CIMMYT. Active promotion of the DTM varieties through collaborative efforts with project partners has been pivotal in ensuring sustained supply and successful adoption thereof by the farming communities in the target ESA countries. Enabling factors have been the project's emphasis on participatory establishment and management of on-farm demonstrations and conducting of field days and various other promotional activities through various media. Leveraging on the wide-network coverage of partner seed companies, their agro-dealers, NGOs, and national partners in the often sparsely rural areas has enabled the scaling efforts of the DTMASS project to achieve greater impact. The products from the project (improved DTM seeds) have reportedly had spillover effects into other areas, regions, and countries that were not originally targeted, mainly through the efforts of the partner seed companies as they seek to expand their markets. These unintended positive consequences have been made possible as a result of increased awareness and enhanced accessibility of the DTM varieties to many smallholders who may have been skeptical about purchasing improved seed due to risks associated with maize varieties that are traditionally known to be vulnerable to abiotic and biotic stresses (CIMMYT 2016). While CIMMYT provides technical backstopping for seed business development and marketing activities as well as overall coordination and management of the DTM scaling strategies, partner seed companies come in handy through the production, promotion, and delivery of DTM products. The National Agricultural Research Systems (NARS), NGOs, and various private associations support country-level scaling initiatives through policy advocacy and implementation as well as diverse promotional activities.

The DTMASS project provides support for production and supply as well as demand generation for the DTM varieties in ESA. The overall aim of the project is to significantly improve access to improved and good quality seed varieties that produce maize which is drought-tolerant, stress-resilient, and high-yielding for smallholder farmers. Emphasis on financial support has since shifted from production and supply toward provision of funding to the seed company partners for demand generation. For instance, in 2017, over 70% of DTMASS sub-grants went to supporting marketing and promotional activities of the partner seed companies, including capacity strengthening for agro-dealer distribution network expansion (CIMMYT 2017). As capacity of the partner SME seed companies continues to grow, they begin to explore new ways for expanding into new territories with the DTM seed. Some few companies have begun supplying foundation seed to fellow seed companies and, in some cases, even exporting into neighboring countries. Table 1 shows the performance of the DTMASS project by 2016.

The number of partner seed companies across the 6 countries has since grown to over 50 in 2017 and the number of DTM seed varieties released to over 90 (see Appendix 1). The quantity of DTM seed produced by the partner seed companies and the number of farming households reached annually have also increased tremendously (Table 2). By September 2017, the DTMASS project had exceeded all its annual targets for key outputs after having worked with 56 partners across the 6 target countries. Production and marketing of over 80 commercial DTM varieties amount to over 12,000 t that were supported by the project in 2017. It is estimated that over 94,000 farming households were directly reached with DTMASS promotional field events conducted by CIMMYT and its partners in 2017 (CIMMYT 2018).

Table 1 DTMASS project achievements (2013–2016)

Country	No. of partner seed companies	No. of seed varieties released	Quantity of DTM seed produced (mt)	Quantity of DTM EGS supported (mt)	# of farming households reached
Ethiopia	6	7	7,920	222	186,454
Kenya	6	12	700	40	38,356
Mozambique	6	17	480	16	11,116
Tanzania	6	13	2,090	12	42,471
Uganda	12	13	700	35	54,264
Zambia	7	13	1,500	65	65,502
Total	**43**	**75**	**13,390**	**390**	**398,163**

Source: CIMMYT – DTMASS Project (2017)
EGS Early generation seed

Table 2 Performance of the DTMASS project by country in 2017

Country		EGS (mt)	Certified DTM seed (mt)	Number of DTM demo plots	Number of DTM field days	Number of agro-dealers trained
Ethiopia	*Target*	137	2,990	160	54	250
	Actual	**214.9**	**3,359**	**203**	**35**	**295**
Kenya	*Target*	31.8	1,000	345	87	100
	Actual	**25.9**	**1,000**	**744**	**48**	**120**
Mozambique	*Target*	11.3	1,228	250	147	54
	Actual	**49**	**1,292**	**168**	**59**	**36**
Tanzania	*Target*	15.2	1,459	515	238	200
	Actual	**33.5**	**2,793**	**598**	**259**	**220**
Uganda	*Target*	39	1,000	955	304	310
	Actual	**40**	**1,000**	**1,010**	**305**	**401**
Zambia	*Target*	50	3,000	1,205	337	0
	Actual	**60**	**3,000**	**2,500**	**658**	**33**
Total	*Target*	148.3	10,677	3,444	1,167	914
	Actual	**423.3**	**12,444**	**5,223**	**1,364**	**1,105**

Source: CIMMYT DTMASS Project (2018)

Capacity building of agro-dealers, which commenced in the second year of the project, came about as a response to the 2015 baseline surveys that revealed low awareness levels for DTM varieties and their characteristics among both farmers and the agro-dealers. The country surveys indicated that most of the dealers lacked knowledge and understanding on how to properly advise farmers on product selection and proper handling and storage of seeds, including the DTM seed. This lack of awareness among agro-dealers has been leading to low uptake of the DTM varieties on the market. The Seed Business Development Unit in CIMMYT collaborated with country-level partners under the DTMASS project to identify and train both formal and informal agro-dealers in raising awareness for DTM varieties, their characteristics, benefits, seed production agronomic practices, seed processing, handling, storage, promotion, and marketing. The trained agro-dealers are expected to use the acquired knowledge and information to improve their service delivery and impart the knowledge to farmers. In 2017 alone, over a thousand agro-dealers were trained in DTM marketing and business promotion skills across the six countries. The developed agro-dealer training curriculum was distributed to partners to enable them to train even more agro-dealers in their respective countries.

Since the inception of the DTMASS project in 2014, CIMMYT has also ramped up efforts to assist the companies to enhance their profitability and product viability through revitalization and digitization of the Seed Road Mapping process. A Seed Road Map (SRM) is a simple but powerful toolkit, based on excel spreadsheet, which assists in the planning of certified seed production and marketing, including promotional aspects. Partner seed company representatives are taken through interactive hands-on training sessions on practical application of the toolkit, which assists them in setting market-based seed production targets in proportionate to their capacities. The SRMs are handy in simplifying the reporting of seed production and marketing figures by the companies since it can organize a company's operations by location, variety, and target market, making it easier for monitoring and evaluation needs. By the end of 2017, all partner seed companies had received the SRM training and toolkit for individual application across the DTM varieties they produce. As the project approaches its end in 2018, the support from CIMMYT has increasingly become skewed toward leveraging promotional activities that accelerate diffusion of the DTM technologies. Thus, there has been more technical and financial support toward establishment of demonstration plots, holding of exhibitions and field days, conducting media promotions, and expansion of agro-dealer distribution networks. New partnerships and alliances continue to be forged with other development organizations that are interested in promoting DTM varieties in their community agricultural development activities.

Another innovation within the DTMASS project entails partnering and working with local (non-seed company) partners to establish and expand the concept of "mega demos," which are meant to showcase multiple DTM varieties against locally favored varieties. The mega demos have proven to be trusted by farmers more than the seed companies' demos that tend to showcase only their products and advertising their proven attributes and benefits. These partners would then train farmers on how to conduct objective evaluations and participatory assessments. While participatory assessments or evaluations (in which farmers get to assess performance of a demo plot themselves) are not an innovation, what is new is the combination of efforts that the DTMASS project is leveraging on during these events. The project then collects data based on the evaluations and leverage of the platform to also raise awareness on the benefits of the DTM varieties in a more practical and convincing manner while at the same time providing greater potential for achieving more effective dissemination of appropriate agronomic practices or crop management techniques. Qualitative socioeconomic data and information on farmers' choices and preferences is also collected to inform future breeding and marketing efforts. This is done as an exit strategy for the project as it winds down support for individual seed company promotional activities, especially the establishment of demonstration plots.

Case Studies

Case Study 1: AgriSeed in Tanzania

AgriSeed in Tanzania is one seed business enterprise that started off in 2010 as a small-scale seed company producing maize seed through donor grants and selling the seed through the government subsidy program. The company continued to struggle until 2015, when it was awarded a grant by CIMMYT under the DTMASS project, which enabled it to produce early generation (foundation) seed for its own new hybrid, WE2112, which has set the company on a massive growth and sustainability path (Kaimenyi 2017). The variety, which is marketed as AgriSeed H12, is the first hybrid variety to be sold by the company. AgriSeed had previously relied on production and marketing of publicly released OPV products that has the disadvantage that they can be licensed to and be produced by any other company. With support from CIMMYT's DTMASS project, the company has been able to transition from the low-value OPVs to high-value hybrids. The company has begun to realize meaningful growth in its seed business through profitable sales of the improved DTM varieties, which are only accessible to few companies in Tanzania, a situation that has given AgriSeed a competitive edge on the market. From merely 3 t produced and sold in 2016 of this improved DTM variety, the company expected the figure to rise to over 50 t in 2017 (Ibid.). The increase is due to enhanced capacity and growing demand from smallholder farmers, thanks to DTMASS project's business promotional support, which enabled the company to set up more than 100 demonstration plots in villages across the southern parts of Tanzania. AgriSeed has benefitted immensely from the technical, financial, and capacity-building support of CIMMYT under the DTMASS project.

Case Study 2: Phoenix Seed Company in Mozambique

In Mozambique, where over 2.5 million smallholder farmers account for over 90% of arable land, the greatest challenge for seed companies has been lack of cost-effective distribution systems in rural areas due to the country's poor road transport infrastructure. Despite the logistical challenges and high transaction costs, Phoenix Seed Company is one emerging seed business enterprise that has managed to take advantage of the vast market opportunity with support of the business development initiatives under the DTMASS project at CIMMYT and other financiers. The support has enabled Phoenix to expand its seed market beyond the borders of its farm into the entire Vanduzi district, Manica Province, and beyond the province into seven of the country's nine provinces within a short space of time. The expansion, which is now targeted at covering all the provinces, the entire country, and beyond the borders of Mozambique into the region, has been aided by the engagement of a growing network of agro-dealers, who stock and sell the company's seeds, thus reducing transaction and transport costs. CIMMYT has played a crucial role in the growth of Phoenix Seeds and a few other SME seed businesses in Mozambique through capacity building of their owners, agronomists, marketing personnel, and agro-dealers. The trainings are augmented by financial and technical support for promotional materials and activities aimed at popularizing the DTM varieties among farmers. During the 2016/2017 season, Phoenix managed to produce and sell over 110 t of seeds and is expected to exceed 1,000 t of mostly ZM523 DTM seed variety by 2018. The company boasts of nearly 700 out-growers and over 300 agro-dealers who are partnering with it in producing, stocking, and retailing these seeds. The initiative has brought huge economic benefits not only to the company, out-growers, and agro-dealers but also to the over 100,000 smallholder farmers who have begun to realize higher yields and improved livelihoods from purchasing and planting the higher-yielding DTM varieties.

Phoenix Seed achieved great success on both supply and demand sides. On the supply side, the company witnessed the success of promoting the DTM OPV,

ZM523 over the years. Large orders of the variety are now being made by both large-scale and small-scale producers as well as the national input distribution scheme. The variety is said to be selling well due to its high-yielding potential (>4.5 t/ha) and ability to withstand the dryer growing conditions. The company's demonstration plots and other promotional activities, augmented by technical assistance from CIMMYT, have gone a long way in creating awareness among farmers about the benefits of DTM varieties. On the demand side, the company managed to issue out 4 t of DTM seed nationwide in 2017 through field days conducted at over 200 demo plots, where small sample (50 g) seed packs were distributed as free handouts to farmers. Also distributed to the farmers during the demo field days are simple agronomic guides, referred to as "Grower Aids." These are one-page infographics with step-by-step guidelines on optimal DTM agronomic practices. Phoenix Seed has also recorded great success with the concept of village-based agents (VBAs), where opinion leaders are trained in technical aspects of seed handling and marketing and then used as distributors of the company's DTM seed in the villages they stay. The VBA concept is reported to have significantly expanded the company's distribution networks and reach.

Case Study 3: Zimbabwe Super Seeds

Zimbabwe Super Seeds (ZSS) is one good example of success stories of incentive-based seed business development for scaling up of improved CSA technologies. The ZSS, formerly known as Zaka Super Seeds, was established in 2012 in Zaka district of Masvingo Province in Zimbabwe as a community cooperative company to produce, process, and market DTM OPV with support from the Seeds and Markets Project (SAMP) coordinated by GRM International, a development organization that designs and implements development projects on behalf of funding organizations through sub-granting of local organizations. A key feature of the company is its strong rural community involvement at the ownership (about 180 farmers are shareholders), production and processing, and utilization of the seed. ZSS has benefited significantly from a strong multi-stakeholder partnership for scaling of its seed business operations. Although the initial focus of ZSS was merely on OPV maize production targeting the resource poor, smallholder farmers in the drought-prone Zaka district, it has since expanded to encompass other drought-tolerant crops like groundnuts, sugar beans, and cowpeas. Within three seasons, the company has managed to spread its tentacles beyond the Zaka district and Masvingo Province in terms of operations and markets. The company produced 26 t of DTM OPV seed in 2012, which increased to 188 t in 2015, while its annual turnover has risen from US$30,000 to US$398,000 during the same period and is expected to reach US$1 million by 2020 (Anandajayasekeram 2016). ZSS currently contracts over 1,000 smallholder farmers in 6 districts across the country to produce maize, groundnuts, sugar bean, and cowpea seeds. With the use of both fixed and mobile processing plants, ZSS procures and process seed produced in each district within that district. The seed produced from the local plant is first supplied to local agro-dealers before surplus is distributed to other districts. The seeds are packed in 2 kg, 5 kg, and 10 kg packs for convenience of purchase by various farmer income groups. As part of its expansion drive, ZSS intends to introduce drought-tolerant quality protein maize (QPM) in 2018 and pro-vitamin A orange maize by 2019 as a way of responding to the nutritional needs of vulnerable households that normally have maize, the staple crop, but lack other protein- and vitamin A-rich foods.

Munyaka et al. (2017) analyzed the performance of the ZSS and found that the community seed business has grown to serve areas beyond the initially targeted single district (Zaka) within a short space of three seasons. The analysis reveals the immense potential for growth and profitability within community seed enterprises if they are managed commercially. For instance, drought-tolerant maize seed produced, processed, and sold by the enterprise rose from 26 metric tons (mt) to 151 mt within the first three cropping seasons of inception. Embarking on maize seed production offered local farmers more income (US$305) per ton than producing commercial grain. The initiative improved availability and affordability of maize seed locally, and the multi-stakeholder partnership approach used by SAMP enhanced its sustainability to the extent that the community seed business now boasts of a vast network of agro-dealers across the whole country. The current promotional activities used by the company include participation in agricultural shows, hosting field days, establishment of demonstration plots, distribution of free sample seed packs, brochures and pamphlets, and branded hats and T-shirts.

The majority of the company's clients are rural general dealer shops with capacity to stock small quantities of the ZSS seed products on consignment. These rural traders, who are mostly financially constrained, are contracted to act as the stockists and outlets for the ZSS seed, and this increases business volumes for them. End users of ZSS products are predominantly the communal smallholder farmers as well as some small-scale commercial farmers. ZSS now boasts of multiple varieties of DTM OPV maize (ZM309, ZM401, ZM521), sugar beans (NUA 45, Cherry, Gloria, Sweet Violet), and cowpeas (CBC2, CBC 3) and is aiming to continue expanding its product portfolio. While the company is targeting to put 1,500 mt of maize seed, 200 mt of sugar beans, 50 mt of groundnuts, and 10 mt of cowpea seeds on the market by 2020, Anandajayasekeram (2016) estimates that ZSS has potential to produce and market up to 10,000 t of DTM OPV seed in Zimbabwe.

In addition to these three companies profiled in the foregoing case studies, there are over 50 other small to medium-sized seed enterprises in ESA countries that are being supported under the DTMASS project and other similar projects to scale up production and marketing of these improved maize varieties. The phenomenal growth in business recorded by most of these companies within the short life span of the project has been instrumental in enhancing availability, accessibility, and affordability of improved and adaptable maize varieties to smallholder farmers in drought-prone regions of the selected ESA countries.

Drivers and Enabling Factors for DTM Scaling

According to Anandajayasekeram (2016), the common drivers that relentlessly push forward the scaling up process include proven models or ideas, vision and leadership, external catalyst, incentives, potential for empowerment, and markets. With regard to proven ideas or models, CIMMYT noted through experiences gained in previous projects like the Drought Tolerant Maize for Africa (DTMA), Improved Maize for African Soils (IMAS), and the Water Efficient Maize for Africa (WEMA) that a model or idea that works well on a pilot basis needs to be tested and promoted under the farmer's conditions. Thus, CIMMYT works with partners to establish and evaluate on-farm trials and demonstration plots in the target areas. The DTMASS project has built on the experiences, successes achieved, and lessons learned from these and other complementary projects, particularly the DTMA, IMAS, and STMA. All of these projects have tremendously contributed to the development of drought- and stress-tolerant improved maize varieties. The preceding projects have not only laid the foundation in terms of development and release of many stress-resilient and nitrogen-use efficient maize varieties but they have also presented opportunities for learning lessons from challenges and experiences encountered in SSA. Thus, the emphasis of DTMASS has exclusively been on the scaling up and enhancing uptake of the improved maize varieties.

Clear vision and leadership are essential for driving the scaling process. A sound vision is required for recognition of the fact that scaling up is necessary, desirable, and feasible. It is often the duty of visionary leaders to drive the scaling process. In DTMASS, CIMMYT is intensively providing oversight while leveraging on already existing public-private partnerships to strengthen the production and delivery of DTM seed in ESA. The coordination role played by CIMMYT has helped the project to achieve remarkable success in all the target countries as reported earlier. External catalysts, in the form of crises such as climate change or pressure from outside actors like donors and partner NGOs, have potential to drive the scaling up process forward. This has been the case in the DTMASS project, where the already existing strong networks of partners in SSA, including national agricultural research institutions, community-based organizations (CBOs), and NGOs, have acted as external catalysts. These partnerships have also had significant influence on national government policies in several countries, creating huge opportunities to scale up and scale out the improved maize varieties under DTMASS.

Incentives are needed for motivation of players and organizations to strive for greater results and impact. These may include rewards, competitions, community demand, peer reviews, and independent evaluations (Anandajayasekeram 2016). Lucrative seed markets and profitability of seed business can be powerful drivers for scaling of DTM varieties. Profit is known to be a very powerful driver for delivering private goods and services. In target areas in each of the six selected countries, the project has ensured that at least two to four prioritized varieties are picked up for profitable scaling up by the key partners. The partners then indicate to CIMMYT the present stage of seed development against each selected variety, along with the nature and amount of support required for effective scaling up and to reach the project targets.

Empowered rural communities have the potential to promote scaling up through declaring their preferences and holding public agencies accountable. They can be a very strong advocacy voice for lobbying political powers to reform policies and regulations for creation of conducive environments for seed production and supply. A good example is where accumulated evidence of farmers/consumers' preference for specific variety(s) through participatory evaluations of on-farm trials has been used to inform decisions about varieties to include in government inputs support programs. Of importance to the rural communities is the possibility for multiple uses of the new variety (e.g., food, feed, fodder, energy).

Conclusion and Recommendations

This chapter is expected to inform climate smart agriculture policy and project interventions for climate risk reduction and resilience among smallholder farming households. In addition to contributing to knowledge on strategies that work in promoting CSA approaches, the chapter has made a case for serious consideration of the role of business (private sector) in facilitating dissemination and adoption of beneficial technologies. Furthermore, it has contributed to making information accessible on available CSA options (improved DTM maize varieties) that have beneficial outcomes for both farmers and business in the context of climate change in ESA. The DTMASS project implemented by CIMMYT from 2013 to 2018 has adopted a business development approach to ensure availability, accessibility, and affordability of improved and adaptable maize seed varieties to smallholder farmers in drought-prone regions of Ethiopia, Kenya, Mozambique, Tanzania, Uganda, and Zambia. The approach has fostered country-level business partnerships and networks that have boosted sustainable production, processing, and marketing (scaling up) of improved DTM seed. The financial and technical support from CIMMYT has boosted the seed production and distribution capacity of SME seed businesses in these countries. Through directly supporting the SMEs' seed production and promotional and marketing activities, the DTMASS project has also enhanced the spread (scaling out) of the improved DTM varieties across various geographical areas. Both scaling up and out rely on giving stakeholders in the maize value chain compelling reasons to continue producing, distributing, and consuming DT maize varieties. These success stories of the SME seed businesses profiled in the case studies presented in this chapter provide a compelling case for future seed business development initiatives across the African continent, where seed industries are yet to realize full potential.

It is important to note that although the DTM varieties are necessary for mitigation of climate change impacts, on their own they are not sufficient for solving the farmers' problems. There is need for the farmers to be assisted to adopt good agronomic practices, including purchase and proper use of external inputs such as fertilizers and agrochemicals, for the desired outcomes to be achieved. The business promotional approach has proven to be effective in nurturing SME seed companies and fostering strong agro-dealer networks for improved supply and affordability of certified DTM seeds. It is critical for CSA interventions to explore the feasibility of private sector-driven, market-based approaches to scaling up of technologies for wider dissemination and adoption. The case of CIMMYT's DTMASS project in ESA provides a compelling case for scaling up.

Scaling up of CSA technologies and practices should be concerned about achieving greater impact through improving the living standards of more people on a lasting basis. It should not be focused on achievements of individual projects although these may be important tools for design and implementation of CSA programs. Scaling up should be about supporting longer-term programs of engagement and building momentum that lasts beyond the program. It should not only be about what governments or donors can do but about getting programs right on the ground, whether with government or external assistance or without. Governments and donors should support and facilitate, rather than hinder, scaling up of CSA technologies like improved DTM varieties. Thus, continuous engagement and dialogue with governments is important to ensure that input subsidy programs implemented in most SSA countries actually contribute to seed market and private development rather than lead to distortion of market forces. The success of any CSA technology, practice, or program will largely be linked to strong government investment and political will. Strong government commitments for implementation of evidence-based policy agendas, dissemination of technologies through robust extension systems, and making resources available to fuel this work, in addition to investment in irrigation development to help manage risk, are invaluable for successful scaling of proven and beneficial CSA technologies and practices.

Cross-References

- Adaptation of Sorghum (*Sorghum bicolor* L. Moench) Crop Yield to Climate Change in Eastern Dryland of Sudan
- Climate Change Adaptation: Adoption of GMOs in Africa
- Drought-Tolerant Crops in Kirinyaga County, Kenya: Climate-Smart Agriculture Adaptation Strategies
- Intensifying Maize Production Under Climate Change Scenarios in Central West Burkina Faso

Appendix 1: DTMASS Project Partners 2017

Country	Partner	Status (public or private or NGO)	Role/responsibility
Multiple	Esoko	Private	Digital data platform provider
Ethiopia			
1	ARARI (Amhara Regional Agricultural Research Institute)	Public	Seed production and promotion
2	EIAR (Ethiopian Institute of Agricultural Research)	Public	Seed production and promotion
3	OARI (Oromia Agricultural Research Institute)	Public	Seed production and promotion
4	SARI (Southern Agricultural Research Institute)	Public	Seed production and promotion
5	MBU (Meki Batu Union)	CBO	Seed production and promotion

(*continued*)

Country	Partner	Status (public or private or NGO)	Role/responsibility
6	TARI (Tigray Agricultural Research Institute)	Public	Seed production and promotion
7	ASE (Amhara Seed Enterprise)	Public	Seed production and promotion
8	ESE (Ethiopian Seed Enterprise)	Public	Seed production and promotion
9	OSE (Oromia Seed Enterprise)	Public	Seed production and promotion
10	SSE (South Seed Enterprise)	Public	Seed production and promotion
11	EVF (Ethiopia Veg and Fruit)	Private	Seed production and promotion
Kenya			
12	STAK (Seed Trade Association of Kenya)	NGO	Seed promotion
13	Kenya Seed Co. (KSC)	Parastatal	Seed production and promotion
14	Dryland Seed Limited (DSL)	Private	Seed production and promotion
15	East African Seed	Private	Seed production and promotion
16	Freshco	Private	Seed production and promotion
17	Western Seed	Private	Seed production and promotion
18	KALRO (Kenya Agricultural and Livestock Research Organization)	Public	Quality control
Mozambique			
19	MOA Seed Department	Public	Seed supply and quality assurance
20	Sementes Nyara	Private	Seed production and promotion
21	Oruwera Seed Co.	Private	Seed production and promotion
22	Dengo Seed	Private	Seed production and promotion
23	Klein Karoo K2 MOZ	Private	Seed production and promotion
24	Phoenix Seed	Private	Seed production and promotion
Tanzania			
25	AgriSeed Tech	Private	Seed production and promotion
26	Aminata	Private	Seed production and promotion
27	IFFA Seed Co	Private	Seed production and promotion

(*continued*)

Country	Partner	Status (public or private or NGO)	Role/responsibility
28	MAMS	Private	Seed production and promotion
29	Meru Agro	Private	Seed production and promotion
30	Suba Agro	Private	Seed production and promotion
31	TOSCI (Tanzania Official Seed Cert Inst.)	Public	Quality control
Uganda			
32	MoA NCSC (National Seed Certification)	Public	Quality control
33	NARO (National Agricultural Research Organisation)	Public	Seed supply and quality assurance
34	Masindi Cooperative	Private	Seed production and promotion
35	EAS (East Africa Seed – UG)	Private	Seed production and promotion
36	Equator Seeds	Private	Seed production and promotion
37	Grow More Seed	Private	Seed production and promotion
38	NASECO	Private	Seed production and promotion
39	FICA	Private	Seed production and promotion
40	Otis Seed	Private	Seed production and promotion
41	Pearl Seeds	Private	Seed production and promotion
42	Simba Seed	Private	Seed production and promotion
43	Victoria Seeds	Private	Seed production and promotion
44	USTA (Uganda Seed Trade Association)	NGO association	Networking and advocacy
Zambia			
45	SCCI (Seed Control and Certification Institute)	Public	Promotion and quality control
46	ZARI (Zambia Agriculture Research Institute)	Pubic	
47	Unity Seed	Private	Seed production and promotion
48	Total Land Care	NGO	Demos/ community-level awareness
49	Kamano Seed	Private	

(*continued*)

Country	Partner	Status (public or private or NGO)	Role/responsibility
			Seed production and promotion
50	Stewards Globe	Private	Seed production and promotion
51	Klein Karoo K2 Zambia	Private	Seed production and promotion
52	Advanta	Private	Seed production and promotion
53	National Agricultural Information Services (NAIS)	Public	Promotion
54	COMACO (Community Markets for Conservation)	Public	Awareness and demonstration
55	CRS (Catholic Relief Services)	NGO	Demos/ community-level awareness

References

Abate T, Shiferaw B, Menkir A, Wegary D, Kebede Y, Tesfaye K, Kassie M, Bogale G, Tadesse B, Keno T (2015) Factors that transformed maize productivity in Ethiopia. Food Sec 7:965–981

Alliance for Green Revolution in Africa (cited as AGRA, 2014) Africa agriculture status report 2014. Climate Change and Smallholder Agriculture in Sub-Saharan Africa. AGRA

Anandajayasekeram P (2016) Scaling up and scalability: concepts, frameworks and assessment. Vuna Research Report, Pretoria

Appadurai AN, Chaudhury M, Dinshaw A, Ginoya N, McGray H, Rangwala L, Srivatsa S (2015) Scaling success: lessons from adaptation pilots in the rain-fed regions of India. World Resource Institute, Washington, DC

Audet-Bélanger G, Gildemacher P, Hoogendoorn C (2016) Malawi study report – final draft. CIMMYT, Texcoco, Mexico

Belt J, Kleijn W, Chibvuma PA, Mudyazvivi E, Gomo M, Mfula C, Mkojera E, Opio M, Zakaria I, Boafo K (2015) Market-based solutions for input supply: making inputs accessible for smallholder farmers in Africa. Kit working papers 5-2015. Development Organisation (SNV), The Hague, Netherlands

Best R, Ferris S, Wheatley C (2015) A Guide to Strengthening Business Development Services in Rural Areas. Catholic Relief Services, Baltimore, USA

Chandy L, Hosono A, Kharas H, Linn J (2013) Getting to scale. Brookings, Washington, DC

Cooley L, Linn J (2014) Taking innovations to scale: methods, applications and lessons. Results for Development and MSI, Washington, DC

Dayo P, Olumuyiwa OJ, Yarama N, Gabriel M, Edet E (2018) Scaling strategies for agricultural innovations in Nigeria. FARA Res Results 2(1):21

Fatunbi AO, Ajayi MT, Obi A, Odularu GO, Adekunle AA (2015) Spreading the gains of agricultural innovations in Africa: a strategy to scale-out and scale-up the IAR4D concept. Forum for agricultural research in Africa (FARA), Accra

Fisher M, Abate T, Lunduka RW, Asnake W, Alemayehu Y, Madulu RB (2015) Drought tolerant maize for farmer adaptation to drought in sub-Saharan Africa: determinants of adoption in Eastern and Southern Africa. Clim Chang 133:283–299

Food and Agriculture Organisation (cited as FAO, 2013). Climate smart agriculture sourcebook. The Food and Agriculture Organisation of the United Nations, Rome

Hartmann A, Linn J (2008) Scaling up: a framework and lessons for development effectiveness from literature and practice, Wolfensohn center working paper no. 5. Brookings, Washington, DC

International Maize and Wheat Improvement Center (CIMMYT, 2016) Stress tolerant maize in Africa (STMA) Project Proposal Narrative Report. CIMMYT. Texcoco, Mexico

International Maize and Wheat Improvement Center (cited as CIMMYT, 2017) Drought-tolerant maize for Africa seed scaling (DTMASS) project annual report for feed the future centrally funded scaling programs for the period: FY16 – October 2015 – September 2016. International Maize and Wheat Improvement Center (CIMMYT)

International Maize and Wheat Improvement Center (cited as CIMMYT, 2018) Drought-tolerant maize for Africa seed scaling (DTMASS) project annual report for feed the future centrally funded scaling programs for the period: FY17 1 October 2016–30 September 2017. International Maize and Wheat Improvement Center (CIMMYT)

Kaimenyi K (2016) Maize seed systems in Africa: understanding the basics. Available at: http://www.cimmyt.org/maize-seed-systems-in-africa-understanding-the-basics/

Kaimenyi K (2017) Tanzanian start-up rises to profitable maize business. Available at: http://www.cimmyt.org/tanzanian-start-up-rises-to-profitable-maize-business/

Kandiwa V, Lweya KB (2017) Gender responsive promotional approaches for improved maize seeds in Africa: guidelines on awareness creation. International Maize and Wheat Improvement Center (CIMMYT), Nairobi, Kenya

Linn J (2012a) Scaling up in agriculture, rural development and nutrition., 2020 focus briefs, no. 19. International Food Policy and Research Institute, Washington, DC

Linn JF (2012b) Scaling up development interventions: a review of UNDP's country programme in Tajikstan. Global economy & development working paper 50. Brookings, Washington DC

Linn JF (2014) Scaling Up Development Impact: a summary of current research, advice and outreach. Short Paper. Brookings Institute. Washington DC

Lunduka RW, Mateva KI, Magorokosho C, Manjeru P (2017) Impact of adoption of drought tolerant maize varieties on total maize production in South Eastern Zimbabwe. Climate and Development, 2017. https://doi.org/10.1080/17565529.2017.1372269

MacRobert J (2009) Seed business management in Africa. International Maize and Wheat Improvement Center (CIMMYT), Harare

Mthetwa C, Mponda L, Maluma E (2010) Agro Business Management Manual for Implementers and Agro-Dealers. (ADAPT) Project, CARE Zambia

Munyaka N, Mazarura U, Mvumi BM (2017) A compelling case for seed enterprises as a tool for rural development in the smallholder farming sector. African J Rural Develop 2(2):199–212

Setimela PS, Magorokosho C, Lunduka RW, Gasura E, Makumbi D, Tarekegne A, Cairns JE, Ndhlela T, Erenstein O, Mwangi W (2017) On-farm yield gains with stress-tolerant maize in Eastern and Southern Africa. Agron J 109(2):406–417

Takim FO (2017) Climate change adaptation options: importance of drought tolerant maize seeds. Policy brief no. 16. UNU-INRA. Available at: https://collections.unu.edu/eserv/UNU:6262/takim_unuinra_policybrief2017_final.pdf

The United States Government (2017) Global Food Security Strategy Technical Guidance scaling for widespread adoption of improved technologies and practices. Feed the Future. U.S. Government's Global Hunger and Food Security Initiative

Innovation Systems to Adapt to Climate Change: Lessons from the Kenyan Coffee and Dairy Sectors

110

Kinfe Asayehegn, Ludovic Temple, Philippe Vaast, and Ana Iglesias

Contents

Introduction: Evidence on the Impacts of Climate Change and its Adaptation 2250
Climate Change Knowledge Asymmetry 2251
The Case Study of the Coffee and Dairy Sectors in Central Kenya 2253
The Need for SSI Approach in Climate Change Adaptation 2254
Analysis of Sources of Knowledge for Climate Change Adaptation 2256
Technological and Institutional Innovations to Adapt to Climate Change in the Coffee and Dairy Sectors 2259
Developing Disease-Resistant Coffee Varieties 2259
Institutional Innovation in the Coffee Sector 2262
Innovation in the Dairy Sector to Adapt to Climate Change in Central Kenya 2264
Technological Innovation in the Dairy Sector 2265
Institutional Innovation in the Dairy Sector 2265
Comparison of the Contributions of Actors in Adaptation to Climate Change in the Coffee and Dairy Sectors of Central Kenya 2266
Conclusions 2268
References 2270

K. Asayehegn (✉)
School of environment, Gender and development Studies, Hawassa University, Hawassa, Ethiopia

Center for International Cooperation in Agricultural Research and Development (CIRAD), Montpellier, France

Department of Agricultural Economics, Universidad Politécnica de Madrid, Madrid, Spain
e-mail: kinfe.asayehegn@gmail.com; mikkinfe@gmail.com; kinfe85@gmail.com

L. Temple · P. Vaast
Center for International Cooperation in Agricultural Research and Development (CIRAD), Montpellier, France

A. Iglesias
Department of Agricultural Economics, Universidad Politécnica de Madrid, Madrid, Spain

W. Leal Filho (ed.), *Handbook of Climate Change Resilience*,
https://doi.org/10.1007/978-3-319-93336-8_25

Abstract

Climate change is one of the most widespread anthropogenic challenges affecting agriculture and agricultural production, where its impacts on economic growth in general are still issues in debate. An adjustment to the actual or expected changes in the agricultural sector using different innovative adaptation strategies has to be among priorities in policy decisions. Research on sectoral systems of innovation, however, has paid little attention to adaptation to climate change, notably in agricultural trade and agribusiness. This chapter, therefore, explores the role of systems of innovation in adapting to climate change, which have methodological, conceptual, and policy implications. It focuses on two case studies in Kenyan agriculture, i.e., the coffee and dairy sectors, which differ in terms of stakeholder and institutional setups. In the coffee sector, the actors' system is highly centralized, and the systems of innovation is oriented toward technology development. In contrast, the dairy sector consists of a diversity of actors, and its systems of innovation is based on institutional building and marketing. The capacity to innovate and adapt, therefore, depends on institutional arrangements in addition to technology development, suggesting that the dairy sector in Kenya could be an example for the coffee sector.

Keywords

Climate change · Coffee · Dairy · Sectoral systems of innovation · Central Kenya

Introduction: Evidence on the Impacts of Climate Change and its Adaptation

Among the most widespread anthropogenic challenges affecting agricultural production are climate change and climate variability. The Intergovernmental Panel on Climate Change (IPCC's) Fifth Assessment Report (AR5) provides indisputable evidence of climate change driven by the increase in global greenhouse gas (GHG) emissions (IPCC 2014). Global GHG emissions have grown since preindustrial times, with a sharp increase of 70% between 1970 and 2004. This caused:

- Minimum and maximum temperatures to increase globally and particularly in the tropics (IPCC 2014).
- Rainfall has increased in variability over the last 60 years (Rosell and Holmer 2007), rainfall patterns (such as onset, duration, and cessation) have changed, and intermittent dry spells have caused critical soil moisture deficit.

The impacts of climate change on economic growth are still an issue of debate. Some consider that climate change could be an opportunity for economic growth, while others (IPCC 2014; Armah et al. 2015; Touzard 2014; Teucher et al. 2016) argue that it will lead to a slow-growing economy or even de-growth. However, this debate must take into account regional differences:

- In the colder regions of the northern hemisphere, climate change could have a positive effect on production.
- In the tropics, it is expected to substantially reduce economic growth (IPCC 2014). In Africa and Latin America, for instance, maize production is expected to decline, which would result in a loss of $2 billion per year as of 2055, compared to current production (IPCC 2014).

An adjustment in response to the actual or expected changes has to be, therefore, among priorities in policy decisions. Farmers' behavioral change toward adaptation and willingness to take action are as important as are policy decisions and technical innovations (World Bank 2016). Technical and innovative adaptation to climate change varies across production systems due to differences in opportunities and determinants. In rain-fed crop production, adaptation, for instance, comprises practices such as adopting drought-resistant varieties or intercropping of different crops (Teucher et al. 2016), while changing breeds and alternative feeding strategies are common in the livestock sector (Seo 2010).

The capacity to adapt to climate change depends on many factors, such as farmers' and other stakeholders' perceptions, market dynamics, private and public investment in R&D, policy, and institutions (Hallegatte 2009). A key factor is the capacity for cooperation between actors in regions and sectors. Touzard (2015), for example, reports that climate change has a systemic impact on both regional vineyards and wine value chains, calling for cooperative solutions among actors at these two levels. Hence, adaptation capacities depend on the type of sector, and actors' interaction and coordination, notably in agriculture.

Climate Change Knowledge Asymmetry

Climatologists have developed different methodologies to analyze and understand the rate of changes in climate, particularly through the uses of predictive models. The future climate trends are, however, uncertain, and hence direct use of the outputs of existing climate models (single information source) developed by scientists may well be insufficient to guide the development of adaptation policies. Most of the evidence produced from models is also at global and regional scales, while the microclimate of local scales is significantly important. It would seem wiser to take into account other sources of information about climate change to develop adaptation policies. This may include the knowledge and information of the technical and social scientists, lifelong farmers' experiences, and other development stakeholders' knowledge systems. There is, however, an active debate among scientists about the inclusion of farmers' knowledge for adaptation planning, with some believing farmers' knowledge of climate change is pivotal while others disregard it (Armah et al. 2015).

Among those who tend to disregard farmers' knowledge, Weber (2006) considers climate change as basically probabilistic and regards it as an issue that is beyond human perception. Farmers' cognition and social institutions are considered of little value (Rodima-Taylor et al. 2012), while the stochastic analysis of historical climate data is considered as the only input necessary for adaptation policies and decisions. Among those who value farmers' knowledge, the contention is that not every adaptation decision requires consideration of long-term historical climate information (Bednarek et al. 2018; Armah et al. 2015; Preston et al. 2015; Cash et al. 2006). Rather they claim that valorizing of local people's knowledge could produce better results.

These two approaches are neither mutually exclusive nor self-exhaustive, but complement each other. This is particularly the case in locations where temperature and rainfall records may be unreliable or unavailable (a situation common to most East African countries) and for which using a purely statistical analysis to determine climate change would be haphazard. For instance, climate change scenarios and models unanimously project an increase in future ground temperature in Eastern Africa. There are however inconsistencies regarding changes in rainfall, with some studies projecting an increase (Haarsma et al. 2005) while others project a decrease (Castellanos et al. 2012). Unreliable regional climate records and poor management of existing meteorological stations contribute to unreliable data results. Perception, on the other hand, is linked to farmers' action and may be hindered by impacts rather than the actual changes. Hence, perception could be distorted by recall bias (Hahn et al. 2009) and hence could yield unreliable climate information and adaptation needs (Armah et al. 2015). Understanding changes both from local perceptions and from historical data is, therefore, fundamental to both climate science and adaptation policy formulation because it helps to combine local and global contexts where farmers, scientists, and policy actors operate (Asayehegn et al. 2017a).

The capacity of climate change to adapt depends on many factors, such as market dynamics, private and public investment in R&D, policy, and institutions (Hallegatte 2009), or even cultural aspects (Dunlap and Brulle 2015). A key factor is the capacity for cooperation between actors in regions and sectors (Boyer 2016). Hence, adaptation capacities depend on the type of sector, and actors' interactions and coordination, notably in agriculture (Soussana 2013). An innovation system, in this context, contemplates the way actors, sectors, and systems interact and coordinate for better adaptation to climate change.

So far, innovation studies have, however, emphasized on the role of innovation in producing economic and social changes (Van Lancker et al. 2016), in particular in each sector separately, where knowledge and technology, the structure of demand, institutions, and firm characteristics are different (Malerba 2002). The concept of the sectoral systems of innovation (SSI) is thus proposed in order to analyze this set of factors, which can strongly determine specific trends in innovation in each sector, with possible effects on economic performance and adaptation to global challenges such as climate change (Malerba 2007). Empirical studies, however, insufficiently take into account sectoral differences in innovation for climate change adaptation and the role that an SSI could play (Touzard 2014).

This chapter therefore provides emphasis on the roles of sectoral systems of innovation (SSI) in the adaptation of sectors to climate change, focusing on the coffee and dairy sectors in Central Kenya. It provides an answer to questions, such as:

1. What characteristics of the systems of innovation are particular to each sector in the process of adaptation to climate change?
2. How do such characteristics affect the adaptation process and competitiveness of the sectors?

The Case Study of the Coffee and Dairy Sectors in Central Kenya

Central Kenya is one of the areas with diversified physical environments and an extreme climate. The area is made up of three agroecological zones each corresponding to a different altitude range (high, mid, and low altitude) and farming system (food crops, coffee, and tea). The different agroecological zones are indeed characterized by a specific climate and weather:

- Cold temperature and higher rainfall in the highlands.
- Average temperature and rainfall in the midlands.
- Hotter temperature and lower rainfall in the lowlands.

Associated with the climate features, the crops cultivated differ according to the agroecological zone:

- Tea and coffee are the main crops at the higher and mid altitude areas, respectively.
- Food crops (maize and beans) dominate in the lowlands.

Coffee, particularly Arabica, grows at an average temperature between 18 °C and 21 °C and at an annual rainfall of 1500–3000 mm (Lin 2007). The lower-altitude, food crop zone of Kenya has a semiarid climate with a high potential for droughts. Rainfall is erratic, uncertain, and unevenly distributed into two seasons. However, both seasons are characterized by a short to very short growing period.

Historically, coffee production was one of the main income-generating activities in Kenya. Following the "Lancaster House Conference" after independence, most of the large-scale coffee farms were sold to local elites (Ratten 1993), and local indigenous people were encouraged to invest in coffee. Due to the expansion of plantations and the attention given to coffee by farmers and the government, the sector grew at an annual rate of 6.6% until 1987 (Thuku 2013). National coffee production has been on an increasing trend for a long time; it increased from 43,778 t in 1963/1964 to 140,000 t in 1987/1988 (Thuku 2013).

Climate change has, however, strongly impacted coffee production of the area (Jaramillo et al. 2014). This is in relation to:

- Long-term rainfall has declined and the temperature has increased (Asayehegn et al. 2017a).
- Previous potential coffee areas have transformed into marginal coffee or food crops (Jaramillo et al. 2014).

- Favorable weather zones have transformed into arid and semiarid zones (Lin 2007).

As a result, coffee plants tend to be water stressed, while new pests and diseases emerge in connection to climate change. This caused, since 1988, coffee production to decline by 62% (Thuku 2013). During this period, coffee production has moved up from the lower altitudes, where it has been replaced by dairy and food crops, for two reasons:

1. Warmer temperatures and erratic rainfall resulted in the transformation of potential areas into semiarid zones, where a 1 °C temperature increase results in a yield decline of 137 kg per hectare (Craparo et al. 2015).
2. Climate change encouraged the infestation of coffee leaf rust (CLR) and coffee berry diseases (CBD), causing the transformation of previous potential coffee-growing areas into marginal coffee or food crop areas (Jaramillo et al. 2014).

Apart from the coffee sector development, the development of the dairy sector in Kenya on the other hand shows three different periods with distinct paradigm shifts in the dairy development and technology. These are (1) the period of steady growth (before 1990), (2) the period of disruption (1991–2002), and (3) the period of revival (since 2003). During the period of steady growth, indigenous smallholder farmers were encouraged to develop dairy production through training, infrastructure development, and service delivery. Annual milk production grew from 75 million liters in 1964 to 392 million liters in 1990. During the period of disruption, dairy production declined from 359 million liters in 1991 to less than 150 million liters in 2002, due to the absence of an efficient market and supply system. During the revival period, the dairy sector experienced a sharp increase in the volume of production, reaching over 4.1 billion liters in 2014 (FAO 2011).

The Need for SSI Approach in Climate Change Adaptation

The concept of SSI basically provides a multidimensional, integrated, and dynamic view of specific economic sectors, including different structural components and a set of actors carrying out market and nonmarket actions (Malerba 2002; Edquist and Chaminade 2006). The key structural components of SSI are the knowledge base and the learning process, the technological base, the specific institutions (rules, norms), and the evolution of demand (Malerba 2007). The agents in the SSI are individuals, firms, and non-firm organizations that are involved at various levels of the production and innovation processes (Malerba 2002). These actors have specific models of interaction, interdependencies, and links, which depend on the evolving structure of each sector (Malerba 2007). Mobilization of the SSI approach is, therefore, important in order to explore conditions of adaptation to climate change in the coffee and dairy sectors, which are similar in terms of farmers' objectives and shifting historical fortunes, but are different in terms of marketing, sociopolitical and technical characteristics, and policy. This approach helps to understand:

- The current climate changes, which are explained by the changes in temperature and rainfall.
- The impacts of the climate change on land and land use change, production, and productivity.
- The adoption of different adaptation strategies by farmers and other stakeholders that adapt/mitigate the changing climate.
- The interactions and relationships between different stakeholders and actors (farmers and their organizations, research institutes, academia, private sectors, policy actors, etc.) in order to take action.
- The potential innovation systems such as technological (improved breeds and varieties, agronomic practices,) and institutional (the contribution and interaction of research centers, policy actors, and other stakeholders in the adaptation process).
- The interaction and impact pathway between the different entities (change in climate, its impacts, adaptation strategies and process, innovation system), which are cause and effect in nature (see Fig. 1).

The first dimension in the above framework illustrates changes in climate, which result from the combined changes in rainfall and temperature. Conceptually, climate change may act as both an impact pathway and an adaptation pathway.

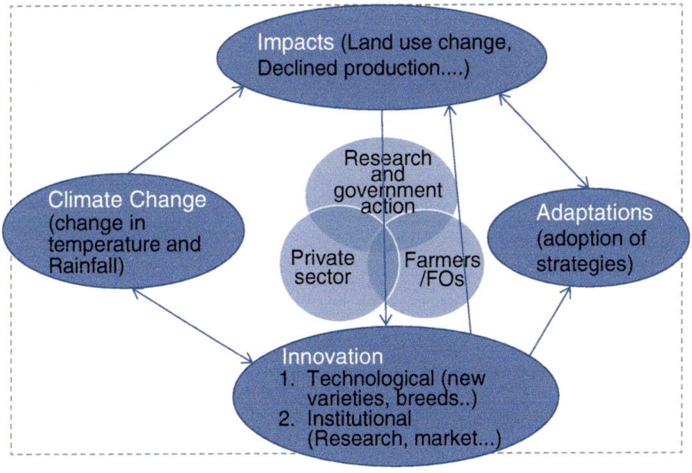

Fig. 1 Conceptual model illustrating interactions between change in climate, climate change impacts, innovation systems, and adaptation to climate change. (Source: Authors' own representation of concepts). Note: some of the arrows above are one-way and others are two-way. The two-way arrows indicate an influence from both sides

- From the impact pathway, climate variability and change may increase the frequency of drought (Fig. 1).
- In the adaptation pathway, adoption of different adaptation strategies minimizes impact of climate change on land and land use change, production, and productivity of the agricultural sector.

The second dimension of the framework (Fig. 1) focuses on the interdependencies in adaptation to climate change between different sectors and their actors. The boundaries of sectors are not fixed but dynamic, which provides a mechanism of emergence, growth, changes, and innovation. For example, over the last 35 years, coffee in Kenya has moved up to higher altitudes in order to adapt to the changing climate, while food crops have been grown at the altitude once reserved for coffee production (Asayehegn et al. 2017b). Land use competition between sectors creates a mechanism of interaction among actors. In the agroforestry coffee systems of Kenya, the emergence of the dairy sector generates new interactions between actors, creating new links between coffee, tea, and dairy systems. In such cases, links and boundaries could be competitive or complementary. The links and interdependencies among actors could be within a sector and/or across sectors. This led to the third dimension in the above framework (Fig. 1), which is the characterization of the innovation and adaptation processes. Innovations to adapt to climate change may be technological, such as the development of new varieties and new breeds, or institutional, such as implementing new rules, norms, or organizations that improve collaborations to reduce the impacts of climate change (Fig. 1). The development of new cultivars of cowpea in the Sahelian West Africa, for example, helped farmers to cope with the climate challenges they experienced (Chhetri et al. 2012). In East and South Africa, scientists developed high-yielding maize varieties, with the objective of reducing vulnerability to drought (Fisher et al. 2015). New agronomic practices are another line of innovation for adaptation. Cropping system diversification, conservation tillage, and new fertilizer management have proved to be sustainable adaptation practices, increasing farmers' income and reducing agronomic risk (Teklewold et al. 2013). Institutional innovation may also exert influence on both the impacts of climate change and adaptation (Fig. 1). For example, the adoption of new varieties depends on both upstream and downstream organizations (dissemination of varieties and credits vs. access to market). New institutions may also improve actors' interactions in providing solutions for adaptation, by improving knowledge-sharing and interorganizational learning, complementary resource development, capacity building, and efficient governance (Asayehegn et al. 2017b; Malerba 2002, 2007).

Analysis of Sources of Knowledge for Climate Change Adaptation

Due to the direct relationship between crop production and climate, agriculture is the sector most affected by climate change (Pelling and Manuel-Navarrete 2011). To understand the changes, adaptations, and resilience of systems depends on different knowledge systems, particularly, farmers' knowledge, scientists' knowledge, and policy makers' and other stakeholders' knowledge at different contexts and scales (Nilsson et al. 2012; Brody et al. 2008). Farmers' knowledge of climate change is mainly based on perception. Farmers have different experiences in understanding of climate change. Their understanding is in general, therefore, based on two sources, i.e., external information on climate and weather change and their own observations from experience and intergenerational knowledge transfer (Armah et al. 2015; García de Jalón et al. 2013). Some farmers might hear about the concept of climate and weather change from external sources in various ways, while others completely depend on their observation to understand changes without any external influence. Those who receive the information may formally get it from meteorological information dissemination, seminars, government and NGO consultations and warning systems, or personal experience sharing and interaction (Asayehegn et al. 2017a, b).

Sector-wise disparities, for instance, in a study by Asayehegn et al. (2017a), revealed that coffee farmers appeared better informed on weather and climate changes than food crop farmers, perhaps as a result of coffee farmers being better off in wealth and social status and hence positing a greater capacity to access information. The procedure to understand farmers' perception of climate change, however, requires developing methodologies like contemplating an understanding if there is a change in the long-term climate (Hahn et al. 2009; Simelton et al. 2013). To analyze such patterns, developing a simple flowchart matrix framework like the one adapted and used by Asayehegn et al. (2017a) could be important to organize the quotes on rainfall and temperature patterns flowing from left to right (Fig. 2).

The first level of the analysis of the flowchart (Fig. 2, left column) establishes whether there is indeed a *change in climate*, relating broadly to rainfall or temperature. The second level (Fig. 2, middle column) identifies *what* is *exactly* changing,

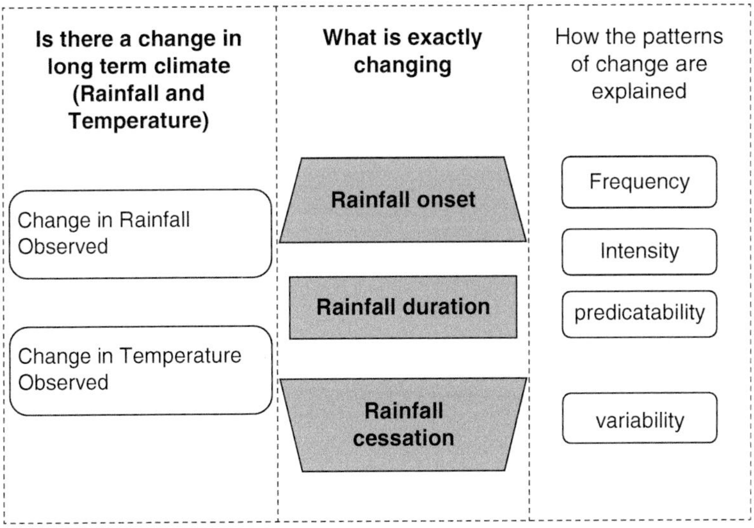

Fig. 2 Analytical flowchart matrix to organize quotes on changes in rainfall and temperature. (Source: Adapted from Simelton et al. 2013)

e.g., changes in onset, duration, or cessation of rainfall. At each level, the number of respondents not perceiving changes is expected also to be noted, and the farmers that perceived no change in temperature or rainfall are expected to be excluded from the next analysis (the pattern). A respondent who perceived a change in at least one of the three categories, i.e., onset, duration, or cessation, is considered for the third level of analysis, which focuses on variables such as frequency, intensity, predictability, and variability of the perceived changes in the case of short rain season temperature, long rain season temperature, or dry season temperature, and finally, how it changed is necessary in the analysis.

Apart from their access and use of official climate information, farmers describe changes in climate by describing their personal experiences and observations of how it used to be in the past and how the rainfall and temperature was during the recent years. Farmers repeatedly explained about the changes in seasons and weather patterns (rainfall and temperature patterns). Analysis of farmers' perceptions could be in relation to three aspects (e.g., Asayehegn et al. 2017a):

1. Change in overall climate
2. Change in rainfall
3. Change in temperature

The case study in Central Kenya shows that nearly 87% of the farmers perceived the overall climate has changed, while 84 and 76% perceived that either the rainfall has declined or the temperature has increased, respectively. The validity of farmers' perceptions could be determined by three indicators that show climate is indeed changing:

1. The trends observed in the pattern of rainfall
2. Trends in temperature
3. Observations on the effects of the changes such as the natural environment

Trend Analysis of Temperature and Rainfall: Meteorological Data Analysis to Understand Climate Change

Historical meteorological record of temperature (daily minimum and maximum) and rainfall (daily mean rainfall) over 35 years taken from representative stations is another source of climate change knowledge which is mostly used by climatologists and other scientists. Understanding of the climate change from an analysis of linear trends in historical meteorological observations considered:

- Minimum and maximum temperature.
- Rainfall in the highland and in low-altitudes zones.
- Rainfall patterns (the starting and finishing of rainfall).

Trend analysis of the long-term climate data clearly shows climate is indeed changing. Two types of changes are detected, i.e., spatial changes, for example, changes between the coffee and food crop zones and temporal changes within each zone.

- Mean minimum and maximum temperature (Fig. 3a, b) shows an increasing trend over time for both the coffee and food crop zones.
- The mean interannual variability of temperature (Fig. 3c) is also increasing.
- Mean annual rainfall (Fig. 4) was higher in the coffee zone compared to food crop zone.
- For rainfall, the trend analysis indicated a weak decrease for the coffee zone (Fig. 4a), while no clear trend was found (Fig. 4b) for the food crop zone.

The mean minimum and maximum temperature trends over time for both the coffee and food crop zones were also confirmed by the Mann-Kendall test and Sen's slope estimator (Table 1). Mean annual rainfall was higher in the coffee zone compared to food crop zone. For rainfall, the trend analysis indicated a decrease for the coffee zone, while no clear trend was found by the Mann-Kendall trend analysis and Sen's slope estimator (Table 1) for the food crops. Besides the trend in annual rainfall, rainfall pattern is also important.

Technological and Institutional Innovations to Adapt to Climate Change in the Coffee and Dairy Sectors

Developing Disease-Resistant Coffee Varieties

The dramatic increase in national coffee production between 1963 and 1987, from 34,000 to 140,000 metric tons, was dependent of two technological innovations. In 1963, local coffee farmers were encouraged and supported to use a new technical system that included fertilizer and cultural practices such as pruning, in order to increase yields (Asayehegn et al. 2017b). The second technological change was the introduction of pesticides in the mid-1960s to prevent frost, coffee berry diseases (CBD), and coffee leaf rust (CLR). Inputs and management supports were provided by the government directly through the cooperatives. In order to further improve production and quality, the Coffee Research Foundation (CRF) developed new disease-resistant varieties during the 1970s, in 1980 delivering a new variety called *Ruiru 11*, which has been disseminated to farmers from 1985 onward (Asayehegn et al. 2017b).

However, this SSI of developing new varieties and disseminating new practices and pesticides failed to increase coffee production or even maintain the level of production. For example, annual coffee production declined from 140,000 to less than 50,000 metric tons between 1988 and 2011, and production per hectare was reduced from 735 to less than 253 kg (Coffee Board of Kenya).

The claim that climate change impacts coffee production is based on the evidence that:

- Analysis of long-term climate data shows that rainfall has declined and that the temperature has increased (Figs. 3 and 4). This caused previous potential coffee-growing areas to be transformed into marginal coffee or food crops.

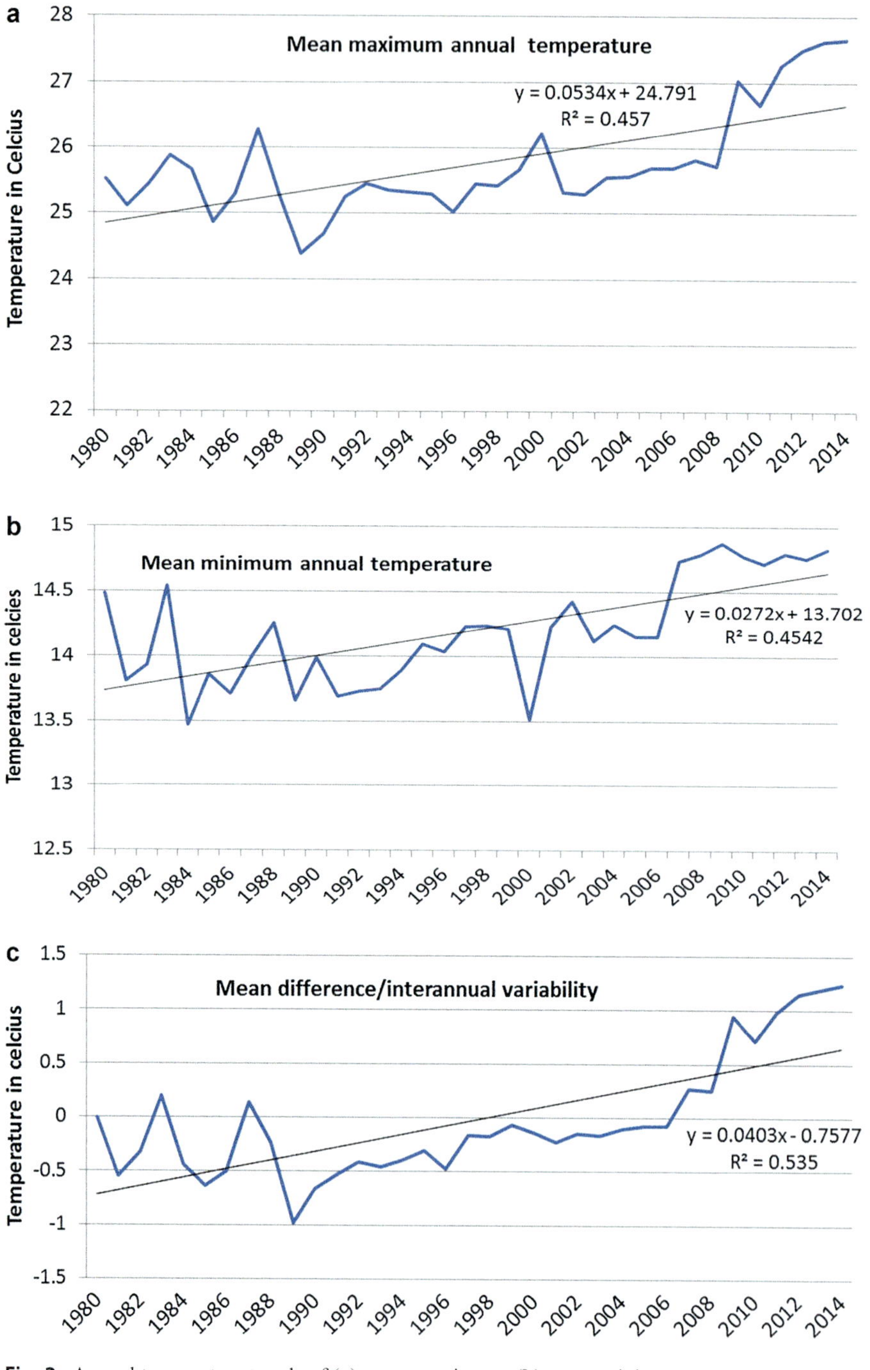

Fig. 3 Annual temperature trends of (**a**) mean maximum, (**b**) mean minimum, and (**c**) interannual mean differences in temperatures of the study area (1980–2014). (Data source: Kenya Meteorological Department)

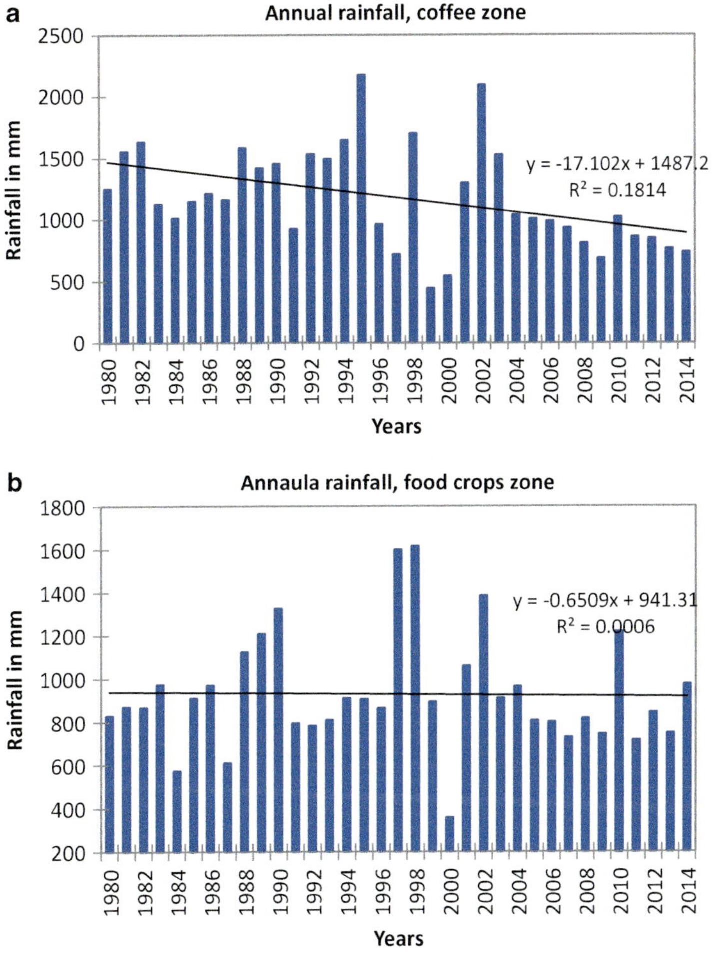

Fig. 4 Annual rainfall trends in millimeters of the coffee zone (**a**) and food crops zone (**b**), for the years of 1980–2014. (Data source: Kenya Meteorological Department)

- Farmers' perceptions of changes indicating seasonal and weather patterns are changed (Fig. 2). These arguments support the interpretation that the changes in coffee production noted above are induced by climate change and that SSIs in the coffee and dairy sectors are able to tackle to the current challenges of climate change.

This converges that:

(i) Institutional conditions and extension services are inadequately taken into account in this dissemination and learning process.

Table 1 Mann-Kendall test of significance

Variable	Mann-Kendall's tau	p-value	Sen's slope	Mean	SD
T max	0.503	<0.0001**	0.043	25.75	0.80
T min	0.509	<0.0001**	0.032	14.19	0.41
T interannual variability	0.592	<0.0001**	0.037	0.033	0.56
Rainfall coffee zone	−0.334	0.0040**	−17.100	1179.00	411.50
Rainfall food crops	−0.052	0.6730	−1.485	925.72	262.61
			Significance level (%): 5		

**indicates statistical significance at 5% level
Data Source: Kenya Meteorological Department

(ii) That climate change modified the conditions of technical innovation, and these were not considered by organizations of the coffee SSI.

In relation to variety selection, farmers are being confronted by two main challenges. First

1. The new varieties, which farmers call *heavy feeders*, mean the new coffee varieties, particularly *Ruiru 11*, require higher doses of fertilizer and frequent watering. This is linked to their physiological characteristics, where the root system is shallow (Asayehegn et al. 2017b).
2. The old varieties are effective for nutrient intake, due to their deep root system, but are highly sensitive to CBD and CLR, where farmers are being forced to invest in purchasing chemicals.

Institutional Innovation in the Coffee Sector

The actors in the coffee sector SSI are classified into three main categories. The first category includes the national and county government organizations which are the direct administrative bodies for the sector. The second category covers research and education institutions, such as the Kenya Agricultural and Livestock Research Organization (KALRO); international research institutions linked to CGIAR (ICRAF, CIAT), ICIPE, CIRAD, or joint research programs and projects with the AU and the EU; etc. These organizations develop research projects for the Kenyan coffee sector. The third category brings together development organizations and community-based organizations such as Danish International Development Agency (DANIDA), AgriProFocus, Swedish International Development Cooperation Agency (SIDA), and United States Agency for International Development (USAID). Figure 5 shows two main points about the coffee sector.

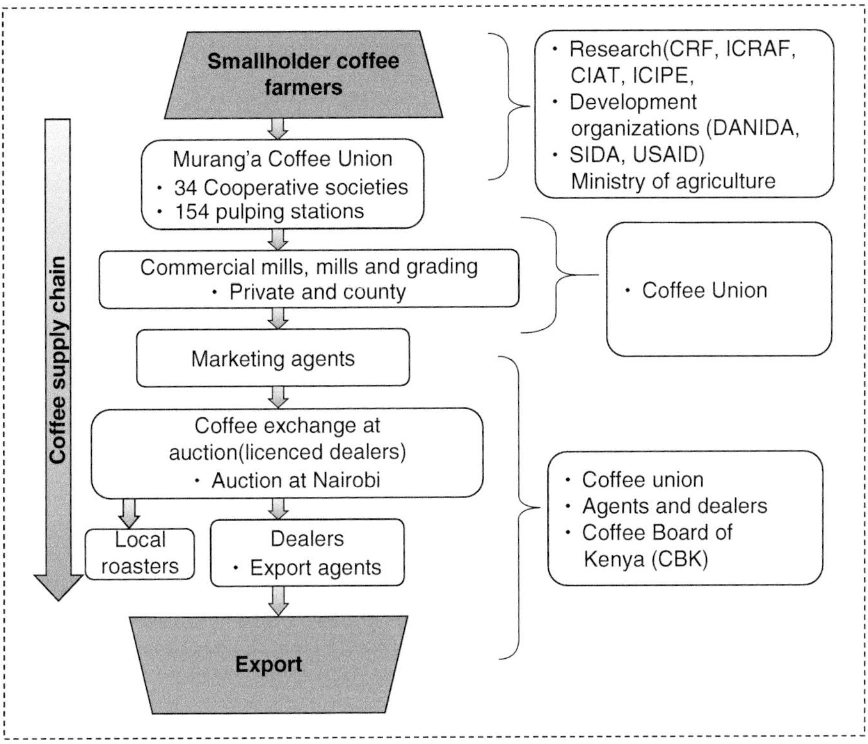

Fig. 5 Smallholder farmers' coffee supply chain and actors' interaction in Kenya. (Source: Own survey data, 2015)

1. The coffee SSI is organized around a long value chain with few actors in the marketing system (an oligopolistic structure) and a focus of the research and development organizations at the production scale.
2. The interactions and collaboration between the different actors in the coffee sector appear weak and dominated by business and administrative links.

The participatory stakeholder analysis of this study revealed that upstream in the coffee value chain, the Coffee Union was the monopoly institution providing financial, administrative, and technical services until coffee liberalization in 1992. It was mandated to supply inputs and control the application of rules and regulations on coffee production and supply. After liberalization, the Coffee Board of Kenya (CBK), which is the regulatory body, became responsible for defining marketing rules. The Coffee Union and CBK have a direct link with the government institutions, Coffee Research Foundation (CRF), milling companies (coffee processing), and auction and export agents (Fig. 5).

Downstream institutions and actors such as societies, local government agents (Ministry of Agriculture, cooperatives), national and international research centers (CRF, KALRO, CGIAR), and community-based development organizations are loosely linked to upstream actors. These unconnected and uncoordinated SSIs create opportunities for private business-oriented actors to have higher input prices. This kind of actors' interaction led coffee farmers in the community to follow four main strategies to adapt to the challenges they encountered.

1. Continue specialization in coffee production by investing in inputs and chemicals.
2. Intercrop food crops and fodder in coffee farms, which reduces coffee production.
3. Abandon coffee management and shift their source of livelihoods to off-farm and nonfarm activities.
4. Completely uproot coffee in order to plant other food crops and fodder for livestock.

Innovation in the Dairy Sector to Adapt to Climate Change in Central Kenya

The SSI in the development of the dairy sector is mainly based on three broad categories of change: technological development, extension and education on best practices for keeping dairy animals, and institutional building for marketing channels (Fig. 6).

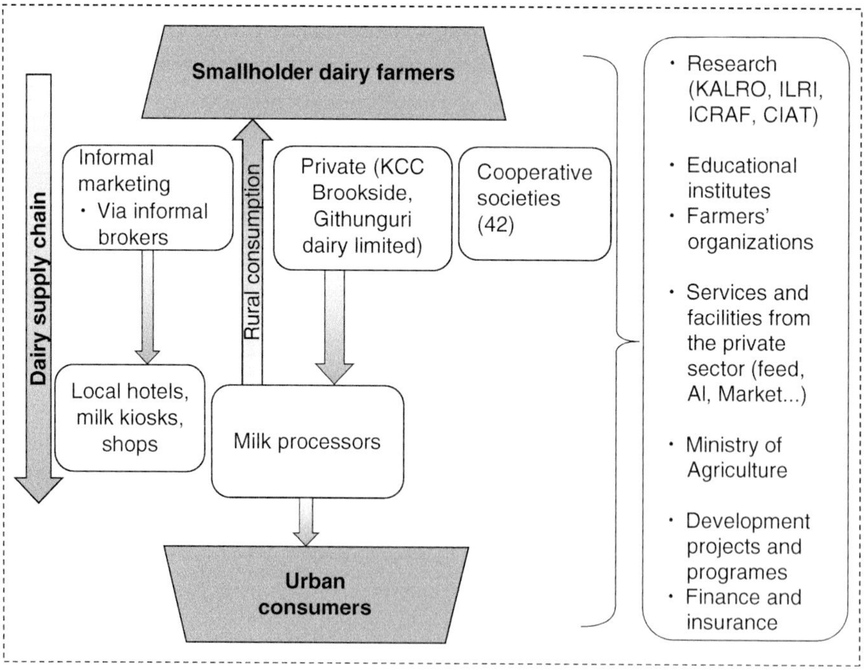

Fig. 6 Supply chain of milk and actors' interaction in Central Kenya. (Source: Own survey data, 2015)

Technological Innovation in the Dairy Sector

The focus of technological innovation is mainly on the genetic improvement of livestock breeds and the sanitary supervision of livestock. It also involves using new grass, shrubs, and commercial feed in order to solve the acute feed shortage linked to climate change. For its part, extension and education is mainly developed by applying a business-oriented private sector approach. The breeding materials, health services, and innovative new feed systems are primarily developed at research centers or directly adapted from abroad.

Originally, the local cow breed in the area was the Zebu, with a medium performance in meat and milk production. As the farmers' objective was mainly to improve milk production, continuous crossing of the best traditional breeds was carried out, leading to higher milk performance crossbreed cows. Artificial insemination of improved breeds is used, and farmers experienced the requirements of the new breeds' in terms of feeding and housing. The other technological innovation in the dairy sector is mainly dedicated to improving safety and milk quality, such as the installation of 35 dairy cooling plants (each with 5,000 liters of capacity) in the milkshed localities.

Institutional Innovation in the Dairy Sector

Information from key informants and group discussions indicated that Kenya Cooperative Creameries (KCC), equivalent to the Coffee Union, was established in 1925 to support the production, marketing, and processing of dairy and dairy products as a monopoly agent. The Kenya Dairy Board (KBD), equivalent to the CBK, was created to regulate the dairy sector. There were essentially three periods in the SSI in the dairy sector.

- During the first period, cattle breeding were fairly well organized and subsidized by the government. Breeding materials and artificial insemination were effectively used to upgrade breeds. KCC continued to be the sole agent for marketing and processing, protected by policy.
- During the second period, the position of cooperatives and the KCC weakened after liberalization of the sector. Farmers' milk delivery to KCC and other cooperatives declined due to irregular payments and delays in response to the liquidation of the KCC. As a consequence of the liberalization of the dairy sector, the services previously delivered by the government stopped. Public breeding and veterinary services were cut back, and artificial insemination services became inadequate. Indeed, the private sector was insufficient and not able to develop the insemination service, as well as the feeding function for the whole sector.
- Around the middle of the 1990s, self-help groups and informal agreements emerged. Deregulation of milk prices created an opportunity for different actors to participate in milk marketing. Three options for milk marketing thus coexisted: The KCC as a government agent, private companies such as Brookside dairy limited, and informal channels (Fig. 6).

During the last period, a new impetus corrected previous administrative and technical failure. Alternative feed sources such as drought-resistant grasses were developed, and new commercial feed companies emerged. Driven by increasing national demand for milk, prices increased, providing farmers with greater motivation to increase their production, in a new institutional context, which allows the stable availability of feed for domestic and commercial opportunities. The KCC was privatized and county cooperatives emerged in a new way.

Five categories of actors played a role in the dairy SSI (Fig. 6):

(i) Government development agencies.
(ii) National research organizations, particularly the Kenya dairy research institute and the Kenya beef research institute, and international research organizations (ICRAF, ILRI, CIAT, CYMMIYT, ICIPE, etc.) or joint research programs with the AU or the EU.
(iii) Development and community-based organizations such as DANIDA, Techno-Serve, SIDA, USAID, and others.
(iv) Organizations from the private sector, such as Brookside, the KCC, Githunguri.
(v) Finance institutions, both public and private, such as banks and microfinance institutions.

Regarding actors' interaction in the dairy SSI, among the government agencies, KDB continues to control the quality of, and to regulate, dairy products from both the cooperatives and private firms. The dairy SSI also includes demonstration and trail fields for higher education institutions such as the University of Nairobi, Kenyatta University, and Egerton University, as important research actors. National and international research organizations participate in collaborative programs on breeding, production, feeding, and marketing. The EADD program was the best example of a program implemented by a consortium of Heifer International, ILRI, dairy cooperatives, TechnoServe, the African breeders' services, and ICRAF, but also private dealers and banks are providing credit to R&D initiatives and marketing organizations, including smallholder cooperatives. This kind of coordinated action opens up the options for farmers in terms of input supply, financial support, and marketing access. Market arrangements and contract agreements are implemented between the county government, private milk processing companies, and dealers. Access to insurance for cows is also one of the agreements included in the package.

Comparison of the Contributions of Actors in Adaptation to Climate Change in the Coffee and Dairy Sectors of Central Kenya

Coffee and dairy SSIs differ in both their structural components and the actors' contribution to innovative and collaborative solutions in each sector. Coffee farmers are less supported in their access to material and inputs, compared to dairy farmers. In the coffee sector, the cooperative societies, the county, and the national government participated in the provision of material and input, although the provisions are not sufficient. In the dairy sector, the cooperatives, financial organizations (banks, credit, and insurance companies), upstream and downstream private firms (input dealers, market agencies), and international research institutes (CGIAR centers, ICIPE) collaborated on input and material supply.

Access to financial and credit services also differed according to each SSI. In the coffee sector, cooperatives and private agents are the main actors, although this is not satisfactory, while dairy farmers are adequately served by a wider range of financial and credit institutions, such as government, cooperatives and farmers' federations, private suppliers and buyers, and pure financial agents (banks and insurance companies). The power imbalance among the actors also affected interactions and performance in the two sectors. For example, coffee was under the "political control" of the county and the national government, while the dairy sector was driven by a more distributed power structure.

Regarding market access and facilitation, the cooperative union was the sole and autonomous organization for processing and selling coffee on behalf of the farmers, but private marketing agents at the auction remained powerful actors. Coffee export marketing was done by auction through an agent hired by the cooperative union. Prices were controlled by the top actors in the chain (exporters), and farmers were price takers, with a payment term of at least 6 months. Consequently, the asymmetry of information in the value chain benefited the actors in the auction and the union, while farmers were disadvantaged and knew hardly anything about quality requirements.

In the dairy sector, cooperative unions and federations, private market agents, development organizations (USAID, TechnoServe, SIDA), and international research institutes were involved in facilitating farmers' access to the market. These actors were more equally distributed throughout the value chain, resulting in more transparency and both competition and cooperation. All the dairy buyers organize and register farmers looking for improvements in production, input service, and marketing.

Finally, our results on the knowledge base and learning process of the two SSIs showed two contrasting situations: in the coffee sector, CRF, the cooperatives and farmers' federations were the primary sources of information and knowledge for coffee farmers, while there was a limited contribution from international research institutes, county and national governments, and other development organizations. In the dairy sector, actors such as cooperative unions and farmers' federations, the national and county governments, international research institutes, financial institutions, private suppliers and buyers, and development organizations were all providers of knowledge and information. The coffee SSI was thus narrower and was separated according to the different stages of the value chain, whereas the dairy SSI contributed to a more complex, diversified, and extended learning process.

The main question here is: what were the consequences of the SSI in the coffee and dairy sectors for adaptation to climate change? Differences in structural arrangements and the interaction of the actors in the coffee and dairy SSI brought about different ways of farmer adaptation to climate change and adaptive capacity. According to results from the FGDs, coffee farmers adapt to the changes either through diversification of enterprises, such as diversifying into food crops and dairy farming, while dairy farmers specialize in investing in external input and feed. This is connected to the functions and contributions of the SSIs of the different sectors.

Conclusions

The long-term rise in temperature evidenced by various studies (Armah et al. 2015; Dhanya and Ramachandran 2015; Kemausuor et al. 2011) confirms the consistency of information obtained from both the historical climate data and local farmers' perception in both the coffee and food crop zones. Discrepancies are, however, observed between farmers' perception and historical data of rainfall at the food crop zone: the analysis of annual records of historical data of rainfall provided no evidence to support farmers' perception of declining trends, other than agreeing about interannual variability and pattern differences.

Why different interpretations are given for a supposedly factual issue and why this matters in the adaptation process could be, therefore, at the center of the climate change research. Such differences could occur for three types of reasons:

1. Rainfall at the food crop zone was erratic and highly variable throughout the three decades (1980–2014), which in turn could have misled farmers into believing there was a steadily decrease.
2. The average trend for rainfall over the last 10 years (2003–2013) was actually declining. Were farmers to give more weight to recent recall, it could well explain the apparent inconsistency between farmers' perception and measured data. Farmers' perception and experiences are affected by emotion. The decreasing trend for the last 10 years may be the source of emotion leading farmers to perceive the long-term rainfall is decreasing.
3. Though there is no clear trend of declining annual rainfall, farmers may actually pay more attention to the trend in rainfall pattern (onset, cessation, duration, frequency, interannual variability) which affects crop harvest. There was evidence that both the perception and historical data showed a variability in the pattern of rainfall. This, therefore, indicates that farmers' perception of annual rainfall may be affected by the pattern of the rainfall, which is a function of production.

The main question here is, what is the impact of such complementarity or discrepancies of the two knowledge sources (farmers' vs. meteorological data) when it comes to adaptation to climate change? Though the farmers' perceptions correlate with the historical data analysis which is mostly used by scientists for policy decisions, there are discrepancies in adaptation needs to climate change. On one hand, farmers for their part seem keen on bringing changes to their agronomic practices.

Policy actors, on the other hand, prefer macro-level long-term investment strategies such as institutional building for vertical and horizontal collaboration among different institutions, capacity enhancement of the financial sector that provides credit and insurance for farmers, improving climate information services through installation and management of meteorological stations, and ensuring the flow of information, multilevel institutional linkage, networking, and formation of policy platforms of different stakeholders.

Macro-level agreements, such as the UNFCCC Paris Agreement, should be designed to encourage debate on how to tackle climate change through the notion of SSI, for both technological innovation and marketing issues. Technological innovation is indeed important, but this is not the only requirement. Enabling an SSI where some technological innovations contribute to adaptation to climate change should be a priority area for action. In the coffee and dairy sectors of Central Kenya, before market liberalization, the input delivery and marketing system was organized through monopoly cooperative agents. The two sectors experienced a decline during the early years of liberalization, but later they tended to take different directions. While the coffee sector has continued to decline, and has entered a full recession, the dairy sector flourished. The main idea of this discussion is, therefore, to understand why these two sectors, and their SSIs, took different trajectories and what these evolutions bring in order to understand and act upon farmers' adaptation strategies to climate change.

Three main reasons contribute to the two sectors taking divergent trajectories.

1. The coffee SSI continued to push technological innovations, such as disease-resistant varieties, following a very top-down process focused on the production stage, whereas the dairy SSI was more driven by the demand side and involved various stakeholders in the learning process.
2. The two sectors followed different value chain policies: in the coffee sector, input and services were left to private firms, with less attention to farmer empowerment, while the marketing of coffee is based on a cooperative monopoly influenced by powerful private export actors at the top of the chain. The dairy sector, on the contrary, was fairly well liberalized, and the cooperatives, private firms, and informal dealers competed equally for service delivery and milk marketing, building diversified collaborations with development organizations. The coffee sector is thus organized along a long vertical supply chain, while the dairy supply chain is shorter and more diversified and complex.
3. The two sectors also differ in how actors interact around innovation. In the coffee SSI, actors are relatively few and focus on supporting production technologies for coffee plantations. In the dairy SSI, many actors interact for different kinds of innovations throughout the supply chain, at the production, collection, marketing, and distribution stages.

Innovating for adaptation to climate change in the coffee and dairy sectors depends on the structure and evolution of each SSI, i.e., the evolving institutional, knowledge, and collaborative environment that can improve a set of innovations, including new varieties and breeds, good agronomic practices, better access to information, input and services, and efficient marketing systems. The coffee sector illustrates that this combination of innovations was lacking, which correlates well with observed difficulties in adapting to climate change. In contrast, the dairy sector shows that this combination of changes was addressed: institutional development, such as creating active and powerful cooperatives, was coupled with technological innovations such as new breeds or milk cooling machines and better access to input, service, and information. We argue that the evolution of the dairy SSI explains why the dairy sector was more resilient to climate change, something confirmed by other studies. For example, cooperative institutions are known catalysts for decreasing production and marketing costs, for developing new markets, and for better access to technical advice. The performance and efficiency of the sector, however, depends on how actors in these cooperatives are involved in interactions and the coproduction of knowledge with multiple actors, private business, community organizations, and public agencies.

References

Armah FA, Yengoh GT, Luginaah I, Chuenpagdee R, Hambati H, Campbell G (2015) Monitored versus experience-based perceptions of environmental change: evidence from coastal Tanzania. J Integr Environ Sci 12(2):119–152

Asayehegn K, Temple L, Iglesias A (2017a) Impact of climate change on the agro-ecological innovation of coffee agroforestry systems in Central Kenya. PhD Thesis dissertation submitted to the SupAgro Montpellier and Polytechnic University Madrid

Asayehegn K, Iglesias A, Bernard T, Philippe P, Temple L (2017b) The Role of systems of innovation in adapting to climate change: the case of the Kenyan coffee and dairy sectors. J Innov Econ Manag 24(3):127

Bednarek AT, Wyborn C, Cvitanovic C, Meyer R, Colvin RM, Addison PFE, Close SL, Curran K, Farooque M, Goldman E, Hart D (2018) Boundary spanning at the science–policy interface: the practitioners' perspectives. Sustain Sci 13:1–9

Boyer J (2016) L'implication des acteurs de la recherche dans les processus d'adaptation au changement climatique : l'exemple des régions viticoles françaises. JIEM, Innovations 51(3):147

Brody SD, Zahran S, Vedlitz A, Grover H (2008) Vulnerability and public perceptions of global United States. Environ Behav 40(1):72–95

Cash D, Adger WN, Berkes F, Garden P, Lebel L, Olsson P, Pritchard L, Young O (2006) Scale and cross-scale dynamics: governance and information in a multilevel world. Ecol Soc 11(2):8

Castellanos EJ, Tucker C, Eakin H, Morales H, Barrera JF, Dıaz R (2012) Assessing the adaptation strategies of farmers facing multiple stressors: lessons from the Coffee and global changes project in Mesoamerica. Environ Sci Policy 26:19

Chhetri N, Chaudhary P, Tiwari PR, Yadaw RB (2012) Institutional and technological innovation: understanding agricultural adaptation to climate change in Nepal. Appl Geogr 33(1):142–150

Craparo ACW, Van Asten PJA, Läderach P, Jassogne LTP, Grab SW (2015) Coffea arabica yields decline in Tanzania due to climate change: global implications. Agric For Meteorol 207:1–10

Dhanya P, Ramachandran A (2015) Farmers' perceptions of climate change and the proposed agriculture adaptation strategies in a semi-arid region of South India. J Integr Environ Sci 8168(September):1–18

Dunlap R, Brulle R (2015) Climate change and society. Oxford University Press, New York

Edquist C, Chaminade C (2006) Industrial policy from a systems-of-innovation perspective. EIB Papers 11:108–132

Fisher M, Abate T, Lunduka RW, Asnake W, Alemayehu Y, Madulu RB (2015) Drought tolerant maize for farmer adaptation to drought in sub-Saharan Africa: determinants of adoption in eastern and southern Africa. Clim Chang 133:283–299

García de Jalón S, Iglesias A, Quiroga S, Bardají I (2013) Exploring public support for climate change adaptation policies in the Mediterranean region: a case study in Southern Spain. Environ Sci Pol 29:1–11

Haarsma RJ, Selten FM, Weber SL, Kliphuis M (2005) Sahel rainfall variability and response to greenhouse warming. Geophys Res Lett 32(April):1–4

Hahn MB, Riederer AM, Foster SO (2009) The livelihood vulnerability index: a pragmatic approach to assessing risks from climate variability and change-a case study in Mozambique. Glob Environ Chang 19(1):74–88

Hallegatte S (2009) Strategies to adapt to an uncertain climate change. Glob Environ Chang 19(2):240–247. Touzard (2015)

Intergovernmental Panel on Climate Change (IPCC) (2014) Summary for policymakers. Working Group I contribution to IPCC fifth assessment report: climate change 2014, Geneva

Jaramillo J, Setamou M, Muchugu E, Olaye AC, Jaramillo A, Mukabana J, Gathara S (2014) Climate change or urbanization ? Impacts on a traditional coffee production system in East Africa over the last 80 years. PLoS One 8:1–11

Kemausuor F, Dwamena E, Bart-plange A, Kyei-baffour N (2011) Farmers' Perception of Climate Change in the Ejura-Sekyedumase District of Ghana. J Agri Biol Sci 6(10):26–38

Lin BB (2007) Agroforestry adaptation and mitigation options for smallholder farmers vulnerable to climate change. Agroecology, ecosystems and sustainability, 1st edn. Taylor and Francis Group, New York. Ratten, 1993

Malerba F (2002) Sectoral Systems of Innovation and Production. Res Policy 31:247–264

Malerba F (2007) Sectoral Systems of Innovation: a framework for linking innovation to the Knowledge Base, structure and dynamics of sectors. Econ Innov New Technol 14(1–2):37–41

Nilsson AE, Åsa G, S & Katarina E. (2012) Knowledge for local climate change adaptation in Sweden: challenges of multilevel governance. Local Environ 17(6–7):751–767. https://doi.org/10.1080/13549839.2012.678316

Pelling M, Manuel-Navarrete D (2011) From resilience to transformation: the adaptive cycle in two Mexican urban centers. Ecol Soc 16(2):11

Preston BL, Mustelin J, Maloney MC (2015) Climate adaptation heuristics and the science/policy divide. Mitig Adapt Strateg Glob Chang 20(3):467–497

Ratten M (1993) Land Reform in Africa : Lessons from Kenya, World Development 71–83

Rodima-Taylor D, Olwig MF, Chhetri N (2012) Adaptation as innovation, innovation as adaptation: an institutional approach to climate change. Appl Geogr 33(1):107–111. https://doi.org/10.1016/j.apgeog.2011.10.011

Rosell S, Holmer B (2007) Rainfall change and its implications for Belg harvest in South Wollo, Ethiopia. Geogr Ann Ser A 89 A(4):287–299

Seo SN (2010) Is an integrated farm more resilient against climate change? A micro-econometric analysis of portfolio diversification in African agriculture. Food Policy 35(1):32–40

Simelton E, Quinn CH, Batisani N, Dougill AJ, Dyer JC, Fraser EDG et al (2013) Is rainfall really changing? Farmers' perceptions, meteorological data, and policy implications. Clim Dev 5(2): 123–138

Soussana JF (ed) (2013) S'adapter au changement climatique, Versailles, QUAE

Teklewold H, Kassie M, Shiferaw B, Köhlin G (2013) Cropping system diversification, conservation tillage and seed adoption in Ethiopia: impacts on household income, agrochemical use and demand for labor. Ecol Econ 93:85–93

Teucher M, Hornetz B, Jätzold R (2016) Increasing livelihood through use of adapted crop varieties-case study from a semiarid region of East Kenya. Agro Sustain Food Syst 40(6):614–632

Thuku GK (2013) Effects of reforms on productivity of coffee in Kenya. Int J Bus Soc Sci 4(15): 196–212

Touzard J-M (2014) Les approches sectorielles de l'innovation. In: Réseau de Recherches sur l'Innovation (ed) Principes d'économie de l'innovation. Peter Lang, Bruxelles, pp 235–245
Touzard J-M (2015) Combining value chain approach and innovation system approach for climate smart agriculture, position & strategic paper Version 3. CSA Booster, Paris
Van Lancker J, Mondelaers K, Wauters E, Van Huylenbroeck G (2016) The organizational innovation system: a systemic framework for radical innovation at the organizational level. Technovation 52(53):40–50
Weber EU (2006) Experience-based and description-based perceptions of long-term risk: why global warming does not scare us. Clim Chang 77:103–120
World Bank (2016) Climate information services, agriculture global practical assistance paper. The World Bank, Washington, DC

Optimizing Outputs of Nigerian Research Institutes to Achieve National Food Security in the Face of Climate Change

111

Samuel Adelowo Olakojo

Contents

Introduction 2274
Interrelationship of Climate Change, Agriculture, Food Security, and Industrialization 2275
Effect of Climate Change on Food Processing, Packaging, Storage, and Food Quality 2275
Optimizing Output of Research Institute to Achieve National Food Security and Industrialization 2276
Important Information for Farmers with Respect to Climate Change and Food Production 2282
Recommendations for National Research Institutes (NARIs) to Note 2283
Recommendations to Government and Policy Makers 2285
Conclusion 2285
References 2286

Abstract

Food security is described as a situation when all people at all times have physical and economic access to sufficient, safe, and nutritious food to meet their dietary needs and food preference for an active healthy life. For food to be produced, the agricultural environment must be conducive to plants, animals, microbes, and people living in it. Climate change is sharply bringing variation in physical, structural, occupational, agricultural, biodiversity, and socioeconomic ways of life. The interaction between man and his environment is grossly responsible for the global warming and climate change. The effect is currently reversing the old order of things with respect to amount and distribution of precipitation, occurrence, and severity of drought, flood, and natural hazards. Earlier knowledge of geographical principles is no longer completely reliable, thereby posing a serious threat to survival of man, live stocks, fisheries, and wild life. Green house

S. A. Olakojo (✉)
Institute of Agricultural Research and Training, Obafemi Awolowo University, Ibadan, Nigeria
e-mail: saolakojo@yahoo.co.uk

W. Leal Filho (ed.), *Handbook of Climate Change Resilience*,
https://doi.org/10.1007/978-3-319-93336-8_26

emission of carbon dioxide, heat and depletion of ozone layer, and carelessness of man in his environment could be blamed for these consequences. Since man cannot bring back to normal the damage done by fiat, he must accept whatever the climate change is offering mankind. The survival and coping strategies should therefore be holistically approached, conscientiously executed, and collectively adhered to by all inhabitants of the planet earth. In this chapter, the effect of climate change on food distribution, expectation from farmers, research institutes, government, and relevant stakeholders are discussed for enhanced food supply, food security, and industrial growth for a better economy. The potentials of some national research institutes in achieving food security are highlighted.

Keywords
Climate change · Food security · Technologies from NARIS · Roles of stakeholders

Introduction

Food security is defined by FAO as physical and economic access to food for all people at all times. However, some modified this definition as "physical and economic access to quality food for all in adequate proportion, nutrition, at affordable prices in sustainable manner for the growth and development of mankind." Some others defined food security as "the outcome of food system processes all along the food chain." Food security was defined by FAO (1996) as "a situation where all people at all times have physical and economic access to sufficient, safe and nutritious food to meet dietary needs and food preferences for an active, healthy life." From these definitions, we could see that access to food, socioeconomically and to the food system (which consists of production, processing, distribution preparation, and consumption) is also very germane; hence these two components ensure food security for a nation.

Climate change: From several literatures consulted, no specific definition of climate change is given. Nevertheless, a lot of scholars described climate change in relation to their discipline or area of interest. For the purpose of this contribution, climate change defined as changes in the global situation that brings about significant and strong variation in different aspect of physical, climatic, occupational, agricultural, and socioeconomic and biodiversity of living organisms occasioned by rise in global temperature, with its consequences on living and nonliving things of the environment.

Research output: This may be simply described as the results, breakthroughs, findings, and technologies that evolved from empirical study of a well-planned research that are reproducible, proven, and socioculturally acceptable and adoptable for the benefit of mankind.

Industrialization: This is a conscientious and articulated effort of technologically growing the economy with a view to mechanizing production operations for efficiency, enhanced output, reduced cost per unit, and adding value to a product to meet international market standard.

Interrelationship of Climate Change, Agriculture, Food Security, and Industrialization

Climate change brings some shifts in food production. In some places, temperature will rise and rainfall will increase; in others, rainfall will decrease. In some places, coastal flooding will reduce the amount of available land for agriculture. Generally speaking, food crops are sensitive to climate change; such changes often affect soil temperature, moisture levels, and organisms in the soil, pests, and disease. Climate change has been predicted to alter production of rice, wheat, maize, soybean, and potatoes which are staples of billions of people across the globe (Pimente 1993). By 2030, according to one scenario, atmospheric carbon dioxide concentration will be double of preindustrial concentration, other greenhouses gases will increase substantially, and temperature in North America and Africa will rise approximately by 2 °C.

The impact of pests will be high and emergence of new strains and species may be problematic. One such example is the outbreak of armyworm this year in subtropical Africa including Nigeria which ravaged hundreds of hectares of maize farms and led to loss of billions of naira.

The current rate of soil erosion also reflects the urgency of stemming soil loss which is presently threatening many crop production environments and food security of our nation. The eastern Nigeria agro-environment is a good example. If it is not checked, it will increase fertilizer need, pollute the soil, increase cost of production, and promote accumulation of chemical residue in major crops being consumed by an average Nigerian. The unpredictability of normal rainy season may lead to a decrease in productivity and agricultural earning may be jeopardized and the agrarian community may be seriously affected.

Effect of Climate Change on Food Processing, Packaging, Storage, and Food Quality

Some of the direct or indirect effects of climate change on food processing, storage, and food quality are highlighted here. (1) Protein energy malnutrition (PEM) and micronutrient deficiencies especially of iron, folic acid, vitamin B12, and iodine. These deficiencies have remain important public health problem (Ashaye 2018) It has been estimated that over 50% of perishable food commodities such as fruits, vegetables, root, and tuber are lost after harvest especially in the hot humid ecologies of tropical Africa due to sharp variations in weather and climate. If African countries are to be food secured, urgent attention should be given to -post-harvest handling of perishable foods. (2) Similarly, excess of carbon that is released into the atmosphere in the form of carbon pollution has direct effect on food. For example, when atmospheric carbon dioxide level rises, plant food quality declines, surge in greenhouse gas also triggers changes in food composition of plant leaves, stem, roots, fruits, and tubers. In the same vein, natural-sugar- to-nutrient ratio is disrupted, while carbon dioxide increases plant carbohydrate content but diminishes other vital nutrient level. It has been estimated that between 1950 and 1990, riboflavin content in crops dropped by margin of 38%. Rising carbon dioxide is also likely to lower the concentration of essential micro- and macroelements such as iron, zinc, calcium, magnesium copper, sulfur, phosphorus, and nitrogen in most plants especially cereals and staple crops. (3) Climate change may also affect quality stored foods due to complex interaction between biotic and abiotic factors on migrating pest and disease pathogen which often result in poor quality of stored products, occasioned by mould and accumulation of mycotoxin (Atanda et al. 2011). Extreme temperature, precipitation, and weather have been earlier described by Ashaye (2018) as drivers of food contamination, while packaging materials that fluctuate as a result of climatic variables are also not good for storing healthy foods (Fasoyiro et al. 2010).

Optimizing Output of Research Institute to Achieve National Food Security and Industrialization

There are more than 20 agricultural research institutes in Nigeria, each of them working on different crop/livestock to develop technologies that will be suitable for farmers. Each of these research institutes is strategically located very close to production belt of mandated crops so that farmers can take maximum advantage of the technologies developed, facilitate easy transfer of technology, and to give feedback to the institutes on the usefulness or otherwise of these technologies. These opportunities are best optimized as follows:

1. **Identify suitable technologies from the relevant institutes**: Establishment of demonstration technologies in farming communities, adopting them, and scaling up these technologies among other farmers. For example, technologies that have direct influence on factors that induce climate change triggered food insecurity and nutritional imbalances. Any technology that promotes efficient use of land, water, inputs, reduces CO_2 emission, reduces heavy use of farm inputs, reduces heat-generating operations, lowers the risk of eco-unfriendly situations, eliminates flood inducing practices, lowers risks from drought, prevents pest, and disease outbreak. Almost all research institutes working on agricultural crops have soil fertility, soil maintenance, and soil health programs as an integral part of their research activities. For example, animal traction is an advantage over mechanical traction by machines which generate heat, pollute the air with emissions and noise during operation.
2. **Selection of ecology-specific planting materials**. Several crop varieties have been developed to address specific production constraint of individual agroecology. For instance, low-nitrogen tolerant maize varieties have been developed and released for nitrogen depleted soil to reduce nitrogen requirement of maize from 120 KgN/ha to as low as 30–36 kg/ha with good yield, drought-tolerant varieties have been released to cope with drought prone agro-environment, early and extra-

early materials have been released to cope with marginal rainfall environment. Some crops were developed specifically for intercropping in farming system practices. Similarly, mid-altitude varieties have been developed and released in high altitudes agro-environments. As soil fertility is declining, it gives an opportunity for parasitic weeds to strive. Striga-resistant varieties maize, cowpea, and soybean have been developed and released for adaptation to striga endemic areas of Nigeria. In the same vein, forest and savannah adapted crop varieties have been released to cope with ecological and climatic dictates of these production environment to maximize yield. Some crop that can be used to mitigate effects of climate change are listed in Table 1.

3. **Close monitoring of evolving pests and diseases**: Climate change, global warming, and its resultant effects also bring along with it mutation of pest and diseases as new strains of pathogen are evolving, breeding season of some pests are elongated, and older chemicals against the known pests and diseases were no longer effective on the newly emerged pest. A very good example is the strain of army worm that broke out this year ravaging maize in Nigeria, tomato Ebola that ravaged tomato farms in the North, and many more. As long as man have no

Table 1 Some crop varieties that can mitigate effects of climate change

Crop	Variety	Outstanding characteristic
Maize	LNTP-W	This is the white version of low N tolerant maize, minimum of 36 kgN/ha fertilizer is sufficient for optimal productivity; promotes good soil health, and reduces per unit cost of production
Maize	LNTP-Y	This is a yellow version with similar characteristics as in white version above
Maize	Ife Maize hyb-06 and many extra-early maturing varieties developed and released by IITA	Drought and Striga pest resistance hybrid maize. This can cope effectively with sharp drought often experienced during production especially when the drought is as a result of sudden climate change with good grain yield
Maize	BR9943-DMR	Stem borer resistance white maize fairly tolerant to devastating effects of armyworm brought about by effects of climate change
Maize	BR9928-DMR	Yellow stem borer resistance maize but tolerant to armyworm as in white version above
Maize	Eruwa-1	Weevil resistant maize variety that stores better in hot humid tropical environment
Rice	TOS6454 × FARO 52	Submergence resistance with long stalk, high grain yield and bred by Africa Rice
Rice	TO6454 × L19	Submergence resistance with short stalk, high yield, and general acceptability

Source: Institute of Agricultural Research and Training, Farmer's Guide, 2014 effective control over these changing temperatures, flooding, drought, destructive wind, hurricanes, tornadoes, etc., they must readjust to the weather and climate by choosing plants and animals that can adapt to such situations. Yvo de Boer (2006) in his UNFCCC report had suggested use of drought-resistant seed and to embrace climate change technologies. For example, early and extra-early varieties to cope with drought situations, flood tolerant genotypes to cope with coastal flooding, pests and disease resistant varieties for pest and disease endemic zone of our production belt becomes very relevant.

In 2014 and 2015, it was an outbreak of maize stripe in many parts of south western and middle belt, in outbreak of bird pest in east and north-eastern region. Farming could only be profitable if the choice of our planting material is suitable and adaptable to production constraints of our environment. Some of these varieties are available with national and international research institutes for farmers to adopt.

4. **Food distribution for nutritional balance**: Choice of food, redistribution of food and proper monitoring of food items, grown, distributed, and consumed in different parts of globe should be of concern to the government at all levels. For example, too much calorie food is consumed by some elites. This results in obesity and inherent cardiac diseases. In some other regions, such as coastal environment, only protein rich food such as fish is consumed, whereas energy food such as starch is lacking while those on protein diet may not be able to do much work as expected for lack of starch, they often resort to taking alcohol to boost energy; on the other hand, elites that consume more of animal protein, drinks a lot of carbonated sugar drinks, and heavy snacks are compelled to register at gym for weekly or daily regular body exercise. Types of food items coming into the country and distributed regionally should consider at socioeconomic status of people in such areas.
5. **Understanding of utilization potential of crops**: The need for farmers to understand the utilization potential of the specific variety of crops to be planted and animal breed to be adopted is important to ensure food security. The release of crop, livestock, and industrial crops (cotton and kenaf, jute and sisal) in Nigeria compelled breeders to carry out nutrient component of such material before release to analyze and ascertain their utilization potential. Farmers and industrial experts should therefore carefully select varieties to be promoted, produced, and marketed based on the nutritional quality, biochemical profile, and the expected end-products of their consumers. For example, maize, cassava, potato, and rice were bred and released for their high beta-carotene, provitamin-A, high starch content, high vegetable oil content, high sucrose content, high fiber content, and many more. All of these varieties are available in the research institutes with these crop mandates.
6. **The need to focus on production of natural food**: Plant vegetable industries may also have a look at some Nigerian oil rich plants for production of vegetable oil as a replacement to bleaching and refining of animal fat. For example, shea butter, soybean, groundnut, Benin seeds, oil palm, sunflower oil, and castor oil

are good alternatives for consumption. A visit to Raw Material Research Institute will provide farmers and industrialists with sufficient information to take a good decision.

7. **Adoption of modern technologies to boost production**: Several proven technologies are available on the shelf of research institutes to enhance higher yield/ha, reduce cost/unit, guarantee sustainable income and are reproducible for adoption by farmers in different parts of the globe. For example, mini yam set technology, yam production in the air without soil, artificial insemination in livestock industries, and use of biotechnology in crop and animal production, shortening of gestation and maturity period of some tree crops by grafting and budding, etc., such approach had been used in shea butter trees, mango, citrus, and other plants of economic importance. All of these technologies enhance higher yield/unit cost, uniformity, reduces gestation period, ease of transfer, and promotes disease/pest resistance. These are some of the comparative advantages over the old or traditional method of doing things in agricultural enterprise. Research institutes in Nigeria have been empowered to breed, develop, and disseminate varieties of crop and agro-raw materials that can precisely fit into industrial needs of each industry. For example, cassava variety TMS 419 was developed by IITA & NCRI for production of starch, while vitamin rich cassava is good for children to boost vitamin A in their body system.
8. **Crop production, storage, and products development**: More than 30% of agricultural products are wasted annually. Some are perishables such as pepper, tomato, mango, plantain, banana, and other fruits that should be preserved or processed for use during the off-season. Similarly, grains such as maize, cowpea are prone to the devastating attack of storage pests such as weevils especially in hot humid environments. Tubers such as yam, cassava, and potato are also wasted during the peak of production. These should normally be converted into flour for use during off-season. Technologies are available in various research institutes as to how to process them, store, and preserve them for further uses. A visit to many institutes such as N.S.P.R.I. Ilorin, FIIRO, Oshodi, IITA, Ibadan, I.A.R.&T., Ibadan, I.A.R., Zaria and NIHORT, Ibadan will equip farmers with information on preservation, storage, and processing methods of these crops to sustain their regular supply, add value, and to bring out better products from them. This will reduce wastages, encourage expanded production programs, and may possibly earn more foreign exchange, if these are processed to meet up with international quality standard. By-products of some of these food crops will serve as raw materials for other industries especially when an integrated program for crop utilization is designed to encourage food security and sustainability.
9. **Aggressive information and technology dissemination**: National Agricultural Extension Research and Liaison Services, Zaria (NAERLS), has the national mandate on information and technology dissemination and each state Agricultural Development Programme ADPs is equally saddled with technology dissemination. Rice/maize center is responsible for carrying out on-farm testing of new

maize and rice varieties before release. Some NGOs such as Justice for Peace and Development Commission of the Roman Catholic is also engaged in providing services among rural farmers. However, a more effective approach is required for disseminating of new technologies and useful information to farmers. For example, information dissemination is better done electrically. Collaboration with GSM service providers and local radio stations and national television is more desirable because of its wider coverage. T.V. drama, radio jingles, and SMS appear more effective having far reaching effect and should be encouraged. These methods have proven very effective in countries like Kenya; in fact, you could sit down in your sitting room and monitor what is going on in your farm using your smart phones. This will not only attract youths to agriculture, but time spent on daily visit to farms may be productively utilized. Some vital information on weather, market, and price forecast can be regularly obtained to enable farmers to plan adequately for farming operations and marketing strategies.

10. **Other general opportunities from research institutes that can enhance food production, food security, and sustainability**
    - Farmers could source information pertaining to cultural and agronomic practices on mandate crops from various research institutes. Information such as soil selection, planting date and geometry, fertilizer recommendation, pests and disease control method, as well harvesting period and storage may be important for new entrants in farming business.
    - Procurement of improved planting materials that are disease and pest resistant, location-specific adaptation, and with specific-end-uses. The concerned institutes are equally ready to render back-up services to farmers on request (may be) for a token. Data on production trend, newly released varieties adoption rate, and utilization potential of many of the varieties could be accessed from these institutes.
    - Opportunity for participatory varietal selection during varietal development before release is available to farmers for them to have a say in the varieties to be released to them. These often result in high adoption and quick access to the new varieties immediately after release. Similarly, information from Management Training Plots (MTPs) usually equips the participating farmers with knowledge of production of such newly released varieties.
    - Opportunity for specialized training, seminar, workshops, and symposium on mandate crops of the relevant institutes. Many of such training are provided almost free or highly subsidized by the government or the institute organizing such workshop.
    - On-shelf technologies could be accessed from these institutes through farmers' guide, video clips, on-farm demonstration, and MTPs.
    - Some research institutes have consultancy outfits that can help train farmers, beginners, and commodity groups, usually with a charge (Table 2).

Table 2 Some proven technologies in national research institutes

Federal Industrial Research Institute, Oshodi	Institute for Agricultural Research, Zaria	Institute of Agricultural Research and Training, Ibadan	National Root Crop Research Institute, Umudike	National Institute of Freshwater Research, New Bussa
Commercial cultivation of indigenous mushroom	Development and release of 44 sorghum varieties, 24 groundnut varieties, 4 sunflower varieties, and SAMCO1-13 cotton, and many maize varieties and hybrids with varied outstanding characteristics	Development and release of over 20 high yielding maize varieties and hybrid IfeMaizehyb-06 and 07 are drought and striga tolerant Three popcorn varieties, three cowpea varieties, and two kenaf varieties for different end users and varied adaptations	Development and release of 46 improved cassava varieties, 19 yam varieties with yield of 27–35 t/ha	Development and release of hybrid fish
Upgrade and standardization of some indigenous fermented food, e.g., *Ugba*	Developed recommendation for irrigation scheduling, quantity of water, irrigation method, water requirement for many crops, and protocols for amelioration of irrigated problems	Developed botanicals for control of cowpea brown blotch disease	Seed production by cutting technique using aeroponic system	Production of zooplankton (*moina)* as natural fish feed for larvae and fry
Processing of dehydrated fruits and vegetables		Developed artificial insemination technique in swine production and use of farm waste in feeding livestock		Development of Kanji solar tent (dries over 100 kg fish in less than 18 h

(*continued*)

Table 2 (continued)

Federal Industrial Research Institute, Oshodi	Institute for Agricultural Research, Zaria	Institute of Agricultural Research and Training, Ibadan	National Root Crop Research Institute, Umudike	National Institute of Freshwater Research, New Bussa
Processing of spices, fruit juice, and fruit nectars		Development of soy-based snacks and foods, pigeon pea- and cowpea-based snacks and food		Development of smoking kilns
Processing of jams and jelly substances		Development of Roselle-based food flavor, drinks, wine, and jam		Fish feed formulation and development of several hand on skill practices in fish farming
Processing of high fiber drinks from indigenous fruits for diabetic		Effective tools for land resources inventory, quality assessment of valley bottom for vegetable production Technology of water harvesting to cope with drought Technology of check mini dams for supplementary irrigation to prevent crop failures		Mechanical boom for control of water hyacinth among others

Source: Websites of the research institutes

Important Information for Farmers with Respect to Climate Change and Food Production

1. Information from meteorological station and bioinformatics should be embraced by farmers. In the developing countries, existing climate can be hard to measure (Mendelsohn 2009), because the weather stations are located mainly in the cities (probably air ports) and not in the villages where agriculture is being practiced.

They are often very far away from each other. Hence, the need to rely more on data from satellite for agricultural production and prediction.

2. Use of appropriate crop varieties which are adaptable ion, disease- and pest-tolerant to production environment. Such variety must be ecology-specific. Beddington et al. (2011) earlier reported the need for positive change and the use of variety of crops. In fact, any variety that can cope with the devastating effects of climate change should as a matter of necessity be adopted.
3. Soil management practices, water management practices, and micro- and macro-element monitoring. Gosain et al. (2016) had suggested that adaptation strategies is one of the best available practices to cope with the induced stresses which may be posed by climate change in India with respect to hydrology.
4. Deployment of suitable and appropriate input *vis-a-vis* timely application, correct dose, and best method of application. Fertilizer application to maize plants for example should include use of organo-mineral fertilizer consisting both inorganic and organic fertilizer rather than use of only mineral fertilizer which render agricultural soil acidic with time and accumulation. Similarly, NPK fertilizer in correct dose depending on soil native component after testing is better than blanket application regardless of soil situation. Under dose will give poor yield, while overdose will affect soil health status. Urea is appropriate while maize is about to tasseling rather than use of urea fertilizer at the seedling stage of production. In the same vein, grain legume requires little or no NPK but single superphosphate because of its root nodulating ability.
5. Subsidy to agricultural sector should focus more on those farmers that can address water, energy, and chemical changes of production environment. Planting methods should be improved and chemicals should be ecofriendly. Jemma et al. (2009) highlighted the implications of climate change for agricultural productivity, to reduce the adverse effects, policy makers at all levels should assist stakeholders in food and agro-raw materials production with policies that can stem the effects, bring relief, and encourage agricultural production sustainability.
6. Study government policy on agriculture and key in at the appropriate time for enhanced food security and sustainability. It should be recalled that Jodie et al. (2009) had predicted that GDP in some developing countries with agriculture as major source of income (specifically Guinea Bissau) will drop by 32.7% by 2080 while reviewing the overall impact of climate change, challenges, adaptation, and mitigation. Whatever individual government is putting in place to mitigate climate change and encourage higher output for enhanced GDP should be adopted.

Recommendations for National Research Institutes (NARIs) to Note

The present global warming, climate change, and its attendant problems cannot be stopped by fiat. In the meantime, mankind is left with option of devising coping strategies that can assist to militate against it and enhance survival on the planet earth. The onus now rests on various research institutes to fashion out relevant research activities that can be of tremendous help for farmers and other stakeholders in the business of food production, distribution, preservation, product development, and food security. The following recommendations among others provide some research direction:

1. Research that focuses on crop and animal production which will reduce flood, heat, disease, and will increase pest resistance. The crop with early or extra-early maturing genotypes. Varieties with low nitrogen tolerant trait to be able to cope with decline in soil natural fertility, crop that are nutritionally good for body healthy living. Crops with good adaptation to drought and which match with existing ecology. Crops that can store for longer period and are resistant naturally to storage pests and diseases. The output of our research should be economically viable and socially acceptable.
2. Research should no longer be to verify the result of some experimental and research findings earlier reported by some workers but should be demand-driven, market-oriented, industrially demanded, and generally acceptable for high adoption. Cutting-edge researches that promote quality, nutrition, are ecofriendly and integrated in nature, allows waste recycling, and avoids environmental pollution need to be encouraged.
3. Research that does not require too much of energy from hydrocarbons, not emitting too much of CO_2 and not generating too much of heat, research that can be powered by renewable energy such as solar and windmill. Technology that allows farmers to operate without necessarily generating heat from energy source should be promoted. National Centre for Agricultural Mechanization should do more to improve animal-driven farm activities to reduce greenhouse emissions. Farm implements that do not dig too deep into the soil that may not permit frequent soil erosion and soil depletion should be encouraged. Outcome of research effort should be sustainable, culturally acceptable, socioeconomically viable and generally affordable to the common man. It must be noted that poor human management of natural resources often bring negative impacts on agricultural environment that affect food production (Olakojo 2015).
4. Technology dissemination: In our contemporary time, findings from our research **must** be scaled up, diffused and must be made accessible to the relevant end users; otherwise the time, resources, funds, and intellectual investment on such will be a waste. Such breakthroughs should not end up on pages of journals for promotion or at best gather dust on our shelves but should be aggressively promoted among end-users for food and nutritional security, sustainability, and industrial growth of national economy. Some of the channels of disseminating our technologies should be in farmers' guides, monthly technology review meeting, books, websites, booklets, radio and TV programs, jingles, mobile phone, newspapers, and training workshops.

Recommendations to Government and Policy Makers

1. Effect of climate change should form an integral part of school curriculum to be taught at all levels of education and agricultural science. It must be made compulsory as a subject in secondary schools, while it forms part of the course content of general studies in universities.
2. Part-time farming may be made compulsory for all civil servants just like some countries in Africa. For example, all major factors of agricultural production (such as land, skill, information, and time) are available to more than 65% of civil servants in Nigeria. They should be put to use for enhanced national food security.
3. Subsidy on agriculture by governments should target farming activities that can stem the effect of global warming and climate change. For example, use of animal traction for land preparation, farm produce processing should be subsidized while discouraging the fossil fuel guzzling farm machines that generate air pollution.
4. Nigeria and many African countries will be food and nutritionally secured, if guarantee markets are available to farmers to encourage sustainable farming business.
5. The activities of middle men should be removed from marketing of agro-materials and food crops by introducing direct sale of food crops. These will encourage production and sustain farmers by keeping them in business.
6. Federal government should legislate against free roaming domestic animals such as cattle, goats, and sheep that often willfully destroy crops and causing clashes between crop and livestock farmers. In advanced countries of the world, live stocks are reared in captivity for proper care and productivity. It appears primitive for us as a nation to completely rely on nomadic way of cattle rearing and livestock husbandry.
7. We cannot over flog the issue of annual decline in budgetary allocation to agriculture which has never enjoyed the 10% international recommendation. Government should do more, now that agriculture is the only trusted sector of the economy to move Nigeria out of recession.
8. National research institutes have been poorly funded. Governments at all levels prefer to patronize the International Research Institutes for what the National Agricultural Research Institutes could conveniently handle.

Conclusion

The issue of food production, security, and sustainability is important for everyone in the face of climate change. Individuals should be an active player in safe-guarding our environment where food and agro-raw materials are produced. Nature had put things right at the onset of creation; onus therefore rests on man to maintain, sustain, and ensure comfortable living and survival on the only planet that support our existence. This generation should also realize that the coming generations will inhabit this same planet where food and essential needs of life and survival are being produced. Therefore, everybody must contribute his or her quota to make living better on the planet earth, for enhanced food security, desirable industrial growth, and for the survival of the national economy.

References

Ashaye OA (2018) Post-harvest handling of food crops in a changing climate: an outlook on research in the Institute of Agricultural research and Training (IAR&T). A lead paper presented in 2018 annual in-house review of IAR&T, aDONAI publishing House, Ibadan, pp 1–29

Atanda SA, Pessu PO, Agod S, Isong IU, Ikotun I (2011) The concepts and problems of post-harvest food in perishable crops. Afr J Food Sci 5(11):603–613

Beddington J, Asaduzzaman M, Clark M, Fernandez A, Guillou M, Jahn M, Erda L, Mamo T, Van Bo N, Nobre CA, Scholes R, Sharma R, Wakhunga J (2011) Achieving food security in the face of climate change: summary for policy makers from the commission on sustainable agriculture and climate change, pp 1–20

FAO (1996) Anchoring agriculture within a Copenhagen Agreement: a policy brief for UNFCCC parties by FAO

Fasoyiro SB, Obatolu VA, Ashaye OA, Lawal OB (2010) Knowledge assessment, improved storage techniques and training of local processors and vendors of soy products on food safety practices in south west Nigeria. J Agric Food Inform 11(4):340–350

Gosain AK, Rao S, Basuray D (2016) Climate change impact assessment on hydrology of Indian River Basin. Curr Sci 90(3):346–353

Jemma G, Betts R, Burke E, Clark R, Camp J, Willett K, Keane J, Page S, Kergna A, Kennan J (2009) Climate and developing country agriculture: an overview of expected impacts, adaptation and mitigation challenges and fund requirements, pp 8–61

Mendelsohn R (2009) The impact of climate change on agriculture in developing countries. J Nat Res Policy Res 1(1):5–19

Olakojo SA (2015) Management of natural aquatic, forest and energy resources. In: Popoola L, Olaniyan O, Alabi R, Olayide O (eds) Ibadan sustainable development summit (ISDS) 2015: the post 2015 sustainable development agenda: whither Africa? University of Ibadan Centre for Sustainable Development, pp 239–259

Pimente (1993) Climate changes and food supply. Forum Appl Res Public Policy 8(4):5460

Sharma R, Wakhunga J (2011) Achieving food security in the face of climate change: summary for policy makers from the commission on sustainable agriculture and climate change, Boddington J (ed), pp 1–20

Wiltshire A (2010) Implications of climate change for agricultural productivity in the early twenty first century. Philos Transr Soc Lond Biol Sci 365(1554):2973–2989

Yvo de Boer (2006) Technology for adaptation to climate change. In: UNCCC report, pp 3–40

Saline Agriculture: A Climate Smart Integrated Approach for Climate Change Resilience in Degraded Land Areas

112

Muhammad Saqib, Javaid Akhtar, Ghulam Abbas, and Hafiz Abdul Wahab

Contents

Introduction 2288
Saline Agriculture 2290
Saline Agriculture Technology: A Case Study 2290
  Experimental Details 2290
  Results 2292
  Discussion 2299
Conclusions and Future Outlook 2303
References 2303

Abstract

Soil and water resources of the world are under increasing strain as a result of climate change and increasing population. Climate change has increased land degradation around the world as a result of increased drought, salinity, temperature rise, sodicity, and rising sea level. Plenty of good-quality water is required to reclaim degraded lands; however such water is not available in most of these areas. The extent of the degraded land areas is significant for food and energy production, particularly for the indigenous people of the degraded land areas. Saline agriculture is an integrated approach, using stress-resistant genetic resources to exploit the production potential of the degraded land and water resources. Saline agriculture involves a careful analysis of a degraded land area, including the soil and water analyses, climatic and market conditions, financial power of the farmers, and local culture. A large variety of genetic resources with

---

M. Saqib (✉) · J. Akhtar · H. A. Wahab
Institute of Soil and Environmental Sciences, University of Agriculture, Faisalabad, Pakistan
e-mail: m.saqib@uaf.edu.pk; drhmsab@yahoo.com

G. Abbas
Department of Environmental Science, COMSATS University Islamabad, Vehari Campus, Vehari, Pakistan

W. Leal Filho (ed.), *Handbook of Climate Change Resilience*,
https://doi.org/10.1007/978-3-319-93336-8_36

diverse functions and stress resistance are available and can be exploited considering the site-specific conditions. This chapter discusses in detail the climate-smart integrated approach of saline agriculture with practical examples and a case study showing its potential for environmental sustainability, ecological rehabilitation, and climate change resilience in degraded land areas.

Keywords
Climate change · Food security · Land degradation · Salinity · Saline agriculture

Introduction

History shows that the prosperity of mankind is closely related to climatic conditions, and a change in these conditions directly impairs the life quality and survival of human societies. Changes in climate have severe implications in all life spheres (FAO 2008). These changes affect food security, health, and ecosystem as well as environmental sustainability. The agricultural sector is also significantly affected by climate change (Lobell et al. 2011), which directly affects food security and life quality. As defined by the UNFCCC (United Nations Framework Convention on Climate Change), climate change means "A change of climate which is attributed, directly or indirectly to human activity that alters the composition of global atmosphere and which is in addition to natural climate variability observed over comparable time periods." Therefore, a shift in average weather conditions of a region or in distribution of these conditions is called climate change (IPCC 2007).

Agriculture, forestry, and fisheries sectors are significantly affected by a change in rainfall, hydrological cycle, seal level, and droughts (Gornall et al. 2010; IPCC 2007). These changes are leading to more dry spells, drought, and land degradation. Similarly, the existing dry land areas are more affected by climate change. At least 75 countries in the word have salt-affected soils to some extent (Qadir et al. 2015). In the last 20 years, the area under salt-affected soils has increased tremendously. With every passing day, around 2000 ha of cropland in irrigated areas is being degraded by salinity and sodicity of different levels (Qadir et al. 2015). This alarming increase in salinity is threatening crop production which should be increased to fulfill the needs of the fast-growing world population. Saline soils are high in soluble salts, and sodic soils are high in exchangeable sodium, whereas saline-sodic soils are high in both soluble salts and exchangeable sodium.

Growth of different plants is reduced under salinized conditions as a result of osmotic effect, ion toxicity, and restricted availability of major plant nutrients (Abbas et al. 2013; Saqib et al. 2013). Higher concentrations of Na^+ and Cl^- are toxic to cells, and, therefore, plants restrict their entry to cytoplasm. For normal functioning of cells, these ions are dumped into vacuole via cellular compartmentalization, which is an important salt tolerance mechanism in many plant species (Munns and Tester 2008). A substantial cytoplasmic level of K^+ is crucial for different cellular functions including cell enlargement, ion homeostasis, and osmoregulation (Schachtman et al. 1997). It is well documented that plants with a potential to maintain higher K^+:Na^+ ratio show more resistance to saline conditions (Abbas et al. 2013; Saqib et al. 2005). Growth and productivity of plants is reduced due to salinity (Abbas et al. 2016; Saqib et al. 2008; Hütsch et al. 2014). The intensity of the damage of salts on plants depends on the growth stage during which exposure to salt happens, salt type and concentration, stress duration, and the species of plant (Lauchli and Grattan 2007; Saqib et al. 2013). Gypsum addition improves the properties of a saline-sodic soil (Ahmad et al. 2016). Similarly, compost and vermicompost reduce the effects of salinity and improve crop growth (Oo et al. 2015). Addition of biochar is also reported to improve the quality of saline soils (Wu et al. 2014) and assist in the reforestation of saline-sodic soils (Drake et al. (2016). Leaf litter decomposition in sodic soils increases carbon and nitrogen contents and microbial activities (Singh et al. 2016).

Due to the decreasing quality of soil and water resources, the use of saline soils and water is becoming inevitable. For growing food crops, these salt-affected soils need to be reclaimed. There are chemical and engineering approaches to reclamation which are not only expensive but also lack sustainability over longer time periods (Qureshi and Barrett-Lennard 1998). In the last century, different chemical amendments and physical and biological interventions have been used for the reclamation of salt-affected soils (Qadir et al. 2007). Among all these, the plant-based technique commonly called as phytoremediation has been found a good practical and cheap technique (Qadir et al. 2015). In this technique, the native source of calcium present in the calcareous salt-affected soils is solubilized with the actions of plant roots, and the calcium made available is utilized in sodium replacement on the exchange soil sites. Chemical amendments are found to be effective only where these are applied (Ilyas et al. 1993); however, phytoremediation has been good even at lower depths of soil depending on the plant species, rooting depth and structure of roots (Arshad et al. 2008). Moreover, the content of macronutrients and organic matter is improved too. Seenivasan et al. (2014) observed in a field experiment that eucalyptus grown on a salt-affected soil caused significant improvement in soil chemical properties (acidity, pH_s; electrical conductivity, EC_e; and sodium adsorption ratio, SAR) and organic matter. The shoot also plays its role in salt uptake and desalinization of soils. However, due to very low concentration of the salt taken up, the role of shoot is not very important even after longer periods of times (Qadir et al. 2007). The plant species show differential responses for phytoremediation of salt-affected soils. The species with higher salinity resistance and more biomass production are considered more suitable for the phytoremediation of such soils.

Under changing climatic conditions, soil salinization is expected to increasingly lead to reduced food and fodder production in the coming years, predominantly in arid and semiarid regions of the world. Considering these facts, we need to consider plant species able to grow on salt-affected soils with high biomass production potential and at the same time have good potential to improve soil properties. It has been observed that tree species are generally more resistant to salinity than crop plants. Moreover, they produce high biomass and cause considerable remediation of the salt-affected lands in arid conditions (Seenivasan et al. 2014; Abbas et al. 2016).

Saline Agriculture

Saline agriculture proposes mixed use of genetic resources of plants, animals, fish and insects, and improved agronomic practices for the profitable and sustainable exploitation of salt-affected soils and water resources (Qureshi and Barrett-Lennard 1998). To fulfill the food, feed, and fiber needs of the growing population, it is necessary to increase the productivity of irrigated salt-affected areas as sufficient good-quality soils for this purpose may not be available (Qadir and Oster 2004). There are established ways to reclaim the salt-affected areas; however they cannot be applied to many saline-sodic soils because of the scarcity of good-quality water, high cost of the inputs, and poor drainage of the soil (Qadir et al. 2008; Qureshi and Barrett-Lennard 1998). The use of salt-resistant plant species including crops and trees along with different organic and inorganic inputs and improved agronomic measures provides another option.

Saline agriculture may comprise of different strategies including site-specific practices and short- and long-term programs of saline agriculture (Jacobsen et al. 2015; Qadir et al. 2008; Qureshi and Barrett-Lennard1998; Rozema and Schat 2013). It starts with the evaluation and appraisal of the existing conditions of soil and water quality, and aims of the land holders. This information helps to devise a strategy and plan for a particular situation as described here:

- Slightly to moderately saline sodic areas may be used for salinity- and drought-resistant crop plants with the help of good agronomic management.
- Highly saline sodic soils may be placed under tree species with good salinity-sodicity tolerance (*Eucalyptus*, *Acacias*, *Tamarix*, etc.). Salt-tolerant grass and forage species like *Atriplex* species, etc. may be grown in between the tree rows.
- Small animals.
- Apiculture, and fishery may also be introduced with time.
- The crops species start providing income in a couple of years; whereas trees may take 5–7 years for giving some profit.

Saline Agriculture Technology: A Case Study

Experimental Details

This case study consists of a field experiment. This field experiment investigates the role of three tree species, *Albizia lebbeck*, *Grewia asiatica*, and *Sesbania sesban*, for the phytoremediation and organic matter improvement of a saline-sodic soil at the Proka Farm, University of Agriculture, in Faisalabad, Pakistan. Small pits were excavated in the field. The dimensions of the pits were

0.60 × 0.15 × 0.15 m (depth × width × length). These pits were planted with healthy seedlings of three tree species. The seedling age was 3 months at the time of transfer. A plant-to-plant distance of 2 m and row-to-row distance of 3 m was maintained. The plants were irrigated weekly during the first month and fortnightly during the rest of the period. The water used for irrigation was saline sodic in nature. The detailed soil and water analyses data are given in Table 1. Other maintenance activities such as manual weeding and hoeing were done occasionally as needed. Randomly, five plants of each tree species were selected and were used for plant and soil data collection.

The data including the plant height, stem circumference, and number of branches and leaves were recorded at 6-month interval for 2 years. The plants were categorized into two groups, i.e., good and poor. A meter rod was used to measure plant height, whereas stem circumference was measured with a measuring tape. The number of branches and leaves was counted manually. Old and young leaves were sampled separately after every 6 months for ionic analyses. These leaves were oven-dried for 72 h at 65 °C and powdered in a grinding mill. The powdered samples were digested in HNO_3 and $HClO_4$ following Ryan et al. (2001). The digested samples were filtered, and the aliquot was analyzed after the required dilution. The concentrations of Na^+ and K^+ were measured by Sherwood-410 flame photometer, and Cl^- was determined with Sherwood-926 chloride analyzer.

At the start of the experiment, soil samples were taken from two depths (0–30 and 30–60 cm) and were analyzed for different properties (Table 1). Later on, every 6 months, soil sampling (from the plant root zone) and analyses were carried out for 2 years. Particle size analysis was done using the hydrometer method (Bouyoucos 1962). USDA textural triangle was followed to find soil textural class. A pH meter (Jenco Model-671P) was used to measure the pH of saturated soil paste, and a conductivity meter (CM-14P; TOA) was used to measure the electrical conductivity (EC_e) of the saturated soil extract. Cations and anions were determined, and sodium adsorption ratio (SAR) was calculated following Richards (1954). Irrigation water was sampled in clean plastic bottles and analyzed following the methods given by Richards (1954).

Table 1 Characteristics of soil and irrigation water used in the experiment

Soil analysis	**pH_s**	**EC_e** (dS m^{-1})	**SAR** (mmol $L^{-1})^{1/2}$	**OM** (%)
Albizzia lebbeck	8.3	13.98	18.2	0.41
Grewia asiatica	8.6	10.5	23	0.38
Sesbania sesban	8.51	15.6	48.4	0.39
Water analysis	**EC** (dS m^{-1})	**SAR** (mmol $L^{-1})^{1/2}$	**RSC** (mmol L^{-1})	–
	3.54	19.5	Nil	–

The values of soil analysis are average of two (0–30 and 30–60 cm) soil depths. *ECe*: electrical conductivity of saturated soil extract, *SAR*: sodium adsorption ratio, *OM*: organic matter, *EC*: electrical conductivity of water, *RSC*: residual sodium carbonate

Results

Albizia lebbeck

Values of the growth parameters differed significantly between good and poor plants over time. A considerable increase in the height of good as well as poor plants of *Albizia lebbeck* was observed (Fig. 1a). For good plants, in the first 6-month data, plant height increased from 347 to 608 cm in 2 years. For poor plants, this increase was from 76 to 164 cm. Likewise, stem circumference increased from 16.8 to 38 cm for good plants and from 4.93 to 13 cm for poor plants (Fig. 1b). Similarly, the number of main branches and leaf numbers on one plant increased significantly, from 54.7 to 180 for good plants and from 29 to 80.7 for poor plants, respectively (Fig. 1c and d).

Plant leaf ionic composition differed significantly for good and poor plants and older and younger leaves (Fig. 2). The ionic analysis showed that older leaves have significantly more Na^+ and Cl^- compared to younger leaves for both good and poor plants. K^+ levels were higher in younger leaves of both good and poor plants, and the comparison of both types of plants showed that

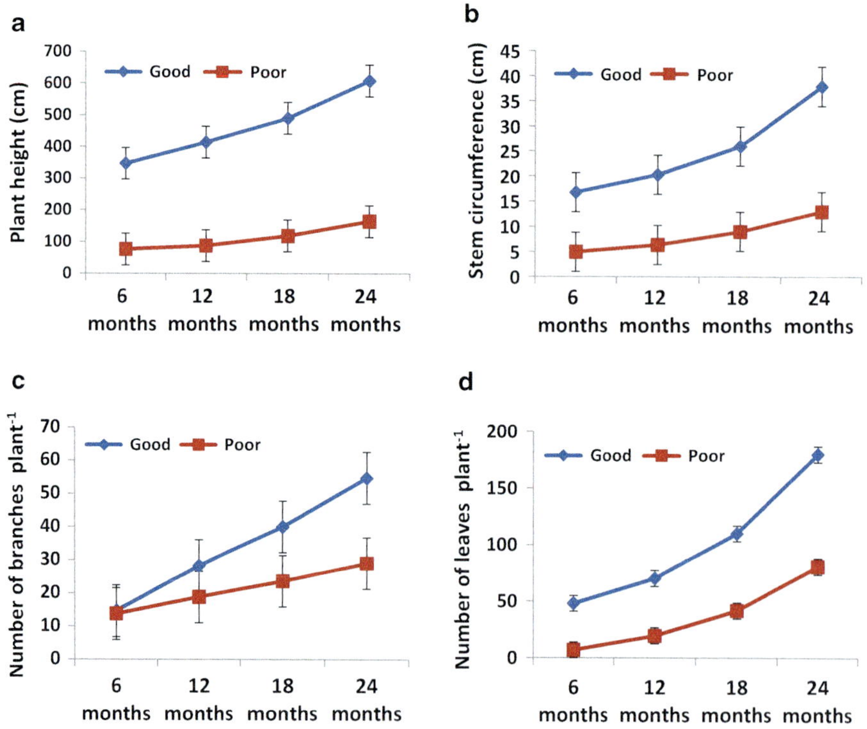

Fig. 1 Growth parameters of *Albizia lebbeck*: (**a**) plant height, (**b**) stem circumference, (**c**) number of branches, and (**d**) number of leaves. Vertical bars indicate the LSD value at $P \leq 0.05$

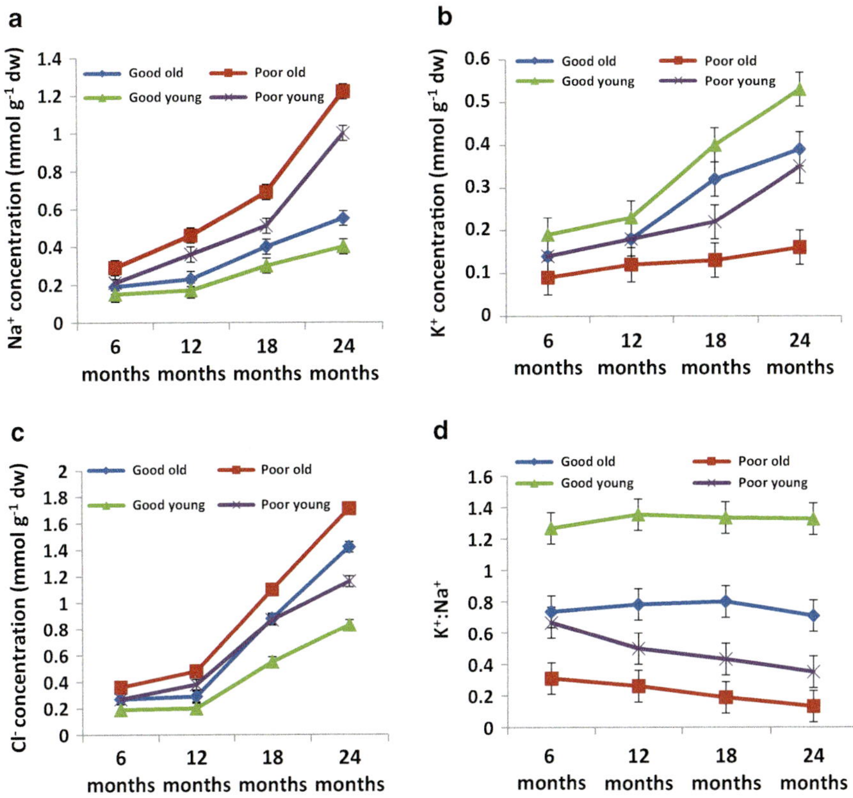

Fig. 2 Ionic parameters of *Albizia lebbeck*: (**a**) sodium, (**b**) potassium, (**c**) chloride concentration, and (**d**) K^+:Na^+ ratio in older (o) and younger (y) leaves. Vertical bars indicate the LSD value at $P \leq 0.05$

this concentration was higher in good plants than in poor plants. K^+:Na^+ ratios were the highest in younger leaves of good plants and were the lowest in older leaves of poor plants. This trend was observed throughout the experimental period.

Data regarding soil analysis showed that with time, there was a significant reduction in salinity and alkalinity indicators (Fig. 3). After 6 months, the values of pH_s, EC_e, and SAR reached 8.2, 12.96, and 17.2 at 0–30 cm depth and 7.98, 10.98, and 19.57 at 30–60 cm depth for good plants. For poor plants, these values were 8.10, 15.09, and 19.4 at 0–30 cm depth and 8.06, 11.31, and 19.77 at 30–60 cm depth. In the next time period, these values further decreased. In the last data, these values reached 7.25, 5.23, and 9.98 at 0–30 cm depth and 7.34, 10.8, and 15.8 at 30–60 cm depth for good plants. For poor plants, these values were 7.47, 11.5, and 14.3 at 0–30 cm depth and 7.54, 11.2, and 16.9 at 30–60 cm depth.

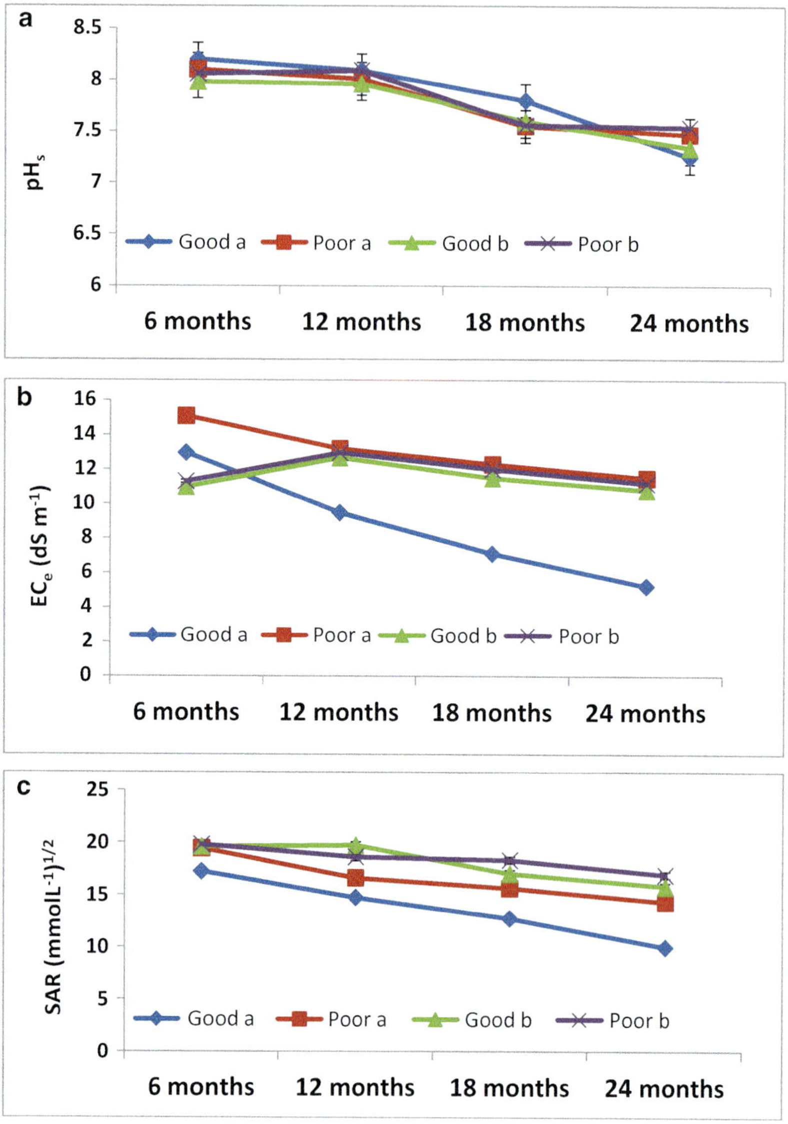

Fig. 3 Soil parameters of *Albizia lebbeck*: (**a**) pH_s, (**b**) EC_e, and (**c**) SAR of 0–30 cm (a) and 30–60 cm (b) soil depth. Vertical bars indicate the LSD value at $P \leq 0.05$

Grewia asiatica

Data regarding plant growth showed significant differences between good and poor plants (Fig. 4). For good plants, plant height increased from 95 to 197 cm between the first- and last 6-month data. For poor plants, height increased from 26.3 to 71.3 cm. Similarly, stem circumference increased from 4.00 to 14.7 cm for good plants and from 1.66 to 4.73 cm for poor plants. Likewise, an increasing trend

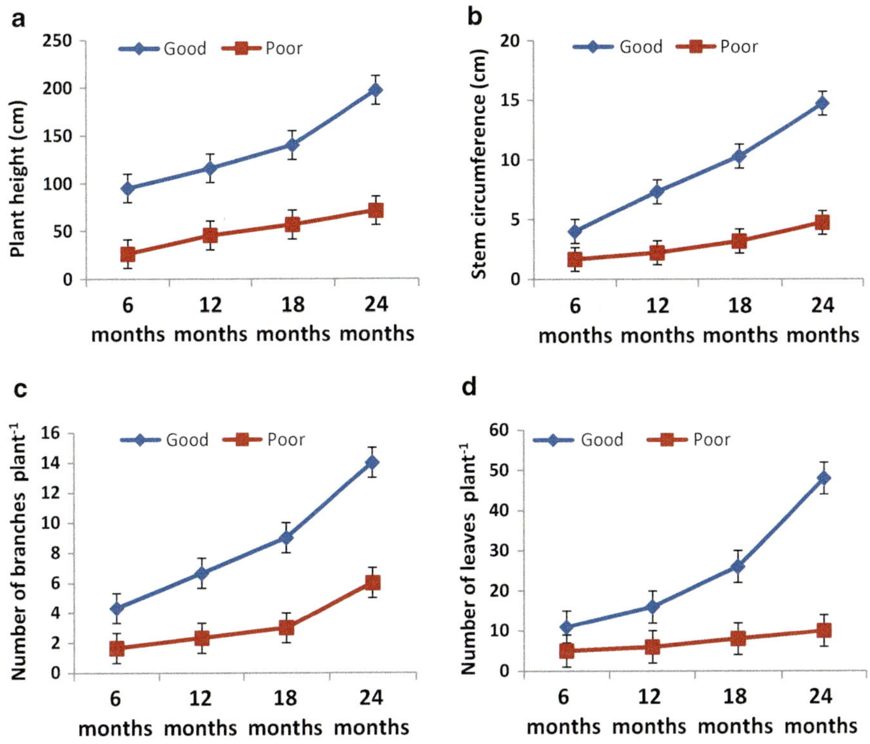

Fig. 4 Growth parameters of *Grewia asiatica*: (**a**) plant height, (**b**) stem circumference, (**c**) number of branches, and (**d**) number of leaves. Vertical bars indicate the LSD value at $P \leq 0.05$

was found in the number of main branches and the number of leaves on one plant, which increased from 4.33 and 11 to 14 and 48 for good plants; whereas, for poor plants, this increase was from 1.66 to 6 and from 5 to 10, respectively.

Leaf ionic composition showed significant differences for the leaf type, plant type, and time of sample collection; however, the interaction was nonsignificant for the three ions determined (Fig. 5). Older leaves had significantly more Na^+ and Cl^- compared to younger leaves for both good and poor plants. K^+ concentration was higher in younger leaves of both good and poor plants, and the comparison of both types of plants showed that this concentration was higher in good plants than poor plants. The K^+:Na^+ ratio was highest in younger leaves of good plants and was lowest in older leaves of poor plants. This trend was observed throughout the experimental period.

The values of alkalinity and salinity indicators of soils were decreased significantly during the study period (Fig. 6). In the first 6 months of the first year, the values of pH_s, EC_e, and SAR were 8.82, 8.29, and 20.2 at 0–30 cm depth and 8.61, 10.14, and 23.3 at 30–60 cm depth for good plants; whereas, for poor plants, these values were 8.87, 10.31, and 26.4 at 0–30 cm depth and 8.67, 10.59, and 27.6 at

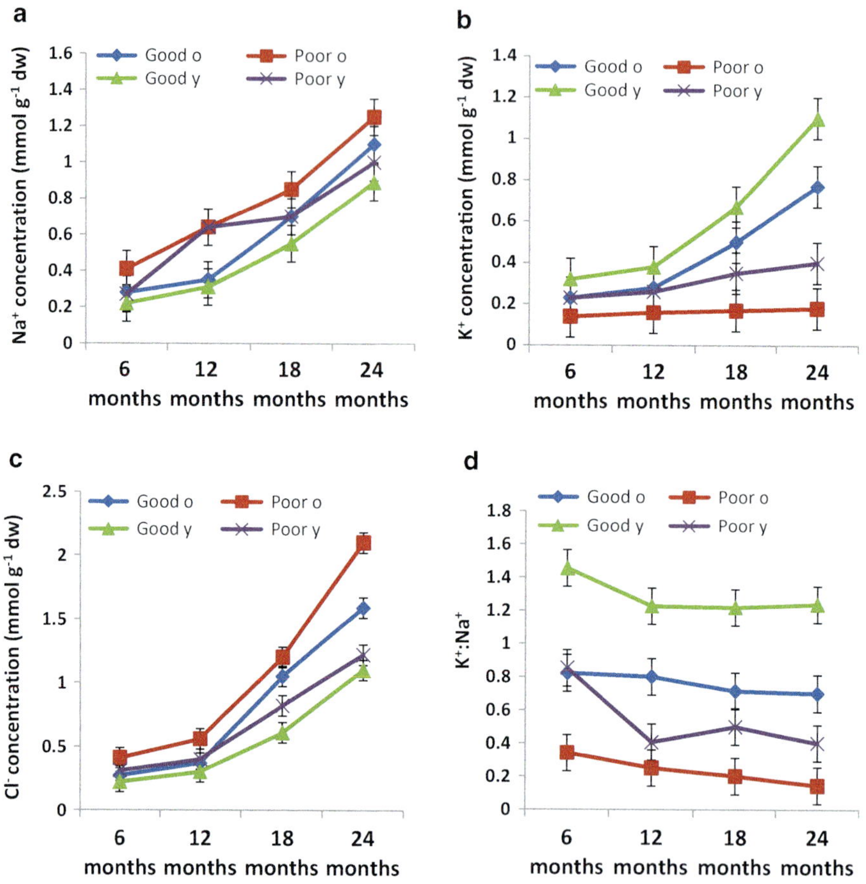

Fig. 5 Ionic parameters of *Grewia asiatica*: (**a**) sodium, (**b**) potassium, (**c**) chloride concentration, and (**d**) K^+:Na^+ ratio in older (o) and younger (y) leaves. Vertical bars indicate the LSD value at $P \leq 0.05$

30–60 cm depth, respectively. In the second year, these values were 8.48, 7.12, and 16.1 at 0–30 cm depth and 7.88, 8.99, and 21 at 30–60 cm depth for good plants; whereas, for poor plants, these values were 8.52, 9, and 24 at 0–30 cm depth and 7.9, 9.65, and 26.3 at 30–60 cm depth, respectively.

Sesbania sesban

Data regarding plant height showed that plant height of *Sesbania sesban* increased significantly with time (Fig. 7). In the first 6 months, plant height of good plants was 412 cm and increased to 837 cm in 2 years. For poor plants, this increase was from 225 to 455 cm. Stem circumference increased from 17.3 to 49 cm and from 10.3 to 25 cm for good and poor plants, respectively. The number of branches

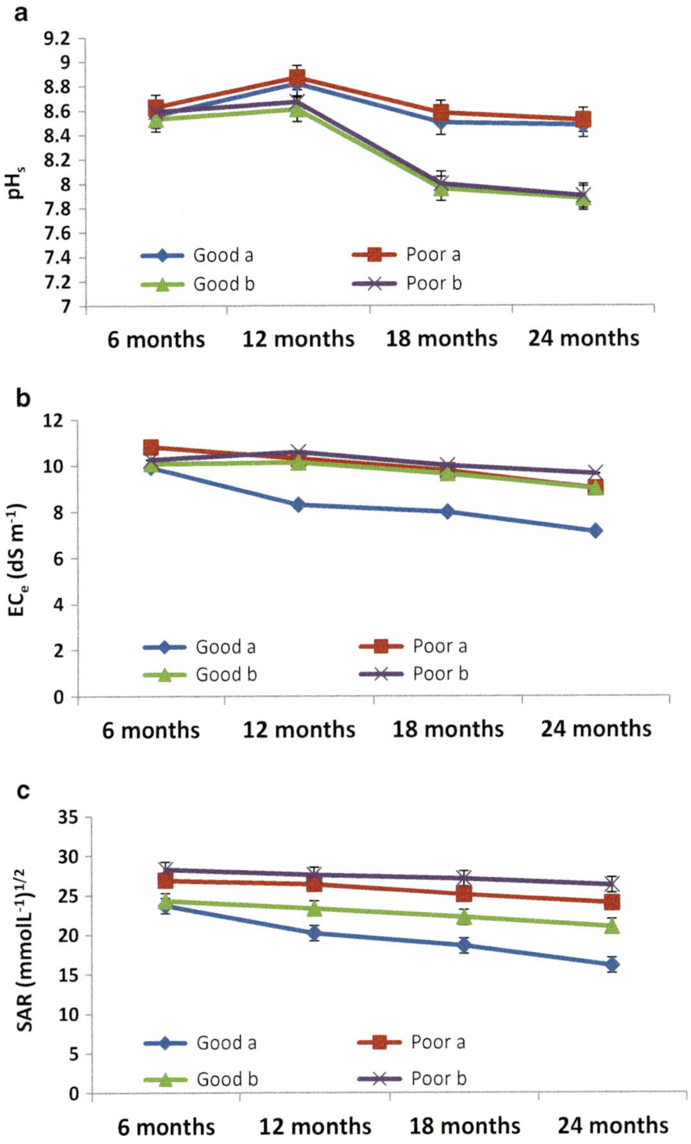

Fig. 6 Soil parameters of *Grewia asiatica*: (**a**) pH_s, (**b**) EC_e and (**c**) SAR of 0–30 cm (a) and 30–60 cm (b) soil depth. Vertical bars indicate the LSD value at $P \leq 0.05$

and the number of leaves increased from 73 and 153 to 244 and 450 for good plants and from 39.7 and 93.3 to 72 and 199 for poor plants, respectively.

There were significant differences between the plant type, time, and leaf type for ionic parameters (Fig. 8). Ionic analysis showed that older leaves had significantly more Na^+ and Cl^- compared to younger leaves for both good and poor plants.

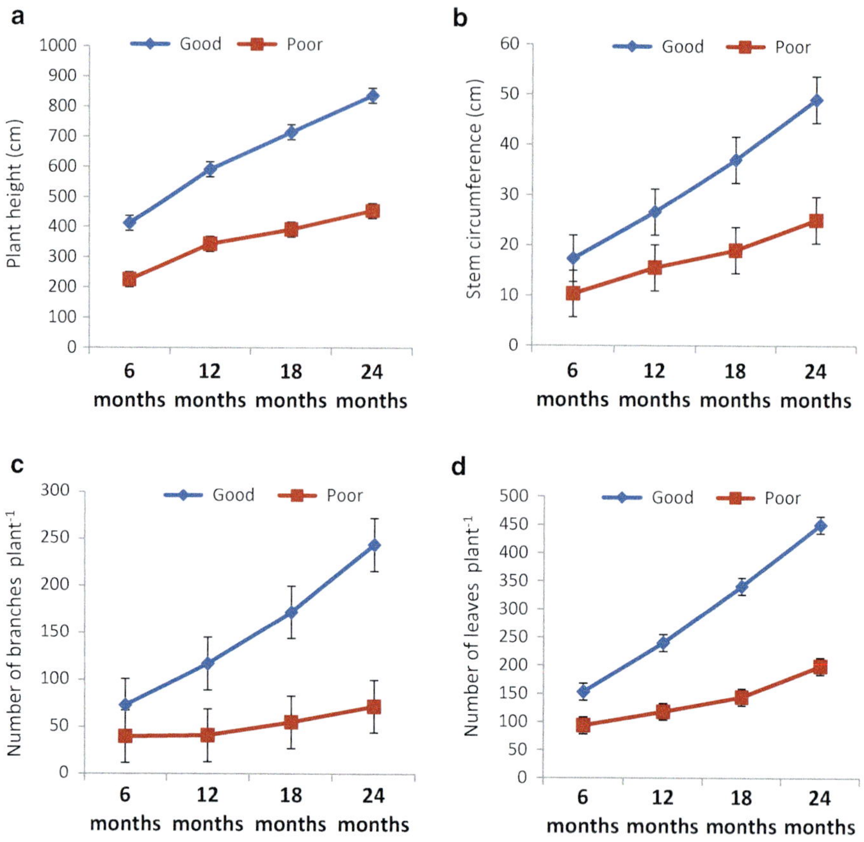

Fig. 7 Growth parameters of *Sesbania sesban*: (**a**) plant height, (**b**) stem circumference, (**c**) number of branches, and (**d**) number of leaves. Vertical bars indicate the LSD value at $P \leq 0.05$

With time, this concentration further increased for both types of leaves. K^+ concentration was higher in younger leaves of both good and poor plants, and the comparison of both types of plants showed that this concentration was higher in good plants than in poor plants. The K^+:Na^+ ratio was highest in younger leaves of good plants and was lowest in older leaves of poor plants. This trend was observed throughout the experimental period.

After 6 months, the values of pH_s, EC_e, and SAR were 8.28, 12.99, and 42.07 at 0–30 cm depth and 8.09, 10.69, and 42.9 at 30–60 cm depth for good plants (Fig. 9). For poor plants, these values were 8.37, 15.65, and 47.7 at 0–30 cm depth and 8.17, 10.54, and 46.3 at 30–60 cm depth. In the last 6-month data, these values were 8.11, 9.98, and 32.5 at 0–30 cm depth and 7.92, 8.5, and 37.8 at 30–60 cm depth for good plants. For poor plants, these values were 8.27, 12.9, and 43.2 at 0–30 cm depth and 8.04, 9.34, and 43 at 30–60 cm depth.

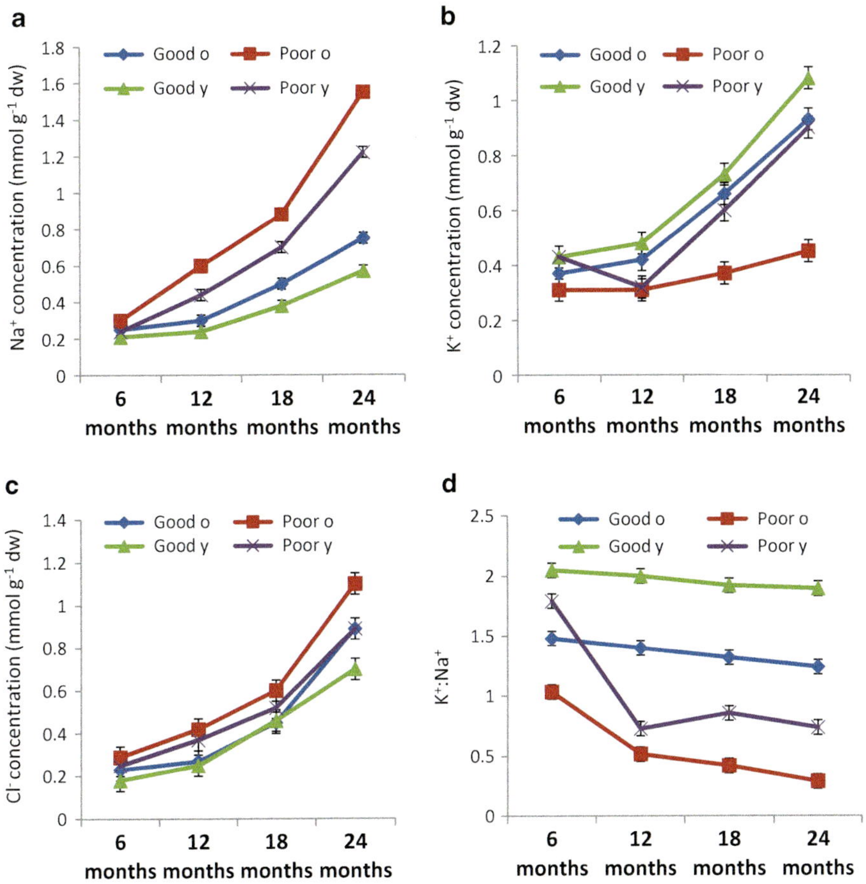

Fig. 8 Ionic parameters of *Sesbania sesban*: (**a**) sodium, (**b**) potassium, (**c**) chloride concentration, and (**d**) K^+:Na^+ ratio in older (o) and younger (y) leaves. Vertical bars indicate the LSD value at $P \leq 0.05$

Discussion

The three tree species grown on the saline-sodic soil showed different growth potentials from one another. Growth indicators such as the plant height, stem circumference, number of branches, and number of leaves slowly increased in the first year, but in the second year and particularly in the last 6 months, these growth parameters increased at a rapid rate. Initially, the plants were not well acclimatized to the harsh environmental conditions, but with the passage of time, they become established and showed good growth performance. Among the three tree species, *Sesbania sesban* produced more plant height, stem circumference, number of branches, and number of leaves than the other two species. This type of growth pattern indicates the differential response of different trees under saline-

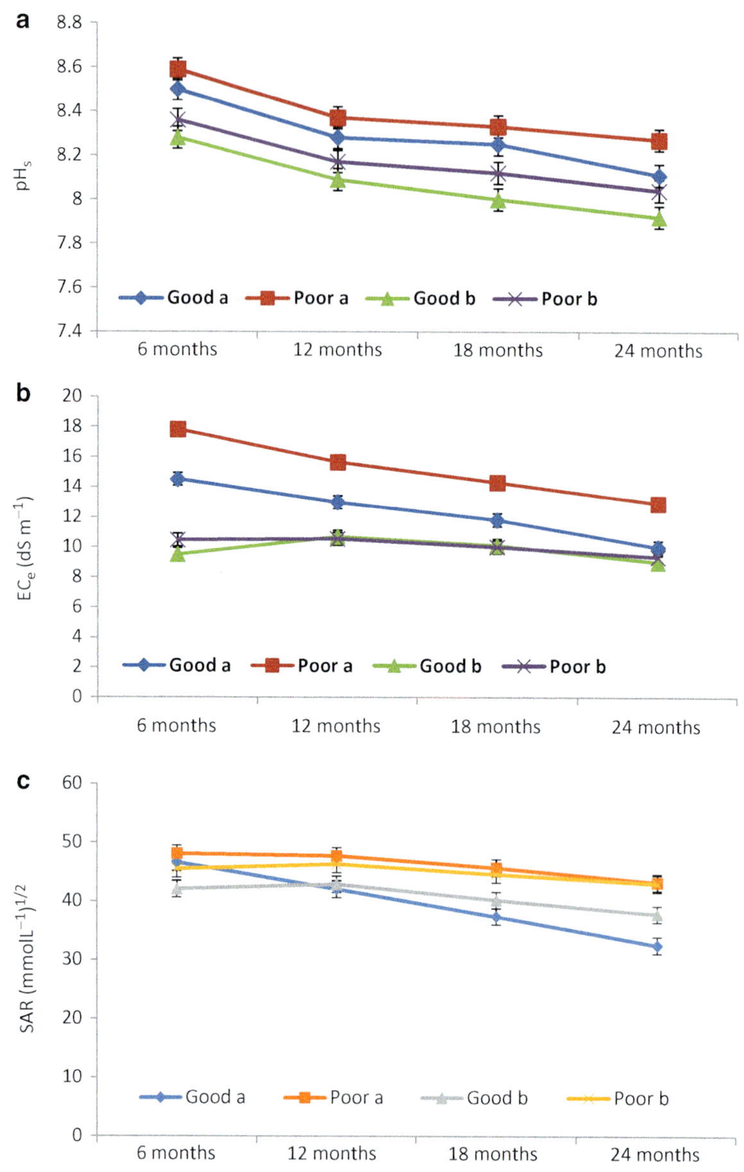

Fig. 9 Soil parameters of *Sesbania sesban*: (**a**) pH_s, (**b**) EC_e and (**c**) SAR of 0–30 cm (a) and 30–60 cm (b) soil depth. Vertical bars indicate the LSD value at $P \leq 0.05$

sodic conditions. Similar results were also found by Abbas et al. (2016) and Ashraf et al. (2008) as they studied different acacia species under salt-affected field conditions having various salinity and sodicity levels. Such growth behavior of plants under salinity stress depends on their growth stages, the type and amount of salts in the root zone, and most importantly the plant species (Lauchli and Grattan 2007; Abbas et al. 2016).

In the present study, we observed that *Sesbania sesban* has greater growth and biomass production potential on salinized barren soils receiving saline-sodic water than the other two species. The salt tolerance potential of *Sesbania sesban* was evaluated by Mahmood et al. (2008) in a pot experiment. The seedlings were irrigated with saline water having various salt concentrations ranging from 5.12 to 16.22 dS m^{-1}. They observed a progressive decrease in growth and nodulation with increasing salinity levels; however the plants survived even at the highest salinity level. This indicates that this tree species has high salinity tolerance potential.

Qureshi et al. (1993) compared the survival and growth performance of different fruit trees including *Grewia asiatica* under saline-sodic field conditions. The survival of fruit tree species varies with the level of salinity and sodicity. *Grewia asiatica* showed 100% survival when the levels of EC and SAR were 10 and 15.1, respectively. The survival was significantly decreased to 50% when EC and SAR were 20.3 and 54.8, respectively.

Varshney and Raj (2000) explored the salt tolerance potential of leguminous tree species, namely, *Albizia lebbeck* and *Leucaena leucocephala*, at germination and early seedling stages. Seeds of both tree species were exposed to sulfate and chloride salt solutions having conductivities of 4, 8, 12, and 16 dS m^{-1}. Although the germination of both species was decreased in both types of salts, chloride salinity was more detrimental for germination and seedling growth. It was noticed that *Albizia lebbeck* responded to salinity in a better way than *Leucaena leucocephala* by producing more seedling length and seedling vigor index. Dagar et al. (2001) conducted a long-term field experiment by growing 30 tree species on a high alkali-saline soil with $pH > 10$. Trees were grown by auger hole technology and were irrigated with sodic water. After 7 years of growth, out of 30 species, only 13 species survived. Moreover, only three surviving species including *Acacia nilotica*, *Tamarix articulata*, and *Prosopis juliflora* produced enough biomass, which make them suitable for these soils. It was further noticed that among many other species, *Albizia lebbeck* (L.) survived the initial 5 years but could not survive afterward due to the high pH of both soil and water. Likewise, Bhushan et al. (2014) found that *Albizia lebbeck* (L.) is highly tolerant to salinity and survived even at 4% salinity level in a pot experiment.

Although these are salinity-tolerant tree species, the growth and biomass production are decreased at higher levels of salts in the growth medium, mainly due to osmotic effect, ionic effect, nutrient deficiency, and oxidative stress (Abbas et al. 2015, 2017). It has been well documented that tree species having relatively higher water contents, limited uptake of toxic ions (Na^+ and Cl^-), and higher activities of antioxidant enzymes have better growth and produce more biomass (Abbas et al. 2015, 2017). Leaf ionic concentration revealed significantly higher Na^+ and Cl^- concentrations in old leaves, whereas lower Na^+ and Cl^- accumulation was observed in younger leaves, which accumulated more K^+. This pattern of ionic concentration relates to the growth performance of the tree species: the species (*Grewia asiatica*) which accumulated higher Na^+ and Cl^- and lower K^+ exhibited lower values of all the growth parameters and the species with lower concentration of deleterious ions (Na^+ and Cl^-), and higher accumulation of K^+ showed better growth performance. Such disturbed ionic concentration results in molecular damage, reduced growth, and plant death in some cases (Zhu 2001). It is very important to maintain a favorable ionic concentration inside the plant for growth and survival under salt-affected conditions (Saqib et al. 2005; Abbas et al. 2013). Higher ratios of K^+:Na^+ were maintained by *Sesbania sesban* compared to the other tree species. Higher K^+ and lower Na^+ or a higher K^+:Na^+ ratio might be responsible for higher growth in *Sesbania sesban* compared to the other tree species. Regarding ionic composition, Shirazi et al. (2001), Nasim et al. (2008), and Abbas et al. (2015, 2017) have also reported similar results.

Growing plants can ameliorate a salt-affected soil. Soil pH_s, EC_e, and SAR slowly decreased over time. Among the three tree species, *Sesbania sesban* caused more reduction in these soil indicators compared to the other two species. Such improvement in the chemical constitution of salt-affected soils with trees had been previously noticed by Basavaraja et al. (2011) and Seenivasan et al. (2014). Abbas et al. (2016) compared the phytoremediation potential of two acacia species under saline-sodic field conditions and concluded that *Acacia ampliceps*, due to more rhizosphere acidification, caused more reduction in pH_s, EC_e, and SAR of soils than *Acacia nilotica*. Plants facilitate the dissolution of Ca^{2+} from calcite in soils leading to phytoremediation of soils. The CO_2 released is converted to H_2CO_3 which helps either to dissolve Ca^{2+} from $CaCO_3$ or dissociation of H_2CO_3 to H^+ and HCO_3^- leads to a reaction of H^+ with $CaCO_3$ resulting in the release of Ca^{2+} ions (Qadir et al. 2007). The release of different organic acids like malic acid, citric acid, and tartaric acid also helps in the release of Ca^{2+} from $CaCO_3$ (Abbas et al. 2016). Exchangeable Na^+ is replaced in exchange sites by Ca^{2+} released from $CaCO_3$ (Qadir et al. 2003). The released Na^+ either leaches down with water or is absorbed by plant roots (Qadir et al. 1997). The rhizosphere acidification potential is higher in leguminous plants carrying out nitrogen fixation, and a release of H^+ ions (Schubert et al. 1990; Tang et al. 2013) and organic acids occurs (Haigang et al. 2013). It has been seen that the phytoremediation potential of a plant is positively correlated with its rhizosphere acidification potential (Martin et al. 2014; Abbas et al. 2016). More rhizosphere acidification by the leguminous tree species, i.e., *Sesbania sesban*, compared to nonleguminous species, might be responsible for a stronger reduction in pH_s and the dissolution of calcite and ultimately a higher reduction in soil SAR. The other possibility of a reduction of soil pH might be Na^+–Ca^{2+} exchange followed by leaching of Na^+ (Qadir et al. 1996).

In line with our findings, Singh and Gill (1992) also observed a significant decrease in pH of saline-sodic soil followed by 20 years of plantation of different tree species including *Albizia lebbeck*, *Eucalyptus tereticornis*, *Acacia nilotica*, *Terminalia arjuna*, and *Prosopis juliflora*. They further observed a significant increase in soil organic carbon and available P and K contents. Likewise, Dagar et al. (2001) and Qureshi et al. (1993) observed a decreasing trend in soil pH of salt-affected soils following the plantation of different tree species. All the tree species added organic matter to soils and increased the carbon sequestration potential of the barren lands. Due to leaching of soluble salts, soil EC_e also decreased, and again the leguminous species was responsible for more reduction in soil EC_e than the other two tree species under consideration. Earlier, Ashraf et al. (2008) and Abbas et al. (2016) reported an increase in organic matter in a saline-sodic soil under acacia plantation. Seenivasan et al. (2014) found that eucalyptus plantations increased organic matter of a saline-sodic soil over a 3-year period of time. This increase in organic matter can be related to more root turnover and litter production with a subsequent decomposition over time and an improvement in soil properties.

Conclusions and Future Outlook

Climate change poses a real challenge to the prosperity and existence of the mankind in this century and beyond. Agriculture, particularly in the dry and degraded land areas, is affected by climate change-induced disturbances leading to poor living conditions of the inhabitants. The situation may be addressed with the use of stress-tolerant plant species which can survive and maintain their production in harsh conditions. It is imperative to identify and domesticate such species and improve their produce quality. The use of such species on barren degraded lands not only improves food provision but also improves the environment and provides an opportunity to earn carbon credits. With the consistent efforts of policy makers, the scientific community, and stakeholders, the menace of climate change and land degradation can be converted into an opportunity.

References

Abbas G, Saqib M, Akhtar J, Basra SMA (2013) Salinity tolerance potential of two acacia species at early seedling stage. Pak J Agric Sci 50:683–688

Abbas G, Saqib M, Akhtar J, Murtaza G, Shahid M (2015) Effect of salinity on rhizosphere acidification and antioxidant activity of two acacia species. Can J For Res 45(1):124–129

Abbas G, Saqib M, Akhtar J, Murtaza G, Shahid M, Hussain A (2016) Relationship between rhizosphere acidification and phytoremediation in two acacia species. J Soils Sediments 16:1392–1399

Abbas G, Saqib M, Akhtar J, Murtaza G (2017) Physiological and biochemical characterization of *Acacia stenophylla* and *Acacia albida* exposed to salinity under hydroponic conditions. Can J For Res 47(9):1293–1301

Ahmad S, Ghafoor A, Akhtar ME, Khan MZ (2016) Implication of gypsum rates to optimize hydraulic conductivity for variable-texture saline-sodic soils reclamation. Land Degrad Dev 27:550–560

Arshad M, Silvestre J, Pinelli E, Kallerhoff J, Kaemmerer M, Tarigo A, Shahid M, Guiresse M, Pradere P, Dumat C (2008) A field study of lead phytoextraction by various scented *Pelargonium* cultivars. Chemosphere 71:2187–2192

Ashraf MY, Shirazi MU, Ashraf M, Sarwar G, Khan MA (2008) Utilization of salt-affected soils by growing some Acacia species. In: Khan MA, Weber DJ (eds) Ecophysiology of high salinity tolerant plants. Springer Science + Business Media, Springer, Dordrecht, pp 289–311

Basavaraja PK, Sharma SD, Dhananjaya BN, Badrinath MS (2011) *Acacia nilotica*: a tree species for amelioration of sodic soils in central dry zone of Karnataka, India. Res J Agric Sci 2:311–316

Bhushan G, Sharma SK, Sagar P, Seth N, Singh AP (2014) Role of arbuscular mycorrhiza fungi on tolerance to salinity of the tree legume *Albizia lebbeck* (L.) inoculated by rhizobium. Indian J Pharm Biol Res 2(1):45–50

Bouyoucos GJ (1962) Hydrometer method improved for making particle size analysis. J Agron 53:464–465

Dagar JC, Singh G, Singh NT (2001) Evaluation of forest and fruit trees used for rehabilitation of semiarid alkali-sodic soils in India. Arid Land Res Manag 15(2):115–133

Drake JA, Cavagnaro TR, Cunningham SC, Jackson WR, Patti AF (2016) Does biochar improve establishment of tree seedlings in saline sodic soils? Land Degrad Dev 27(1):52–59

Food and Agriculture Organization (FAO) (2008) Climate change and food security: a framework document. Rome. http://www.fao.org/forestry/15538-079b31d45081fe9c3dbc6ff34de4807e4.pdf

Gornall J, Betts R, Burke E, Clark R, Camp J, Willett K, Wiltshire A (2010) Implications of climate change for agricultural productivity in the early twenty-first century. Philos Trans R Soc Lond Ser B Biol Sci 365:2973–2989

Haigang L, Zhang F, Rengel Z, Shen J (2013) Rhizosphere properties in monocropping and intercropping systems between faba bean (*Vicia faba* L.) and maize (*Zea mays* L.) grown in a calcareous soil. Crop Pasture Sci 6:976–984

Hütsch BW, Saqib M, Osthushenrich T, Schubert S (2014) Invertase activity limits grain yield of corn (*Zea mays* L.) under salt stress. J Plant Nutr Soil Sci 177:278–286

Ilyas M, Miller RW, Qureshi RH (1993) Hydraulic conductivity of saline-sodic soil after gypsum application and cropping. Soil Sci Soc Am J 57:1580–1585

Intergovernmental Panel on Climate Change (IPCC) (2007) In: Solomon S, Qin D, Manning M, Chen Z, Marquis M, Averyt KB, Tignor M, Miller HL (eds) Climate change 2007: the physical science basis. Contribution of working group I to the fourth assessment report of the Intergovernmental Panel on Climate Change. Cambridge University Press, Cambridge, UK

Jacobsen SE, Sørensen M, Pedersen SM, Weiner J (2015) Using our agrobiodiversity: plant-based solutions to feed the world. Agron Sustain Dev 35:1217–1235

Lauchli A, Grattan SR (2007) Plant growth and development under salinity stress. In: Jenks MA, Hasegawa PM, Jain SM (eds) Advances in molecular breeding towards drought and salt tolerant crops. Springer, Dordrecht, pp 285–315

Lobell D, Schlenker W, Costa-Roberts J (2011) Climate trends and global crop production since 1980. Science 333:616–620

Mahmood A, Athar M, Qadri R, Mahmood N (2008) Effect of NaCl salinity on growth, nodulation and total nitrogen content in *Sesbania sesban*. Agric Conspec Sci 73(3):137–141

Martin BC, George SJ, Price CA, Ryan MH, Tibbett M (2014) The role of root exuded low molecular weight organic anions in facilitating petroleum hydrocarbon degradation: current knowledge and future directions. Sci Total Environ 472:642–653

Munns R, Tester M (2008) Mechanisms of salinity tolerance. Annu Rev Plant Biol 59:651–681

Nasim M, Qureshi RH, Aziz T, Saqib M, Nawaz S, Sahi ST, Pervaiz S (2008) Growth and ionic composition of salt-stressed *Eucalyptus camaldulensis* and *Eucalyptus tereticornis*. Pak J Bot 40(2):799–805

Oo AN, Iwai CB, Saenjan P (2015) Soil properties and maize growth in saline and nonsaline soils using cassava-industrial waste compost and vermicompost with or without earthworms. Land Degrad Dev 26(3):300–310

Qadir M, Oster JD (2004) Crop and irrigation management strategies for saline-sodic soils and waters aimed at environmentally sustainable agriculture. Sci Total Environ 323:1–19

Qadir M, Qureshi RH, Ahmad N (1996) Reclamation of a saline-sodic soil by gypsum and *Leptochloa fusca*. Geoderma 74:207–217

Qadir M, Qureshi RH, Ahmad N (1997) Nutrient availability in a calcareous saline-sodic soil during vegetative bioremediation. Arid Soil Res Rehabil 11:343–352

Qadir M, Steffens D, Yan F, Schubert S (2003) Sodium removal from a calcareous saline-sodic soil through leaching and plant uptake during phytoremediation. Land Degrad Dev 14:301–307
Qadir M, Oster JD, Schubert S, Noble AD, Sahrawat KL (2007) Phytoremediation of sodic and saline-sodic soils. Adv Agron 96:197–247
Qadir M, Tubeileh A, Akhtar J, Larbi A, Minhas PS, Khan MA (2008) Productivity enhancement of salt-affected environments through crop diversification. Land Degrad Dev 19:429–453
Qadir M, Noble AD, Karajeh F, George B (2015) Potential business opportunities from saline water and salt-affected land resources. Resource recovery and reuse series, vol 5. International Water Management Institute (IWMI). CGIAR Research Program on Water, Land and Ecosystems (WLE), Colombo. https://doi.org/10.5337/2015.206. 29p
Qureshi RH, Barrett-Lennard EG (1998) Saline agriculture for irrigated land in Pakistan: a handbook. ACIAR, Canberra
Qureshi RH, Nawaz S, Mahmood T (1993) Performance of selected tree species under saline-sodic field conditions in Pakistan. In: Lieth H, Al Masoom A (eds) Towards the rational use of high salinity tolerant plants, vol 1. Kluwer, Dordrecht, pp 259–269
Richards LA (1954) Diagnosis and improvement of saline and alkali soils. Handbook, vol 60. USDA Agric., Washington, DC
Rozema J, Schat H (2013) Salt tolerance of halophytes, research questions reviewed in the perspective of saline agriculture. Environ Exp Bot 92:83–95
Ryan J, Estefan G, Rashid A (2001) Soil and plant analysis laboratory manual. ICARDA, Aleppo
Saqib M, Zorb C, Rengel Z, Schubert S (2005) Na^+ exclusion and salt resistance of wheat (*Triticum aestivum* L.) are improved by the expression of endogenous vacuolar Na^+/H^+ antiporters in roots and shoots. Plant Sci 169:959–965
Saqib M, Akhtar J, Qureshi RH (2008) Sodicity intensifies the effect of salinity on grain yield and yield components of wheat. J Plant Nutr 31:689–701
Saqib M, Akhtar J, Abbas G, Nasim M (2013) Salinity and drought interaction in wheat (*Triticum aestivum* L.) is affected by the genotype and plant growth stage. Acta Physiol Plant 35:2761–2768
Schachtman DP, Kumar R, Schroeder JI, Marsh EL (1997) Molecular and functional characterization of a novel low-affinity cation transporter (LCTI) in higher plants. Proc Natl Acad Sci 94:11079–11084
Schubert E, Schubert S, Mengel K (1990) Effect of low pH of the root medium on proton release, growth, and nutrient uptake of field beans (*Vicia faba*). Plant Soil 124:239–244
Seenivasan R, Prasath V, Mohanraj R (2014) Restoration of sodic soils involving chemical and biological amendments and phytoremediation by *Eucalyptus camaldulensis* in a semiarid region. Environ Geochem Health 37(3):575–586
Shirazi MU, Khanzada B, Alam SM, Ansari R, Mujtaba SM, Ali M, Khan MA (2001) Seasonal nutrient variation in two acacia species growing under saline environment. Pak J Biol Sci 4(5):514–517
Singh G, Gill HS (1992) Ameliorative effect of tree species on characteristics of sodic soils at Karnal. Indian J Agric Sci 62(2):144–146
Singh K, Trivedi P, Singh G, Singh B, Patra DD (2016) Effect of different leaf litters on carbon, nitrogen and microbial activities of sodic soils. Land Degrad Dev 27:1215–1226
Tang C, Weligama C, Sale P (2013) subsurface soil acidification: its possible causes and management options, In: Xu JM, Sparks D (eds) *Molecular Environmental Soil Science at Critical Zone Interfaces*, Springer, Dordrecht, pp 389–412
Varshney UK, Raj K (2000) Effect of chloride and sulphate salinity on germination and seedling development of two tropical leguminous tree species. Indian Forester 126(8):856–860
Wu Y, Xu G, Shao HB (2014) Furfural and its biochar improve the general properties of a saline soil. Solid Earth 5(2):665–671
Zhu JK (2001) Plant soil tolerance. Trends Plant Sci 6:66–71

Discussion with Farmers in Southwestern Nigeria on Climate Change and Agriculture

113

Obasanjo Joseph Oyedele

Contents

Background and Problem 2308
Literature Review 2309
Participation Model 2311
Methodology 2312
Findings 2314
One–Way Model of Communication 2314
Dynamics of Responses from Women, Youths, and Educated Farmers 2315
Raise Awareness, Inform, and Warn Farmers 2316
Dynamics of Responses from Women, Youths, and Educated Farmers 2316
Warnings Overshadowed Knowledge, Attitude, and Practice 2317
Age and Educational Dimensions on Credibility of Sources 2319
Timeliness and Complexity of Messages 2320
Gender, Age, and Education Dynamics on Timeliness and Complexity of Messages 2320
Conclusion 2321
References 2321

Abstract

Africa is the worst hit by the crisis of food insecurity because of the effects of climate change on agriculture among farmers in the continent. Global discourse on sustainable management of climate change in developing countries is on access to climate change information and practice of climate-friendly farming strategies by farmers. However, new studies have established that empirical conversation on media communication of climate change in developing countries should move from determining level of awareness to promoting inclusive climate change communication wherein stakeholders will dialogue for sustainable

O. J. Oyedele (✉)
Department of Mass Communication, Faculty of Social and Management Sciences, Bowen University Iwo, Osun State, Nigeria
e-mail: obasanjo.joseph@gmail.com

W. Leal Filho (ed.), *Handbook of Climate Change Resilience*,
https://doi.org/10.1007/978-3-319-93336-8_39

climate change management. This chapter examines how climate change communication to farmers in Southwestern Nigeria fulfills the empirical recommendation on two-way climate change communication exchange. Interviews and focus group discussion sessions among farmers in Osun, Oyo, and Ondo states reveal that agricultural information about climate change mitigation and adaptation follows a one-way communication; warnings and weather forecasts overshadow building of positive knowledge, attitude, and practice. An appropriate feedback and participatory mechanism is needed, and farmers need to be active stakeholders in climate change communication for sustainable agricultural development in Nigeria.

Keywords
Climate change communication · Food security · Two-way communication · Farmers · and Sustainable agricultural development

Background and Problem

Africa is the worst hit by the crisis of food insecurity because of the effects of climate change on agriculture among farmers in the continent. The concept of food security extends beyond access to food by citizens in any country; it covers constant availability of food in enough quantity, stability in supply, and improved quality of food for people to lead a healthy lifestyle and be able to guarantee food provision for future generations. In Nigeria, statistics show a downward trend in local food production (growth in food production is less than 1%) at a time that population growth rate is 3.2%. Also, 60% of the population are below the poverty line, and the country is only above 13 of 104 countries on the 2015 Global Hunger Index. In 2012, it ranked 153rd out of 187 countries on the United Nations Development Programme's (UNDP) Human Development Index, and if affordability, availability, quality, and safety indices of the Global Food Security Index are considered, the country takes number 91 out of 109 assessed (Oyedele 2017; Metu et al. 2016). Between 2014 and 2016, 233 million people were reportedly undernourished. Farmers are prominently relied on to halt this negative trend by producing food in large quantity and quality for the continent to achieve food security by 2050 (Food and Agriculture Organization (FAO) of the United Nations, International Fund for Agricultural Development (IFAD), and World Food Programme (WFP) 2015; Liz et al. 2014). The decreasing rate of food production in Nigeria in the face of high population growth rate is at the heart of food insecurity, and the agricultural economics' strategy for removing this imbalance is to raise agricultural productivity and make food available and accessible through distribution (Oladejo 2013).

However, this drive is threatened by the inability of farmers in the country to meet local demand for food and the effects of climate change on agricultural development (FAO et al. 2015; Liz et al. 2014). Global discourse on sustainable management of climate change in developing countries is on access to climate change information and practice of climate-friendly farming strategies by farmers. Empirical findings show and predict a prevalence of negative effects of climate change on food security and sustainable agricultural, environmental, and economic development of many developing countries, especially in rural communities. In Nigeria, existing poor knowledge, attitude, and practices among farmers exist parallel to government's reactive actions to climate change management (FAO 2008; British Broadcasting Corporation (BBC) World Service Trust 2010; Oladipo 2010; Centre for Research on Environmental Decisions (CRED) 2009).

Some studies have recognized governments, scientists, and the mass media (representing the public) as the major players in climate change communication, following the rationalist-instrumental, one-way model of communication wherein the scientists identify research problems and find their solutions, communicate these findings to government for policy implementation, and use the media to raise awareness, inform, and mobilize the public for actions (Weingart et al. 2000). However, new studies have established that empirical conversation on communication of climate change to farmers in developing countries should move from determining level of awareness to promoting two-way participatory climate change communication wherein scientists, policy makers, development partners, researchers, and farmers will promote knowledge, encourage dialogue and appropriate attitude, integrate media, design strategies for responding to climate change, and influence behavioral change (Moser 2010; FAO 2008; BBC World Service Trust 2010; Oladipo 2010; CRED 2009).

This modification in empirical conversation supports Wilcox's (2003) and World Bank's conception of participation as "a process where individuals, groups, and organisations choose to take an active role in making decisions that affect them" and as "a process through which stakeholders influence and share control over development initiatives and the decisions and resources which affect them" (Usadolo and Caldwel 2016: 2). This chapter examines the extent to which climate change communication to and with farmers in Southwestern Nigeria has fulfilled the two-way development communication model established in literature.

Literature Review

The global concern about increased climate change vulnerability in this century stems from observable negative impacts of climate change on the ecosystems, economies, and communities of the world. Scientific evidence shows that mortality, illnesses, destruction of natural resources, loss of source of income and food, food insecurity, poor air quality, water- and food-borne contamination, increased pest and disease invasion, loss of farmland, reduced harvest, extreme heat, collapsed infrastructures, increase in average air and ocean temperatures, rising sea level, melting of snow and ice, and other dangerous economic and social effects are plaguing the world. These negative trends have necessitated the renewed call for climate change adaptation and increased resilience worldwide (Berry et al. 2009). Environmental challenges did shape and are currently shaping the structures of development in both developed and developing countries, and studies have concluded that most of today's problems are products of unplanned and uncontrolled urge to use environmental resources for industrial and economic activities. This anthropogenic narrative about sources of climate change are increasingly blurring the strong natural narratives that dominated climate change discourses of different stakeholders in most countries of the world (IPCC 2007; Guti et al. 2012; Ho and Lin 2012; Shishi et al. 2015).

Specifically, in Nigeria, flooding; urban decay; industrial pollution; desertification; drought; deforestation; water pollution; oil pollution; loss of biodiversity; increased temperature; heat and intensity of sunshine; poor yield of crops and animals; starvation; death of crops, man, and animals; loss of soil nutrients; and soil and coastal erosion are among the prominent environmental challenges associated with climate change. Various national policies (Vision 2010 and Vision 20: 2020), legislative actions, sectoral programs, and cooperation with local and international organizations are targeted at achieving the global mandate of the Millennium Development Goals, Sustainable Development Goals, and Agenda 21. Furthermore, international organizations such as World Bank, International Atomic Energy Agency(IAEA), International Fund for Agricultural Development (IFAD), Canadian International Development Agency (CIDA), United Nations Development Programme (UNDP), United Nations Industrial Development Organization (UNIDO), United Nations Environment Programme (UNEP), and Agricultural Development Company Limited support the country in developing environmental policies and programs, and they offer financial and technical assistance as Nigeria strives to achieve sustainable climate change mitigation and adaptation (Olagunju 2015).

Various national development plans of the Nigerian government show a deep concern for the impact of climate change on the nation's socioeconomic development and planned actions to combat its negative effects. Especially in the current national development plan called Vision 20:2020, there is a greater understanding that the nation's prosperity and future development could be greatly threatened by climate change. This has led the nation to formulate principles ranging from having a strategic response compatible and consistent with the nation's development priorities, addressing climate change, and using energy to drive high economic growth with a focus on sustainable development framework to placing priority on mitigation and adaptation as policy response and coping strategy and ensuring a permeation of climate change policy in other policies. They also include the cooperation of all ministries, agencies, sectors, communities, and stakeholders, ensuring that national response produces greater and concrete entrepreneurial opportunities (Nachmany et al. 2015). The extent to which these principles and policies have been actualized is not the focus of this study.

From the Fourth Assessment Report of the Intergovernmental Panel on Climate Change, man-made factors that cause and aggravate effects of climate change have been on the increase since the mid-twentieth century. There is a growing belief that the current geological age is a product of man's quest for modern lifestyles, a persistent urge to consume the ecosystem with increased greenhouse gas emission. Central to the efforts at understanding this negative trend and proposing relevant solutions are the media which present the discourses of all stakeholders on climate change management. That is why mass, digital, and/or social media are pivotal to the climate change debate.

People do not become stakeholders in a case or a matter unless they are involved or influenced by that issue or case. The umbrella called "stakeholders" covers different broad groups involved in policy making, policy implementation, and the beneficiary of the development project. When these people are classified under policy and program designers and implementers, they include decision makers, public administration officers at all levels of government, and local leaders who make decisions. There are farmers, landowners, forest managers, rural businesses, actors in the food chain, environmental interest groups, actors in research and innovation, and rural community groups. The last group encompasses individuals concerned about rural and national development (European Network for Rural Development 2015).

Why is stakeholder participation very important? It is because policies target the particular and original needs of both the policy makers and the end users/beneficiaries who are involved in a two-way communication process. When beneficiaries are not included right from the conception of the policies and programs, there are threats to the successful implementation of the projects and programs. Also, the inclusion of all stakeholders, especially the potential beneficiaries of the project, ensures that development is not a one-off, top-down flow, but a cycle involving engagement, feedback, and a process of continuation and sustainability (European Network for Rural Development 2015).

Participation Model

Among popular participation theories, Sherry Arnstein's ladder of citizen participation has been prominently used since 1969. Arnstein's conception of citizen participation is premised on citizen power, power sharing, and power redistribution. The hallmark of citizen participation is when the have-nots join in determining how information is shared, goals and policies are set, tax resources are allocated, programs are operated, and benefits like contracts and patronage are parceled out. The ladder of citizen participation has eight rungs (manipulation, therapy, informing, consultation, placation, partnership, delegated power, and citizen control) divided into three stages: nonparticipation, tokenism, and citizen power. Movement on the ladder from its base to the top shows an increase in the level of inclusion and citizen participation in public policy. Therefore, on the first two rungs (manipulation and therapy), citizens are nonparticipants since their inclusion, contribution, and opinions are not even needed at all. Propaganda reigns supreme, and there is no real power and function for citizens' inclusion in governance. The next three rungs are classified as the stage of tokenism. The stage starts with one-way information provision from government offices, agencies, and bureaucracies (media inclusive) without any channel for feedback and negotiation. Consultation which is the fourth rung has a semblance of two-way information flow, but no assurance that citizens'

Fig. 1 Arnstein's ladder of citizen participation (1969)

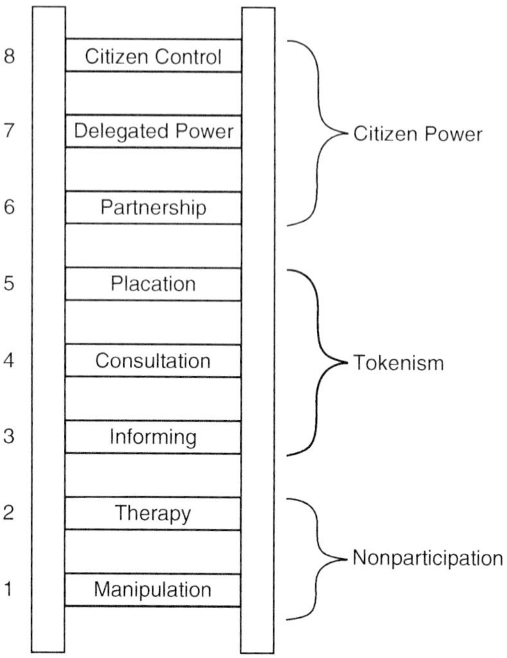

informed contribution will be considered and implemented. At the last lap of tokenism, some form of power starts building up for citizens as they are members of committees and boards though their participation is subjected to the outplay of number and overbearing tendencies of the government actors. At the last stage of the ladder, partnership, delegated power, and citizen control show real inclusive machinery and participation as joint policy boards, planning committees, and accountability mechanisms confer appreciable level of dominant authority over plans, implementation drives, and ownership of development agenda (Brooks and Harris 2008; Carpentier 2016, 2011; Arnstein, 1969) (Fig. 1).

Methodology

The chapter was conducted in Osun, Oyo, and Ondo states in Southwestern Nigeria among farmers who were captured under the Agricultural Development Programme of each state. Nigeria has six geopolitical divisions (South-West, South-South, South-East, North-West, North-East, and North-Central). The selected states are three of the six states in the South-West. Oyo State is in the Southwestern Nigeria on latitude 70.8 N and 7°.9S and longitude 3°.8E and 4°.0 W. The state is bounded on the north by Kwara State, Osun State on the east, Ogun State on the south, and Republic of Benin on the west. The state has a combination of tropical rainforest and derived savannah. The state produces cocoa, cotton, grains, palm oil, tobacco, maize, and yam as its major agricultural crops. Ondo State was created with six other states in 1976 with its capital at Akure. It lies between latitudes 4°3N and 6°.0S and 5°.4E and 8°.1 W. On the north, the state is bounded by Ekiti and Kogi states, on the east by Edo State, west by Osun and Ogun states, and south by the Atlantic Ocean. The state is noted for the production of palm oil, cocoa, cassava, and timber (Oyedele 2017).

Osun State was created from Oyo State in 1991. The state is located between longitudes 5°.4E and 4°W and latitudes 8°.1 N and 6°.5S. Agriculture is the predominant occupation in the three states as crop production is the main source of income for about 75% of its population. Farmers in the states produce food and cash crops such as yam, maize, cocoyam, cassava, beans, cocoa, kola nut, and palm oil. Though there are many ethnic tribes in the states, it is predominantly populated by the Yoruba-speaking people. Low level of socioeconomic activities, poverty, low purchasing power, and inadequate basic infrastructure are the major characteristics of most rural areas in the states, and these are also the communities where raw materials and food are produced (Oyedele 2017).

Most of the farmers were above 35 years; it was only in Ondo State where three of the farmers were youths, others were old men and women involved in both food crop and cash crop production. Most farmers in Osun and Oyo states were not literate in their mother tongue, though more than half of the participants in Ondo States attained above secondary school education. It is important to note that there were more female farmers among the FGD participants than male farmers. This was because more women farmers were available when the researcher visited their communities and not because they were housewives. The interviewees were farmer-opinion leaders as identified under the sampling technique, and they were purposively selected because of their influence among farmers in their communities. Note that in each community, women were among those interviewed, especially at Owena in Ondo State and Ibapon-Adindi in Oyo State where the wife of the leader of the community was interviewed. All the farmers were married, there was a mixture of nuclear and extended family types, and all of them belonged to one informal social network or another. Literature arguably lists high level of illiteracy, unemployment, poverty, low standard of living, and near absence of basic facilities such as electricity, modern roads, pipe-borne water, hospital, and functional schools as the development issues in rural Nigeria where over 65% of citizens (who are predominantly farmers) live (Asemah et al. 2013). Therefore, most of the farmers relied on battery-powered transistor radio sets and occasional visits by agricultural extension agents for information.

The multistage sampling procedure was adopted. The first stage involved the adoption of the balloting strategy in picking Osun, Ondo, and Oyo states from the six states in the geopolitical zone through the simple random sampling technique. In the next stage, the existing stratification of states under the Agricultural Development Programme (ADP) in Nigeria into agricultural zones, blocks, and cells was followed. From the list of zones, blocks and cells were followed, and the simple random sampling technique was used to pick Iwo and Ife/Ijesha zones in Osun State, Ibapon/Ibarapa and Ogbomoso zones in Oyo State, and Akure and Owena zones in Ondo State. From the selected zones, the same sampling technique was adopted in picking Iwo and Ife South as blocks and Oloba Ogundiran and Omifunfun as cells in Osun

State and Ogbomoso and Ido as blocks and Ibapon-Adindi and Koguo as cells in Oyo State, while Idanre and Akure North were picked as blocks, with Owena and Ilu Aabo as cells in Ondo State (Oyedele 2017).

To recruit farmers from their cells to participate in the focus group discussion (FGD) and in-depth interview (IDI) sessions, the available and purposive sampling techniques were employed, respectively. Realizing that most of the farmers would be on their farms for many hours every day, the researcher had to recruit male and female farmers who were at home during the researcher's visit for the FGD session. For the interview, the researcher purposively selected only farmer-opinion leaders or representatives. These were people other farmers rely on for leadership and interpersonal information provisioning, sensitization, and education. They were also farmers who attend seminars, workshops, and ADP meetings with government and agricultural extension units and other non-governmental organizations. More than 15 farmers in each location participated in the focus group discussion session giving an approximate of 90 farmers who participated in all the FGD sessions held. Interviews were conducted among 18 purposively selected farmer-opinion leaders in the three states (six farmers in each state). The researcher visited the Ministry of Agriculture in each state to get approval and to also ascertain villages and farmers captured under the Agricultural Development Programme. It is important to note that the sessions were conducted very early in the morning (8–9 o'clock) on each day because the farmers were fond of returning from their farmlands to their villages late in the evening. Also, each FGD session took an approximate of 1 h, and each interview was for about 30 min (Oyedele 2017).

Findings

The findings are presented and discussed under the sections below. Under each heading, the researcher first presents the positions of women, youths, and educated participants before the responses of other groups of the participants. The focus is to present distinguished and representative responses of women, youths, and educated participants in order to understand the varying degrees of influence and perception.

One-Way Model of Communication

Farmers in the three states had access to climate change information, and this was mostly through radio and television stations owned by government. There were daily releases of information mostly as weather forecasts after the routine newscasts, and there were also specific agricultural programs through which climate change messages were disseminated. Most of the farmers observed during the FGD and interview sessions that when rain will fall, when to expect disasters, when to clear the land and make ridges, when to plant crops, and when to apply fertilizer and other chemicals were the usual mass media offerings on climate change.

Dynamics of Responses from Women, Youths, and Educated Farmers

Most of the women, youths, and literate farmers agreed that climate change communication followed a one-way, top-down information dissemination model. Their opinions were represented by the submissions of Mrs. Racheal Orodokun, wife of the village Baale or head at Ibapon-Adindi in Ogbomoso, Oyo State, those of Mr. Ikuejafo Ololade who was a young farmer and a graduate, and those of Queen Josephine Akiitan who echoed the position of the farmers in the area:

> We hear such information on Ajilete FM and Parrot FM.Also, because we do not always tune it to stations and we are not always at home, it is those who are around who also listen to radio that will hear. These days, what is happening to me is beyond listening to radio.

> There was a time I heard on radio that torrential rainfall will lead to flooding in some parts of Nigeria, it really happened as predicted. Those people experienced flooding and their homes and farmlands were destroyed. That was a time that we did not even enjoy rainfall here at all. Radio was saying it many times everyday and we actually heard that it led to loss of lives, but there was no rain here.

Furthermore, at Owena in Ondo State, Mr. Dare Ayodele and Mr. Ikuejafo Ololade who were young and educated and Queen Josephine Akiitan (wife of the village head) submitted during the interview and FGD sessions, respectively, that:

> Government media stations are very important in this regard. OSRC (Ondo State Radio Corporation), Positive FM, Oba lagbe/Farmers are kings (a radio and television agricultural programme on agricultural development on OSRC), "Farmers Congress" and Adaba FM do air programmes and news on climate change.

> They say it every morning. It is everyday on OSRC (Ondo State Radio Corporation). They will say that rain will fall in this place, there will harmattan in this place, rainfall will be heavy in this place and it will not fall at all in this place. They will say areas that the thing will affect and areas that will not be affected. But all these things are human wisdom they use to compare how it was ten years before now and fifteen years to come. It works sometimes and fails sometimes.

The opinions of these critical stakeholders were unanimously supported by FGD participant numbers 2, 3, 6, 9, 4, and 13 who were older members in the communities. However, at Iwo in Osun State, Mr. Wakili Aremu and Mr. Asimiyu Adeniyi expressed a different opinion:

> We rarely have such except for 'obalagbe' that they run in Ilesha in unique FM, they just talk about a crop or honey; they don't really address any issue on climatic change.we only hear radio broadcasting that the government have released money for fertilizer, drug and so on, but we only hear them, it does not get to us because it is the same government that use the money for their selfish interest. It is the educated farmers that get the incentives.

The submissions of these farmers on their access to climate change information and available media platforms for engagement on climate change show that the existing model of climate change communication to farmers has been one-way. Though farmers were getting information daily, frequent release of information gave no opportunity for the farmers to explain what they were encountering, the failure or otherwise of the messages disseminated, and the strategies they were also employing in managing the effects of climate change. Farmers had no opportunity to discuss their challenges with government and other stakeholders, and there was no means of exchanging information on the effects of climate change with policy makers. This is contrary to the position of African Climate Policy Centre (2014) on cross-sectoral, cross-regional, multilevel, and multi-stakeholder involvement in climate change governance. It does follow the one-way, non-inclusive model of participation designed by Mwazi and Ndokosho (2011), but it does not move beyond the first stage on Arnstein's ladder of participation (1969) since farmers as citizens are only passive receivers of climate change information, but are seriously yearning to be active participants in the climate change communication process.

Raise Awareness, Inform, and Warn Farmers

The mass media have performed well in their responsibility of raising awareness, informing farmers, and warning them about the dangerous effects of climate change on agricultural production. According to the information provided by farmers during data collection, there were some programs on radio and television, such as a debate on whether climate was changing or not, and the focus was to raise awareness and warn farmers to avoid imminent flooding, and gradual reduction in the length of day was also mentioned in the news.

Dynamics of Responses from Women, Youths, and Educated Farmers

Mrs. Racheal Orodokun observed:

> By the time it is relayed by the radio, the farmer will have gone to the farm. And the farmer cannot take the radio to the farm. Assuming it was in the morning like this that the programme is being done, the farmers will be hearing. They said the weather has changed, do not plant maize at this particular time, and do not apply fertilizer to the farm at this time. The farmer must know all these things but the kind of period in which they say it, the farmer has gone to the farm.

At Ibapon-Adindi, Ogbomoso, Oyo State, Pa. Ojetola Moses, one of the farmer-opinion leaders interviewed, said it was uncommon for climate change information from the mass media to include new strategies farmers could adopt for sustainable mitigation and adaptation and that announcement and warnings were prioritized over other components of climate change management.

They will say that rain will fall in this place, there will harmattan in this place, rainfall will be heavy in this place and it will not fall at all in this place. They will say areas that the thing will affect and areas that will not be affected.

Farmers do not understand, this weather is from God and it is not clear to us. It does not come from human beings, there is no person that will like to work and fail to reap, we should abide by what God has said. We do not have control over this change in climatic conditions. Who will go to God to plead our case? The extension officers and the mass media cannot change what God has planned to do.

From the observations of the farmers above, it is clear that the warnings and information provisioning have not translated to understanding. Although they got warnings on when to plant and when not to, when rain would fall, and the kind of negative effects that could accompany such, the farmers bemoaned their lack of understanding of what climate change is and what they could do to manage it. This is also evident from the conclusion of the farmers that God was punishing them with the negative effects of climate change on their farms. Daily weather forecasts, news stories, and programs could not translate to understanding, not to mention knowledge, attitude, and behavior change or modification of practices in relation to climate change. The findings of the nationwide chapter conducted by the BBC World Service Trust (2010) support the position that awareness raising, information provisioning, and warnings have been the focus of climate change communication in Africa (and Nigeria especially). The more reason that Moser (2010), CRED (2009), and BBC World Service Trust (2010) recommended a shift of focus to inclusive stakeholder participation using dialogue, inclusive design, formulation, implementation, and evaluation of policies and programs on climate change governance. This inclusive model, supported by the European Network for Rural Development (2015), places beneficiaries of the policies and programs (in this case farmers in Nigeria) as one of the crucial stakeholders that must not be ignored right from the beginning to the end of the policy formulation and implementation structure. Failure to do so, the European Network for Rural Development (2015) observes, is a threat to the successful implementation of the governance structure.

Warnings Overshadowed Knowledge, Attitude, and Practice

In behavior change communication, there is an expectation that access to and accessibility of development messages will produce desired knowledge and lead to the formation of positive attitudes and the adoption of recommended practices or modification of behavior. However, when there is a problem in the communication for development process, such an expectation may be jeopardized. This explains the lamentation of farmers that participated in this study. Most of them observed that information on climate change management was not common, "what we hear is announcement and warning; there is no clear information on what we can do to manage it by ourselves." The failure of information overload and the negative effect of superimposing warnings over knowledge production, formation of the right attitudes, and eventual adoption of promoted climate-friendly farming strategies became apparent when after the instructions and recommendations, the farmers observed that:

> sometimes it is dangerous to adhere because if they predict that rain will fall in March, we may not see anything until June. So many farmers might have started making ridges and planting as early as March with no rain. This results in failure of crops and there is no hope for the farmers. If rain starts to fall in July and you plant yam it can never yield as expected. It is only good for cash crops and not arable crops.

The warnings also failed in some instances as explained by Elder Abolarin Taye, one of the farmers at Omifunfun in Osun State:

> There was a time I heard on radio that torrential rainfall will lead to flooding in some parts of Nigeria, it really happened as predicted. Those people experienced flooding and their homes and farmlands were destroyed. That was a time that we did not even enjoy rainfall here at all. Radio was saying it many times everyday and we actually heard that it led to loss of lives, but there was no rain here.

To clearly show that their knowledge about climate change was poor, most of them did not know anything else about climate change apart from weather and rainfall. For farmers it is only rain that indicates that there is a climate change. This poor knowledge could be responsible for their conclusion that God is behind climate change and that He (God) was using it as a form of punishment for their sins. FGD participant numbers 5, 4, 9, 7, 14, 8, 2, 6, 10, and 12 at Owena in Ondo State and numbers 1, 2, 3, 5, 7, 10, 15, 12, and 8 and numbers 4, 2, 7, 6, 14, 5, 8, and 13, at Ibapon-Adindi and Koguo, respectively, in Oyo State echoed the submission of most of the farmers:

> Unknown sins are also a reason for climate change, like ritual killings and kidnappings. All these things were not happening in the days of our fathers. We know this is from God because it was a big punishment for us this year and we are begging God to help us.

As a result of this conclusion, their attitude to climate change management was to offer sacrifices and appease their gods as their forefathers used to. Here is Mr. Akintiyibo Adesoji, Mr. Ojetola Moses, Pa. Lateef Jimoh, and Asimiyu Adeniyi on this:

> During the days of our fathers,once they notice there is no rain; they make sacrifices to the god of the sky and then rain will fall after the sacrifices. Now that religion has come things have not been going on well, aside this, it is a sign that there is so much sin in the land. In addition to what has been said, the climate has changed, but only God knows why.

According to the farmers, the only solution to this "spiritual" problem was to continue to plant in anticipation that God would grant their prayers and "release" rain. Therefore, Pa Tuki Joshua, Mr. Oliver Ngbegun at Ilu Aabo, and Queen Josephine Akiitan at Owena, respectively, in Ondo State and Mr. Jelili Onasiji at

Koguo in Oyo State offered this advice to farmers facing negative effects of climate change on their farms:

> What can be done about it is, you see, they said our fathers say a proverb that "The farmer is foolish, it is God that makes it result in wisdom." We can't relent because of that. If there is something to be planted on a time schedule we have known previously, we will keep planting it. God may make one to result in wisdom or the reverse may happen. We will endure it because we are not God. We do not know what do to do about it. Whatever we should plant in August, we should plant it. Whatever we should plant in June, we should plant it. The rest remains in God's hands.

This spiritual dimension to the farmers' understanding of climate change has clear misconceptions and cannot lead to or promote the right attitudes and practices for sustainable climate change management. This poor level of understanding and its accompanying spiritual interpretation of effects of climate change are prominent in the findings of the BBC World Service Trust (2010) and Oladipo (2010) and Moser (2010). These studies and the findings of Oyedele (2015) have recommended a shift from concentrating on using mass media to raise fear appeals through warnings. This is because fear appeals can be counterproductive. Instead, research and practice on climate change communication should focus on building knowledge, encouraging positive attitude, and engendering good, climate-friendly practices among people.

Age and Educational Dimensions on Credibility of Sources

Another big factor that undermined the success of the climate change information through the mass media was the credibility of the sources. According to the farmers, before their adoption of media messages (predictions and warnings) on climate change, they would ask this question: how many times have we received information or predictions from this source and it was reliable? In most cases, their level of trust in the radio and television stations releasing climate change warnings and predictions determined their decision to adopt the recommendations. Because they expected the predicted heavy, persistent, and torrential rainfall and this failed in some instances, most of them would continue with the indigenous, agelong knowledge and practices of their forefathers that stood the test of time and credibility. They did not understand why these predictions were failing and their crops had been destroyed because of this delayed rainfall. Mr. Dare Ayodele, an educated youth at Owena in Ondo State observed:

> Concerning what they tell us about climate change, some of the information come true while some do not. It is a game of probability. If we should look at our fathers who are older farmers, we cannot rubbish their experience in this farming business. Before formal education became a popular phenomenon, they have been farmers planting in the mud and other types of soil. There are many younger farmers who can never beat these old farmers when it comes to farming. So, their knowledge, practices and beliefs are still very relevant. During the days of our fathers, once they notice there is no rain; they make sacrifices to the god of the sky and then rain will fall after the sacrifices. Now that religion has come things have not been going on well, aside this, it is a sign that there is so much sin in the land.

Timeliness and Complexity of Messages

As rural dwellers and farmers, the participants submitted that they spent most of the hours during the day on their farms; and they were busy with manual farm operation. Apart from the fact that this prevented them from getting any climate change information on television stations and other sources during this period, it was even very difficult for them to combine this manual operation with radio listening. Good and beneficial information on changing weather and climatic conditions, warnings on when to and when not to plant, and when not to apply fertilizer to the farm was relayed at a period that the farmers have gone to their farms. Pa Lateef Jimoh of Koguo village in Oyo State said:

> It is not everyone that hears the news. It is not everyone that has the time to listen to the news. You know, we are farmers; we go to the farm early. By the time the news is being said, the farmer would have gone to the farm. At night, it is not everyone that has television. It is on the television that I hear and see it almost every day that they talk about the weather. It is not everyone that has access to television. So, I might have information and it could be negligible. Only a few individuals can hear it.

Gender, Age, and Education Dynamics on Timeliness and Complexity of Messages

This issue of timeliness is compounded with the complexity associated with climate change. As a scientific concept, climate science and its communication are complex because the phenomenon is new with difficult terms, issues, explanations, and calculations that are known and better understood by experts in climatology, geography, and data science. Therefore, farmers, most of whom did not have formal education, would find it difficult to understand when such complex messages are relayed on the mass media without any opportunity for clear explanations, feedbacks, and clarifications. This position is explained further by Mr. Akintiyibo Adesoji (educated youth) of Omifunfun in Osun State, Mr. Ikuejafo Ololade (educated youth) of Owena in Ondo State, and Mrs. Racheal Orodokun of Ibapon-Adindi in Oyo State:

> There is a possibility that climate change is beyond the understanding of many farmers because it is something new. They see it as something big, hard and complex for them to understand. I can say this is what is affecting farmers in this area because information on climate change is new, making what they already know to be obsolete. There is a challenge when they want to process their established knowledge with latest, modern information on climate change. They see new things happening as predicted by the media, it may cofound some of them. There is confusion because they start thinking whether to follow their personal knowledge and experience or what learned experts release and advise through various media sources.

On credibility, timeliness, and complexity of climate change messages, Hasan (2013) advises that source credibility, accurate timing of message dissemination, and communication of simple messages should be promoted. In the contributions of

Harvey et al. (2012), Nwabueze et al. (2015), and Kibue et al. (2015), appropriate format, medium, and language must be employed for successful promotion and communication of climate change messages on mitigation and adaptation. The advice is important because the media are relied upon by all stakeholders and are central to farmers' understanding of climate change because of their agenda setting functions and effects in the society. Therefore, the media must therefore be involved, planned for, and selected based on the needs of the farmers.

Conclusion

The chapter has revealed that mass media are used to pass climate change information to farmers with no meaningful interaction or exchange of ideas between farmers and other stakeholders involved in the climate change discourses. Most especially, women and educated youths lamented inappropriate or odd time such messages are relayed. The credibility and timeliness of climate change information dissemination were stressed by these groups of the audiences as significant to the understanding and actions of farmers. Most of the older men still connected effects of climate change with the supernatural, an indication of poor awareness and knowledge, which could presume a practice of ill-motivated and unfriendly farming strategies. The fear appeals that characterized this information dissemination in form of warnings overshadowed the required knowledge building and promotion of sustainable mitigation and adaptation practices. Especially for women and educated youths who could see through the weaknesses inherent in the media offering, it was hard to believe and practice the strategies recommended by and in the media. In Nigeria, recognition is given to governments, scientists, and the mass media (representing the public) as the major players in climate change communication and food security value chain. This lopsided, one-way, and non-inclusive model of climate change communication and climate change governance is a threat to sustainable food security and development. This chapter has explained that the contributions of all stakeholders (farmers, government, scientists or researchers, non-governmental organizations, mass media, traditional and religious leaders, opinion leaders, etc.) are central to the achievement of the objectives of inclusive growth and development in the agricultural sector in Nigeria. Therefore, agricultural extension officers, media practitioners, and farmers should be engaged as active policy actors in the national climate change communication campaigns to promote inclusive growth and sustainable agricultural development in Nigeria.

References

African Climate Policy Centre (2014) African climate change strategy, www.un.org/en/africa/osaa/pdf/au/cap_draft_auclimatestrategy_2015.pdf, pp 1–40

Arnstein, S. (1969). 'A ladder of citizen participation'. J Am Inst Plann, Vol. 35, pp. 216–224

Asemah ES, Anum V, Edegoh LO (2013) Radio as a tool for rural development in Nigeria: prospects and challenges. AFRREV IJAH 2(1):17–35

BBC World Service Trust (2010) Africa talks climate: the public understanding of climate change in ten countries. BBC World Service Trust, London, pp 1–19

Berry P, Clarke K, Pajot M, Hutton D, Verret M (2009) The role of risk perception and health communication in adapting to the health impacts of climate change in Canada. Report for the climate change impacts and adaptation program, Earth Science Sector, Natural Resources Canada, Ottawa

Brooks R, Harris G (2008) Citizen participation, NEPA, and land-use planning in northern New York, USA. Environ Pract 10(4):140–151

Carpentier N (2016) Beyond the ladder of participation: an analytical toolkit for the critical analysis of participatory media processes. Javnost The Public 23(1):70–88. https://doi.org/10.1080/13183222.2016.1149760

Carpentier N (2011) Media and participation: a site of ideological-democratic struggle. Intellect, Bristol, pp 1–408

Centre for Research on Environmental Decisions (CRED) (2009) The psychology of climate change communication: a guide for scientists, journalists, educators, political aides, and the interested public. CRED, New York, pp 1–48

European Network for Rural Development (2015) Improving stakeholder involvement. EU rural review no 19. Publications office of the European Union, Luxembourg, pp 1–56

Food and Agriculture Organisation (2008) Climate change adaptation and mitigation in the food and agriculture sector. Technical background document from the expert consultation, Rome, March 5–7, 2008, pp 1–17

Food and Agriculture Organization of the United Nations (FAO), International Fund for Agricultural Development (IFAD) and World Food Program (WFP) (2015) The state of food insecurity in the world 2015. Meeting the 2015 international hunger targets: taking stock of uneven progress. Rome, FAO. Retrieved on 19 July 2016 from www.fao.org/3/a-i4646e.pdf

Guti B, Aji MM, Magaji G (2012) Environmental impact of natural resources exploitation in Nigeria and the way forward. J Appl Technol Environ Sanitat 2(2):95–102

Harvey B, Ensor J, Carlile L, Garside B, Patterson Z, Naess LO (2012) Climate change communication and social learning-review and strategy development for CCAFS. In: Working paper No. 22. CGIAR research program on climate change, agriculture and food security (CCAFS), Copenhagen, Denmark. Retrieved 12 Feb 2016 from www.ccafs.cgiar.org

Hasan S (2013) Mass communication: principles and concepts, 2nd edn. CBS Publishers and Distributors, New Delhi, pp 1–456

Ho YH, Lin CY (2012) An empirical study on Taiwanese logistics companies' attitudes toward environmental management. Adv Manage Appl Econom 2(4):223–241

Intergovernmental Panel on Climate Change (IPCC) (2007) In Metz B, Davidson OR, Bosch PR, Dave R, Meyer LA (eds) Contribution of working group III to the fourth assessment report of the intergovernmental panel on climate change, 2007. Cambridge University Press, Cambridge, UK and New York, pp 1–841

Kibue GZW, Pan G, Joseph S, Xiaoyu L, Jufeng Z, Zhang X, Li L (2015) More than two decades of climate change alarm: farmers' knowledge, attitudes and perceptions. Afr J Agric Res 10(27):2617–2625

Liz M, Malcolm MAS, Williams R (2014) Agricultural production under climate change: the potential impacts of shifting regional water balances in the US. Invited paper prepared for presentation at the agricultural & applied economics association's 2014 AAEA Annual Meeting, Minneapolis, MN, July 27–29, 2014, pp 1–60

Metu AG, Okeyika KO, Maduka OD (2016) Achieving sustainable food security in Nigeria: challenges and way forward. In: Third international conference on African development issues, Ogun State, Nigeria. Covenant University Press, pp 182–187

Moser SC (2010) Communicating climate change: history, challenges, process and future directions. Wiley, vol 1 January/February, pp 31–53, www.wires.wiley.com/climatechange. Accessed 19 May 2016

Mwazi FN, Ndokosho J (2011) Effective communication of climate change by extension agents. Agricola 6:39–43

Nachmany M, Fankhauser S, Davidova J, Kingsmill N, Tucker L, Hitomi R, Schleifer P, Setzer J, Amelia S, Singleton CS, Sundaresan J, Townsend T (2015) Climate change legislation in Nigeria: an excerpt from the 2015 global climate legislation study, a review of climate change legislation in 99 countries. Grantham Research Institute on Climate Change and Environment. www.lse.ac.uk/Granthaminstitute/legislation/

Nwabueze C, Nnaemeka F, Umeora D (2015) Nigerian newspapers' coverage of climate change issues. Eur Sci J 11(17):171–184

Oladejo JA (2013) Comparative analysis of factors influencing quantity of maize marketed among agricultural household in Oyo and Osun states, Nigeria. Eur Centre Res Train Develop 1(1):1–16

Oladipo E (2010) Towards enhancing the adaptive capacity of Nigeria: a review of the country's state of preparedness for climate change adaptation. Report submitted to Heinrich Boll Foundation, Nigeria, pp 1–44

Olagunju TE (2015) Drought, desertification and the Nigerian environment: a review. J Ecol Nat Environ 7(7):196–209

Oyedele OJ (2017) Climate change information sources and knowledge, attitudes and practices of farmers in South-Western Nigeria. An unpublished PhD thesis, University of Ibadan, Ibadan, Nigeria, pp 1–173

Oyedele OJ (2015) Climate change communication research: trends and implications. Afr J Sustain Develop 5(1):239–255

Shishi KP, Sashidharan RN, Nazry Y, Jeannot AK (2015) An integrated model of the likelihood and extent of adoption of green practices in small and medium sized logistics firms. Am J Econom 5(2):251–258

Usadolo SE, Caldwel M (2016) A stakeholder approach to community participation in a rural development project. SAGE Open 6(1):1–9

Weingart P, Engels A, Pansegrau P (2000) Risks of communication: discourses on climate change in science politics, and the mass media. Public Understand Sci 9:261–283

Wilcox D (2003) The guide to effective participation. http://www.partnerships.org.uk/guid/index.htm

Cross-Disciplinary Drivers: Benefit to Smallholder Farmers and to Achieve SDGs by Various Means

114

Ijaz Rasool Noorka and J. S. (Pat) Heslop-Harrison

Contents

Introduction ... 2326
Genetic Diversification ... 2327
Environmental Stresses ... 2327
Climate Proofing and Domestication ... 2328
Genetic Manipulation ... 2328
Education ... 2328
Small Landholder's Options to Survive ... 2329
Agriculture and Human Welfare ... 2329
Water Is a Key Unit for the Productivity of Field Crops ... 2330
Genetic Basis of Drought Tolerance ... 2330
Variation: The Basis of Plant Breeding ... 2331
Biotechnology ... 2331
Protoplast and Cell-Fusion Technologies ... 2332
Future Global Food Security ... 2332
Conclusion ... 2333
References ... 2333

Abstract

Development of policies to mitigate climate change, adaptation, and migration needs cross-disciplinary work to understand the drivers and motivations to address the adversities of climate change. Interdisciplinary or multisector solutions are essential so that the mobility of people will help to achieve the United

I. R. Noorka (✉)
Department of Plant Breeding and Genetics College of Agriculture, University of Sargodha, Sargodha, Pakistan
e-mail: ijazphd@yahoo.com

J. S. P. Heslop-Harrison
Department of Genetics and Genome Biology, University of Leicester, Leicester, UK
e-mail: phh4@leicester.ac.uk

W. Leal Filho (ed.), *Handbook of Climate Change Resilience*,
https://doi.org/10.1007/978-3-319-93336-8_40

Nations Sustainable Development Goals (SDGs) including "1 no poverty," "2 zero hunger," and sustainability in food production for all. It has been considered that farming underpins the economies and stability of most of the poorest countries in the world. However, in Western countries, people – especially the young – do not want to be working long, hard, hours on farms, often with uncertain production and visible environmental degradation. This is one of the many drivers of increased urbanization and associated migration. The chapter present that there is dire need to develop viable agricultural systems, including appropriate genetics approaches, agronomy, technology, and logistics, to meet the aspirations of the rural populations of developing countries. Exploitation of the genetic diversity of conventional and nonconventional crops is the basic for the adaptation to climate change and provides crops with secure yields delivered with low inputs because of their optimal genetics. It was also concluded that the lack of coordination among all agricultural approaches to ensure food security. Due to the over exploitation of genetic diversity, the biodiversity is going to loss. It is also suggested that participatory approaches to conserve genetic diversity of crop plant have the potential to save the endangered plant genetic resources to ensure food security.

Keywords
Agriculture · Environment · Genetic · Diversity · Food security

Introduction

The man's life is closely associated with the diverse developmental phases of its own and plants life. It should be the food, fiber, fuel, livelihood, energy, sourcing, physical, reproductive, as well as mental health (WHO 2017). The formulation and implementation of viable policies regarding migration, labor mobility, and sustainable production has paved the path for the motivational approach (Brown 1999). The participatory plant breeding and food security approaches will attract the people to be a part of interdisciplinary or multisector solutions to accomplish the major food issues raised by the United Nations Sustainable Development Goals (SDGs), environmental degradation, and longer lifespan of the flora and fauna to maintain our ecosystem (Bruinsma 2009).

The small landholder farmers are the basic units,can bring a change to develop a viable agricultural systems. It is recognized that our farming system underpins the economies and stability of most of the poorest countries in the world (Noorka et al. 2017). However, as in Western countries, people – especially the young – do not want to work long, hard hours on farms, often with uncertain production and visible environmental degradation. This is one of the many drivers of increased urbanization and associated migration (Borlaug and Dowswell 2005). It is necessary to develop viable agricultural systems, including appropriate plant breeding, genetics, agronomy, technology, and logistics, to meet the aspirations of the rural populations of developing countries (Araus et al. 2002; Asins 2002). Higher level education gives the young, who are often the most mobile, but also the most vulnerable, a route to improve their lives whether through (1) their own education, (2) policies driven by well-educated people, or (3) dissemination of knowledge through local, well-educated actors. In the face of the alarming situation of the abrupt food security, this chapter emphasized that nowadays in food production system, the major role is being played by small farm holders, but they are facing a lot of problems. Some of these problems are administrative and the most of them can be overcome by employing the cross-disciplinary drivers to benefit smallholder farmer communities. The crucial steps will be the participatory agricultural approaches to conserve genetic resources, grow more food, carry out better agricultural marketing, overcome environmental stresses, climate proofing of the crops, and gene manipulation to minimize the risk of plant diseases. The purpose of this chapter is to present some topics about genetics, crop improvement with biodiversity, better sustainability, agricultural technology, and education as a proposal to benefit smallholder farmer communities and to achieve the SDGs.

Genetic Diversification

Exploitation of the genetic diversity of conventional and nonconventional crops are basics for adaptation to climate change and providing crops with secure yields delivered with low inputs because of their optimal genetics (Amorim et al. 2003; Athar and Ashraf 2005). It is true that climatic change and poor farm management has resulted in unpredictability with drought, precipitation, loss of soil, and heat waves-environmental services. However, a well-managed, agricultural ecosystem can deliver healthy change to ensure food security (Dekkers and Hospital 2002; Anhalt et al. 2009).

Environmental Stresses

A series of environmental stresses have played its role in most of globe particularly in Africa. These stresses have not only threatened food security but also affected the cropping pattern, shrinking in vegetative and reproductive production stages of the crops (Achard et al. 2006). The changes in both water and temperature affect the photosynthesis ability, roots' behavior, crop fruiting, and harvesting (Chowdhry et al. 1999; Chaves et al. 2009). Furthermore, the postharvest losses often amount to more than 40%, exacerbated when crops must be harvested and transported to urban populations (USDA 2012; Ray et al. 2013; Sendhil et al. 2018). Food that is not eaten after harvest means that the land, water, nutrients, and farmers' efforts are entirely lost. Specific sources of adaptation underpinning the environmental stresses may be addressed by the improvement of genetic resources (Blum 2009; Banziger and Araus 2009). Under the prevailing environmental conditions, global agricultural scientists have realized that plant genetic resources must be captured and stored in in-situ and ex-situ states, including gene bank, pollen banks, DNA libraries, and biorepositories (Bauer et al. 2005; Batool et al. 2013). The time has come to create the state of diversity by proper preservation of genetic resources including obsolete varieties, wild relatives by the proper incorporation of local and exotic germplasm, using well-supported and robust methods to measure, document, and disseminate the genetics available (Braun et al. 2013; Dahiya et al. 2017).

Climate Proofing and Domestication

The initiation of climate proofing, domestication, introduction, hybridization, evaluation, implementation, and international collaboration will pave the path towards the road of food security and sustainable agriculture (Dzanku et al. 2015). The development of the unified database and extension of data exchange program among the plant breeders, botanists, physiologists, pathologists, agronomists, economists, and plant scientists will check the associated risk of narrowing the genetic base of current commercial and domestic cultivars and climate change. The process must also involve at all levels the stakeholders among smallholder farmers and leaders in villages (Fereres and Soriano 2007; Carvalho et al. 2004). The conservation of indigenous varieties to combat biotic and abiotic stresses will provide yield stability by cutting down the cost of production.

Genetic Manipulation

The process of plant breeding and genetic manipulation will accelerate and unfolds the neglecting and orphan plant attributes for genetic improvement. Since with the beginning of agriculture, the natural genetic variability has been exploited in many folds to meet the subsistence food, feed, fiber, fuel, and even pharmaceutical (the 5F) needs. In the mid-1960s, the most of south Asian developing countries has already experienced the introduction of genetic manipulation in the shape of green revolution by the introduction of higher yielder and nitrogenous responsive dwarf hybrids in mostly consumed crops like wheat and rice. Today, we are at the dawn of advancement of agricultural and related science and technology but still wondering how to feed the globe by 2050 without destroying the planet; how to meet the challenging food prices when the per capita availability of food and water will worsen not only in agriculture but also in nonagriculture sectors (Iqbal et al. 2010; Rasheed et al. 2018); and how to meet the rightful aspirations of people to have a rich and varied diet (often with more meat) while having time for leisure and pleasure (Habash et al. 2009). It is therefore more pertinent to behave our genetic resources and ultimately, the agriculture not only as a food machine but also a big source of employment and savior of culture (Ferrio et al. 2007).

Education

The educational level of the migrant is increasing over the time; however, the migrant with tertiary education accounts the lion share in economic impact. In many cases, their movement deprives the country of origin of their skills and a return on the huge investment in that training. In the context of the global deliberations on financing for farm and nonfarm infrastructure, research and development of agricultural produce has diversified the employment statures of the country. The capacity-building projects in higher education funded generously by the advanced nations around the globe have developed the grassroots level change in education, working directly with teachers, mentors, administrative staffs, and even the students. The idea is to explore the development of their skills, knowledge, and experience that leads to a bottom-up approach to change in higher education institutions and to support the modernization, accessibility, and internationalization of tertiary education (Ray et al. 2013; Sendhil et al. 2018). The rural livelihoods literature revealed that rural households do not receive a significant proportion of their farm and nonfarm incomes (Kashiwagi et al. 2006). It has also been seen that female-headed households tend to have larger farm and nonfarm incomes, income shares compared to male-headed families. The results also showed that the gender composition of households, age, education, and access to credit, electricity, and markets are important determinants of multiple nonfarm activities and nonfarm income (Weber et al. 2012; Samal et al. 2017).

Small Landholder's Options to Survive

Agriculture is critical to the well-being of the world's population. Due to unplanned population and limited land resources, the farming community owned land fragmentation is a continuous process due to prevailing inheritance laws. Day by day the number of small land holders is increasing, and their economic position is going downward. However, smallholders and farmers have the right to a high standard of living, like the "rich" they see among some urban dwellers that rely on the farms of the poor for their nutrition. It must be recognized that technological change is allowing farmers and their children to reach their aspirations, and the policies adopted can have a major role in determination migration and labor movement patterns (Liniger et al. 2017). It is important that agricultural and genetic scientists have input into the vision for farms of the future, providing safe, nutritious food in adequate, stable, and sustainable quantities, while preserving the environment and meeting challenges of water and climate change (Koning and van Ittersum 2009).

Agriculture and Human Welfare

The age which is now passing away is largely the age of physical science, with its inventions and discoveries, which have given us power over the forces of nature. It is to be hoped that when this war between physical and biological sciences will be over, then there is an age of biological science (Shirsath et al. 2017). Within the biological sciences, agriculture not only implies the food, fuel, and fiber production but also it refined the food production process by employing all the available disease management approaches by integrated disease management including cultural, biological, and chemical control with the main objective to keep the disease incidence below economic threshold level (McKenzie and Williams 2015). This path leads towards the plentiful food production, good income with low expenditure. The process of welfare starts with the initiation of good farm income particularly in small farm holders (Ortiz et al. 2009).

Water Is a Key Unit for the Productivity of Field Crops

Productivity of crop is not solely affected by its genetic but also by environmental factors. Abiotic as well as biotic factors affect the yield obtained; among abiotic stress, drought is at the top of the list. Drought actually refers to the severe condition of shortage of water even at the critical stages of development of a crop. Globally, drought is a leading threat to limit the yield of field crops (Yang et al. 2001). However, due to the shortage of supplemental irrigation, the plant modifies itself for the following traits:

- Drought tolerance
  Drought tolerance is an ability of the plant to withstand water stress condition as delineated by Mohamed et al. (2000) and Liniger et al. (2017), which is characterized by unpredictable weather, long dry and hot spells, and refers to the ability of a plant to withstand drought condition without affecting the economic yield up to greater extent (Banziger and Araus 2009). Drought tolerance may be in a way that plant escape from the drought or may withstand water stress through different morphological, physiological, and biochemical mechanisms.
- Drought escaping
  Drought escaping is the ability of a plant to escape from the onset of drought by increasing growth rate for a longer time (Chaves et al. 2009).
- Drought avoidance
  Drought avoidance is related to use the judicious use of irrigation water with the minimum loss of water and maximum water uptake (Abdulmalik et al. 2017). The plant develops the deep root system for the avoidance of drought effects (Noorka and Heslop-Harrison 2015).

Genetic Basis of Drought Tolerance

Some genes are drought influenced and produce different types of water stress-related proteins and enzymes. A lot of research is being done on the genetic basis but still very little has been revealed about drought responsive genes, and their role is not properly determined. Expression of Abscisic acid (ABA) genes, protein and carbohydrate synthesis are actually the molecular basis of water stress tolerance (Yang et al. 2001; Zhou 2010). Seedling stage of the crop plant is junction stage and is more vulnerable to drought stress. Stress inducible expression gene, that is dehydration tolerant HVA1, helps the wheat plant to grow at seedling stage under water-stressed conditions. It actually codes for a protein that is late embryo abundant (LEA-III) and has 11 amino acids in it (Noorka and Heslop-Harrison 2015).

Variation: The Basis of Plant Breeding

Plants differ in many ways. It may be safely generalized that no two plants are exactly alike, even though we may limit our observations to a single species like corn. Upon casual examination, we may be impressed with the similarity of the plants within a field of hybrid corn. There is a constancy of features among the corn plants: the development of the stalk; the size, shape, and arrangement of the leaves on the stalk; the termination of the stalk in a tassel; the formation of ear shoots at nodes midway of the stalk; and many other developmental features. By these characteristics, we recognize the corn plant and distinguish it from a plant of wheat, or cotton, or potato.

There are variations in the size, shape, color, composition, and developmental stages among plants matters particularly when responding to different intensities of an environmental stresses. Environmental variations may be observed by comparing plants in a genetically uniform population. Two genetically identical bean seeds, one large and one small, will produce seeding plants differing in size because the large seed has more stored nutrients to start the seedling plant (Noorka and Heslop-Harrison 2014). Two plants of wheat from the same pure strain will differ in development and yield if one is severely infected with a rust pathogen while the other is free from the pathogen. The variations in growth and development of genetically uniform plants described above are caused by variations in the environmental stresses to which the plants are exposed. Environmental effects on a plant are not transmitted to the plant's progeny, and thus selection in a genetically uniform population does not isolate strains that differ in response to environmental stress.

Biotechnology

Biotechnology already has delivered many benefits to society such as new drug vaccines, diagnostic tools, and exciting research techniques to fight human diseases and genetic disorders (Ashraf 2010). Similar biotechnology techniques with plants will benefit society in many ways, including providing a better understanding of plant growth and development, a more stable production of food and fiber, a reduction in crop production inputs, greater profits for farmers, greater international competitiveness, and a higher quality end product (Abdulmalik et al. 2017).

The benefits from this powerful technology will be enormous. However, powerful technologies also provide risks. Biotechnology is expensive, increasing pesticide usage, reducing the labor force in agriculture, and manipulating the genetics of nature. In addition, growing concerns over food safety and environmental quality may become attached to biotechnology (Adebayo et al. 2017).

Protoplast and Cell-Fusion Technologies

Plants and most microbial cells are characterized by a distinct outer wall or exoskeleton which gives the shape characteristic to the cell or organism. Immediately within the cell wall is the living membrane, or plasma membrane, retaining all the cellular components such as nuclei, mitochondria, vesicles, etc.

These protoplasts are extremely fragile but can be maintained in isolation for variable periods of time. Isolated protoplasts cannot propagate themselves as such, requiring first the regeneration of a cell wall before regaining reproductive capacity. Protoplasts from different strains can sometimes be persuaded to fuse and so overcome the natural sexual mating barriers. However, the range of protoplast fusions is severely limited by the need for DNA compatibility between the strains concerned. Fusion of protoplasts can be enhanced by treatment with the chemical polyethylene glycol, which, under optimum conditions, can lead to extremely high frequencies of recombinant formation which can be increased still further by ultraviolet irradiation of the parental protoplast preparations. Protoplast fusion can also occur with human or animal cell types. Protoplasts will also be an important part of genetic engineering in facilitating recombinant DNA transfer (Kashiwagi et al. 2006). Fusion may provide a method of reassorting whole groups of genes between different strains of macro- and microorganisms.

The genomics transformation has been guiding vegetation sciences to open new open doors that are not too far off (Agren and Franklin 2003). The exploration subjects to give us a look at where plant science will lead us to in the coming decade will be as follows:

- People grow crops according to their own needs which cause the reduction of crops species and their varieties. People just grow those which they need and discard other; by discarding, many species become extinct. Many traits are lost by this practice.
- Many new varieties replace the old ones; by this, the old varieties will vanish, so the many good traits in them are also lost.
- Many varieties have weedy ancestors but farmers remove them.
- Groundwork is also the basic cause of genetic erosion such as constructing new roads, colonies, railways, factories, etc.

Future Global Food Security

More than half of the world's migrants continue to live in G20 and oil-rich Gulf Cooperation Council (GCC) countries in search of their better future. Migration is a universal phenomenon having an old history even on the onset of the humanity, and every year, millions of people move around the world for their economic and social life, bringing enrichment to their own lives and those of others (Jat et al. 2017; Jat et al. 2018). This migration pattern has explored the strong binding and relationships between nations to build interfaith harmony, new opportunities for employers and employees, farm and nonfarm incomes, and even the employment diversification.

Conclusion

This chapter revealed and encircled the biotic and abiotic stresses that are depleting our crop genetic resources. The multidisciplinary drivers are used in agriculture and its associated welfare. It is true that climate change and poor farm management has resulted in unpredictability with drought, precipitation, loss of soil, and heat waves – environmental services – that a well-managed agricultural ecosystem can deliver. The chapter also concluded that global researchers need to develop a viable and new world order overhaul present agricultural systems in combination with brown biotechnology (soil), green biotechnology (gene therapy and genetic tools), red and white biotechnology (livestock for both meat and milk purposes), blue biotechnology (marine) coupled with agronomical approaches, and extending extension services, outreaching services to all stakeholders in the process of food production to achieve the SDGs according to the needs of underprivileged populations. The abrupt climate changes have shaken farm management and unskilled agriculture has no future in forthcoming agriculture system which can be covered by cross-disciplinary drivers to benefit small farm holders and poor farmer communities with the best possible agricultural technology and sustainable modern education system.

References

Abdulmalik RO, Menkir A, Meseka S, Unachukwu N, Ado S, Olarewaju JD, Aba DA, Hearne S, Crossa J, Gedil M (2017) Genetic gains in grain yield of a maize population improved through marker assisted recurrent selection under stress and non-stress conditions in West Africa. Front Plant Sci 8:841

Achard P, Cheng H, De Grauwe L, Decat J, Schoutteten H, Moritz T (2006) Integration of plant responses to environmentally activated phyto hormonal signals. Science 311:91–94

Adebayo MA, Menkir A, Hearne S, Kolawole AO (2017) Gene action controlling normalized difference vegetation index in crosses of elite maize (*Zea mays* L.) inbred lines. Cereal Res Commun 45(4):675–686

Agren GI, Franklin O (2003) Root: shoot ratios, optimization and nitrogen productivity. Ann Bot 92:795–800

Amorim EP, de Souza CC, Almeida MJC, MeloSereno BF, Neto JFB (2003) Genetic variability in sweet corn using molecular markers. Maydica 48:177–181

Anhalt UCM, Heslop-Harrison JS, Piepho HP, Byrne S, Barth S (2009) Quantitative trait loci mapping for biomass yield traits in a Lolium inbred line derived F2 population. Euphytica 170:99–107. https://doi.org/10.1007/s10681-009-9957-9

Araus JI, Slater GA, Reynolds MP, Toyo C (2002) Plant breeding and drought in C3 cereals, what should we breed for? Ann Bot 89:925–940

Ashraf M (2010) Inducing drought tolerance in plants; recent advances. Biotechnol Adv 28:169–183

Asins MJ (2002) Present and future of Quantitative traits loci analysis in plant breeding. Plant Breed 121:281–291

Athar HR, Ashraf M (2005) Photosynthesis under drought stress. In: Pessarakli M (ed) Hand book photosynthesis, 2nd edn. C.R.C. Press, New York, pp 795–810

Banziger M, Araus JL (2009) Recent advances in breeding maize for drought and salinity stress tolerance. In: Jenks MA, Hasegawa PM, Mohan S (eds) Advances in molecular breeding toward drought and salt tolerant crops. Springer, Dordrecht, pp 587–601

Batool A, Noorka IR, Afzal M, Syed AH (2013) Estimation of heterosis, heterobeltiosis and potence ratio over environments among pre and post green revolution spring wheat. J Basic Appl Sci 9:36–43

Bauer I, Drinic SM, Filipovic M, Konstanti K (2005) Genetic characterization of early maturing maize hybrids (*Zea mays* L.) obtained by protein and RAPD markers. Genetika 37(3):235–243

Blum A (2009) Effective use of water (EUW) and not water-use efficiency (WUE) is the target of crop yield improvement under drought stress. Field Crop Res 112:119–123

Borlaug NE, Dowswell CR (2005) Feeding a world of ten billion people: a 21st century challenge. In: Proceedings of In the wake of double helix: from the green revolution to the gene revolution 27–31st May 2003 Bologna

Braun HJ, Tadessa W, Morgounov AI, Akin B, Keser M, Kaya Y, Sharma RC, Rajaram S, Singh M, Baum M, van Ginkel M (2013) Breeding progress for yield in winter wheat genotypes targeted to irrigated environments of the CWANA region. Euphytica 194:177–185

Brown LR (1999) Feeding nine billion. In: Starke L (ed) State of the world 1999. W.W. Norton and Co., New York

Bruinsma J (2009) The resource outlook to 2050: by how much do land, water and crop yields need to increase by 2050? FAO expert meeting on how to feed the world in 2050. Food and Agriculture Organization, Rome

Carvalho VP, Ruas CF, Ferreira JM, Moreira RMP, Ruas PM (2004) Genetic diversity among maize landraces assessed by RAPD markers. Genet Mol Biol 27:228–236

Chaves MM, Flexas J, Pinheiro C (2009) Photosynthesis under drought and salt stress: regulation mechanisms from whole plant to cell. Ann Bot 103:551–560

Chowdhry MA, Rasool I, Khaliq I, Mehmood T, Gilliani MM (1999) Genetics of some metric traits in spring wheat under normal and drought environments. RACHIS 18(1):34–39

Dahiya S, Punia SS, Singh J, Kakraliya SK, Singh B, Jat HS, Malik R (2017) Yield and yield attributes as affected by different sowing dates and different maturity classes cultivar on direct seeded rice. Chem Sci Rev Lett 6(21):149–152

Dekkers JCM, Hospital F (2002) The use of molecular genetics in the improvement of agricultural populations. Nat Rev 3:22–32

Dzanku FM, Jirström M, Marstorp H (2015) Yield gap-based poverty gaps in rural sub-Saharan Africa. World Dev 67:336–362

Fereres E, Soriano MA (2007) Deficit irrigation for reducing agricultural water use. J Exp Bot 58(2):147–159

Ferrio JP, Mateo MA, Bort J, Voltas J, Abdahla O, Araus JL (2007) Relationships of grain sigma13C and sigma 18O with wheat physiology and yield under water limited conditions. Ann Appl Biol 150:207–215

Habash DZ, Kekel Z, Nachit M (2009) Gonomic approaches for designing durum wheat ready for climate change with a focus on drought. J Exp Bot 60:2805–2815

Iqbal MA, Bonder G, Heng LK, Eeitzinger J, Hassan A (2010) Assessing yield optimization and water reduction potential for the summer growth and spring sown maize in Pakistan. Agric Water Manag 97(5):731–737

Jat ML, Singh B, Stirling C, Jat HS, Tetarwal JP, Jat RK, Singh R, Lopez-Ridaura S, Shirsath PB (2017) Soil processes and wheat cropping under emerging climate change scenarios in South Asia. Adv Agron 148:111–171

Jat RK, Singh P, Jat ML, Dia M, Sidhu HS, Jat SL, Bijarniya D, Jat HS, Parihar CM, Kumar U, Lopez-Ridaura S (2018) Heat stress and yield stability of wheat genotypes under different sowing dates across agro-ecosystems in India. Field Crop Res 218:33–50

Kashiwagi J, Krishnamurthy L, Upadhyaya DH, Krishna H, Chandea S, Vadez V, Serraj R (2006) Genetic variability of drought avoidance root traits in the mini core germplasm collection of chickpea (*Cicer arietinum* L). Euphytica 146:213–222

Koning N, van Ittersum MK (2009) Will the world have enough to eat? CurrOpin Environ Sustain 1:77–82

Liniger H, Mekdaschi S, Rima, Moll P, Zander U (2017) Making sense of research for sustainable land management. Centre for Development and Environment (CDE), University of Bern and Helmholtz-Centre for Environmental Research GmbH – UFZ, Bern/Leipzig

McKenzie FC, Williams J (2015) Sustainable food production: constraints, challenges and choices by 2050. Food Security 7:221–233

Mohamed MAH, Harris PJC, Henderson J (2000) In vitro selection and characterization of a drought tolerant clone of *Tagetes minuta*. Plant Sci 159:213–222

Noorka IR, Heslop-Harrison JS(P) (2014) Water and crops: molecular biologists, physiologists, and plant breeders approach in the context of evergreen revolution. In: Hand book of plant and crop physiology. CRC Press/Taylor and Francis, Boca Raton, pp 967–978

Noorka IR, Heslop-Harrison JS(P) (2015) Agriculture and climate change in Southeast Asia and the Middle East: breeding, climate change adaptation, agronomy and water security submitted. In: Handbook of Climate Change Adaptation [110652]. Springer, Berlin/Heidelberg. https://doi.org/10.1007/978-3-642-40455-9_74-1

Noorka IR, Taufiqullah, Heslop-Harrison JS(P), Schwarzacher T (2017) The agriculture-nutrition-health nexus at the cost of water availability in maize diverse genotypes to ensure food security. Int J Water Reso Arid Environ 6(2):242–251

Ortiz R, Taba S, Tovar VHC, Mezzalama M, Xu Y, Yang J, Crouch JH (2009) Conserving and enhancing maize genetic resources as public goods. A perspective from CIMMYT. Crop Sci 50:13–28

Rasheed A, Ogbonnaya FC, Lagudah ES, Appels R, Zhonghu H (2018) The goat grass genome's role in wheat improvement. Nat Plants 4:56–58

Ray DK, Mueller ND, West PC, Foley JA (2013) Yield trends are insufficient to double global crop production by 2050. PLoS One 8:e66428. https://doi.org/10.1371/journal.pone.0066428

Samal SK, Rao KK, Poonia SP, Kumar R, Mishra JS, Prakash V, Mondal S, Dwivedi SK, Bhatt BP, Naik SK, Choubey AK, Kumar V, Malik RK, McDonald A (2017) Evaluation of long-term conservation agriculture and crop intensification in rice-wheat rotation of Indo-Gangetic Plains of South Asia: carbon dynamics and productivity. Eur J Agron 90:198–208

Sendhil R, Balaji SJ, Ramasundaram P, Anuj Kumar, Satyavir Singh, R Chatrath GP Singh (2018) Doubling farmers income by 2022: trends, challenges, pathway and strategies. Research Bulletin No: 40, ICAR-Indian Institute of Wheat and Barley Research, Karnal, pp 1–54

Shirsath PB, Aggarwal PK, Thornton PK, Dunnett A (2017) Prioritizing climate-smart agricultural land use options at a regional scale. Agric Syst 151:174–183

USDA (2012) USDA agricultural projections to 2021. U.S. Department of Agriculture, Longterm Projections Report OCE-2012-1, pp 102

Weber VS, Atlin GN, Crossa J, Hickey JM, Grudloyma P, Jannick J-L, Sorrels M, Ramen B, Cairns JE, Tarekegne A, Semagn K, Beyene Y, Grudloyma P, Technow F, Riedelsheimer C, Melchinger AE (2012) Effectiveness of genomic prediction of maize hybrid performance in different breeding populations and environments. Gene Genet Genomic Nat Genet 44(7): 803–807

World Health Organization (2017) Inheriting a sustainable world? Atlas on children's health and the environment. Geneva 17. https://www.unicef.org/publications/files/UNICEF_Clear_the_Air_for_Children_30_Oct_2016.pdf. 16

Yang J, Zhang J, Wang Z, Zhu Q, Liu L (2001) Water deficit induced senescence and its relationship to the remobilization of pre-stored carbon in wheat during grain filling. Agron J 93:196–206

Zhou M (2010) Improvement of plant water logging tolerance. Water logging signaling and tolerance in plants. Springer, Berlin

Climate-Resilient Agricultural Practices in Different Agro-ecological Zones of Bangladesh

115

Muhammad Abdur Rahaman, Mohammad Mahbubur Rahman, and Md. Shahadat Hossain

Contents

Introduction 2338
Agroecosystem Versus Agro-ecological Zones 2339
Study Area 2339
Methodology 2341
Result and Discussion 2343
  Impact of Climate Change on Agriculture 2343
Resilient Agricultural Options 2348
  Shyamnagar Upazila (Ganges Tidal Floodplain AEZ) 2348
  Tanore Upazila (High Ganges River Floodplain, Level Barind Tract, and High Barind Tract AEZ) 2353
  Rangamati Sadar Upazila (Northern and Eastern Hills AEZ) 2354
  Chilmari Upazila (Active Tista Floodplain, Tista Meander Floodplain, and Active Brahmaputra-Jamuna Floodplain AEZ) 2355
  Khaliajuri (Old Brahmaputra Floodplain and Sylhet Basin AEZ) 2357
Reshaping Policy to Promote Climate-Resilient Agriculture in Bangladesh 2359
Conclusion 2360
Cross-References 2361
References 2361

M. A. Rahaman
Climate Change Adaptation, Mitigation, Experiment and Training (CAMET) Park, Noakhali, Bangladesh
e-mail: rana.bries@gmail.com

M. M. Rahman (✉)
Network on Climate Change, Bangladesh (NCC,B) Trust, Dhaka, Bangladesh
e-mail: mmrahman.ju39@gmail.com

M. S. Hossain
Institute of Water Modeling, Dhaka, Bangladesh
e-mail: shahadat.research@gmail.com

W. Leal Filho (ed.), *Handbook of Climate Change Resilience*,
https://doi.org/10.1007/978-3-319-93336-8_42

Abstract

Bangladesh is one of the most vulnerable countries to climate change due to its unique geographical position, the dominance of floodplains, low elevation, high population density, high levels of poverty, and overwhelming dependence on nature, its resources, and services. Increasing temperatures, irregular rainfall, drought, and cyclones are adversely affecting agricultural production, in turn creating a high risk to the food security of Bangladesh's large population. Large-scale climate-resilient practices (structural and nonstructural) are being implemented in different agro-ecological zones (AEZs) of Bangladesh, which have the potential to reduce the vulnerability and risks associated with climate change and contribute to sustainable agricultural development. This chapter explores the spontaneous and planned resilient practices and their possible contribution to food security in different AEZs of Bangladesh. We systematically classify and characterize agrarian adaptation options to climate change. To this end, first, we assess the impacts of climate change on the agriculture sector in Bangladesh. In addition, we analyze the determinants of farmer's choices between alternative adaptation measures available in different AEZs. Finally, we identify the gaps in the implementation of those practices and the way forward with policy recommendations.

Keywords

Climate change · Agriculture · Agro-ecological zones · Climate resilient · Practices · Bangladesh

Introduction

Bangladesh is one of the most vulnerable countries to the impacts of climate change in the world. According to the Global Climate Risk Index (GCRI), in 2017, Bangladesh was the sixth most climate-vulnerable country in the world (Kreft et al. 2017), though during 2010 it was the most climate-vulnerable country (Harmeling 2009). The Global Climate Risk Index (GCRI) 2010, covering the period 1990–2008, estimates that, on an average, 8241 people died each year in Bangladesh, while the cost of damages was around US $ 1.2 billion per year, and loss of gross domestic product (GDP) was 1.81% during the period (Harmeling 2009). Agriculture is the most vulnerable sector of Bangladesh due to climate-induced disasters which restrict the national food security. Different agro-ecological zones (AEZ) of Bangladesh suffer from different climate-induced disasters that affect agricultural production. Farmers introduced climate-resilient practices, through a large number of agricultural innovations by their own initiative and with the financial and technical support of various government and nongovernment organizations. It appears important at this stage to understand what climate-resilient agricultural practices are anticipated, based on current climate-induced disasters and knowledge of farmers, in each of the AEZs of the country.

Agroecosystem Versus Agro-ecological Zones

Though agroecosystems and agro-ecological zones are interrelated, agroecosystems are disaggregated into smaller crop suitability space based on some geographical and climatic characteristics such as land type, soil structure, soil texture, rainfall pattern, annual quantity of rainfall, incoming solar radiation, hydrometeorological disaster type, etc. Agroecosystems are an assembly of mutually interacting organisms and their environment, in which materials related to crop production are interchanged in a largely cyclical manner. An ecosystem has physical, chemical, and biological characteristics, along with energy sources and pathways of energy and materials interchange (Banglapedia 2014a). Bangladesh contains four main types of ecosystems: coastal and marine ecosystems, inland freshwater ecosystems, terrestrial forest ecosystems, and man-made ecosystems (Daniels 2003). The AEZs of Bangladesh are determined based on physiography, hydrology, cropping pattern, season, soil types, and tidal activity. There are 30 AEZs in Bangladesh. These 30 AEZs are subdivided into 88 agro-ecological subregions, which cover the whole country (Table 1). Again, these are divided into 535 agro-ecological units (Banglapedia 2014b).

Study Area

The study was conducted in five sub-districts of four agroecosystem as well as 10 agro-ecological zones based on different agroclimatic vulnerability. Shyamnagar Upazila in the Satkhira district, Tanore Upazila in the Rajshahi district, Khaliajuri Upazila in the Netrokona district, Rangamati Sadar Upazila in the Rangamati Hill district, and Chilmari Upazila in the Kurigram district are selected as the study area (Fig. 1), which covers the Ganges Tidal Floodplain for the Shyamnagar AEZ; the High Ganges River Floodplain, the Level Barind Tract, and High Barind Tract for the Tanore AEZ; the Northern and Eastern Hills for the Rangamati Sadar AEZ; the Active Tista Floodplain, Tista Meander Floodplain, and Active Brahmaputra-Jamuna Floodplain for the Chilmari AEZ; and the Old Brahmaputra and Sylhet Basin for the Khaliajuri AEZ (Table 2).

Shyamnagar Upazila (An administrative subunit of districts locally named Upazila or Thana) in the Satkhira district is located between 21°36′ and 22°24′ N and between 89°00′ and 89°19′ E and is a coastal settlement adjacent to the Bay of Bengal and the largest mangrove forest of the Sundarbans (Banglapedia 2015a). Tanore Upazila in the Rajshahi district is located between 24°29′ and 24°43′ N and between 88°24′ and 88°38′ E and is an area highly vulnerable to drought and displays Barind zone characteristic (Banglapedia 2014c).

Table 1 Agro-ecological zones of Bangladesh. (Source: BARC 1988)

Agro-ecological zones (AEZs)	Districts covered
1. Old Himalayan Piedmont Plain	Dinajpur, Panchagarh, Thakurgaon
2. Active Tista Floodplain	Gaibandha, Kurigram, Lalmonirhat, Nilphamari, Rangpur
3. Tista Meander Floodplain	Bogra, Dinajpur, Gaibandha, Joypurhat, Kurigram, Lalmonirhat, Naogaon, Nilphamari, Panchagarh, Rangpur, Rajshahi
4. Karatoya-Bangali Floodplain	Bogra, Pabna, Sirajganj
5. Lower Atrai Basin	Bogra, Naogaon, Natore, Rajshahi, Sirajganj
6. Lower Punarbhaba Floodplain	Naogaon, Nawabganj
7. Active Brahmaputra-Jamuna Floodplain	Bogra, Chandpur, Gaibandha, Jamalpur, Kurigram, Manikganj, Pabna, Sirajganj, Tangail
8. Young Brahmaputra and Jamuna Floodplain	Dhaka, Gazipur, Jamalpur, Kishoreganj, Manikganj, Munshiganj, Mymensingh, Narayanganj, Narsingdi, Netrokona, Sherpur, Tangail
9. Old Brahmaputra Floodplain	Gazipur, Jamalpur, Kishoreganj, Mymensingh, Narayanganj, Narsingdi, Netrokona, Sherpur, Tangail
10. Active Ganges Floodplain	Chandpur, Dhaka, Faridpur, Rajbari, Kushtia, Madaripur, Manikganj, Munshiganj, Natore, Chapai Nawabganj, Pabna, Rajshahi, Shariatpur
11. High Ganges River Floodplain	Bagerhat, Chuadanga, Jessore, Jhenaidah, Khulna, Kushtia, Magura, Meherpur, Naogaon, Narail, Natore, Chapai Nawabganj, Pabna, Rajshahi, Satkhira
12. Low Ganges River Floodplain	Bagerhat, Barisal, Dhaka, Faridpur, Rajbari, Gopalganj, Jessore, Khulna, Kushtia, Madaripur, Magura, Manikganj, Munshiganj, Narail, Natore, Pabna, Pirojpur, Shariatpur, Sirajganj
13. Ganges Tidal Floodplain.	Bagerhat, Barguna, Barisal, Bhola, Jhalakati, Khulna, Madaripur, Patuakhali, Pirojpur, Satkhira, Shariatpur
14. Gopalganj-Khulna Beels	Bagerhat, Barisal, Gopalganj, Jessore, Khulna, Madaripur, Narail, Pirojpur
15. Arial Beel	Dhaka, Munshiganj
16. Middle Meghna River Floodplain	Brahmanbaria, Chandpur, Comilla, Kishoreganj, Munshiganj, Narayanganj, Narsingdi
17. Lower Meghna River Floodplain	Chandpur, Lakshmipur, Noakhali
18. Young Meghna Estuarine Floodplain	Barisal, Bhola, Chandpur, Chittagong, Feni, Lakshmipur, Noakhali, Patuakhali
19. Old Meghna Estuarine Floodplain	Barisal, Brahmanbaria, Chandpur, Comilla, Dhaka, Feni, Gopalganj, Habiganj, Kishoreganj, Lakshmipur, Madaripur, Munshiganj, Narayanganj, Noakhali, Shariatpur
20. Eastern Surma-Kushiyara Floodplain	Habiganj, Moulvibazar, Sunamganj, Sylhet
21. Sylhet Basin	Brahmanbaria, Habiganj, Kishoreganj, Netrokona, Sunamganj

(*continued*)

Table 1 (continued)

Agro-ecological zones (AEZs)	Districts covered
22. Northern and Eastern Piedmont Plain	Brahmanbaria, Comilla, Habiganj, Jamalpur, Moulvibazar, Mymensingh, Netrokona, Sherpur, Sylhet
23. Chittagong Coastal Plain	Chittagong, Cox's Bazar, Feni
24. St Martin's Coral Island	Cox's Bazar
25. Level Barind Tract	Bogra, Dinajpur, Gaibandha, Joypurhat, Naogaon, Natore, Chapai Nawabganj, Rajshahi, Sirajganj
26. High Barind Tract	Naogaon, Chapai Nawabganj, Rajshahi
27. North-eastern Barind Tract	Bogra, Dinajpur, Gaibandha, Joypurhat, Rangpur
28. Madhupur Tract	Dhaka, Gazipur, Jamalpur, Kishoreganj, Mymensingh, Narayanganj, Narsingdi, Tangail
29. Northern and Eastern Hills	Bandarban, Brahmanbaria, Chittagong, Comilla, Cox's Bazar, Feni, Habiganj, Jamalpur, Khagrachhari, Moulvibazar, Mymensingh, Netrokona, Rangamati, Sherpur, Sunamganj, Sylhet
30. Akhaura Terrace	Brahmanbaria, Habiganj

Rangamati Sadar Upazila in the Rangamati Hill district is located between 22°30′ and 22°49′ N and between 92°04′ and 92°22′ E longitudes and is occupied by hills (Banglapedia 2015b). Chilmari Upazila in the Kurigram district is located between 25°26′ and 25°40′ N and between 89°38′ and 89°48′ E and is highly vulnerable to upstream floods and riverbank erosion. The area is adjacent to the entry point of the trans-boundary river Padma (Banglapedia 2014d). Finally, Khaliajuri Upazila in the Netrokona district is located between 24°36′ and 24°50′ N and between 90°00′ and 90°16′ E and is situated in the haor basin of Bangladesh (Banglapedia 2014e).

Methodology

The chapter aimed to increase the AEZ-based climate-resilient knowledge on agricultural interventions through field data collection and accumulated data from other secondary sources. Available scientific literature related to the impact of climate-induced sudden and slow onset disaster in agriculture has been reviewed and superimposed on a selected AEZ database on a geographical information system (GIS)-based platform to identify AEZ-based climate hotspots. The chapter incorporated both qualitative and quantitative information from secondary sources reviewing available literature related to climate-resilient agricultural practices in different agro-ecological zones of Bangladesh and primary information collection through household survey (HHS), focus group discussions (FGD), key informants interviews (KII), case studies, making use of the AEZ database and exploring climate-resilient agro-based innovations through ground truthing using case studies and field observation, and consultation with different stakeholders including the Department of Agriculture, local government institutions, farmers, livestock

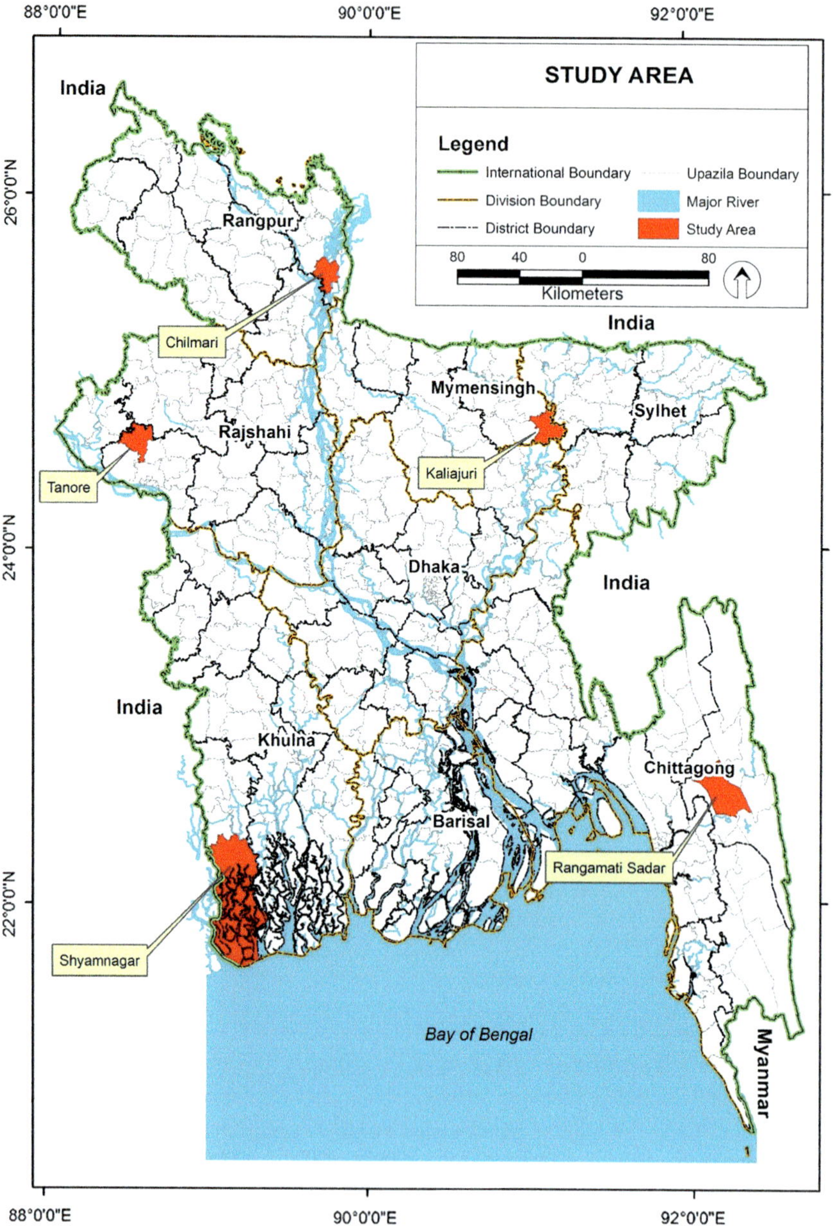

Fig. 1 Study area

department, the fisheries department, and NGOs involved in climate-resilient agricultural promotional activities in flood-, drought-, and salinity-prone areas and in hill ecosystems, as illustrated in Tables 2 and 3 and Fig. 1.

Table 2 Study area

Districts	Upazilas	Agro-ecological zones (AEZs)
Satkhira	Shyamnagar	Ganges Tidal Floodplain (13)
Rajshahi	Tanore	High Ganges River Floodplain (11), Level Barind Tract (25), and High Barind Tract (26)
Rangamati	Rangamati Sadar	Northern and Eastern Hills (29)
Kurigram	Chilmari	Active Tista Floodplain (2), Tista Meander Floodplain (3), and Active Brahmaputra-Jamuna Floodplain (7)
Netrokona	Khaliajuri	Old Brahmaputra Floodplain (9) and Sylhet Basin (21)

Table 3 Primary data collection tool and sample size for the target people

Sl.	Data collection tool	Target people	Sample size
1	HHS	Climate-vulnerable people	100 in each study area
2	FGD	Climate-vulnerable people	
3	KII	Department of Agriculture, local government institutions, Department of Livestock, Department of Fisheries, NGOs	8 in each sub-district
4	Case studies	Best practice documentation on resilient interventions related to agriculture, livestock, fisheries	3 in each sub-district

Result and Discussion

Impact of Climate Change on Agriculture

Due to climate change, precipitation patterns changed in Bangladesh, causing more intense rainfall on rainy days and more dry days in the summer season (in a year). As a result, agricultural production, the critical sector in Bangladesh, is suffering. The impacts of climate change have significantly reduced agricultural production in flood-, drought-, and salinity-prone areas. Climate-induced hazards related to agricultural damage in our study area are illustrated in Table 4.

Shyamnagar (Ganges Tidal Floodplain AEZ)

The coastal Shyamnagar Upazila AEZ, which is in the Satkhira district, is adjacent to the Bay of Bengal and the mangrove forest of the Sundarbans. Being in the nearby Bay of Bengal, the Upazila, like the other Upazilas of the Ganges Tidal Floodplain AEZ, is highly vulnerable to salinity intrusion into agricultural land (Fig. 2), cyclone, storm surge and tidal surge, the impact of

Table 4 Existing climatic hazards in the study areas. (Source: Authors)

District	Upazila	Climatic Hazards
Satkhira	Shyamnagar	Salinity intrusion, tidal surge, cyclone and storm surge
Rajshahi	Tanore	Drought, storm/hailstorm
Rangamati	Rangamati Sadar	Landslide, flash flood
Kurigram	Chilmari	Flood, riverbank erosion, storm/hailstorm
Netrokona	Khaliajuri	Flash flood, storm/hailstorm

sea level rise, drainage congestion and flooding, and storm surges and tidal inundation. Every year, about 500 ha of land are converted to saline land. In 2000, the level of salinity was about 23.93 dS/m, and in 2009 about 28.64 dS/m in the Shyamnagar Upazila (Islam et al. 2015).

Rangamati Sadar

The Rangamati Sadar Upazila is an ecologically and economically constrained area due to the prevalence of hilly ecosystem. The Upazila is characterized by climatic hazards including flash floods (Fig. 3), landslides, and medium-level droughts. Climatic hazards result in increased ecosystem fragmentation, depletion of wildlife, the decline in forest resources, and changes in the livelihood patterns of indigenous people in Rangamati (Huda 2013).

Chilmari

Floods, flash floods, and riverbank erosion are the most common climate-induced disasters for the Chilmari Upazila in the Kurigram district (Fig. 4). Climate change is contributing to a reduction of cultivable land through erosion and agricultural land loss due to the seasonal floods in the area. In the last 10 years, sandstorms and the sand wave (the storm which washes away the sand with strong wave) have been observed in the Chilmari Upazila, as related by respondents during the community consultation. Climate-induced disasters restrict the accessibility of common property resources sectors like forestry and fisheries, because of their depletion, due to cold waves, dense fog, groundwater depletion, heavy rainfall, and untimely and irregular rainfall in these areas (Rahaman 2018).

Tanore

The Tanore Upazila in the Rajshahi district, which is located in the Barind Tract region, is a highly drought-prone area (Fig. 5). The area also suffers from cold waves during winter. Farmers face frequent extreme weather events in the area, like in other areas of the High Ganges River Floodplain, Level Barind Tract, and High Barind Tract. The impacts were noticed mostly in the form of erratic rainfall and distribution patterns, changes in temperature, droughts, depletion of

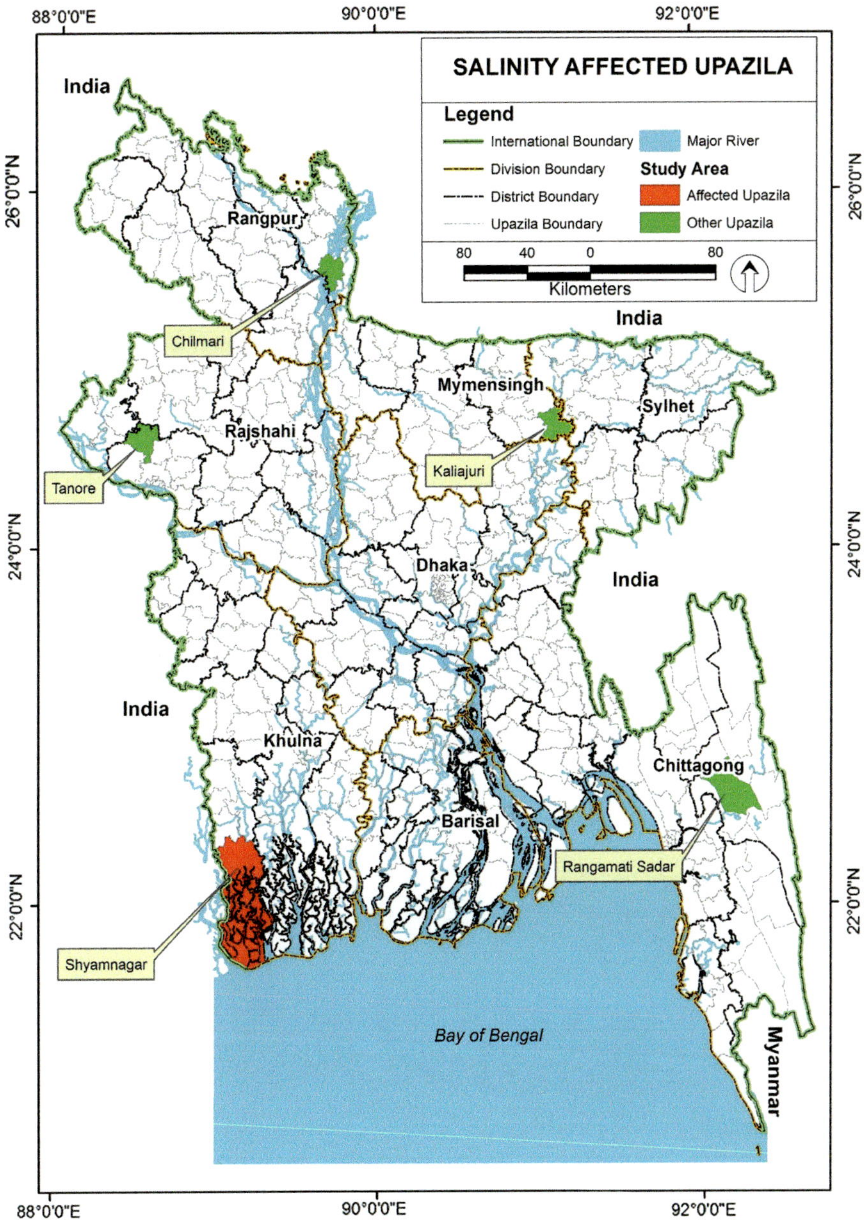

Fig. 2 Salinity affected Upazila in the overall study area

groundwater reserves, more pest and disease problems, and water scarcity. Some of these are of course only indirectly linked to climate change and variability (Nagothu et al. 2014).

Khaliajuri

Like other areas of the Old Brahmaputra and Sylhet Basin AEZs in the haor basin, flash floods (Fig. 6) are a common climate-induced disaster in Khaliajuri. Along with flash floods, the high-intensity wave action which is locally known as *afal* is an

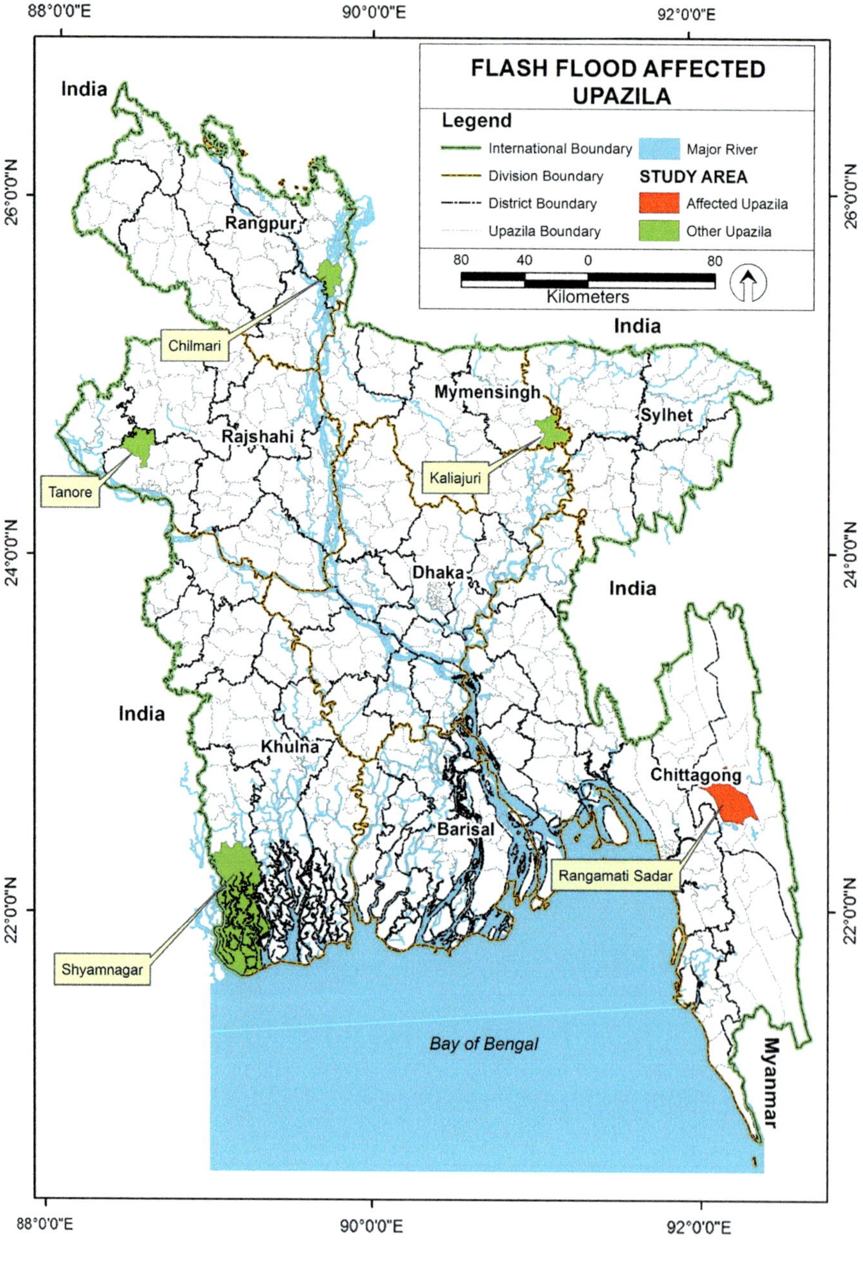

Fig. 3 Flash flood-affected Upazila in the overall study area

extreme climate-induced event in the area. During summer, water crises also hamper irrigation in the area, which restricts crop production. Each year, flash floods damage huge crop surfaces in the area.

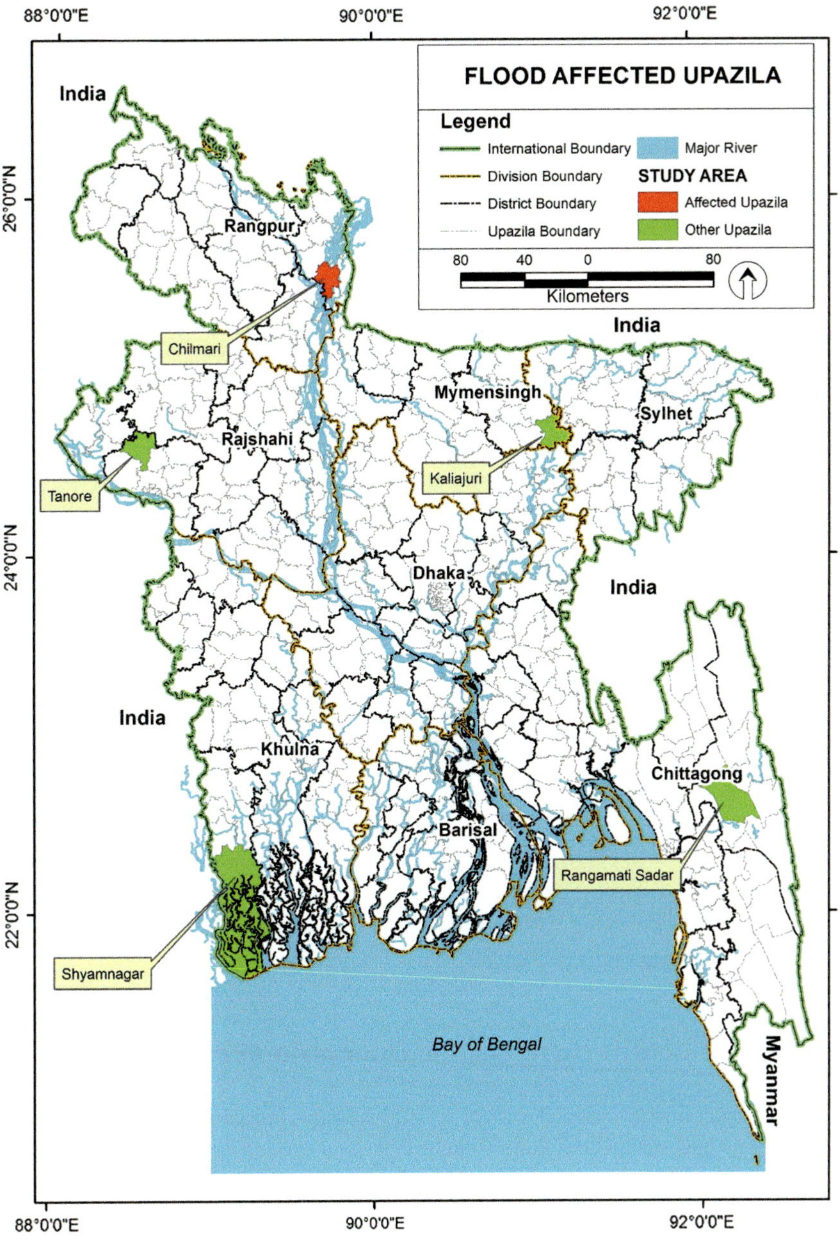

Fig. 4 Flood affected Upazila in the overall study area

Resilient Agricultural Options

Agriculture is the principal livelihood option of Bangladesh. But it is the most vulnerable to the increasing frequency and intensity of extreme events such as floods, cyclones, storm surges, hailstorms, erratic and heavy rainfall, and salinity intrusion (Mainuddin et al. 2011). Climate change and climate variability are affecting the land use patterns, crop systems, productivity, and optimum agriculture output (GOB 2009). Many forms of adaptation practices such as hard and soft adaptation are being implemented throughout the world to reduce the loss and damage from extreme climate events and climate variability (ADB 2011). Many agricultural adaptation options are being practiced in Bangladesh to adapt to climate-induced agricultural disasters like salinity, flood, waterlogging, drought, etc. To ensure climate-resilient agricultural development in Bangladesh, many structural and nonstructural soft and hard interventions are being practiced all over the country in different agro-ecological zones. Some of these innovations are devised by local communities through their indigenous knowledge, and some are planned interventions promoted by different government and nongovernment organizations. Nelson et al. (2007) and Alam et al. (2013) mention that floating bed crops/vegetables in the south-central and southern areas, plant bed raising, and dyke cropping at the shrimp *gher* (Water bodies which are artificially generated through raising dyke around the lowland) is an old but effective practice used nowadays.

Adaptation procedures can be autonomous, planned, or natural (Chambwera et al. 2014). Autonomous activities are those which are undertaken by private actors, prompted by climate change-induced market or welfare changes. Planned actions are those which are carried out by both private and public actors. These actions mainly include deliberate policy decisions based on the awareness that conditions have changed or are about to change and that action is required to return to, maintain, or achieve the desired state (Carter et al. 1994). Natural actions appear within the ecosystem as a reaction to climate change (Chambwera et al. 2014). The agriculture sector adaptation in Bangladesh is triggered by autonomous, planned, and natural adaptation in different agro-ecological zones by the government and nongovernment sector. The following are the key resilient agricultural practices in the Ganges Tidal Floodplain, High Ganges River Floodplain, Level Barind Tract, High Barind Tract, Northern and Eastern Hills, Active Tista Floodplain, Tista Meander Floodplain, and Active Brahmaputra-Jamuna Floodplain AEZs of Bangladesh (Table 5).

Shyamnagar Upazila (Ganges Tidal Floodplain AEZ)

To cope with climate-induced disasters like salinity intrusion, tidal surges, and waterlogging and ensure resilient livelihoods, multidimensional resilient options are being implemented in the Shyamnagar Upazila. Our research indicates that, at present, 31.3% (Fig. 7) of the farmers of the Shyamnagar Upazila have adopted alternative land use practices like shrimp farming instead of crop production

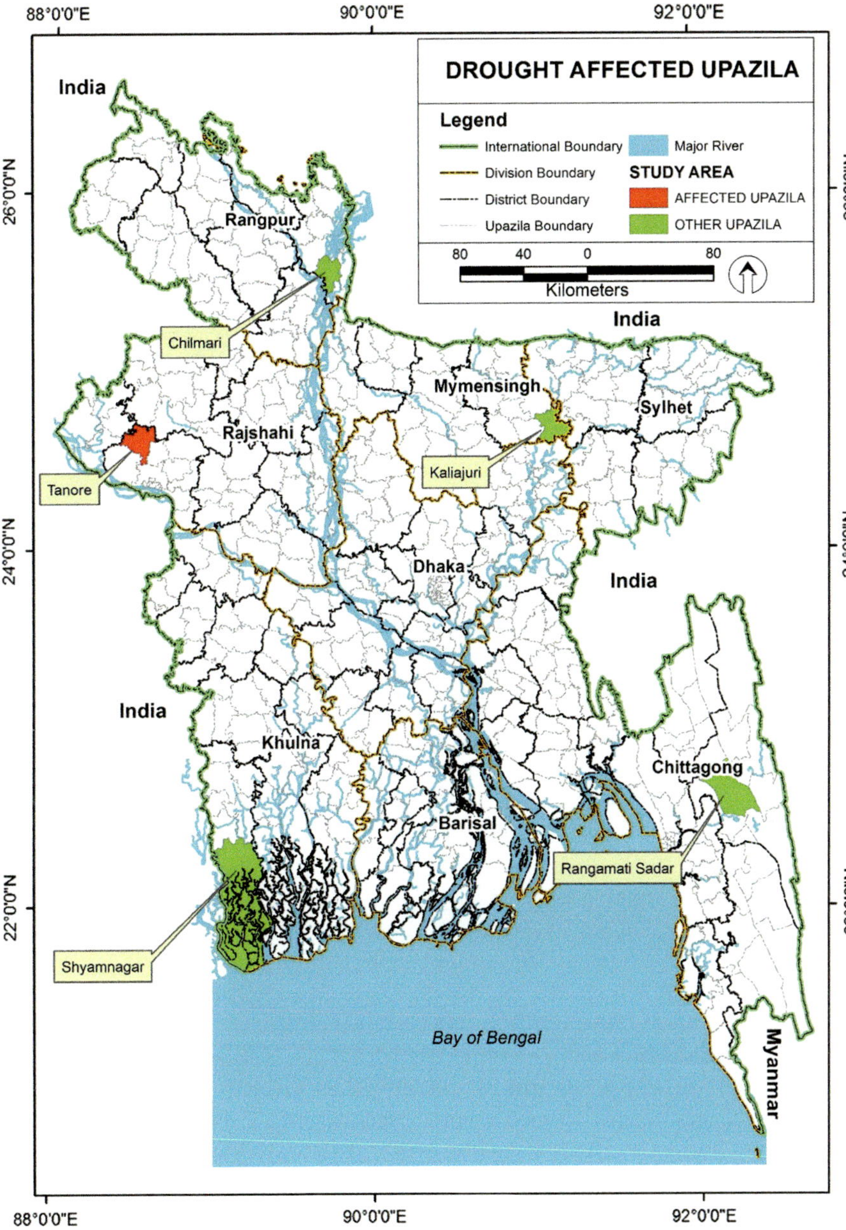

Fig. 5 Drought affected Upazila in the overall study area

(Field Survey 2017). Though the use of land for crop farming is decreasing gradually, there are some alternative options for agriculture emerging, such as embankment cropping, plantation of mangrove trees, cultivation of saline-tolerant grass,

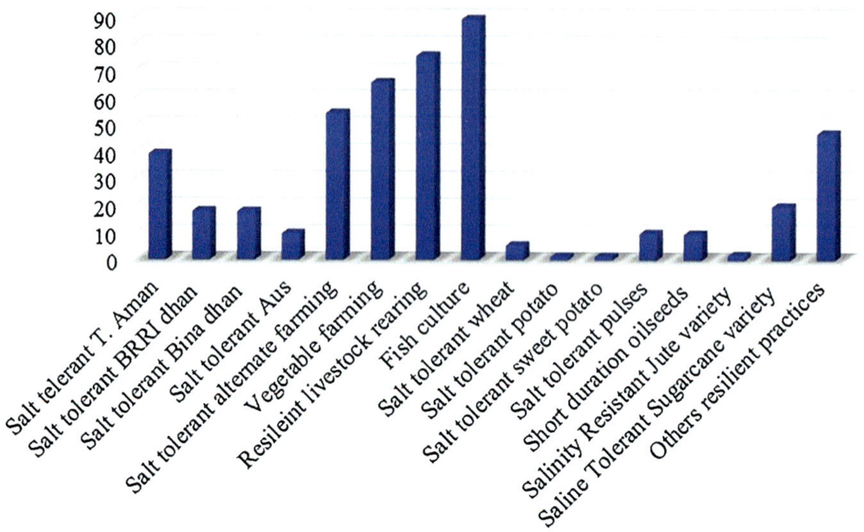

Fig. 6 Climate-resilient agro-based practices in Shyamnagar Upazila. (Source: Field Survey 2017)

using flood- and saline-tolerant rice, etc. The major climate-resilient agro-based practices of the Ganges Tidal Floodplain AEZ are illustrated in Table 6.

Farmers of the Shyamnagar Upazila under the Ganges Tidal Floodplain AEZ have introduced a number of climate-resilient practices in different agro-based farming, like the introduction of salinity-tolerant rice varieties such as T. Aman: BR-22 and BR-23; Bina shail flash flood; BRRI dhan 33, 56, 57, and 62; Bina dhan 7 and 16; BRRI dhan 47, 61, 67; Bina dhan-8 and 10; T. Aman: BRRI dhan 40, 41, 53, 54; and Aus: BRRI dhan 65; T. Aman: BR-22 and BR-23; cage fishing, mele (reed) cultivation, floating dhap (The practice of growing vegetable saplings on water bed, locally known as "dhap") cultivation, shifting planting time, short-duration rice varieties, integrated farming, crab patenting, semi-scavenger housing for goat, duck, and hen rearing, net fishing, dyke farming, salt-tolerant wheat like Bijoy, BARI Gom-25, BAU-1059 line; salt-tolerant potato: BARI Alo-22 CIP Clone −88.163; salt-tolerant sweet potato: BARI Mishti Alo-8,9, salt- and heat-tolerant pulses: BARI Mug-2,3,4,5,6, BM-01, BM-08 BARI Falon-1, BARI Sola-9; short-duration and salt-tolerant oilseeds: BARI Sharisha-14,15; BARI Chinabadam-9, BINA China badam-1, BINA China badam-2, BARI Soyabean-6 BARI Til-2,3,4; salt-resistant jute varieties by Bangladesh Jute Research Institute (BJRI): (i) HC-2, (ii) HC 95, (iii) CVL 1; salt-tolerant sugarcane varieties by Bangladesh Sugarcane Research Institute (BSRI): (i) ISWARDI-40 (Table 5).

Figure 6 illustrates that 39.2% of farmers use salt-tolerant T. Aman varieties BR-22, BR 23, BR 22, and Bina shail; 18.1% of farmers use salt-tolerant BRRI dhan varieties BRRI 33, BRRI 56, BRRI 57, BRRI 62, BRRI 40, BRRI 41, BRRI 53, and BRRI 54; 17.9% of farmers use salt-tolerant BINA dhan varieties BINA 7, BINA 16, BINA 8, and BINA 10; and only 9.8% of farmers use the salt-tolerant Aus variety

Table 5 Climate-resilient agricultural practices in different AEZs. (Source: Authors)

Agro-ecological zones (AEZs)	Study area	Climate stress	Adaptation options
Ganges Tidal Floodplain (13)	Shyamnagar	Salinity intrusion, tidal surge, cyclone, and storm surge	Shifting planting time, short-duration rice varieties, increasing the height of the mud wall, using flood-tolerant rice varieties, floating bed agriculture, improved flood warning system and communication, T. Aman: BR-22 and BR-23 (rice varieties developed by the Bangladesh Rice Research Institute (BRRI)); Bina shail flash flood, BRRI dhan 33, 56, 57, and 62 (rice varieties developed by the Bangladesh Rice Research Institute (BRRI)); Bina dhan 7 and 16 (rice varieties developed by Bangladesh Institute of Nuclear Agriculture (BINA))
High Ganges River Floodplain (11) Level Barind Tract (25) High Barind Tract (26)	Tanore	Drought, storm/ hailstorm	Short-duration rice varieties, surface water-based irrigation, shifting planting time, drought-tolerant rice varieties, rainwater harvesting, (wheat) BRRI dhan 56, 57, and 66
Northern and Eastern Hills (29)	Rangamati Sadar	Landslide, flash flood	Replantation, early crop harvesting, short-duration varieties, shifting planting time
Active Tista Floodplain (2) Tista Meander Floodplain (3) Active Brahmaputra-Jamuna Floodplain (7)	Chilmari	Flood, riverbank erosion, storm/ hailstorm	Rice-prawn/shrimp farming, cultivating saline-tolerant varieties, rainwater harvesting, desalinization, Boro, BRRI dhan 47, 61, 67 and Bina dhan-8 and 10; T. Aman, BRRI dhan 40, 41, 53, 54; and Aus, BRRI dhan 65, (wheat) Bijoy, BARI Gom-25, BAU-1059 line
Old Brahmaputra Floodplain (9) Sylhet Basin (21)	Khaliajuri	Flash flood, storm/ hailstorm	Excavating the canals of surrounding fields, digging drains, excess water is withdrawn by pump from the field, BRRI dhan 51 and 52, Bina dhan-11 and 12

BRRI 65. On the other hand, 54.3% of farmers practice alternate cropping including mele (reed) cultivation and salt-tolerant grass farming in the study area (Fig. 6). Salt-tolerant grass (fodder) farming is increasing day by day, as found during the community consultation. Resilient livestock farming, including semi-scavenger housing for livestock (hen, duck, sheep, and goat), is practiced by 75.6% of households to protect the livestock from climate-induced salinity and flood-oriented

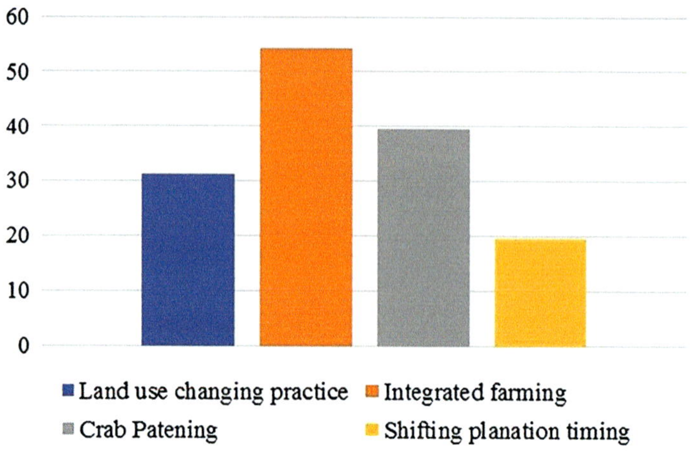

Fig. 7 Alternate climate-resilient practices in Shyamnagar Upazila. (Source: Field Survey 2017)

Table 6 Agro-based resilient practices in Ganges Tidal Floodplain AEZ (Shyamnagar). (Source: Authors)

Resilient sector	Options
Salt-tolerant T. Aman	BR-22 and BR-23, Bina shail
Salt-tolerant BRRI dhan	33, 56, 57, and 62 BRRI dhan 40, 41, 53, 54
Salt-tolerant Bina dhan	7 and 16 Bina dhan-8 and 10
Salt-tolerant Aus	BRRI dhan 65
Salt-tolerant alternate farming	Mele (reed) cultivation, salt-tolerant grass farming
Vegetable farming	Floating dhap cultivation, dyke farming, homestead farming
Resilient livestock rearing	Semi-scavenger housing for goat, semi-scavenger housing for duck, semi-scavenger housing for hen, semi-scavenger housing for sheep
Fish culture	Net fishing, cage fishing
Salt-tolerant wheat	Bijoy, BARI Gom-25, BAU-1059
Salt-tolerant potato	BARI Alo-22, CIP Clone -88,163
Salt-tolerant sweet potato	BARI Mishti Alo-8,9
Salt-tolerant pulses	BARI Mug- 2,3,4,5,6, BM-01, BM-08 BARI Falon- 1, BARI Sola-9
Short-duration oilseeds	BARI Sharisha-14,15 BARI Chinabadam –9, BINA China badam-1, BINA China badam-2, BARI Soyabean-6 BARI Til-2,3,4
Salinity-resistant jute variety	HC-2, HC 95, CVL 1
Saline-tolerant sugarcane variety	ISWARDI-40
Others resilient practices	Land use changing practice, integrated farming, crab patenting, shifting plantation timing

diseases. Fish culture is the common practice of coastal Bangladesh, but due to sea level rise, tidal inundation, storm surges, and flood, this practice is being restricted for the recent time (Field Survey 2017). Most of the farmers stopped fish culture in open pond, and now 89.1% of farmers are practicing net fish in their *gher* to protect shrimp and other fishes from flood and tidal inundation and some of the homestead level cage fishing; crab farming is observed in the study area which is being practiced by almost 89% of farmers (Fig. 6).

Vegetable farming is restricted by high salinity, waterlogging, and flood in the study area. But in the last 20 years, through the initiative of government and nongovernment organizations, some climate-resilient practices like homestead farming, dyke farming, *gher/dhap* farming, or integrated farming have been introduced in the study area, and 65.7% of farmers are practicing such farming for their household needs (Fig. 6). Commercial vegetable farming is not possible in the study area, as found from the community survey. Salt-tolerant wheat, jute, sugarcane, oilseed, pulse, etc. cultivation are practiced at a small scale.

Land use changing practice, integrated farming, crab patenting, and shifting plantation timing are introduced as climate-resilient practices in the study area. 31.3% of crop producers are farming shrimp instead of crop in their paddy fields, and some farmers changed crop calendar following seasonal crop suitability in the study area (Fig. 7).

Tanore Upazila (High Ganges River Floodplain, Level Barind Tract, and High Barind Tract AEZ)

Tanore (in the High Ganges River Floodplain, Level Barind Tract, and High Barind Tract AEZ) is vulnerable to drought, groundwater depletion, and hailstorms. Drought and groundwater depletion were the most damaging disasters for agricultural crop production, according to respondents during community consultations. The farmers of Tanore Upazila under the High Ganges River Floodplain, Level Barind Tract, and High Barind Tract AEZ are involved in climate-resilient agricultural practices like short-duration rice varieties, surface water-based irrigation, solar irrigation, rainwater harvesting for irrigation, canal re-excavation, shifting planting time, and use of drought-tolerant rice varieties like BRRI dhan 56, 57, and 66, short duration: BRRI dhan 33, 56, 57, and 62; Bina dhan 7, 16, and 66, and drought-tolerant wheat varieties like BARI Gom-26; salt- and heat-tolerant pulses: BARI Mug-2,3,4,5,6, BM-01, BM-08, BARI Falon-1, BARI Sola-9 oilseeds (Table 7). A small portion of farmers introduced conservation farming and mulching as experimental farming practices.

Field consultations revealed that 21.3% of farmers are practicing short-duration rice varieties of BRRI dhan 33, 56, 57, and 62 and Bina dhan 7, 16, and 66 and that 49.9% of farmers are practicing drought-tolerant rice varieties BRRI dhan 56, 57, and 66 as climate-resilient practices in the study area (Fig. 8). Although being a drought-prone area, wheat is a suitable crop for the area, and 11.6% of farmers are cultivating drought-tolerant wheat varieties like BARI Gom-26 in the area. Groundwater depletion is the major crisis in the drought-prone Tanore as well as the High Ganges River Floodplain,

Table 7 Agro-based resilient practices in High Ganges River Floodplain, Level Barind Tract, and High Barind Tract AEZs of Tanore Upazila. (Source: Authors)

Resilient sector	Options
Short-duration rice varieties	BRRI dhan 33, 56, 57, 62; Bina dhan 7, 16, 66
Drought-tolerant rice varieties	BRRI dhan 56,57, 66
Drought-tolerant wheat farming	BARI Gom-26
Heat-tolerant pulses	BARI Mug- 2,3,4,5,6, BM-01, BM-08 BARI Falon-1, BARI Sola-9
Other resilient practices	Conservation farming, mulching, shifting planting time

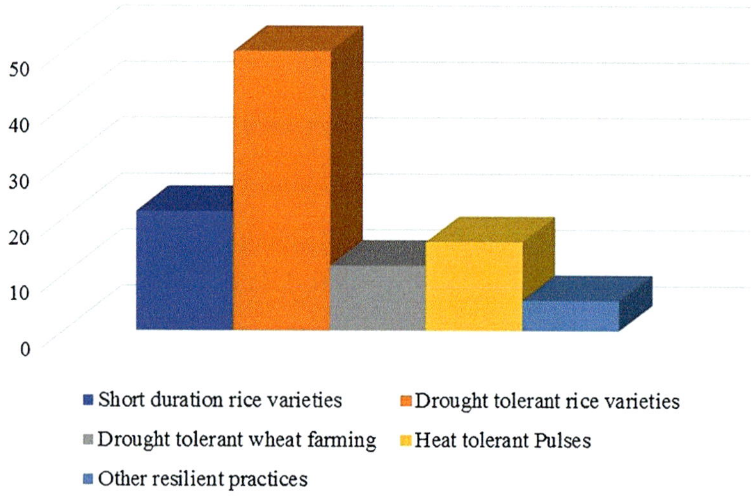

Fig. 8 Climate-resilient agro-based practices in Tanore Upazila. (Source: Field Survey 2017)

Level Barind Tract, and High Barind Tract AEZ, and inhabitants of the study area are managing water for irrigation and domestic purposes through surface water-based irrigation (17%), solar irrigation (14%), rainwater harvesting for irrigation (6%), and canal re-excavation (9%), and the intensity of water management options are shown in Fig. 9 (Field Survey 2017). A very small number of farmers are using conservation farming, mulching, and shifting planting time to adapt to drought in the study area.

Rangamati Sadar Upazila (Northern and Eastern Hills AEZ)

Hill ecosystems are not suitable for farming. Previously, Jhum was the only crop production option for the hill people. But due to climate change, Jhum farming is restricted. In recent times, through the initiatives of government and nongovernment organizations, the hill farmers have started practicing climate-

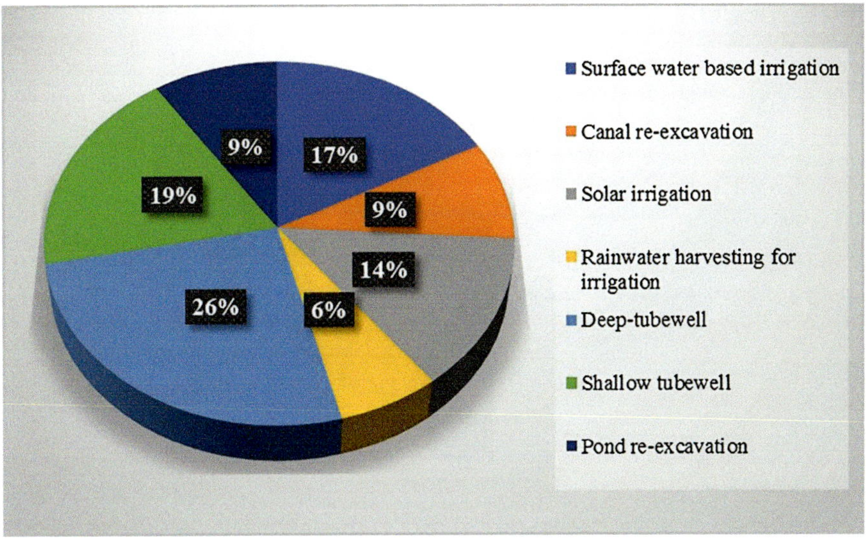

Fig. 9 Drought-resilient water management options in Tanore Upazila. (Source: Field Survey 2017

resilient zero tillage farming, terrace farming, rainwater harvesting for irrigation, replantation, early crop harvesting, short-duration varieties, and shifting planting time.

Figure 10 shows that zero tillage farming is the most practiced climate-resilient agricultural system in Rangamati Sadar as well as Northern and Eastern Hills AEZ. It is practiced by 54.9% of the total farmers in the study area. Terrace farming is practiced by 39.2% of farmers to adapt to landslide and flash flood in the hill areas. During summer, not only irrigation water but also drinking water crises affect the hill dwellers, and the farmers harvest rainwater for agricultural irrigation; 40.5% of farmers introduced the water system as climate-resilient irrigation system (Fig. 10). Short-duration crop varieties are adopted by 23.4% of the farmers to harvest early in the rainy season. Crop calendars changes as well as shifting planting time are practiced by 37.3% of total farmers. Mulching is the most prominent water conservation agricultural technology, and 28.6% of farmers are using it as a water conservation technique as well as a climate-resilient farming practice in hill tracts areas (Fig. 10).

Chilmari Upazila (Active Tista Floodplain, Tista Meander Floodplain, and Active Brahmaputra-Jamuna Floodplain AEZ)

Chilmari is affected by flash floods, regular floods, and riverbank erosion. Access to agriculture and natural resources including forestry and fisheries are restricted in the area. The farmers are implementing some climate-resilient practices. Flash flood- and submergence-tolerant rice varieties such as BRRI dhan 51 and 52; Bina dhan-11

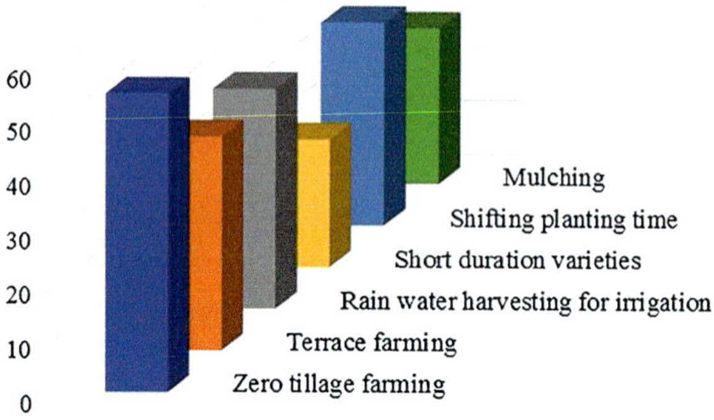

Fig. 10 Climate-resilient agro-based practices in Rangamati Sadar Upazila. (Source: Field Survey 2017)

and 12, T. Aman BR-22 and BR-23, Bina shail flash flood, BRRI dhan 33, 56, 57, and 62, Bina dhan 7 and 16, BRRI dhan 47, 61, and 67; Bina dhan-8 and 10, T. Aman BRRI dhan 40, 41, 53, and 54, Aus BRRI dhan 65, T. Aman BR-22 and BR-23; floating dhap cultivation, shifting planting time, semi-scavenger housing for goat, duck, and hen rearing, net fishing, dyke farming, embankment cropping, roadside plantation, agroforestry, homestead farming, *SORJAN*, dyke farming, etc., are the most innovative climate-resilient practices in these AEZs.

Figure 11 illustrates that homestead farming is the most widely used resilient agro-based practice in the study area of the Active Tista Floodplain, Tista Meander Floodplain, and Active Brahmaputra-Jamuna Floodplain AEZ; it is practiced by 72.4% of households of the Chilmari Upazila to meet their household vegetable demand around the year, especially in the rainy season when floods hit the area. The most used resilient agro-based interventions are using flash flood- and submergence-tolerant rice varieties, including BRRI 51, BRRI 52, BRRI 33, BRRI 56, BRRI 57, BRRI 62, BRRI 40, BRRI 41, BRRI 53, BRRI 54, BRRI 47, BRRI 61, BRRI 67, BINA 11, BINA 12, BINA 7, BINA 16, BINA 8, BINA 10, BR 22, BR 23, Aus BRRI 65, and T. Aman BR 22 and BR-23, which are being used by 47.1% of farmers of the study area (Fig. 11). Agroforestry is another popular and widely practiced climate-resilient intervention in the study area, being practiced by 39.4% of farmers for fuelwood, fruits, and timber (Fig. 11). 24.6% of households also practice roadside plantation as climate-resilient intervention; 11.8% are involved in embankment cropping; 11.4% are practicing plantation time shifting as well as crop calendar changes to harvest early crops before the floods (Fig. 11). 32.6% of households are involved in semi-scavenger housing for goat, semi scavenger housing for duck, and semi-scavenger housing for hen rearing as climate-resilient practice for livestock rearing in the study area (Fig. 11). Net fishing is found as climate-resilient fish culture at the household level as the resilient fish culture which is being practiced by only 5% of households in their pond (Fig. 11). Fish culture is hardly an intervention

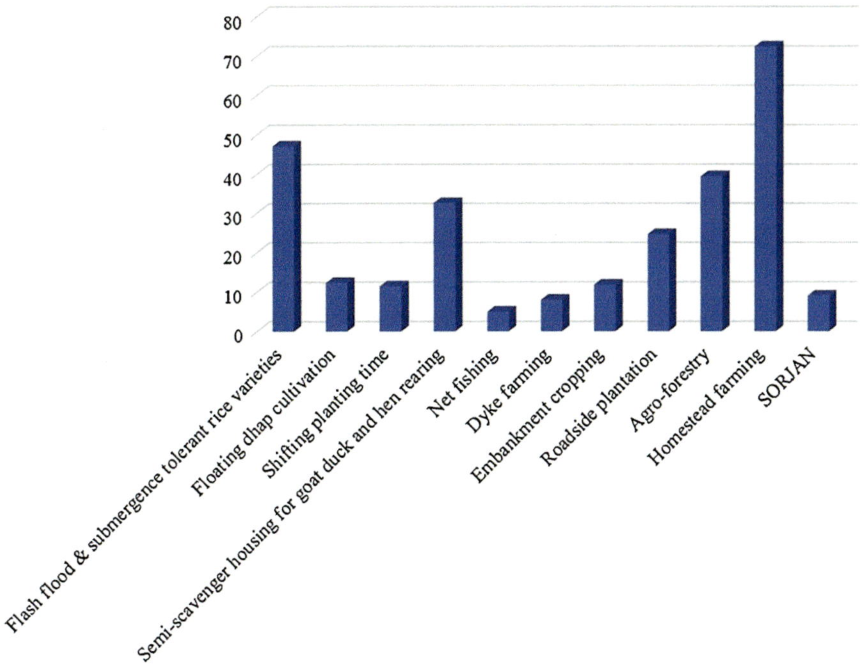

Fig. 11 Climate-resilient agro-based practices in Chilmari Upazila. (Source: Field Survey 2017)

in the study area because during the rainy season, all the waterbodies become submerged and on the other hand, during winter and summer, the area suffers from water deficit. During the field study in 2017, it was found that dyke farming and *SORJAN* farming, as a climate-resilient vegetable, are being practiced by 8% and 9% of households in the study area, respectively (Fig. 11).

Khaliajuri (Old Brahmaputra Floodplain and Sylhet Basin AEZ)

The area is flash flood prone. In response, the farmers of of Khaliajuri Upazila under the Old Brahmaputra Floodplain and Sylhet Basin AEZ have introduced excavation of the canals of surrounding field, *SORJAN*, submersible dykes, digging drains, withdrawal of excess water from the fields by pump, BRRI dhan 51 and 52; Bina dhan-11 and 12, submersion-tolerant rice varieties BRRI dhan 51 and 52, Bina dhan-11 and 12, T. Aman BR-22 and BR-23; Bina shail flash flood, BRRI dhan 33, 56, 57, and 62, Bina dhan 7 and 16, BRRI dhan 47, 61, and 67, Bina dhan-8 and 10, T. Aman BRRI dhan 40, 41, 53, and 54, Aus BRRI dhan 65, T. Aman BR-22 and BR-23, floating dhap cultivation, shifting planting time, semi-scavenger housing for goat, duck, and hen rearing, net fishing, dyke farming, embankment cropping, roadside plantation, agroforestry, and homestead farming as climate-resilient

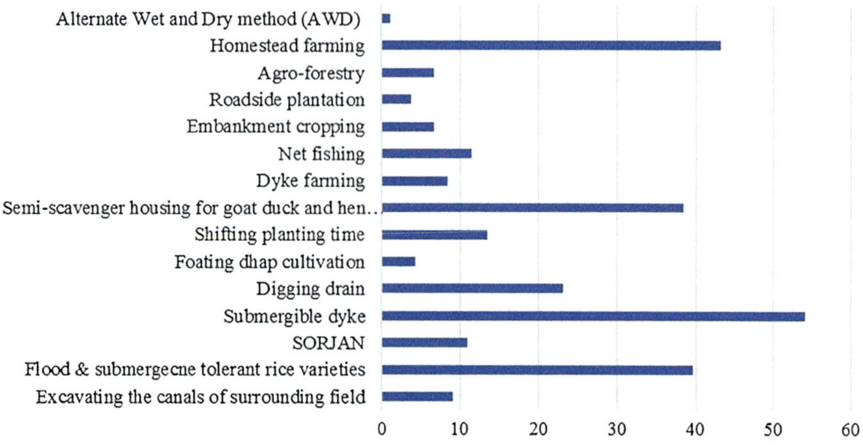

Fig. 12 Climate-resilient agro-based practices in Khaliajuri Upazila. (Source: Field Survey 2017)

agricultural practices. The alternate wet and dry (AWD) method toward conserving irrigation water is introduced as an experimental irrigation technology for the water-stricken summer season, in order to provide adequate water to irrigate all the Boro lands in haors, in the absence of reliable and economically exploitable groundwater aquifer systems.

The AWD is a newly tested climate-resilient water management option in the Khaliajuri haor area, which at present is used by only 1.2% of farmers to cultivate rabi crops during winter when the haor becomes dry (Fig. 12). In this area, the submersible dyke is the most widely used resilient intervention for the farmers to mitigate floods during the rainy season. Submersible dykes support multi-dimensional services for the haor dwellers in cropping and communication during summer and rainy season. It was found that 44.1% of farmers depend on submersible dykes for farming (Fig. 12). Homestead farming is a common practice, which has become a major climate-resilient farming practice for the haor people where 43.2% of households are involved in homestead farming in order to restore rural agro-based economy and cope with climatic stress in terms of vegetable supply and food security (Fig. 12) (Field Survey, 2017). Rice cultivation is difficult and somewhat restricted in the study area due to regular flash floods, regular floods, and submergence. The people of the study area use flood- and submergence-tolerant rice varieties. 39.6% of farmers in Khaliajuri Upazila are practicing this with the varieties BRRI dhan 51 and 52; Bina dhan-11 and 12, BRRI dhan 51 and 52; Bina dhan-11 and 12, T. Aman BR-22 and BR-23; Bina shail; BRRI dhan 33, 56, 57, and 62; Bina dhan 7 and 16, BRRI dhan 47, 61, and 67; Bina dhan-8 and 10, T. Aman BRRI dhan 40, 41, 53, and 54, Aus BRRI dhan 65, and T. Aman BR-22 and BR-23. Semi-scavenger housing for hen, duck, and goat rearing is another climate-resilient household level intervention found in the study area, which is being practiced by 38.4% of households. 23.2% of farmers dig drains around the agricultural land during the rainy season to drain flooded water, which is also a flood mitigation intervention for the paddy farmers. *SORJAN* (11% of farmers), floating dhap cultivation (4.3% of farmers), and dyke farming (8.4% of farmers) are climate-resilient vegetable farming practices in the area. A small number of households are involved in net fishing and climate-resilient fish cultures, in order to meet the households' needs; only 11.5% of households use this practice in the study area (Fig. 12). Agroforestry, embankment cropping, and homestead farming are other climate-resilient farming practices used by 6.7%, 3.8%, and 43.2% of farmers, respectively. During the field survey, 13.5% of farmers were found to change crop calendars in order to cope with early flash floods (Fig. 12).

Reshaping Policy to Promote Climate-Resilient Agriculture in Bangladesh

According to the official data from the Soil Resource Development Institute (SRDI) (Rahman and Hasan 2013), in 2010, crop agriculture land covered 9.5 million hectares in Bangladesh, though the Bangladesh Bureau of Statistics (BBS 2016) estimated that 8.52 million hectares were under crop in 2010–2011 and the Department of Agricultural Extension (DAE) stated in the Krishi (agriculture) Diary (DAE 2011) that the crop area is about 9.1 million hectares. Moreover, due to population growth, the share of land per capita is shrinking every year, which is threatening the land resource base for agriculture, forest, and wetlands. For example, in 1983–1984 (BBS 1985), there were 21,442 thousand acres of total cultivable crop area, which dropped to 19,133 thousand acres in 1994–1995 (BBS 1997). On average, Bangladesh is losing nearly 82,000 ha of arable land each year. This is mainly due to the conversion of land into urban, peri-urban, industrial uses. Agriculture is also responsible for some conversions (e.g., Chakaria Sundarbans of Cox's Bazar district and its adjacent areas). Agricultural land has decreased to 12,742,274 ha with the loss of 23,391 ha in 2000, and again with the loss of 56,537 ha in 2010, it reached 12,176,904 ha (Rahaman 2016). The annual loss of agricultural land during the study period (1976–2010) was 33,140 ha. The availability of agricultural land was in a decreasing trend, with much higher rates during the period from 2000 to 2010 (Rahaman 2016). Agriculture accounts for 20% of Bangladesh's gross domestic product and employs 65% of the country's labor force. However, more than 50% of Bangladesh's arable land has been affected by saltwater intrusion, submergence, or drought (IFC 2011). Bangladesh urgently needs support in developing a climate-resilient agriculture if its people are to survive and prosper in the long term. Climate change is affecting the country in many ways. For instance, rising sea levels are causing some agricultural land in coastal areas to become more saline, reducing both the quality and quantity of the products available. The impact of climate change on agriculture is undeniable and will most certainly worsen if climate-resilient farming practices are not adopted. In southern districts, where land is only centimeters higher than the brackish estuarine water, large swathes of agricultural land are becoming arid. Crop yields are shrinking because of increased salinity due to rising water levels in the Bay of Bengal. In coastal areas, coconut and betel nut trees do not yield half of what they did two decades ago, while banana groves are dying out in their hundreds. At the same time, vegetables sold in the urban markets of Dhaka, Khulna, and Rajshahi are deemed tasteless and fetch low prices compared to produce from salt-free regions. In a country where almost 80% of the population lives in rural areas (Hossain 2001), this is bad news. Bangladesh needs support for climate-resilient agriculture. Support for sustainable climate-resilient agriculture is the key trigger to enabling farmers to increase food security and adapt. In order to address this, farmers must raise their vegetable beds, maintain the soil's moisture by covering the seed-beds (and the manure around plants) with straw and leaves to prevent excessive evaporation and erosion, increase the amount of organic material in the soil, and modify cropping patterns.

The National Agriculture Policy was finalized in 2013, and it has also taken cognizance of the Environment Policy of 1992, Forest Policy of 1994, Fisheries Policy of 1998, Agricultural Land Use Policy of 2001, National Jute Policy of 2002, Livestock Resources Policy and Action Plan 2005, National Livestock Development Policy of 2007, National Food Policy of 2008, and National Poultry Development Policy of 2008. One of the stated objectives of the Policy is to make Bangladesh's agriculture sustainable and climate resilient such as to deal with climate change-related threats. It has put an emphasis on research, extension services, technology transfers, and information in order to make this happen. But there are no specific action plans based on agro-ecological zones.

Although the National Food Policy Plan of Action (2008–2015) has recognized the threats of climate change and emphasized the development of long- and short-term forecasts or the development of climate change-related early warning system (EWS) to deal with risks to food production, the policy does not reflect agro-ecological zone-based climate-resilient action plans, research, and strategies. From this point of view, reshaping agricultural policies and strategies in climate-smart agriculture in Bangladesh is urgently needed.

Conclusion

Bangladesh is one of the most climate-vulnerable countries in the world due to the combination of frequent natural disasters, high population density, and low resilience to economic shocks. Bangladesh has a primarily agrarian economy. Agriculture is the single largest producing sector of the economy since it makes up nearly 30% of the country's GDP and employs around 60% of the total labor power. The operation of this sector has a significant impact on major macroeconomic objectives like job creation, poverty mitigation, human resources development, and food security (Chowdhury and Chowdhury 2011). However, the agriculture sector is largely impacted by the adverse impact of climate change-related extreme events including cyclone, flooding, drought, salinity intrusion, etc. As a result, the food security of the people, especially the people living in the climate change hotspots, is in a precarious condition. Despite all these climatic stressors, people are trying to cope with the adverse situation, both through their traditional knowledge of agrarian adaptation and through newly emerged technologies in the agricultural sector. These adaptation options vary with geographic location, extent and type of climatic impacts, etc. The people of Bangladesh are using those practices since they are well-known adaptation options for those climate-induced disasters and extreme events. They are trying their level best to build their resilience in combating climate change impacts in the agricultural domain. These efforts need to be more organized and developed, based on improved technologies existing in the world. The government should consider recognizing the local and indigenous agrarian adaptation practices in its relevant policies and should move forward with new strategies for better implementation of those adaptation techniques.

Cross-References

- ▶ Assessment of Information on Successful Climate-Smart Agricultural Practices/ Innovations in Tanzania
- ▶ Drought-Tolerant Crops in Kirinyaga County, Kenya: Climate-Smart Agriculture Adaptation Strategies
- ▶ Effect of Mulching on Soil Temperature and Moisture for Potato Production in Agro-ecological Zones of Central Highlands of Kenya
- ▶ Local Knowledge of Climate Change Among Arable Farmers in Selected Locations in Southwestern Nigeria

References

Alam M, Ahammad R, Nandy P, Rahman S (2013) Coastal livelihood adaptation in changing climate: Bangladesh experience of napa priority project implementation. In: Shaw R, Mallick F, Islam A (eds) Climate change adaptation actions in Bangladesh. Springer, Tokyo, pp 253–276

Asian Development Bank (ADB) (2011) Adapting to climate change: strengthening the climate resilience of water sector infrastructure in Khulna, Bangladesh. © Asian Development Bank. http://hdl.handle.net/11540/987. License: CC BY 3.0 IGO

Bangladesh Bureau of Statistics (BBS) (1985) Agricultural Census of Bangladesh. Ministry of Planning, Bangladesh Bureau of Statistics, GOB, Dhaka

Bangladesh Bureau of Statistics (BBS) (1997) Agricultural Census of Bangladesh. Ministry of Planning, Bangladesh Bureau of Statistics, GOB, Dhaka

Bangladesh Bureau of Statistics (BBS) (2016) Yearbook of Agricultural Statistics-2016. Ministry of PlanningBangladesh Bureau of Statistics, GOB, Dhaka

Banglapedia (2014a) Agroecosystem. Banglapedia, National Encyclopedia of Bangladesh. Asiatic Society of Bangladesh, Dhaka

Banglapedia (2014b) Agroecological Zone. Banglapedia, National Encyclopedia of Bangladesh. Asiatic Society of Bangladesh, Dhaka

Banglapedia (2014c) Tanore Upazila. Banglapedia, National Encyclopedia of Bangladesh. Asiatic Society of Bangladesh, Dhaka

Banglapedia (2014d) Chilmari Upazila. Banglapedia, National Encyclopedia of Bangladesh. Asiatic Society of Bangladesh, Dhaka

Banglapedia (2014e) Khaliajuri Upazila. Banglapedia, National Encyclopedia of Bangladesh. Asiatic Society of Bangladesh, Dhaka

Banglapedia (2015a) Shyamnagar Upazila. Banglapedia, National Encyclopedia of Bangladesh. Asiatic Society of Bangladesh, Dhaka

Banglapedia (2015b) Rangamati Sadar Upazila. Banglapedia, National Encyclopedia of Bangladesh. Asiatic Society of Bangladesh, Dhaka

BARC (Bangladesh Agriculture Research Council) (1988) Land resources appraisal of Bangladesh for agricultural development. BARC, Dhaka

Carter TR, Parry ML, Nishioka S, Harasawa H (1994) Technical guidelines for assessing climate change impacts and adaptations. Climate change 1995: impacts, adaptations and mitigation of climate change: scientific-technical analyses. In: Contribution of Working Group II to the Second Assessment Report of the Intergovernmental Panel on Climate Change. Cambridge University Press, Cambridge

Chambwera M, Heal G, Dubeux C, Hallegatte S, Leclerc L, Markandya A, McCarl BA, Mechler R, Neumann JE (2014) Economics of adaptation. In: Field CB, Barros VR, Dokken DJ, Mach KJ, Mastrandrea MD, Bilir TE, Chatterjee M, Ebi KL, Estrada YO, Genova RC, Girma B, Kissel ES, Levy AN, MacCracken S, Mastrandrea PR, White LL (eds) Climate change 2014: impacts, adaptation, and vulnerability. Part A: global and sectoral aspects. Contribution of Working Group II to the Fifth Assessment Report of the Intergovernmental Panel on Climate Change. Cambridge University Press, Cambridge and New York, pp 945–977

Chowdhury TA, Chowdhury SS (2011) Performance evaluation of agricultural banks in Bangladesh. Int J Bus Manag 6(4):75–89. https://doi.org/10.5539/ijbm.v6n4p75

Daniels RJR (2003) Ecosystem diversity in Bangladesh: status and trends. A Report for the Bangladesh National Conservation Strategy and Action Plan, Dhaka. IUCN, Dhaka

Department of Agriculture Extension (DAE) (2011) Krishi (Agriculture) Diary. Ministry of Agriculture, Department of Agriculture Extension, GOB, Dhaka

Field Survey (2017) Field survey for the study titled climate-resilient agricultural practices in different agro-ecological zones of Bangladesh. Bangladesh, Authors

GOB (Government of Bangladesh) (2009) Bangladesh climate change strategy and action plan 2008. Ministry of Environment and Forests, Government of the People's Republic of Bangladesh, Dhaka

Harmeling S (2009) Global climate risk index 2010: who is the most vulnerable? Weather-related loss events since 1990 and how Copenhagen needs to respond. Germanwatch, Bonn

Hossain MZ (2001) Farmer's view on soil organic matter depletion and its management in Bangladesh. Nutr Cycl Agroecosyst 61:197–204

Huda MN (2013) Understanding indigenous people's perception on climate change and climatic hazards: a case study of Chakma indigenous communities in Rangamati Sadar Upazila of Rangamati District, Bangladesh. Nat Haz 65:2147–2159

International Finance Corporation (IFC) (2011) IFC promotes climate-resilient agriculture in Bangladesh, helping ensure food security. World Bank, IFC Research Group, Dhaka

Islam MA, Shitangsu PK, Hassan MZ (2015) Agricultural vulnerability in Bangladesh to climate change induced sea level rise and options for adaptation: a study of a coastal Upazila. Journal of Agriculture and Environment for International Development (JAEID) 109(1):19–39

Kreft S, Eckstein A, Melchior I (2017) Global climate risk index 2017: who suffers most from extreme weather events? Weather-related loss events in 2015 and 1996 to 2015. Germanwatch, Bonn

Mainuddin K, Rahman A, Islam N, Quasem S (2011) Planning and costing agriculture's adaptation to climate change in the salinity-prone cropping system of Bangladesh. International Institute for Environment and Development (IIED), London

Nagothu S, Nemes A, Biswas JC, Sarker MH (2014) Climate change impacts, vulnerability and adaptation: sustaining rice production in Bangladesh. Bioforsk Rapport

Nelson DR, Adger WN, Brown K (2007) Adaptation to environmental change: contributions of a resilience framework. Annu Rev Environ Resour 32:395–419

Rahaman MA (2016) Assessment of agricultural land loss and food security in Bangladesh. Social Science Research Council, Planning Commission, GOB

Rahaman MA (2018) Climate-induced disaster, livelihood and food security nexus: overview of marginalized communities of northern Bangladesh in patterns and effects of climate change on the food security and livelihoods of marginalised communities in the north of Bangladesh. German NETZ, Dhaka (Unpublished)

Rahman MT, Hasan MN (2013) Assessment of shifting of agricultural land to non-agricultural land in Bangladesh. Ministry of Agriculture, Soil Resource Development Institute, GOB, Dhaka

Agro-met Services and Farmer Responsiveness to Advisories: Implications for Climate Smart Agriculture

116

Ramkumar Bendapudi, Nitin Kumbhar, Prithviraj Gaikwad, and Crispino Lobo

Contents

Context 2366
Objectives 2367
Conceptual Framework 2367
WOTR Agro-Met Set up 2368
Components of WOTR Advisory System and Interactions Between Key Actors 2369
Contractual Relationships: Actors and Governance Arrangements 2369
Technical Relationships 2371
Uptake of Agro-Advisories by the Farmers: Implications for Improving Service Relationship 2372
Profile of the Sample Farmers in the Area 2373
Major Crops Grown by the Sample Farmers 2373
Major Climate Risks as Perceived by Farmers 2373
Sources of Information Accessed by Farmers in the Study Area 2373
Response of Farmer's to WOTR Agro-Met Advisory Information 2375
Conclusions: Implications for Making Advisory Systems Demand Driven 2380
Cross-References 2381
References 2381

Abstract

Agro-met services delivered through SMS by Watershed Organization Trust (WOTR) provides crop and locale-specific agro-advisories based on weather forecasts and the particular crop growth stage in order to reduce risks and improve agriculture productivity despite local climatic variations. The objective of the

R. Bendapudi (✉) · C. Lobo
Watershed Organisation Trust (WOTR), Pune, India
e-mail: rbendapudi@yahoo.com; cslobo1@gmail.com

N. Kumbhar · P. Gaikwad
Watershed Organisation Trust (WOTR), Pune, India

WOTR Centre for Resilience Studies (W-CReS), Pune, India
e-mail: nitin.kumbhar@wotr.org.in; prithviraj.gaikwad@wotr.org.in

W. Leal Filho (ed.), *Handbook of Climate Change Resilience*,
https://doi.org/10.1007/978-3-319-93336-8_47

chapter is to understand the nature and importance of different components of WOTR advisory system and the response of farmers to the agro-met advisory services.

An important aspect for developing a farmer-responsive agro-advisory system is the willingness of diverse stakeholders to collaborate (farmers, NGOs, Research institutions, Government institutions, private organizations) and converging of respective strengths. It was found that the content of advisories needs to be relevant to the location and to the specific crop to make an advisory system demand driven that is based on farmers' needs.

Keywords
Agro-advisory system · Farmer response · Onion and Pearl millet

Context

Birner et al. (2009) defined "agricultural advisory services are defined as the entire set of organizations that support and facilitate people engaged in agricultural production to solve problems and to obtain information, skills, and technologies to improve their livelihoods and well-being."

Farmers and other rural actors can benefit from numerous types of advisory services defined as much by their content (technical, economic, social, and environmental) as by the way they are provided (disseminating information and techniques, reinforcing the learning process, or accompanying actors) (SanneChipeta 2006).

There have been many initiatives to provide weather-based crop advisories in India. The National Centre for Medium Range Weather Forecasting (NCMRWF) under the Ministry of Earth Sciences (MoES), Government of India, in collaboration with India Meteorological Department (IMD), Indian Council of Agricultural Research, and State Agricultural Universities had been providing Agrometeorological Advisory Services (AAS) to the farming community based on location-specific medium-range weather forecast to the districts under different agro-climatic zones (now called Integrated Agrometeorological Advisory Service) (Maini and Rathore 2011).

There is a large interest and takeoff of similar private and public programs on advisory systems in India (Mittal et al. 2010), and it is important to understand the impact of these interventions. Policies to improve access to ICT in rural areas need to focus as much on content and education as infrastructure. Local content creation needs to be linked to institutional innovations to provide farmer-responsive extension services (Anderson 2007).

The present study examines the case of agro-met services delivered through SMS by Watershed Organization Trust (WOTR) in Sangamner block located in Ahmednagar district of Maharashtra. WOTR's knowledge-embedded service to farmers provides crop and locale-specific agro-advisories based on weather forecasts and the particular crop growth stage in order to reduce risks and improve agriculture productivity despite local climatic variations.

Objectives

The specific objectives of the chapter are to:

(i) Understand the nature and importance of different components of advisory system and their interactions in the context of WOTR's agro-met advisory system.
(ii) Identify the different knowledge systems available to and accessed by the farmers in the area.
(iii) Understand farmers response (adoption or non-adoption) to different types of agro-met advisories.

Conceptual Framework

We use the framework of advisory system used by Faure et al. (2011) (Fig. 1) to analyze the case of agro-met advisory system that was developed by Watershed Organisation Trust (WOTR), a not-for-profit organization. The framework by Faure et al. (2011) was adapted from Gadrey (1994) and Birner et al. (2009), which depicts the actors providing the advisory service, the interactions among them, and the factors influencing the service.

The operation of an advisory system is explained by interlinked components of governance, financing mechanisms, skills and qualifications of advisors and managers of service providers, and method by which advice is provided, which is characterized by the service relationship between the advisor and the farmer (Birner

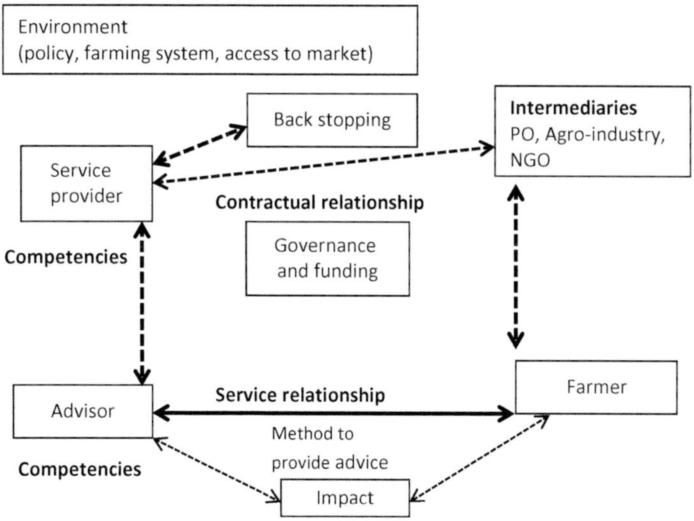

Fig. 1 Advisory system used by Faure et al. (2011).

et al. 2009; Faure et al. 2011, 2016). According to Birner et al. (2009), quality of advisory services is indicated by the "(i) accuracy and relevance of the content of the advice, (ii) timeliness and outreach of the advice, (iii) quality of the partnerships established and the feedback effects created, and (iv) efficiency of service delivery and other economic performance indicators."

The present study mainly focuses on the *service relationship* between advisor and farmer. Service relationship was divided into technical relationship (content of the service, the tools used to provide the service) and contractual relationship (organization of work, price of service, planning of advisory activities) (Faure et al. 2011). In the context of WOTR agro-met advisory system, it is of relevance to understand the response of the farmer, who is the final beneficiary or client of the advisory system, which in turn has implications on the content of advisory, outreach, and the effectiveness of the service delivery system.

WOTR Agro-Met Set up

The integrated agro-meteorology service evolved by WOTR consists of four components that are interlinked as follows: (i) weather awareness, local weather data acquisition, and short range weather forecasts; (ii) crafting of agro-advisories, dissemination, and feedback gathering; (iii) an automated content management system (ACMS) for agro-meteorological advisory generation and dissemination; and (iv) on-site capacity building, knowledge and technology transfer, and engagement with local institutions (Lobo et al. 2017).

WOTR, in collaboration with the India Meteorological Department (IMD), has installed 103 Automated Weather Stations (AWS) in Jalna, Ahmednagar, Aurangabad, Pune, Nashik, Dhule, and Nandurbar districts of Maharashtra. Altogether, they make for a high density grid ranging 3–5 km between the weather stations, the reason being to provide a sufficiently dense network of high quality data points to enable the IMD to better calibrate their weather models, given the diversity of topographies and agro-climatic ecologies present in the project district. Of these, 92 AWSs are telemetrically linked to WOTR's servers and send on an hourly basis, weather information round-the-clock. This data is subject to quality control at WOTR and then forwarded to the IMD. IMD sends WOTR daily, 3-day weather forecasts for respective villages.

Based on these short-term (3-day) local weather forecasts, agricultural experts from WOTR prepare agro-advisories with the help of crop-weather calendars which have been prepared with inputs from the Central Research Institute for Dryland Agriculture (CRIDA) and two state agricultural universities (the Mahatma Phule Agriculture University (MPKV) at Rahuri and the Vasantrao Naik Marathwada Agriculture University (VNMAU) at Parbhani) with whom WOTR has a knowledge-sharing collaboration. These advisories that are crop and locale-specific include integrated nutrient-water-pest-and diseases management recommendations that stress organic and environmentally sustainable interventions. These advisories are issued in the local language at least twice a week in the summer months and more frequently during the agricultural season, as required, thus alerting farmers and giving them enough time to implement suggested measures. The advisories are disseminated through SMSs to mobile phones, wallpapers that are put up at prominent places in the project villages, and by word-of-mouth.

Components of WOTR Advisory System and Interactions Between Key Actors

Contractual Relationships: Actors and Governance Arrangements

WOTR has been engaging with rural communities and farmers since nearly 25 years now. In the early year of this millennium, it became evident that agriculture, especially rainfed farming systems, was in distress. Climate variability was beginning to affect crop productivity and weather-induced losses began increasing; adverse market conditions and declining soil health resulted in low returns and increasing costs of cultivation. This led WOTR to launch an initiative to offer a combo of services that included timely weather forecasts, weather responsive crop management advisories, information on prevailing market prices of key crops, as well as on-farm extension and technical services.

With a view to developing an approach that can be scaled up in a cost-effective manner and contribute to improve the country's efforts in the field of applied agrometeorology, WOTR entered into a collaborative partnership with key developmental, scientific, and academic institutions, namely, the India Meteorological Department (IMD), the Central Research Institute for Dryland Agriculture (CRIDA), the Mahatma Phule Agriculture University (MPKV), and the Vasantrao Naik Marathwada Agriculture University (VNMAU) (Table 1). The specific objective of this initiative is to provide farmers locale- and crop-specific weather-based crop management and assist them in using this information for agricultural decision-making. This effort was financially and technically supported by the Swiss Agency for Development and Cooperation (SDC) and the National Bank for Agriculture and Rural Development (NABARD).

The advantage of such an arrangement involving stakeholders working across scales (from the local, national to the international), having complementary strengths and working towards a common objective, is that it makes possible a pooling and sharing of valuable resources and expertise across domains and institutional boundaries and facilitates mutual learning, cogeneration of practical knowledge and technology transfer. More importantly, such a partnership opens up possibilities to channel insights and experiences from the ground to decision and policy makers which can contribute to improving the quality and effectiveness of the agrometeorological services provided by state and national entities.

Even though the collaborative partnerships between the various institutions were within a formal contractual framework, the trust (developed over a period of interaction and personal relationships) and service motivation were a common binding force for the collaboration (Table 1).

Table 1 Key actors in the WOTR agro-advisory system, roles and relationships

Actor	Type of organization	Roles	Type of relationships (Actor linkages with key facilitator – WOTR)
Watershed organisation trust (WOTR)	Nongovernment organization	Conceptualized and facilitated the entire agro-met advisory system – Design and framework of the advisory system; community mobilization, awareness, generating, establishing AWSs and providing weather data to IMD for processing, crafting advisories, delivery of advisories and services to farmers; feedback from end users	Key facilitator
India meteorological department (IMD), Ministry of Earth Sciences, government of India	Central government	Establish transmission linkages between WOTR, the AWSs, and IMD; expertise in analyzing meteorological data and providing short-term weather forecast for respective villages; hand-holding support for management of AWS (maintenance, calibration); training of WOTR personnel on applied meteorology	Trust, service, contractual – Memorandum of understanding
Mahatma PhuleAgriculture university (MPKV), Rahuri	Agriculture university	Support to developing crop-weather calendars and crop management practices	Trust, service, contractual – Memorandum of understanding
Vasantrao Naik Marathwada agriculture university (VNMKV), Parbhani	Agriculture university	Support to developing crop-weather calendars and crop management practices	Trust, service, contractual – Memorandum of understanding
Central research Institute for Dryland Agriculture (CRIDA), Indian Council of Agricultural Research (ICAR)	Research	Provide inputs on cropping/ production systems management; contingency crop planning	Trust, service, contractual – Memorandum of understanding
AWS supplier	Private company	Supply of automated weather stations; first-level training on installation and operations/ maintenance	Business

(continued)

Table 1 (continued)

Actor	Type of organization	Roles	Type of relationships (Actor linkages with key facilitator – WOTR)
Network service provider	Private company	Bulk SMS services	Business
Farmers	End users	Users of advisories	Service, trust

Source: Authors field notes.

At the farmer level, the relationship of WOTR is still based on service and trust (social capital developed over a period of interactions with farmers) with no formal contractual relationship. Over a period of time, from point of financial sustainability of the advisory system, it is also important to transition into business contracts, which could also ensure quality advisories and accountability.

Technical Relationships

Capacity Building, Structuring Advisories

The content of crop- and location-specific advisories was prepared in a knowledge-sharing collaborative mode under contractual arrangements between WOTR, state agricultural universities, and central research institute (as summarized in Table 1). But it is equally important to develop awareness among the farmers about the advisory system and develop local human capital at the village level for improved adoption of advisories sent to them.

Agro-Met trainings were conducted in all villages to enable communities to understand the purpose and utility of the weather stations and how to protect and maintain them. Weekly visits of "para-agriculture workers" (*Wasundhara-Sevak*) often accompanied by WOTR subject matter specialist) are made to the villages to discuss about the automatic weather station and various aspects of its uses. This helped to get community become inquisitive and keen to opt for the new initiative.

Agricultural professionals provided extension, technology, and training support at the farm and village level, established "demonstration plots," undertake Farmer Field Schools and engaged with farmers' groups, local governance institutions, and government departments (Lobo et al. 2017). To ensure community participation and engagement on this component, exposure visits were arranged to nearby villages which already had experience of using the Agro-Met Advisory services. Farmers met other groups of farmers and sought their feedback.

Tools Used for Advisory Dissemination

Advisories were disseminated through mobile telecommunication networks (via SMSs) directly to individual farmers, by announcements over the village public address system (loudspeakers), and weekly posters that are put up at prominent places in some villages and on walls/blackboards (Lobo et al. 2017).

Feedback Mechanism in Place to Monitor

In order to develop effective advisories, there is need to get farmers' feedback about their utility, impact, and how they have applied them, if at all. WOTR has put in place a system for regular feedback collection which involves collection of on-field data and farmer's observations. For a particular season, three farmers per key crop were identified in selected villages. Feedback against the key advisories was collected from these selected farmers in a prescribed register on a 15 days interval basis by para agronomist, which is then reviewed by an agronomist, who is in-charge of a cluster of villages. This validated data is then entered into software developed by WOTR and uploaded into the system. This software helps collate, assess, and analyses the feedback received from the farmers.

The agro-met advisory feedback system is necessary to assess the impact of the disseminated agro-advisories and to understand the problems faced by farmers in implementing them. The feedback system also helps to update WOTR's knowledge bank; refine or modify advisories generally prescribed for specific problems; generate better targeted and effective crop management practices; and better understand the challenges faced by farmers and how they respond to and adapt to climate-induced risks and extreme events.

Advisors

The advisors included a team of qualified subject matter specialists from WOTR and Agriculture Universities and "para-agronomists." The subject-matter specialists were agriculture graduates and postgraduates who trained the para-agronomists and farmers through Farmer Field Schools and on-field demonstrations.

The para-agronomists were local literate rural youth with education up to diploma or higher secondary school education. They were trained along with selected farmers on various crop management practices, importance of agro-advisory system, and also the process of feedback on the advisories.

These advisors are the key connecting link in the advisory system that combines contractual relationships and technical relationships. They play a two-way role: that of influencing the farmer's adoption of the different crop advisories sent through SMS and at the same time gathering feedback regarding farmer's informational needs and developing or refining content of advices that meet the farmer requirements.

Uptake of Agro-Advisories by the Farmers: Implications for Improving Service Relationship

A short feedback survey of the farmers in Sangamner block of Ahmednagar district was undertaken to understand how farmers perceive the advisories at different crop growth stages and their action (follow-up or not following up) on the advisories.

Sangamner block of Ahmednagar district comes under the Western Maharashtra Scarcity Zone. This area suffers from the twin problems of low productivity and high instability as a result of inadequate and unpredictable rainfall. The average annual rainfall is less than 500 mm and ranges from 350 mm to 450 mm. High temperatures and high wind velocity result in high potential evaporation values leading to moisture deficit. In Sangamner, kharif is an important cropping season where pearl millet, pulses, groundnut, maize, soybean, and tomato are major crops. Onion in this region is predominantly grown in shallow to medium type of soil during both kharif (June–September) and rabi seasons (October–February).

The feedback on WOTR advisories was taken from 120 randomly selected farmers receiving SMS in Kharif (rainy) season of 2014–2015. Farmer interviews were conducted through structured questionnaires to collect information on farmers' response to the advisories on various crop management practices (adoption or non-adoption).

Descriptive statistics cross classification tables and measures of association based on chi-square statistics and Cramer V were used to understand differences in adoption of advisories. The rank-based quotient (RBQ) was used to understand farmer's preferences regarding the modes of communication for getting agro-met information.

Profile of the Sample Farmers in the Area

Most of the farmers were small and medium farmers (average land ownership of 1.2 h) with some basic education (mean education of about 10 years) (Table 2).

Major Crops Grown by the Sample Farmers

The major crops in the kharif (rainy) season are onion (103 plots) and pearl millet (70 plots) with an average cultivated plot area of 0.53 and 0.51 acres, respectively (Table 3).

The average price of onion in Ahmednagar district (in 2015) was Rs. 25 per kg, which is higher than that of pearl millet (Rs. 16 per kg). This is also one reason for farmers' preference to cultivate onion.

Major Climate Risks as Perceived by Farmers

Late onset of monsoon and dry spells were perceived as the major climate risks by 100% and 89% of the respondents, respectively. Majority also considered other risks associated with agriculture such as diseases and pests (Table 4).

Sources of Information Accessed by Farmers in the Study Area

The traditional ICTs included TV, radio, newspapers, other farmers, government agricultural extension services, traders, input dealers, seed companies, and relatives.

Table 2 Characteristics of sample farmers in the area

	Mean (N = 120)	Minimum	Maximum	Standard deviation
Family size (no.)	4.5	4.0	11.0	1.6
Age of household head (Years)	41.6	24	66	10.6
Education of household head (Years)	9.6	0	17	3.1
Total land owned (Hectares)	1.2	0.1	6.4	0.9
Total number of large ruminants	2	0.0	19.0	2.3
Annual household income (USD[a])	1741	309	5868	999

Source: Field survey data
[a]One USD is equal to 68 INR (2018)

Table 3 Crops grown in Kharif 2014–2015 (rainy season) among the sample households

Crop	No. of plots	Mean (Ha)
Pearl millet	70	0.51
Sorghum	2	0.50
Onion	103	0.53
Groundnut	10	0.42
Tomato	6	0.28
Marigold	1	0.40
Green peas	2	0.30

Source: Field survey data

These have been an important tool since past several decades to disseminate scientific and technical agricultural knowledge to farmers and also leading to improved adoption of technologies (Mittal and Mamta 2012).

Majority of the sample farmers (about 88% of the respondents) indicated that SMS service from WOTR was one of the major sources of crop advisories. Agricultural information through television was the second most important source of information (as indicated by 72.5% of the sample farmers). Other progressive farmers were also considered as important sources of information (by about 55% of farmers) (Table 5).

The sample farmers were asked to rank their preferences regarding the modes of communication for getting information on weather and crop management practices.

It was found that SMS through mobile phone was ranked first followed by television programs, which was ranked second (Table 6). Gowda and Dixit (2015) found that farmers with higher education level showed better comprehension of advisories, acted upon the advisories more promptly and shared the information with fellow farmers more often than those with lower education level.

There are also other web-based services such as iKisan, an online informational resource for farmers by private entity, Nagarjuna Fertilizers and Chemicals Ltd. (NFCL) (Meera Shaik et al. 2004). In the state of Maharashtra, the Department of Agriculture, Government of Maharashtra, provides wide range of agricultural information and services through its web portal. Other such advisory services include

Table 4 Major risks as perceived by farmers

Risk	No. of farmers	Percent
Climate risks		
Late onset of monsoon	120	100.0
Dry spells	107	89.2
Unseasonal precipitation	33	27.5
Other agriculture related risks		
Diseases	112	93.3
Pests	112	93.3

Source: Field survey data

Table 5 Major sources of information (n = 120)

	No. of farmers	% of total sample farmers
TV	87	72.5
Radio	22	18.3
Agricultural extension officer	51	42.5
Kisan call center	35	29.2
Progressive farmers	66	55.0
SMS WOTR	106	88.3
SMS government	17	14.2

Source: Field survey data

Table 6 Farmers preferences regarding modes of communications

	Mean rank	RBQ	Rank
SMS through mobile	1.2	94	I
TV program	2.5	62	II
Wallpaper	3.1	48	III
Newspaper	3.2	46	IV

Source: Field survey data

"Kisan Call Centres (KCCs)," an initiative of Ministry of Agriculture & Cooperation and Farmers welfare, Government of India.

Response of Farmer's to WOTR Agro-Met Advisory Information

The total number of plots under cultivation among the 120 sample farmers was 195 in the rainy season (kharif) of 2014–2015. Out of these, 70 plots were under pearl millet and 103 plots were under onion. The type of advisories sent to the farmers included location-specific and crop-specific information on weather, land preparation, seed varieties, seed treatment, sowing time, irrigation, FYM application, DAP application, urea application, complex fertilizers, micronutrient application, pesticides, hand weeding, weedicides, and harvesting. Farmers may choose parts rather than the whole package of practices. So, many farmers adopt pieces of the package rather the entire package (Al-Karablieh et al. 2009). This was noticed in the

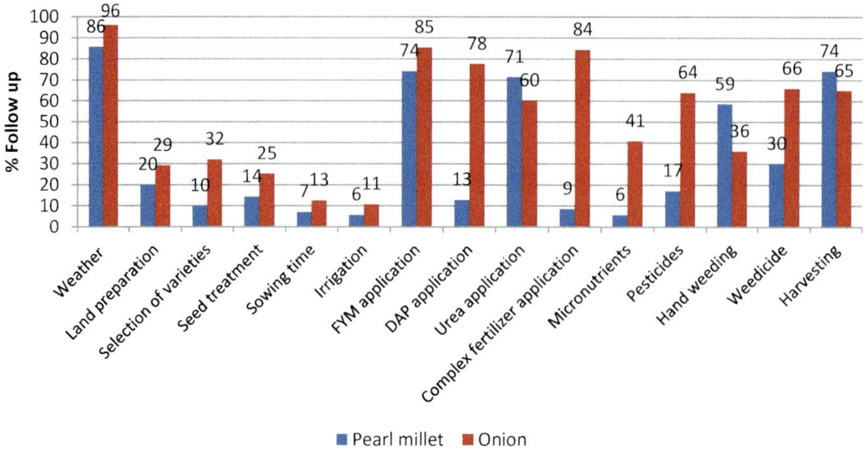

Fig. 2 Positive follow-up on the crop advisories in case of pearl millet onion crops.

respondent farmers' in the area. Not all farmers followed up on all the advisories. Some crop advisories received greater positive responses in terms of adoption as compared to others (Fig. 2). This also varied with the type of crop, in this case: pearl millet and onion.

Weather Advisories

The weather advisories included a 3-day forecast of rainfall, temperature, wind speed, and humidity. The aim was to help the farmers to plan intercultural operations such as weeding, fertilizer application, pest and disease management, and harvesting. It was found that weather-related advisories were followed up in the case of both pearl millet (86% of plots) and onion crops (96% of plots) (Fig. 2). Chi-square test was used to test if there was any association between the type of crop grown and the follow-up action on weather advisories (Table 7). The chi-square test indicated significant association at 1% significant level.

It was noticed that farmers were not flexible to change their crop preference. For example, short duration crop was recommended instead of onion due to dry spells. But the farmers ended up resowing the same crop repeatedly instead of sowing an alternative crop (pearl millet, pulses – green gram, black gram, and cowpea).

Seed

The advisories on seed advisories included seed treatment and time of sowing. The follow-up was low in both pearl millet and onion crops. In the case of pearl millet, farmers were advised to treat the seed with bio-fertilizers to improve germination percentage. For onions, advisory was for the seedlings to be treated with fungicides and pesticides. Since the seeds were pretreated at the time of purchase itself in both the major crops, namely, pearl millet and onion, there was no necessity to follow-up on the advisory.

Table 7 Relationship between type of crops (pearl millet or onion) and follow-up on advisories (crop management practice-wise)

Crop type	Pearl millet (Total no. of plots = 70)		Onion (Total no. of plots = 103)				
Management practice	Follow-up	Not followed	Follow-up	Not followed	Chi Square	P-value	Cramers V
Weather	60	10	99	4	6.06[a]	0.01	0.18
Land preparation	14	56	30	73	1.83	0.17	0.10
Seed	14	56	33	70	3.05	0.08	0.13
Selection of varieties	7	63	33	70	11.39[a]	0.00	0.25
Seed treatment	10	60	26	76	3.52	0.06	0.14
Sowing time	5	65	13	90	1.36	0.24	0.08
Irrigation	4	66	11	92	1.30	0.25	0.08
FYM application	52	18	88	15	3.36	0.06	0.14
DAP application	9	61	80	23	70.08[a]	0.00	0.63
Urea application	50	20	62	41	2.30	0.13	0.11
Complex fertilizer application	6	64	87	16	96.56[a]	0.00	0.74
Micronutrients	4	66	42	61	26.25[a]	0.00	0.39
Pesticides	12	58	66	37	37.08[a]	0.00	0.46
Hand weeding	41	29	37	66	8.63[a]	0.00	0.22
Weedicide	21	49	68	35	21.64[a]	0.00	0.34
Harvesting	52	18	67	39	1.65	0.19	0.10

Source: Authors calculation based on field survey data
[a]Significant at 1% probability level

Based on the 3-day weather forecast, farmers were advised on appropriate sowing times. Most often, farmers persisted with their normal sowing dates and resisted changing sowing dates. This could be because of village to village variation in onset of monsoon and type of soils. Farmers having clay soils go for sowing even with low rainfall as these soils have higher field capacity and thereby have ability to retain water for a longer time (volumetric soil moisture content remaining at field capacity is about 45–50%). Conversely, farmers with sandy soils wait for adequate rainfall due to lower field capacity before they start sowing. Farmers having access to irrigation tend to grow their crops on normal date of sowing.

Even though the follow-up on recommended specific variety was low (in 10% of plots in case of pearl millet and 32% of plots in onion), chi-square test indicated significant association (at 1% level) between the crop type and follow-up on advisory (refer to Table 7). One reason for farmers unable to follow-up on recommended seed variety was that these varieties were not easily available to farmers in adequate quantity in the market. Thereby, the farmers opted for other varieties in the market from private companies.

Irrigation

The advisory for irrigation was based on the phenological growth stage of the crop and rainfall forecast (during prolonged dry spells). For pear millet, the critical stages for irrigation were tillering, flowering, and grain formation. In onion, the critical stages included seedling stage, vegetative, bulb formation, and development stages. There was very low follow-up on the irrigation advisories in both types of crops. Farmers were irrigating as per the prevailing local conditions using their respective experiences. Type of soil again plays an important role in the farmer's decision on timing of irrigation. Sandy soils have lower water retention capacity and require more frequent irrigations during dry spells as compared to clayey soils. Another reason could be that farmers have water sharing arrangements with neighboring farmers possessing groundwater source such as tube well. Sometimes 8–10 farmers share the same irrigation source in which case it could take about 12–15 days for a farmer to get his turn to irrigate and hence the irrigation advisory timing does not match with the situation.

Farm Yard Manure (FYM) Application

There was high follow-up on advisories regarding application of FYM in 74% and 85% of the plots belonging to pearl millet and onion crops, respectively. However, there was no significant association between the type of crops and follow-up since farmers adopted the recommendations irrespective of type of crop. The advice on FYM Application is given at the time of land preparation such that FYM is incorporated into the soil before last harrowing so that volatization losses of ammonia are reduced.

Fertilizer Application

Fertilizers applied included urea, DAP, complex fertilizers, and micronutrients. The advisories for urea application were sent at the time of transplantation, 30 days and 60 days after transplantation in the case of onion. For pearl millet, the advisories for urea application were sent at the time of sowing and 35 days after sowing.

The follow-up on urea advisory was in 71% and 60% of plots for pearl millet and onion crops, respectively. Urea is one of the major sources of nitrogen fertilizer and easily available in the market. Chi-square test indicated no significant association between the type of crop and adoption of advisories.

The advisories for DAP/complex fertilizer application was at the time of transplantation in the case of onion and for pearl millet, it was at the time of sowing as basal dose. In the case of DAP and complex fertilizers, the adoption was high in onion (in almost 78% and 84% of plots, respectively) as compared to pearl millet (in about 13% and 9% plots, respectively). Accordingly, the chi-square test indicated significant association (at 1% level) between the type of crop and adoption of the advisory (Table 7). The Cramer V, a measure to compare the strength of association between any two cross-classification tables, was 0.63 and 0.74 in the case of DAP and complex fertilizers, respectively indicating strong associations between crop type and adoption of advisories. Adoption of DAP and complex fertilizer advisory was higher in onion crop since nutrient requirement is more in onion as compared to pearl millet. Farmers also invested more in onions due to higher expected returns. Some farmers could not purchase DAP due to nonavailability during peak demand period and instead applied complex fertilizers such as 19:19:19 and 10:26:26.

For micronutrient advisories, even though adoption of advisory was on lower side in both types of crops, it was relatively high in onion crop (in 41% of plots) as compared to pearl millet (6% of plots). Chi-square test indicated a significant association between adoption of advisory and type of crop. In general, adoption of advisories on micronutrients might be less due to relatively poor awareness about the importance and role of micronutrients among the farmers. Some progressive farmers followed the advisory due to awareness created by agriculture service center sand private dealers.

Pesticide Application

Onion crop is infested by major pests like thrips, aphids, onion fly, etc. while pearl millet is infested by stem borer and blister beetle. The adoption of advisory was high in case of onion (almost in 64% of plots) as compared to pearl millet (in only 17% of plots). This could be due to higher pest occurrence in onion as well as higher economic importance of the crop as compared to pearl millet. The chi-square test indicated significant association (at 1% level) between the type of crop and adoption of advisory. The Cramers V was 0.46 indicating reasonably strong relationship between the two variables.

Weeding

Advisories on weeding included hand weeding and use of weedicides. Adoption of advisories on hand weeding was more in case of pearl millet (in 59% of plots) as compared to onion crop (36% of plots). The chi-square test indicated significant association (at 1% level). In case of advisories on application of weedicides, the adoption of advisories was high in case of onion crop (in 66% of plots) as compared to pearl millet (in only 30% of plots). This indicates that farmers are willing to spend more on onion crop, which fetches higher returns. There was a significant association between the adoption of advisory and crop type. Onion is very succulent and close growing crop, which makes it susceptible to damage during hand weeding operations, which in turn increase infestation of pest and disease.

Harvesting

The follow-up on advisories was high in both pearl millet and onion crops in 74% of plots and 65% of plots, respectively. This was mainly by way of taking precautionary measures during light rains. Farmers harvested the crops before the event of rainfall and thereby avoided losses. Farmers could avoid risk of loss from pest incidence and reduced quality of grain in pearl millet and rotting in onion. Also it becomes difficult to harvest onion once rains occur. There was no significant association between type of crop and adoption of advisory. At the time of harvesting, losses due to unseasonal rainfall is very high and also affect the quality of produce. Sometimes farmers lose their whole produce/income within 1 h due to unseasonal rainfall. Therefore, adoption of advisories at the time of harvest stage of crop was higher in both types of crops.

Therefore, as indicated by Chandra Babu et al. (2012), it is important for extension programs to consider tailoring the delivery of agricultural information to the different information search behaviors of farmers. The quality of information, timeliness of information, and trustworthiness of information are the three important aspects that have to be delivered to the farmers to meet their needs and expectations (Mittal and Tripathi 2009).

The results also corroborate with observation of Weiss et al. (2000) that even though information needed for diverse groups of end users growing the same crops maybe similar, the differences in implementation of the information occur due to differences in human and financial resources available, and methods of information dissemination. Mittal and Mamta (2012) found that although farmers indicated that they would like more information delivered to them via mobile, they were not proactively seeking it out.

Conclusions: Implications for Making Advisory Systems Demand Driven

Setting up an advisory system that is location specific and responsive to individual famers needs, involves interactions between diverse actors (farmers, NGOs, Research institutions, Government institutions, private organizations). One of the key factors for developing such a system is willingness of diverse institutions to collaborate and convergence of respective strengths. Even though formal contracts provide framework for working together, it is also important to develop trust and accountability among the different actors for smooth delivery of services as well as the ability to innovate and respond to changes on the ground.

It also requires certain amount of resources to improve capacities, especially those of advisors, both subject-matter specialists and local para-agronomists, who directly engage with the farmers (client) and form the critical link in the advisory system.

In the context of uptake/follow-up of advisories by the farmers, there were differences depending on the type of crop grown. In the present case, the major kharif crops for which advisories were sent were pearl millet and onion. Weather advisories had good uptake irrespective of the type of crop. Among the advisories pertaining to fertilizer application, there was high follow-up in case of advisories for application of DAP and complex fertilizers in case of onion and less in pearl millet. In general, farmers tend to follow on advisories more in high-return crops (such as onion) as compared to cereal crops such as pearl millet.

Some of the advisories were not adopted irrespective of the type of crop such as irrigation scheduling. In such cases, content of advisories can be upgraded by taking into account the location or plot specific characteristics (such as soil characteristics, water availability, etc.) and thereby increasing the rate of adoption of advisories.

Sometime other issues such as poor mobile connectivity (quality and coverage) and low literacy levels affect the uptake of advisories. It is important to recognize that farmers generally adopt new technologies and practices when these are built on their knowledge and skill sets and demonstrate usefulness of specific technologies.

Extreme weather conditions (such as drought) could also negatively affect the demand for advisories. Crop failures would directly impact on the number of farmers following up on the advisories. Such risks have to be taken into account, especially in cases where agro-advisory services are chargeable. Some farmers are willing to pay if the information is customized to their specific soil, crop, and weather conditions.

The above findings can help in refining the content of specific advisories and thereby making the system demand driven based on farmer's needs. This would help in increasing the effectiveness of the advisory system. Such feedback would also help in making the whole agro-advisory system sensitive and responsive to changes and requirements on the ground.

Cross-References

▶ Climate Change Adaptation Through Science-Farmer-Policy Dialogue in Mali
▶ Climate Smart Adaptations in the African Tropics: Scaling Weather Information for Decision Support Outcomes in Nigeria Savannahs
▶ Educating Farmers in Rural Areas on Climate Change Adaptation for Sustainability in Nigeria
▶ Enhancing Resilience of Livelihoods and Production Systems to Climate Variability and Other Related Risks in Africa
▶ Scaling Up Climate-Smart Agricultural (CSA) Solutions for Smallholder Cereals and Livestock Farmers in Zambia

References

Al-Karablieh EK, Al-Rimawi AS, Hunaiti DA (2009) Logit models for identifying the factors that influence the adoption of barley production technologies in low rainfall areas. Jordan J Agric Sci 5(3):251–265

Anderson JR (2007) Agricultural advisory services, Background paper for the world development report 2008

Birner R, Davis K, Pender J, Nkonya E, jayasekeram PA, Ekboir J, AdielMbabu DJS, Horna D, Benin S, Cohen M (2009) From best practice to best fit: a framework for designing and analyzing pluralistic agricultural advisory services worldwide. J Agric Educ Ext 15(4):341–355. https://doi.org/10.1080/13892240903309595

Chandra Babu S, Glendenning CJ, Asenso-Okyere K, Govindarajan S (2012) Farmers' information needs and search behaviors, case study in Tamil Nadu, India. IFPRI discussion paper 01165. Eastern and Southern Africa Regional Office, International Food Policy Research Institute, Washington, DC

Faure G, Rebuffel P, Violas D (2011) Systemic evaluation of advisory services to family farms in West Africa. J Agric Educ Ext 17(4):325–339

Faure G, Davis KE, Ragasa C, Franzel S, Babu SC (2016) Framework to assess performance and impact of pluralistic agricultural extension systems the best-fit framework revisited IFPRI discussion paper 01567 November 2016. Development Strategy and Governance Division, Washington, DC

Gadrey J (1994) Les relations de Service Dans le Secteur Marchand. In: Bandt J, Gadey J (eds) Relations de Service, Marchés de Services. CNRS Editions, Paris, pp 23–41
Gowda CMJ, Dixit S (2015) Influence of farmers educational level on comprehending, acting upon and sharing of agro advisories. J Agric Rural Dev Trop Subtrop 116(2):167–172
Lobo C, Chattopadhyay N, Rao KV (2017) Making smallholder farming climate-smart integrated agrometeorological services. Econ Pol Wkly LII(1):53
Maini P, Rathore LS (2011) Economic impact assessment of the agrometeorological advisory Service of India. Curr Sci 101(10):1296
Meera Shaik N, Jhamtani A, Rao DUM (2004) Information and communication technology in agricultural development: a comparative analysis of three projects from India. The Agricultural Research and Extension Network, Department for International Development (DFID), London. Network Paper No. 135
Mittal S, Mamta M (2012) How Mobile phones contribute to growth of small farmers? Evidence from India. Q J Int Agric 51(3):227–244
Mittal S, Tripathi G (2009) Role of mobile phone technology in improving small farm productivity. Agric Econ Res Rev 22(Conference Number):451–459
Mittal S, Gandhi S, Tripathi G (2010) Socio economic impact of Mobile phones on Indian agriculture Surabhi Mittal Sanjay Gandhi Gaurav Tripathi. ICRIER working paper 246. International Council for Research on International Economic Relations, New Delhi
SanneChipeta (2006) Neuchâtel Group. Pub. Swiss Center for Agricultural Extension and Rural Development (AGRIDEA). Department for International Cooperation, Lindau
Weiss A, Van Crowder L, Bernardi M (2000) Communicating agrometeorological information to farming communities. Agric For Meteorol 103(2000):185–196

Vulnerability and Adaptation Levels of the Construction Industry in Kenya to Climate Change

117

Onkangi N. Ruth, Mwangi Peter Njiiri, Erick Maklago, and Ondari Lilian

Contents

Introduction 2384
Impacts of Climate Change on Infrastructure 2387
Assessment of Climate Change Impacts on Infrastructure 2388
    Vulnerability 2389
    Adaptation 2390
    Reactive Adaptation: Adapting as the Changes Manifest Themselves (Losses) 2390
Methodology 2391
Results 2391
    Vulnerability Assessment 2392
    Adaptation Assessment 2394
Conclusion 2397
Recommendations 2397
Cross-References 2398
References 2399

Abstract

Climate change is increasingly becoming a threat to future development plans, sustainability of existing infrastructure, and biodiversity conservation. Assessment of vulnerability levels, adaptation measures, models, and climate

O. N. Ruth (✉) · E. Maklago
Research and Business Development, National Construction Authority, Nairobi, Kenya
e-mail: konkangi@gmail.com; e.wesonga@nca.go.ke

M. P. Njiiri
Kenya Wildlife Service, Nairobi, Kenya
e-mail: pnjiiri@kws.go.ke

O. Lilian
University of Toledo, Toledo, OH, USA

Clayco Construction, Inc., Chicago, IL, USA
e-mail: nyanchyond@gmail.com

W. Leal Filho (ed.), *Handbook of Climate Change Resilience*,
https://doi.org/10.1007/978-3-319-93336-8_65

change variables have majorly focused on biological systems. Besides biodiversity, the construction industry faces equally high threats from climatic changes with enormous carbon price-related risks.

Construction in the developing nations is gaining momentum and is a development indicator as well as a major contributor to GDP. However, infrastructure is very vulnerable structurally and financially to extreme weather conditions and events. This calls for a paradigm shift in conducting business in the construction industry in developing countries. Awareness of climate change among engineers, architects, quantity surveyors, construction managers, and contractors in the industry in Kenya was evaluated in this chapter. Further, using vulnerability and adaptation indicators in the assessment tool, the extent of integration of climate knowledge in planning and implementing infrastructure projects in Kenya was evaluated. The findings indicate that there is great need to initiate early warning systems, incorporate rigorous risk assessments to determine infrastructure vulnerability levels, integrate adaptation measures and strengths for infrastructure and buildings to continue functioning in a changing planet, and avert retrogression and mark-timing development wise.

Keywords

Climate change · Construction · Awareness · Mitigation · Adaptation · Vulnerability

Introduction

Climate change is a major global challenge whose effects are likely to be felt for years even if its causality is completely or suddenly withdrawn now. While endeavoring to lower greenhouse gas emissions in a bid to combat climate change effects on progeny, the effects of the already altered climate are here. Our mitigation efforts in Kenya though at best feeble hold potential for possible synergy and will bear fruit in the future. Nonetheless, there is little influence in changing the here and now hence the need for incorporating adaptation measures in new and existing infrastructure. The infrastructure and building industry permeate every sector through provision of much needed housing and connectivity that is indispensable for growth and development.

Kenya's climate change mitigation and adaptation efforts have been invested in biodiversity and human health. The agriculture sector has had more attention especially in arid and semiarid lands (ASALs) with Makueni County Government taking the lead. Climate change affects every being and thing that is influenced by weather conditions. Infrastructure though man-made to meet human needs is embedded in the ecosystem and is therefore subject to climatic influences. For continued functioning amid variations in temperature, wind strength, and water availability, infrastructure "must be designed to meet climate standards, such as 10 year precipitation conditions, low stream flows, and high and low temperatures" (Wilbanks and Fernandez 2014).

Nevertheless, the attention given to climate change infrastructure-wise is focused on mitigating GHGs, while adaptation and building resilience of infrastructure to climate change is not given the attention it deserves. Although Kenya as a developing nation has less obligations on cutting GHG emissions under the Kyoto Protocol (1997), it remains very vulnerable to the negative effects of climate change hence the urgent need to adapt and enhance resilience besides mitigating climate change.

According to the National Climate Change Response Strategy Paper, "the damage caused by the eight-month 1997/1998 El-Niño rains to the country's transport and telecommunication infrastructure was estimated at one billion US Dollars." The Response Strategy Paper further recognizes physical infrastructure as a priority area in need of immediate action and calls for climate proofing of infrastructure in Kenya. To this effect, there is a National Adaptation Plan (NAP) reviewed every 5 years. Overall, the country has addressed climate change by enacting laws and policies to guide its reforms and mainstreaming into sectors. These include the Climate Change Act (2016) which empowers the National Climate Change Action Plan (NCCAP) "to identify strategic areas of national infrastructure requiring climate proofing" (National Climate Change Action Plan 2013–2017). The National Climate Change Response Strategy recognizes the threat of climate change to infrastructure (NCCRS 2010).

Kenya's climate change adaptation strategy acknowledges the high growth in infrastructure and the sector's requirement to incorporate low-carbon technologies as well as be climatically resilient. Per the strategic plan, climatic changes expected in the country include continued increase in temperature, higher number of hot days and nights, consequently a decrease in cold days and nights, and increased precipitation more so in highlands and coastal regions as wet areas are forecasted to get more wet and dry areas even drier. The plan further recognizes increase of GHGs with population growth as well as increase in consumption as the economy expands "from 59 million tons of carbon dioxide equivalent (MtCO2e) in 2010 to 102 MtCO2e in 2030."

Besides instituting and promoting mitigation measures for the anticipated and experienced negative climatic changes, there is dire need for adaptation to cope with the already altered conditions since the effects of this global phenomenon have already caught up with us. Loss of infrastructure and lowered functional performance as well as collapsing of buildings due to weather-related interruptions have been widely observed. Anecdotally, waste management is a major problem in Kenya, mostly in cities. Flooding in Nairobi and other cities in Kenya is a major problem after every downpour, due to clogged drainage systems with plastics and metal cans. Poor solid waste management and overburdened sewers are thus observed to be an amplifier of negative impacts and improper functioning of infrastructure arising from climate change. In harmony with the Cancun Agreement (United Nations 2010), adaptation and mitigation needs to march in lockstep in the infrastructure sector as well.

Infrastructure developed is posed to be impacted by climate change both directly (through weather and climate) and indirectly through material costs, utility,

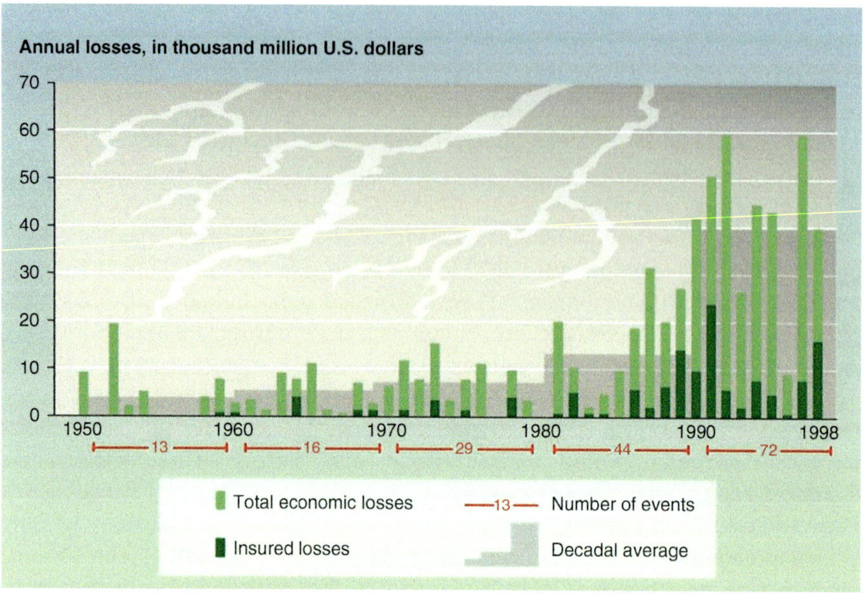

Fig. 1 Global costs of extreme weather events from 1950 to 1998. Climate change and weather-related interruptions have proven to be costly. (Source: UNEP)

maintenance, insurance, and legislation in form of carbon tax (Camilleri 2000). Impacts of climate change can be quantified by considering climatic changes at regional level and effects on the expected functional performance of infrastructure.

Buildings appear to be standing alone in the mitigation and adaptation wave. The aforementioned policy documents address infrastructure solely and fail to loop in buildings, despite buildings being established to be the places where people spend 90% of their time (Brager 2013). The number and impact of collapsing of buildings has been on the rise over the years in Kenya. Globally buildings are responsible for one third of GHG emissions. IPCC's Fourth Assessment Report quantified building-related GHG emissions to be approximately 8.6 million metric tons CO_2 in 2004 (IPCC AR4 WGII 2007 as quoted in UNEP 2009). Consequently, the construction industry largely contributes to climate change, and climate change in return largely impacts the industry functionally and financially. Climate change and weather-related interruptions have proven to be costly as illustrated in Fig. 1. Therefore, mitigating impacts and building climate resilience are necessary for the sector's survival and performance as well as aversion/minimization of expenses incurred.

Vulnerability of small and large and simple and complex infrastructures to functional failures is an underlying concern that is growing with every investment, population growth, and worsening climate conditions. However, it is also worth noting that vulnerabilities and risks associated with climate change are heightened where infrastructure is exposed to a series of weather stresses, not just a one-time climatic or weather-related impact. Therefore, for adequate risk management, innovative interventions need to be adopted for structural and operational aspects of infrastructure (Wilbanks and Fernandez 2014). These interventions can be adopted and tailored to local prevailing conditions by first understanding the vulnerabilities and adaptation levels of the industry at local and national level.

This research acknowledges that impacts of climatic changes and varying weather conditions transcend the physical infrastructure itself to the costs incurred in retrofitting, cleanups, repair, compensation, environmental impacts, derailed project outcomes due to disruptions, health and safety, and time overruns due to extensive interdependencies (inter-sector, interregional) in the industry that lead to unexpected cascading impacts and reduction in economic outputs.

Research literature in the construction industry in Kenya has not given major attention on climate change assessments unlike the agriculture and biodiversity conservation sectors. This chapter therefore is breaking a new ground in climate change vulnerability and adaptation assessment in the construction sector in Kenya thus serving as a start to addressing concerns by stakeholders and policymakers regarding the construction industry and climate change.

Impacts of Climate Change on Infrastructure

Climate change challenges developed and developing nations to rethink about infrastructure development in both urban and rural settings. Fortunately, developing nations still have 66% of infrastructure to build, which presents an opportunity to do it right and sustainably; otherwise they will attract retrogression and all manner of losses in the long run. This also highlights the crucial role developing nations have in sustainable resource consumption and production.

Changes associated with climate changes are forecasted to contribute to degradation of infrastructure and building materials (Rosenzweig et al. 2011a; Wilby 2007). Water availability is likely to affect demand and supply especially in prolonged dry spells, and this can affect construction projects as they are heavily reliant on water. Increased precipitation will affect storm water drainages, load, wastewater treatment systems, and road networks causing project delays and reducing the lifetime of infrastructure (Cela et al. 2010; Kirshen et al. 2008; Nelson et al. 2009). Equipment and machinery will be impacted by extreme temperatures requiring regular maintenance, reducing their lifetimes; increased precipitation will increase strain on sewer flows, culverts, and bridges and cause pollution on waterways, affecting waterway transportation systems and infrastructure (Major et al. 2014).

Cascading of failures due to linkages of infrastructure will affect the national economy and health hence the need to increase coping abilities of the masses and supporting infrastructure networks. Table 1 gives a summary of anticipated impacts with weather variations.

Table 1 Summary of expected impacts on infrastructure with weather variations. (Source: Andrey and Mills 2003; World Bank 2015)

Climatic exposure	Potential impact
Increased temperature	High energy use for cooling Material fatigue and enhanced aging Decreased comfort and severe health challenges Buckling of rail tracks and roads Thermal expansion and strains causing cracking/buckling, slab curl, and blow ups in joint designs Softening and bleeding Increased oxidation/hardening of asphalt binder, crack, and joint sealants Increased span and strength of urban heat island effects associated with paved surfaces
Increased sea level rise	Flooding High tides damage embankments Damage and inaccessibility to low-lying infrastructure Soil subsidence risks Increased water table, surface flooding damage to walls and foundations
Increased precipitation	Overflowing sewers Water intrusion Opportunity for increased hydropower generation Increased inundation and decreased retreating rates of flash floods hence buildup and risk in eroding of infrastructure like railway construction Damage to foundations and basements Landslides and mudslides Obliteration of buildings and infrastructure Slope failure Increased sedimentation of drainages Erosion and washout Lengthy construction duration
Decreased precipitation and droughts	Sinking of infrastructure due to decreased water table
Increased wind strengths	Opportunity for energy generation Damage to overhead power lines, rail transport, and cranes Increased threat to structural integrity of lengthy bridges

Assessment of Climate Change Impacts on Infrastructure

Various developed countries and corporates like World Bank have assessed the resilience of their infrastructure to climate change and stepped up adaptation measures through retrofitting for existing infrastructure and incorporating climate change in design and construction of new infrastructure (World Bank 2015). Exarro, a mega coal company in South Africa, used downscaled models to assess climate change impacts on the company's operations and employees' residential locations. Similarly, Vale, a Brazilian company, conducted vulnerability assessment on water availability and biodiversity (Nelson and Schuchard 2011).

To integrate climate into building design in Eastern Africa, the UN HABITAT has developed material on "Sustainable Building Design for Tropical Climates Principles and Applications for Eastern Africa" (Butera et al. 2015). The uptake of the information in this material is slow though crucial.

European Investment Bank (EIB) has developed strategies which address climate change adaptation. The bank also supports adaptation projects in water resource management.

European Bank for Reconstruction and Development (EBRD) has developed a methodology for establishing risks posed by climate change and related impacts. Pilot Program for Climate Resilience (PPCR) is another program that the European Bank for Reconstruction and Development is also supporting. Pilot Program for Climate Resilience gives pro-adaptation technical assistance to investment projects. Further, it has developed and espoused the ISO Guide 64 which fosters environmental aspects in product standards. This provides standardization of environmental aspects even to non-environmental professionals (European Commission 2007).

Vulnerability

All infrastructures are of great value and importance to everyday operations, both financially as an investment and as a critical element for a vibrant economy. However, vulnerability and loss for non-operation or disruption can be localized or widened, depending on the sector and type of infrastructure (HM Government 2011).

According to the Commission to the European Parliament 2007 report: "The location of an infrastructure, together with the adaptive capacity of local businesses, governments and communities, usually determines an asset's climate sensitivity and vulnerability. This vulnerability to various climate parameters is also strongly sector-specific and closely linked to the technology used for construction and operation."

Vulnerability of the infrastructure more so the transport sector is both through the flooding and heat avenues. Increased temperature will exert an energy demand for cooling homes and offices, and cold temperatures will draw a demand for heating. Heat can reduce the life of tires and cause roads and pavements made of concrete to "explode or buckle." Under similar circumstances, asphalt roads can soften and are reported to have an even shorter lifetime. Steel rails are susceptible to excessive heat which can distort rail lines and cause them to buckle tampering with the alignment resulting in derailments (Mehrotra et al. 2011).

Infrastructure more so in urban areas is interdependent especially through the water and wastewater systems as well as the drainage system. Infrastructure to infrastructure ties are also very present (water or non-water); therefore, weather-related disruptions whose genesis is hydrologic – floods and intense precipitation – can reverberate through the entire system. Therefore, management practices in one infrastructure will definitely impact another (Wilbanks and Fernandez 2014).

Vulnerability is exacerbated by the aging of many built infrastructures and growing populace hence overburdening existing infrastructure, sewer lines, and bridges (Wilbanks and Fernandez 2014). Development pressures to build on vulnerable areas are mounting with every population increase and appetite for monetary gain

(Titus et al. 2009). Lack of coordination and political fragmentation hinders management for climate change response (Leichenko et al. 2010). There lacks "local capacity to translate climate science into public policy" (Krause 2011 as quoted by Wilbanks and Fernandez 2014).

Infrastructure is not just forecasted to be negatively influenced by climate change or cause climate change. It also can reduce climate change risks if resilience is built into the structures through assimilation of a range of adaptive behaviors: "physical/capital equipment adaptations; technology adaptations; and institutional adaptations; self-initiated 'autonomous' adaptations; and 'planned' adaptations related to changes in external signals, requirements" (ibid.).

Adaptation

Adaptation can be planned adaptation/proactive adaptation or reactive adaptation. Proactive adaptation is recommendable as it reduces system stressors and sensitivity/vulnerability. Reactive adaptation responds to change as it manifests itself. Proactive adaptation can be achieved through material availability, client's demand for resilient infrastructure/sustainable buildings, and lifetime infrastructure service. The population in the urban areas is projected to rise by more than two thirds by 2050. Approximately 90% of the increase will be in urban areas of Africa and Asia (UN 2014). Therefore, there is a huge remaining opportunity to build climate-resilient infrastructure in the developing countries and avert retrofitting costs besides other losses associated with climate change.

Disregarding climate change attracts serious risks. But adapting in the wrong way also attracts greater risks. "As long as climate risk analysis is fully integrated in the project cycle, starting from the upstream stages of planning at the national, river basin, regional, and power-pool levels, and in pre-feasibility studies of individual investments, climate risks can be significantly mitigated in a cost-effective manner" (Cervigni et al. 2015). Robust adaptation strategies have been suggested to counter uncertainties, for example, the mini-maxi criterion which has a high level of risk aversion (ibid.). Integrated models developed and criterions adapted can enable adaptation and building resilience into climate change.

Reactive Adaptation: Adapting as the Changes Manifest Themselves (Losses)

Adaptation measures suggested include leak detection, water conservation, management interventions in buildings and water systems including "low-flow toilet," and water reuse in landscape irrigation (Rosenzweig et al. 2007). Urban heat islands can be reduced through green infrastructure – tree planting, green roofs, and permeable pavements (Stone et al. 2010; Hamin and Gurran 2009) – and solid waste management and removal of debris to avoid clogging of drainages and encouraging of inundation (Mehrotra et al. 2011).

Managing regional convergence of infrastructure is a risk management strategy. Implementing appropriate adaptation strategies and initiating early warning and monitoring systems will enable risk management of disruptive impacts associated with climate change. This can be drafted through assessment of infrastructure interdependencies, revising building codes, material, and engineering standards for infrastructure as well as initiating transformational changes where necessary. Nairobi and other Kenyan metropolis can learn from developed world cities like Boston, which have gone a step further to require developers of new projects to incorporate effects of climate change in their projects. The Public Works Department of Boston has further evaluated impacts of heat on road durability (Wilbanks and Fernandez 2014).

Developing or adopting integrated assessment models (IAMs) to examine climate change scenarios will pivot risk management. IAMs are tested and tried approaches that incorporate GHGs and respective mitigation measures. They enable risk quantification for societies and economies. *Envision system* by the Institute for Sustainable Infrastructure is a tool that is very useful in rating resilience of infrastructure, which is yet to be employed in Kenya. The tool incorporates multihazard resilience, infrastructure asset management planning, business continuity, and ecosystem protection (Wilbanks and Fernandez 2014).

Adapting to climate change through value creation and value protection has yielded enormous business opportunities. Robust technologies, designs, and materials have been developed that improve infrastructure resilience, strengthen adaptation strategies, and enhance conservation of resources.

Methodology

Researchers assessed adaptation and vulnerability after adopting some criteria of respective indicators as illustrated in the Fig. 2. A descriptive survey tool of structured questionnaires assessing the indicators was administered to contractors, engineers, architects, quantity surveyors, construction firms, project managers, and construction managers. Desktop study, observation, and assessment of infrastructure more so where failures were reported yielded information to complement the feedback from questionnaires on vulnerability and adaptation levels. These findings gathered from the expert community were analyzed using SPSS, and the evaluation forms the bulk of this chapter.

Results

Two hundred seventy-two questionnaires were issued, and 194 duly filled in questionnaires were collected, giving a response rate of 71%. The awareness survey revealed that 86.7% of the respondents reported to have noticed climate change through observation or learnt through the media or while undertaking professional training in school. None of the respondents denied climate change as a fable or a fallacy. Increased precipitation, stronger winds, higher temperatures, and higher

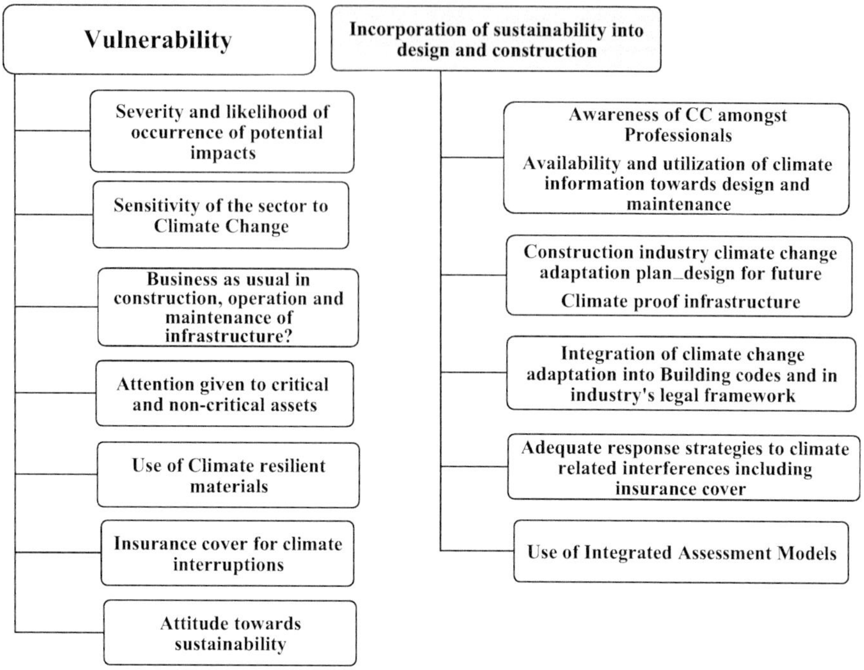

Fig. 2 Study's adopted criteria of respective indicators for assessment of vulnerability and adaptation. (Source: Research Data, 2016)

tides were widely reported to be the noted climatic changes. As for the general feel on climate change and infrastructure, 33% of the respondents expressed fear about climate change, 46% were hopeful, and 6.7% expressed concern/worry, while 13.3% expressed confusion about climate change.

Vulnerability Assessment

It was worth noting that confidence on the suitability of existing infrastructure for climatic changes in Kenya was only ascertained by 27% of the respondents as illustrated in Fig. 4. This can be ascertained by an assessment on climate resilience. Losses of over KShs. 600 million were reported to have been incurred and directly attributed to weather interruptions.

Awareness Among Professionals

There was no correlation between awareness levels and profession, so no deduction whether engineers are more aware of climatic impacts on infrastructure than architects or quantity surveyors or vice versa. There was also no correlation between a profession and switch to use of climate-resilient materials, as well as attitude toward use of sustainable technologies.

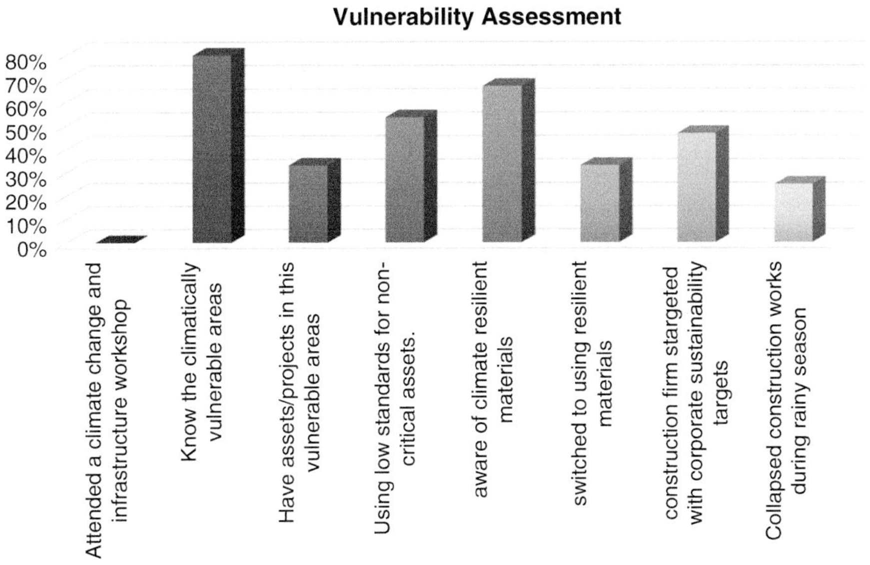

Fig. 3 Results for key indicators on vulnerability levels of the construction industry in Kenya to climate change. (Source: Research Data, 2016)

It was very noticeable that none of the respondents has ever attended a climate change and its impacts on infrastructure workshop. Besides noticing climatic changes, none had been apprised particularly on climate change and its impact on infrastructure, building for the future, and retrofitting to increase resilience.

Eighty percent of the respondents knew the climatically vulnerable areas, and 33% had assets/projects in these vulnerable areas. Awareness on vulnerable areas may have guided in investment on assets in these areas. But this could be misleading as there are engineering interventions that can enable anchorage of assets in vulnerable areas. This can be averted by investing in research and development as well as disseminating knowledge on new technologies, materials, and designs that can thrive in vulnerable areas.

A summary of these results on assessment of critical vulnerability indicators is illustrated in Fig. 3.

Confidence Level on Key Assets

The respondents' confidence on key roads' ability to handle percent increase in flows due to intense rainfall was 58%. A higher figure of 83% was the reported confidence for pavements withstanding more prolonged exposure to high intensity heat. The confidence level dropped to 59% for rain intensity affecting slope of most buildings as shown in Fig. 4. Probably due to increased building collapses in rainy season, 83% of the participants expressed need for marking and maintaining of bridges with high-risk value.

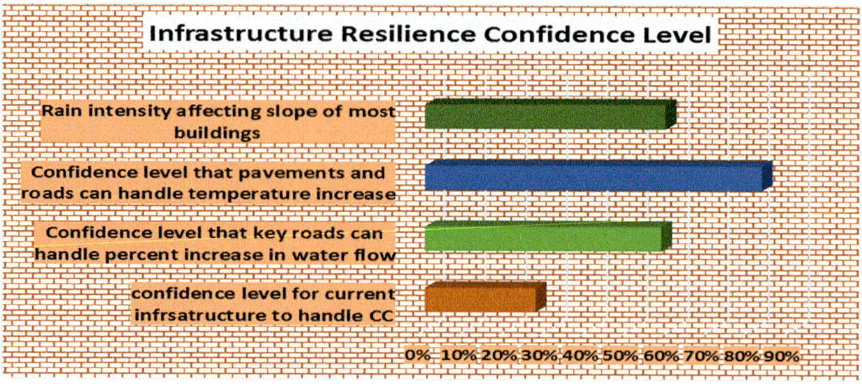

Fig. 4 Results on assessment of perception and confidence level of participants on infrastructure, handling certain weather extremities. (Source: Research Data, 2016)

Attention to Critical and Non-critical Assets

53.4% of the respondents reported using low standards for non-critical assets, and 75% of the developers and contractors have *not* insured against infrastructure collapse.

Infrastructure Failures Attributed to Climate Change

Eighty percent of the respondents reported having experienced weather-related disruptions in their construction projects. Of infrastructure projects that collapsed, 25% happened during heavy rains. Increased inundation beyond the usual flood plains and low retreating rates of flood waters led to the collapse of the KShs. 80 million (800,000USD) bridge in Galana as shown in Fig. 5. Increased inundation also tested the structural integrity of Kenya's most recent and celebrated standard gauge railway flagship project as shown in Fig. 6.

Adaptation Assessment

Adaptation in Kenya is majorly reactive not proactive. Respondents who experienced weather-related interruptions in their projects implemented measures after construction not before at design phase.

Professionals Considering Adaptation Measures in Design

Only 50% of the respondents reported to use models that incorporated green building design. 93.3% stated they would consider adaptation measures despite the costs, and achievement of zero energy is a goal to 73.7%. 66.7% of the respondents reported to be aware of climate-resilient materials, and only half of these had switched to using these materials and replace existing materials where possible. 58.3% reported to consider rainwater load, storm damage, indoor climate, wind speed, water use efficiency more so in arid areas, and material recycling and

Fig. 5 Experienced climate change impacts in infrastructure: collapsed Galana bridge following increased inundation of river Tana due to heavy precipitation. (Source: State Department of Public Works-Kenya 2015)

Fig. 6 Experienced climate change impacts in infrastructure; a washed-out section of SGR following heavy precipitation. (Source: Muasya 2016)

reuse. Rainwater load was the most considered climatic change in any construction project. And aqua planning on roads was given attention by 58.7%.

Perception and Uptake of Sustainable Methods

A significant number of the respondents (86.7%) felt sustainable building methods should be used even if capital costs are greater. However, only 46.6% of the construction firms targeted had corporate sustainability targets! 91.7% expressed that there is inadequate awareness on sustainable construction and climate change within the industry in Kenya.

Compliance to Legal Framework on Climate Change and Sustainable Development

Compliance to Kenyan environmental regulations as a minimum requirement to infrastructure resilient is as low as only 33% of the respondents stated to be guided by Environment Management and Coordination Act 1999, Environmental Impact Assessment/Environmental Audit regulations 2003, Building Codes, and Kenya Bureau of Standards for materials. 66.7% of these felt the regulations had big effect on the industry. The rest ignored these regulations citing varied reasons including corruption as well as uncoordinated or conflicting legal/policies thus opting to ignore them.

A summary of the results on adaptation levels is illustrated in Fig. 7.

Respondents, however, voiced that the legal framework governing the industry needs to incorporate climate change especially the building code, to disseminate knowledge of disaster management techniques, to raise level of awareness on impacts of climate change and adaptation measures, to map vulnerable areas, and to develop regional climate change models and availing them to construction industry stakeholders. Further, climate and meteorological (met) data was reported to be costly. Thus, many industry practitioners opted not to include it when designing for infrastructure to cut costs.

All the respondents called for improved early warning systems, and 91.7% expressed confidence that early adapters to climate change in the construction

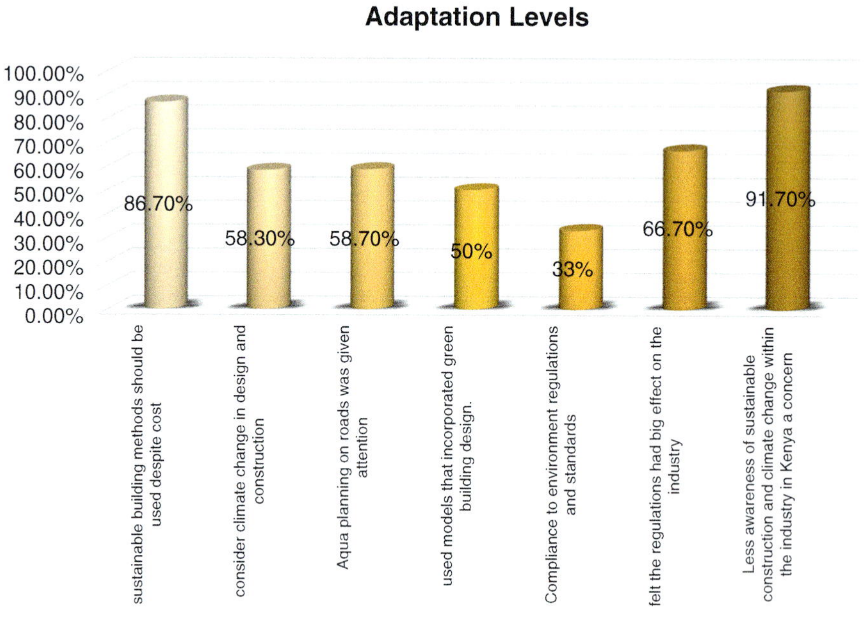

Fig. 7 Results on adaptation levels of the construction industry in Kenya to climate change. (Source: Research Data, 2016)

industry will leverage a competitive advantage over laggards. It was equally expressed that tenants will prefer green buildings though with higher rents but lower operating costs.

There is a high consensus among industry stakeholders on climate changes and strong evidence on impacts of climate change on infrastructure in Kenya. Evidently there is strong will among stakeholders to mitigate, adapt to climate change, and ensure the industry thrives in the changing planet. This goodwill needs to be nurtured and supported by the government as the major developer and beneficiary of investment for the people of Kenya.

While there is the will and "spirit" to incorporate sustainability in the industry, the "flesh" of implementation is weak due to costs, lack of popularity, and enforcement. Where people are aware of climate change and sustainable technologies, shifting to these materials is slow, further lagging adaptation. Therefore, mainstreaming climate change into the policies, materials standards, and laws governing the infrastructure sector is crucial. With business as usual and more infrastructure to build, developing countries are thus likely to remain vulnerable to climate change, attract retrogression, and reinforce poverty and inequalities as well contribute to the historical build of GHGs levels in the atmosphere thus affecting progressive reduction of GHG to safe levels as envisioned in international agreements. As country capabilities improve, the differentiated responsibilities in the Kyoto Protocol (1997) need to be made "common."

Conclusion

From the findings, it is evident that the construction industry in Kenya is very vulnerable to climate change due to low awareness and weak linkages between policy formulation and institutional framework for implementation. Regulations governing infrastructure have not adequately captured climate change and guided its mainstreaming into the sector. Adaptation to climate change in the construction industry is also extremely low, and enforcement of even the common climate change adaptation mechanisms among industry players is also wanting. Generally, expensive met data and lack of integrated climate change assessment models to support decision-making have led to more of building in the "dark" and future retrofitting costs. Considering the effect of interdependencies in infrastructure, equal attention is required in all development projects as opposed to apportioning more attention to "critical assets," while those deemed as non-critical assets are given less attention. This largely contributes to increasing vulnerability levels particularly in the developing countries like Kenya.

Recommendations

The government through construction industry regulating bodies needs to proactively adapt by:

1. Capacity building professionals and improve awareness levels of climate change among contractors, artisans, and developers. The government should also raise the floor and set precedence: establish a pattern by assessing and building resilience into state infrastructure.
2. Greater involvement and commitment from finance institutions: climate finance/green finance to encourage climate-resilient construction.
3. Evaluate adequacy and review policies, building codes, and laws to integrate CC and achieve adaptation objectives. Accordingly, have a climate change (CC) response strategy/policy focusing on infrastructure.
4. Initiate early warning and adaptation systems as well as investing in integrated assessment models for Kenya to guide planned adaptation.
5. Map vulnerable and critical areas, and advise proper engineering interventions for construction in these sites. This will enable developing of site-level adaptation strategies and enhance climate resilience of infrastructure operations in their surrounding regions.
6. Incentivize low-carbon investments like green buildings.
7. Green infrastructure should be secured in road expansion and rail construction.
8. Develop local carbon pricing and carbon footprinting.
9. Improve on met data disclosure and transparency.
10. Conduct vulnerability assessments at local/regional level as well as inter-sector and interregional collaboration in conservation to manage interdependencies.
11. Improve climate science base, and mainstream it into the infrastructure sector and academic training curriculum.
12. Use the investment in R&D locally to spur innovation and resilience of materials and technologies, in addition to making visible sustainable technologies and giving incentives to adopt sustainable approaches in construction.
13. Advocate for equal attention on critical and noncritical assets.
14. Improve waste management systems more so on solid waste through cleaning and campaigns against littering. And expand capacity of sewer lines to manage load with population growth and increased volumes of storm water. New urban areas have separate storm water drains and sewer lines.
15. Reverse perverse incentives on more profit for "unwise" construction and increase incentives for smart buildings and green ratings.
16. Incorporate climate change adaptation and mitigation measures to Environmental Impact Assessment (EIA)/Environmental Social Impact Assessment (ESIA) and reporting.

Cross-References

▶ Building Resilience of Urban Ecosystems and Communities to Sea-Level Rise: Jamaica Bay, New York City
▶ From Vulnerability Assessments to Low/No Regret Resilience Planning in Rural Contexts
▶ Resilience to Extreme Events in the City of La Paz, Mexico

- Strengthening Climate Change Adaptation in the Cities of West Africa: Policy Implications for Urban Resilience
- Urban Climate Change Vulnerability, Responses, and Policies in Qatar: An Assessment
- Urbanization, Climate Linked Water Vulnerability as Impediments to Gender Equality: A Case Study of Delhi, India
- Vulnerability of Uganda's Electricity Sector to Climate Change: An Integrated Systems Analysis

References

Andrey J, Mills N (2003) Climate change and the Canadian transportation system: vulnerabilities and adaptations. Dep Geogr Publ Ser Univ Waterloo 55:235–279

Brager S (2013) Benefits of improving occupant comfort and well-being in buildings. In: Proceedings of the 4th international Holcim forum for sustainable construction: the economy of sustainable construction, pp 181–194

Butera F, Adhikari R, Aste N (2015) Sustainable building design for tropical climates. UN-Habitat-UNON-United Nations Offices, Nairobi, Publishing Services Section

Camilleri T (2000) Implications of climate change for the construction sector: houses. BRANZ, Porirua

Cela M, Hulsey J, Titus G (2010) South Florida. In: Titus JG, Trescott DL, Hudgens DE (eds) The likelihood of shore protection along the Atlantic Coast of the United States, Volume 2: New England and the Southeast. Report to the U.S. Environmental Protection Agency, Washington, DC

Cervigni R, Liden R, Neumann J, Strzepek K (2015) Enhancing the climate resilience of Africa's infrastructure: the power and water sectors. World Bank Publications, Washington, DC

Climate Change Act (2016) (Ken)

European Commission (2007) Communication from the Commission to the European Parliament, the Council, the European Economic and Social Committee and the Committee of the Regions. COM (2007), 569, Brussels

Hamin M, Gurran N (2009) Urban form and climate change: balancing adaptation and mitigation in the U.S. and Australia. Habitat Int 33(3):238–245

HM Government (2011) Department for Environment, & Rural Affairs. Climate resilient infrastructure: preparing infrastructure for a changing climate (Vol. 8065). The Stationery Office

IPCC AR4 WGII (2007) Climate change 2007 – impacts, adaptation and vulnerability. Contribution of Working Group II to the Fourth Assessment Report of the Intergovernmental Panel on Climate Change, 2007. Glossary available online http://www.ipcc.ch/pdf/glossary/ar4-wg2.pdf

Kirshen P, Knee K, Ruth M (2008) Climate change and coastal flooding in Metro Boston: impacts and adaptation strategies. Clim Change 90:453–473

Kyoto Protocol (1997) United Nations framework convention on climate change. Kyoto Protocol, Kyoto, 19

Leichenko RM, O'Brien KL, Solecki WD (2010) Climate change and the global financial crisis: a case of double exposure. Annals of the Association of American Geographers, 100(4):963–972

Major D, Bader D, Leichenko R, Johnson K, Linkin M (2014) Projecting future insured coastal flooding damages with climate change

Mehrotra S et al (2011) Climate change and urban transportation systems. In: Rosenzweig C, Solecki D, Hammer A, Mehrotra S (eds) First assessment report on climate change and cities (ARC3). Cambridge University Press, New York

Muasya P (2016, 21 November) Floods wash away part of Standard Gauge Railway (SGR) project. Standard Media Group. Retrieved from https://www.standardmedia.co.ke/article/2000224139/floods-wash-away-part-of-standard-gauge-railway-sgr-project
National Climate Change Action Plan (2013–2017) (Ken)
National Climate Change Response Strategy (2010) (Ken)
Nelson J, Schuchard R (2011) Adapting to climate change: a guide for the mining industry. Business for Social Responsibility
Nelson C et al (2009) Forecasting the combined effects of urbanization and climate change of stream ecosystems: from impacts to management options. J Appl Ecol 46(1):154–163
Rosenzweig C, Major C, Demong K (2007) Managing climate change risks in New York City's water system: assessment and adaptation planning. Mitig Adapt Strateg Glob Chang 12(8):1391–1409
Rosenzweig C et al (eds) (2011a) *First assessment of research on Climate change in cities*, Cambridge University Press
Rosenzweig C, Solecki W, DeGaetano A (eds) (2011b) Responding to climate change in New York State: the ClimAID integrated assessment for effective climate change adaptation. Technical report. New York State Energy Research and Development Authority (NYSERDA), Albany. Chapter 9
Stone J, Hess J, Howard F (2010) Urban form and extreme heat events: are sprawling cities more vulnerable to climate change than compact cities? Environ Health Perspect 118(10):1425–1428
Titus JG, Hudgens DE, Trescott DL, Craghan M, Nuckols WH, Hershner CH, O'Connell JF (2009) State and local governments plan for development of most land vulnerable to rising sea level along the US Atlantic coast. Environmental Research Letters, 4(4), 044008
UNEP SBCI (2009) Buildings and climate change: summary for decision-makers. United Nations Environmental Programme, Sustainable Buildings and Climate Initiative, Paris, pp 1–62
United Nations (2010) Report of the Conference of the Parties on its sixteenth session, vol 31, Cancun
United Nations (2014) World urbanization prospects: the 2014 revision, highlights. Department of Economic and Social Affairs. Population Division, United Nations
Wilbanks TJ, Fernandez S (2014) Climate change and infrastructure, urban systems, and vulnerabilities: technical report for the US Department of Energy in Support of the National Climate Assessment. Island Press
Wilby L (2007) A review of climate change impacts on the built environment. Built Environ 33(1):31
World Bank (2015) Moving toward climate-resilient transport the World Bank's experience from building adaptation into programs. World Bank Group. Washington DC

Drought-Tolerant Crops in Kirinyaga County, Kenya: Climate-Smart Agriculture Adaptation Strategies

118

Peterson N. M. Njeru, Jayne Mugwe, Monicah Mucheru-Muna, Immaculate Maina, Stephen K. Kimani, and David K. Lelgut

Contents

Introduction 2402
Materials and Methods 2404
  Site Description 2404
  Soil Physical Characteristics of Kirinyaga West County 2404
  Experimental Design 2404
  Data Analysis 2406
Results 2406
  Water Harvesting Technologies Available in Ndia West Division 2406
  Farmer's Evaluation on Treatment Performance 2407
  Treatment Score by Gender 2407
  Field Experiment Results 2409
  Treatment Effect at all Levels of Interactions 2409
Discussions 2411
  Farmer's Evaluation on Treatment Performance 2411
Conclusion 2412
Limitation 2413
References 2413

P. N. M. Njeru (✉) · S. K. Kimani · D. K. Lelgut
KALRO- Muguga South, Nairobi, Kenya
e-mail: njerupeterson@yahoo.com; njerupeterson1@gmail.com; Peterson.Njeru@kalro.org; 2skimani@gmail.com; David.Lelgut@kalro.org

J. Mugwe · M. Mucheru-Muna
Kenyatta University, Nairobi, Kenya
e-mail: jaynemugwe@yahoo.com; moniquechiku@yahoo.com

I. Maina
KALRO Headquarters, Nairobi, Kenya
e-mail: Lusike.Wasilwa@kalro.org

W. Leal Filho (ed.), *Handbook of Climate Change Resilience*,
https://doi.org/10.1007/978-3-319-93336-8_80

Abstract
Soil fertility degradation remains the major biophysical cause of declining per capita crop production on smallholder farms in Central Kenya highlands. A study was conducted to compare farmers' perception and biophysical data on selected water harvesting and integrated soil fertility management technologies on sorghum (*Sorghum bicolor* (L.) Moench) and cowpea (*Vigna unguiculata* L.) production in Kirinyaga County of Kenya. Three hundred and seventy-one smallholder farmers were invited to evaluate 36 plots laid out in partially balanced incomplete block design (PBIBD) replicated three times. The treatment which was ranked best overall rated as "good" by the farmers was farmers' practice with a mean score of 2.78 and yielding 3.5 t/ha under sorghum alone plus external soil amendment of 40 Kg P/ha + 20 Kg N/ha. This treatment was closely followed by water harvesting technologies tied ridges and contour furrows overall rated as "good" by the farmers. These being under cultivation of sorghum alone plus external soil amendment of 40 Kg P/ha + 20 Kg N/ha + manure 2.5 t/ha and 40 Kg P/ha + 40 Kg N/ha + manure 5 t/ha both with a mean score of (2.7) and yielding (3.0 t/ha) and (2.9 t/ha), respectively. Generally, all experiment controls were overall scored as "poor" yielding as low as 0.3 t/ha to 0.6 t/ha. Therefore, integrating organic and inorganic inputs on sorghum production under farmers' practice could be an alternative option contributing to food security in central highland of Kenya.

Keywords
Food security · Water harvesting · Integrated soil fertility management · Central Kenya

Introduction

Smallholder farms in central highlands of Kenya are characterized by unreliable rainfall distribution and declining soil fertility that are unsuitable for sustainable rain-fed agriculture in semiarid lands (SALs) (Miriti 2011). Approximately 83% of Kenya's land surface is classified as arid and semiarid lands (ASALs) that are characterized by low and erratic rainfall (100–900 mm per annum) which is not suitable for sustainable rain-fed agriculture. Agricultural production is highly affected by the high variability of rainfall onset in short and long rain season, distribution and frequent droughts that usually occur during the growing season, often resulting in depressed yields and persistent crop failures (Keating et al. 1992; Miriti et al. 2012).

Agricultural intensification requires the maintenance of soil physical, chemical, and biological quality. The loss of nutrients through plant nutrient mining, removal of crop residues, erosion, leaching or volatilization, and the deterioration of soil physical properties can independently or interactively result in yield reduction

(Bielders et al. 2002). Appropriate technologies need to be employed on the basis of their ability to maintain soil fertility (mulching; e.g., Bationo et al. 2000), to conserve the soil against soil losses by wind erosion (mulching or ridging) and to improve soil physical conditions in the topsoil. This suggests that only low-input technologies may be currently suitable and need to be adopted through a known crop intensification technologies that could be enhanced in arid and semiarid lands (ASALs) of Kenya. Therefore, more research is needed to find out the comparison of scientific research as compared to farmers' perceptions toward these technologies.

Soil fertility degradation is a major biophysical cause of declining per capita crop production and food security on smallholder farms in Central Kenya (Bationo et al. 2004: Kimani et al. 2007). The soil fertility decline is a result of a combination of processes such as high rates of soil erosion, nutrient leaching, removal of crop residues, continuous cultivation of the land without adequate fertilization, and fallowing (Njeru et al. 2011; Okalebo et al. 2006). The average annual loss in soil nutrients of 42 kg nitrogen (N), 3 kg phosphorus (P), and 29 kg potassium (K) ha^{-1} in Kenya is among the greatest in Africa (Smaling et al. 1997). The rising cost of inputs has led to many smallholder farmers reducing or abandoning the use of chemical fertilizer altogether in Central Kenya (Gachimbi 2002). The Kenyan ASALs are also experiencing low crop production due to a combination of biophysical problems such as factors like low rainfall, surface sealing, unavailability of high-quality manure, declining soil fertility due to continuous cultivation, and crust formation that reduces soil water availability to crops (Gicheru 2002; Gitau 2004). Several recent studies have yielded little evidence on occurrence of dry spells to increase the frequency of rainwater use efficiency in ASALs of Africa (Stroosnijder 2009). This has been contributed by mixed crop-livestock systems being currently projected to see reduction in crop production throughout most East Africa regions due to climate change by 2050 (Thornton et al. 2010).

Therefore, food security situation is expected to continue deteriorating and could worsen in future if water harvesting and integrated soil fertility technologies are not taken up quickly particularly in Kirinyaga County. Traditional crops such as sorghum and cowpea are considered as the crop for the future that can contribute to food security (Fongod et al. 2012). Improving agricultural productivity is crucial for resolving food crises, enhancing food security, and accelerating pro-poor growth. Most food security research and development programs tend to focus high and medium potential on promoting technologies for a limited number of major crops such as maize and beans in high potential areas and neglected the high-valued traditional crops which are drought tolerant and provide local nutrition in the vulnerable areas. Yet, sorghum and cowpea are locally important for food and household nutrition and provide income opportunities for the most vulnerable people and women in particular. These premium crops have potential to diversify the farming systems, adapt to spread risks, and are more resilient to climatic variations and climate change. This study assessed the farmer's evaluation on selected water harvesting and integrated soil fertility management technologies on sorghum and cowpea productivity in Ndia West division of Kirinyaga west subcounty in central highland of Kenya. This study was conducted to compare farmers' perception and biophysical data on selected water harvesting and integrated soil fertility management technologies on sorghum and cowpea production in Kirinyaga County of Kenya.

Materials and Methods

Site Description

The study was conducted in Ndia West division of Kirinyaga County which represents an area of declining potential occupying total area of 1437 km^2. The County lies between latitudes 0^0 1^1 and 0° 40^1 south and longitudes 37° and 38^1 east at an altitude of 1480 meters to 6800 meters above the sea level. The total population of the County is 509,157 individuals out of which 30.2% are considered food poor. The population density is 309 people per km^2. The County has about 97,970 farm families working in the agricultural sector occupying about 96,938 farm holdings with an average farm size of 1.25 ha per family (Government of Kenya 2007). There are four major agroecological zones (AEZs) in the County (LH1, UM1, UM2, and UM3), but the study was done in UM3, with maize-beans, horticulture, French beans, dairy, coffee, and banana production being the major crops. It receives a mean rainfall between 900 and 2700 mm per annum and has temperatures of between 14 °C and 30 °C. The soils in this County are volcanic which are known as andisols favorable for maize crop production. In addition, the County has two rain seasons – long and short rains between March and June and July and December, respectively.

Figure 1 shows Ndia West division of Kirinyaga County indicating the location of the study site in Central Kenya. The figure also shows the level of poverty in these areas, and this indicates the need for employing water harvesting and integrated soil fertility management technologies in the area.

Soil Physical Characteristics of Kirinyaga West County

The results of the three types of soil texture such as sand (13%), silt (78%), and clay (10%) have indicated that the soil types are silty loams in Kirinyaga County.

Experimental Design

The treatment were arranged in a factorial structure each treatment being a combination of one of the three levels of water harvesting techniques, cropping systems were two levels and the soil fertility amendment options were six levels thus giving a total of 36 treatments.

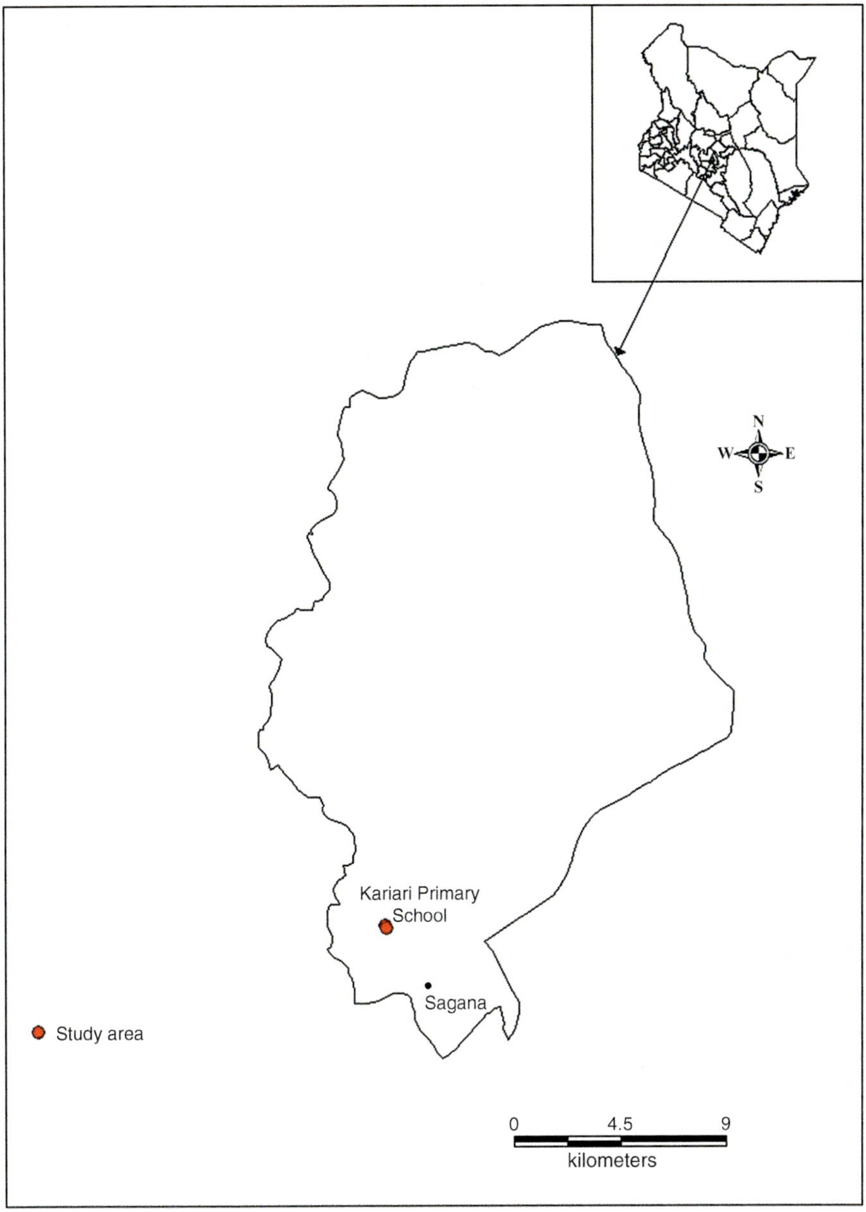

Fig. 1 Map of Kenya showing the study area in Kirinyaga County of Central Kenya

- **Water harvesting techniques (three levels)**
  - Tied ridges
  - Contour furrows
  - Conventional tillage/farmers' practice

- **Cropping systems (two levels)**
  - Sole sorghum (Gadam)
  - Sorghum and cowpea (M66) intercrop
- **Soil fertility amendment options (six levels)**
  - Control
  - 40 Kg P/ha + 40 Kg N/ha
  - 40 Kg P/ha + 20 Kg N/ha
  - 40 Kg P/ha + 40 Kg N/ha + manure 5 t/ha
  - 40 Kg P/ha + 20 Kg N/ha + manure 2.5 t/ha
  - Manure 5 t/ha

They were laid out in a partially balanced incomplete block design (PBIBD) with six incomplete blocks per replicate each containing six treatments, replicated three times making a total of 108 plots. Treatments were assigned to blocks randomly. The plot size was 6 m × 4 m. The dryland sorghum (Gadam) and cowpea (M66) varieties were used as the test crops. Then at the end of the short rain 2011 season, smallholder farmers were invited for a field day to evaluate each plot by scoring in a scale of good, fair, and poor according to their own observation on crop performance, and this was compared with scientific data collected on crop productivity. They were all given equal opportunity to evaluate 108 plots in the field experiment. They were also asked the kind of water harvesting and soil fertility management they used in their farms.

Data Analysis

Social data was coded and analyzed by the use of SPSS version 17. Data was analyzed by the use of descriptive analysis where frequencies of scores for each treatment were computed. Dependency tests were also conducted to find out if there was a relationship between gender and the treatment score. The difference between treatment scores and gender was declared significant at $P \leq 0.05$. The biophysical data on crop yield was analyzed using statistical Analysis of Variance (ANOVA) using SAS version 8. Differences between treatment effects were declared significant at $P \leq 0.05$.

Results

Water Harvesting Technologies Available in Ndia West Division

Table 1 shows water harvesting technologies which were in use on-farm in Ndia West division, Kirinyaga County. Tied ridges and convectional tillage were the widely used means of water harvesting technologies where 34.7% of respondents

Table 1 Household (%) practicing water harvesting technologies in Ndia West division

Household distribution of different water harvesting technologies practiced	(%)
Tied ridges	34.7
Contour furrow	28.2
Convectional tillage	34.7
Others	2.4
Total	**100**

(N = 371)

used them on their farms while 2.4% of respondents did not use any of the above technologies (Table 1).

Farmer's Evaluation on Treatment Performance

In Ndia West division, the farmers' criteria for distinguishing plots was on a scale of good, fair, and poor that included crop yield and performance. The findings (Table 2) underscore the value of taking into consideration the visual and morphological crop characteristics used by farmers as a key criterion for scientific crop evaluation and development during short rains 2011.

Table 2 shows that treatment under farmers' practice with sorghum alone plus soil amendment of 40 Kg P /ha + 20 Kg N /ha attracted the highest preference of farmers who rated it as "good" with a mean score of 2.78 and was ranked number 1 out of 36 treatments. This was followed by tied ridges and contour furrows under sorghum cropping system plus the same soil amendment of 40 Kg P /ha + 20 Kg N /ha + manure 2.5 t/ha both with a mean score of 2.7. Also contour furrow under sorghum cropping system with optimal application of fertilizers and manure was also ranked under "good" category. The results show that all the technologies that ranked "good" included a combination of fertilizers and manure or standalone application.

The results further indicate that the overall rating of the top eight treatments was rated in the "good" category by the smallholder farmers. This was contrary to the experiment controls which were rated as "poor" treatment. The results further indicated that the experiment controls with tied ridges were rated in the "fair" category.

Treatment Score by Gender

There was no significant difference ($P \geq 0.05$) regarding scoring by gender in all the 36 treatments of experiment which were ranked in the scale of good, fair, and poor. However, there was a highly significant difference ($p < 0.001$) on rating of treatments by smallholder farmers in Ndia West division, Kirinyaga West County.

Table 2 **Farmers' rating on water harvesting**, cropping systems, and ISFM technologies in Ndia West division

Water harvesting	Cropping systems	Soil fertility management regimes	Mean rank	Overall rating
Farmers' practice	Sole crop	40 Kg P/ha + 20 Kg N/ha	2893.87	Good
Tied ridges	Sole crop	40 Kg P/ha + 20 Kg N/ha + manure 2.5 t/ha	2763.26	Good
Tied ridges	Sole crop	40 Kg P/ha + 40 Kg N/ha + manure 5 t/ha	2767.22	Good
Contour furrows	Sole crop	40 Kg P/ha + 20 Kg N/ha + manure 2.5 t/ha	2753.38	Good
Contour furrows	Intercrop	40 Kg P/ha + 20 Kg N/ha + manure 2.5 t/ha	2667.23	Good
Farmers' practice	Sole crop	Manure 5 t/ha	2641.8	Good
Tied ridges	Intercrop	40 Kg P/ha + 20 Kg N/ha + manure 2.5 t/ha	2590.86	Good
Tied ridges	Sole crop	40 Kg P/ha + 40 Kg N/ha	2480.37	Good
Farmers' practice	Sole crop	40 Kg P/ha + 20 Kg N/ha + manure 2.5 t/ha	2443.79	Fair
Tied ridges	Intercrop	40 Kg P/ha + 40 Kg N/ha + manure 5 t/ha	2462.85	Fair
Contour furrows	Sole crop	40 Kg P/ha + 40 Kg N/ha	2403.52	Fair
Farmers' practice	Sole crop	40 Kg P/ha + 40 Kg N/ha	2384.7	Fair
Tied ridges	Sole crop	40 Kg P/ha + 20 Kg N/ha	2397.72	Fair
Tied ridges	Intercrop	40 Kg P/ha + 40 Kg N/ha	2310.8	Fair
Contour furrows	Intercrop	40 Kg P/ha + 20 Kg N/ha	2368.56	Fair
Tied ridges	Sole crop	Manure5t/ha	2338.03	Fair
Farmers' practice	Intercrop	40 Kg P/ha + 40 Kg N/ha	2280.98	Fair
Farmers' practice	Intercrop	40 Kg P/ha + 20 Kg N/ha	2243.88	Fair
Farmers' practice	Intercrop	40 Kg P/ha + 40 Kg N/ha + manure 5 t/ha	2256.27	Fair
Contour furrows	Sole crop	Manure 5 t/ha	2208.08	Fair
Contour furrows	Intercrop	40 Kg P/ha + 40 Kg N/ha + manure 5 t/ha	2180.3	Fair
Contour furrows	Sole crop	40 Kg P/ha + 20 Kg N/ha	2155.78	Fair
Farmers' practice	Intercrop	Manure 5 t/ha	2139.53	Fair
Contour furrows	Sole crop	40 Kg P/ha + 40 Kg N/ha + manure 5 t/ha	2123.15	Fair

(*continued*)

Table 2 (continued)

Water harvesting	Cropping systems	Soil fertility management regimes	Mean rank	Overall rating
Contour furrows	Intercrop	40 Kg P/ha + 40 Kg N/ha	2107.65	Fair
Tied ridges	Intercrop	40 Kg P/ha + 20 Kg N/ha	2098.91	Fair
Farmers' practice	Intercrop	40 Kg P/ha + 20 Kg N/ha + manure 2.5 t/ha	1740.38	Fair
Contour furrows	Intercrop	Manure 5 t/ha	1687.24	Fair
Tied ridges	Intercrop	Manure 5 t/ha	1686.05	Fair
Farmers' practice	Sole crop	40 Kg P/ha + 40 Kg N/ha + manure 5 t/ha	1385.2	Fair
Tied ridges	Sole crop	Control	1412.16	Fair
Tied ridges	Intercrop	Control	1160.19	Fair
Contour furrows	Sole crop	Control	1053.38	Poor
Contour furrows	Intercrop	Control	865.29	Poor
Farmers' practice	Sole crop	Control	677.2	Poor
Farmers' practice	Intercrop	Control	534.38	Poor

(N = 371), Test statistics: Kruskal-H test; Chi-square = 1237.5; d.f = 35; $p = 0.000$

Field Experiment Results

The results (Table 3) underscore the scientific crop evaluation from the field experiment during short rains 2011. The results further indicate the effect of various water harvesting and integrated soil fertility management options for sorghum and cowpea productivity on biomass and grain yield.

The results (Table 2) indifca three types of water harvesting, two cropping system and six fertility amendment levels but only fertility levels that differed significantly from one another ($p = 0.0001$) in terms of sorghum grain yield. The three levels of water harvesting and the two cropping systems did not differ significantly in terms of grain yield among themselves ($p = 0.8413$ and $p = 0.7168$, respectively). The total dry matter amount varied significantly among levels of cropping system and fertilizer application ($p = 0.0216$ and 0.0001, respectively). However the total dry matter amount did not vary significantly across water harvesting methods ($p = 0.5743$). The sorghum biomass was significantly different among cropping system ($p = 0.0020$), while water harvesting and fertility levels did not differ significantly ($p = 0.3930$ and 0.0698).

Treatment Effect at all Levels of Interactions

The results further indicated that sorghum without manure application did not differ significantly in yield production with plots that did not receive fertilizer application.

Table 3 The effects of water harvesting, cropping system, and soil fertility regimes on sorghum yields in Ndia West division

Water harvesting method	Cropping system	Soil fertility management regimes	Total dry matter T/Ha	Grain yield T/ha
Farmer practice	Sole crop	40 Kg P/ha +20 Kg N/ha	7.0	3.5
Tied ridges	Sole crop	40 Kg P/ha + 20 Kg N/ha + manure 2.5 t/ha	7.1	3.0
Contour furrows	Sole crop	40 Kg P/ha +40 Kg N/ha	6.4	2.9
Tied ridges	Sole crop	40 Kg P/ha +40 Kg N/ha	6.3	2.9
Contour furrows	Intercrop	40 Kg P/ha +40 Kg N/ha	6.0	2.9
Farmer practice	Sole crop	Manure 5 t/ha	6.6	2.8
Farmer practice	Sole crop	40 Kg P/ha +40 Kg N/ha	7.8	2.7
Contour furrows	Sole crop	40 Kg P/ha +20 Kg N/ha + manure 2.5 t/ha	6.1	2.7
Farmer practice	Intercrop	40 Kg P/ha +40 Kg N/ha	4.9	2.7
Tied ridges	Intercrop	40 Kg P/ha + 20 Kg N/ha + manure 2.5 t/ha	5.8	2.6
Tied ridges	Intercrop	40 Kg P/ha +40 Kg N/ha + manure 5 t/ha	5.6	2.6
Contour furrows	Intercrop	40 Kg P/ha +20 Kg N/ha	4.9	2.6
Farmer practice	Intercrop	Manure 5 t/ha	6.2	2.5
Contour furrows	Sole crop	40 Kg P/ha +40 Kg N/ha + manure 5 t/ha	5.9	2.5
Tied ridges	Sole crop	Manure 5 t/ha	6.7	2.4
Contour furrows	Intercrop	40 Kg P/ha + 20 Kg N/ha + manure 2.5 t/ha	5.7	2.4
Tied ridges	Intercrop	40 Kg P/ha +20 Kg N/ha	5.7	2.4
Contour furrows	Sole crop	Manure 5 t/ha	4.9	2.4
Farmer practice	Intercrop	40 Kg P/ha + 20 Kg N/ha + manure 2.5 t/ha	6.0	2.3
Contour furrows	Intercrop	40 Kg P/ha +40 Kg N/ha + manure 5 t/ha	7.9	2.3
Contour furrows	Intercrop	Manure 5 t/ha	5.4	2.3
Farmer practice	Intercrop	40 Kg P/ha +20 Kg N/ha	6.2	2.2
Farmer practice	Sole crop	40 Kg P/ha + 20 Kg N/ha + manure 2.5 t/ha	6.2	2.1
Tied ridges	Sole crop	40 Kg P/ha +20 Kg N/ha	5.8	2.1
Tied ridges	Intercrop	40 Kg P/ha +40 Kg N/ha	5.2	2.1
Farmer practice	Sole crop	40 Kg P/ha +40 Kg N/ha + manure 5 t/ha	5.1	2.1
Tied ridges	Intercrop	Manure 5 t/ha	5.0	2.1
Contour furrows	Sole crop	40 Kg P/ha +20 Kg N/ha	6.0	2.0
Tied ridges	Sole crop	40Kg P/ha +40Kg N/ha + manure 5 t/ha	5.9	2.0
Farmer practice	Intercrop	40 Kg P/ha +40 Kg N/ha + manure 5 t/ha	5.7	2.0

(*continued*)

Table 3 (continued)

Water harvesting method	Cropping system	Soil fertility management regimes	Total dry matter T/Ha	Grain yield T/ha
Tied ridges	Sole crop	Control	1.7	0.6
Tied ridges	Intercrop	Control	1.5	0.6
Contour furrows	Sole crop	Control	1.9	0.6
Contour furrows	Intercrop	Control	2.4	0.5
Farmer practice	Sole crop	Control	1.4	0.4
Farmer practice	Intercrop	Control	1.0	0.3
Means			5.6	2.3
CV			19.0	22.9
LSD			1.96	0.98

However, plots that received fertilizer and no manure gave slightly higher sorghum yield as compared to plots that received manure and no fertilizer (Table 2). The highest sorghum yield (3.5 t/ha) was recorded from farmer practice in sole sorghum cropping system with external nutrient replenishment of 40 Kg P/ha + 20 Kg N/ha, followed by 3.0 t/ ha under tied ridges, sole sorghum cropping system with combination of half rate of fertilizer and manure. In the third place, there were three treatments (2.9 t/ha) under tied ridges and contour furrow, sole cropping system in both water harvesting and one intercrop under contour furrows with all under maximum nutrient replenishment of 40 Kg P/ha + 20 Kg N/ha. However the four treatments yield did not differ significantly from one another ($p < 0.05$). The lowest sorghum yield (< 2.0 t/ha) was observed in treatments regarded as "control" with neither fertilizer nor manure regardless of other intervention (water harvesting method or cropping system). Total dry matter was more in plots with farmer practice, sole cropping, and maximum (40 Kg N/ha) N fertilizer application but no manure (7.7 t/ha), followed by 7.1 t/ha from plots under tied ridges, sole cropping with half rate (20 Kg N/ha) fertilizer, and 2.5 t/ha manure. Others that did well in total dry matter production were treatments under farmer practice, sole crop, and intercrops with half rate of (20 Kg N/ha) N fertilizers and (2.5 t/ha) manure (6.9 t/ha). All these top producers did not differ significantly from one another ($p < 0.05$).

Discussions

Farmer's Evaluation on Treatment Performance

The results (Table 1) indicated that 34.7% of the respondents both used tied ridges and convectional tillage as their main land preparation methods followed by contour furrows 28.2%. This was because of lack of knowledge on water harvesting methods they use when growing sweet potatoes and tobacco in the area. The consistently high preference (Table 2) by farmers on overall rating of treatments as "good" on farmers' practice, tied ridges, and contour furrow under sorghum alone with organic and inorganic inputs at half dose application of nitrogen was an indication that farms in Kirinyaga West County require minimal nutrient replenishment. Studies by Mugendi et al. (2010) and Gachimbi (2002) have also reported that farms in Central Kenya require nutrient replenishment every season from manures, fertilizers, and also return of crop residue in their farms. It has also been reported by Njeru et al. (2010) and Mairura et al. (2007) that soil fertility can be accessed through visual observation of crop performance and yield. The results further indicate that water harvesting technologies played a major role in moisture conservation where most of the technologies were highly ranked by the farmers. This is in agreement with what Miriti et al. (2012) has further found that farmer perception of soil fertility is closely related to the soil's water holding capacity.

The results (Table 2) from the farmers' responses show that the best four treatments of tied ridges, contour furrow, and farmers' practice under intercrop of sorghum and cowpea were generally ranked fairly and they were also dominated by sole cropping system (Table 3) from the field experimental results. This could be because of nutrient competition since cowpeas are heavy nutrient miners as they are associated with interspecific competition in mixed stands. The same results have been reported by Katsaruware and Manyanhaire (2009) that crop yield reduction can be experienced in intercrops where they are associated with interspecific/nutrient competition in mixed stands and the absence of interspecific competition in the monocrops. The results further indicate that probably intercropping sorghum with cowpea depressed sorghum yields influencing farmers to rank them in fair category. This outcome for sorghum (Tables 2 and 3) could be in line with reports for maize from Kenya (Nadar 1984) and in Tanzania (Jensen et al. 2003) where maize grain yield reduction of 46–57% and 9% occurred when maize was intercropped with cowpea due to the competition for moisture between the two crops. Alternatively due to slow mineralization of manure which could graduate to good category with given number of seasons (Lekasi et al., 2003). The results by Miriti (2011) reported that cowpea was a nutrient competitor for maize production in semiarid areas of Eastern Kenya. The results (Tables 2 and 3) had a very clear trend on farmer's perception and crop yield that all those treatments regarded as "controls" were poorly ranked by the farmers and had lower crop yields regardless of water harvesting and cropping system. This is in line with cultivation in the same piece of land for a continuous cultivation results to nutrient depletion and requires nutrient replenishment (Mugwe et al. 2009; Miriti et al. 2003). Land degradation as a result of various biophysical factors contributing to reduction in agricultural productivity has contributed to farmers being discouraged from adoption of these water conservation structures due to labor shortage and land tenure uncertainty (Demelash and Stahr 2010).

Conclusion

The results reported in the study demonstrate that smallholder farmers' knowledge can provide a consistent treatment evaluation on crop yield. There was clear evidence from the study that no difference was noted in terms of scoring of treatments by gender in

Ndia West division, Kirinyaga West County. Therefore, both genders could be used by agricultural extension services and researchers for evaluation of other related scientific work in this study area. The results have demonstrated that Kirinyaga West County is characterized by low and erratic rainfall and generally fragile ecosystems which are not suitable for sustainable rain-fed agriculture. The results have indicated that there is a need to incorporate various water harvesting technologies and minimum soil nutrient supplements during seasons with low rainfall distribution regardless of water harvesting technology employed in semiarid areas. This also suggests that only low-input technologies are currently suitable and need to be adopted through a known crop intensification technologies that could be enhanced in these areas of central highland of Kenya. The results have also demonstrated a very clear message to smallholder farmer, extension services, and other stakeholders that there is a need for water harvesting technologies and nutrient replenishment on-farm every season for increased sorghum and cowpea productivity in central highland of Kenya.

Limitation

During the research period, there was a big challenge on rainfall distribution where the crop did not perform very well during the first season of the experiment in the County. It was within this season where it was reported with a lot of stem borer infestation, but they were controllable by spraying pesticides.

References

Bationo A, Kimetu J, Ikeera S, Kimani S, Mugendi D, Odendo M, Silver M, Swift MJ, Sanging N et al (2004) The African network for soil biology and fertility: new challenges and opportunities. In: Bationo A (ed) Managing nutrient cycles to sustain soil fertility in sub-Saharan Africa. Academy Science Publishers and Tropical Soil Biology and Fertility Institute of CIAT, Nairobi, pp 1–23

Bielders CL, Michels K, Bationo A et al (2002) On-farm evaluation of ridging and residue management options in a Sahelian millet cowpea intercrop. Soil quality changes. Soil Use Mgmt 18:216–222

Demelash M, Stahr K (2010) Assessment of integrated soil and water conservation measures on key soil properties in south Gonder, north-western highlands of Ethiopia. Soil Sci Evntal Mgmt 1:164–176

Fongod AGN, Mih AM, Nkwatoh TN et al (2012) Morphological and agronomic characterization of different accessions of sweet potatoes (*Ipomoea batatas*) in Cameroon. Inter Res J Agric Sci Soil Sci 2:254–245

Gachimbi LN (2002) Technical report of soil survey and sampling results: Embu-Mbeere Countys, Kenya. LUCID working paper series number: 9. KARI-kabete, Nairobi, Kenya

Gicheru PT (2002) The effect of different soil management practices on crust formation, soil moisture conservation and crop growth in Machang'a area, Mbeere County, Kenya. PhD thesis. University of Nairobi, Kenya

Gitau AN (2004) Mechanical behaviour of hardsetting Luvisal soil. PhD thesis, University of Nairobi, Kenya

GOK (2007) The Kenya vision 2030; a competitative and prosperous Kenya. Government of Kenya. Ministry of planning and national development, Nairobi

Jensen JR, Bernhard RH, Hansen S, McDonagh J, MØberg JP, Nielsen NE, Nordbo E et al (2003) Productivity in maize based cropping systems under various soil-water-nutrient management strategies in semi-arid Alfisol environment in East Africa. Agric Water Mgt 59:217–237

Katsaruware RD, Manyanhaire IO (2009) Maize-cowpea intercropping and weed agronomy and soil science: Madison, WI, USA. Suppression in leaf stripped and detasselled maize in Zimbabwe. Electron J Environ Agric Food Chem 8(11):1218–1226

Keating BA, Wafula BM, Watiki JM et al (1992) Exploring strategies for increased productivity. The case for maize in semi-arid eastern Kenya. In: Proceeding of symposium held in Nairobi, Kenya, 10–11 December 1990. Goanna Pry Ltd, Canberra, pp 90–99

Kimani SK, Esilaba AO, Odera MM, Kimenye L, Vanlauwe BJ, Batiano A et al (2007) The effect of organic sources of nutrients on maize yield in three Countys of Central Kenya. In: Batiano A (ed) Advances in integrated soil fertility management in sub-Sahara Africa, challenges and opportunities. Springer, Dordrecht, pp 283–288

Lekasi JK, Tanner JC, Kimani SK, Harris PJC et al (2003) Cattle manure quality in Maragwa County, Central Kenya: effect of management practices and development of simple method of assessment. Agric Ecos Environ 94:289–298

Mairura FS, Mugendi DN, Mwanje JI, Ramisch JJ, Mbugua PK, Chianu JN et al (2007) Integrating scientific and farmers' evaluation of soil quality indicators in Central Kenya. Geoderma 139:134–143

Miriti JM (2011) Effect of conservation tillage on water and nitrogen use efficiency in maize-cowpea cropping systems in semi-arid eastern Kenya. PhD Thesis. University of Nairobi, Kenya

Miriti JM, Kironchi G, Esilaba AO, Heng LK, Gachene CKK, Mwangi DM et al (2012) Yield and water use efficiencies of maize and cowpea as affected by tillage and cropping systems in semi-arid Eastern Kenya. Agri water Mgt 115:148–155

Miriti JM, Thomas DB, Gachene CCK et al (2003) Soil chemical properties, sunflower growth and yields as affected by tillage, compost and mulch application. East Afr Agric J 68(3):159–163

Mugendi E, Gitonga N, Cheruiyot R, Maingi J et al (2010) Biological nitrogen fixation by Promiscous Sobean (Glycine max L. Merril) in the central highlands of Kenya: response to soil Amendends. World J Agric Sci 6:381–387

Mugwe J, Mugendi D, Mucheru-Muna M, Merckx R, Chianu J, Vanlauwe B et al (2009) Determinants of the decision to adopt integrated soil fertility management practices by smallholder farmers in the central highlands of Kenya. Exp Agric 45:61–75

Nadar HM (1984) Intercropping and intercrop component interaction under varying rainfall conditions in eastern Kenya. II: maize/cowpea intercrop. East Afri Agri J 44:176–181

Njeru PNM, Maina I, Miruka M, Amboga JS, Gitari J, Rono B, Mahasi M, Murithi F et al (2011) Soil fertility characterization of smallholder farms under group and individual management in Central Kenya. Proceedings on held at Maputo, Mozambique on 10th to 13th October. Afri Crop Sci Conf Prod 10:1–4

Njeru PNM, Hjalten J, Kimani SK et al (2010) A book on the effect of push-pull management and different fertilization strategies on maize crop yield in Central Kenya. Lambert Acadamic publishers, Saarbrücken SSBN 978-3-8433-6376-1

Okalebo JR, Othieno CO, Woomer PL, Karanja NK, Semoka JRM, Bekunda MA, Mugendi DN, Muasya RM, Bationo A, Mukhwana EJ et al (2006) Available technologies to replenish soil fertility in East Africa. Nutr Cycl Agroecosyst 76:153–170

Smaling EMA, Nandwa SM, Jansen BH et al (1997) Soil fertity is at stake. In: Replenishing soil fertility in Africa, Special publications no. 51. American Society of Agronomy, Crop Science Society of America, Madison

Stroosnijder L (2009) Modifying land management in order to improve efficiency of water use in the African highlands. Soil Till Res 103:247–256

Thornton PK, Jones PG, Alagarswamy G, Andresen J, Herrero M et al (2010) Adopting to climate change: agricultural system and household impact in East Africa. Agric Syst 103:73–82

Effect of Mulching on Soil Temperature and Moisture for Potato Production in Agro-ecological Zones of Central Highlands of Kenya

119

Eva N. Gacheru, Charles K. K. Gachene, Patrick T. Gicheru, and Lieven Claessens

Contents

Introduction 2417
Materials and Methods 2419
  Study Sites Description 2419
  Study Design and Description 2422
  Data Collection 2422
  Potato Growth and Yield 2423
Results and Discussion 2424
  Gravimetric Soil Moisture as Influenced by Mulching Rates 2424
  Soil Temperature (°C) at 12 cm Depth as Influenced by Mulching Rates Across the Three Different AEZs 2426
  Effects of Mulching Rates on Soil Temperature at 20 cm Depth Across Different AEZs 2430
  Potato Emergence (%) as Influenced by Mulching Rates 2432
  Effect of Mulch Rates on Potato Yield Across Different AEZs 2434

E. N. Gacheru (✉)
Kenya Agriculture and Livestock Research Organisation, Nairobi, Kenya
e-mail: evagacheru@yahoo.com; Evah.gacheru@kalro.org

C. K. K. Gachene
Department of Land Resource Management and Agricultural Technology (LARMAT), University of Nairobi, Nairobi, Kenya

Kenya Agriculture and Livestock Research Organisation, Nairobi, Kenya
e-mail: gachenecharles@gmail.com

P. T. Gicheru
Kenya Agriculture and Livestock Research Organisation, Embu, Kenya
e-mail: ptgicheru@yahoo.com

L. Claessens
International Institute of Tropical Agriculture (IITA), Arusha, Tanzania, Kenya
e-mail: I.claessens@Cgiar.org

W. Leal Filho (ed.), *Handbook of Climate Change Resilience*,
https://doi.org/10.1007/978-3-319-93336-8_86

Conclusion 2435
Limitations 2436
Cross-References 2436
References 2436

Abstract

In Kenya, potato plays a significant role in the livelihood of smallholder farmers, serving both as a food security and cash crop. Sensitive to both heat and moisture stress, in the tropics the future of potato production is threatened by global warming and the associated changes in temperatures and rainfall variability. A study to determine the potential of mulching as an adaptation strategy to climate change in potato production system was undertaken along an elevation gradient of varying agroecological zones (AEZs) in the south eastern slopes of Mt. Kenya in Embu County. The study sites were Mbuvori, representing the lower highland humid (LH1), at 1900–2100 m.a.s.l.; Nembure, the upper midland subhumid (UM2), at 1400–1600 m.a.s.l.; and Karurumo, the lower midlands semihumid (LM3), at 1070–1280 m.a.s.l. AEZs. Dry maize stover was applied at the rate of 0, 4, 8, and 12 t ha^{-1}. The experiment was laid out in a RCBD and was replicated three times per site. Soil temperature measurements were taken at depths of 12 and 20 cm at 8.00 to 9.00 and 14.00 to 15.00 h and gravimetric soil moisture determined at 0–15 cm depth. Potato emergence and yield were also monitored. The results showed that application of 12 t ha^{-1} of maize stover significantly improved soil moisture, at Karurumo and during period of less rainfall in the other two sites. Mulching rates interacted well with soil temperature, whereby with increasing quantities of mulch, the temperature gap between the hot and drier AEZs sites and the cooler zones decreased. At 12 cm depth, mulching reduced afternoon soil temperatures at Nembure to levels that were comparable to those observed in the cool and moist Mbuvori site and was negatively correlated to increasing mulch quantities. At 20 cm depth, soil temperature in plots applied with 12 t ha^{-1} mulching rate, at Karurumo, was comparable to that observed at Nembure, while the difference between the two sites in bare soil was as high as 5.5 °C. Mulching reduced the gap between soil temperature observed between Karurumo and Mbuvori. The gap between the two sites bare soil plots ranged between 5.0 °C and 7.8 °C while that beneath the 12 t ha^{-1} was 2.3 °C. At Karurumo, afternoon soil temperature at both 12 cm and 20 cm depths was negatively correlated to increasing mulching rate ($p < 0.05$). Mulching enhanced tuber emergence at Karurumo. Application of 8 and 12 t ha^{-1} resulted in increased tuber yield at both Nembure and Karurumo. This study shows that mulching can be a reliable adaptation strategy to climate change in smallholder potato production system of Kenya.

Keywords

Climate change · Adaptation · Mulching · Soil temperature · Soil moisture · Potato yield · Emergence · AEZs

Introduction

Potato (*Solanum tuberosum* L.) plays a significant role in the global food supply and is considered as the most important noncereal food crop, in terms of production and volume (Cromme et al. 2010). In Kenya, potato is ranked as the second most important staple food after maize and serves as both a food security and income-earning crop (Gildemacher et al. 2009; Lung'aho et al. 2006). Smallholder farmers account for approximately 90% of the national potato production (MoA 2008) with an annual farm gate earning of five billion Kenya shillings (MoA 2009). In Kenya, potato production is typical tropical country and is mainly concentrated at the mid altitude and the highlands from approximately 1200 to 3000 m. a. s. l. (Lutaladio et al. 1995). The major production areas include counties around the slopes of Mt. Kenya, the Aberdare ranges; the Rift valley; and Cherangani hills (MoA 2009; Kaguongo et al. 2008). Most of the potato crop is grown under rainfed conditions (Gildemacher et al. 2009). Potato is mainly grown twice a year, during the long rain and the short rain seasons (Lung'aho et al. 2006).

In the tropics, high temperature plays a critical role in determining the growth and potato production areas (Caldiz et al. 2002; Haverkort 1990; Manrique 1992). The rate of development from shoot sprouting, tuber initiation, and bulking depends on soil temperature within the root zone, the optimum ranging from 15 °C to 23 °C (Struik 2007). Under high-temperature conditions, tuber formation is delayed and can even be inhibited at temperatures above 30 °C (Kar and Kumar 2007). The potato crop is shallow rooted with most of the roots concentrated in the top 30 cm of the soil; consequently it is sensitive to moisture stress (Lynch et al. 1995; Yuan et al. 2003). Climate change and the associated rise in temperatures and changes in rainfall amounts and variability pose a serious threat to potato production more so in the vulnerable rainfed production systems in Kenya.

Kenya is already experiencing the impact of climate change as manifested by the changes in weather patterns (Omambia et al. 2009). Unfavorable weather conditions have been acknowledged as one of the major contributors to the declining yield in potato in Kenya (Kaguongo et al. 2008). Even in the highlands, farmers are already experiencing the dire effects of climate change. For instance, in eastern Kenya, lowland temperature belts have shifted upward by 100 m and the midland zone by 60 m in the period between 1981 and 2000 compared to 1937–1968, with an accompanying temperature rise of 0.5 °C and 0.35 °C, respectively (Jaetzold et al. 2006). In the central highlands, the main food surplus area that is supposed to receive at least 500 mm of rainfall during the main crop growing seasons, March, April, May, June (MAMJ), has been shrinking by about 15 km per decade, and the amount of rainfall declined by 100 mm since 1960 (Funk et al. 2010).

The potential of adaptation to reduce the negative impacts of climate change has been reported, such as in studies conducted by Hijmans (2003), Parry et al. (2005), and Rosenzweig and Parry (1994). On global cereals production, Rosenzweig and Parry (1994) and Parry et al. (2005) analyzed the impact of farm-level adaptation strategies and reported a 3% reduction on yield losses associated with climate change. On potato production, Hijmans (2003) analyzed the adaptation potential of strategies and projected 40% restoration of yield losses associated with climate change. These studies focused on global-scale adaptation measures and have limitations of application at local levels. In their studies, Rosenzweig and Parry (1994), Parry et al. (2005), and Hijmans (2003) indicated disparities between developed and developing countries and acknowledged the need for localized studies. Developing countries have different and unique circumstances, both at national and local level (Fischer et al. 2002). While temperate countries can take advantage of the longer thermal growing periods, in the tropics, soil moisture availability and heat stress may be the main determinants of crop production especially for temperate crops like potato.

Farming practices such as mulching the soil surface can modify the hydrothermal properties of the soil and, in the process, create favorable conditions for crop growth. Surface mulching has been widely used to reduce evaporation and to ameliorate soil temperature (Cook et al. 2006). Compared to bare soil, crop residue mulches have a high light-reflective capacity (albedo) and combined with their low thermal conductivity impede solar radiation entry into the soil and in process reduce soil heating (Blanco-Canqui and Lal 2009; Chen et al. 2007). Several investigators have reported that the thermal conditions under mulch differ from that of bare soils, the daytime soil temperature under mulch cooler than that of bare soil (Bhat and Khero 2006; Cook et al. 2006; Mahmood et al. 2002). The magnitude of soil temperature modification is also influenced by the quantity of mulch applied (Blanco-Canqui and Lal 2009; Blanco-Canqui et al. 2006; Rees and Chow 2005).

The rainfall amount received per season is not the key limiting factor for crop yield, rather the fraction that is available for crop consumption. In sub-Saharan Africa, water management is a big challenge, with most of the seasonal rainfall getting lost through nonproductive means such as evaporation and surface runoff, with only a paltry 15–30% utilized by crops (Falkenmark and Rockstrom 2006). Evaporation accounts for 25–50% of these losses (Abouziena et al. 2014; Ramakrishna et al. 2006). The effect of mulching on soil moisture conservation has been reported in several studies (Bhat and Khero 2006; Blanco-Canqui and Lal 2009; Cook et al. 2006; Mulumba and Lal 2008; Rees and Chow 2005). Some of these studies have shown that mulching rate determines the amount of moisture retained in the soil but differs in the optimum rate of mulching required (Blanco-Canqui et al. 2006; Pervaiz et al. 2009; Rees and Chow 2005; Rees et al. 2002). Mulching rates exceeding certain amount could impact negatively on crop growth and development (Wang et al. 2005). In Kenya and in smallholder farm holdings such as the ones found in the Kenya highlands, maize stover is the most abundant crop residue (Methu et al. 2001) but has numerous competing uses such as fuelwood and dry season feed for livestock (Nyaata et al. 2000; Methu et al. 2001). This study aimed at determining the effect of mulching rates on soil moisture, soil temperature, and potato yield under varying agroecological conditions.

Materials and Methods

Study Sites Description

The study was carried out in Embu County, on the southeastern slopes of Mt. Kenya. Embu County is located between latitudes 0°8′ and 0°35′ south and longitudes 37°19′ and 37°42′ east (Jaetzold et al. 2006). The study had two variables: (1) temperature belts represented by different EZs and (2) mulching rates. Embu County was selected because it is covered by different agroecological zones ranging from the high altitude tea dairy lower highland humid zone (LH1) to the livestock-millet lower midland semiarid (LM5) agroecological zone (AEZ) within a short distance (Fig. 1). Rainfall is bimodal with the main cropping season or the long rain (LR) season starting from March through June (MAMJ) and the short rain (SR) season from October to December (OND). The climate gradient is based on three levels of climatic conditions in form of temperature belts and rainfall variation. The experiments were carried out in three farmers' fields that represented three different temperature belts (AEZs). The three AEZs included the humid lower highland (LH1) zone at Mbuvori, the subhumid upper midland (UM2) zone at Nembure, and the semihumid lower midland (LM3) zone at Karurumo (Fig. 1).

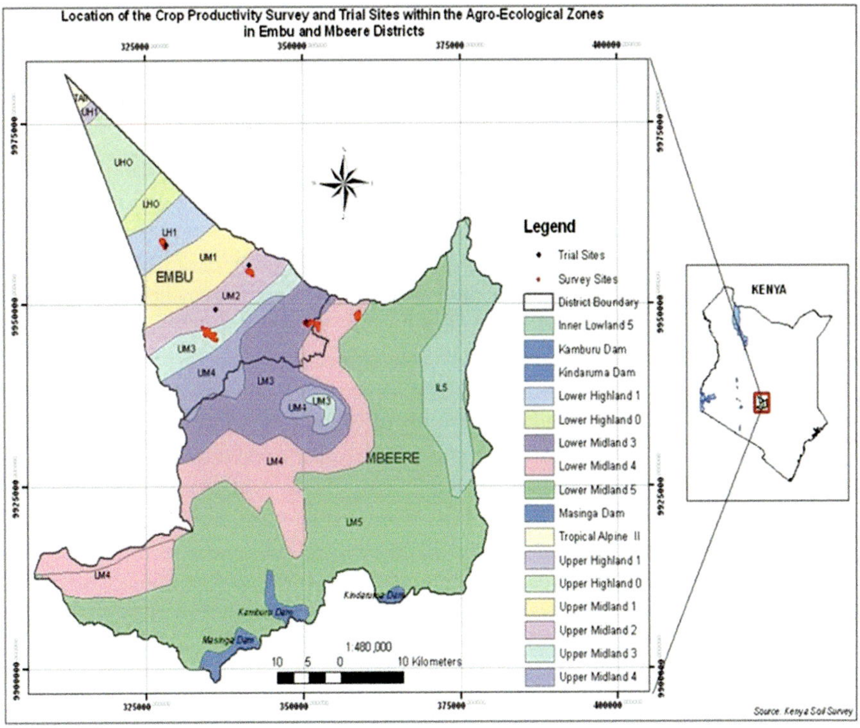

Fig. 1 Study sites and AEZ

Table 1 Study sites elevation and respective agroecological zones characteristics

Site	Elevation (m)	Agroecological zone	Average mean temperature (°C)	Average annual rainfall (mm)	Altitude	Soil types
Mbuvori	1788*	LH1	15.8–17.8	1750–2000	1900–2100	Andosols
Nembure	1517	UM2	18.9–20.1	1200–1500	1400–1600	Nitisols
Karurumo	1235	LM3	21.0–22.0	900–1100	1070–1280	Ferralsols

Adapted from Jaetzold et al. 2006
Note: Mbuvori is located in a valley bottom, hence the lower elevation* compared to the altitude for the LH1 AEZ

Table 2 Monthly mean air temperature (°C) at study sites and the Embu Meteorological Centre during the SR09, LR10, and SR10 seasons

Sites/agroecological zones	Seasons									
	Short rains 2009			Long rains 2010				Short rains 2010		
	Oct	Nov	Dec	Mar	April	May	June	Oct	Nov	Dec
Mbuvori (LH1)	18.1	20.1	18.5	–	19.7	19.3	17.2	18.9	18.3	17.9
Nembure (UM2)	22.0	21.1	20.7	20.2	22.3	21.9	19.9	22.3	20.4	20.5
Embu met. (UM2)	20.9	20.4	19.9	20.5	21.0	20.4	19.0	21.1	19.9	19.8
Karurumo (LM3)	23.9	24.5	23.9	24.3	25.8	25.0	24.1	23.5	21.8	23.2

The details of the study sites location, dominant soil types, and climatic characteristics of the respective AEZs are presented in Table 1. Whereas soil types varied per site, the differences were not factored in as variable in this study.

The sites mean monthly ambient temperature records and rainfall distribution during the study period are shown in Table 2 and Fig. 2, respectively. Mbuvori was cooler than Nembure and Karurumo by 2.6 °C and 5.3 °C, respectively. An inter-season variation on air temperature was observed, particularly during the two SR seasons in November. Compared to the same period, the SR10 was cooler than the SR09 season by 1.3 °C and the month of November 2010 cooler by 1.8 °C, 0.7 °C, and 2.7 °C at Mbuvori, Nembure, and Karurumo, respectively (Table 2). During the SR09 and LR10 seasons, rainfall amount followed an altitudinal gradient, the highest amount recorded at the LH1 (Mbuvori) site and the lowest at the LM3 (Karurumo) AEZ (Fig. 2). During the SR09 season, the total rainfall recorded at Mbuvori was 533 mm, 473.9 mm at Nembure, and 330 mm at Karurumo. During the LR10 season, 361 mm was recorded at Karurumo compared to 796 mm and 979 mm at Nembure and Mbuvori, respectively. At Karurumo, rainfall was poorly distributed, with 49% falling in the last 3 days of March and the first week of April, after which a dry spell of 19 days was experienced. During the SR10 season, the highest amount of rainfall was recorded at Karurumo (453.1) and the least amount at Nembure (281.4 mm). At both Nembure and Karurumo, the SR10 season rainfall

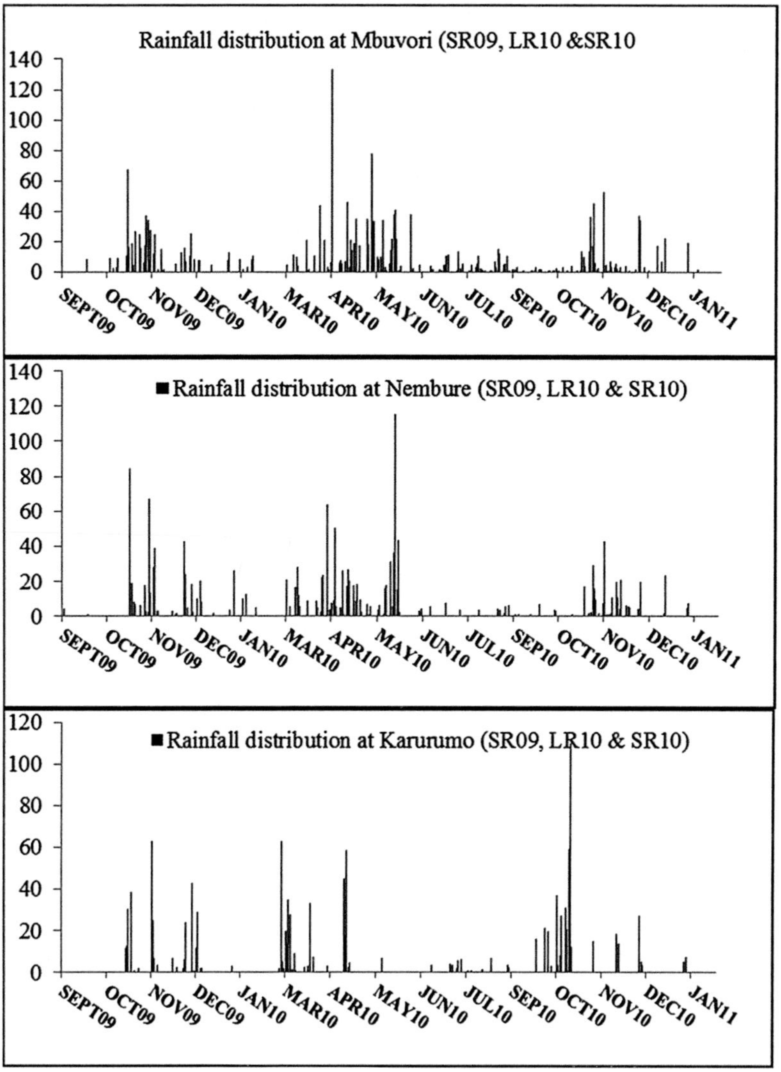

Fig. 2 Rainfall distribution at Mbuvori (LH1), Nembure (UM2) and Karurumo (LM3) during the SR09, LR010 SR10 seasons

was poorly distributed. Nembure experienced drought during the first 17 days of October and Karurumo 19 days. At Nembure, December, which coincided with the bulking period of potato, was also dry, the amount of rainfall recorded less by 47% of the amount recorded in 2009. At Karurumo, though the highest amount of rainfall was recorded during the SR10 season, it was poorly distributed with 324.6 mm falling during the month of November in 5 consecutive days and 110 mm of this amount falling in 1 day (Fig. 2).

Plate 1 Freshly applied dry maize stover mulch

Study Design and Description

The experiment was undertaken for four seasons (i.e., two LR and two SR seasons) in three different AEZs. The treatments consisted of four mulching rates (0, 4, 8, and 12 t ha^{-1}) of dry maize stover in each sites. In each season, dry maize stover applied in the three sites was sourced from one farm after maize crop harvest. Moisture content of the materials was determined and was applied on dry weight bases. The potato variety ASANTE was used as a test crop. Land preparation and opening of planting furrows were undertaken using a hand hoe. Fertilizer, diammonium phosphate (DAP), was spread evenly within the furrow at a rate of 500 kg ha^{-1}. Potato seed were placed in the furrow at a spacing of 30 cm, covered with the soil, which was heaped in the middle of the furrow to make a ridge. In the mulching treatments, dry un-chopped maize stover (stalk, leaves, and husks) was spread evenly on the soil surface, immediately after planting (Plate 1). The experiment was established in a randomized complete block design, and each treatment was replicated 3 times in each site. The plot size was 4.5 m wide and 5 m long. The plots consisted of six rows, 5 m long spaced 75 cm apart. The plots were separated from each other by a 1 m path.

Data Collection

Soil Moisture Determination

Soil moisture determination was initiated during the second season of the study. Soil was sampled at a depth of 0–15 cm, from four random spots per plot using an Edelman soil sampling auger, and placed in a plastic bucket. The soil from the four spots was thoroughly mixed and a subsample placed in a polythene paper bag. Moisture determination was done at the regional research center KALRO-Embu soil laboratory using gravimetric method as outlined by Anderson and Igram (1993).

Soil Temperature Measurements

Data collection on soil temperature was also started during the second season (SR09). Soil temperature at 20 cm depth was measured using modified dry bulb soil thermometer. In each plot, one thermometer was installed on the potato planted ridge (Plate 2b). Data was collected twice weekly both in the morning (8.00–9.00 h) and in the afternoon (15.00–16.00 h) in all the three sites for 12 weeks. At crop harvest, the thermometers were removed and reinstalled at planting of the next season's crop. At the 12 cm depth measurements were taken using the Hanna soil temperature and conductivity meter (Plate 2a). At both Mbuvori and Nembure, readings were taken twice weekly and once every 2 weeks at Karurumo. Since most of the potato growth stages happens below ground, it was not possible to determine the actual growth period. The number of days from planting to a specific growth period was based on description of potato growth periods and duration by Loon 1981. The data was averaged in triplicates; weeks 1–4 (emergence), weeks 5–8 (tuber initiation), and weeks 9–12 (tuber bulking) per site.

Potato Growth and Yield

Data on potato tuber emergence was collected from the four middle rows every 2 days until maximum emergence was observed in each plot. The counts were

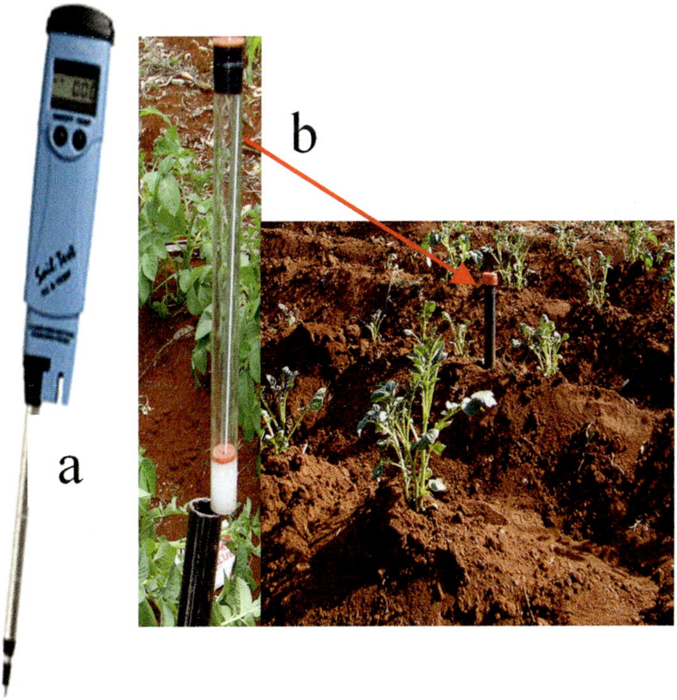

Plate 2 (**a**) Hanna's soil tester, (**b**) modified dry bulb thermometer

transformed to percentages based on the total number of tubers planted in the four rows. Fresh tuber yield was determined from a net plot of 9 m^2.

Statistical Analysis

Data was subjected to analysis of variance for the parameters studied (soil moisture, soil temperature, potato emergence and yield) using Mstat-C software version 1.4 (Michigan State University, USA). When results were statically significant ($P \leq 0.05$), means were separated using the Student-Newman Keels test using the same software, while linear regression analysis was done using Genstat Edition 3 (www.vsni.co.uk) software.

Results and Discussion

Gravimetric Soil Moisture as Influenced by Mulching Rates

The high moisture contents observed at Mbuvori masked the effects of mulching in the drier sites. Consequently, each site's soil moisture data was analyzed separately (Tables 3, 4, and 5). At Mbuvori, mulching caused no significant effect on soil moisture during the SR09 and LR10 seasons (Table 3). The nonsignificant effects of mulching could be attributed to rainfall events prior to the dates of soil sampling and the high and well-distributed rainfall during those two seasons (Fig. 2). For instance, during the SR09 season, 38 mm of rainfall was recorded 1 day prior to the 50 DAP soil sampling date. During the SR10 season, soil moisture conserved at 54 DAP under the 8 and 12 t ha^{-1} mulching rates was significantly higher than bare soil by 1.3–1.9% (Table 3). The significant effects of mulching during the SR10 season was attributed to the low rainfall amounts recorded during that season, which was 50% and 70% less the amount recorded during the SR09 and LR10 seasons, respectively. The nonsignificant difference between 8 and 12 t ha^{-1} in this site corroborates with Mulumba and Lal (2008) observation that mulching beyond

Table 3 Effect of mulching rates on soil gravimetric soil moisture (%) at Mbuvori during the SR09, LR10, and SR10 seasons

	SR09		LR10		SR10		
Mulching rate (t ha^{-1})	30 DAP	50 DAP	42 DAP	65 DAP	18 DAP	54 DAP	72 DAP
0	40.0	38.9	40.4	37.9	19.0	25.5b	20.3
4	41.6	40.3	41.9	38.0	19.4	25.3b	20.2
8	41.4	40.2	41.6	37.8	19.1	26.8a	20.5
12	42.4	42.2	42.9	38.1	20.3	27.4a	20.5
F prob	ns	ns	ns	ns	ns	0.002	ns
s.e.	0.62	1.12	0.58	0.91	0.69	0.24	0.19
CV (%)	2.6	2.9	2.3	4.2	6.2	1.6	3.2

Means followed by the same letter within a column are not significantly different at $p = 0.05$

Table 4 Effect of mulching rates on gravimetric soil moisture (%) at Nembure during the SR09, LR10, and SR10 seasons

	SR09		LR10		SR10		
Mulching rate ($t\ ha^{-1}$)	30 DAP	50 DAP	42 DAP	65 DAP	18 DAP	54 DAP	72 DAP
0	36.0	28.9	29.9b	31.0	9.4	25.70b	16.62
4	38.2	27.8	35.1ab	37.4	12.6	27.48a	18.47
8	37.8	32.2	34.9ab	33.4	11.0	27.48a	18.47
12	37.5	33.0	37.7a	34.2	10.9	27.48a	18.63
F prob	ns	ns	0.04	ns	ns	0.002	ns
s.e.	0.581	2.337	1.452	2.451	0.97	0.21	0.19
CV (%)	2.7	13.3	7.3	12.5	15.3	1.4	1.8

Means followed by the same letter within a column are not significantly different at $p = 0.05$

Table 5 Effect of mulching rates on gravimetric soil moisture (%) at Karurumo during the SR09, LR10, and SR10

	SR 2009		LR 2010		SR 2010		
Mulching rate ($t\ ha^{-1}$)	30 DAP	50 DAP	42 DAP	65 DAP	18 DAP	54 DAP	72 DAP
0	30.4	24.6b	23.4b	24.6b	5.1	18.5	17.5
4	31.7	27.2ab	24.7ab	27.2a	4.5	23.5	17.5
8	32.2	28.1ab	24.7ab	27.2a	5.2	24.4	17.3
12	33.0	30.3a	25.9a	28.5a	4.7	24.6	17.6
F prob	ns	0.01	0.01	0.01	ns	ns	ns
s.e.	0.96	0.78	10.49	0.53	0.53	1.96	0.20
CV	26	4.9	3.4	3.4	19.0	16.8	2.0

Means followed by the same letter within a column are not significantly different at $p = 0.05$
DAP days after planting

8 $t\ ha^{-1}$ rates made no significant difference in moisture conservation. Blanco-Canqui et al. (2006) found no significant difference in moisture conservation between 5 and 10 $t\ ha^{-1}$ rates. The nonsignificant effects of the 4 $t\ ha^{-1}$ mulching concurs with Blanco-Canqui et al. (2006), who also observed no difference between 3.75 $t\ ha^{-1}$ and bare soil.

At Nembure, mulching showed no significant effect ($p = 0.05$) in soil moisture conservation, during the SR09 season (Table 4). Similar to the situation at Mbuvori, rainfall events prior to the date of soil sampling could have masked the effects of mulching on soil moisture. Three and five days prior to soil sampling at 30 DAP of the SR09 season, 34.4 and 84.6 mm, respectively, of rainfall were recorded. During the LR10 season, mulching with 12 $t\ ha^{-1}$ of maize stover, at 42 DAP, conserved 7.8% more moisture compared to bare soil (Table 4). At 65 DAP, the difference in soil moisture content between mulched treatments and bare soil was not significant. The previous week was exceptionally wet (a total of 250 mm) falling daily until 3 days prior to the date of sampling. At this date, the potato crop had also developed a canopy, which could have shaded the soil surface even in the non-mulched plots. According to Bristow (1988) and Midmore et al. (1986a), the effect of mulching on the crop's environment diminishes with progressive development of the canopy. During the SR10 season, soil moisture was significantly influenced by mulching at 54 DAP, irrespective of rates. At 18 DAP and 72 DAP, soil moisture content was very low irrespective of mulching rate and was not significantly different (Table 4). This was attributed to the extended drought spells prior to the dates of soil sampling, the low and poorly distributed rainfall. October 2010 was rather dry. Prior to the date of soil sampling, an extended dry spell lasting 17 days was observed. Also prior to the 72 DAP sampling date, an extended dry spell of 9 days was experienced (Fig. 2). The lack of moisture conservation response to mulching during these two dates corroborates Ji and Unger (2001), Mulumba and Lal (2008), and Ramakrishna et al. (2006), observations that the effectiveness of mulching in moisture conservation diminishes with increasing drought period.

At Karurumo, contrary to the results in the other two sites, soil moisture was significantly higher in mulched treatments, during the SR09 and LR10 seasons (Table 5). At 50 DAP of the SR09 season and 42 DAP during the LR10 seasons, soil moisture beneath 12 t ha^{-1} of maize stover was higher by 5.7% and 2.5%, respectively. Soil moisture beneath the 4 and 8 t ha^{-1} was not significantly different from bare soil. The implication of this result is that for effective moisture conservation in the drier and hotter Karurumo site; the optimum rate of mulching should be above 8 t ha^{-1}. By the 65th DAP of the LR10 season, soil moisture was significantly higher in all mulching treatments irrespective of rates and ranged between 2.7% and 4.2% (Table 5). The effectiveness of the lower mulching rates in moisture conservation was likely due to self-shading by the crop canopy. In this site, potato growth in bare soil was not vigorous enough to form a thick canopy for self-shading indicating the importance of mulching in hot and dry climates. During the SR10 season, mulching caused no significant effect on soil moisture at this site. This was associated with the poorly distributed rainfall and the extended dry spells prior to dates of soil sampling (Fig. 2). At 18 DAP, this site had not received any rainfall, hence the very low moisture content.

Soil Temperature (°C) at 12 cm Depth as Influenced by Mulching Rates Across the Three Different AEZs

The effect of mulching on soil temperature at 12 cm soil depth across the three sites indicated no significant effects of mulching at 08.00–09.00 h (data not presented). Comparison of the afternoon soil temperature (15.00–16.00 h) between Mbuvori and Nembure showed a significant ($p = 0.003$) site by mulching rate interaction (Table 6a). At Mbuvori, application of 8 and 12 t ha^{-1} of mulching reduced soil temperature to levels that were not significantly different from those observed at Nembure. While the difference in soil temperature in bare soil plots between the two sites was 4.2 °C, the gap between the two sites in 8 and 12 t ha^{-1} mulched plots was reduced to as low as 1.3 °C to 2.6 °C and was not significantly different (Table 6). During the LR10 season, the difference in mean air temperature between the two sites was 2.6 °C. The reduction in soil temperature by 8 and 12 t ha^{-1}, in this site, is an indicator that with global warming as projected in future scenarios, mulching has the potential of reducing the negative impacts. While Karurumo was hotter than Nembure, there were no interaction effects between these two sites (Table 6b and c). During the LR10 season, average soil temperature in the two sites was significantly depressed by the 8 and 12 t ha^{-1} mulching rate and reduced by 1.1 °C and 2.6 °C, respectively. During the SR10 season, soil temperatures beneath 8 and 4 t ha^{-1} was not significantly different from bare soil. Mulching with 12 t ha^{-1}depressed soil temperature by 3.0 °C (Table 6c). The nonsignificant effects of the 8 and 4 t ha^{-1} during the SR10 season could be due low and poorly distributed rainfall in the two sites that resulted to dry soil conditions hence low thermal capacity.

Across, season analysis at Nembure showed no significant season by mulching rate interaction (Table 7). Averaged soil temperature showed significant (all the three crop growth periods at p <0.001) effect of mulching (Table 7). During the emergence period, mulching with 12 t ha^{-1} reduced soil temperature by 3.5 °C compared to bare soil and was significantly lower from 8 to 4 t ha^{-1} rates (Table 7a). At tuber initiation period, soil temperature beneath both 8 and 12 t ha^{-1} was lower by 1.7 °C and 2.6 °C, respectively (Table 7b). The same trend was observed at tuber bulking stage (Table 7c). As crop growth progressed, the gap between bare soil and mulched plots reduced. During the early period of potato growth, tuber emergence (EM), a negative correlation between mulching rates and soil temperature was observed at p = 0.003 and 0.001 during the LR10 and SR10 seasons, respectively (Fig. 3). Soil temperature decreased with increasing mulching rates, with every 2 t ha^{-1} depressing soil temperature by 0.6 °C and 0.4 °C during the LR10 and SR10 seasons, respectively. The optimum mulching rate was 10 t ha^{-1} which reduced soil temperatures to 23.3 °C and 25.5 °C during the LR10 and SR10 seasons, respectively. The higher temperature at optimum mulching rate, during the SR10 season, was attributed to the dry weather conditions. During the first and fourth WAP

Table 6 Effects of mulching rates on soil temperature (°C at 12 cm depth) across maximum Mbuvori (LH1), Nembure (UM2), and Karurumo (LM3), during the LR10 and SR10 seasons

Mulch rate (t ha^{-1})	(a) Average(4–6 WAP) Mbuvori and Nembure, LR10			(b) Average (3 and 6 WAP) Nembure and Karurumo, LR10			(c) Average (3 and 6 WAP) Nembure and Karurumo, SR10		
	LH1	UM2	Mean	UM2	LM3	Mean	UM2	LM3	Mean
12	20.1b	22.7b	22.4	22.6	24.4	23.5c	24.4	24.6	24.5b
8	22.6b	23.9b	23.2	23.3	24.8	24.1bc	25.5	25.1	25.3b
4	22.0b	25.5a	23.7	24.1	25.1	24.6 b	25.1	25.3	25.2b
0	22.2b	26.5a	24.4	24.7	27.4	26.1a	28.1	26.8	27.5a
P	0.003			<0.001			<0.001		
s.e	0.43						0.28		
CV	3.1%			2.4			0.24		

Numbers followed by the same letter in the same column or row (comparison between sites) are not significantly different from each other

WAP weeks after planting

Table 7 Effect of mulching rates on soil temperature (°C) at 12 cm depth across the SR09, LR10, and SR10 during different potato growing periods at Nembure

Mulch rate ($t\ ha^{-1}$)	(a) Emergence			(b) Tuber initiation			(c) Tuber bulking			
	LR10	SR10	Mean	LR10	SR10	Mean	SR09	LR10	SR10	Mean
12	23.2	25.2	24.2c	22.2	23.1	22.6c	22.9	23.0	21.9	22.3c
8	24.2	26.1	25.1b	23.4	23.7	23.5bc	22.3	23.0	23.4	22.9bc
4	24.5	26.4	25.4b	24.4	24.4	24.4ab	22.9	23.4	24.3	23.5b
0	27.1	28.3	27.7a	24.6	25.9	25.2a	24.4	26.5	23.4	24.8a
P			<0.001	ns		<0.001				<0.001
s.e			0.275			0.24				0.33
CV	2.4							4.2		

Numbers followed by the same letter in the same column are not significantly different from each other

of the LR10 season, a total of 62 mm and 91 mm of rainfall was recorded, respectively, and was distributed in 5 days in each week. On the other hand, the SR10 season was dry and recorded a total of 6.5 and 21 mm during the first and fourth WAP, respectively. The second and third WAP were dry with only a paltry 1 mm received during the third week after planting (Fig. 3). During the LR10 season, the relationship between soil temperature and mulching quantities was not significant at both tuber initiation and bulking. However during the SR10 season, the rate of soil cooling at both tuber initiation and bulking growth stages was negatively correlated to mulching rates $p = <0.001$ and $p = 0.008$, respectively (Fig. 3b and c). For every 2 $t\ ha^{-1}$ mulching rate, a reduction of 0.9 °C observed. The optimum mulching rate was about 10 $t\ ha^{-1}$ which reduced soil temperatures from 26.2 °C to 22.7 °C (Fig. 3c). During the LR10 season, application of 8–9 $t\ ha^{-1}$ decreased soil temperatures from 23.5 °C to 22.3 °C which, though in the upper limit, is within the optimum soil temperature for potato tuber growth. According to Struik 2007, the optimum soil temperature for tuber growth varies from 15 °C to 23 °C. Due to drought conditions experienced during the SR10 season, the optimum soil temperature for potato growth was not achieved even with the application of the optimum mulching rate (Fig. 3a). The influence of rainfall on soil temperature responses to mulching rates observed in this site concurs with Zhang et al. (2009) who reported a relationship between soil temperature response to mulching and rainfall. Moist soils have a higher thermal capacity and have therefore lower temperature than dry soils (Abu-Hamdeh and Reeder 2000; Sauer et al. 1998).

At Karurumo, across-season analysis indicated no significant season by mulching rate interaction on soil temperatures at 8.00–9.00 h. Based on seasonal average, soil temperature was significantly decreased by mulching irrespective of the quantity applied (Table 8a). A negative correlation between soil temperatures and mulching rates at $p < 0.001$ and $p = 0.03$ was observed during the LR10 and SR10 seasons, respectively (Fig. 4a). For every 2 $t\ ha^{-1}$ of mulch applied, soil temperatures at 8.00 to 9.00 h decreased by 0.3 °c and 0.14 °C during the LR10 and SR10 seasons, respectively (Fig. 4a). At 15.00–16.00 h, a season by mulching rate interaction was observed, the highest difference between treatments observed during the SR09 and the LR10

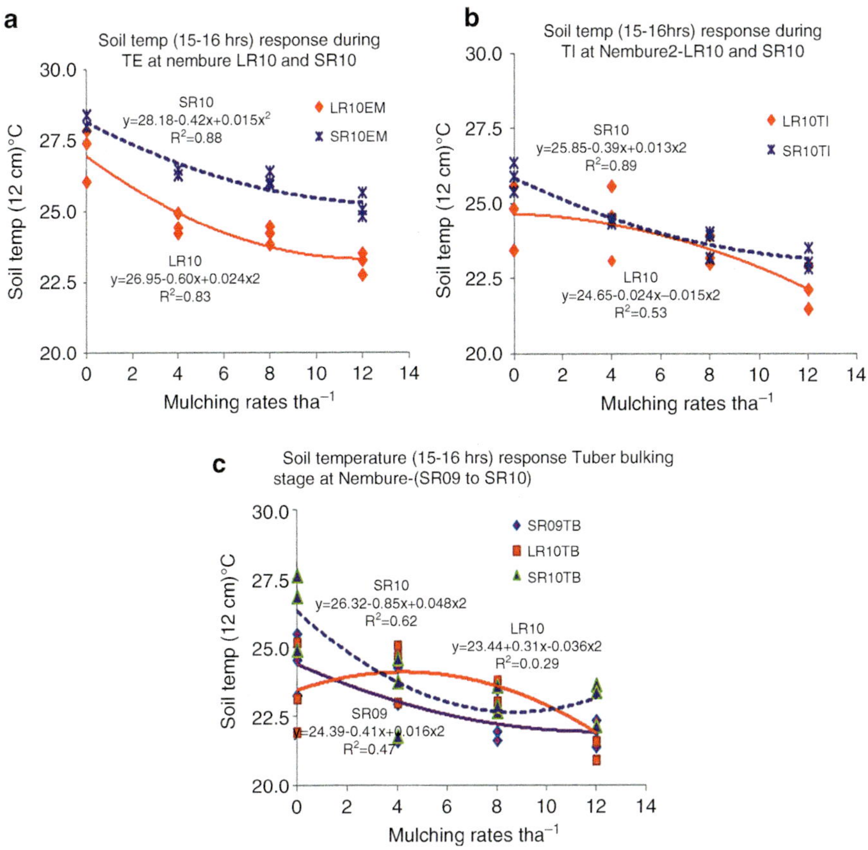

Fig. 3 Soil temperature (12 cm depth) response to mulching (**a**) at tuber emergence (EM), (**b**) initiation (TI), and (**c**) bulking (TB), during the SR09, LR10, and SR10 seasons at Nembure (UM2)

seasons compared to the SR10 season (Table 8). Application of 12 t ha^{-1} mulch decreased soil temperatures by 5.3 °C, 4.0 °C, and 2.5 °C during the SR09, LR10, and SR10 seasons, respectively (Table 8). During the SR09 season, soil temperature was negatively correlated ($p < 0.001$) to mulching rate and decreased by 0.7 °C for every 2 t ha^{-1} of mulch applied, down to a low of 23.8 °C (Fig. 4b). During the LR10 and SR10 seasons, soil temperature was negatively correlated $p = 0.005$ and $p = 0.002$, respectively. Application of 10 t ha^{-1} of maize stover reduced soil temperature to 22.3 °C and 20.5 °C during the LR10 and SR10 seasons, respectively (Fig. 4b). These temperature levels are within the optimum range for potato production. The negative correlation between soil temperatures and increasing quantity of mulching at both the Nembure and Karurumo is in agreement with observations by Blanco-Canqui and Lal (2009), Chen et al. (2007), Rees and Chow (2005), and Wang et al. (2005) that soil temperature suppression increased with increasing mulching rates. The reduction of soil temperature to optimum potato crop growing conditions in these sites indicates the potential of mulching as an adaptation strategy.

Table 8 Effect of mulching rates on soil temperature (°C) at 12 cm depth, across the SR09, LR10, and SR10 seasons at Karurumo (LM3)

Mulch rate	(a) Soil temperature at 8.00–9.00 h				(b) Soil temperature at 15.00–16.00 h			
(t ha^{-1})	SR09	LR10	SR10	Mean	SR09	LR10	SR10	Mean
12	21.3	22.3	20.5	21.4b				
12	21.3	22.3	20.5	21.4b	23.4f	24.4de	22.9f	23.6
8	21.9	22.3	20.7	21.6b	24.5de	24.8d	23.5ef	24.2
4	21.9	22.4	20.7	21.6b	26.3c	25.1d	23.6ef	25.0
0	22.7	23.4	21.3	22.5a	28.7a	27.4b	25.4d	27.1
Mean	21.9b	22.6a	20.8c					
p	<0.001			<0.001	<0.001			
CV	1.1				1.9			

Means followed by the same letter on the same column and row (comparisons of treatments across seasons) are not significantly ($p = 0.05$) different from each other

Effects of Mulching Rates on Soil Temperature at 20 cm Depth Across Different AEZs

A significant site by mulching rate interaction was observed across the three sites (Table 9). At Mbuvori and Nembure, mulching caused no significant effects ($p \leq 0.05$) on soil temperature at 20 cm depth. At Mbuvori, the non-significant effects of mulching are in agreement with Rees and Chow, (2005) who also observed no significant effects by varying rates of mulch application for 4 consecutive years. The lower ambient air temperature, the well-distributed rainfall, and soil type could have all contributed to high soil moisture and in turn nonresponse to mulching. Nullet et al. (1990) stated that differences between soil and air temperature are influenced by rainfall, the soil temperature in wet areas approaching air temperature. Moist soil has a higher thermal capacity and therefore requires higher energy to warm up (Rees and chow 2005; Sauer et al. 1998). Though not a variable in this study, the dominant soil type in this site is an andosol which has a high water holding capacity. At Nembure, the nonsignificant effects of mulching at 20 cm depth are contrary to the results for 12 cm depth (Tables 6 and 7). This could be due to the relatively cooler climatic conditions, compared to Karurumo, and the soil depth. According to Li et al. (2012), the magnitude of mulching effects on soil temperature is higher near the soil surface and decreases with depth. In the hotter and drier Karurumo site, mulching reduced the gap between soil temperatures at this site and the relatively cooler Nembure and Mbuvori. At Karurumo, while bare soil was warmer than at Nembure by 1.3–5.5 °C, the difference in 12 t ha^{-1} mulching between the two sites ranged between 0.63 °C and 0.3 °C and was not significantly different. The gap in soil temperature in bare soil between Mbuvori and Karurumo was 7.8 °C, while the differences in mulched plots ranged between 4.43 °C to 4.6 °C and 2.13 °C to 2.4 °C, in plots covered with mulch rates of 4 t ha^{-1} and 12 t ha^{-1} of mulch,

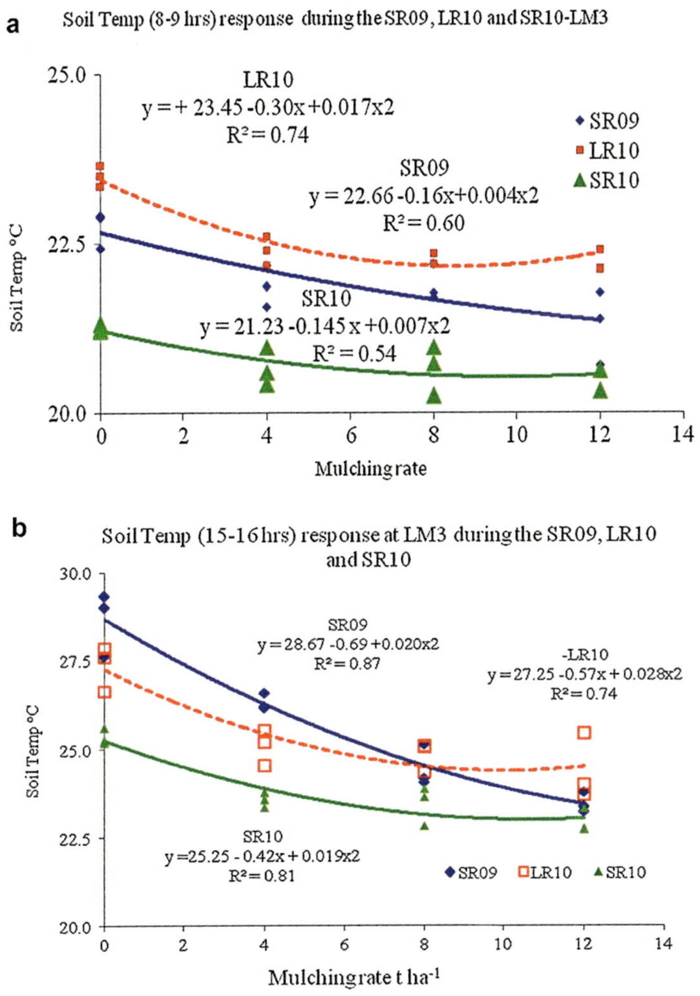

Fig. 4 Influence of mulching rates on soil temperature (°C at 12 cm depth) across season at Karurumo (**a**) at 08.00–09.00 h and (**b**) at 15.00–16.00 h

respectively (Table 9). During the latter part of the crop growing period (i.e., tuber bulking), soil temperatures beneath 12 t ha^{-1} of mulching rate at the Karurumo were comparable to those observed at both Nembure and Mbuvori (Table 9d). Such a high drop in soil temperatures under mulching conditions later in the crop growing period is attributed to the closed crop canopy as has been reported by other researchers (Cook et al. 2006; Midmore et al. 1986a; Sauer et al. 1998).

During the crop emergence/vegetative (EM) growth stage, soil temperature at 15.00–16.00 h at Karurumo was negatively correlated ($p < 0.001$) with increasing mulching rate (Fig. 5a). The same trend ($p = 0.016$) was also observed at tuber

Table 9 Effect of mulching rates on maximum soil temperature (°C) at 20 cm depth across AEZs during the SR09 and LR10 rain seasons

Mulch rate (t ha^{-1})	(a) Soil temp. (°C) at EM, LR10				(b) Soil temp. (°C) at TI, LR10			
	LH1	UM2	LM3	Mean	LH1	UM2	LM3	Mean
12	20.8e	22.9d	23.2 cd	22.3	20.6d	23.1c	23.4bc	22.3
8	20.9e	22.7d	24.6bc	22.9	20.5d	23.0c	24.7bc	22.9
4	21.1e	23.7 cd	25.7b	23.2	20.6d	23.5bc	25.0b	22.8
0	20.9e	23.1 cd	28.6a	24.5	20.6d	23.4bc	28.1a	24.0
P		0.000		0.000		0.000		
s.e		0.41				0.38		
CV		3.1%				2.8%		
Mulching rate t ha^{-1}	(c) Soil temp. (°C) at TB, SR09				(d) Soil temp. (°C) at TB, LR10			
	LH1	UM2	LM3	Mean	LH1	UM2	LM3	Mean
12	19.7d	21.7 cd	22.0c	21.1	21.2bc	21.8bc	22.3bc	21.7
8	21.8 cd	21.9 cd	23.8b	22.5	20.8c	22.0bc	23.2b	22.0
4	20.1 cd	21.8 cd	24.4b	22.1	20.5c	21.9bc	23.1b	21.9
0	21.3 cd	21.9 cd	26.3a	22.9	20.8c	21.2bc	24.8a	22.3
P		0.003				0.05		
s.e		0.48				0.46		
CV		3.7				2.6		

Means followed by the same letter with a column are not significant. Means with the same letter with a row per crop growth and season are not significantly different
LH1 Mbuvori, *UM2* Nembure, *LM3* Karurumo, *EM* tuber emergency, *TI* Tuber initiation, *TB* tuber bulking

initiation crop growth (TI) stage (Fig. 5b). The negative correlation between soil temperature and the amount of mulch applied concurs with observations made by Blanco-Canqui and Lal (2009), Chen et al. (2007) and Wang et al. (2005) that soil temperature suppression increased with increasing mulch rates. During these two crop growth periods, soil temperature at Karurumo declined by close to 1 °C for every 2 t ha^{-1} of mulch applied and by 0.3 °C latter in the season (Fig. 5c). The optimum rate was 10 t ha^{-1} and was effective in reducing soil temperatures to 23.5 °C which is close to the soil optimum range for potato growth. This result proves that in the relatively hot and drier Karurumo site, mulching reduces soil temperature from levels that are close to the optimum levels for potato growth.

Potato Emergence (%) as Influenced by Mulching Rates

At Mbuvori, tuber emergence was not significantly influenced by mulch (Fig. 6a). At Nembure, the percentage of emerged plants was negatively correlated to increasing mulching rates (p <0.001) during the early stages of emergence (Fig. 6b). However, with time the difference between treatments diminished and was not significantly different (Fig. 6b). The delay in potato emergence in high mulching rate plots at Nembure could be attributed to the relatively lower soil temperatures in

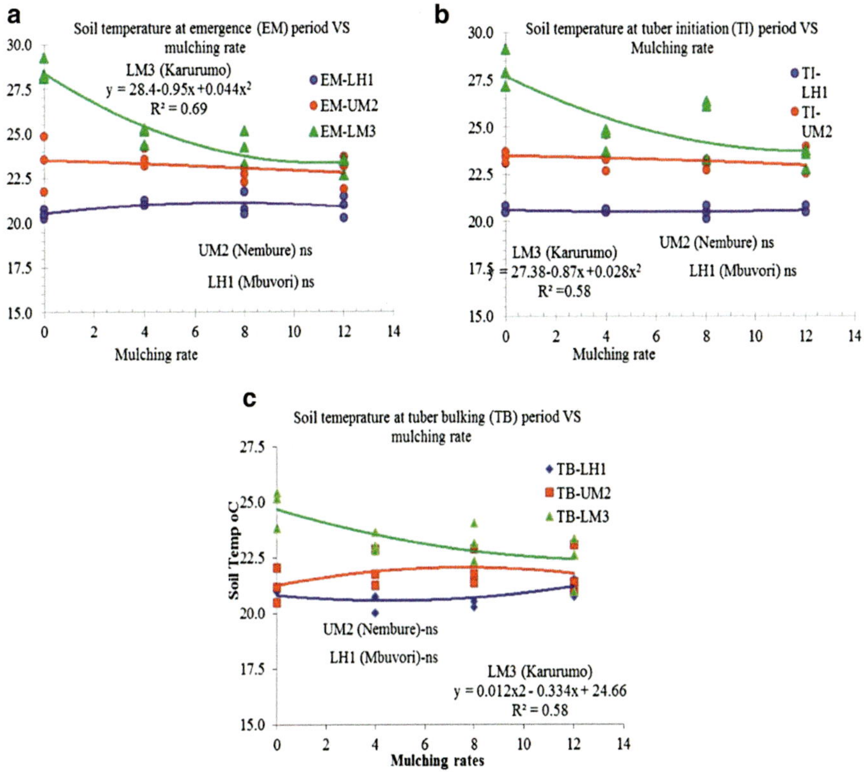

Fig. 5 Soil temperature (°C at 20 cm depth) response to mulching rates at 15.00–16.00 h during tuber emergence (EM), tuber initiation, and tuber bulking (TB) during the LR10 season

high mulching rates plots and the thick surface through which potato had to penetrate. Blanco-Canqui et al. (2006) also associated the delay in the emergence of maize and wheat with reduced soil temperatures. In the hotter and drier Karurumo site, a positive relationship between mulching rates and the percentage of emerged tubers was observed. At 31 DAP of the SR09 season, though not significantly different, there was tendency toward a positive correlation ($p = 0.06$) between the percentage of emerged plants with increasing mulch quantity. As the period from the date of planting to emergence progressed (61 DAP), a significant ($p = 0.007$) positive relationship between mulching rates and tuber emergence was observed (Fig. 6c). The high percentage of emerged tubers in this site could be due to the reduction in soil temperatures and moisture conservation. This result is in agreement with Midmore et al. (1986a) and Mahmood et al. (2002) observation of hastened potato emergence by mulch in hotter climates. They also associated the early tuber emergence with temperature reduction due to mulching. In addition to the improved thermal conditions, moisture conservation could also have contributed to the faster potato emergence in mulched plots.

Effect of Mulch Rates on Potato Yield Across Different AEZs

During the LR09 and SR09 seasons, a significant site by mulching rate interaction was observed (Table 10). During the LR09 and SR09, application of 4–12 t ha^{-1} of mulch at Karurumo resulted in high tuber yield that were not significantly different from yield observed in relatively cooler Nembure (Table 10a and b). During the LR10 season, the yield obtained from the non-mulched treatments at Nembure was 85% significantly higher than Karurumo, but mulching irrespective of rate reduced that gap to nonsignificant levels (Table 10a). During the SR09 season, mulching irrespective of the rate resulted in increased total tuber yield at the UM2 site that were also comparable to that observed at Mbuvori (Table 10). The yield increase with even the low mulching is similar to the observation by Rees and Chow (2005), who reported an 11–14% yield increase by applying 2.25–4.5 t ha^{-1} of hay. Application of 12 t ha^{-1} rate increased yield at both UM2 and LM3 and was not significantly different from the yield observed

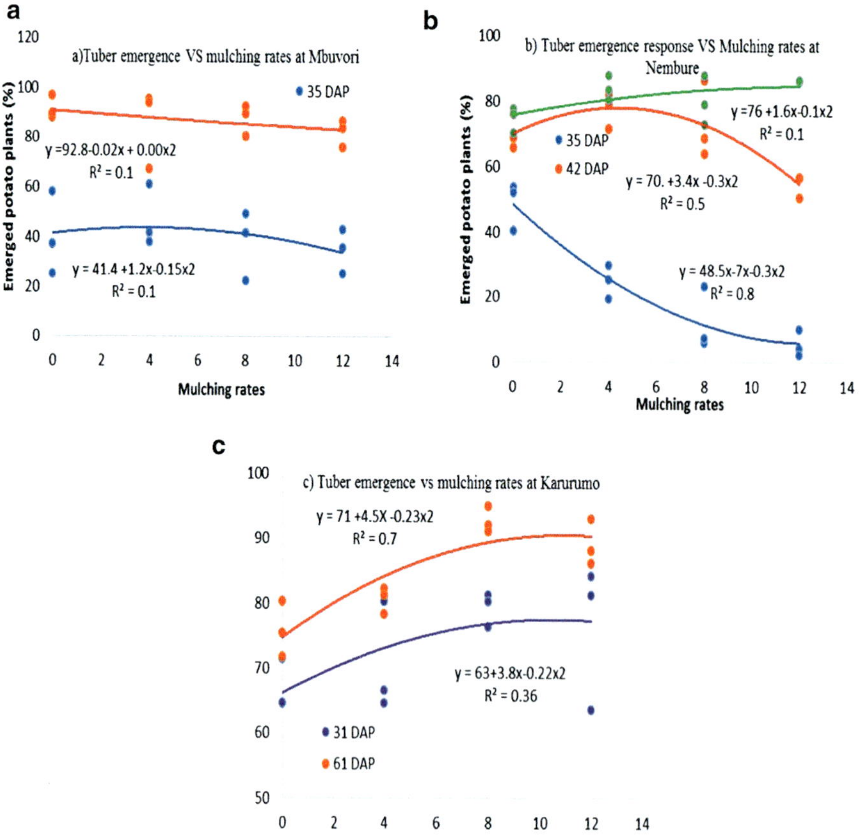

Fig. 6 Tuber emergence as influenced by mulching rates at (**a**) Mbuvori, (**b**) Nembure, and (**c**) Karurumo

Table 10 Total tuber yield (t ha^{-1}) as influenced by mulching rates at Mbuvori, Nembure, and Karurumo during the LR09 to SR10 season

Mulch rate (t ha^{-1})	Sites				Sites			
	LH1	UM2	LM3	Mean	LH1	UM2	LM3	Mean
	(a) LR09				**(b) SR09**			
0	23.0ab	17.2d	9.3 e	18.4	28.0a	15.6b	4.1c	18.4
4	30.6a	21.9 cd	21.3 cd	24.6	26.5a	19.0ab	14.8b	24.6
8	31.9 a	24.4bc	24.0bc	26.8	27.4a	22.5ab	18.1ab	26.8
12	27.2abc	23.5bc	23.6bc	24.8	28.7a	22.4ab	20.4ab	24.8
Mean	29.6	21.8	19.53		27.6	19.9	14.4	
	$p = 0.002$ s.e. = 1.45 CV =10.6				$p = 0.04$ s.e. 2.2 CV =18.3			
	(c) LR10				**(d) SR10**			
0	23.2	15.3	16.3	18.4b	20.0	4.4	10.0	10.0b
4	25.9	14.3	22.9	21.0a	20.9	8.3	13.4	13.4a
8	24.3	17.2	22.4	21.3a	23.9	8.2	14.9	14.7a
12	25.8	17.9	21.6	21.8a	20.9	8.3	15.4	14.9a
	TRT, $p = 0.015$ s.e = 0.72 CV = 10.5				TRT, $p = 0.0002$ s.e = 0.83 CV = 18.9			

Means followed by the same letter with a column are not significant. Means with the same letter with a row per season (treatment effects across sites) are not significantly different

at Mbuvori. Comparison of yield obtained from bare soil treatments, Nembure and LM3 during the SR09 season, indicated a 3.8 times higher at the UM2 zone. However application mulching rates as low as 4 t ha^{-1} reduced that gap to nonsignificant levels (Table 10b). The yield advantage of mulching was more significant in the hotter and drier environment. Application of 8–12 t ha^{-1} at the LM3 AEZ increased yield 4.4 and 5.7 times, respectively, compared to bare soil and was not significantly different from Nembure and Mbuvori (Table 10b). In irrigated potato Kar and Kumar (2007) observed a 24–42% yield increase and associated it with both moisture conservation and temperature moderation. The high yields observed in the Nembure and Karurumo were due to the soil temperature reductions in those two sites. This result shows that even with temperature rise as high as 5 °C yield losses associated with climate change can be arrested by mulching. The increased yield in the hotter LM3 AEZ is contrary to Midmore et al. (1986b) who reported less yield response to mulching in hot climatic conditions. In this study, during the LR09 season while application of 12 t ha^{-1} by 1.4-folds in the Um2 AEZ, the yield in the hotter LM3 AEZ was 2.5-folds (Table 10).

Conclusion

This study indicates that mulching was effective in conserving soil moisture and slowed the drying process. For effective moisture conservation, a mulching rate in the range of 8 t ha^{-1} and above will be required. With increasing length of the drought period, the effectiveness of mulch to conserve moisture diminishes. At both the UM2 and LM3 AEZs, mulching at 10 t ha^{-1} depressed soil temperature at 12 cm depth and at 20 cm at LM3, to within optimum levels for potato tuber growth and development. Rainfall and in turn soil moisture influenced the response of soil temperature to mulching the extent to which soil temperature decrease determined by the duration of the dry spell or rainfall amounts. In the relative hotter and dry LM3 AEZs, mulching with 8–12 t ha^{-1} rates reduced soil temperatures to levels that were comparable to those observed in the relatively cooler and wetter UM2 AEZ. Mulching at the 8–12 t ha^{-1} at the UM2 AEZ depressed soil temperatures to levels that were comparable to the cool and humid LH1 AEZ. Mulching enhanced tuber emergence under hot climatic conditions and was associated with the reduced soil temperature by mulching. In the LM3 AEZ, mulching with as low as 4 t ha^{-1} more than doubled potato yield. These results show that mulching can be used as an adaptation strategy to climate change in potato farming systems of central Kenya.

Limitations

The results are limited to the current climatic conditions in the study sites and are not necessarily representation of future climate change scenarios but can be used to infer possible outcomes. Under prolonged drought periods, mulching will not be effective in decreasing soil temperature and the associated yield losses.

Cross-References

- ▶ Ecotourism as an Adaptation Strategy for Mitigating Climate Change Impacts on Local Communities Around Protected Areas in Ghana
- ▶ Water Harvesting Technology for Enhancing Food Security Livelihood: The Case of Northern Katsina State, Nigeria

Acknowledgment This work was funded by the Germany Ministry of Economic cooperation (BMZ) in partnership with the International Potato Center.

References

Abouziena HF, El-Saeid HM, Ahmed A, Amin AAE (2014) Water loss by weeds: a review. Int J ChemTech Res 7(1):323–336

Abu-Hamdeh NH, Reeder RC (2000) Soil thermal conductivity: effects of density, moisture, salt concentration, and organic matter. Soil Sci Soc Am J 64:1285–1290

Anderson JM, Igram JSL (eds) (1993) Tropical soil biology and fertility; A handbook of methods, 2nd edn. CAB International, Wallington. ISBN085198210

Bhat R, Khero KL (2006) Effect of tillage and mode of straw application on soil erosion in the submontane tract of Punjab, India. Soil Tillage Res 88:107–115

Blanco-Canqui H, Lal R (2007) Impact of long term wheat straw management on soil hydraulic properties under No-tillage. Am Soc Agron 71(4):1166–1173

Blanco-Canqui H, Lal R (2009) Crop residue removal impacts on soil productivity and environmental quality. Crit Rev Plant Sci 28(3):139–163
Blanco-Canqui H, Lal R, Post WM, Owen LB (2006) Changes in long-term no-till corn growth and yield under different rates of Stover mulch. Agron J 98:1128–1136
Bristow KL (1988) The role of mulch and its architecture in modifying soil temperature. Aust J Soil Res 26(2):269–280
Caldiz DO, Gaspari FJ, Moreno Kiernan A, Struik PC (2002) Agro-ecological zoning at the regional level: spatio-temporal variation in potential yield of the potato crop in the Argentinian Patagonia. Agric Ecosyst Environ 88(1):3–10
Chen SY, Zhang XY, Pei D, Sun HY, Chen SL (2007) Effects of straw mulching on soil temperature, evaporation and yield of winter wheat: field experiments on the North China Plain. Ann Appl Biol 150(3):261–268
Cook HF, Valdes GSB, Lee H (2006) Effects on rainfall interception, soil physical characteristics and temperature under *Zea mays* L. Soil Tillage Res 91:227–235
Cromme N, Prakash AB, Lutalado N, Ezeta E (2010) Strengthening potato value chains; Technical and policy options for developing countries. FAO, Rome, p 148. ISBN: 9789251066270
Falkenmark M, Rockstrom J (2006) The new blue and green water paradigm: breaking new ground for water resources planning and management. J Water Resour Plan Manag 132:129–132
Fischer G, Shah M, van Velthuizen H (2002) Climate change and agricultural vulnerability. IIASA, Laxenburg, 150pp
Funk C, Eilerts G, Davenport F, Michaelsen J (2010) A climate trend analysis of Kenya-August 2010, USGS Fact Sheet 2010-3074. http://pubs.usgs.gov/fs/2010/3074/pdf/fs2010-3074.pdf
Gildemacher P, Kaguongo W, Ortiz O, Tesfaye A, Woldegiorgis G, Wagoire W, Kakuhenzire R, Kinyae P, Nyongesa M, Struik P, Leeuwis C (2009) Improving potato production in Kenya, Uganda and Ethiopia: a system diagnosis. Potato Res 52(2):173–205
Haverkort AJ (1990) Ecology of potato cropping systems in relation to latitude and altitude. Agric Syst 321:251–272
Hijmans RJ (2003) The effects of climate change on global potato production. Am J Potato Res 80:271–280
Jaetzold R, Schmidt H, Hornet ZB, Shisanya CA (2006) Farm management handbook of Kenya. Natural conditions and farm information, 2nd edn, vol 11/C. Eastern Province. Ministry of Agriculture/GTZ, Nairobi
Ji S, Unger PW (2001) Soil water accumulation under different precipitation, potential evaporation and straw mulch conditions. Soil Sci Soc Am J 65:442–448
Kaguongo W, Gildemacher P, Demo P, Wagoire W, Kinyae P, Andrade J, Forbes G, Fuglie K, Thiele G (2008) Farmer practices and adoption of improved potato varieties in Kenya and Uganda. Social sciences working paper 2008-5. International Potato center (CIP), Lima, 85p
Kar G, Kumar A (2007) Effects of irrigation and straw mulch on water use and tuber yield of potato in eastern India. Agric Water Manag 94(1–3):109–116
Li R, Hou XQ, Han QF, Yang BP (2012) Effect of rainfall harvesting and mulching technologies on soil water, temperature and maize yield in Loess Plateau region of China. Soil Res 50:105–113
Lung'aho C, Nderitu SKN, Kabira JN, El-Bedewy R, Olanya OM, Walingo A (2006) Yield performance and release of four late blight tolerant potato varieties in Kenya. J Agron 5:57–61
Lutaladio NB, Ewell PT, Kidanemarian H (1995) Advances in potato research in Eastern and central Africa. In: Akoroda MO, Ekanayake IJ (eds) Root crops and poverty alleviation: Proceedings of the sixth triennial symposium of the International Society for Tropical Root Crops-Africa Branch (ISTRC-AB), ISTRC-AB, Lilongwe, October 22–28, 1995
Lynch MDR, Foroud N, Kozub GC, Farries BC (1995) The effect of moisture stress at three growth stages on the yield, components of yield and processing quality of eight potato varieties. Am Potato J 72:375–386
Mahmood MM, Farooq K, Hussain A, Sher R (2002) Effect of mulching on growth and yield of potato. Asian J Plant Sci 1(2):132–133

Manrique LA (1992) Potato production in the tropics: crop requirements. J Plant Nutr 15:2679–2726
Methu JN, Owen E, Abate AL, Tanner JC (2001) Botanical and nutritional composition of maize Stover intakes and feed selection by dairy cattle. Livest Prod Sci 71(2–3):87–96
Midmore DJ, Roca J, Berrios D (1986a) Potato (*Solanum* spp.) in the hot tropics II. Soil temperature and moisture modification by mulch in contrasting environments. Field Crop Res 15:97–108
Midmore DJ, Roca J, Berrios D (1986b) Potato (*Solanum* spp.) in the hot tropics III. Influence of mulch on weed growth, crop development, and yield in contrasting environments. Field Crop Res 15:109–124
MoA (2008) National policy on potato industry: policy reforms to revitalize the potato industry. Ministry of Agriculture, Nairobi
MoA (2009) National potato task force report; final report. 45 pages. Ministry of Agriculture, Nairobi
Mulumba LN, Lal R (2008) Mulching effects on selected soil physical properties. Soil Tillage Res 98:106–111
Nullet D, Ikawa H, Kilham P (1990) Local differences in soil temperature and soil moisture regimes on a mountain slope, Hawaii. Geoderma 47:171–184
Nyaata OZ, Dorward PT, Keatinge JDH, O'Neill MK (2000) Availability and use of dry season feed resources on smallholder dairy farms in central Kenya. Agrofor Syst 50:315–331
Omambia AN, Shemsanga C, Li Y (2009) Combating climate change in Kenya: efforts, challenges and opportunities. Rep Opin 1(6):65–76
Parry M, Rosenzweig C, Livermore M (2005) Climate change, global food supply and risk of hunger. Philos Trans R Soc B 360:2125–2138
Pervaiz MA, Iqbal M, Shahzad K, Hassan AU (2009) Effect of mulch on soil physical properties and N, P, K concentration in maize (*Zea mays* L.) shoots under two tillage systems. Int J Agric Biol 11:119–124
Ramakrishna A, Tam HM, Wani SP, Long TD (2006) Effect of mulch on soil temperature, moisture, weed infestation and yield of groundnut in northern Vietnam. Field Crop Res 95:115–125
Rees HW, Chow TL (2005) Impact of incorporated hay mulch on selected physical, chemical and microbiological properties of a loam soil in Atlantic Canada after five years of continuous potato production. Can J Soil Sci 85:161–172
Rees HW, Chow TL, Loro PJ, Lavoie J, Monteith JO, Blaaux A (2002) Hay mulching to reduce runoff and soil loss under intensive potato production in northwestern New Brunswick, Canada. Can J Sci 82(2):249–258
Rosenzweig C, Parry LM (1994) Potential impact of climate change on world food supply. Nature 367:133–138
Sauer TJ, Hatfield JL, Prueger JH, Norman JM (1998) Surface energy balance of a corn residue-covered field. Agric For Meteorol 89:155–168
Struik PC (2007) Responses of the potato plant to temperature. In: Bradshaw DV, Gebhardt C, Govers F, Mackerron DKL, Taylor MA, Ross HA (eds) Potato biology and biotechnology. Elsevier Science B.V, Amsterdam, pp 367–393
Van Loon CD (1981) The effect of water stress on potato growth development and yield. Am Potato J 58:51–69
Wang XL, Li FM, Jia Y, Shi WQ (2005) Increasing potato yield with additional water and increased soil temperature. Agric Water Manag 78:181–194
Yuan B, Nishiyama S, Kang Y (2003) Effects of different irrigation regimes on the growth and yield of drip-irrigated potato. Agric Water Manag 63:153–167
Zhang S, Lovdahl L, Grip H, Tong Y, Yan-Jun X, Wang Q (2009) Effect of mulching and catch cropping on soil temperature, soil moisture and wheat yield on the Loess Plateau of China. Soil and Till. Res. 102:78–86

Ecosystem-Based Adaptation in Tigray, Northern Ethiopia: A Systematic Review of Interventions, Impacts, and Challenges

120

Mulubrhan Balehegn, Mitiku Haile, Chao Fu, and Wu Liang

Contents

Introduction 2441
Materials and Methods 2441
  Description of the Study Area 2441
Methods 2443
  Selection of Studies and Records 2443
  Evaluation of EBA Interventions 2443
Results and Discussions 2444
  Types of EBA Interventions 2444
Impact of Interventions 2444
  Impacts of Biological Rehabilitation 2446
  Impact of Conservation Agriculture Interventions 2457
  Impact of Water Harvesting Interventions 2460
  Impacts of Soil and Water Conservation 2460
  Impact of Soil Fertility Improvement Interventions 2464
  Impact of Integrated Watershed Management Interventions 2464
Challenges with EBA Interventions and Future Directions 2469
  Challenges on Mobilizing Public Support 2469

M. Balehegn (✉)
Department of Animal, Rangeland and Wildlife Sciences, Mekelle University, Mekelle, Ethiopia
e-mail: mulubrhan.balehegn@mu.edu.et; mu.gebremikael@ufl.edu

M. Haile
Department of Land Resource Management and Environmental Protection, Mekelle University, Mekelle, Ethiopia

C. Fu
Key Laboratory of Ecosystem Network Observation and Modeling, Institute of Geographic Sciences and Natural Resources Research, Chinese Academy of Sciences, Beijing, China

W. Liang
Research Center for World Geography and Resources, Institute of Geographic Sciences and Natural Resources Research, Chinese Academy of Sciences, Beijing, China

W. Leal Filho (ed.), *Handbook of Climate Change Resilience*,
https://doi.org/10.1007/978-3-319-93336-8_117

Conclusion 2475
Cross-References 2476
References 2477

Abstract
The Tigray regional state in the northern part of Ethiopia has been severely affected by centuries of land degradation and climate change-induced recurrent drought and extreme weather variability. Government and people have, therefore, steadily implemented internationally recognized (Example at Rio 20 for Innovative Hunger Solutions and recently Gold Medal Winner at the World Future Council award for Best policy in Combating Desertification and Land Degradation: The World's Best Policies https://www.worldfuturecouncil.org/p/champions/) land rehabilitation and ecosystem-based adaptation (EBA) programs. Despite the international recognition for the successes, the specifics of the different interventions implemented, the impacts observed, challenges encountered have not been well articulated and communicated, in a way that would enable other similar regions to learn from the experiences of Tigray region of Ethiopia. In this chapter, we reviewed 170 publications on 30 EBA interventions in 400 sites in Tigray. Interventions fall in either of the following categories: Soil and Water Conservation (SWC) (62.69%), Biological Rehabilitation (BR) (18.41%), Water Harvesting and Production (WHP) (7.46%), Soil Fertility Improvement (6.22%), Conservation Agriculture (3.48%), and Integrated Watershed Management (1.74%). While many studies reported impressive biophysical changes, e.g., decrease in runoff and increased sediment deposition (in 63.93% of cases), farmer-relevant impacts such as improvements in crop and livestock yields and income/livelihoods have been quantified only in 8.46%, 1.6%, and 4.16% of cases, respectively, implying that implementers are failing to communicate impacts, causing a missed opportunity for fast dissemination and mobilization in other similar places. The most successful interventions include exclosures and variety of soil and water conservation measures (mainly stone bunds), while those that showed limited positive impacts are water harvesting and production schemes. Popular support and capitalization on locally available resources are the most cited reasons for success of interventions, while overambitious and myopic project planning, top-down approach, limited technical skills are common challenges encountered. In this chapter, we summarized one of the most successful cases of EBA and land rehabilitation intervention in the world, identified strengths, weaknesses, challenges, and suggested solutions in a way that could enable other communities learn from the EBA experiences of Tigrian farmers and government.

Keywords
EBA · Land rehabilitation · Tigray · Northern Ethiopia · Land degradation

Introduction

Northern Ethiopia has historically been under extreme demographic influence, civil war, and climate change that increased the occurrence of drought and extreme weather variability and concomitant environmental degradation (Conway 2000). Perhaps nowhere in the world is land degradation problem more manifest than in the marginal highlands of northern Ethiopia (Hengsdijk et al. 2005), with a huge cost and economic implication (Haregeweyn et al. 2008a). Some of the worst human calamities caused by drought and subsequent famines have been reported in the northern Ethiopian regions of Tigray, Amhara, and Afar, so much so that the region has been associated with famine and misery in popular imagination. Such challenges have made people and government in northern Ethiopia to implement steady ecosystem-based adaptation (EBA) and community-based land rehabilitation and conservation programs, with impressive and internationally recognized success (Bewket 2007). The interventions implemented in northern Ethiopia utilize biodiversity and ecosystem services to support climate change adaptation and enhance environmental rehabilitation, making them typical cases of EBA (Munang et al. 2013).

Nonetheless, despite northern Ethiopia being repeatedly recognized for its efforts-impacts of its interventions (https://www.worldfuturecouncil.org/p/champions/), information on what actually has been changed and what has been achieved is only available in the form of results of dispersed studies that evaluate the impact of different interventions on wide variety of environmental and social variables (Gebremeskel et al. 2017). So far, impacts brought about through different interventions, the challenges encountered, and possible future solutions have not been analyzed and articulated in a way that would enable sharing of experiences with other communities (Gebremeskel et al. 2017). Therefore, studies undertaken on different EBA interventions in the Tigray regional state with the objective of identifying understanding the EBA interventions, their impacts, challenges encountered in implementation, and recommendations for better successful implementation in the future. Moreover, the study contributes towards evidence on effectiveness of EBA that is generally limited worldwide and concentrated only in developed countries (Doswald et al. 2014) and provides an analysis of different factors that affect the effectiveness of EBA interventions based on cases from northern Ethiopia.

Materials and Methods

Description of the Study Area

Tigray regional state is one of the nine regional states in Ethiopia located at the northern most extreme of the country (Fig. 1) between 12°15′ and 14°50′N and between 36° 27′ and 39° 59′E with an area of 80,000 km^2. It is surrounded by Sudan in the west, Eritrea in the north, and the Ethiopian regions of Amhara and Afar in the

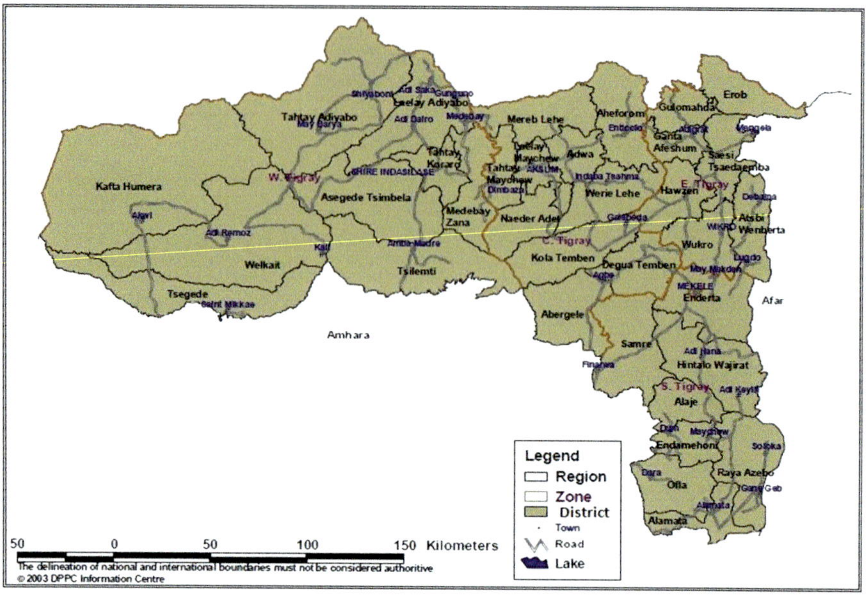

Fig. 1 The Tigray regional state showing its administrative units

south and east, respectively. Tigray is characterized by undulating terrain and steep slopes with altitude varying from 500 to 4000 m.a.s.l. (Gebremeskel et al. 2017). The agroecology is semi-arid with distinct dry and wet seasons and rainfall ranging 200–1000 mm and with an average annual temperature of 18 °C (Hagos et al. 1999).

The land use types dominant in Tigray include croplands, exclosures, remnant forests, villages, and built-up areas. Thirteen major soil types are identified in Tigray: *cambisols*, *rendzinas*, *lithosols*, *acrisols*, *fluvisols*, *luvisols*, *regosols*, *nitosols*, arenosols, vertisols, xerosols, solonchaks, and andosol (Nyssen et al. 2008a). The geological formations consist of Precambrian metavolcanics and Mesozoic sedimentary rocks such as Adigrat sandstone, Antalo limestone, Agula shales, and Amba Aradam sandstone, which in turn are intruded by Cenozoic dolerite dykes/sills (Haregeweyn et al. 2008b).

Tigray has a population of 4.4 million growing at 3% annually (CSA 2010). The farming system is dominated by small-scale rainfed agriculture, which utilizes traditional crop and livestock production technologies (Hagos et al. 2016). Approximately 90% of the population depends on the centuries-old plow-based subsistence cultivation, currently having 1.2 ha average landholdings per household (Pender and Gebremedhin 2007). Agriculture contributes to 60% of the regional total gross domestic product (Hagos et al. 2016). Dominant cereal crops grown are *tef* (*Eragrostis tef*), barley (*Hordeum vulgare* L.), wheat (*Triticum sp.*), sorghum (*Sorghum bicolor*), and maize (*Zea mays*), accompanied by leguminous crops such as field peas (*Pisum sativum*), chickpeas (*Cicer arietinum*), and horse bean (*Vicia faba*). Gesho (*Rhamnus prinoides*). Cattle, sheep, goat, equines, beehives, and poultry are dominantly found in livestock types in Tigray.

Methods

Selection of Studies and Records

Published studies were searched from the Web of Science and Google Scholar Databases using relevant key words. Moreover, gray literature or office and field reports by different organizations working in the EBA issues in the Tigray, Masters/MA, PhD level unpublished theses and dissertations have also been included. Though there was no restriction in the years for publication of studies and reports, most of the 170 identified studies fall within the years 1997–2017. Only studies or peer reviewed publications that have been published in journals indexed by the Web of Science, Scopus, African Journals OnLine, and other legitimate indexing services were included, and those published in the so-called predatory journals have been excluded, as their quality have been repeatedly questioned (Balehegn 2017a).

Evaluation of EBA Interventions

EBA interventions were first categorized into any of the six categories namely: (1) Biological Rehabilitation (BR): Those that intend to enhance or improve the biological potential of degraded grazing lands, farmlands, etc.; (2) Conservation Agriculture (CA): Types of agriculture or farming that tried to maximize the conservation of moisture for improved yield; (3) Integrated Watershed Management (IWM) catchment or watershed level approaches that implement combinations of many interventions for an overall ecological and livelihoods improvement; (4) Soil Fertility Improvement (SFI), which included variety of interventions that aim to improve the fertility status of degraded farmlands; (5) Soil and Water Conservation (SWC), interventions aimed at conserving soil and water or protecting soil from erosion; and (6) Water Harvesting and Production (WHP), interventions that intend to conserve water from loss or extract more water for agricultural and other uses. The observed impacts of EBA interventions were also categorized into 11 general categories (Fig. 5) namely: (1) Enhanced drought and climate change adaptation (EDCCA), (2) Enhanced soil characteristics (ESC), (3) Enhanced vegetation (EV), (4) Improved carbon stocks (ICS), (5) Improved crop yields (ICY), (6) Improved income and livelihoods (IIaL), (7) Improved livestock productivity (ILsP), (8) Improved wildlife diversity (IWD), (9) Improved water harvesting and use efficiency (IWHUE), (10) Reduced runoff and increased sediment deposition (RRISD), and (11) Others. Moreover, for most of the EBA interventions, main challenges encountered during the implementation, solutions recommended or implemented have also been identified (Table 8).

Results and Discussions

Types of EBA Interventions

A total of 30 types of EBA interventions including SWC ($n = 16$) followed by WHE ($n = 10$), BR ($n = 10$), CA (n = 7) and SFI ($n = 3$) have been reported. The typology of the different interventions is given in Fig. 2. The EBA interventions according to the percentage of cases reported (total number = 402) are SWC (62.69%), BR (18.41%), WHP (7.46%), SFI (6.22%), CA (3.48%), and IWM (1.74%) (Fig. 3). The larger diversity of SWC interventions is because of the diverse agro-ecological setting in northern Ethiopia that requires different solutions. Moreover, soil erosion and land degradation have always been the most important environmental problems (Hurni 1988) with serious economic consequences such as, for example, causing a loss of 3.4 million Euros per year, just from the erosion caused loss of N and P in Tigray (Haregeweyn et al. 2008a). As a result therefore, SWC interventions have been the commonest types of EBA interventions in northern Ethiopia, with about 522, 600 ha of land already covered by some form of SWC interventions, mainly stone bunds from 1991 to 2002 (Nyssen et al. 2007). Exclosures are the second most commonly implemented and studied interventions. Currently, there are about three million hectares of land under exclosure management all over Ethiopia (Lemenih et al. 2014), with most of it (around 1.2 million hectares) of exclosures being in Tigray (Tetemke et al. 2017).

The description, purposes, and sites for implementation of the different EBA interventions is given in Table 1. Some examples graphical representations of EBA interventions are also given in Fig. 4.

Impact of Interventions

The most commonly reported impact is reduction in runoff and increase in sediment deposition (461 cases) followed by improved crop yield (61 cases) and enhanced vegetation (49 cases) (Fig. 3). Other impacts, reported to a lesser extent also included impacts on income, livestock productivity, soil fertility, carbon stocks and others (Fig. 3).

Apart from limited number of studies (30 cases), which tried to quantify and demonstrate the economic or livelihood impact of interventions, most studies demonstrated only biophysical impacts on soil, vegetation and water. This is probably because of the methodological difficulties of differentiating the impact of the EBA interventions on the livelihoods and income (Haregeweyn et al. 2015) but is critical problem because, the emphasis on biophysical impacts with little consideration on

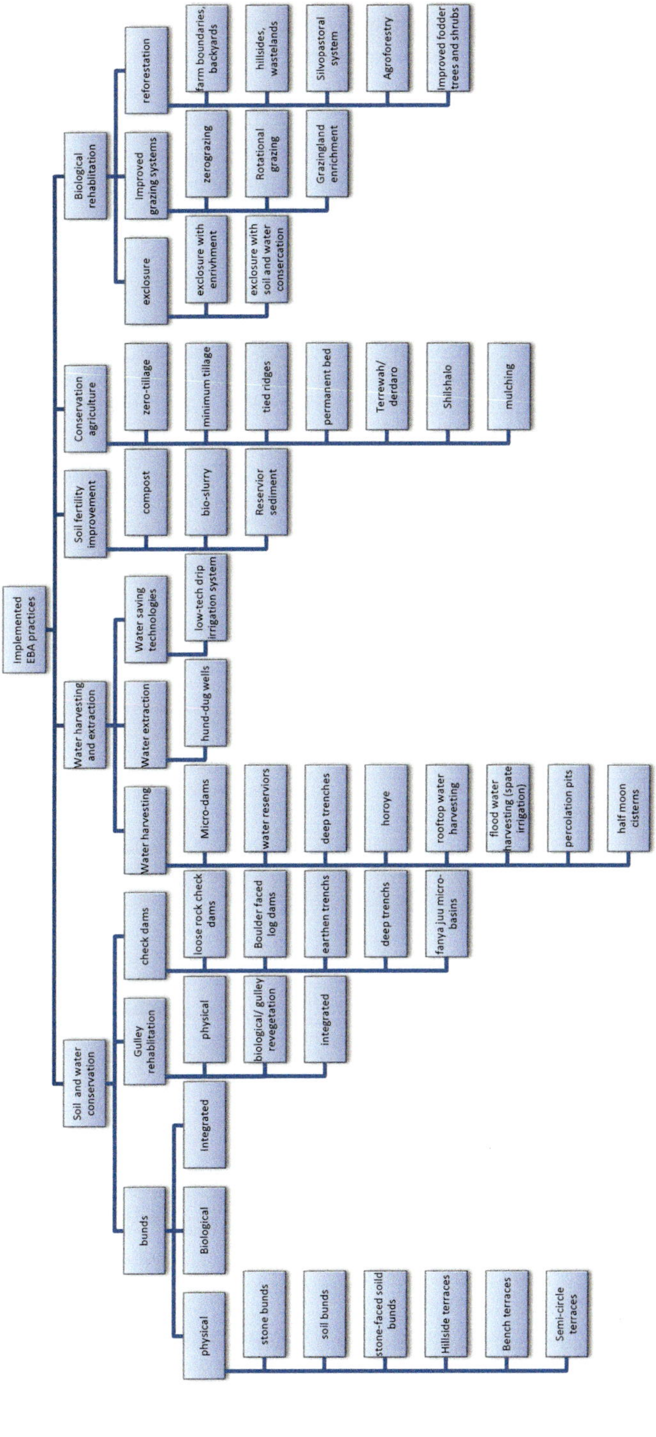

Fig. 2 Typology of ecosystem-based adaptation techniques implemented in northern Ethiopia

the social and economic benefits of interventions makes it difficult to provide evidence for convincing policy makers in expanding the implementation of the EBA practices and popularizing the practices among local communities (Awulachew et al. 2005). Even when it is currently difficult to quantify economic and livelihood benefits, as ecological improvement will take time to manifest as livelihood improvement, it is important that modeling studies be undertaken to generate evidence for livelihood impacts of interventions, unless and otherwise it will be difficult to convince farmers (Balana et al. 2010).

Impacts of Biological Rehabilitation

Positive impacts as a result of the most common biological rehabilitation interventions – exclosures – include improvements in: litter accumulation (1802–2108.57%), total organic carbon (15–64%), available Nitrogen (187–5125%); soil phosphorus (290–1150%), and other soil chemical characteristics (Table 2). Similarly, exclosures have resulted into improved vegetation attributes including: herbaceous cover (329.27%), herbaceous species diversity (31–50%), woody species cover (436%), woody species diversity (50–81%), and other vegetation attributes including species richness, vegetation density, basal cover and ground cover, and bird and mammal species diversity (Table 2). Reduction in soil erosion (46–79%), runoff (83–95), and

Fig. 3 Percentage cases of studied interventions. *SWC* Soil and water conservation, *WHP* Water harvesting and production, *SFI* Soil fertility improvement, *EA* Ecological Agriculture, *BR* Biological rehabilitation, *IWM* Integrated watershed management

Table 1 EBA interventions implemented in northern Ethiopia and their descriptions

EBA/Subtype	Description of the intervention	Objectives or aims	Sites or areas applied	Sample references
Biological rehabilitation				
Exclosures	Areas closed off from human and livestock disturbance; may or may not be accompanied with enrichment or planting of tree species or soil and water conservation practices	To promote natural regeneration of degraded grazing lands and hillsides	Degraded hillsides	Aerts et al. (2009)
Zero-grazing	A system of grazing where livestock are restricted in a small space (usually in backyards or barns) and fodder is cut and carried to where animals are, instead of animals freely grazing on grazing lands	To prevent overgrazing and destruction of conservation structures and young seedlings by livestock	Degraded grazing lands	Gebreyohannes and Hailemariam (2011)
Silvopastures	Planting or enhancing fodder or other multipurpose trees in pasturelands, farm boundaries, backyards, etc.	To improve livestock feed availability while providing other multipurpose benefits, e.g., fuel wood	Grazing land, farmland boundaries, wastelands, backyards	Balehegn (2017b)
On farm agroforestry	The use, planting and enhancement of trees on farmlands or farmland boundaries	Improve soil fertility, serve as wind break and prevention of pests while providing multiple benefits	Farmlands and farmland boundaries	Ernstberger (2017)
Reforestation/ enrichment	Planting of seedling of exotic or indigenous tree species or reseeding of seeds of trees and herbaceous plants into degraded lands	For rehabilitating degraded area, recovering plant cover	Degraded hillsides, degraded rangelands	Meaza et al. (2016)

(*continued*)

Table 1 (continued)

EBA/Subtype	Description of the intervention	Objectives or aims	Sites or areas applied	Sample references
Conservation agriculture				
Zero-tillage	Not tilling the land before sowing, may involve weeding or removal of unnecessary vegetation	To reduce moisture loss and soil disturbance due to excessive plowing	Drought-affected farmlands	McHugh et al. (2007)
Minimum tillage	Also called conservation plowing or noninversion tillage (e.g., reduced plowing which minimizes the disruption of soil structure. The frequency of plowing is reduced to less than the 3–12 times practiced traditionally for different crops in northern Ethiopia	To improve soil physical characteristics and moisture content	Drought-affected farmlands	Tesfahunegn (2015)
Terrewah	Traditional plowing followed by making every 1.5–2 m contour furrows. Furrows are made at 2–4 m intervals along the contour the same day after planting, especially practiced for teff (*Eragrostis teff*)	To conserver moisture on farm and prevent runoff	Farmlands	Araya et al. (2016a)
Shilshalo	A traditional plowing where contour furrows are created within the standing crop (mostly sorghum and maize) during second weeding operation	To conserve moisture on farm and prevent runoff, reduce weed infestation	Sorghum and maize fields	Nyssen et al. (2010)
Derdero	Bed furrows prepared along contour using the	Conserving water on farm	Farmlands	Nyssen et al. (2010)

(*continued*)

Table 1 (continued)

EBA/Subtype	Description of the intervention	Objectives or aims	Sites or areas applied	Sample references
	traditional *marasha (Plowsahre)* at the last tillage operation or after farmers broadcast seeds over the farmland			
Tied ridges	Small earthen *ridges*, 15–20 cm high, with an upslope furrow which accommodates runoff from a catchment strip between the *ridges*	Conserve water on farm, reduce amount of moisture lost due to excessive plowing	Farmlands	Grum et al. (2017a)
Crop residue mulching	Leaving crop-residue on the farm, after harvest so that they serve as mulch against evaporation of moisture	To conserve residual moisture	Drought-affected farmlands	Araya and Stroosnijder (2010)
Integrated watershed management	A holistic watershed or catchment development that integrates various interventions with in a catchment while utilizing synergies and reducing competitiveness among interventions	Overall ecological rehabilitation, improvement of yields, and adaptation to natural challenges	Degraded watersheds	Alemayehu et al. (2009)
Soil fertility improvement				
Compost	Decaying domestic organic wastes, weeds, and plant materials to use them for fertilizing farmlands	Improve fertility of degraded farmlands with low cost input	Degraded or infertile farmlands	Edwards (2007a)
Bioslurry	A byproduct of an aerobic fermentation of livestock excreta or dung in a biogas digester	Used as a fertilizer while the biogas used as a source of fuel	Farmlands	Edwards (2012)

(continued)

Table 1 (continued)

EBA/Subtype	Description of the intervention	Objectives or aims	Sites or areas applied	Sample references
Reservoir sediment	Sediment harvested from the bottom of reservoirs when they dry up during the dry season and applied to farmlands	To utilize nutrients leached from upper soil and stored in reservoirs for land reclamation	Degraded farmlands	Girmay et al. (2012)
Soil and water conservation				
Earthen trenches	Or simply trenches are ditches dug on the path of an erosion or a catchment	To trap sediment, reduce runoff or erosion	On high erosion areas (grazing lands, hillsides)	Taye et al. (2014)
Traditional *Daget*	Traditionally established by farmers as untilled strip of about 2 m wide at a lower plot limit in a slopping farmland. The grass in this strip is allowed to grow year after year	The strip is intended to reduced runoff velocity and allows water to infiltrate and sediment to be trapped	At the edge or lower strip of farmlands	Nyssen et al. (2000)
Stone bunds/ terraces	Stone structures constructed along a contour in a slopping land (usually hillsides and slopping farmlands)	To reduce erosion or runoff	On hillsides, sloppy farmlands	Nyssen et al. (2008b)
Soil bunds	Soil-based structures constructed along a contour in a slopping land (usually only on slightly sloped farmlands)	To reduce erosion and runoff	On a slightly slopping farmland	Teshome et al. (2013)
Gabion check dams	Check dams constructed using rocks and reinforced by gabion or wire mesh for increased strength or resistance	To control gulley expansion, trap sediment, and conserve water	On gullies, streams and eroded areas	Mekonnen et al. (2015)

(*continued*)

Table 1 (continued)

EBA/Subtype	Description of the intervention	Objectives or aims	Sites or areas applied	Sample references
Boulder-faced log dams	Dams constructed by embedding two parallel logs with a spacing that varies between 5 and 25 cm, lifting the logs approximately 0.50–1 m above bed and facing their upper side with spaced boulders (0.3–0.7 m across) that rested on coarse bed load	To counter the destruction of gabion dams caused by abrasion in heavy currents	In rivers and streams with heavy currents	Nyssen et al. (2017)
Loose rock check dams	Check dams constructed across small and medium sized streams using only rocks and also mud or soil	To control gulley expansion or rehabilitate gullies	On gullies	Nyssen et al. (2004b)
Water harvesting and production				
Horeye	Trapezoidal ponds with their floors and sides sometimes covered with plastic, riprap, or compacted to prevent seepage	To harvest water at household level from small catchments	On wasteland with catchment	Gebremeskel et al. (2017)
Micro-dam reservoirs	Small dams built in catchments and valleys to reserve water during the rainy season for using it in the dry season	Harvesting and reserving water for irrigation and livestock use	On catchments and valley bottoms	Berhane et al. (2016a, b)
Percolation pond	Pits and ditches of various sizes and shapes, dug or constructed for the sole purpose of allowing runoff to percolate	Increase ground water availability in downstream areas	On wastelands, degraded areas upstream	Grum et al. (2017b)
Deep trench	Deep trenches at lower part of a catchment or across the slop of a catchment dug to harvest water and silt during	To control runoff, trap silt, and enhance groundwater recharge processes	On road sides, degraded areas, farmland sides	Woldearegay et al. (2015)

(*continued*)

Table 1 (continued)

EBA/Subtype	Description of the intervention	Objectives or aims	Sites or areas applied	Sample references
Hand-dug wells	Wells dug deep into the ground to access and collect ground water	To provide water for domestic use and irrigation	On backyards, near farmlands in the lower section of catchments	Woldearegay and Van Steenbergen (2015)
Roof water harvesting	A simple system and structure to harvest water from roofs and store it in underground or above ground concrete tankers or any other form of water storage facility	To harvest rain water that would otherwise be lost	In urban and peri-urban areas	Ham (2008)
Spate irrigation	The diversion of seasonal floods to irrigate farmlands	To utilize excessive flood water for growing crops	Dryland farming systems	Hiben and Tesfa-alem (2014)
Low cost family drip irrigation	Low cost family-based drip irrigation systems that use small plastic tubes and water conserving structures to save water	Saving water	On backyard and highly fertile soils	Waktola (2007)

improvement in other soil and water variables such as rain percolation, and sediment accumulation are also some of the impacts of exclosures that have been measured in various studies (Table 2).

Though limited in numbers, economic-related impacts have also been quantified for exclosures. These include improvement in yield of economically important products such as Frank incense from *Boswellia paprifera* tree (43.9%), improvement in livestock productivity (20%), and overall increase in net value of land as a result of conversion to exclosures (28%). Though numerical quantification is lacking, exclosures have also improved the income of poor families through the production of honey in exclosures and sell of grass (Meaza et al. 2016). In fact, a study by Babulo et al. (2009) indicated that products from exclosures including (honey, fuel wood, grass, etc.) account for 15% of the overall average forest environmental income in Tigray.

Fig. 4 Different EBA interventions implemented in Tigray (Top row right to left: Gulley treatment, stone bunds, integrated gulley treatment, zero-grazing (cut and carry from exclosures). Middle row: Reservoirs, hand-dug wells, and river diversion). Bottom row: Treated gulley, exclosure enriched with drought-tolerant halophytes, earth sided deep trenches, the derdero system of plowing)

Table 2 Impacts of biological rehabilitation interventions in northern Ethiopia (Sampled studies showing impacts on various variables)

Observed impact	Untreated	Treated	Percentage change	Description of the study setting	Sampled references
Exclosures					
Litter biomass	17.5 (g/m^2)	386.5 (g/m^2)	2108.57%	Old exclosure compared with freely grazed land; values average for two sites	Descheemaeker et al. (2009a)
Available organic matter	9 g/m^2	148.8 g/m^2	1553.33%	Average value for exclosures of different ages, comparison with unprotected grazing land	Descheemaeker et al. (2006)
Carbon stock	7.76 mg ha^{-1}	22.29 mg ha^{-1}	187.24%	Average value for exclosures of different ages, comparison with unprotected grazing land	Solomon et al. (2017)
Soil organic carbon	54 mg C ha^{-1}	74 mg C ha^{-1}	37.04%	0–40 cm soil depth, comparison with cultivable land	Girmay and Singh (2012)
Available total nitrogen	0.2 g/m^2	4.32 g/m^2	2060.00%	Average value for exclosures of different ages, comparison with unprotected grazing land	Descheemaeker et al. (2006)
Available phosphorus	0.02 g/m^2	0.078 g/m^2	290.00%	Average value for exclosures of different ages, comparison with unprotected grazing land	Descheemaeker et al. (2006)
Soil potassium	0.06 (g/m^2)	2.995 (g/m^2)	4891.67%	Exclosure compared with freely grazed land; values average for two sites	Descheemaeker et al. (2009a)
Soil calcium	1.7	70.65	4055.88%	Exclosure compared with freely grazed land; values average for two sites	Descheemaeker et al. (2009a)
Soil magnesium	0.12 (g/m^2)	3.85 (g/m^2)	3108.33%	Exclosure compared with freely grazed land; values average for two sites	Descheemaeker et al. (2009a)
Soil sodium	0.0035 (g/m^2)	0.29 (g/m^2)	8185.71%	Exclosure compared with freely grazed land; values average for two sites	Descheemaeker et al. (2009a)
Soil pH	6.25	7.2	–	Seven-year exclosure average value for soil depth 0–50 cm, comparison with free grazing land	Mekuria et al. (2017)
Mycorrhiyzal colonization	1.763*105/g soil	4.39*105/g soil	149.01%	Average value for exclosures of different ages, comparison with unprotected grazing land	Birhane et al. (2017)
Woody plant canopy cover	6%	32.16%	436.00%	Average value for 5-year and 10-year-old exclosure and comparison with freely grazing land	Mekuria et al. (2007)

Woody species diversity	2.97	5.4	81.82%	Exclosures of age 5–10 years compared with adjacent grazing lands	Yayneshet et al. (2009)
Ground cover	7%	44.83%	540.43%	Average value for 5-year and 10-year-old exclosure and comparison with freely grazing land	Mekuria et al. (2007)
Above ground biomass	2 mg/ha	15 mg/ha	650.00%	Average value for 20-years-old exclosure compared with adjacent grazing land	Mekuria and Veldkamp (2012)
Herbaceous species diversity	2.70	4.00	48.15%	Exclosures of age 5–10 years compared with adjacent grazing lands	Yayneshet et al. (2009)
Herbaceous basal cover	0.41%	1.76%	329.27%	Ten years exclosure compared with freely grazed land, average values from three locations or sites	Abesha (2014)
Common bird species with forests	5%	20%	300.00%	Exclosures compared with free grazing land	Aerts et al. (2008)
Number of large mammals	0	24		29 year exclosure compared with freely grazed land	Yami et al. (2007)
Runoff	6 times higher		−83%	Cultivated land compared with exclosure	Girmay et al. (2009)
Soil loss	68.03 t/ha/year	36.7 t/ha/year	−46.05%	Average value for 5-year and 10-year old exclosure and comparison with freely grazing land	Mekuria et al. (2007)
Evapotranspiration	358 mm	549 mm	53.35%	Exclosures from three sites with exclosures 5–20 years old	Descheemaeker et al. (2009b)
Frankincense yield from *Boswelia papirifera*	288 g/tree/year	414.58 g/tree/year	43.95%	Exclosures compared to areas open for livestock and people	Tilahun et al. (2011)
Livestock fodder		5000 kg/ha more		Exclosures compared to freely grazed land	Asres (2012)
Shannon Wiener Diversity Index (H′) for honey bee flora	0.45	0.875	94.4%%	Exclosures compared to freely grazed land. Values are average for dry and wet seasons	Teklay (2011)
Livestock water productivity		20% higher water productivity	20%	Creating exclosure on 40% of the land in a study site	Descheemaeker et al. (2010)
Net present value		28% higher (837 US $)	28%	Net present value exclosures compared with wheat farms	Mekuria et al. (2011)

(*continued*)

Table 2 (continued)

Observed impact	Untreated	Treated	Percentage change	Description of the study setting	Sampled references
Silvopastures					
Livestock productivity		Replace 50% concentrate	50%	Freed from silvopastoral fodder trees helped replace 50% concentrate feed while improving goat productivity	Balehegn et al. (2015)
Total nitrogen	1.6 mg/ha	3.5 mg/ha	118.75%	Silvopasture compared with rainfed farmland	Gelaw et al. (2015a)
Total organic carbon	16.1 mg/ha	39.1 mg/ha	142.86%	Silvopasture compared with rainfed farmland	Gelaw et al. (2015a)
Agroforestry					
Total nitrogen	0.34 g/kg	0.655 g/kg	92.65%	Agroforestry trees on farmland compared with rainfed farmland with not trees	Gelaw et al. (2015a)
Soil organic carbon	3.2 g/kg	5.95 g/kg	85.94%	Agroforestry trees on farmland compared with rainfed farmland with not trees	Gelaw et al. (2015a)
Soil quality indices (SQI)	0.47 g/kg	0.58 g/kg	23.40%	Agroforestry trees on farmland compared with rainfed farmland with not trees	Gelaw et al. (2015b)
Density of *Feidheria albida*	12.15 trees/ha	13.56 trees/ha	11.60%	On farm agroforestry compared with exclosures, values are mean for two sites	Noulekoun et al. (2017)
Reforestation					
Income from grasses		1381/ha/yr	–	Average value from four sites and 3 years	Meaza et al. (2016)
Income from timber		1722/ha/yr	–	Average value from four sites and 3 years	Meaza et al. (2016)
Erosion		15% less	–	A modeling study	Hengsdijk et al. (2005)
Soil carbon	10.7 kg/mg soil	18.0 kg/mg soil	68.22%	Restricted grazing lands compared with free grazing traditional land	Rimhanen et al. (2016)

All these improvements observed as a result of exclosure are mainly due to removal of overgrazing that not only cause removal of vegetation but also physically disturb soil structure making soil unsuitable for natural recruitment, ultimately contributing to land degradation. In the specific case of northern Ethiopia, traditional free-grazing system has contributed to the steady degradation of land that has taken place in the highlands of Ethiopia for centuries (Mekuria et al. 2007; Taddese 2001) and has resulted in continuous decline in the availability of livestock feed and other ecosystem services from natural rangelands (Gebremedhin et al. 2004). Free-grazing system did not only result in increased land degradation but also has limited the effectiveness of human endeavor in rehabilitating degraded areas through the physical destruction of soil and water conservation structures (Meshesha et al. 2012). Many communities in northern Ethiopia have therefore implemented the exclosure intervention and have voluntarily established traditional local by-laws which helped in enforcing strict protection of exclosures from livestock and human disturbance (Yami et al. 2013). Economic and ecological benefits accrued from exclosures, however, depend on the age of exclosures, where at least 7 years of exclosure is needed before significant improvements in soil physical and chemical characteristics can be detected (Mekuria et al. 2017). Therefore, it is important for planners and communities to understand that benefits may not be realized with in short period of time.

Similar to exclosures, other biological rehabilitation interventions including silvopastures, agroforestry, and plantations or reforestation have resulted in improvements such as increase in soil nitrogen, soil carbon, vegetation attributes, and farmland and livestock productivity, and even income (Table 2). These interventions act by increasing the number of multipurpose trees planted on farms, farm boundaries, wastelands, hillsides, and other similar areas. Specific multipurpose trees such as *Ficus thonningii* (Balehegn 2017b), *Feidherbia albida* (Gelaw et al. 2015a), *Acacia etbica* (Yayneshet et al. 2008) introduced in agroforestry and silvopastoral systems have been observed to result in ecological and economic improvements such as improved soil fertility (Gelaw et al. 2014), livestock productivity (Balehegn et al. 2014), and income (Meaza et al. 2016).

Impact of Conservation Agriculture Interventions

The different quantified positive impacts of conservation agriculture include reduction in runoff by *Derdaro* (49–82%), mulching (64.44%), permanent beds (60.49%), *Shilshalo* (41.65%), *tied ridges* (56%) and combination of tied ridges, straw mulch, and effective microorganisms (80.85%) (Table 3). Reduction in runoff has also resulted in the expected reduction in soil loss from farmlands at rates of 53–78.5% for *Derdero*, 78.89% for permanent beds, 21–61.03% for *Shilshalo*, and 87% for combination of *Terrrewah*, straw mulch, and effective microorganisms (Table 3). The different conservation agriculture interventions and their combinations have also resulted in reduction in loss of

Table 3 Impact of conservation agriculture interventions in Northern Ethiopia

Interventions	Impacts	Untreated	Treated	Percentage change	Description of the study setting	References
Derdaro	Runoff	91.6 mm	46.3 mm	−49.45%	Derdero compared with conservation tillage	Araya et al. (2011)
	Soil loss	30 t/ha	14 t/ha	−53.33	Terrewah compared with traditional plowing over 9 year average	Araya et al. (2016b)
	Crop yield	1.51 t/ha	2.03 t/ha	34.43%	Derdaro compared with conventional plowing, grain yield is average for 9 years for teff, wheat, barley, and grass pea)	Araya et al. (2016a, b)
	Straw yield	3.54 t/ha	4.81 t/ha	35.8%	Derdaro compared with conventional plowing, straw yield is average for 9 years for wheat and teff	Araya et al. (2016a, b)
Mulching	Runoff	67.2 mm	20.5 mm	−69.49%	Mulching compared with unmulched control. Values are average for 4 years	Araya and Stroosnijder (2010)
	Barley yield	1262.25 kg/ha	1271.33 kg/ha	0.71%	Mulching compared with unmulched control. Values are average for 4 years	Araya and Stroosnijder (2010)
Permanent bed	Runoff	653 m^3/ha	255 m^3/ha	−60.94%	Permanent bed compared with traditional plowing	Gebreegziabher et al. (2009)
	Soil loss		81% less	−81%	Permanent beds compared with conventional plowing	Nyssen et al. (2010)
Terewah	Runoff		11% less	−11%	Terewah compared with conventional tillage	Nyssen et al. (2010)
	Soil loss	19.5t/ha	7.6 t/ha	−61.03%	Terrewah compared with traditional plowing	Gebreegziabher et al. (2009)
	Crop yield	1.51 t/ha	1.78 t/ha	17.88%	Terrewah compared with conventional plowing, grain yield is average for 9 years for teff, wheat, barley, and grass pea	Araya et al. (2016a, b)
	Straw yield	3.54 t/ha	4.51 t/ha	27.4%	Terrewah compared with conventional plowing, straw yield is average for 9 years for wheat and teff	Araya et al. (2016a, b)

Shilshalo	Sorghum yield	1.4 mg/ha	1.78 mg/ha	21.35%	Tied ridges compared with flat tillage	Brhane et al. (2006)
	Sorghum stover yield	5.29 mg/ha	7.02 mg/ha	32.7%	Tied ridges compared with flat tillage	Brhane et al. (2006)
Terrewah, straw mulch, and effective microorganism	Soil loss	90.6 kg/ha	11.7 kg/ha	−87.07%	Combination of treatments compared with no treatment	Grum et al. (2017a)
	Total N loss	364.1 g/ha	98.8 g/ha	−72.86%	Combination of treatments compared with no treatment	Grum et al. (2017a)
	Total P loss	259.2 g/ha	62.8 g/ha	−75.77%	Combination of treatments compared with no treatment	Grum et al. (2017a)
Tied ridges	Runoff	67.2 mm	29.5 mm	−56.10%	Tied ridges compared with conventional tillage. Values are average for 4 years	Araya and Stroosnijder (2010)
	Barley yield	1262.25 kg/ha	1784.5 kg/ha	41.37%	Tied ridges compared with conventional tillage. Values are average for 4 years	Araya and Stroosnijder (2010)
	Soil water content		45.5% higher	45.5%	Tied ridges compared with flat tillage	Brhane et al. (2006)
	Sorghum grain yield	1.4 mg/ha	2.52 mg/ha	80%	Tied ridges compared with flat tillage	Brhane et al. (2006)
	Soghum stover yield	5.29 mg/ha	10.31 mg/ha	94.89%	Tied ridges compared with flat tillage	Brhane et al. (2006)
Tied ridges, straw mulch, and effective microorganism	Runoff	4.7 mm	0.9 mm	−80.85%	Combination of treatments compared with no treatment	Grum et al. (2017a)

soil nutrients including nitrogen and phosphorus. All these positive impacts have eventually resulted in an increase in yields of different crops including *Derdaro* (34–48%), *Shilshalo* (17.88–21.35%), and tied ridges (80–94.89%) (Table 3).

These improvements are achieved because the simple interventions of changing how land is tilled resulted in artificially made micro-basins that conserve soil moisture and reduce the loss of soil and nutrients from farmlands (Brhane et al. 2006). For instance, a simple tide ridging during plowing resulted in an increase of soil water content by 45.5% (Brhane et al. 2006). Similar improvements in water balance, crop productivity, and reduction in runoff and soil loss, due to conservation agriculture interventions, have also been observed elsewhere (McHugh et al. 2007).

Impact of Water Harvesting Interventions

Various water harvesting interventions including micro-dam reservoirs, river diversion, ground water production, and various catchment-level water harvesting structures including deep trenches, hand-dug wells (Woldearegay and Van Steenbergen 2015) have improved the available water for agriculture and domestic use, while reducing runoff and soil loss. The increased water availability for agriculture has inevitably resulted in an increase in income of up to 50% (micro-dam reservoirs), 50% (river diversion), and 50% (ground water development schemes) (Table 4). Catchment level integrated water harvesting schemes have also resulted in decrease in runoff by 43% and catchment level sediment yield by 54.5% (Table 3). Other reported impacts include increase in crop yields ranging from 71% to 233.3% for various crop types as a result of implementation of spate irrigation schemes (Table 3). Spate irrigation's positive impact is particularly important because there is about 9265.95 million meter^3 of flood water and 661853.6 ha of arable and 695,000 ha of communal land that can be irrigated using spate irrigation (Yazew 2015).

Impacts of Soil and Water Conservation

The most common types of soil and water conservation interventions (stone bunds), have resulted in improvements of: soil organic matter content (10–100.1%), soil nitrogen (5.71–100%), available phosphorus (1.4–31%), yields of various crops (4.78–25%), value of crop productivity (2.53%), net return (30%), and reductions in runoff (22%) and soil loss 64.9% (Table 5).

Other improvements observed include reductions in gulley head retreat rate (100%), runoff (11.11–55.56%), and number of destroyed dams (2.82%), due to interventions of subsurface geomembrane dams, check dams with vegetation and boulder-faced check dams, respectively (Table 5).

Table 4 Impact of water harvesting and production technologies

Interventions	Measured impact	Before intervention	After intervention	Percentage difference	Description of study setting	References
Micro-dam reservoirs (MDR)	Sediment deposition	0 t/ha	19 t/ha/year		Average value from 11 MDR	Tamene et al. (2006)
	Income		50% more income	50%	Irrigators compared with rainfed cultivators	Gebregziabher et al. (2009)
River diversion	Income		50 more income	50%	Irrigators compared with rainfed cultivators	Gebregziabher et al. (2009)
Ground water	Income		50% more income	50%	Irrigators compared with rainfed cultivators	Gebregziabher et al. (2009)
Catchment level water harvesting interventions	Runoff		43% reduction	−43%	Average value for two catchments	Grum et al. (2017b)
	Sediment deposition		54.5% reduction	−54.5%	Average value for two catchments	Grum et al. (2017b)
Roof water harvesting	Water harvested		5000 Lt per household		Harvested water in normal year, 2500 Lt in drier year	Helvetas (2017)
Spate irrigation	Wheat yield	4 ton/ha	13 ton/ha	225%	Yield compared with and without intervention	Van den ham (Ham 2008)
	Barley yield	7 ton/ha	12 ton/ha	71.4%	Yield compared with and without intervention	Ham (2008)
	Teff yield	3 ton/ha	6 ton/ha	100%	Yield compared with and without intervention	Ham (2008)
	Haricot bean yield	6 ton/ha	15 ton/ha	150%	Yield compared with and without intervention	Ham (2008)
	Maize yield	3 ton/ha	10 ton/ha	233.3%	Yield compared with and without intervention	Ham (2008)
Roadside water harvesting	Soil moisture content		100% increase	100%	Soil moisture around water harvesting trenches compared before and after intervention	Woldearegay et al. (2015)

Table 5 Impacts of soil and water conservation interventions

Intervention	Measured impact	Before	After	% change	Description of experimental settings	References
Stone bunds on hillside	Soil organic carbon	1.72%	2.34%	36.05%	Terraced hillsides compared with nonterraced hillsides	Hishe et al. (2017)
Stone bunds	Soil organic matter	2.01%	4.04%	100.99%	Terraced hillsides compared with nonterraced hillsides	Hishe et al. (2017)
	Soil organic carbon	0.63%	0.70%	11.11%	Terraced farmlands compared with nonterraced farmlands	Hishe et al. (2017)
	Soil organic matter	1.09%	1.20%	10.09%	Terraced farmlands compared with nonterraced farmlands	Hishe et al. (2017)
	Total nitrogen	0.05%	0.1%	100%	Terraced farmlands compared with nonterraced farmlands	Hishe et al. (2017)
	Total nitrogen	0.14%	0.19%	5.71%	Terraced hillsides compared with nonterraced hillsides	Hishe et al. (2017)
	Available phosphorus	0.71 ppm	0.72 ppm	1.40%	Terraced hillsides compared with nonterraced hillsides	Hishe et al. (2017)
	Available phosphorus	0.45 ppm	0.56 ppm	24.44%	Terraced farmlands compared with nonterraced farmlands	Hishe et al. (2017)
	Plant available phosphorus	10.69 mg/kg	14.07 mg/kg	31.61%	Soil in nonterraced land compared with soil in terrace bunches	Vagen et al. (1999)
	Reduced runoff	1.11	1.36	–22.52%	Interventions compared with untreated land, values are average for three land use types	Taye et al. (2013)
	Soil loss	57 t/ha/yr	20 t/ha/yr	–64.91%	Terraced farmland compared with nonterraced farmland	Gebremichael et al. (2005)
	Sediment deposition	–	59 t/ha/yr		Sediment deposition measured behind the stone bunds	Gebremichael et al. (2005)
	Sediment accumulation		58 t/ha/yr		An interdisciplinary comparison of terraced and nonterraced land	Nyssen et al. (2007)

	Cereal yields	632 kg/ha	683 kg/ha	8.07%	Terraced farmlands compared with nonterraced farmlands	Vancampenhout et al. (2006)
	Teff yield	501 kg/ha	556 kg/ha	10.98%	Terraced farmlands compared with nonterraced farmlands	Vancampenhout et al. (2006)
	Cicer arietinum yield	335 kg/ha	351 kg/ha	4.78%	Terraced farmlands compared with nonterraced farmlands	Vancampenhout et al. (2006)
	Grain yields	0.58 t/ha/yr	0.65 t/ha/yr	12.07%	An interdisciplinary comparison of terraced and nonterraced land	Nyssen et al. (2007)
	Value of crop productivity	1614 birr/ha	2026 birr/ha	25.53%	Terraced farmland yield compared with nonterraced one in a modeling study	Kassie et al. (2008)
	Net return over 30 year planning horizon		30% more	30%	Terraced farms compared with nonterraced ones	Gebremedhin et al. (1999)
Sub-surface geomembrane dam	Gulley head retreat rate	0.34 m/ year	0 m/year	100%	Treated area s head cut retreat compared with an average value from former study (Frankl et al. 2012)	Frankl et al. (2016)
Trenches		1.11	1.93	73.87%	Trenches compared with untreated land, values are average for three land use types	Taye et al. (2013)
Sandstone check dam with vegetation	Reduction in runoff discharge	18%	8%	−55.56%	Sandstone check dam without vegetation compared with the one with vegetation	Guyassa et al. (2017)
Limestone check dam with vegetation	Reduced runoff	9%	8%	−11.11	Limestone check dam without vegetation compared with the one with vegetation	Guyassa et al. (2017)
Boulder-faced log dams	Destroyed dams	39%	37.9%	−2.82%	Durability of boulder-faced log dams compared with that of looser rock check dams from another study (Nyssen et al. 2004b)	Nyssen et al. (2017)

Impact of Soil Fertility Improvement Interventions

The different soil fertility improvement interventions have resulted in various positive impacts. These include improved grain yields of 10.39–103.08% (use of compost), 63.7–71.99% (use of bioslurry), and 125.76% (use of reservoir sediment) (Table 6). The use of compost and bioslurry is not only a cheap source of fertilizers but also that their use helps in management and removal of domestic and farmland wastes, which would otherwise be source of health concerns in rural areas (Beyene 2011). Moreover, plants that are normally considered noxious weeds, such as *Parthenium hysterophorus*, *Datura stramonium*, and *Argemone mexicana*, have been have been used for composting resulting in both improved soil fertility while at the same time removing weeds (Tedla 2010).

Some EBA soil fertility management options may not be able to improve crop productivity as compared to the inorganic fertilizers, where crop yields have been observed to decline as inorganic fertilizers are replaced by composts (Table 6). However, even though inorganic fertilizers cause a boost in crop yields in the short term, their prolonged and intensive use causes decline of agro-biodiversity and over all soil and ecological health (Hadgu et al. 2009). Therefore, owing to the desirable long-term impacts they have, as compared to inorganic fertilizers, and the significant contribution in crop yields, there is a potential for the use of EBA practices in soil fertility improvement such as composting and bioslurry.

Impact of Integrated Watershed Management Interventions

Observed impacts of IWM interventions include reductions in: runoff (7.9–94%), erosion (89%), soil loss (32–37%), sediment yield (77%), and percentage bareland (75%) and increments in: sediment deposition (22.41%), percentage irrigated land (3077%), forest cover (142.3–202.4%), bush land (72%), and yield of various crops (60–100%) (Table 7). The runoff and erosion rates observed by the reviewed studies have shown a tremendous improvement from the 42 tons/ha that has been recorded before the start of the integrated watershed management interventions (Hurni 1988). Most of the positive improvements reported for northern Ethiopia are similar to the impacts of IWM reported elsewhere (Mekonen and Tesfahunegn 2011). A qualitative assessment by Michael and Waters-Bayer (2007) also reported similar observations of increased vegetation cover, more water infiltrated into the soil, reduced siltation, and increased crop yields. Generally speaking, because of the various integrated interventions, the effect of recurrent drought and extreme land degradation is effectively offset that the Tigray regional state is now more productive and greener than it used to be some 145 years ago (Nyssen et al. 2014).

Table 6 Impact of soil fertility improvement interventions

Intervention	Measured impact	Before treatment (kg/ha)	After treatment (kg/ha)	Percentage change	Description of study setting	References
Compost	Grain yield	1200	2437	103.08%	Farmlands with compost compared with untreated farmlands; results are 7 years' average for seven cereal crops	Edwards (2007a, b)
	Grain yield	1812	2437	34.49%	Farmlands with compost compared with farmlands treated with inorganic fertilizers; results are 7 years' average for seven cereal crops	Edwards (2007a, b)
	Teff yield	872	1113	27.63%	Compost applied at 6.4 t/ha/year compared with nonfertilized land	Tedla (2010)
	Barley yield	2173	2950	35.75%	Compost applied at 6.4 t/ha/ year compared with nonfertilized land	Tedla (2010)
	Faba bean yield	3334	4230	26.87%	Compost applied at 6.4 t/ha/ year compared with nonfertilized land	Tedla (2010)
	Faba bean yield	3832	4230	10.39%	Compost applied at 6.4 t/ha/ year compared with a land fertilized by inorganic fertilizer at standard rates	Tedla (2010)
	Teff yield	1120	1113	−0.625%	Compost applied at 6.4 t/ha/ year compared with a land fertilized by inorganic fertilizer at standard rates	Tedla (2010)
	Barley yield	3025	2950	−2.48%	Compost applied at 6.4 t/ha/ year compared with a land fertilized by inorganic fertilizer at standard rates	Tedla (2010)
	Straw yield	2477	4073	64.43%	Farmlands with compost compared with untreated farmlands; results are 7 years' average for seven cereal crops	Edwards (2007a, b)

(*continued*)

Table 6 (continued)

Intervention	Measured impact	Before treatment (kg/ha)	After treatment (kg/ha)	Percentage change	Description of study setting	References
Bioslurry	Wheat yield	1711	2800	63.70%	Bioslurry-fertilized land compared with nonfertilized land	Beyene (2011)
	Barley yield	1528	2628	71.99%	Bioslurry-fertilized land compared with nonfertilized land	Beyene (2011)
	Barley straw	2417	4056	67.81%	Bioslurry-fertilized land compared with nonfertilized land	Beyene (2011)
	Wheat yield	4607	2800	−39.22%	Bioslurry-fertilized land compared with land fertilized by in-organic fertilizer	Beyene (2011)
	Wheat straw	4707	4267	−9.35	Bioslurry-fertilized land compared with land fertilized by in-organic fertilizer	Beyene (2011)
	Wheat straw	3732	4267	14.34%	Bioslurry-fertilized land compared with nonfertilized land	Beyene (2011)
Fine reservoir sediment	Garlic yield	2721	6143	125.76%	Land fertilized with fine reservoir sediment applied at 15–30 cm deep compared with untreated land	Girmay et al. (2012)

Table 7 Impacts of integrated watershed management interventions

Measured impact	Before intervention	After intervention	Percentage change	Description of study	References
Runoff		27% reduction	−27%	A modeling study measuring changes before and after intervention	Haregeweyn et al. (2012)
Runoff from cultivated land	156.66	37.39	−76.13%	Measurement taken before and after intervention	Alem et al. (2015)
Runoff from Bushland	169.54 mm/year	156.01 mm/year	−7.98%	Measurement taken before and after intervention	Alem et al. (2015)
Runoff from bareland	188.40 mm/year	10.93 mm/year	−94.19%	Measurement taken before and after intervention	Alem et al. (2015)
Runoff from exclosure	188.40 mm/year	110.55 mm/year	−41.32%	Measurement taken before and after intervention	Alem et al. (2015)
Sheet and rill erosion		89% reduction	−89%	A modeling study measuring changes before and after intervention	Haregeweyn et al. (2012)
Soil loss by sheet and rill erosion		78% less	−78%	Sediment deposition measured at different conservation structure	Nyssen et al. (2008b)
Soil loss		32% reduction	−32%	Analysis of landscape photographs in 1975 and 2006	Munro et al. (2008)
Soil loss	14.3t/ha/yr	9 t/ha/yr	−37.06%	Land use and management were mapped and analyzed for 2000 and 2006	Nyssen et al. (2009)
Sediment deposition	5.8 t/ha/yr	7.1 t/ha/yr	22.41%	Land use and management were mapped and analyzed for 2000 and 2006	Nyssen et al. (2009)

(*continued*)

Table 7 (continued)

Measured impact	Before intervention	After intervention	Percentage change	Description of study	References
Sediment yield	8.5 t/ha/yr	1.9 t/ha/yr	−77.65%	Land use and management were mapped and analyzed for 2000 and 2006	Nyssen et al. (2009)
Area of irrigated land	7 ha	222.4 ha	3,077%	An analysis of aerial photograph in one watershed	Alemayehu et al. (2009)
Forest cover	32.4 ha	98 ha	202.4%	An analysis of aerial photograph in one watershed	Alemayehu et al. (2009)
Forestland	2.6%	6.3%	142.30%	Analysis of areal landscape photographs (1972, 1984/1986, and 2000)	Mûelenaere et al. (2014)
Bareland	32%	8%	−75%	Analysis of areal landscape photographs (1972, 1984/1986, and 2000),	Mûelenaere et al. (2014)
Bush land	25%	43%	72%	Analysis of areal landscape photographs (1972, 1984/1986 and 2000),	Mûelenaere et al. (2014)
Teff yield	0.3 t/ha	0.6 t/ha	100%	Yields compared between 1997 (before intervention) and 2004 (after intervention)	Alemayehu et al. (2009)
Wheat yield	0.5 t/ha	0.8 t/ha	60%	Yields compared between 1997 (before intervention) and 2004 (after intervention)	Alemayehu et al. (2009)
Barley yield	0.45 t/ha	0.75 t/ha	66.67%	Yields compared between 1997 (before intervention) and 2004 (after intervention)	Alemayehu et al. (2009)

Challenges with EBA Interventions and Future Directions

Though the changes brought about by the different EBA interventions are impressive and have been confirmed and reconfirmed by many studies, certain issues threaten the sustainability of the positive benefits of the interventions exist. The most important challenges include challenges of mobilizing public support, due to limited immediate economic benefits from interventions and a misguided approach for recruiting popular support and specific technical challenges with some interventions. While the challenge with the approach of recruiting popular support is explained below, specific technical challenges and recommended solutions are outlined in Table 8.

Challenges on Mobilizing Public Support

In a review of soil and water conservation interventions in Tigray, Gebremeskel et al. (2017) concluded that voluntary popular participation in the form of free labor days (mid to late 1980s) and then work for food program in the 1990s, accompanied by integration of disciplines (social, technical, and institutional) were most important reasons for the observed impressive success. Similarly, Nyssen et al. (2004a) indicated that farmers participate voluntarily, because in highly degraded areas like northern Ethiopia, farmers are left with no other alternatives, except improved land management or rehabilitation.

However, Segers et al. (2008a) argued that local farmers' mass participation is more of a support to the political agenda or conformity to popular practice than an understanding of or anticipation of the financial benefits to be accrued from participating in such schemes. In other words, it means that farmers accept interventions even knowing fully that they are not suitable for them, perhaps, why some very ambitious projects like the "*horeye*" water harvesting, despite obviously destined to fail, did not face any opposition by farmers. The Tigray regional administration was mainly successful in gaining popular support for its land rehabilitation projects by tapping into the common history and legacies of solidarity during the armed struggle of the Tigrian People's Liberation Front (TPLF) against the "*Dergue*" military regime, than by convincing farmers of the economic benefits of the projects (Segers et al. 2008a). It has been demonstrated earlier in this chapter, the number of studies that showed economic or livelihood benefits is also very limited (Fig. 5). This makes it difficult to make economic arguments for further dissemination of successful interventions. It is, however, also important to note that economic or livelihood benefits need time to manifest and that implementers need to subsidize farmers' efforts until economic incentives can attract voluntary local engagement (Gebremedhin et al. 1999). Secure land tenure rights can also reinforce private incentives to make long-term investments in soil conservation (Gebremedhin and Swinton 2003). Moreover, it is important to raise the Tigrean masses awareness that the interventions are ultimately meant to increase income and productivity, and are

Table 8 Main challenges of important EBA interventions in Tigray regional state and recommended solutions

Types of EBA interventions	Main drawbacks, limitations, challenges and side effects	Causes of problems	References	Proposed or Recommended solutions	References
All interventions	Failure to make economic arguments for the interventions	Most of the observed impacts need time to be translated into economic or livelihood benefits	Teshome et al. (2013)	Capitalizing on those interventions that would provide immediate capital return	Teshome et al. (2013)
Exclosures	Lack of economic incentives for farmers	In many places, procedures on economic utilization of grass and other biomass from exclosures not set	Gebremeskel et al. (2017)	Developing guidelines on grass sharing, seasonal grazing etc., expanding the practice of honey bee rearing in exclosures	Gebregziabher et al. (2017)
Exclosures	Complete exclusion of livestock may result in encroachment of invasive bushes and weeds	Because grazing provides services like breaking seed dormancy, nutrient cycling and removal of old outgrowth	Stavi et al. (2016)	Prescribed moderate grazing	Yayneshet et al. (2009)
Zero-grazing	Low level of adoption	Shortage of livestock feed, low level of awareness, culture of keeping large number of livestock; unprotected communal areas, limited water resources for forage development	Yami et al. (2011, Gebreyohannes and Hailemariam (2011)	Breed improvement accompanied by destocking, participation of local communities during planning and implementation	Yami et al. (2011, Gebreyohannes and Hailemariam (2011)
Tree plantation/ enrichment	The survival rate of seedlings very low	Lack of immediate benefits that make investment in tree nursing very limited: Wrong choice of tree species	Reubens et al. (2011, Aerts et al. (2007)	Improved nursing practices (watering, sheltering and improved planting conditions), providing incentives for tree grower: Identification and selection of drought tolerant high survival potential species	Reubens et al. (2011, Balehegn et al. (2015)

Agroforestry (trees on farm)	Tree shade crops, resource depletion, and barrier for cultivation	Failure to choose the right tree species for the right land, lack of knowledge tree management	Ernstberger (2017)	Awareness creation on the long term and short term benefits of agroforestry	
Silvopasture	Low level of adoption	Farm size, negative interaction, water scarcity, obstruction during plowing and incapability to fence tree seedlings	Hachoofwe (2012)	Choosing drought tolerant, multipurpose, adaptive and productive tree species	Balehegn (2017b), Balehegn et al. (2015)
Microdam reservoirs	Water leakage/seepage, insufficient inflow, structural damages, spillway erosion problems. siltation, lack of proper water management practices,	Insufficient or improper understanding of the geological and edaphic characteristics of sites before embarking on construction	Berhane et al. (2016a)	Undertaking detailed feasibility study for future construction. Installation of a cut-off wall at the upstream toe of the dam for those with leakage; utilization of the water from big diameter shallow hand dug wells from the leakage zone downstream	Berhane et al. (2016b)
	Eutrophication and a high occurrence of blooms of toxic *cyanobacteria*:		Dejenie et al. (2008)	Understanding the mechanisms of eutrophication for developing management options	Dejenie et al. (2008)
Check dams	Crop desiccation, collapse of dams, etc.	Soil piping	Frankl et al. (2016)	Application of Geomembrane dams	Frankl et al. (2016)
Gabion check dams	Gabion failure in ephemeral streams with coarse bed load, which abrades at the chute step: Reported problems of abrasion and collapse in torrents with coarse bed load and (2) the cost of gabion baskets	Abrasion of wires by bed-load movement over stepped chutes	Nyssen et al. (2017)	The use of boulder faced check dams in areas with strong streams	Nyssen et al. (2017)

(*continued*)

Table 8 (continued)

Types of EBA interventions	Main drawbacks, limitations, challenges and side effects	Causes of problems	References	Proposed or Recommended solutions	References
Horeye	Water seepage, inability to hold water, early dry up	Mostly abandoned because there was no consultation of farmers, pure top down approach	Segers et al. (2008a, b)	Developing construction guidelines relevant to specific topography, soil or geology. Implementing small scale water harvesting in consultation to users	Gebremeskel et al. (2017)
Shallow ground water irrigation	Well design and construction problems, high costs of pumps and oil, depletion of groundwater due to excessive withdrawal, emergence of conflicts due to competition for resources,	Lack of technical competence on the side of farmers and experts	Woldearegay and Van Steenbergen (2015)	Technical training and capacity building on site selection, construction and designing, management and utilization of water	Woldearegay and Van Steenbergen (2015)
Spate irrigation	Technical inefficiency and failures (e.g., huge amount of sediment deposition, salinity), significant water loss due to inefficient canal system, water resources conflicts	Limited technical competence, lack of involvement of beneficiaries at different states of projects	Hiben and Tesfa-alem (2014), Kidane (2009)	Development and dissemination of design guidelines and a manual, managing conflicts, following a participatory approach	Hiben and Tesfa-alem (2014)

Stone bunds/ terraces	Decline in effectiveness of sediment trapping, with time	With time, stone bunds are filled and reach the point where they no more prevent soil loss	Taye et al. (2015), Nyssen et al. (2007)	Periodic maintenance of bunds or implementing self-sustaining bunds that grow in height with time (e.g. cactus or other plant hedges)	Taye et al. (2015), Nefzaoui and El Mourid (2010)
	Harbors rodents causing crop damage	Porous designs serve as shelters for small rodents	Meheretu et al. (2014, 2015)	Developing a design that deters rodents e.g. integrating branches from spiny trees into the bund	Meheretu et al. (2014, 2015)
	Cost of building almost equal to the benefit induced by the increase in crop yield	Building is too labor intensive, especially where stone is not readily available	Nyssen et al. (2007, 2008b)	Current programs of food for work, safety net. In areas where farmers cannot see incentive in building stone bunds, government should subsize	Gebremedhin et al. (1999)
	Sometimes aggravates sheet and rill erosion on farms	In areas where stone is a limited resource, excessive removal of stone and gravel from farms causes sheet and rill erosion	Nyssen et al. (2007)	Utilizing alternative bund building materials, soil or vegetation bunds, transporting stone from other areas	
	Take land out of production	In low productive areas the increase in productivity or soil protection is offset by the land taken out of production	Nyssen et al. (2007)		
Bioslurry compost	Shortage of feed stock or dung	Most dung in rural areas is used as a fuel	Kelebe et al. (2017)	The use of alternative feed stock such as human excreta	Kelebe et al. (2017), Beyene (2011)

(*continued*)

Table 8 (continued)

Types of EBA interventions	Main drawbacks, limitations, challenges and side effects	Causes of problems	References	Proposed or Recommended solutions	References
Lose rock check dams	Frequent collapse, close to 39%	Technical inefficiencies in the construction	Nyssen et al. (2004b)	Development of technical construction guidelines for different types of geologic and edaphic formations, where catchment areas are large or there are steep slopes, repair dams as soon as partial collapse starts use biological gulley control methods	Nyssen et al. (2004b)
Small scale irrigation	Not adopted well	Myopic project planning, inadequate engineering studies, unsound designs, too short a time for implementation, and lack of full farmer involvement at different levels of project planning and implementation	Aberra (2004)	Participatory approaches to project planning, design and management. Skills training to farmers and technicians	Aberra (2004)
Roof top water harvesting	Not adopted well	Lack of knowledge base, technical skills, awareness and extension in urban areas	Biazin et al. (2012)	Awareness creation on the technology, provision of training and technical skills building	Biazin et al. (2012)

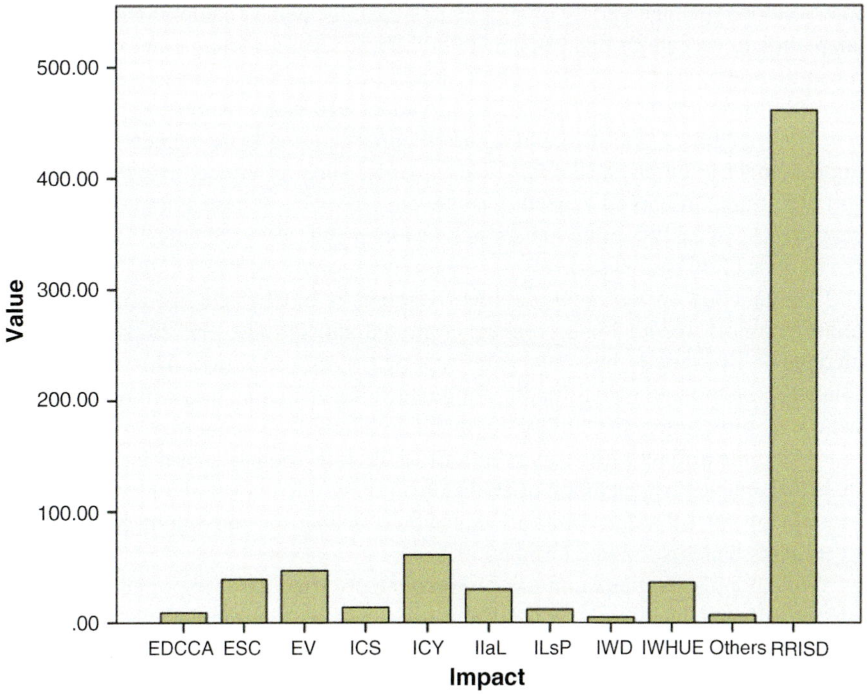

Fig. 5 Number of cases of evidences for different positive impacts. *EDCCA* Enhanced drought and climate change adaptation, *ESC* Enhanced soil characteristics, *EV* Enhanced vegetation, *ICS* Improved carbon stocks, *ICY* Improved crop yields, *IIaL* Improved income and livelihoods, *ILsP* Improved livestock productivity, *IWD* Improved wildlife diversity, *IWHUE* Improved water harvesting and use efficiency, *RRISD* Reduced runoff and increased sediment deposition

not orders that they have to blindly follow, as such a thinking would improve the household level adoption of these technologies, which is currently either lacking or very limited (Segers et al. 2008a).

Conclusion

In this chapter, we critically reviewed studies on decades of extensive EBA adaptation practices, implemented in one of, hitherto, the most ecologically fragile, hunger, and poverty-ridden corners of the world that has been repeatedly plagued by recurrent drought and extreme weather variability. The review includes 170 publications on 30 different types of EBA interventions in 400 sites in Tigray. Reviewed studies indicated that Tigrian farmers, together with the Tigray State Government of Ethiopia, their donors and development partners have demonstrated that through dedicated, evidence-based interventions, it is not only possible to beat the problems of land degradation, recurrent drought, extreme weather variability but also turn around the challenges and convert land into highly stable productive ecosystems.

Interventions spanned variety of simple, low cost, locally available ecosystem-based adaptation interventions that can generally fall into six categories, namely, biological rehabilitation, conservation agriculture, integrated watershed management, soil fertility improvement, soil and water conservation, and water harvesting and production interventions. Quantified and reported impacts of the EBA included in general, improvement in biophysical ecological variables such as reduced runoff, increased soil deposition, improved vegetation cover etc., (63.93% of cases), improvements in crop (8.46% of cases), improvement in livestock yields (1.6% of cases), and income/livelihoods (4.16% of cases).

Despite wide spread reported improvement in biophysical variables such as reduced runoff, reduced soil erosion, improved soil fertility, etc., improvement in income or livelihoods has only been reported 4.16% of cases. This is probably caused by methodological difficulty for quantifying economic or livelihood impacts, which in turn limits the ability to make livelihood or economic argument for promoting the different EBA interventions. Therefore, there is a need for developing tools that enable the generation of evidence on economic impact of interventions and capitalizing on EBA interventions that would enable the maximum economic or livelihoods benefits to local farmers.

While popular participation has been repeatedly cited as an important reason for successful implementation of the EBA interventions, household level or private adoption of the interventions still remains very limited. Almost all practices are implemented in communal or government land. This puts a question as to the motive of farmers in participating in the implementation of interventions. Though, many indicated that farmers participate voluntarily, by sometimes offering free labor days, there is evidence that this is mainly behavior of conformity, rather than a need-based voluntary involvement. This might explain why most of the interventions are lacking from private land or household-based interventions. This is important because, without popular household-based adoption, the sustainability of the interventions could be under question. Therefore, it is important to raise awareness on the short and long term economic or livelihood benefits of the EBA interventions, so that individual farmers can adopt them.

Finally, the types of EBA and land rehabilitation interventions, challenges faced and solutions implemented give important lessons for replicating the success in northern Ethiopia to similar places elsewhere, with similar challenges.

Cross-References

- Area Exclosure as a Strategy for Climate Change Mitigation: Case Study From Tigray Region, Northern Ethiopia
- Building Resilience to Climate Change: Water Stewardship in Rainfed Agrarian Villages in Maharashtra, India
- Climate Change, Vulnerability, and Adaption Under the Small Farming Households of Konso Community, Southern Ethiopia

- Ecosystem-Based Adaptation and Gender Perspectives From a Participatory Vulnerability Assessment in Mountainous Rural Vietnam
- Effect of Mulching on Soil Temperature and Moisture for Potato Production in Agro-Ecological Zones of Central Highlands of Kenya
- Environmental Sustainability and Systems Thinking: A Foundation for More Effective Climate Policy
- Fertilization Strategies Based on Climate Information to Enhance Food Security Through Improved Dryland Cereals Production
- Forest Cover Change and Its Impacts on Ecosystem Services in Katimok Forest Reserve, Baringo County, Kenya
- Resilience in Climate Stressed Environment Through Water Grabbing
- Water Harvesting Technology for Enhancing Food Security Livelihood: The Case of Northern Katsina State, Nigeria
- Water Management as a Means for Climate Change Adaptation and Sustainable Development

References

Aberra Y (2004) Problems of the solution: intervention into small-scale irrigation for drought proofing in the Mekele Plateau of northern Ethiopia. Geogr J 170:226–237

Abesha GA (2014) Herbaceous vegetation restoration potential and soil physical condition in a mountain grazing land of eastern Tigray, Ethiopia. J Agric Environ Int Dev (JAEID) 108:81–106

Aerts R, Negussie A, Maes W, November E, Hermy M, Muys B (2007) Restoration of dry Afromontane Forest using pioneer shrubs as nurse-plants for *Olea europaea* ssp. *cuspidata*. Restor Ecol 15:129–138

Aerts R, Lerouge F, November E, Lens L, Hermy M, Muys B (2008) Land rehabilitation and the conservation of birds in a degraded Afromontane landscape in northern Ethiopia. Biodivers Conserv 17:53–69

Aerts R, Nyssen J, Haile M (2009) On the difference between "exclosures" and "enclosures" in ecology and the environment. J Arid Environ 73:762–763

Alem T, Tsegazeab G, Micheale G, Atinkut M (2015) Characterization and impact assessment of water harvesting techniques: a case study of Abreha Weatsbeha watershed, Tigray, Ethiopia. Department of Land Resources Management and Environmental Protection, Mekelle University, Ethiopia

Alemayehu F, Taha N, Nyssen J, Girma A, Zenebe A, Behailu M, Deckers S, Poesen J (2009) The impacts of watershed management on land use and land cover dynamics in eastern Tigray (Ethiopia). Resour Conserv Recycl 53:192–198

Araya A, Stroosnijder L (2010) Effects of tied ridges and mulch on barley (*Hordeum vulgare*) rainwater use efficiency and production in northern Ethiopia. Agric Water Manag 97:841–847

Araya T, Cornelis W, Nyssen J, Govaerts B, Bauer H, Gebreegziabher T, Oicha T, Raes D, Sayre KD, Haile M (2011) Effects of conservation agriculture on runoff, soil loss and crop yield under rainfed conditions in Tigray, northern Ethiopia. Soil Use Manag 27:404–414

Araya T, Nyssen J, Govaerts B, Baudron F, Carpentier L, Bauer H, Lanckriet S, Deckers JA, Cornelis W (2016a) Restoring cropland productivity and profitability in northern Ethiopian drylands after nine years of resource-conserving agriculture. Exp Agric 52:165–187

Araya T, Nyssen J, Govaerts B, Deckers JA, Sommer R, Bauer H, Gebrehiwot K, Cornelis W (2016b) Seven years resource-conserving agriculture effect on soil quality and crop productivity in the Ethiopian drylands. Soil Tillage Res 163:99–109

Asres HG (2012) Effect of exclosure on environment and its socio economic contributions to local people: in the case study of halla exclosure, Tigray, Ethiopia. Norwegian University of Life Sciences, Ås

Awulachew SB, Merrey D, Kamara A, van Koppen B, Penning de Vries F, Boelee E (2005) Experiences and opportunities for promoting small-scale/micro irrigation and rainwater harvesting for food security in Ethiopia. IWMI, Colombo, Sri Lanka

Babulo B, Muys B, Nega F, Tollens E, Nyssen J, Deckers J, Mathijs E (2009) The economic contribution of forest resource use to rural livelihoods in Tigray, northern Ethiopia. Forest Policy Econ 11:109–117

Balana BB, Mathijs E, Muys B (2010) Assessing the sustainability of forest management: an application of multi-criteria decision analysis to community forests in northern Ethiopia. J Environ Manag 91:1294–1304

Balehegn M (2017a) Increased publication in predatory journals by developing countries' institutions: what it entails? And what can be done? Int Inf Libr Rev 49:97–100

Balehegn M (2017b) Silvopasture using indigenous fodder trees and shrubs: the underexploited synergy between climate change adaptation and mitigation in the livestock sector. In: Leal Filho W, Belay S, Kalangu J, Menas W, Munishi P, Musiyiwa K (eds) Climate change adaptation in Africa: fostering resilience and capacity to adapt. Springer International Publishing, Cham

Balehegn M, Eik L, Tesfay Y (2014) Replacing commercial concentrate by *Ficus thonningii* improved productivity of goats in Ethiopia. Trop Anim Health Prod:1–6

Balehegn M, Eik LO, Tesfay Y (2015) Silvopastoral system based on *Ficus thonningii* : an adaptation to climate change in northern Ethiopia. Afr J Range Forage Sci 32:183–191

Berhane G, Gebreyohannes T, Martens K, Walraevens K (2016a) Overview of micro-dam reservoirs (MDR) in Tigray (northern Ethiopia): challenges and benefits. J Afr Earth Sci 123:210–222

Berhane G, Kebede S, Gebreyohannes T, Martens K, van Camp M, Walraevens K (2016b) An integrated approach for detection and delineation of leakage path from Micro-Dam Reservoir (MDR): a case study from Arato MDR, northern Ethiopia. Bull Eng Geol Environ 75:193–210

Bewket W (2007) Soil and water conservation intervention with conventional technologies in northwestern highlands of Ethiopia: acceptance and adoption by farmers. Land Use Policy 24:404–416

Beyene G (2011) Bio-slurry – is it a fertilizer in the Making? SNV Netherlands Development Organisation http://www.snv.org/public/cms/sites/default/files/explore/download/et_bioslurry_-_is_it_a_fertiliser_in_the_making.pdf

Biazin B, Sterk G, Temesgen M, Abdulkedir A, Stroosnijder L (2012) Rainwater harvesting and management in rainfed agricultural systems in sub-Saharan Africa – a review. Phys Chem Earth 47/48:139–151

Birhane E, Gebremedihin KM, Tadesse T, Hailemariam M, Solomon N (2017) Exclosures restored the density and root colonization of arbuscular mycorrhizal fungi in Tigray, northern Ethiopia. Ecol Process 6:33

Brhane G, Wortmann CS, Mamo M, Gebrekidan H, Belay A (2006) Micro-basin tillage for grain sorghum production in semiarid areas of northern Ethiopia. Agron J 98:124–128

Conway D (2000) Some aspects of climate variability in the north east Ethiopian highlands-Wollo and Tigray. SINET: Ethiop J Sci 23:139–161

CSA (2010) The 2007 population and housing census of Ethiopia

Dejenie T, Asmelash T, de Meester L, Mulugeta A, Gebrekidan A, Risch S, Pals A, van der Gucht K, Vyverman W, Nyssen J (2008) Limnological and ecological characteristics of tropical highland reservoirs in Tigray, northern Ethiopia. Hydrobiologia 610:193–209

Descheemaeker K, Muys B, Nyssen J, Poesen J, Raes D, Haile M, Deckers JA (2006) Litter production and organic matter accumulation in exclosures of the Tigray highlands, Ethiopia. For Ecol Manag 233:21–35

Descheemaeker K, Muys B, Nyssen J, Sauwens W, Haile M, Poesen J, Raes D, Deckers JA (2009a) Humus form development during forest restoration in exclosures of the Tigray highlands, northern Ethiopia. Restor Ecol 17:280–289

Descheemaeker K, Raes D, Nyssen J, Poesen J, Haile M, Deckers JA (2009b) Changes in water flows and water productivity upon vegetation regeneration on degraded hillslopes in northern Ethiopia: a water balance modelling exercise. Rangel J 31:237–249

Descheemaeker K, Mapedza E, Amede T, Ayalneh W (2010) Effects of integrated watershed management on livestock-water productivity in water scarce areas in Ethiopia. Phys Chem Earth 35:723–729

Doswald N, Munroe R, Roe D, Giuliani A, Castelli I, Stephens J, Möller I, Spencer T, Vira B, Reid H (2014) Effectiveness of ecosystem-based approaches for adaptation: review of the evidence-base. Climate Dev 6:185–201

Edwards S (2007a) The impact of compost use on crop yields in Tigray, Ethiopia. Institute for Sustainable Development (ISD). In the proceedings of the international conference on organic agriculture and food security. FAO, Rome. Obtainable at: ftp://ftp.fao.org/paia/organicag/ofs/02-Edwards. pdf

Edwards S (2007b) Role of organic agriculture in preventing and reversing land degradation. In: Climate and land degradation. Springer, Berlin

Edwards MS (2012) Biogas-bioslurry: a package for narrowing gender disparity in the rural households-the case of Hintalo Wajirat and Ofla Woredas, Tigray Region

Ernstberger J (2017) Perceived multifunctionality of agroforestry trees in northern Ethiopia. MSc.: A case study of the perceived functions and associated personal values of trees for farming households in Tigray. Thesis. Swedish University of Agricultural Sciences. Alnarp, Sweden.

Frankl A, Poesen J, Deckers J, Haile M, Nyssen J (2012) Gully head retreat rates in the semi-arid highlands of northern Ethiopia. Geomorphology 173:185–195

Frankl A, Deckers J, Moulaert L, van Damme A, Haile M, Poesen J, Nyssen J (2016) Integrated solutions for combating gully erosion in areas prone to soil piping: innovations from the drylands of northern Ethiopia. Land Degrad Dev 27:1797–1804

Gebreegziabher T, Nyssen J, Govaerts B, Getnet F, Behailu M, Haile M, Deckers J (2009) Contour furrows for in situ soil and water conservation, Tigray, northern Ethiopia. Soil Tillage Res 103:257–264

Gebregziabher G, Namara RE, Holden ST (2009) Poverty reduction with irrigation investment: an empirical case study from Tigray, Ethiopia. Agric Water Manag 96:1837–1843

Gebregziabher D, Soltani A, Hofstad O (2017) Equity in the distribution of values of outputs from exclosures in Tigray, Ethiopia. J Arid Environ 146:75–85

Gebremedhin B, Swinton SM (2003) Investment in soil conservation in northern Ethiopia: the role of land tenure security and public programs. Agric Econ 29:69–84

Gebremedhin B, Swinton SM, Tilahun Y (1999) Effects of stone terraces on crop yields and farm profitability: results of on-farm research in Tigray, northern Ethiopia. J Soil Water Conserv 54:568–573

Gebremedhin B, Pender J, Tesfay G (2004) Collective action for grazing land management in crop–livestock mixed systems in the highlands of northern Ethiopia. Agric Syst 82:273–290

Gebremeskel G, Gebremicael T, Girmay A (2017) Economic and environmental rehabilitation through soil and water conservation, the case of Tigray in northern Ethiopia. J Arid Environ 151:113–124

Gebremichael D, Nyssen J, Poesen J, Deckers J, Haile M, Govers G, Moeyersons J (2005) Effectiveness of stone bunds in controlling soil erosion on cropland in the Tigray highlands, northern Ethiopia. Soil Use Manag 21:287–297

Gebreyohannes G, Hailemariam G (2011) Challenges, opportunities and available good practices related to zero grazing in Tigray and Hararghe, Ethiopia. Drylands Coordination Group (DCG) Report

Gelaw AM, Singh B, Lal R (2014) Soil organic carbon and total nitrogen stocks under different land uses in a semi-arid watershed in Tigray, northern Ethiopia. Agric Ecosyst Environ 188:256–263

Gelaw AM, Singh B, Lal R (2015a) Organic carbon and nitrogen associated with soil aggregates and particle sizes under different land uses in Tigray, northern Ethiopia. Land Degrad Dev 26:690–700
Gelaw AM, Singh BR, Lal R (2015b) Soil quality indices for evaluating smallholder agricultural land uses in northern Ethiopia. Sustainability 7:2322–2337
Girmay G, Singh BR (2012) Changes in soil organic carbon stocks and soil quality: land-use system effects in northern Ethiopia. Acta Agric Scand Sect B Soil Plant Sci 62:519–530
Girmay G, Singh B, Nyssen J, Borrosen T (2009) Runoff and sediment-associated nutrient losses under different land uses in Tigray, northern Ethiopia. J Hydrol 376:70–80
Girmay G, Nyssen J, Poesen J, Bauer H, Merckx R, Haile M, Deckers J (2012) Land reclamation using reservoir sediments in Tigray, northern Ethiopia. Soil Use Manag 28:113–119
Grum B, Assefa D, Hessel R, Woldearegay K, Kessler A, Ritsema C, Geissen V (2017a) Effect of in situ water harvesting techniques on soil and nutrient losses in semi-arid northern Ethiopia. Land Degrad Dev 28:1016–1027
Grum B, Woldearegay K, Hessel R, Baartman JE, Abdulkadir M, Yazew E, Kessler A, Ritsema CJ, Geissen V (2017b) Assessing the effect of water harvesting techniques on event-based hydrological responses and sediment yield at a catchment scale in northern Ethiopia using the Limburg Soil Erosion Model (LISEM). Catena 159:20–34
Guyassa E, Frankl A, Zenebe A, Poesen J, Nyssen J (2017) Effects of check dams on runoff characteristics along gully reaches, the case of northern Ethiopia. J Hydrol 545:299–309
Hachoofwe EM (2012) Local ecological knowledge of trees on farms, constraints and opportunities for further integration in Tigray Region, northern Ethiopia: a case study of smallholder farmers in Abreha Wa Atsbeha and Adi gudom. Bangor University
Hadgu KM, Rossing WAH, Kooistra L, van Bruggen AHC (2009) Spatial variation in biodiversity, soil degradation and productivity in agricultural landscapes in the highlands of Tigray, northern Ethiopia. Food Sec 1:83–97
Hagos F, Pender J, Gebreselassie N (1999) Land degradation in the highlands of Tigray and strategies for sustainable land management. International Livestock Research Institute, Addis Ababa
Hagos EY, Schultz B, Depeweg H (2016) Reservoir operation in view of effective utilization of limited water in semi-arid areas the case of Gumsalasa earthen dam irrigation scheme in Tigray, Ethiopia. Irrig Drain 65:294–307
Ham J (2008) Dodota Spate irrigation system Ethiopia: a case study of Spate irrigation management and livelihood options
Haregeweyn N, Poesen J, Deckers JA, Nyssen J, Haile M, Govers G, Verstraeten G, Moeyersons J (2008a) Sediment-bound nutrient export from micro-dam catchments in northern Ethiopia. Land Degrad Dev 19:136–152
Haregeweyn N, Poesen J, Nyssen J, Govers G, Verstraeten G, de Vente J, Deckers J, Moeyersons J, Haile M (2008b) Sediment yield variability in northern Ethiopia: a quantitative analysis of its controlling factors. Catena 75:65–76
Haregeweyn N, Berhe A, Tsunekawa A, Tsubo M, Meshesha DT (2012) Integrated watershed management as an effective approach to curb land degradation: a case study of the Enabered watershed in northern Ethiopia. Environ Manag 50:1219–1233
Haregeweyn N, Tsunekawa A, Nyssen J, Poesen J, Tsubo M, Meshesha DT, Schutt B, Adgo E, Tegegne F (2015) Soil erosion and conservation in Ethiopia a review. Prog Phys Geogr 39:750–774
HELVETAS (2017) Rural roof water harvesting initiative in Tigray [Online]. Available: https://ethiopia.helvetas.org/en/projects/rrwhi/. Accessed 10 Jan 2017
Hengsdijk H, Meijerink GW, Mosugu ME (2005) Modeling the effect of three soil and water conservation practices in Tigray, Ethiopia. Agric Ecosyst Environ 105:29–40
Hiben MG, Tesfa-Alem G (2014) Spate irrigation in Tigray: the challenges and suggested ways to overcome them. In: Flood-based Farming for food security and adaption to climate change in Ethiopia: potential and challenges, International Water Management Institute (IWMI). Colombo, Sri Lanka, pp 137–148

Hishe S, Lyimo J, Bewket W (2017) Soil and water conservation effects on soil properties in the middle Silluh Valley, northern Ethiopia. Int Soil Water Conserv Res 5:231–240

Hurni H (1988) Degradation and conservation of the resources in the Ethiopian highlands. Mt Res Dev 8:123–130

Kassie M, Pender J, Yesuf M, Kohlin G, Bluffstone R, Mulugeta E (2008) Estimating returns to soil conservation adoption in the northern Ethiopian highlands. Agric Econ 38:213–232

Kelebe HE, Ayimut KM, Berhe GH, Hintsa K (2017) Determinants for adoption decision of small scale biogas technology by rural households in Tigray, Ethiopia. Energy Econ 66:272–278

Kidane H (2009) Community spate irrigation in Raya Valley: the case of three spate irrigation systems. Master's thesis, Addis Ababa University, Addis Ababa, Ethiopia

Lemenih M, Kassa H, Ababa A (2014) Re-greening Ethiopia: history, challenges and lessons. Forests 5:1896–1909

Mchugh OV, Steenhuis TS, Abebe B, Fernandes EC (2007) Performance of in situ rainwater conservation tillage techniques on dry spell mitigation and erosion control in the drought-prone north Wello zone of the Ethiopian highlands. Soil Tillage Res 97:19–36

Meaza H, Tsegaye D, Nyssen J (2016) Allocation of degraded hillsides to landless farmers and improved livelihoods in Tigray, Ethiopia. Norsk Geografisk Tidsskrift-norwegian J Geogr 70:1–12

Meheretu Y, Sluydts V, Welegerima K, Bauer H, Teferi M, Yirga G, Mulungu L, Haile M, Nyssen J, Deckers J (2014) Rodent abundance, stone bund density and its effects on crop damage in the Tigray highlands, Ethiopia. Crop Prot 55:61–67

Meheretu Y, Welegerima K, Sluydts V, Bauer H, Gebrehiwot K, Deckers J, Makundi R, Leirs H (2015) Reproduction and survival of rodents in crop fields: the effects of rainfall, crop stage and stone-bund density. Wildl Res 42:158–164

Mekonen K, Tesfahunegn GB (2011) Impact assessment of soil and water conservation measures at Medego watershed in Tigray, northern Ethiopia. Maejo Int J Sci Technol 5:312–330

Mekonnen M, Keesstra S, Baartman J, Ritsema C, Melesse A (2015) Evaluating sediment storage dams: structural off-site sediment trapping measures in northwest Ethiopia. Cuadernos de Investigación Geográfica 41:7–22

Mekuria W, Veldkamp E (2012) Restoration of native vegetation following exclosure establishment on communal grazing lands in Tigray, Ethiopia. Appl Veg Sci 15:71–83

Mekuria W, Veldkamp E, Haile M, Nyssen J, Muys B, Gebrehiwot K (2007) Effectiveness of exclosures to restore degraded soils as a result of overgrazing in Tigray, Ethiopia. J Arid Environ 69:270–284

Mekuria W, Veldkamp E, Tilahun M, Olschewski R (2011) Economic valuation of land restoration: the case of exclosures established on communal grazing lands in Tigray, Ethiopia. Land Degrad Dev 22:334–344

Mekuria W, Langan S, Noble A, Johnston R (2017) Soil restoration after seven years of exclosure management in northwestern Ethiopia. Land Degrad Dev 28:1287–1297

Meshesha DT, Tsunekawa A, Tsubo M (2012) Continuing land degradation: cause–effect in Ethiopia's Central Rift Valley. Land Degrad Dev 23:130–143

Michael YG, Waters-Bayer A (2007) Trees are our backbone: integrating environment and local development in Tigray region of Ethiopia. Iied, London

Mûelenaere S, Frankl A, Haile M, Poesen J, Deckers J, Munro N, Veraverbeke S, Nyssen J (2014) Historical landscape photographs for calibration of Landsat land use/cover in the northern Ethiopian highlands. Land Degrad Dev 25:319–335

Munang R, Thiaw I, Alverson K, Mumba M, Liu J, Rivington M (2013) Climate change and ecosystem-based adaptation: a new pragmatic approach to buffering climate change impacts. Curr Opin Environ Sustain 5:67–71

Munro RN, Deckers J, Haile M, Grove A, Poesen J, Nyssen J (2008) Soil landscapes, land cover change and erosion features of the Central Plateau region of Tigrai, Ethiopia: photo-monitoring with an interval of 30 years. Catena 75:55–64

Nefzaoui A, El Mourid M (2010) Cactus pear for soil and water conservation in arid and semi-arid lands. In: Improved utilization of cactus pear for food, feed, soil and water conservation and other products in Africa, Cactus Net Newsletter. Mekelle, Ethiopia p 117–126

Noulekoun F, Birhane E, Chude S, Zenebe A (2017) Characterization of *Faidherbia albida* (Del.) A. Chev. population in agroforestry parklands in the highlands of northern Ethiopia: impact of conservation, environmental factors and human disturbances. Agrofor Syst 91:123–135

Nyssen J, Haile M, Moeyersons J, Poesen J, Deckers JA (2000) Soil and water conservation in Tigray (northern Ethiopia): the traditional daget technique and its integration with introduced techniques. Land Degrad Dev 11:199–208

Nyssen J, Haile M, Moeyersons J, Poesen J, Deckers J (2004a) Environmental policy in Ethiopia: a rejoinder to Keeley and Scoones. J Mod Afr Stud 42:137–147

Nyssen J, Veyretpicot M, Poesen J, Moeyersons J, Haile M, Deckers JA, Govers G (2004b) The effectiveness of loose rock check dams for gully control in Tigray, northern Ethiopia. Soil Use Manag 20:55–64

Nyssen J, Poesen J, Gebremichael D, Vancampenhout K, Daes M, Yihdego G, Govers G, Leirs H, Moeyersons J, Naudts J (2007) Interdisciplinary on-site evaluation of stone bunds to control soil erosion on cropland in northern Ethiopia. Soil Tillage Res 94:151–163

Nyssen J, Naudts J, de Geyndt K, Haile M, Poesen J, Moeyersons J, Deckers J (2008a) Soils and land use in the Tigray highlands (northern Ethiopia). Land Degrad Dev 19:257–274

Nyssen J, Poesen J, Moeyersons J, Haile M, Deckers J (2008b) Dynamics of soil erosion rates and controlling factors in the northern Ethiopian highlands–towards a sediment budget. Earth Surf Process Landf 33:695–711

Nyssen J, Clymans W, Poesen J, Vandecasteele I, de Baets S, Haregeweyn N, Naudts J, Hadera A, Moeyersons J, Haile M (2009) How soil conservation affects the catchment sediment budget–a comprehensive study in the north Ethiopian highlands. Earth Surf Process Landf 34:1216–1233

Nyssen J, Govaerts B, Araya T, Cornelis WM, Bauer H, Haile M, Sayre K, Deckers J (2010) The use of the marasha ard plough for conservation agriculture in Northern Ethiopia. Agron Sustain Dev 31:287–297

Nyssen J, Frankl A, Haile M, Hurni H, Descheemaeker K, Crummey D, Ritler A, Portner B, Nievergelt B, Moeyersons J (2014) Environmental conditions and human drivers for changes to north Ethiopian mountain landscapes over 145years. Sci Total Environ 485:164–179

Nyssen J, Gebreselassie S, Assefa R, Deckers JA, Zenebe A, Poesen J, Frankl A (2017) Boulder-faced log dams as an alternative for gabion check dams in first-order ephemeral streams with coarse bed load in Ethiopia. J Hydraul Eng 143:05016005

Pender J, Gebremedhin B (2007) Determinants of agricultural and land management practices and impacts on crop production and household income in the highlands of Tigray, Ethiopia. J Afr Econ 17:395–450

Reubens B, Moeremans C, Poesen J, Nyssen J, Tewoldeberhan S, Franzel S, Deckers J, Orwa C, Muys B (2011) Tree species selection for land rehabilitation in Ethiopia: from fragmented knowledge to an integrated multi-criteria decision approach. Agrofor Syst:1–28

Rimhanen K, Ketoja E, Yli-Halla M, Kahiluoto H (2016) Ethiopian agriculture has greater potential for carbon sequestration than previously estimated. Glob Chang Biol 22:3739–3749

Segers K, Dessein J, Hagberg S, Develtere P, Haile M, Deckers J (2008a) Be like bees–the politics of mobilizing farmers for development in Tigray, Ethiopia. Afr Aff 108:91–109

Segers K, Dessein J, Nyssen J, Haile M, Deckers J (2008b) Developers and farmers intertwining interventions: the case of rainwater harvesting and food-for-work in Degua Temben, Tigray, Ethiopia. Int J Agric Sustain 6:173–182

Solomon N, Birhane E, Tadesse T, Treydte AC, Meles K (2017) Carbon stocks and sequestration potential of dry forests under community management in Tigray, Ethiopia. Ecol Process 6:20

Stavi I, Argaman E, Zaady E (2016) Positive impact of moderate stubble grazing on soil quality and organic carbon pool in dryland wheat agro-pastoral systems. Catena 146:94–99

Taddese G (2001) Land degradation: a challenge to Ethiopia. Environ Manag 27:815–824

Tamene L, Park SJ, Dikau R, Vlek PLG (2006) Reservoir siltation in the semi-arid highlands of northern Ethiopia: sediment yield–catchment area relationship and a semi-quantitative approach for predicting sediment yield. Earth Surf Process Landf 31:1364–1383

Taye G, Poesen J, Wesemael BV, Vanmaercke M, Teka D, Deckers J, Goosse T, Maetens W, Nyssen J, Hallet V (2013) Effects of land use, slope gradient, and soil and water conservation structures on runoff and soil loss in semi-arid northern Ethiopia. Phys Geogr 34:236–259

Taye G, Poesen J, Vanwesemael B, Vanmaercke M, Teka D, Deckers J, Goosse T, Maetens W, Nyssen J, Hallet V (2014) Effectiveness of soil and water conservation structures in reducing runoff and soil loss for different land use and slope gradients: case study from northern Ethiopia. EGU General Assembly conference abstracts

Taye G, Poesen J, Vanmaercke M, van Wesemael B, Martens L, Teka D, Nyssen J, Deckers J, Vanacker V, Haregeweyn N (2015) Evolution of the effectiveness of stone bunds and trenches in reducing runoff and soil loss in the semi-arid Ethiopian highlands. Z Geomorphol 59:477–493

Tedla HA (2010) The effect of compost on soil fertility enhancement and yield increment under smallholder farming: a case of Tahtai Maichew District-Tigray Region, Ethiopia. Inst. für Bodenkunde und Standortslehre, Stuttgart

Teklay A (2011) Seasonal availability of common bee flora in relation to land use and colony performance in Gergera Watershed Atsbi Wemberta District, Eastern Zone of Tigray, Ethiopia. MSc Thesis, Hawassa University, Hawassa, Ethiopia

Tesfahunegn GB (2015) Short-term effects of tillage practices on soil properties under Tef [Eragrostis tef (Zucc. Trotter)] crop in northern Ethiopia. Agric Water Manag 148:241–249

Teshome A, Rolker D, de Graaff J (2013) Financial viability of soil and water conservation technologies in northwestern Ethiopian highlands. Appl Geogr 37:139–149

Tetemke BA, Gebremeskel D, Takenaka K (2017) Ecological networking to scale-up and sustain ecological restoration in the dry lands of northern Ethiopia. 日本森林学会大会発表データベース 第 128 回日本森林学会大会. 日本森林学会, 819

Tilahun M, Muys B, Mathijs E, Kleinn C, Olschewski R, Gebrehiwot K (2011) Frankincense yield assessment and modeling in closed and grazed *Boswellia papyrifera* woodlands of Tigray, northern Ethiopia. J Arid Environ 75:695–702

Vagen T, Tilahun Y, Esser K (1999) Effects of stone terracing on available phosphorus and yields on highly eroded slopes in Tigray, Ethiopia. J Sustain Agric 15:61–74

Vancampenhout K, Nyssen J, Gebremichael D, Deckers JA, Poesen J, HAILE M, Moeyersons J (2006) Stone bunds for soil conservation in the northern Ethiopian highlands: impacts on soil fertility and crop yield. Soil Tillage Res 90:1–15

Waktola DK (2007) Rainwater harvesting in Ethiopia: capturing realities and harnessing opportunities. Rainwater Urban Design 2007:484

Woldearegay K, van Steenbergen F (2015) Shallow groundwater irrigation in Tigray, northern Ethiopia: practices and issues. In: Lollino G, Arattano M, Rinaldi M, Giustolisi O, Marechal J-C, Grant GE (eds) Engineering geology for society and territory - volume 3: river basins, reservoir sedimentation and water resources. Springer International Publishing, Cham

Woldearegay K, van Steenbergen F, Perez MA, Grum B, van Beusekom M (2015) Water harvesting from roads: climate resilience in Tigray, Ethiopia. IRF Examiner. 12 (Winter 2017): 1–7

Yami M, Gebrehiwot K, Stien M, Mekuria W (2007) Impact of area enclosures on density and diversity of large wild mammals: the case of May Ba'ati, Douga Tembien district, Central Tigray, Ethiopia. East Afr J Sci 1:55–68

Yami M, Vogl C, Hauser M (2011) Informal institutions as mechanisms to address challenges in communal grazing land management in Tigray, Ethiopia. Int J Sustain Dev World Ecol 18:78–87

Yami M, Mekuria W, Hauser M (2013) The effectiveness of village bylaws in sustainable management of community-managed exclosures in northern Ethiopia. Sustain Sci 8:73–86

Yayneshet T, Eik LO, Moe SR (2008) Feeding *Acacia etbaica* and *Dichrostachys cinerea* fruits to smallholder goats in northern Ethiopia improves their performance during the dry season. Livest Sci 119:31–41

Yayneshet T, Eik LO, Moe SR (2009) The effects of exclosures in restoring degraded semi-arid vegetation in communal grazing lands in northern Ethiopia. J Arid Environ 73:542–549

Yazew E (2015) Flood-based farming and the second five years growth and transformation plan of Tigray. Mekelle University, Mekelle, Tigray, Ethiopia

Seasonal Climate Forecasts Among Farmers in Nigeria: Assessment of the Usability

121

Glory Ikponmwosa Edwards and Dominic Kniveton

Contents

Introduction 2486
Background 2486
Research Objective and Questions 2488
Literature Review 2488
The Value of SCF 2488
Usability of SCF 2489
Factors Affecting Usability of SCF 2490
Production of SCF in Nigeria 2493
Conceptual Framework 2494
Research Methodology 2495
Agro-Climate of Nigeria 2495
Study Areas and Description 2497
Data Collection 2497
Data Processing and Analysis 2498
Ethical Considerations 2498
Results 2499
Characteristics of Respondents 2499
Forecast Information Farmers Need 2499
Access to Forecast and Timeliness of Forecast 2502
Value of SCF to Farmers Relative to Other Kinds of Climate Information 2503
Correlations of Some Questionnaire Items 2504
Discussion 2504
Relevance of Forecast Information to Farmers Decision-Making Needs 2504
Accessibility and Timeliness of SCF 2505

G. I. Edwards (✉)
Climate Change, Development and Policy, School of Global Studies,
University of Sussex, Falmer, UK
e-mail: edwardsglory@gmail.com

D. Kniveton
Climate Science and Society (Geography),
University of Sussex, Falmer, UK
e-mail: d.r.kniveton@sussex.ac.uk

W. Leal Filho (ed.), *Handbook of Climate Change Resilience*,
https://doi.org/10.1007/978-3-319-93336-8_133

Reliability of Farmers on SCF ... 2506
Opportunities for Usability of SCF ... 2507
Constraints to Usability of SCF ... 2508
Conclusion ... 2509
Summary of the Research ... 2509
Significance of Study ... 2510
Limitations of Study ... 2510
Suggestions for Further Study ... 2510
References ... 2511

Abstract

Despite the improved skill and availability of climate information, there is still a persistent gap between climate information production and its uptake among farmers in tropical countries to inform decision-making. In this study, the usability of seasonal climate forecasts (SCF) was assessed among farmers from three states, representative of different agroecological zones in Nigeria. The objective of the study was to establish how SCF Fit, comprising of SCF relevance, accessibility, timeliness and reliability, can alter SCF usability. Questionnaires were administered to a total of 60 respondents, 20 from each state to assess the usability of seasonal climate forecasts. In addition, semi-structured interviews were conducted for the Nigerian Meteorological Agency (NIMET) Heads of the three states. This study drawing insights from farmers and producers of climate information found that usability is enhanced when users' decision-making needs correspond with the components of the SCF and when SCF is available and accessed in optimal time; which was found to be beginning of the year to beginning of the season. The research found a positive relationship between the factors of relevance, accessibility, timeliness, and reliability and a correlation of these with uptake and application of SCFs. In order for forecasts to be useful to farmers in decision-making, they need to both provide relevant information and be delivered through several sources on timely basis.

Keywords

Seasonal climate forecasts · Usability · Farmers · Fit · Adaptation

Introduction

Background

It is widely projected that as the planet warms, there will be an increase in changes in frequency and intensity of weather events and shifts, changes in climate variability, and shifts in the mean statistics of climate variables such as temperature and precipitation around the world (Intergovernmental Panel on Climate Change

2012). Adaptation to climate change impacts in particular local adaptation efforts is therefore necessary. Major interventions to support local adaptation efforts have focused on providing climate information and financial support (Agrawal and Pern 2008). At the local level, climate information is needed by users in order to make robust adaptation decisions. Especially for farmers who practice rainfed-agriculture, there is a need for information on rainfall and temperature to know when to plan their farming activities.

The term "seasonal climate forecasts" (SCF) refers to predictions on seasonal timescales, which is the climate that can be expected in the coming months, up to 3–6 months, provided ahead of the season (Weisheimer and Palmer 2014). It is a rapidly growing field of research and plays a crucial element in informing climate change adaptation strategy. It involves the ability of climate models to predict the variability in the climate system for the coming months. Farmers require forecasts on seasonal time scales to help with decisions on which crops to plant, when to plant, when to apply herbicides/fertilizers, and when to conserve water for irrigation. Such farming decisions need to be taken based on timely, skillful forecasts to reduce farmer's vulnerability to climate variability while increasing productivity. The ability to provide timely, skillful climate forecasts has improved over the years to supply farmers with needed climate information.

The potential for SCF to reduce the vulnerability of agriculture to climate variability lies in the uptake and application of SCF by the farmers. Studies have shown that despite the improved skill and availability of SCF, there is still a persistent gap between SCF production and its uptake among farmers in tropical countries to inform decision-making (Vogel and O'Brien 2006; Roudier et al. 2014). This gap is commonly called the "usability gap." The Oxford English Dictionary defines usability as the degree to which something is able or fit to be used (OED, n.). The International Standard Organization 9241 standard Part 11 defines it as "the extent to which a product can be used by specified users to achieve specified goals with effectiveness, efficiency and satisfaction in a specified context of use" (ISO 1997). This definition emphasizes that the usability of a product is based not only on the features of the product itself but also by the specific circumstances in which the product is used.

In climate science, the usability gap exists because scientists/climate information producers believe that all they produce is useful but a lot of the climate information remains unused for decision-making. A range of organizational, socioeconomic, and behavioral factors have been studied from different approaches and in different contexts on how they affect the usability of SCF. This research focuses on the perception of Fit, a factor that encompasses perception of SCF relevance, accessibility, timeliness, and reliability of SCF. After production of climate information, it can go from useful to usable as it is disseminated and communicated to approach specific users' perceived needs to support decision-making for adaptation to climate variability (Lemos et al. 2012). Drawing on the conceptual model by Lemos et al. (2012), the usability of SCF produced by the Nigerian Meteorological Agency is assessed among crop farmers from three agroecological zones of Nigeria.

Research Objective and Questions

The objective of this to assess whether and how the factor of Fit comprising of SCF relevance, accessibility, timeliness, and reliability can enhance SCF usability among farmers. The research questions are given below:

1. **How does SCF match farmers' information needs for decision-making?**
   Is the information beneficial to the farmers? Are farmers putting it into use? How satisfied are the farmers with information received?
2. **How and when do farmers obtain SCF?**
   What are the forms of communication, accessibility, and what is the timing of forecast receipt for relevant decision-making?
3. **What is the value of SCF to farmers relative to other kinds of climate information?**
   Do farmers see SCF as reliable? Do they use it in decision-making relative to other kinds of climate information?
4. **How do the factors of relevance, accessibility, timeliness, and reliability affect the usability of SCF?**

Literature Review

The Value of SCF

Climate predictions are necessary for decision-making for adaptation despite inherent uncertainties (Dessai et al. 2009). The probabilistic nature of seasonal forecasts does not make it invalid but provides estimates of seasonal-mean statistics of the weather expected representing these as anomalies from mean conditions of the past. For example, SCF can predict that the coming months/season will have wetter-than-normal conditions or drier-than-normal conditions. Such information carries great value for government, commercial and financial planning, and across sectors such as water management, agriculture, and disaster risk reduction, especially where they are reliable and skillful.

SCF in Individuals' Decision-Making

The potential benefits of SCF are observable in areas of great climate variability and where there is the dependence of economic activities on the climate conditions, such as the agriculture sector. Where SCF is present, it reduces the uncertainty that comes with climate variability thereby equipping the decision-maker to make more informed economic decisions. SCF can be used to avoid/reduce the impacts of climate change or to maximize favorable conditions. Meza et al. (2008) mention two ways in which SCF are of value to users: by reducing the spread of possible outcomes for the coming months in comparison with the mean statistics of climatology and by conveying shifts in the central tendency of climatic outcomes. To communicate uncertainty in SCF, forecasters usually use probabilities to measure the likelihood of an occurrence of an event so that SCF does not present a false sense of certainty. The difference between probabilistic and uncertainty is that with probabilistic information, the unknowns are assumed to be quantifiable while uncertainty includes both unknowns that we know about and also unknowns that what we don't know about (Kniveton et al. 2015a). Where users lack a proper understanding of uncertainty, trust in the forecast can be lost and failure to act based on forecast ensues. Proper communication of uncertainty helps decision-makers make robust decisions.

Accordingly, farmers must consider climate information to develop agricultural adaptation techniques and make economic decisions. When SCF is used, there is evidence of increased agricultural yield, an example of up to 9.4% yield increase in Zimbabwe (Patt and Gwata 2002). Also, the economic value of SCF can instigate decision-making with economic implications, include taking insurance against a poor yield, or taking out a loan for investment, expecting high yields, or farm management decisions on what to plant and when to plant.

SCF in Policy-Making

There is a causal relationship between seasonal climate and crop production which has a contributing relationship with food shortages and rise in food prices (Bellemare 2015). Drought in East Africa has been associated with strong El Nino warming, determining the seasonal climate of below-normal rainfall (Funk et al. 2016). In 2011, the drought led to a food crisis in Somalia and Ethiopia and a drop in GDP with farmers being the worst hit. Since climate variability can have macroeconomic implications like these, governments and institutions need SCF for effective policy formulation for agricultural planning. Farmers can be equipped and prepared with help from governments and institutions by providing resilient seed varieties, storage reserves, irrigation, crop insurance and tax, or subsidies policy especially in preparation for the next planting season.

An example of the beneficial use of SCF in policy-making is in the application of SCF information for grain procurement, importation, and storage at the national level in the Philippines (Mina and Reyes 2009). The Inter-Agency Committee on Rice and Corn (IACRC) meets quarterly to assess the supply/demand situation for rice and corn based on SCF. The Agency recommends importation (timing, volume, and where to distribute to), if necessary, to the government through the Department of Agriculture. Importation is resorted to only during a foreseen production shortfall. The linking of SCF to policy-making can benefit and enhance decisions for both individuals and Government.

Usability of SCF

Prior studies have focused on investigating if climate science information is used by society in decision-making (Patt and Gwata 2002; Vogel and O'Brien 2006; Roudier et al. 2014). These studies and more have established that there is a gap between climate information producers and the users which is seen in the usage of climate

Table 1 Some empirical literature on the use of SCF among farmers in Africa

Authors	Research	Use of SCF
Blench (1999)	Farmers in Southern Africa	Often not used
Broad and Agrawala (2000)	Famine prevention in Ethiopia	Underutilized
Ingram et al. (2002)	Farmers in Burkina Faso	Limited usage but interested in more
Patt and Gwata (2002)	Farmers in Zimbabwe	Not directly useful. When used, it led to 9.4% yield
Luseno et al. (2003)	Pastoralists in southern	Less commonly believed or
	Ethiopia and northern Kenya	Received relative to indigenous methods
Vogel and O'Brien (2006)	Farmers in southern Africa	Very limited usage, constraints are many
Cash et al. (2006)	Southern Africa	Not well used
Roudier et al. (2014)	Small holder farmers in Senegal	Used to some extent

Source: Modified and updated from Dilling and Lemos (2011)

information in practice. This gap is commonly referred to as the "climate information usability gap" (Lemos et al. 2012). Table 1 below summarizes the empirical literature on the use of SCF in agriculture in Africa. It is evident from these examples that for several reasons intrinsically relevant climate information is not maximized for decision-making.

Factors Affecting Usability of SCF

Empirical studies have been carried out on the factors affecting the usability of SCF are discussed below. These are reviewed in three subsections below.

Characteristics of SCF

As shown in Table 1, a lot of the climate information remains unused and does not benefit the user in decision-making. So, what makes SCF usable? In a broader sense, what makes science usable? Dilling and Lemos (2011) argue that for science to be considered usable, it must directly contribute to solving a problem or be used in policy formulation. In software development, usable products are said to be designed incorporating product features known to be beneficial to users in particular contexts of use (Bevan 2009). Hansen (2002) also argues that usable science is beneficial science that contributes to problem-solving.

As to the character of the science information itself, Cash et al. (2002) list credibility, legitimacy, and salience as the three quality criteria that must characterize science for it to be usable. Credibility means that the scientific information is considered accurate and valid and that it meets the thresholds of "scientific plausibility and technical adequacy" (Cash et al. 2002: 4). The legitimacy of science information refers to the fairness of the process of information production that it follows required values while salience signifies that science information is relevant to users for decision-making. The three quality criteria – credibility, salience, legitimacy – must be present and not one without the other. However, there are compromises in trying to improve on one characteristic. For instance, in trying to improve on salience by involving stakeholders in science production, credibility could be reduced because more nonprofessionals in climate science are involved in the production process.

Since credibility is of the scientific process, individuals are often unable to independently evaluate the credibility of scientific information but can be asked how reliable they consider the scientific information to be. This makes reliability a better evaluator of science information usability. Millner and Washington (2011) suggest that forecast reliability is proportional to the forecast adoption rates and therefore can be used as a metric for evaluation.

This proposition that usable science is proportional to forecast adoption rates is limiting and should not be generalized. As it is with people, even in the availability of usable science with all the characteristics of credibility, salience, legitimacy, timeliness, reliability, resolution, there could still be low uptake and application. Users may choose to ignore SCF due to reluctance to change their ways, due to lack of understanding of how SCF fits their decision-making, or even due to lack of capacity to respond to forecasts. This leads us to the factor of users' perceptions which is discussed in the next subsection.

Users' Perceptions

The complexity of the users requires approaching research in the usability of SCF as highly contextual, specific to users' decision-making context. But approaching SCF production in this manner is not always feasible due to cost (von Hippel and Katz 2002). Kirchhoff et al. (2013) propose a model which broadens the use of scientific information by creating a typology of users as a function of proactivity for obtaining and using information and users' decision environment – early adopters, tentative users, proactive users, and laggards. This typology reflects varying users' perception of risk. These perceptions vary across people, cultures, time, and experience and influence the type of climate information sought and used in the decision.

The understanding of the perceptions and attitudes of users allow for upscaling of climate information and application to a wider audience. The suggestion is also made of the use of tool-kits for users to tailor to their needs by application of the tools (von Hippel and Katz 2002). In this way, it is purported that climate information can be produced for heterogeneous users. This seems to be user-friendly and allowing to reach a wider audience but it does not seem practicable and may end up further making science unusable by target audience where interpretation for application in decision-making fails. In response to this, usable science should be amenable to users' needs. Since users are always making decisions that change with their changing needs, for science to be usable, it must be designed to be amenable to interpretation to suit present and evolving needs of the users. Not only are the needs of users changing but their physical and socioeconomic environment and circumstances are constantly changing and can affect the application of scientific information. In addition to flexibility, the uncertainty of SCF requires SCF to possess robustness under varying future climatic conditions to allow low-regrets options (Ranger 2013). That is options that have low-cost relative to their benefits both today and under a wide range of plausible future climates.

Users' perception also influences the perception of forecast value. If users perceive information as reliable and relevant, they are likely to use and apply SCF to decision-making (Cash et al. 2006). Several factors, in turn, can enhance the perception of SCF as reliable and relevant. For instance, where an evaluation is made for past SCF and they are found to closely match observation, farmers increase their trust in the skill of the forecast. In addition, dissemination of intraseasonal forecasts ensured continuous contact between farmers and SCF producers and thus built trust in SCF.

Users' perception is also affected by the interplay between traditional knowledge, personal experience, and scientific information (Lemos 2008). Traditional knowledge involves the use of traditional forecasting methods and indicators such as birds, animal, and plant observations and the work of diviners and rainfall makers (Luseno et al. 2003; Musembi and Cheruiyot 2016; Mpandeli et al. 2017). It also encompasses the knowledge and practices that are acquired by local people over time through society–nature relationships, community practices, and experiences over generations (Kniveton et al. 2015b). Personal experience and traditional knowledge interact to affect the perception of users in the adoption of SCF.

Communication and Knowledge Exchange

Information communication and dissemination pathways impact on the uptake and application of science-based knowledge to support decision-making (Cash et al. 2003; Jacobs et al. 2005; Buizer et al. 2016). The ways in which SCF is translated and communicated to farmers, the usability of SCF can be enhanced or hindered. Features of forecast communication such as forecast timeliness and forecast presentation have been found to influence usability (Ingram et al. 2002; Hansen 2002). For SCF to be of value to the end users, strategies for its dissemination, uptake and usage by the end users must be incorporated in the production chain (Troccoli 2010). Uptake and usage here mean that usable science does not end at dissemination but involves knowledge exchange and is found in the decision-making of users. Prokopy et al. (2017) regard usable science as science that has these certain characteristics – directly reflects expressed fundamental needs of the end-user, is understandable to users, is available at the times and places it is needed, and is available and accessible to the user community. The authors point out the factors of relevance, availability, accessibility, and understanding of forecasts. Understanding is a key to application and ensures that SCF does not end in dissemination paths but is taken up guaranteeing knowledge exchange.

Boundary organizations such as agricultural extension workers can serve as efficient intermediaries between climate forecasters and farmers. There is clearly a positive role played by boundary organizations in enhancing the usability of climate information (Kirchhoff et al. 2015). Boundary organizations can influence users to perceive climate information as credible. By stabilizing the interface between producers and users of climate information, they improve two-way interaction which has its benefits (Kirchhoff et al. 2013). However, boundary organizations can also lead to poor uptake of SCF where the boundary organization activities are not productive in carrying out bridging and mediation between science and decision-making by users. Kirchhoff et al. (2015) present a framework on the idea of boundary chains where several boundary organizations work together to improve usability by the synergy of more efficient production of usable science and more effective interaction between producers and users of climate information.

In this chapter, a synthesis has been made of the literature on factors affecting scientific information usability. SCF usability is highly contextual phenomena and so the factors mentioned above are not exhaustive but rather indicative of the rich literature that has been produced on SCF usability among farmers.

Production of SCF in Nigeria

In Nigeria, the Nigerian Meteorological Agency (NIMET) is the primary producer of meteorological data used for climate information in the country. These data are collected from weather-observation stations located across the nation in airports and other remote stations. At present, there are 224 rainfall stations, 167 agromet stations, 42 synoptic stations, 5 upper air stations, 13 climate stations, 3 marine stations, 1 ozone station, and 1 background pollution monitoring station (NIMET 2017). The density of the stations over much of the nation is inadequate and rural areas are left out largely. The country should have 9000 stations based on the World Meteorological Organization (WMO) standard of a station at every 100 miles (WMO 2007). The data collected from these stations are processed at a central station in the Federal Capital Territory and resulting information is produced which is used across a variety of sectors. Three climate-sensitive sectors where the focus is placed on to provide with climate information are marine, aviation, and agriculture sectors (Personal communication, Festus Omoregbei, NIMET). NIMET produces several publications or periodicals to disseminate climate information – Daily Weather Forecasts, DekadelAgromet Bulletins, Quarterly Weather Bulletins, Monthly Drought and Flood Monitor Bulletin, Monitoring for Environment and Security in Africa (MESA) Land Thema Bulletin, Annual Climate Review, Marinemet Bulletin, Seasonal Rainfall Predictions, Weekly Climate Review, and NIMet Digest (NIMET 2017).

This research focuses on the seasonal rainfall prediction (SRP). Ahead of each rainy season, NIMET releases the SRP which represents the major SCF produced in Nigeria. The SRP contains predictions of the onset and cessation dates of the rainy season, the length of growing season, the amount of rainfall for the year and temperature forecasts, the seasonal amount of rainfall, little dry season and dry spells, malaria severity forecast, etc. These are presented in tables, in maps, bar charts, and text. The information is based on the El-Nino Southern Oscillation (ENSO), Sea surface temperatures anomalies, and the rain-bearing weather systems in Nigeria. The SRP publication also contains advisories generated from the forecast for climate-sensitive sectors such as water resources management, disaster risk management, tourism, health, communication, transportation sectors, etc. The SRP also contains the evaluation of the last year's forecast to test the skill of the forecast. Since 2011, SRP booklet has been printed for dissemination and it is available online for free download (NIMET 2017).

Conceptual Framework

Apart from the factors discussed in section "Factors Affecting Usability of SCF," different frameworks and models can also be found in literature which assists in addressing the usability gap problem. Some examples are those of Kirchhoff et al. (2015) on boundary organizations; Allen and Stephens (2016) on multiple stakeholder engagements; Kniveton et al. (2015a) on access, understanding, and appropriate application; and Lemos et al. (2012) on the perception of fit, interplay, and interaction. Lemos et al. 2012 theorize that perception of fit, interplay, and interaction are what link the path between usefulness and usability of climate information (Fig. 1). The presence of these makes information usable whereas the absence of these constricts the transition gap between producers and users.

Interplay refers to how new knowledge relates to other kinds of knowledge that are currently used by users while *Interaction* is an interaction between information producers and users whether it is one-way or two-way communication affects the usability of climate information.

Fit is described as the perception of the information by users as relevant; the character of the information such as reliability and uncertainty and the ability of the users to deploy knowledge. Forms of communication such as accessibility, timeliness, format, information flow, and knowledge exchange are part of users' perception of fit. Lemos et al. (2012) argue that Fit is not static but evolves and is closely linked with interplay and interaction.

This research focuses on the Fit variable analyzing how the perception of SCF relevance, accessibility, timeliness, and reliability influence SCF usability as SCF flows from producers to users. After production of climate information, it can go from useful to usable as it is disseminated, translated, and communicated to approach users' perceived needs (Lemos et al. 2012). Some authors believe that the focus of efforts to increase resilience against climate change and variability should be shifted away from the improvement of forecast skill and dissemination which encompass the Fit variable (Luseno et al. 2003; Vogel and O'Brien 2006). However, in Nigeria where the study is carried out, the availability and application of agricultural information including climate information remain a key constraint to agricultural development (Collier and Decon 2014; Otitoju and Enete 2016). Availability and accessibility of timely climate information relevant to farmers' decision-making can contribute to addressing the challenge of agricultural development. Therefore, Fit is of paramount importance and is central to SCF usability while the

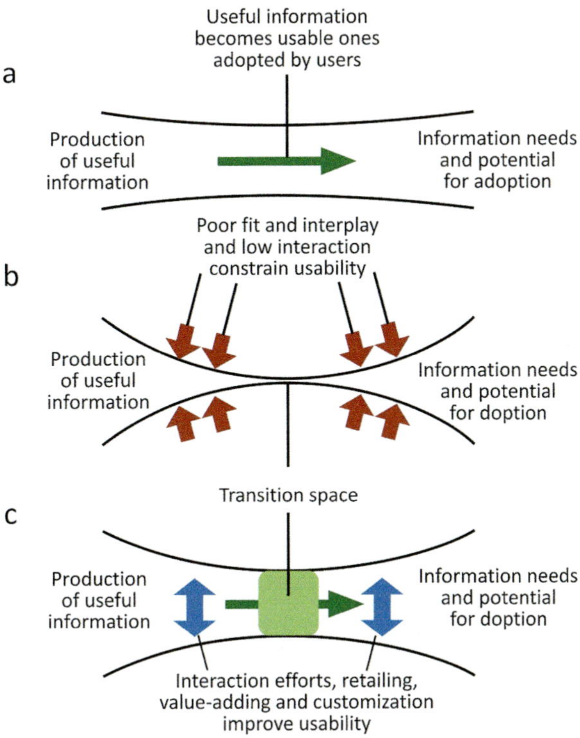

Fig. 1 The conceptual model for usability of climate information. (**a**) Information moves from useful to usable by being deployed by users in decision-making. (**b**) From a producer's perspective, all information produced is potentially useful. Information needs fit, interplay, and interaction to be usable. (**c**) Interaction, retailing, customization, and value-adding improve fit. (Source: Lemos et al. 2012)

other components of the model – interplay and interaction – impact on Fit to enhance or constrict usability of SCF. Fit remains central while whatever alters users' perception of Fit acts to impact usability.

Research Methodology

Agro-Climate of Nigeria

Nigeria is in West Africa between latitude 4–14°N and longitudes 2–15°E. The climate varies from south to north, transitioning from humid in the south to sub-humid in the central parts unto semi-arid in the North. This variation in climate creates an ecologically diverse landscape with marked seasonal differences. Figure 2 below shows the map of Nigeria and indicates the study areas.

The rainy season usually begins around March or April in the southern parts (4°–8°N), commencing at about May/June in the northern parts (8°–14°N) and ending at about October. In Nigeria, "seasonal rainfall" refers to precipitation during the 3-month period (July, August, and September) during which 80% of annual rains typically fall (Fig. 3).

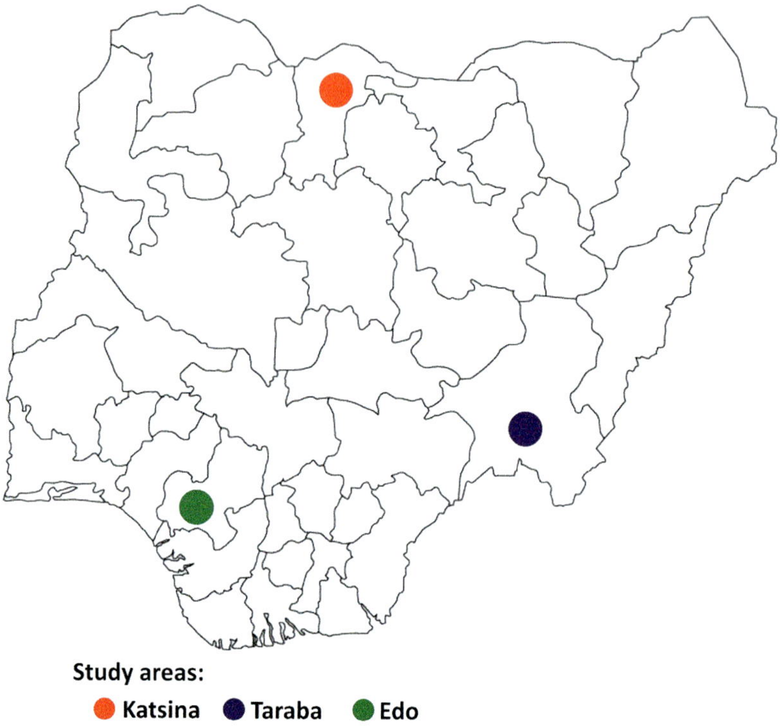

Fig. 2 Map of Nigeria showing study areas

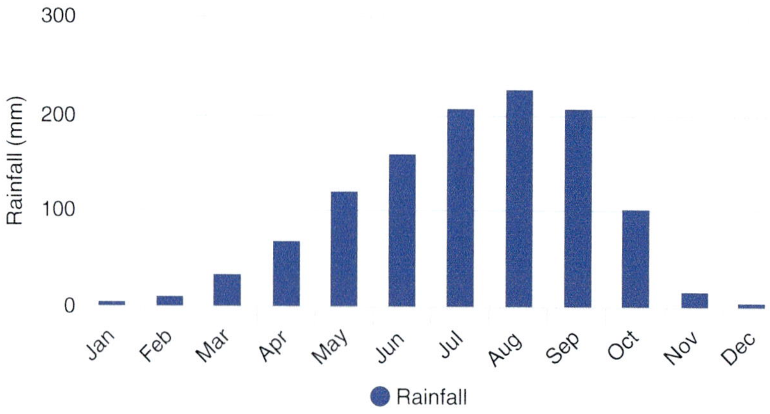

Fig. 3 Average monthly rainfall for Nigeria from 1991–2015 showing the 3-month period (July, August, and September) of peak rainfall. (Source: Climatic Research Unit (CRU) data set from Climate Change Knowledge Portal (World Bank 2017))

Study Areas and Description

The study was conducted in three states: Edo, Taraba, and Katsina states. Study areas were selected from the different agroecological zones for proper representativeness of Nigeria. Major features of the selected three study areas are listed in Table 2. In Edo State, Oredo, Orhionmwon, and Ikpoba-Okha Local Government Areas (LGA) were visited. In Katsina, sites selected were Kankara, Katsina, Funtua, and Kaita LGAs. In Taraba State, two LGA were selected, Lau and Gassol. All sites are among the major food crops producing LGA in their respective states.

Data Collection

Data for this study were collected from purposely selected LGA from each state through NIMET. The LGA had previous contact with NIMET and are located close to the provincial capital of the states where NIMET state stations are located. All the LGA are farming communities and the respondents are crop farmers. The farms were selected randomly from these LGA. This sampling approach ensured that data came from farms that had received SRP from NIMET, thus validating the assumption that the participants had exposure to SRP. Also, all farmers had a reliance on rainfall primarily for their farming, validating all farmers need for SCF.

Table 2 Major climatic and agricultural characteristics of the three study areas

State/longitude and latitude	Agroecological zone/prevalent climate	Predicted seasonal rainfall amount 2017 (mm)	Average annual temperatures (°C)	Main crops farmed	Highlight of SCF for 2017
Katsina/12.51° N and 7.61° E	Sudan savannah/semi-arid	576	30.8	Millet, groundnut, and beans	Early onset and early cessation date, i.e., short length of growing season
Taraba/8.00° N and 10.26° E	Woodland/semi humid	965	26.8	Rice, guinea corn, groundnut	Early onset and dry spells in middle of rainy season
Edo/6.54° N and 5.90° E	Rain forest/humid	2132	24.5	Cassava, maize, yam and water melon	Extended rains of 3–8 days

Source: NIMET (2017)

The research was conducted in July, the middle of the rainy season which is also the growing season. Data collection spanned 2 weeks in the month of July. The study used a nonprobability technique and a cross-sectional survey design, which constituted a collection of data from a stratified population of farmers at a single point in time. The method is important in assessing the prevalence of a phenomenon, problem, attitude, or issue by taking a cross-section of the target population. Therefore, more than one LGA were visited in each state. A structured questionnaire was administered to farmers to collect information.

In addition, semi-structured interviews were conducted. Three people, the NIMET Heads from each of the study states were interviewed in total. The three interviews were audio recorded to acquire information on state-specific activities. Interview questions included questions on communication channels used to disseminate SCF, availability of any institutional support offered by NIMET to farmers, and opportunities and barriers to SCF usability.

The third source of data came from the existing SRP for 2017 produced by NIMET. The information from the SRP booklet combined with the findings from the semi-structured interview provided information on the activities of NIMET. While the findings from the questionnaire respondents provided information on the SCF users – the farmers.

Data Processing and Analysis

Data from the questionnaires were analyzed using Statistical Package for Social Sciences (SPSS version 23). Descriptive statistics were used to analyze data. Correlations of some quantitative items of the questionnaire were carried out also.

The semi-structured interviews were transcribed for trend and content analysis. The purpose was to detect any trends in inferences about SCF usability in Nigeria. Descriptive statistics were used for comparison purposes on variables of interest to explain phenomena. The data were analyzed using the content analysis technique. This is a systematic and replicable technique that compresses many words into fewer content categories (Hsieh and Shannon 2005). It can be a useful technique for allowing researchers to discover and describe important messages or focus of individual or group. Qualitatively, content analysis can involve any kind of analysis where communication content (speech, written text, interviews, and images) is categorized and classified.

All three interviews were transcribed from the recording device and then the responses were coded to each area of questioning. The information was then reviewed and categorized into major theme areas by highlighting key words and phrases.

Ethical Considerations

To adhere to the significance of the rule of ethics when conducting a research, the researcher obtained consent from participants to participate in this research and ensured their anonymity, confidentiality, and respect for different personalities throughout the process. The data do not contain any references to individuals. Participants were all given information sheets explaining the research and giving clear options to join or not.

Results

Characteristics of Respondents

Important demographic characteristics of respondents in the study include age, sex, and gender from the three different agro-zones in Nigeria. Sixty-eight percent of the respondents were male while the other 32% were females. Most of the respondents are between the ages of 25–64 years. Only one respondent is 65 years or older (See Fig. 4 below). Results also show that 45% of respondents farmed more than one crop while the other 55% farmed only one crop. Differences were found across the three agroecological zones in the type of crops farmed and in the number of crops mentioned per farmer. Whereas cassava, maize, yam, and water melon were the most mentioned crops farmed by Edo farmers; Taraba farmers planted mostly rice, guinea corn, and groundnut while Katsina farmers mostly planted millet, groundnut, and beans. Also, in Edo State, 17 out of 20 farmers; in Taraba, 6 out of 20 farmers; and in Katsina, only 4 farmers out of 20 farmed more than one crop.

The study found that 42% of the respondents are smallholder farmers, that is who own two/less than two hectares of land (Fig. 5 below). Farmers who own between 2 and 5 hectares of land follow closely at a percentage of 35% of respondents.

Forecast Information Farmers Need

Farmers selected all the forecast information useful to them and results to show that they are more interested in the timing of rainfall as per onset and cessation dates and

Fig. 4 Age of survey respondents

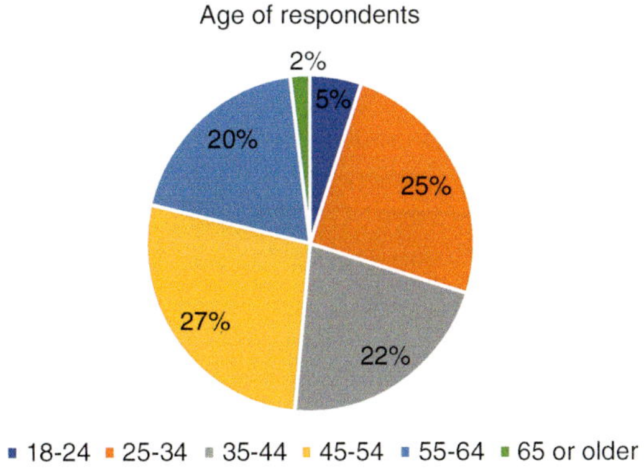

Fig. 5 Area of farm in hectares of land owned by respondents

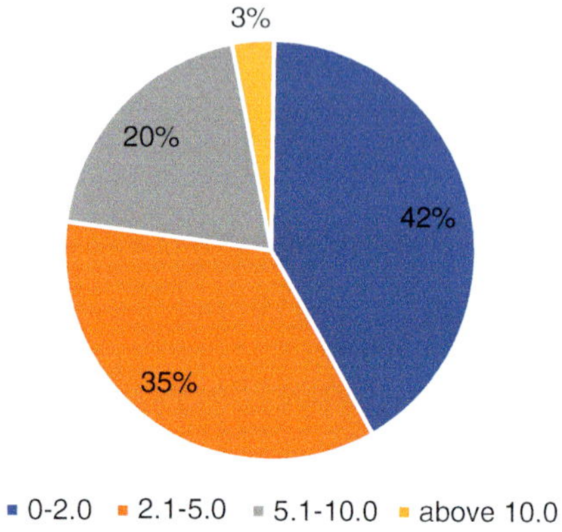

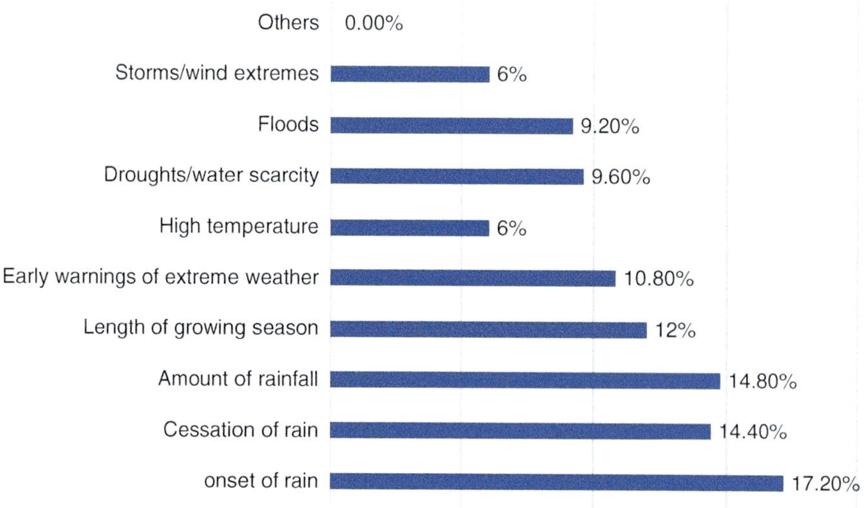

Fig. 6 Forecast information most useful to respondents

amount of rainfall than in other features of the SCF followed by explicit early warning of extreme weather which may lead to poor yield (Fig. 6). Mention of extreme events such as high temperature and wind extremes gathered the least count (15). Floods and droughts had intermediate counts relative to other forecast information with a percentage contribution of 9.6% and 6% respectively. Climate variability features were of more hierarchical importance than extreme events.

SCF help farmers make decisions on their agricultural activities. The results of this survey show that most farmers apply SCF in determining when to start planting. This response was closely followed by the decision of which crop to plant and the need to use inputs such as fertilizers (Fig. 7). The decision to plan irrigation was mentioned only 14 times showing that it is not done by most of the respondents.

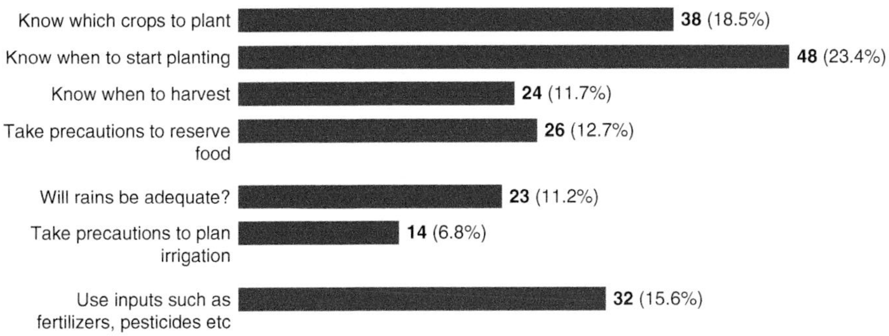

Fig. 7 Ways in which forecast is useful to farmers for making farming decisions

Table 3 Respondents decisions and adaptation strategies in response to SCF highlight for the season

State	Highlight of SCF for 2017	Decision taken	Count
Edo	Extended rains of 3–8 days	Early planting	8
		Delay planting till after extended rains	2
		Continuous farming	2
Taraba	Early onset and dry spells in middle of rainy season	Early planting	17
		Between 4–11 dry spell days recorded by all respondents	
		More water for irrigation	1
		Early fertilizer application	2
		Use of herbicides	2
		Early harvest to enable planting again in the same season	1
Katsina	Early onset and early cessation date, i.e., short length of growing season	Expansion of farm	2
		Early maturing crops planted	4
		Early planting	2
		To know what to plant	6
		To know when to plant	2

The options chosen for the question "the ways in which forecast is useful to farmers?" closely match the adaptation decisions taken in response to the SCF highlight for each state (Table 3). The most mentioned response to SCF highlight was the decision to plant early. Other decisions include the choice of what to plant, early harvesting, and fertilizer application. Respondents from Taraba recorded dry spell days experienced as predicted by the SCF. While 87% of total respondents were aware of SCF highlight, 13% said they are not aware. Out of the 87%, not all took adaptation actions based on the highlight. Seventy three percent of total respondents have already taken or plan to take the adaptation actions which are presented in Table 3 below while 14% of respondents although they had knowledge of SCF didn't make any decisions based on it. Two respondents gave answers in reconsideration of their plans after becoming aware of SCF highlight. "Now that I know, I will plan when to harvest, dry, stack and store," and the other similarly said, "Now that I know, I will store and preserve for next season." Both farmers are in Edo State where SCF highlight was of extended rains of 3–8 days. SCF highlight was also applied in crop selection. A farmer from Katsina wrote that he chose to plant beans instead of guinea corn while another recorded "I decided to plant millet."

Access to Forecast and Timeliness of Forecast

Farmers were asked questions on the ways through which they received SCF, the time they received SCF, and if they paid for SCF or not. Findings are presented in the subsections following.

Ways Through Which Farmers Receive SCF

Results from questionnaire showed that farmers receive SCF through diverse ways. Most mention was made of Radio/TV. As this was a multiple response question, farmers could choose one or more ways in which they accessed SCF. Thirty percent of respondents chose Farmers' Association, while 16% accessed SCF direct from NIMET. The lowest count was recorded for internet source (9%). Radio/TV was the highest mentioned way in which SCF was received (45%). One of the interviewees mentioned that farmers came direct to NIMET office to obtain SCF especially when SCF had not gotten to them and the season was beginning to draw near. From the interview, radio jingles were a repeatedly mentioned way of SCF dissemination.

For state-level access of SCF, mention of Radio/TV broadcasts varied among states with the highest count recorded for Katsina, closely followed by Edo State and then Taraba State (Table 4). Direct from the internet had the lowest count in all three states. Whereas 10–11 counts in Katsina and Edo States were recorded for Farmers' Association, only 3 counts were recorded for same in Taraba State.

Timing of Forecast Dissemination

Sixty six percent of respondents mention that they received SCF at the beginning of the year while 31% received at the beginning of the season and the remaining 3% late in the season. Most farmers, therefore, receive the SCF, months before the start of the rainy season and only 3% (2 farmers) received SCF late after the season had already begun.

Table 4 State-level access of SCF by respondents

State	NIMET	Internet	Farmers' Association	Radio/TV
Edo	2	5	10	13
Taraba	7	2	3	8
Katsina	4	0	11	15
Total	13	7	24	36

Value of SCF to Farmers Relative to Other Kinds of Climate Information

The perceived value of SCF is reflected in farmers' choice of NIMET as the source of their climate information and as the most reliable and most used climate information.

Source of Climate Information

This was a multiple response question to pick all applicable sources of climate information. The options included both direct sources and intermediate sources as shown in Fig. 8 below. The highest count goes to NIMET as the source of climate information. However, if investigated further, it is likely that the count is more than this for NIMET as it is not clear from the results who organizes workshops and from whom ADP receives information from. Interviewee data shows that ADP is one of the agencies who receive climate information from farmers and acting as a boundary organization they communicate information to farmers through workshops. Therefore, an overlapping in results is likely to be present since ADP can be an organizer of workshops or NIMET or any other institution. Personal communication also had the next higher count after NIMET as a source of climate information while mention was made of the internet under the other sources option. Data also showed that 82% of farmer respondents do not pay for SCF while the other 18% said they pay for SCF.

Climate Information Perceived to Be Most Reliable

Farmers chose the most reliable and useful source of climate information in this order: NIMET > traditional knowledge > personal experience. The detailed results of most reliable and used climate information; state-by-state are presented in Table 5 below. The reliance on science-based forecast did not vary considerably among

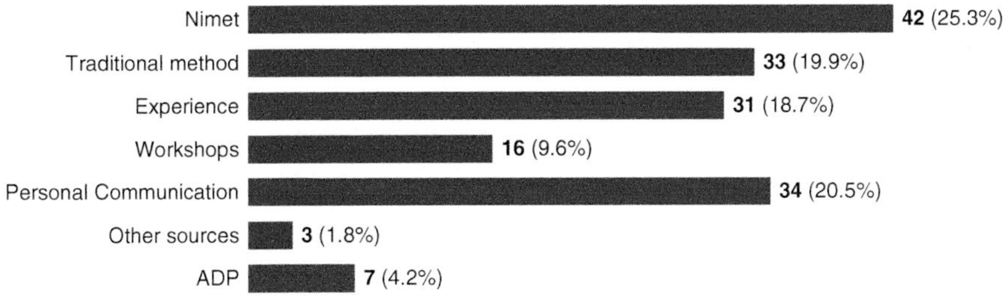

Fig. 8 Respondents' source of climate information

Table 5 Climate information most reliable and used by respondents

State	Traditional knowledge	Personal experience	NIMET	Total
Edo	6	3	11	20
Taraba	5	6	9	20
Katsina	8	3	9	20
Total	19	12	29	60

Table 6 Correlations of items representing reliability, timeliness, relevance, and usability of SCF

Research Items	1	2	3	4	5
1	Which climate information is most reliable and used by you?				
2	When do you receive seasonal rainfall prediction?	0.363**			
	Sig	0.005			
3	How useful is the forecast to you?	0.348**	0.313*		
	Sig	0.006	0.016		
4	Are you aware of the 2017 seasonal highlight prediction for your state?	0.045	0.129	−0.162	
	Sig	0.733	0.331	0.218	
5	Have you made a farming decision based on this prediction?	0.329*	0.312*	0.249	0.255*
	Sig	0.010	0.016	0.055	0.049

states. While traditional knowledge was the next reliable climate information with Edo and Katsina States leading, personal experience was the next reliable climate information in Taraba State.

Correlations of Some Questionnaire Items

Correlational analyses were used to examine the relationship between reliability, timeliness, relevance, and use of SCF using the items as shown in Table 6 below. Results indicate significant correlations between the type of climate information most reliable to the farmers and time of receipt of SCF ($r = 0.363$, $p = 0.005$, $p < 0.01$). Statistically, significant correlation occurred between science-based information from NIMET as information most reliable to the farmers and degree of usefulness of the forecast ($r = 0.348, p = 0.006, p < 0.01$). Also between the type of climate information most reliable to the farmers and farmers' decision-making in response to SCF highlight ($r = 0.329$, p $= 0.01, p < 0.05$).

Discussion

Relevance of Forecast Information to Farmers Decision-Making Needs

A consistency is found in these set of responses: the hierarchical choice of information needs from users' point of view shown in Fig. 6, ways in which forecast is useful to farmers for making farming decisions (Fig. 7), and farmer's decisions and adaptation strategies in response to SCF highlight for the season (Table 3). Farmers are most interested in onset and cessation of rains and they apply SCF information to decide majorly when to plant and which crops to plant. Adaptation strategies in response to SCF highlight of the year also reflected decisions involving timing of planting (e.g., early planting) and use of early maturing varieties.

Findings show that the respondents, due to using SCF, practiced crop selections. Instead of planting guinea corn, a farmer chose to plant beans. Another chose to plant groundnut and another gave priority to the planting of millet. Crop selection is important because different crops thrive in different climatic and soil conditions. Therefore, matching crops to existing and/or predicted climatic situations is needed. By using a simulation of farmer decision-making, another research found that a farmer using SCF has more diversified crop selections, which results in rises in average agricultural income compared to a "one-crop all the time" farmer (Gunda et al. (2017). Crop selection using SCF brings about higher productivity and ensures that climatic conditions are harnessed to the farmers' favor.

The use of early maturing varieties was mentioned by farmers as an adaptation strategy in response to SCF highlight for early cessation of rainfall. It was found to be an adaptation strategy also in addressing late-season drought and therefore reducing risk of production losses in parts of West Africa (Badu-Apraku et al. 2013) and in North-West Bangladesh (Malabayabas et al. 2014). Farmers in Kenya responded to SCF predicting early cessation of rains with similar action – using early-maturing crop varieties (Kniveton et al. 2015b). Consequently, this is an effective strategy among farmers.

Other decisions made by farmers include the timing of farming operations such as to delay planting or to carry out early harvesting. Timing of farming activities is everything and can determine whether the farmer has a good yield or not. Especially for rain-fed agriculture, too early or too late planting is a production risk. In addition to planting time, the type, amount, and timing of fertilizers and herbicides application is associated with the knowledge of rainfall since herbicides and fertilizers require the right soil temperature, air temperature, and the right amount of water (Iizumi et al. 2013). Fertilizer application must be timed in relation to rainfall as this is critical to determining the risk of gaseous loss and effectiveness of fertilizers (Iizumi et al. 2013). These results are consistent with findings on the importance of timing of planting and use of herbicides/fertilizers (Ingram et al. 2002; Hansen and Indeje 2004).

Accessibility and Timeliness of SCF

Different dissemination strategies are used to send out SCF to farmers by NIMET and farmers receive it from one or multiple sources. Since a diversity of farmers make the audience, there would be differences in information access facilities and in the time devoted to digesting the SCF. Understanding the commonest method through which information is received by farmers is important for widespread information dissemination. It can serve as an effective way to disseminate information to a large portion of the community in a shortest possible time. The commonest method found in this study is Radio/TV. Similar studies in Africa found that farmers use more of Radio/TV than printed material or internet sources (Ingram et al. 2002; Roncoli et al. 2009). That farmers reported that they obtained SCF from Farmers' Association gives credence to the role of networks and boundary organizations, Farmers' Association, and cooperatives, in communicating climate information to farmers (Cash et al. 2003; Jacobs et al. 2005). Roncoli et al. (2009) found similarly in Burkina Faso that the many networks to which farmers belong accelerate the spread of SCF and its uptake.

Timing of forecast issuance is critical to be of benefit to farmers' decision-making. Forecasts can be made, but if does not get to the users' hands in time, getting "off the shelf," it loses its relevance. Especially for agricultural purposes, timely forecasts are important so that farmers can take full advantage of weather conditions. Correlations found in this study suggests that farmers who rely on science-based information from NIMET get SCF early in the year or at the beginning of the planting season. Sixty six percent of respondents received SCF at the beginning of the year while 31% received at the beginning of the season and just 3% said they had received forecast late in the season this year. From findings, the time of forecast receipt for relevant decision-making is at the beginning of the year and to a lesser extent, beginning of the season. Similarly, Ingram et al. 2002 report that farmers in Botswana requested that SCF arrive 1–2 months before the rainy season before they make irrevocable decisions like what to plant. In that study when SCF arrived late, farmers made decisions based on the previous season's conditions without waiting on SCF.

Reliability of Farmers on SCF

The farmer is the one that translates the SCF into risk and decides what steps to take to avert unfavorable climate situation or to maximize predicted desirable seasonal climate. For a farmer to make decisions based on SCF means that the farmer considers the information to be reliable. Results show that NIMET is the most mentioned source of climate information followed by personal communication. The role of personal communication in the dissemination of SCF is exemplified from this research. In Roncoli et al. (2009), interviews conducted found that almost all respondents who received the forecasts by workshop or other means shared them with others by direct individual conversation or through village meetings and religious gatherings.

Traditional method and experience were the next sources of climate information mentioned more after NIMET and personal communication (Fig. 8 above). Although NIMET remained the chief source of climate information, respondents still reflected a tendency to seek autonomy in decision-making by the choice of traditional knowledge by 32% of respondents and personal experience by 20% of the respondents. In essence, about half of the respondent population (52%) still hold on to other kinds of knowledge. This indicates that an interplay between traditional knowledge and experience with science-based knowledge can exist. Traditional knowledge and local practices can be a barrier to usability of climate information Where traditional methods fail, farmers move on quickly to science-based information but where the methods work, farmers are not so keen on the use of science-based forecasts. Luseno et al. (2003) found in a study with Ethiopian and Kenyan pastoralists that due to the failed skill of traditional knowledge the farmers had abandoned it for science-based forecast whereas Musembi and Cheruiyot (2016) found that Kenyan farmers' reliability on traditional methods was absolute and hindered use of science-based forecasts. However, the high responses from traditional knowledge/experience from this study alongside application of science-based SCF reflects that interplay between traditional knowledge and science-based knowledge can be positive and should be seen as a complement and not as a barrier. When SCF complements traditional knowledge and experience, it can enhance and create positive synergies, thereby, enhancing the value of SCF.

On inquiring from farmers on the source of their climate information, farmers gave multiple responses for sources. Farmers consult a wide range of information sources and so this presents an opportunity for wider dissemination of SCF through these multiple sources.

Farmers were required to choose most reliable source of climate information between science-based forecast from NIMET, traditional knowledge, and experience. Science-based forecast from NIMET had the highest count in all three states with almost half of the responses (29 out of 60–48%) (Table 5). Traditional knowledge followed with 32% of responses and personal experience with 20% of total responses. From the statistical analysis carried out, respondents who rely on science-based information from NIMET and the information to be useful, get SCF early at timing relevant for decision-making, and make decisions based on SCF.

We summarize the opportunities that enhance and the barriers that constrain SCF usability in Nigeria below from farmers' responses and interviewee responses too.

Opportunities for Usability of SCF

This research found that timeliness as an aspect of information is an opportunity for increased usability of SCF. Timing is a core element of farmers' decisions and adaptation actions. Also, significant positive relationship was found between relevance, reliability, accessibility, and timeliness in the usability of SCF by farmers in decision-making as described in sections "Relevance of Forecast Information to Farmers Decision-Making Needs," "Accessibility and Timeliness of SCF," and "Reliability of Farmers on SCF."

Evaluation is important for testing the skill of forecast. NIMET's SCF contained evaluation of the previous year's forecast, therefore the skill of forecast can be checked by farmers. This could lend itself to increasing reliability on SCF.

Floods due to extended rains are destroying farm investments and causing more awareness of the shift in mean statistics of historical climate. The 2012 floods in Nigeria served as a trigger for the use of SCF. "People used to say, 'If it is not sun, it's rain so I don't need a forecast' but increase in floods (extreme events) are making more people aware of the need for forecasts" (statement by interviewee). Collaboration between institutions was stated by interviewees as an opportunity to enhance usability in the aspects of awareness creation and information dissemination. An institutional collaboration that enhanced usability was mentioned by one of the interviewees and restated by the others. Last year, downscaled SCF was produced for Katsina State to all LGAs containing more details at the request of the state government and this led to a high yield. Lemos et al. (2012) mention information customization as a factor that widens the usability gap. Institutional support such as these take SCF further faster to the target audience with relevant components for their decision-making.

All interviewees mention that they have Memorandum of Understanding (MoU) and existing relationships with Government Agencies such as Ministry of Agriculture, ADP, and Ministry of Water Resources. In particular, Katsina State Government collaboration with NIMET to produce a downscaled SCF led to increased crop yield in 2016. However, besides this, there is no evidence of any positive results that have come out of MoU between NIMET and other institutions. These relationships and collaborations can, if strengthened, critically accelerate usability of SCF.

Constraints to Usability of SCF

Scientist–stakeholder interactions as a key component in enhancing usability of SCF has been mentioned in key literature (Cash et al. 2002; Lemos et al. 2012, 2014). In the case of this research, from interviewees comments, no evidence of feedback from farmers was found. This is a constraint to information Fit development and could hinder usability of SCF. Fit, rather than operating independently, is interconnected with interaction between stakeholder and scientists (Lemos et al. 2012). Fit can be improved by strengthening the interaction levels between NIMET and farmers through government institutions and the networks people belong too.

SCF is produced in English language only and not in the local dialects of farmers. Interviewees confirmed that translation is made into local dialect or in Nigerian pidgin when disseminating SCF via radio. This could account for the high response rate of access to SCF through radio/TV. Interviewee response also raised the concern about farmers paying for SCF. From the study, 18% of farmer respondents say they pay for forecast while the majority does not pay. Promoting the uptake of climate information in decision-making requires not only an appreciation of SCF characteristics but an understanding of the economic and social factors that could play a role in deterring uptake. Although the SRP booklet is free, middlemen sometimes hoard it and sell it at a price. In response to these, NIMET plans to disseminate SRP directly through the weather stations and Agro-met centers scattered across the states.

Conclusion

Summary of the Research

Climate variability affects farmers, especially those who rely on rain-fed agriculture. The emergence of skillful and credible SCF provides farmers with the ability to make farming decisions to avert unfavorable climate conditions or to maximize good climate conditions.

In this research, investigation was made on the usability of SCF provided by NIMET, in the form of seasonal rainfall prediction. The study focused on the *Fit* of SCF to farmers' decision-making needs. The aspects of relevance, accessibility, timeliness, and reliability of the SCF were analyzed to find out the opportunities for and barriers against usability of SCF in Nigeria. The study areas were three states – Edo, Katsina, and Taraba chosen from three distinct agroecological zones of Nigeria. By using a questionnaire survey, 20 respondents from each state were asked questions related to relevance, accessibility, timeliness, and reliability of SCF relative to other sources of climate information. Data was also collected through semi-structured interviews conducted for NIMET Heads of the three states.

Farmer respondents have multiple sources of climate information with NIMET and personal communication taking the lead. As to the routes of receipt of SCF, radio/TV was the most mentioned route. Responses of Farmers' Association as the access route of SCF followed radio/TV closely. This shows the importance of the many networks that farmers belong to in enhancing the fit of SCF information. Through sharing of information by personal communication with other farmers and Farmers' Association meetings, dissemination of SCF can go further faster for relevant decision-making at optimal time. The role of other institutions such as ADP and the state governments was mentioned in the dissemination of SCF and as a source of SCF.

SCF match respondents' information needs and is used by farmers for decision-making. The prominent decisions taken include that of when to plant and what to plant, and the use of fertilizers and herbicides while the forecast information most useful to farmers are the onset of rain, cessation of rain, and amount of rainfall. Adaptation actions taken by farmers or decisions made include early planting, planting of early maturing crops, delay in planting, and the use of fertilizer and herbicides in response to the respective SCF highlight they received and based on the crops they farm. We found that SCF was applied to crop production planning and production input planning and to farming operations. Timing is central to these farming decisions taken. Therefore, to obtain maximum value of SCF, farmers need to access SCF on time before making decisions. Some of these decisions need to be taken before the growing season. Time for forecast receipt for relevant decision-making was found to be beginning of the year before the start of the rainy season and to a lesser extent beginning of the season.

The results show that access to relevant and timely climate information enhances reliability on SCF as a precondition for usability of SCF in decision-making among farmers in Nigeria. This research has established that farmers across the three agroecological zones of Nigeria receive SCF through various sources and that those who relied on science-based information from NIMET find it useful and apply it in decision-making. The positive relationship between interplay and interaction in altering Fit was also indicated from the interviews with NIMET Heads. The study agrees with Kirchhoff (2010) and Lemos et al. (2012) that rather than operating independently, Fit, interplay, and interaction are interconnected and shape each other to widen or constrain the transition gap to the usability of climate information.

Significance of Study

The findings are relevant for policy-making in Nigeria in support of agricultural development and for adaptation to climate variability. This study highlights the importance of knowing the key decisions farmers make and in turn the kind of climate information which can best support these decisions. The research identifies the pathways where SCF can best support adaptation among farmers. The study is useful to NIMET in understanding how to improve SCF to fit users' needs for decision-making. Also, knowing the sort of decisions farmers make in response to SCF is an indication of areas where value addition can be done for increased SCF usability. NIMET can also use information on sources and access points of SCF to more effectively disseminate SCF to farmers.

A unique output of the research is the focus on the Fit component of the conceptual model for usability shown in Fig. 1 above. Theoretically, there is an extension of the Fit usability model by highlighting relevance, accessibility, timeliness, and reliability. This evaluation of usability of SCF can be scaled up for use in several contexts. Another significance is the Nigerian demographic context on which the research is based. This explains the subject of SCF usability from a new paradigm in the literature.

Limitations of Study

One limitation of this study is the use of nonprobability sampling technique. Farmers were picked from selected LGAs which NIMET interviewees confirmed had access to SCF. The composite nature of the respondents, from different sites/LGAs with different crops they farm, different socioeconomic status may confound the study analyses. However, all respondents had several characteristics in common including reliance on rain for agriculture, all were farmers of food crops from food-producing LGA close to the provincial capital of the state.

Suggestions for Further Study

There are unexplored factors that would influence SCF usability among farmers in Nigeria. For instance, the relationship between crop type farmed and usability of

SCF was not explored in this study. Also, it would be interesting to conduct this research in more states in Nigeria to include other agroecological zones. This presents an opportunity to scale up the research and could uncover more enhancers and barriers to climate information usability.

References

Agrawal A, Pern N (2008) Climate Adaptation, Local Institutions and Rural Livelihoods. IFPRI Working Paper W08I-6. School of Natural Resources and Environment. Michigan, USA: University of Michigan

Allen E, Stephens JC (2016) Enhancing the usability of climate information and models through stakeholder engagement. In: Information, models, and sustainability. Springer International Publishing, Cham, pp 121–135

Badu-Apraku B, Oyekunle M, Menkir A, Obeng-Antwi K, Yallou CG, Usman IS, Alidu H (2013) Comparative performance of early-maturing maize cultivars developed in three eras under drought stress and well-watered environments in West Africa. Crop Sci 53(4):1298–1311

Bellemare MF (2015) Rising food prices, food price volatility, and social unrest. Am J Agric Econ 97(1):1–21

Bevan N (2009) Extending quality in use to provide a framework for usability measurement. In International Conference on Human Centered Design (pp. 13–22). Springer, Berlin, Heidelberg

Blench R (1999) Seasonal climate forecasting: who can use it and how should it be disseminated? Nat Resour Perspect 47:1–4

Broad K, Agrawala S (2000) The Ethiopia food crisis – uses and limits of climate forecasts. Science 289:1693–1694

Buizer J, Jacobs K, Cash D (2016) Making short-term climate forecasts useful: linking science and action. Proc Natl Acad Sci 113(17):4597–4602

Cash D, Clark W, Alcock F, Dickson N, Eckley N, Jager J (2002) Salience, credibility, legitimacy and boundaries: linking research, assessment and decision making. Harvard University, John F. Kennedy School of Government

Cash DW, Clark WC, Alcock F, Dickson NM, Eckley N, Guston DH, Jäger J, Mitchell RB (2003) Knowledge systems for sustainable development. Proc Natl Acad Sci 100(14):8086–8091

Cash DW, Borck JC, Patt AG (2006) Countering the loading-dock approach to linking science and decision making: comparative analysis of El Niño/Southern Oscillation (ENSO) forecasting systems. Sci Technol Hum Values 31(4):465–494

Collier P, Dercon S (2014) African agriculture in 50 years: smallholders in a rapidly changing world? World Dev 63:92–101

Dessai S, Hulme M, Lempert R, Pielke R (2009) Climate prediction: A limit to adaptation? In W. Adger, I. Lorenzoni, K. O'Brien (Eds.), Adapting to Climate Change: Thresholds, Values, Governance (pp. 64–78). Cambridge: Cambridge University Press. https://doi.org/10.1017/CBO9780511596667.006

Dilling L, Lemos MC (2011) Creating usable science: opportunities and constraints for climate knowledge use and their implications for science policy. Glob Environ Chang 21(2):680–689

Funk C, Harrison L, Shukla S, Korecha D, Magadzire T, Husak G, Galu G, Hoell A (2016) Assessing the contributions of local and east Pacific warming to the 2015 droughts in Ethiopia and Southern Africa. Bull Am Meteorol Soc 97(12):575–580

Gunda T, Bazuin JT, Nay J, Yeung KL (2017) Impact of seasonal forecast use on agricultural income in a system with varying crop costs and returns: an empirically-grounded simulation. Environ Res Lett 12(3):034001

Hansen JW (2002) Realizing the potential benefits of climate prediction to agriculture: issues, approaches, challenges. Agric Syst 74(3):309–330

Hansen JW, Indeje M (2004) Linking dynamic seasonal climate forecasts with crop simulation for maize yield prediction in semi-arid Kenya. Agric For Meteorol 125(1):143–157

Hsieh HF, Shannon SE (2005) Three approaches to qualitative content analysis. Qual Health Res 15(9):1277–1288

Iizumi T, Sakuma H, Yokozawa M, Luo JJ, Challinor AJ, Brown ME, Sakurai G, Yamagata T (2013) Prediction of seasonal climate-induced variations in global food production. Nat Clim Chang 3:904–908

Ingram KT, Roncoli MC, Kirshen PH (2002) Opportunities and constraints for farmers in West Africa to use seasonal precipitation forecasts with Burkina Faso as a case study. Agric Syst 74:331–349

Intergovernmental Panel on Climate Change (IPCC) (2012) Managing the risks of extreme events and disasters to advance climate change adaptation. In: Field CB, Barros V, Stocker TF et al (eds) A special report of working groups I and II of the Intergovernmental Panel on Climate Change. Cambridge University Press, Cambridge

International Standard Organization (1997) ISO 9241-11: ergonomic requirements for office work with visual display terminals. Part II guidelines for specifying and measuring usability. International Standard Organization, Geneva

Jacobs K, Garfin G, Lenart M (2005) More than just talk: connecting science and decision-making. Environ Sci Policy Sustain Dev 47(9):6–21

Kirchhoff CJ, Lemos MC, Dessai S (2013) Actionable knowledge for environmental decision making: broadening the usability of climate science. Annu Rev Environ Resour 38:393

Kirchhoff CJ, Lemos MC, Kalafatis S (2015) Creating synergy with boundary chains: can they improve usability of climate information? Clim Risk Manag 9:77–85

Kniveton D, Visman E, Daron J, Mead N, Leathes B (2015a) A practical guide on how weather and climate information can support livelihood and local government decision making: an example from the adaptation consortium in Kenya. Draft

Kniveton D, Visman E, Tall A, Diop M, Ewbank R, Njoroge E, Pearson L (2015b) Dealing with uncertainty: integrating local and scientific knowledge of the climate and weather. Disasters 39 (s1):s35

Lemos MC (2008) What influences innovation adoption by water managers? Climate information use in Brazil and the US. J Am Water Resour Assoc 44:1388–1396

Lemos MC, Kirchhoff CJ, Ramprasad V (2012) Narrowing the climate information usability gap. Nat Clim Chang 2(11):789–794

Lemos MC, Kirchhoff CJ, Kalafatis SE, Scavia D, Rood RB (2014) Moving climate information off the shelf: boundary chains and the role of RISAs as adaptive organizations. Weather Clim Soc 6(2):273–285

Luseno WK, McPeak JG, Barrett CB, Little PD, Gebru G (2003) Assessing the value of climate forecast information for pastoralists: evidence from southern Ethiopia and northern Kenya. World Dev 31(9):1477–1494

Malabayabas AJB, Kajisa K, Mazid MA, Palis FG, Johnson DE (2014) Impacts of direct-seeded and early-maturing varieties of rice on mitigating seasonal hunger for farming communities in northwest Bangladesh. Int J Agric Sustain 12(4):459–470

Meza FJ, Hansen JW, Osgood D (2008) Economic value of seasonal climate forecasts for agriculture: review of ex-ante assessments and recommendations for future research. J Appl Meteorol Climatol 47(5):1269–1286

Millner A, Washington R (2011) What determines perceived value of seasonal climate forecasts? A theoretical analysis. Glob Environ Chang 21(1):209–218

Mina CD, Reyes CM (2009) Profitable use of SCF in a policy context: the case of rice stockholding in the Philippines (No. 2009-09). PIDS discussion paper series

Mpandeli S, Maponya P, Liphadzi S, Backeberg G (2017) Indigenous knowledge systems for managing climate change in South Africa. Indigenous knowledge systems and climate change management in Africa, p 255

Musembi DK, Cheruiyot HK (2016) A Case for validation of indigenous knowledge in forecasting rainfall among the kamba community of makueni county. Lower Eastern Kenya. J. Meteorol 9

Nigerian Meteorological Agency NIMET (2017) http://www.nimet.gov.ng

Otitoju MA, Enete AA (2016) Climate change adaptation: uncovering constraints to the use of adaptation strategies among food crop farmers in south-west, Nigeria using principal component analysis (PCA). Cogent Food Agric 2(1):1178692

Patt A, Gwata C (2002) Effective seasonal climate forecast applications: examining constraints for subsistence farmers in Zimbabwe. Glob Environ Chang 12(3):185–195

Prokopy LS, Carlton JS, Haigh T, Lemos MC, Mase AS, Widhalm M (2017) Useful to usable: developing usable climate science for agriculture. Clim Risk Manag 15:1–7

Ranger N (2013) Topic guide. Adaptation: Decision-making under uncertainty (Evidence on Demand Report). London: Department for International Development

Roncoli C, Jost C, Kirshen P, Sanon M, Ingram KT, eaWoodin M, Somé L, Ouattara F, Sanfo BJ, Sia C, Yaka P (2009) From accessing to assessing forecasts: an end-to-end study of participatory climate forecast dissemination in Burkina Faso (West Africa). Clim Chang 92(3–4):433

Roudier P, Muller B, d'Aquino P, Roncoli C, Soumaré MA, Batté L, Sultan B (2014) The role of climate forecasts in smallholder agriculture: lessons from participatory research in two communities in Senegal. Clim Risk Manag 2:42–55

Troccoli A (2010) Seasonal climate forecasting. Meteorol Appl 17(3):251–268

Vogel C, O'Brien K (2006) Who can eat information? Examining the effectiveness of seasonal climate forecasts and regional climate-risk management strategies. Clim Res 33(1):111–122

Von Hippel E, Katz R (2002) Shifting innovation to users via toolkits. Manag Sci 48(7):821–833

Weisheimer A, Palmer TN (2014) On the reliability of seasonal climate forecasts. J R Soc Interface 11(96):20131162

World Bank Group (2017) Average monthly rainfall for Nigeria from 1991–2015. Climatic Research Unit (CRU) dataset from Climate Change Knowledge Portal. http://sdwebx.worldbank.org/climateportal/index.cfm?page=country_historical_climate&ThisCCode=NGA

World Meteorological Organization (2007) Climate information for adaptation and development needs. WMO-No. 1025

Asphalt Making Potential of Pyrolytic Bitumen from Waste Rubber Tyres: An Adaptive Measure to Climate Change

122

J. G. Akinbomi, S. O. Asifat, A. Ajao, and O. Oladeji

Contents

Introduction 2516
Building Resilience to Climate Change 2517
Research Gap and Objective 2518
Significance of the Study 2519
Scope of the Study 2519
Limitation of the Study 2520
Theoretical Framework of the Study 2520
Methodology 2521
Experimental Materials 2521
Experimental Procedure 2523
Bitumen Extraction and Asphalt Concrete Production 2524
Results and Discussions 2524
Results Presentation 2524
Statistical Analysis of the Results 2526
Conclusion 2527
Cross-References 2527
References 2528

J. G. Akinbomi (✉)
Department of Chemical and Polymer Engineering, Faculty of Engineering, Lagos State University, Lagos, Nigeria
e-mail: Julius.Akinbomi@lasu.edu.ng

S. O. Asifat
Lopek Engineering and Construction Ltd, Lagos, Nigeria
e-mail: semitecivil@gmail.com

A. Ajao · O. Oladeji
Amsol Bio Company, Lagos, Nigeria
e-mail: ajaoayodele1980@gmail.com; oladejioluwsola5000@gmail.com

W. Leal Filho (ed.), *Handbook of Climate Change Resilience*,
https://doi.org/10.1007/978-3-319-93336-8_145

Abstract

The aftermath of man's continuous depletion of the planet's natural resource, as well as his inappropriate waste disposal, has manifested over time into severe weather and environmental conditions, which could snowball into an epic catastrophe, if not checked. Therefore, the aim of this study is to develop a viable gas-fired pyrolysis process for extracting bitumen from waste rubber tyres. The pyrolysis system was composed of gas-fired furnace, heavy oil condenser, two cyclones for light oil condensation, scrubber for gas cleaning, and gas storage bag. The extracted bitumen was obtained from the pyrolysis of 9 kg of shredded waste rubber tyres. The bitumen was tested for its asphalt-making potential to verify its suitability as a replacement for the petroleum bitumen commonly used in making asphalt. The performance tests on the asphalt concrete indicated values of 3150 N, 2.6 mm, 3.1%, and 77.4% for Marshall stability, flow value, percent air void in the mixture, and voids filled with bitumen, respectively. When compared with the standard specification of the Nigerian asphalt concrete, all the three properties, except Marshall stability, passed the requirement. In general, results from the study indicated that on further quality improvement, waste tyre bitumen could be used as a substitute to petroleum bitumen.

Keywords

Asphalt · Resource-conservation · Wastes · Climate change · Pyrolytic bitumen

Introduction

It is obvious that human activities, which result in increased concentration of atmospheric greenhouse gases, are mainly responsible for the climate change being experienced globally (Azar et al. 2006; Cook et al. 2013; Dodman 2009). The increased concentration of atmospheric greenhouse gases has caused more reflected infrared heat radiation from the earth to be trapped by the greenhouse gases, resulting in significant global earth warming. The increase in global average temperature gives rise to the melting of polar ice sheets and glaciers as well as expansion of water molecules in the seas, rivers, and oceans, which eventually lead to increased water level. These effects, when combined with strong hurricane, usually result in flooding and erosion of infrastructure in low-lying coastal areas. In some regions, the increased global temperature causes extreme heat waves, severe drought, famine, and wild fires which endanger biodiversity.

The disastrous effects of climate change, as illustrated in Fig. 1, usually impose heavy costs on human environment, society, and economy. For example, when floods, extreme heat, as well as, wild fires occur; property and infrastructure are damaged, while there may also be widespread water- and heat-related diseases (Epstein 2005; Haines et al. 2006; Patz et al. 2005). The cost implications of these effects on human environment, society, and economy cannot be understated. It is

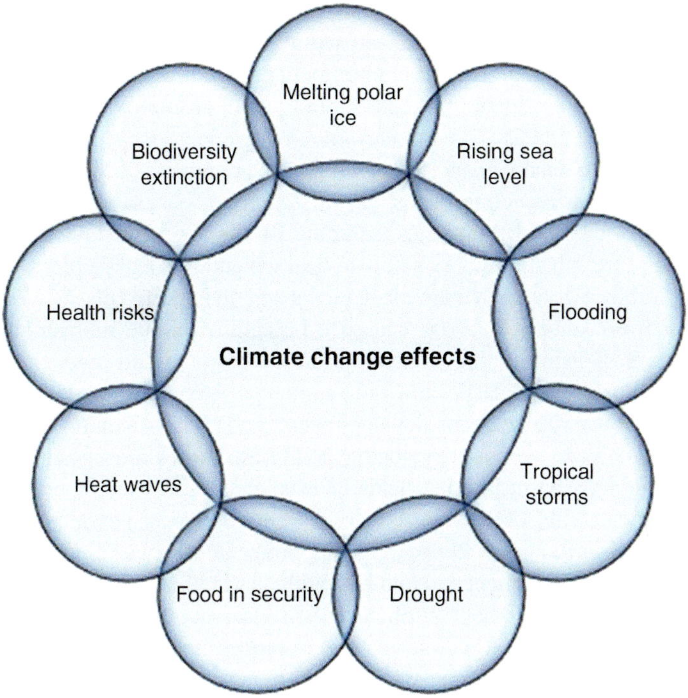

Fig. 1 Climate change effects

therefore necessary to address the climate change phenomenon by building climate change resilience in the social, economic, and environmental systems (Adger et al. 2005).

Building Resilience to Climate Change

Climate change resilience has been defined as the capacity of a system (social, economic, or environmental) to anticipate, prepare for, withstand, and respond to hazardous events or disturbances related to climate change (Folke 2006; Holling 1973; IPCC 2014; Satterthwaite 2013; Sharifi and Yamagamata 2016; Smit and Wandel 2006). According to Holling (1973), natural ecosystem would act to mitigate any climate-induced disturbance to its ecological equilibrium, and also attempt to return to its prior state of equilibrium. However, the socio-ecological systems could also stabilize around a multitude of possible states when subjected to disturbances induced by climate change. The behavior of socio-ecological systems to either return to its previous state or to adapt to a new state could be managed, integrated, and harnessed for improved sustainability of the systems.

Climate change resilience can be attained through various approaches, actions, and strategies that reduce climate change impacts, including, among others, efficient waste management system. Greenhouse gas emissions can be drastically reduced by efficient management of huge amount of waste generated through human activities. The waste is managed to protect the environment and the health of the people, while at the same time provide valuable materials for the economic improvement of the society. Waste management practices are not uniform among countries. Although most developed nations have made giant strides in the development of waste management technologies such as recycling plants, gasification, pyrolysis, sanitary landfill, anaerobic digestion, some of the countries including Britain, the United States, and Japan, among others, are still struggling on how to deal with nonbiodegradable waste such as plastic and rubber wastes. For example, China's ban on imported plastic wastes which took effect from January 1, 2018, caused a buildup of plastic rubbish at recycling plants around Britain as the country used to ship around two thirds (about 500,000 tons) of its used plastics to China annually (Telegraph-reporters 2018).

Regarding waste management, developing nations still have a number of issues that need to be resolved. In most developing countries, wastes are usually dumped in unsanitary landfills, burnt in open fields or even thrown in water channels, thereby causing nonavailability of clean water, untold environmental pollution and adverse health effects (Murillo et al. 2006; Williams et al. 1990). In addition to impacting negatively on the environment and human health, land filling, littering, and burning of wastes are activities that do not contribute to the conservation of natural resources used in the production of the wastes. Since natural resources are exhaustible in nature, their quantities are declining gradually with their exploitation. Besides, the financial and environmental costs of extracting the natural resources from their underground deposits are alarming. It is, therefore, essential to preserve the natural resources, especially, when there are alternative materials that can be used as substitutes for them.

Research Gap and Objective

In view of the aforementioned concerns, experts are exploring the application of wastes, including rubber tyre wastes, as replacements for the depleting natural resources. Many researchers have explored the potentials of using waste rubber tyres for different applications including, for artificial reefs, as aggregates in asphaltic concrete, as a fuel in cement kilns, and for the production of activated carbon (Corinaldesi et al. 2011; Corinaldesi and Moriconi 2004; Eldin and Senouci 1999; Lee and Park 1993; Presti 2013; Teng et al. 1995; Topcu 1995; Trouzine et al. 2012). Besides, value-added products such as fuel gases, liquid oil, and carbon black have also been recovered from waste materials through pyrolysis process (Gungor et al. 2015; Kalargaris and Tian 2017; Zabaniotou and Stavropoulos 2003). It is a process that has since been employed as an environmental friendly tool for treatment and adding value to waste. Although, many researchers have worked on the possibilities of using waste rubber tyres for various purposes, there is still dearth of adequate literature on the application of extracted bitumen from waste tyres for asphalt mix production. The study aims to extend the existing knowledge on the usage of waste-based pyrolytic bitumen as a substitute for the petroleum-based bitumen. The focus in the study is, therefore, the development of a viable gas-fired pyrolysis process for bitumen production from waste rubber tyres, as a way of building climate change resilience by the socioeconomic system in Nigeria.

Significance of the Study

The significance of the study could be related to the benefits associated with the production of pyrolytic bitumen from waste rubber tyres for asphalt concrete. Besides its benefit of being used as an instrument for conserving the earth's natural resources, in this case, fossil fuels, valuable energy resources, including combustible gas, fuel oil, and activated charcoal, among others, could also be got from the process. Moreover, the study will be of great importance to relevant agencies that are interested in making policies to deal with problems associated with climate change. For example, in Nigeria, apart from the challenge of waste disposal facing the country, there is also the problem of acute power shortage due to the vulnerability of the sources of electricity to climate change and sabotage. There are two main types of power generating plant systems operating in Nigeria, including hydroelectric and fossil fuel power plants (Oyewo et al. 2018). Nigeria needs to reduce her dependence on fossil fuels, as this can lead to reduction in greenhouse gas emission. In the meantime, climate change effects, including shifting raining patterns and droughts, often affect the power generation from the hydroelectric power plant as the climate change effects affect the volume of water in the dam. Therefore, in order to prevent climate change from compromising the practicability of using hydropower as an energy source, other sources of power generation need to be explored. A gas-fired pyrolysis process could be designed for households or community for recovering valuable resources (bitumen, diesel oil, gasoline, gas for power production, and activated carbon) from wastes. The study, therefore, intends to discourage people from landfilling or ocean dumping of waste rubber tyres, but instead embraces the technique of resource recovery from waste rubber tyres.

Scope of the Study

The study intends to develop a pilot scale pyrolysis reactor system that would be used for the production of bitumen from waste rubber tyres. The activities involved in the study included collecting waste rubber tyres from Nigerian dumpsites, fabricating the pilot scale reactor system using local materials, initiating pyrolysis process, extracting bitumen from the heavy oil produced, using bitumen for asphalt mix production, and conducting performance test on the asphalt mix. Although, during the pyrolysis of the waste tyres, the gas produced was used in a gas generator to power 80 Watt electric bulb for 3 hours, the scope of the study did not cover the gas and char produced, material and energy balance, air pollution control, as well as carbon footing of the pyrolysis process. All these activities will be carried out in future studies.

Limitation of the Study

The most obvious limitation of the study was the inefficient design of the air-fuel intake port which regulated the ratio of air to fuel that would produce desired burning flame in the gas-fired furnace. Intermittently, large yellow flame was observed during the experiment, which might be responsible for loss of valuable time and heat energy. This clearly showed that the reactor design was not efficient and that it must be improved upon in future related research.

Theoretical Framework of the Study

The concept of utilization of pyrolytic bitumen from waste rubber tyres in asphalt concrete mix is based on the premise that the application of the pyrolytic bitumen will prevent the excessive use of petroleum bitumen for the same purpose, thereby, serving as a means of conserving the earth's natural resources, in addition to being an environment-friendly approach of waste management. Generally, rubber tyres, such as car and truck tyres, are made of multiple polymeric materials, including natural rubber (14–27 wt. %), synthetic rubber (in the form of styrene-butadiene rubber and polybutadiene) (14–27 wt. %), fillers (carbon black and silica) (26–28 wt.%), plasticizers (oil and resins), chemical additives (sulfur) (5–6 wt. %), reinforcement metal (16.5–25 wt. %), and textiles (5.5 wt. %) (Danon and Gorgens 2015). Natural rubber from rubber trees provides high mechanical resistance and thermal stability in tyres while synthetic rubber, which is a petrochemical by-product, enables tyres to resist deformation at high temperature. Fillers provide strength and distinctive black color to tyres. The plasticizers enhance durability and rolling resistance of tyres; the presence of sulfur improves the tyres' flexibility and heat resistance, while reinforcement metal and textiles provide high degree of rigidity and strength to tyres (Williams and Besler 1995).

The polymeric materials in tyres do not decompose easily, thereby making rubber tyres durable and nonbiodegradable. The nonbiodegradability of rubber tyres, however, poses great challenge to the disposal of the spent rubber tyres through landfilling and ocean dumping of the rubber tyres (Adhikari et al. 2000). Meanwhile, other properties such as high heating value and volatile content of rubber tyres make pyrolysis process suitable as a means of disposing spent rubber tyres. Tyre pyrolysis is a clean process with little emission when performed properly, and could therefore be used as a technique for building resilience against climate change (Aylón et al. 2010; Bandyopadhyay et al. 2008). Pyrolysis is a process that involves heating a material to a temperature range of 250–500 °C, in the absence of oxygen, so that the material can be volatilized and decomposed to oil, gas, and char (Martínez et al. 2013). The yields, distribution, and physical and chemical properties of pyrolytic products (noncondensable gas, condensable liquid oil, and char) are influenced by pyrolysis conditions including temperature, heating rate, particle size, residence time, feedstock composition, carrier gas type and flow rate, presence of catalysts, reactor type, and atmospheric pressure (Teng et al. 1995).

Temperature is an important process variable in pyrolysis since it controls the cracking reaction of the polymer chain molecules. When temperature in a rubber tyre increases during pyrolysis, the vibration of the molecules inside the rubber tyre will be greater and the molecules tend to evaporate away from the surface of the tyre. Some of the tyre molecules vaporize as noncondensable gas, while other components of the vaporized molecules are condensed into liquid oil. The nonvolatile components of the pyrolyltic products remain in the reactor as solid char. The pyrolytic products can be used in various applications such as fuels, activated carbon, bitumen, and asphalt concrete productions, among others. Bitumen is a by-product of fractional distillation of crude petroleum. It is used as a binder in asphalt concrete for the construction of flexible pavements and also helps to improve the strength and life of road pavement. The asphalt concrete is a controlled mixture of bitumen, coarse sand, fine sand and filler. The coarse aggregates are the materials retained on an 8-mesh ASTM sieve. The function of the coarse aggregates is to give stability to the asphalt concrete. Fine aggregates are materials between 8-mesh and 200-mesh. They add to the stability of the asphalt concrete and reduce voids in the coarse aggregates. Fillers are the materials passing a 200-mesh ASTM sieve. They act as the final void filling in asphalt concrete (Bunger et al. 1979).

Methodology

The research methodology was based on the research objectives of the study, which include developing a viable gas-fired pyrolysis process for extracting bitumen from waste rubber tyres, extracting bitumen from the pyrolytic oil mixture to make asphaltic concrete, and testing the asphalt making potential of the extracted bitumen. The time frame for the research was a period of 6 months from August 2017 to January 2018. The period covered the time for feedstock and material collection (1 month), design and fabrication of pyrolysis system (3 months), designing and conducting the experiment (1 month), asphalt mix production, as well as quality test analysis of products (1 month).

Experimental Materials

The pyrolysis system comprised of gas-fired furnace, pyrolysis reactor, heavy oil condenser, two cyclones for light oil condensation, scrubber for gas cleaning, and gas storage bag (Fig. 2). The pyrolysis reactor was a cylindrical steel vessel with thickness, internal diameter and capacity of 12 mm, 250 mm, and 25 kg of shredded polymer, respectively. The reactor vessel was closed with two pairs of flanges (top and bottom). There was a hole in the center of the reactor vessel which served as the gas exit. The gas-fired furnace was made in such a way that it would provide uniform heat to the pyrolysis reactor. Solar-power temperature sensor was inserted in the reactor vessel to measure the temperature in the pyrolysis reactor (Fig. 3). Regarding the feedstock, waste rubber tyres were the only raw material used for the pyrolysis

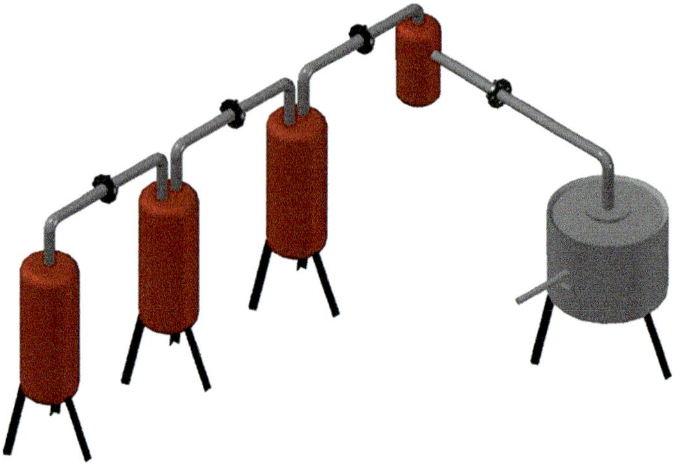

Fig. 2 Pyrolysis system setup

Fig. 3 Gas-fired furnace with attached solar-powered temperature sensor

process during the research work. The waste tyre samples were collected from dumpsites around Lagos State University, Ojo, Lagos, Nigeria. Reinforced steel was removed from the waste tyres before they were shredded manually into chips, washed with clean water, and then spread under the sun for drying. Chemical balance, measuring cylinder, and stop clock were also used during the pyrolysis experiment for measuring feedstock weight, cooling water volume, and progressing reaction time, respectively. Empty containers were also prepared for the collection of condensed liquid oil and char.

Experimental Procedure

The main reactor containing 9 kg of the dried shredded tyres was placed inside liquefied petroleum (LPG) gas-fired furnace for batch pyrolysis process. Before starting the pyrolysis process, Between 20 and 30 litres of clean water were measured into each of the two cyclones used for light oil condensation. The 9 kg of shredded tyres was pyrolyzed at a temperature range between 236 °C and 488 °C. The temperature was maintained for about 5 hours (residence time) at 488 °C. At the end of the pyrolysis experiment, the noncondensable gas was collected in a gas storage bag (Fig. 4) and utilized for the lighting of 80 Watt electric bulb. The condensable liquid oil was collected in two different containers; the first container was used for heavy oil collection, while the second was used for light oil collection. About 4 liters of heavy oil and 1 liter of light oil were collected. The furnace was left to cool down to room temperature (25 °C) before the char left in the reactor vessel was removed.

Fig. 4 A gas storage bag filled with noncondensable pyrolysis gas

Bitumen Extraction and Asphalt Concrete Production

Extraction of bitumen from the pyrolytic heavy oil was made possible by the addition of concentrated sulfuric acid (25 cl) to the heavy oil (2 l). The acid was required to heat up the heavy oil and speed up the rate of the chemical reaction so that bitumen could settle down from the oil, making filtration possible. The mixture was then left for about 48 hours to allow the reaction to take place. Part of the extracted bitumen was sent to the MEDSON Nigeria Limited for laboratory analysis and quality test. The other part of the extracted bitumen was also sent to the same company for asphalt concrete production. Quality test was performed on the asphalt concrete to determine the viability of using the extracted bitumen for making road asphalt design.

Results and Discussions

Results Presentation

The results obtained from the quality tests performed on the extracted bitumen and the asphalt concrete are presented in Tables 1 and 2, while the results of the physico-chemical analysis performed on the light oil got from the pyrolysis of the waste rubber tyres are presented in Table 3. Comparison of the results of the properties of the extracted bitumen with standard specifications by Federal Ministry of Works and Housing in Nigeria (FMW 2007) indicated that all the values of the extracted bitumen properties (specific gravity, penetration, flash point, softening point, ductility, loss of heating, and solubility in trichloroethylene) fell within the specified standard values for 80–100 bitumen grades. However, it could be observed that the penetration value of the extracted bitumen (93 mm) was above the grade of 60–70 penetration value stated in the specification for tropical areas, like the study

Table 1 Quality test on extracted bitumen

S/N	Property	Test method	Unit	Measured values	Standard limits (FMW 2007)[a]
1	Specific gravity at 15 °C and 25 °C	ASTMD70	°C	1.02	1.00–1.05
2	Penetration at 15 °C	ASTMD5	mm	93	80–100 grade
3	Flash point	ASTMD92	°C	235	≥225
4	Softening point	ASTMD36	°C	47	42–52
5	Ductility at 25 °C	ASTMD113	cm	150	≥100
6	Loss of heating	ASTMD6	%	0.5	0.5
7	Solubility in trichloroethylene	ASTMD2042	%	99	99

[a]FMW – Federal Ministry of Works in Nigeria

Table 2 Quality test on compacted asphalt concrete

S/N	Test Property	Unit	Measured values	Standard limits (FMW)[a]
1	Marshall stability	N	3150	≥3500
2	Flow value	mm	2.55	2–4
3	Percent air voids in the mix	%	3.1	3–5
4	Voids filled with bitumen	%	77.4	75–82

[a]FMW – Federal Ministry of Works in Nigeria

Table 3 Physicochemical analysis of used-tyre oil and standard diesel oil

S/N	Fuel property	Unit	Diesel fuel	Used-tyre oil
1	Density at 20 °C	Kg/m^3	845	936
2	Kinetic viscosity at 40 °C	mm^2/sec	3.4	3.6
3	Flash point	°C	60	52
4	Diesel index		55.2	42.8
5	Heat value	MJ/kg	46	43.22
6	Sulfur amount	%	0.1	0.904
7	Centane number		53	45

area (Nigeria). This indicated that the extracted bitumen might be softer than the standard petroleum bitumen, and the softness might be due to the effect of the concentrated acid (H_2SO_4), which was used to speed up the extraction.

Regarding asphalt concrete, the results of performance tests on the compacted asphalt concrete indicated values of 3150 N, 2.55 mm, 3.1%, and 77.4% for Marshall stability, flow value, percent air void in the mixture, and voids filled with bitumen, respectively. When compared with the asphalt standard specification (FMW 2007), all the three properties, except Marshall stability, passed the requirement. The value of the Marshall stability, though very close to the standard, failed the requirement of FMW. This failure might be due to the quality and amount of the asphalt components (bitumen, filler, fine, and coarse aggregates). For example, poor quality of coarse aggregates in asphalt concrete may affect the stability of the concrete, since the aggregate is known to give the strength required for adequate stability of the concrete. Therefore, based on the low stability value of the asphalt concrete, further quality improvement of the extracted bitumen is required for it to be a good substitute for petroleum bitumen.

For comparative analysis, the properties of the light oil obtained from the waste tyre pyrolysis were compared to the properties of the standard diesel oil; the results showed that the light oil, upon quality improvement, can also be used as a replacement for the normal diesel oil. One of the properties that need improvement is the low cetane number (CN) of the light oil (45), since higher cetane oil will have shorter ignition delay periods than lower cetane oil. Cetane number is an indication of the combustion speed of the diesel fuel and compression needed for ignition.

Statistical Analysis of the Results

The Minitab paired t-test was used to analyze if there was significant mean difference between the property values of the pyrolytic bitumen (experimental group) and the standard bitumen (control group). For the bitumen data (Table 4), seven properties, including specific gravity, penetration, flash point, softening point, ductility, loss of heating, and solubility in trichloroethylene, were examined. The mean difference between the experimental and standard values was 9.00, while the standard deviation and standard error of the mean across the difference row were 18.45 and 6.97, respectively. The 95% CI mean difference row showed a 95% confidence interval (95% CI) for the mean difference of −8.06 to 26.06. From the paired t-test output for bitumen, the t-value is 1.29, and the associated *p*-value is 0.244. Before paired t-test is conducted, it is a common practice to choose α-level = 0.05 as the criterion p-value to reject or accept the hypotheses. The hypotheses are H0 (the null hypothesis), the mean difference = 0, and H1 (the alternative hypothesis), the mean difference ≠ 0. From the Minitab paired t-test output, it could be observed that *p*-value was 0.244 and was greater than α-level = 0.05. As the p-value was greater than 0.05, it could be concluded that there was no statistical difference between the properties of pyrolytic bitumen and the standard bitumen.

The Minitab paired t-test was also used to analyze if there was significant mean difference between the property values of the asphalt concrete (experimental group) and the standard asphalt (control group). For the asphalt data (Table 5), four properties, including Marshall stability, flow value, percent air void in asphalt mix, and voids filled with bitumen, were examined. The mean difference between the experimental and standard values was −88.1, while the standard deviation and standard error of the mean across the difference row were 174.6 and 87.3, respectively. The 95% CI mean difference row showed a 95% confidence interval (95% CI) for the mean difference of −365.9 to 189.7. From the paired *t*-test output for asphalt, the *t*-value was −1.01 and the associated *p*-value was 0.387. As the *p*-value was greater than 0.05, it could be concluded that the difference between the properties of pyrolytic asphalt and the standard asphalt was not statistically significant.

Table 4 Minitab paired t-test output for significance difference between pyrolytic bitumen and standard bitumen

Bitumen		Paired T-test and CI: Exptal, Control				
Experimental Values (Exptal)	Standard values (Control)	Paired t-test for Exptal, Control				
1.02	1.03		N	MEAN	STDEV	SEMEAN
93	90	Exptal	7	89.4	84.2	31.8
235	225	Control	7	80.4	77.0	29.1
47	47	Difference	7	9.00	18.45	6.97
150	100					
0.50	0.50	95% CI mean difference: (−8.06, 26.06)				
99.00	99	T-test of mean difference = 0 (vs ≠ 0): *T*-value = 1.29 *P*-value = 0.244				

Table 5 Minitab paired *t*-test output for significance difference between pyrolytic asphalt concrete and standard asphalt concrete

Asphalt		Paired T-test and CI: exptal and control				
Experimental values (exptal)	Standard values (control)	Paired t-test for exptal and control				
3150	3500		N	MEAN	STDEV	SEMEAN
93	3	Exptal	4	808	1562	781
235	4	Control	4	896	1736	868
47	78.50	Difference	4	−88.1	174.6	87.3
		95% CI for mean difference: (−365.9, 189.7)				
		T-test of mean difference = 0 (vs ≠ 0): *T*-value = −1.01 *P*-value = 0.387				

Conclusion

As a way of building climate change resilience for the socioeconomic system in Nigeria, the study aims to extend the existing knowledge on the usage of waste-based pyrolytic bitumen as a substitute for the petroleum-based bitumen'. The waste-based pyrolytic bitumen was tested for its asphalt-making potential to verify its suitability as a replacement for the petroleum bitumen, and the quality of the extracted bitumen was comparable to the standard petroleum bitumen. Moreover, the feasibility of using the pyrolytic bitumen as a binder in the asphalt mix production was confirmed, as there was no significant difference between the properties of asphalt made from pyrolytic bitumen and standard bitumen. However, further quality improvement of the extracted bitumen is still required as evident by the low Marshall stability of 3150 N of the asphalt mix made from the pyrolytic bitumen.

Other area that needs to be given attention in future studies is how to improve the efficiency of the pyrolysis system, especially, the design of the air-fuel intake port which regulated the ratio of air to fuel that would produce desired burning flame in the gas-fired furnace. This would help to optimize the input variables (time, energy, gas, and feedstock) to achieve quality products. Future studies could also cover areas such as material and energy balance, carbon footing of the pyrolysis process, as well as the utilization of the gas and char produced.

Cross-References

▶ Biodiversity, Ecosystem Degradation, and Climate Change Effects on Livelihoods in the Bitumen Area of Nigeria
▶ Climate Change and Coastal Resilience in Nigeria
▶ Climate Change Education for Sustainable Development: Lessons for Nigeria

- Climate Resilience in African Coastal Areas: Scaling up Institutional Capabilities in the Niger Delta Region
- Leveraging Best Practices for Climate Adaptation in the West African Sahel: The Emergence of Global Alliance for Resilience
- Strengthening Climate Change Adaptation in the Cities of West Africa: Policy Implications for Urban Resilience
- Vulnerability of Uganda's Electricity Sector to Climate Change: An Integrated Systems Analysis

References

Adger WN, Arnell NW, Tompkins EL (2005) Successful adaptation to climate change across scales. Glob Environ Chang 15:77–86

Adhikari B, De D, Marti S (2000) Reclamation and recycling of waste rubber. Prog Polym Sci 25:909–948

Aylón E, Fernández-Colino A, Murillo R, Navarro MV, Garcia T, Mastral AM (2010) Valorisation of waste tyre by pyrolysis in a moving bed reactor. Waste Manag 30:1220–1224

Azar C, Lindgren K, Larson ED, Mollersten K (2006) Carbon capture and storage from fossil fuels and biomass-costs and potential role in stabilising the atmosphere. Clim Chang 74:47–79

Bandyopadhyay S et al (2008) An overview of rubber recycling. Prog Rubber Plast Recycl Technol 24:73–112

Bunger J, Thomas K, Dorrence S (1979) Compound types and properties of Utah and Athabasca tar sand bitumens. Fuel 58:183–195

Cook J et al (2013) Quantifying the consensus on anthropogenic global warming in the scientific literature. Environ Res Lett 8:1–7

Corinaldesi V, Moriconi G (2004) Durable fiber reinforced selfcompacting concrete. Cem Concr Res 34:249–254

Corinaldesi V, Mazzoli A, Moriconi G (2011) Mechanical behaviour and thermal conductivity of mortars containing waste rubber particles. Mater Des 32:1646–1650

Danon B, Gorgens J (2015) Determining rubber composition of waste tyres using devolatilisation kinetics. Thermochim Acta 621:56–60

Dodman D (2009) Blaming cities for climate change? An analysis of urban greenhouse gas emissions inventories. Environ Urban 21:185–201

Eldin NN, Senouci AB (1999) Rubber-tyre particles as concrete aggregate. J Mater Civ Eng 5:478–496

Epstein PR (2005) Climate change and human health. N Engl J Med 353:1433–1436

FMW (2007) Pavement and materials design in highway manual part I: Design. Volume 3, Federal Ministry of Works, Abuja, Nigeria

Folke C (2006) Resilience: the emergence of a perspective for social-ecological systems analyses. Glob Environ Chang 16:253–267

Gungor C, Serin H, Ozcanli M, Serin S, Aydin K (2015) Engine performance and emission characteristics of plastic oil produced from waste polyethylene and its blends with diesel fuel. Int J Green Energy 12:98–105

Haines A, Kovats RS, Campbell-Lendrum D, Corvalan C (2006) Climate change and human health: impacts, vulnerability and public health. Public Health 120:585–596

Holling CS (1973) Resilience and stability of ecological systems. Annu Rev Ecol Syst 4:1–23

IPCC (2014) Intergovernmental Panel on Climate Change. Climate: impacts, adaptation and vulnerability. Contribution of Working Group II to the fifth assessment report of the Intergovernmental Panel on Climate Change. IPCC, Geneva

Kalargaris SGI, Tian G (2017) Combustion, performance and emission analysis of a DI diesel engine using plastic pyrolysis oil. Fuel Process Technol 157:108–115

Lee BI, Park BS (1993) Mechanical properties of carbon-fiber-reinforce- polymer impregnated cement composite. Cem Concr Compos 15:153–163

Martínez JD, Puy N, Murillo R, García T, Navarro MV, Mastral AM (2013) Waste tire pyrolysis – a review. Renew Sust Energ Rev 23:179–213

Murillo R, Aylón E, Navarro MV, Callén MS, Aranda A, Mastral AM (2006) The application of thermal processes to valorise waste tyre. Fuel Process Technol 87:143–147

Oyewo A, Aghahosseini A, Bogdanov D, Breyer C (2018) Pathways to a fully sustainable electricity supply for Nigeria in the mid-term future. Energy Convers Manag 178:44–64

Patz JA, Campbell-Lendrum D, Holloway T, Foley JA (2005) Impact of regional climate change on human health. Nature 438:310–317

Presti DL (2013) Recycled tyre rubber modified bitumens for road asphalt mixture; a literature review. Constr Build Mater 49:863–881

Satterthwaite D (2013) The political underpinnings of cities accumulated resilience to climate change. Environ Urban 25:381–391

Sharifi A, Yamagamata Y (2016) Principles and criteria for assessing urban energy resilience. A literature review. Renew Sust Energ Rev 60:1654–1677

Smit B, Wandel J (2006) Adaptation, adaptive capacity and vulnerability. Glob Environ Chang 16:282–292

Telegraph-reporters (2018) Plastic waste already building up in UK following China's ban. The Telegraph. www.telegraph.co.uk. Accessed 16 Jan 2018

Teng H, Serio MA, Wójtowicz MA, Bassilakis R, Solomon PR (1995) Reprocessing of used tires into activated carbon and other products. Ind Eng Chem Res 34:3102–3111

Topcu IB (1995) The properties of rubberized concrete. Cem Concr Res 25:304–310

Trouzine H, Bekhiti M, Asroun A (2012) Effects of scrap tyre rubber fibre on swelling behavior of two clayey soils in Algeria. Geosynth Int 19:124–132

Williams PT, Besler S, Taylor DT (1990) The pyrolysis of scrap automotive tires. Fuel 69:1474–1482

Williams PT, Besler S (1995) Pyrolysis-thermogravimetric analysis of tires and tyre components. Fuel 74:1277–1283

Zabaniotou AA, Stavropoulos G (2003) Pyrolysis of used automobile tires and residual char utilization. J Anal Appl Pyrolysis 70:711–722

Local Knowledge of Climate Change Among Arable Farmers in Selected Locations in Southwestern Nigeria

123

David Olabanjo Sanni

Contents

Introduction 2532
Research Questions 2533
Literature Review of Climate Change in Nigeria 2533
  Change and Variability in Rainfall 2533
  Change and Variability in Temperature 2535
Materials and Method 2535
  Study Area 2535
  Data Collection and Methodology 2537
Results and Discussion 2537
  Demographic and Psychographic Elements of the Selected Farmers 2537
  Indigenous Knowledge of Climate Change 2538
  Sources of Information on Climate Change Among Farmers in Selected Locations in Southwestern Nigeria 2539
  Indigenous Knowledge on the Seasonal Rainfall Total in Southwestern Nigeria 2540
  Indigenous Knowledge on the Night- and Daytime Temperatures in Southwestern Nigeria 2541
Analyses in Climate Variables 2542
  Annual Rainfall in Southwestern Nigeria (1962–2012) 2542
  Annual Temperature Trend in Southwestern Nigeria (1962–2012) 2543
Response to the Changing Climate (Adaptation/Coping Strategies) by the Farmers 2544
  Adaptation Strategies to Climate Change in Southwestern Nigeria 2545
Conclusion 2545
References 2546

D. O. Sanni (✉)
Department of Geography, Obafemi Awolowo University,
Ile Ife, Nigeria
e-mail: olabanjosanni@gmail.com

W. Leal Filho (ed.), *Handbook of Climate Change Resilience*,
https://doi.org/10.1007/978-3-319-93336-8_147

Abstract

One of the essential steps to manage and adapt to the changing climate is having adequate knowledge on the changes in the climatic pattern. This chapter analyzed the responses of rural arable farmers to the change observed in climate in Southwestern Nigeria, which include their source of information, level of awareness, and perception of the amount of rainfall and temperature, leading them to take decisions on how to combat and/or adapt to the changes. Rural arable farmers in four states within the Southwestern Nigeria made up the study population (through random selection). Findings show the climate is no longer stable but fluctuating. Farmers in the area under study have a high perception about climate change; ironically a large percentage of the farmers are yet to put into practice the adaptation methods for reasons such as high cost, religious belief, attachment to tradition, low economic status, and others. Sources of information include educated children, mass media, the Internet, and majorly personal observation. It was concluded that adapting agricultural practices to climate change, regular localized observation of climate variables and the local media disseminating updated adaptations strategies and weather reports would go a long way to reduce the impact of climate change on the society.

Keywords

Local knowledge · Climate variation · Arable farmers · Southwestern Nigeria · Perception

Introduction

The IPCC (Intergovernmental Panel on Climate Change) has reported that the global climate is not as it used to be, which strongly indicates climate change, which is heavily anthropogenic (IPCC 2014). The impact of the climate change will be felt in places where the dependence on water resources is high and the adaptive capacity is low (IFAD 2009; Abiodun 2011). Africa as a continent is well exposed to the severity of climate change. According to the IPCC (2007a) by 2020, 75–200 million people living in Africa will be exposed to the vagaries associated with climate change, which include shortage of water supply and reduced agricultural production. These are threats to food security and may jeopardize the achievement of the Sustainable Development Goals (SDGs) in the continent.

Notwithstanding the impending catastrophe, the level of indigenous awareness of climate change among rural farmers is relatively insufficient especially in the sub-Saharan region of Africa. Perceptions of climate change and the various threats it poses to farmers are strongly dependent on values, trust of public opinions, and personal experience. Given that perception of and response to climate change is largely rooted in experience and accumulated knowledge, farmers possess the right knowledge of climate change through firsthand personal experience of how climate variations impact farming practices (Vedwan and Rhoades 2001). The experiences they gather over time shape their perceptions and ultimately determine their response and decision-making. The value an individual assign to the risk associated with change is dependent on local knowledge.

Most of the rural farmers in the tropical rain forest region feel "secure" since there are no immediate or visible dangers associated to climate change on their crop yield in contrast to the tropical Sudan and Sahel regions. In the dryer regions, the impact of the dwindling rainfall amount is relatively known as presented by the dry lands, increasing presence of sand dunes, dying animals, etc. That notwithstanding, there is a nexus between the indigenous perception and how farmers respond to the changing climate. Previous studies (NEST 2004; Speranza 2010; IPCC 2014) have reported the level of knowledge of climate change among farmers at the national level and regional level, which is not specific to a microlevel, making it irrelevant for crop cultivation at the local level. It is necessary to carry out regional studies than having a generalized conclusion that the farmers' awareness level is high. Once the indigenous perception to climate change is known, it increases the adaptive capacity of the local farmers and helps reduce the impact of the observed climate changes on the farmers' crop at the local level (Broadhead and Howard 2009).

There are differences in the severity of the climate change on a localized level as a result of the global warming. This necessitates a more in-depth knowledge of the indigenous perception at a regional or localized level.

This chapter identified the sources of information and examined the farmers' knowledge on changes in rainfall and temperature patterns between 1962 and 2012; this will determine the level of the indigenous knowledge of climate change among the farmers in selected location in Southwestern Nigeria.

Research Questions

The chapter provides answers relating to the major sources of information, perception/indigenous knowledge of climate change, and the coping methods adopted by the arable farmers against climate change.

Literature Review of Climate Change in Nigeria

Change and Variability in Rainfall

Rainfall is the main climatic factor in agriculture (Matthew and Oladokun 2009). Several studies have examined rainfall trends in tropical Africa, which include Nigeria (Oladipo 1995; Odekunle and Adejuwon 2007; Balogun 2009; Matthew and Oladokun 2009). The rainfall onset has been described as the period at the beginning of the rainy season when rainfall distribution becomes adequate for crop development and rainfall retreat is the period toward the end of the rainy season when rainfall distribution may no longer sustain crop growth (Odekunle et al. 2005).

The growing season has been defined as the length of the period between the onset and the retreat (Odekunle and Adejuwon 2007). Studies (Odekunle and Adejuwon 2007; Balogun 2009) have shown that the rainfall onset and retreat dates for southern Nigeria had remained between March and November, respectively, making the growing season last for 8 months. However, Matthew and Oladokun (2009) observed that since 1968 there has been a significant decline in both the amount and frequency of rainfall, and the rainfall onset period have progressively been delayed.

The rainfall change/variability over climatic zone in Nigeria has been investigated. It was indicated that in Nigeria, the rainfall pattern has changed in the past four to five decades (Oladipo 1995; Ologunorisa and Tersoo 2006). The gradual decline of the rainfall in most part of the country started after a period of high rainfall amount, which signifies a wet period, from the early 1960s to the early 1980s. This indicates a period of change of climate. Odekunle et al. (2008) likewise showed that the Sahelian region in the northern part has recently exhibited a remarkable change in its climate as it is tending toward a wetter condition compared to the twentieth century. Matthew and Oladokun (2009) observed that there is no regular pattern associated with rainfall values except for some dry and wet years, which alternates every 2–3 years, which indicates that rainfall is highly variable in both time and space over Nigeria. Matthew and Oladokun (2009) posit that there was a sharp increase in rainfall from 2005 in the southern part and there was no evidence of returning to its original value of the 1960s. The decadal rainfall trends show that Ibadan and Port Harcourt recorded increases in rainfall since 2005 with an average of about 130 mm and 200 m per annum, respectively, while Jos witnessed stationary rainfall amount of about 130 mm.

A reasonable conclusion as observed by Balogun (2009) and Oluleye (2009) and Odekunle (2010) is that there has been deviation of rainfall from their normal mean values. The shift can be attributed to the shift of climate pattern that was observed in the 1970s, with an increase in the Sahel, Sudan, Guinea, and the Mangrove swamp, but a little increase in the rain forest zone. The annual rainfall of the Sahelian zone has reduced by 100–150 mm during the early 1960s and late 1980s (Olaniran 1983). Matthew and Oladokun (2009) have established that the decadal rainfall trend in Jos reduced slightly from 1976 to 1998. Sokoto and Maiduguri recorded decline in rainfall of about 5 mm per decade from 1958 and rise again to about 70 mm/annum in 2007. Between 1941 and 2000, the annual rainfall and temperature of coastal areas like Port Harcourt increased significantly by about 2–4 mm but reduced by 2–8 mm across most part of Nigeria (Building Nigeria Response to Climate change; BNRCC 2011).

In Makurdi, located in the middle belt area of Nigeria, it was noted that there has being a continuous reduction in the annual rainfall received between 1970s and 2000s with a little break in the mid-1990s to year 2000. The period of high rainfall coincides with the time of the highest frequency of rainfall and attendant flood frequencies, but the city also shows a decline for rainfall received which made the town drier but was still receiving less frequent unprecedented heavy rainfall (Olgunorisa and Tersoo 2006). Odekunle et al. (2008) and Speranza (2010) opined that the annual rainfall pattern in Nigerian weather stations has a seemingly random and strong cyclic variation that corresponds to the El-Nino/Southern oscillations (ENSO) cycle. Most of the southern stations like Ibadan, Calabar, and Enugu show a positive trend with Kano (which is a northern station), while most northern stations (Jos, Nguru, Minna, etc.) and a southern station (Ikeja) show a negative trend in rainfall.

Nieuwolt (1982) observed that the principal controlling element for tropical agriculture has been recognized to be rainfall. This is so because temperature in the tropics is relatively uniform, and seasonal variations are small. Changes were also observed in weather stations close to the Sahel region of West Africa, showing the period of changes in climate was between 1965 and 1970 (Balogun 2009; Oluleye 2009). To date, simulation studies of future rainfall patterns have been limited by a lack of reliable data, and such have provided only *"best-guess"* estimates with little or no information on uncertainties that result from choices in model structure, parameter values, and scaling techniques (Fischer et al. 2002).

Change and Variability in Temperature

It has been established that there will be an increase in both minimum and maximum temperature in the twenty-first century in Nigeria (Adejuwon 2004; Adejuwon and Odekunle 2006; Adesina and Odekunle 2011). The warmness was interrupted by a short cooling season in the mean temperature in 1993 and 1994. According to the report by the Nigeria Environmental Study/Action Team (NEST 2004) in the National Adaptation Strategy and Plan of Action on Climate Change for Nigeria (NASPA-CCN), there was evidences of long-term temperature increases in most parts of the country. The main exception was in Jos area, where a slight cooling was recorded. The most significant increases were recorded in the extreme northeast, extreme northwest, and extreme southwest, where average temperatures rose by 1.4–1.9 °C. According to Adepitan et al. (2017), Jackson et al. (2012) and Adebayo et al. (2017), Ibadan has witnessed a significant increase in temperature in the last four decades – 1973–2008 – which is gradually turning the area to a derived savannah.

Materials and Method

Study Area

The area is Southwestern Nigeria, which consists of Ekiti, Lagos, Ogun, Ondo, Osun, and Oyo States. It lies approximately between longitudes 3°E and 7°E and latitudes 4°N and 9°N (Fig. 1). The area occupies about 180,205 km^2 (MEFRN 2003).

The climate of Southwestern Nigeria is humid and tropical in nature, characterized by wet and dry seasons (Ojo 1977). The wet season lasts from March to

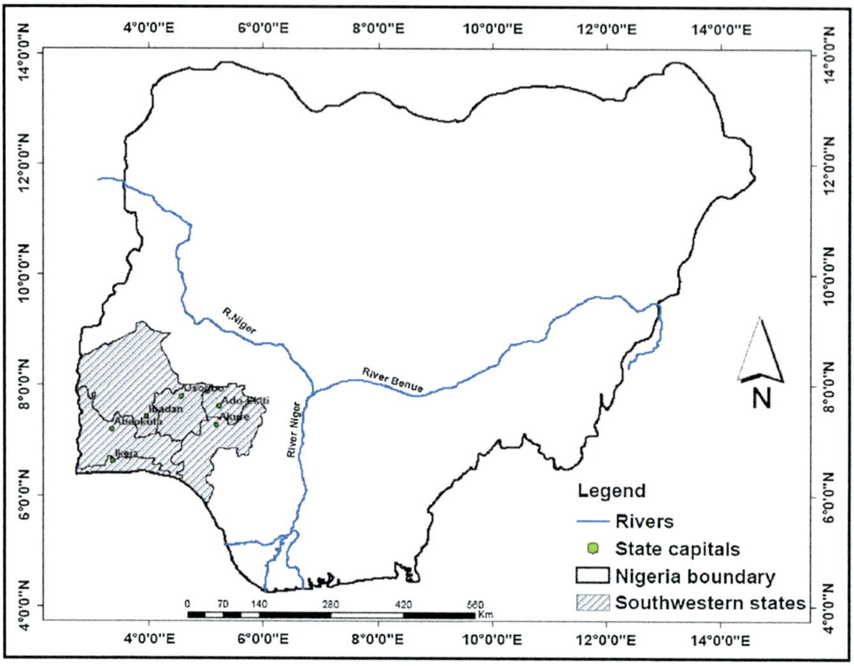

Fig. 1 The study area

October in the southern part (Mangrove/ rainforest) and April to October in the northern fringes (Southern Guinea savannah). The area experiences double rainfall maxima with July being the first maximum and September the second. The period when the rainfall amount reduces occurs in August. This phenomenon of reduced August rainfall is known as the "*Little Dry Season*" (LDS) or August Break (Odekunle and Adejuwon 2007). Climate in Southwestern Nigeria is determined by the influence of the three air masses, namely: the tropical continental (cT), tropical maritime (mT), and the easterlies (E). The equatorial easterlies (E) originating from the eastern part of the continent, which is unstable sometimes, undercut the two air masses and flow in the upper atmosphere giving rise to line squalls (Iloeje 1981; Odekunle 2010).

The region enjoys luxuriant vegetation of high forest zone (evergreen rain forest) in the south and savannah on the northern fringe. The northern part of the area is a transitional zone between the rainforest and the savannah (Filani and Alokan 1993; Emielu 2000).

The soils are mostly ferralitic and ferruginous tropical type of the interior and the alluvial types along the coast. The ferralitic or the forest soils are deeply weathered that are especially good for producing cash crops such as cocoa, kola nut, rubber, and palm produce (Emielu 2000). Notable food crops grown are yam, cassava, vegetable, maize, banana, and millet.

Data Collection and Methodology

This chapter used primary data. These were obtained using questionnaire administration on adult farmers in selected locations in Southwestern Nigeria. The questionnaire ascertains the farmers' awareness and response to climate change. Other data obtained using the questionnaires was to ascertain the farmer's observation of the behavior of climate in the past and at present.

The multistage sampling procedure was used in selecting respondents for the chapter. The respondents are farmers residing in communities close to towns/cities with climate stations with long time data in Southwestern Nigeria (stations: Ibadan, Osogbo, Ondo, and Ikeja – at least 50 years of daily rainfall and temperature data) and cultivate staple food crops like maize, cassava, vegetables, and sorghum. The climate data duration is from 1962 to 2012, sourced from the archives of the Nigeria Meteorological Agency (NIMET), Oshodi, Lagos. This period is good enough to show any variation of climate. Rainfall (and to an extent temperature) is recognized as the principal controlling element for tropical agriculture (Nieuwolt 1982); hence, rainfall and temperature data only were used in this study.

The selected towns/cities are Ibadan (Oyo State), Ede (Osun State), Ondo town (Ondo State), and Ijebu-Igbo (Ogun State). From locations close to the aforementioned towns/cities, five communities with an estimated population of at least 100 farmers were purposively selected. Twenty (20) communities in all were sampled for the chapter. In each of the communities, 10% of the farmers were purposively selected for questionnaire administration. The farmers were selected using a systematic random sampling procedure to select any adult farmer available. Overall, two hundred (200) questionnaires were administered. Descriptive and quantitative methods like percentages, frequencies, tables, and charts were employed to determine the indigenous knowledge of climate change, their source of information, and their observations of changes in some of the climatic parameters (temperature and rainfall). The climate data were analyzed using descriptive second-order polynomial trend analysis (Odekunle et al. 2008). The values of rainfall and temperature were plotted as graphs with trend lines fitted into them to arrive at the objective division of the period under study. The reason for trend analysis is to determine the changes over time (temporal) of extreme climate parameters. It allows for observation of climate behavior while examining inter-annual and inter-decadal variability of the climate parameters. The data was subjected to the bivariate linear regression analysis (BLRA) to test for the significance of the trend statistically (upward and downward movements).

Results and Discussion

Demographic and Psychographic Elements of the Selected Farmers

Most (45%) of the respondents (selected farmers) are within the age bracket of 45–60 years. This implies that most of the farmers are fit to work for a longer period, bear higher risk, are more innovative, and have a better capacity to cope with daily challenges (active stage in life) (Hansen et al. 2012). A high percentage of the farmers are literate to at least primary school education (57%). This makes it easier for them to seek out the cause of the changing climate and open to improved coping strategies. The percentage of the married is relatively high (63.9%). The abundance of married farmers has a large implication for subsistence labor supply and creates responsible farmers (Nnamerenwa et al. 2017). Each farmer household consists of at least 6–7 persons (moderate sized); this is of high benefit for the provision of family labor and reduction in the cost of hiring laborers. It also helps in other household business (Ali and Erenstein 2017). Likewise, a major alternative source of employment among the farmers is business (56%) and civil service (43%). This is what they resort to when there is a short fall in crop yield.

Indigenous Knowledge of Climate Change

Table 1 presents the rural farmers awareness of climate change. The majority of the respondents (99%) agreed that there have been changes in climate over the years in their locality. Only 1% was of the opinion that there had been no change in climate at all in their area.

Majority of the farmers (95%) indicated that they started noticing the changes in climate less than 20 years ago (Table 1). This observation further confirmed the results obtained in earlier studies (e.g., NEST 2004; Adejuwon 2006; Ologunorisa and Tersoo 2006; Odekunle and Adejuwon 2007; IPCC 2007b) which states that the southern part of Nigeria has been affected by climate change and extreme events. Only 4% of the respondents indicated that they started noticing the changes in the atmospheric conditions more than 20 years ago, while 1.0% was of the opinion that there is no conspicuous change. The visible effect of climate change is reduced crop yield, drying soils, flash flooding, late onset of rainy

Table 1 Climate change awareness among farmers in Southwestern Nigeria

Variable	Categories	Frequency	Percentage
Changes in climatic conditions	Yes	198	99
	No	2	1.0
	Total	200	100
Duration	Less than 20 years	190	95
	20 years and above	8	40
	No change at all	2	1.0
	Total	200	100

Source: Authors' fieldwork

season, early rainfall retreat, and shorten length of growing season. All this evidence is worthy to state that farmers have a good knowledge of climate change over the years.

Sources of Information on Climate Change Among Farmers in Selected Locations in Southwestern Nigeria

Most of the respondents claimed to have obtained their information about climate change from different sources. Figure 2 shows the specific sources of information on climate change among the rural farmers. The results showed that the largest proportion of the rural farmers (66.5%) personally observed the change themselves (and to some extent their educated children), while those that obtained their information from radio ranked next (30%). Next are the agricultural extension workers (which sadly had only 1.5%), then television (0.5%), and the Internet (0.5%), while the rest 1% insists that there have been no changes at all on the climate over the years. The agricultural extension workers that ought to be close to the rural communities for the best and latest techniques have not been doing so in the innermost rural areas of Southwestern Nigeria as attested to in the results. In contrast, the high percentage of those that have their information from the radio indicates that the media/press has been active in carrying out their role of informing the society on the changing climate, which may also help in informing the society on the latest methods of adapting the agricultural practices to climate change.

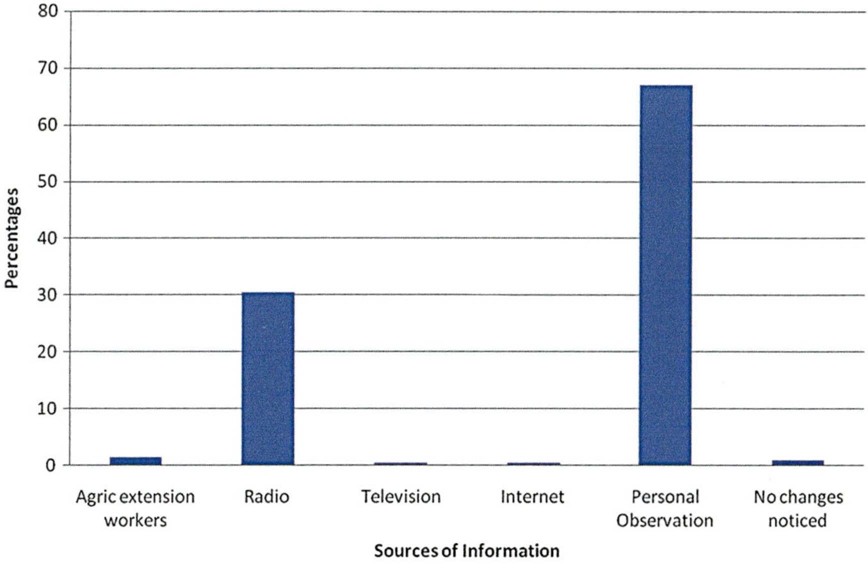

Fig. 2 Farmers' sources of information on climate change. (Source: Authors' fieldwork)

Key: Individual perception of climate change

Questions	Response for coding
Have you noticed any changes in the climatic conditions in your community?	Yes ()…1 No ()…2
If yes, what is your source of your information?	Agricultural extension workers ()…1 Radio ()…2 Television ()…3 Internet ()…4 Personal observation ()…5 No changes noticed ()…6

Indigenous Knowledge on the Seasonal Rainfall Total in Southwestern Nigeria

Table 2 shows the results of the farmer's perception of changes in the seasonal rainfall total in their communities. The result indicated that majority of the farmers (71%) are of the opinion that the seasonal rainfall has been on the increase over the past 10 years and above, 20% believed it has been declining, while the remaining 6.5% agreed that it's been fluctuating. About 83% of the rural farmers observed that the seasonal rainfall total has been stable in the past 30 years and above but started fluctuating in the last 20 years and 13% are of the opinion that it has been on the increase in the last 30 years and above. These results agree with studies like Oladipo (1995), Odekunle and Adejuwon (2007), Odekunle and Eludoyin (2008) which states that rainfall was on the increase within area and all around the country since the mid-1980s. The rainfall mostly occurred over a relatively short period, with heavy storms that are quite destructive. It further leads to flash flooding that cause unquantifiable havoc in materials, food products, and even life!

Table 2 Perception of farmers on the changes in seasonal rainfall totals

	Seasonal rainfall (30 years and above)		Seasonal rainfall (20 years and above)		Seasonal rainfall (10 years and above)	
Timing	Frequency	Percentage	Frequency	Percentage	Frequency	Percentage
Increasing over time	26	13	46	23.0	142	71.0
Decreasing over time	0	0	2	1.0	40	20.0
Fluctuating	8	4.0	138	69.0	13	6.5
Stable	166	83.0	14	7.0	5	2.5
Total	200	100.0	200	100.0	200	100.0

Indigenous Knowledge on the Night- and Daytime Temperatures in Southwestern Nigeria

A large percentage of the farmers (about 92% and 93%) observed that the night and daytime temperatures within the past 10 and 20 years and above have been on the increase, respectively (Tables 3 and 4). The farmers are of the opinion that the daytime intensity of solar radiation has increased compared to the past. Nowadays, most of the farmers sleep outside during the nighttime to reduce the effects of the increasing nighttime temperature. 93% and 94.5% of the farmers, respectively, are of the opinion that the night- and daytime temperatures have been stable 30 years and above. The increases in temperatures signpost climate change. Most of the farmers stated that they feel more uncomfortable sleeping at night, which necessitates sleeping outdoors in recent times due to the increasing nighttime temperature unlike the past when they can sleep in the night with ease. In the area, cases of increased heat-related human diseases such as heat rashes and measles are on the increase, while the wilting of trees (flowers and leaves) are observed, which is attributable to increasing temperature.

Table 3 Farmers' perception of changes in nighttime temperature

	Nighttime temperature (30 years and above)		Nighttime temperature (20 years and above)		Nighttime temperature (10 years and above)	
Timing	Frequency	Percentage	Frequency	Percentage	Frequency	Percentage
Increasing over time	1	0.5	36	18.0	184	92.0
Decreasing over time	0	0	0	0	2	1.0
Fluctuating	13	6.5	36	4.0	10	5.0
Stable	186	93.0	156	100.0	4	2.0
Total	200	100.0	200	100.0	200	100.0

Source: Authors' fieldwork

Table 4 Farmers' perception of changes in daytime temperature

	Daytime temperature (30 years and above)		Daytime temperature (20 years and above)		Daytime temperature (10 years and above)	
Timing	Frequency	Percentage	Frequency	Percentage	Frequency	Percentage
Increasing over time	2	1.0	40	20.0	186	93.0
Decreasing over time	0	0	1	0.5	1	0.5
Fluctuating	9	4.5	152	76.0	10	5.0
Stable	189	94.5	7	3.5	3	1.5
Total	200	100.0	200	100.0	200	100.0

Source: Authors' fieldwork

Analyses in Climate Variables

Annual Rainfall in Southwestern Nigeria (1962–2012)

The annual rainfall data of 1962–2012 were analyzed for the climate stations (Osogbo, Ibadan, Ikeja, and Ondo). It was observed that the rainfall amount decreases for all the stations from the early 1960s (1962) to the early 1980s (1982–1983), and afterward, there was a significant increase for rainfall recorded from 1983 to 1992/1993. This was followed by another decline in the late 1990s especially in the Osogbo station. This continues until mid-2000s when the rainfall total suddenly starts rising again. The last decade has witnessed several occurrences of flooding in most of the urban areas within Southwestern Nigeria. The most noticeable increase is in the Ibadan station where there was an increase in rainfall total in the early 2000s (2004–2012); this coincides with the period when flooding was a common occurrence in the town. The Ikeja stations also witness a lowering in the rainfall total from the year 2000 to 2012. The Ondo station experienced a significant increase in rainfall total especially from 1999 to 2007.

The regression output shows a statistically negative downward trend at the 95% confidence level for the Ikeja and Osogbo stations. These also include the spatial average for the whole station. The result of the BLRA is shown in Table 5. The correlation values are as follows: −0.54 and − 0.35 for Ikeja and Osogbo stations, respectively, and − 0.38 for the spatial average. It is only in Ibadan station that the observed trend is not statistically significant (with correlation value of −0.2). The result of the second-order polynomial was entirely different for the Ondo station; the trend indicates a consistent increase in the rainfall amount throughout the period under study, but is not statistically significant.

Table 5 Results of the rainfall data analysis (1962–2012)

Dependent variable	Explanatory variable	Unstandardized beta coefficient (B) (constant)	Correlation values
Ondo annual rainfall totals	Years (1960–2012)	1593.9	0.09
Ikeja annual rainfall totals	Years (1962–1987)	1924.4	−0.54[a]
	Years (1987–2012)	1557.8	−0.03[b]
Osogbo annual rainfall totals	Years (1962–1983)	1429.7	−0.35[a]
	Years (1983–2012)	1351.5	0.06[b]
Ibadan annual rainfall totals	Years (1962–1983)	1368.5	−0.2[b]
	Years (1983–2012)	1235.1	0.23[a]
Annual rainfall total	Years (1960–1983)	1682	−0.38[b]
Spatial average	Years (1983–2012)	1421	0.41[a]

[a]Correlation is significant at the 0.05 level (1 tailed)
[b]Correlation is significant at the 0.01 level (1 tailed)

Annual Temperature Trend in Southwestern Nigeria (1962–2012)

The annual temperature data of 1962–2012 were analyzed for the climate stations (Osogbo, Ibadan, Ikeja, and Ondo). The results of Ondo station show an increase in temperature from 1962 to 2002/2003 and afterward a noticeable downward trend from 2006 to 2007, then a noticeable increase until 2012. For Ibadan station, the increase lasted from 1962 to 1994/1995, then a downward trend until 2005. From then it rises again until the end of the period under study. The Osogbo station experienced an increase in air temperature throughout the period under study. The Ikeja station shows an upward trend throughout the period (1962–2012). The trend of mean air temperature shows increasing tendency in most of the individual stations after the early 1990s and/or year 2000–2009, from its initial upward tendency since 1960s. The average annual temperature increase ranges from 0.2 °C to 0.4 °C in the area.

To test for the statistical significance, the trends (both the upward and downward limbs) were subjected to bivariate linear regression analysis (BLRA). The results obtained are shown in Table 6. The results indicated that for Ondo and Ibadan, the upward trend was statistically significant at the 99% confidence level with a correlation value of 0.69 and 0.78, respectively. The regression output for Osogbo and Ikeja stations shows a statistically significant positive trend at 99% and 95% confidence level, respectively. Their correlation values are 0.51 and 0.66, respectively. It was observed that in all the stations, there were increasing tendency in the air temperatures.

Table 6 Results of the regression outputs of the trends for the mean annual temperature in Southwestern Nigeria

Dependent variable	Explanatory variable	Unstandardized beta coefficient (B) (constant)	Correlation values
Ondo mean annual temperature totals	Years (1960–2003) Years (2003–2012)	26.2 27.7	0.69^b 0.70^a
Ikeja mean annual temperature	Years (1962–2012)	26.5	0.66^b
Osogbo mean annual temperature	Years (1962–2012)	26.2	0.51^a
Ibadan mean annual temperature	Years (1962–1983) Years (1983–2012)	26.2 27.8	0.78^b 0.50^a
Spatial average Mean annual temperature	Years (1960–1983) Years (1983–2012)	26.2 27.7	0.73^b 0.75^a

[a]Correlation is significant at the 0.05 level (1 – tailed)
[b]Correlation is significant at the 0.01 level (1 – tailed)

The downward trend shows that there is a statistically significant negative trend at the 95% confidence level in Ibadan and Ondo and the spatial average. The correlation values are 0.50, 0.70, and 0.75, respectively. A difference pattern was noticed in Ikeja and Osogbo stations where after subjecting it to the BLRA, there was a statistically significant positive trend for the 99% confidence level. The Ikeja station has a correlation value of 0.66; the Osogbo station which records a steady increase in the trend line has a statistically significant positive trend at the 95% confidence level with a correlation value of 0.51.

Response to the Changing Climate (Adaptation/Coping Strategies) by the Farmers

Using the Likert scale on the questionnaire, all the farmers (100%) agree that there is the need for adaptation of the various agricultural practices to climate change to avoid further loss of crops. The scale indicates if any improved agricultural practices proposed to the communities for adoption would have a wide acceptance (Table 7). The majority (99%) of the farmers strongly agree that steps taken so far against climate change have not been effective; they are of the view that more efforts need to be put in to curtail the impact of climate change in their communities. The farmers are of the opinion that improved agricultural practices will help them improve their crop yields. The farmers are worried that the best of their efforts are not yielding much result hence their desire for more improved practices. Finance is a major determinant to having improved yields. This is obvious from the results obtained as 97.5% of the farmers are of the view that the cost of improved agricultural practices is higher than the traditional practices. The initial high capital requirement makes the farmers to have some initial fear on the acceptance of the improved agricultural practices due to its costly nature, but this will be overcome once they see its efficacy when any pilot project is conducted.

Table 7 Response to improved agricultural practices in the future

S/N	Variables	SA %	A %	D %	SD %
1	Need to adapt agricultural practices to climate change	59	41	0	0
2	Steps taken against climate change have not been effective	54.5	45	0	0.5
3	Improved agricultural practices will help to improve yield	55.5	44.5	0	0
4	Finance is a major hindrance to better agricultural practices	56	44	0	0
5	The cost of improved agricultural practices is higher than traditional practices	50	47.5	0	2.5
6	The agricultural extension workers are doing a good job for the farmers	6.5	27.5	1.5	64.5
7	Planting of improved seeds provide higher yields	53.5	43	1.5	2

Source: Author's fieldwork

Key: *SA* strongly agree; *A* agree; *SD* strongly disagree; *D* disagree

The activity of the government through its agricultural extension workers is not noticeable in most of the communities where the respondents live in. Majority of the farmers (about 66%) are of the opinion that the agricultural extension workers carry out their duty below expectations, because they are hardly seen in their communities, especially in the farms. However, about 34% of the farmers believed the agricultural extension workers are doing their duty well. The agricultural extension workers ought to the harbingers of the latest techniques on farming, by sourcing for the latest technologies and transferring it to the farmers in a language best understood by them. 96.5% supported the view that improved seeds when planted give high yield. This is believed to be effective in reducing the negative effects of the dry spells being experienced in their communities due to the late onset and early retreat of rainfall. The improved varieties are drought-resistant, fast-germinating, resistant to disease, and high-yielding crop varieties. These entire boosters would make introduction of better varieties be easily accepted by the farmers.

Adaptation Strategies to Climate Change in Southwestern Nigeria

The current adaptation methods adopted against climate change differ from farmer to farmer. Some of the farmers combine more than one method as a coping strategy. The planting of high-yielding and fast-germinating seeds by the farmers is the most used adaptation method against climate change (99%) and followed by mulching (91.5%). Other adaptation strategies used include the use of chemical as pesticide and organic fertilizers (85%), practicing mixed farming (76.5%), quick processing of harvested crops (18%), migration from communities (16.5%), and the least which is planting of cover crops to reduce the effect of temperature and flooding (13.5%) (Fig. 3, Table 8).

Conclusion

These indigenous knowledge and perceptions help to inform the policy makers on the need to implement the national and international climate adaptation policy (as designed by each country and the IPCC). Indigenous knowledge about the changing climate is a necessary tool in combating the impact that easily affects the food security of any nation. The rural farmers in the Southwestern region of Nigeria have shown that they are well informed about climate change (changes observed indicate climate change, which poses a threat to food security). They are very observant of the changes in the climatic parameters especially rainfall and temperature but have not taken adaptation strategies due to the high cost of some of the adaptation methods. Some of the farmers also indicated that they believe that the variations noticed are due to the "*immorality*" of the people; hence, they might not be able to stop or reduce it (through adaptation methods) except by prayers to their deity for forgiveness. The high level of indigenous awareness of climate change has helped to increase the farmers' adaptive capacity.

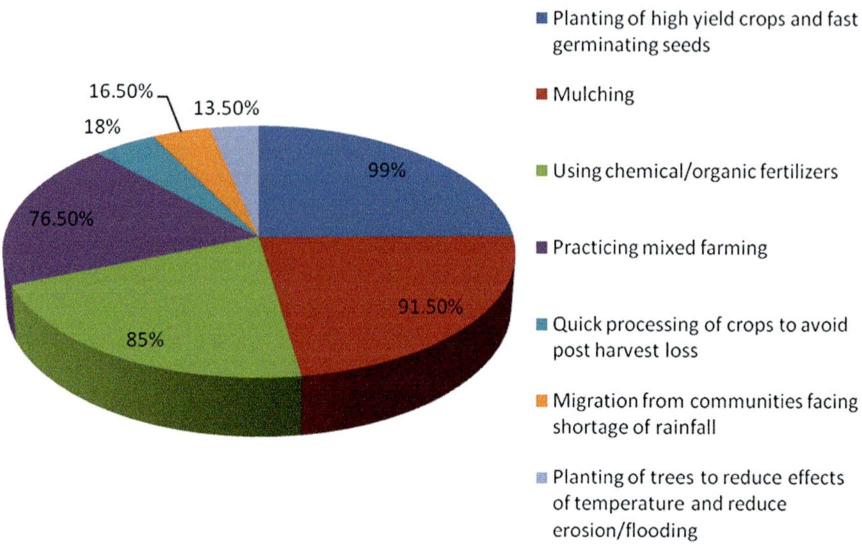

Fig. 3 Major adaptation/coping strategies to climate change by selected farmers in Southwestern Nigeria

Table 8 Adaptation strategies adopted

S/N	Climate change adaptation measures	Percentages (%)
1	Planting of high-yield crops and fast-germinating seeds	99
2	Mulching	91.5
3	Using chemical/organic fertilizers	85
4	Practicing mixed farming	76.5
5	Quick processing of crops to avoid postharvest loss	18
6	Migration from communities facing shortage of rainfall	16.5
7	Planting of trees to reduce effects of temperature and reduce erosion/flooding	13.5

Source: Author's fieldwork

References

Abiodun EO (2011) Effects of climate change on food and human security in Nigeria, global food insecurity rethinking agricultural and rural development paradigm

Adejuwon JO (2004) Vulnerability in Nigeria: a national-level assessment. In: Leary N, Conde C, Kulkarni J, Nyong A, Pulhin J (eds) Climate change and vulnerability. EARTHSCAN, London, pp 198–217

Adejuwon JO (2006) Food crop production in Nigeria II: potential effects of climate change. Clim Res 32:229–245

Adejuwon JO, Odekunle TO (2006) Variability and the severity of the "Little dry season" in Southwestern Nigeria. J Am Meteorol Soc 19:483–493

Adesina FA, Odekunle TO (2011) Climate change and adaptation in Nigeria: some background to Nigeria's response – III. In: International conference on environmental and agriculture engineering IPCBEE, vol 15. IACSIT Press, Singapore

Adebayo F, Balogun I.A, Adediji AT, Akande O, Abdulkareem SB (2017) Assessment of Urban heat island over Ibadan metropolis using landsat and MODIS. International Journal of Environment and Bioenergy 12(1):62–87

Adepitan, JO, Falayi, EO and Ogunsanwo, FO (2017) Confirmation of climate change in Southwestern Nigeria through analysis of rainfall and temperature variations over the region. Covenant Journal of Physical & Life Sciences (CJPL) Vol. 5 No. 1

Ali A, Erenstein O (2017) Assessing farmer use of climate change adaptation practices and impacts on food security and poverty in Pakistan. Clim Risk Manag 16:183–194

Balogun IA (2009) Assessing climate change: the Nigerian rainfall perspective and its implications. In: AT Salami, IE Ofoezie, OO Awotoye (eds) Your Planet needs you – Unite to combat climate change, vol 2. Proceedings of the 2nd annual conference of the Institute of Ecology and Environmental studies, OAU, Ile –Ife, pp 186–192

Broadhead L, Howard S (2009) Deepening the debate over 'sustainable science': indigenous perspectives as a guide on the journey. Sustain Dev 19(5):301–311

Building Nigeria's Response to Climate Change (BNRCC) Project (2011) National adaptation strategy and plan of action on climate change for Nigeria (NASPA-CCN). Prepared for the federal ministry of environment special climate change unit

Emielu SA (2000) Senior secondary geography, Revised edn. Geographical bureau Nig. Ltd, Ilorin

Filani MN, Alokan O (1993) Osun state. In: Udo RK, Mamman AB (eds) Nigeria: giants in the tropics. Gbajumo publishers, Lagos

Fischer G, Shah M, van Velthuizen H (2002) Climate change and agricultural vulnerability. International Institute for Applied Systems Analysis, Report prepared under UN institutional contract agreement, 11–13 for World Summit on Sustainable Development, Laxenburg

Hansen J, Sato M, Ruedy R (2012) Perception of climate change. Proc Natl Acad Sci USA 109: E-2415–E-2423

IFAD (2009) Enabling poor rural people to overcome poverty. International Fund for Agricultural Development. www.ifad.org/operations/projects/regions/pa/factsheets/ng.pdf

Iloeje NP (1981) A new geography of Nigeria. Longman, Lagos, Nigeria, p 59

International Panel on Climate Change-IPCC (2007a) Climate change 2007: synthesis report. Contributions of Working Groups I, II, and III to the fourth assessment report of the intergovernmental panel on climate change. IPCC, Geneva

International Panel on Climate Change-IPCC (2007b) Climate change (2007). In: Solomon S, Qin D, Manning M, Chen Z, Marquis M, Averyt KB, Tignor M, Miller HL (eds) The physical science basis. Contribution of working group I to the fourth assessment report of the intergovernmental panel on climate change. Cambridge University Press, Cambridge, p 996

International Panel on Climate Change-IPCC (2014) Climate change 2014: IPCC report on climate impacts: "It's time to get serious". Synthesis report, contributions of working group I, II, and III to the fifth assessment report of the Intergovernmental Panel on Climate Change, Geneva

Jackson OI, Isienyi NC, Osudiala CS, Odofin BT, Adeyemi AA, Odeleye OA and Amoo VO (2012) Analysis of temperature trends of Ibadan, Nigeria over the period of 1965–2013. Journal of Forestry Research and Management 9:61–72

Matthew OJ, Oladokun DO (2009) The trends of spatial rainfall variability/change over Nigeria during the last 50 years. In: Salami AT, Ofozie IE, Awotoye OO (eds) Your planet needs you-Unite to combat climate change, vol 2. Proceedings of the 2nd annual conference of the Institute of ecology and environmental studies. OAU, Ile –Ife, pp 14–21

MEFRN (2003) Nigeria's first national communication under the United Nations framework convention on climate change. Landmark publication, Lagos

Nieuwolt S (1982) Tropical climatology: an Introduction to the climates of the low latitudes. Wiley, New York

Nigerian Environmental Study Team (NEST) (2004) Regional climate modelling and climate scenarios development in support of vulnerability and adaptation studies: outcome of regional climate modelling effects over Nigeria. NEST, Ibadan

Nnamerenwa GC, Jessie GF, Nwatu C (2017) Analysis of price differentials in the marketing of plantain in Abia state, Nigeria. In: Proceeding of the 1st International conference of postgraduate students association of Michael Okpara University of Agriculture Umudike in conjunction with postgraduate school Michael Okpara University of Agriculture Umudike, Abia State

Odekunle TO (2010) Trends in the extreme daily amount and the frequency of rainy days in Sokoto, Sudano-Sahelian ecological zone of Nigeria. Ife Res Publ Geogr 9(1):1–14. Geography Department, Obafemi Awolowo University, Ile –Ife, Nigeria

Odekunle TO, Adejuwon SA (2007) Assessing changes in the rainfall regime in Nigeria between 1961 and 2004. GeoJournal 70(2–3):145–159

Odekunle TO, Eludoyin AO (2008) Sea surface temperature patterns in the Gulf of Guinea: their implications for the spatio-temporal variability of precipitation in West Africa. Int J Climatol. https://doi.org/10.1002/joc.1656. Available online on www.interscience.wiley.com. Accessed 8 Dec 2015

Odekunle TO, Balogun EE, Ogunkoya OO (2005) On the prediction of rainfall onset and retreat dates in Nigeria. Theor Appl Climatol 81:101–112

Odekunle TO, Andrew O, Aremu SO (2008) Toward a wetter Sudano- Sahelian ecological zone in the twenty-first century, Nigeria. Weather 63:66–70

Ojo O (1977) The climates of West Africa. Heinemann publishers, London

Oladipo EO (1995) An indication of abrupt change of rainfall and its potential impact on energy development in Nigeria. In: Umolu JC (ed) Global climate change: impact on energy development. DAMITECH Nigeria Limited, Lagos

Olaniran OJ (1983) The onset of the rains and the start of the growing season in Nigeria. Niger Geogr J 26(1 &2):81–88

Ologunorisa TE, Tersoo T (2006) The changing rainfall pattern and its implication of flood frequency in Makurdi, Northern Nigeria. J Appl Sci Environ Manag 10(3):97–102

Oluleye A (2009) Change detection in rainfall characteristics across climatic zones in Nigeria. J Metrol Climatol Sci 7:6–10

Speranza CI (2010) Resilient adaptation to climate change in African agriculture. DIE, Bonn. (Studies / Deutsches Institut für Entwicklungspolitik; 54) ISBN 978-3-88985-489-6. Deutsches Institut für Entwicklungspolitik

Vedwan N, Rhoades RE (2001) Climate change in the Western Himalayas of India: a study of local perception and response. Clim Res 19:109–117 Inter-Research

Adaptation of Sorghum (*Sorghum bicolor* L. Moench) Crop Yield to Climate Change in Eastern Dryland of Sudan 124

Haitham R. Elramlawi, Hassan I. Mohammed, Ali W. Elamin, Omer A. Abdallah, and Abdel Aziz A. M. Taha

Contents

Introduction 2551
Climate Change 2552
Dryland Farming 2553
Dryland Farming in Sudan 2554
Sorghum Cultivation and Production in Gadarif State 2555
Problems Related to Conventional Tillage in Dryland Farming of Gadarif 2556
Soil and Water Management in Dryland Farming 2556
Objectives of the Study 2559
Methodology 2559
  Site Description 2559
  Experimental Design and Methods 2560
  Measurements, Calculations, and Statistical Data Analysis 2560

H. R. Elramlawi (✉)
Center of Dryland Farming Research and Studies (CDFRS), Faculty of Agricultural and Environmental Sciences, University of Gadarif, Gadarif, Sudan
e-mail: elramlawi@gmail.com

H. I. Mohammed
Department of Agricultural Engineering, College of Agricultural Studies, Sudan University of Science and Technology, Khartoum, Sudan
e-mail: hassantal@gmail.com

A. W. Elamin · O. A. Abdallah
Department of Agricultural Engineering, Faculty of Agriculture, University of Khartoum, Khartoum, Sudan
e-mail: aliwidaa59@gmail.com; omerabdallah@gmail.com

A. A. A. M. Taha
Department of Agricultural Engineering, University of Gadarif, Gadarif, Sudan
e-mail: azizabdallah@gmail.com

W. Leal Filho (ed.), *Handbook of Climate Change Resilience*,
https://doi.org/10.1007/978-3-319-93336-8_157

Results 2561
Variation in Annual Rainfall 2561
Grain Yields, WUE, Water Use, Soil Water in Responses to Treatments 2562
Discussion 2566
Conclusion 2568
References 2569

Abstract

Dryland sorghum (*Sorghum bicolor* L. Moench) crop production in semiarid areas of eastern Sudan is vulnerable to climate change through the direct impacts of sporadic and highly varied rainfall accompanied by high temperature conditions. Continuous practice of conventional tillage (WLD) in clay soil using wild-level disc harrow over more than 70 years had created a hard subsurface layer that prevents both adequate storage of water and deep penetration of crop roots. Thus, sorghum crop experiences frequent water stress periods during its life cycle due to unreliable water source (rainfall) and deteriorated reservoir (soil). Investigating the capability of conservation tillage (CT) practices and in-situ water harvesting (ISWH) techniques to achieve sustainable and adaptable sorghum yield is the main objective of this study.

The behavior of sorghum grain yield (GY), water use efficiency (WUE), water use (ET), and soil water status at sowing and at harvest (SWS and SWH, respectively) were the main parameters that explored under CT and ISWH: WLD, as a control treatment, no-till (ZT-M) and contour chisel plowing (CCP-M), both with residue mulching, as CT practices, and contour ridge-furrow (CRF-M), and contour tied-ridging (CTR-M CT), both with residue mulching, as ISWH techniques. The research was carried out at a Pilot Farm of University of Gadarif, Twawa area, during three consecutive rainy seasons (2011–2013). The experiment was set-up using a randomized complete block design (RCBD) with four replications.

Mean GY ($Kg.ha^{-1}$) and WUE ($Kg.ha^{-1}. mm^{-1}$) in response to CT and ISWH treatments were about 1895 and 5.0, 2184 and 6.2, 2198 and 6.0, 1945 and 5.2, and 2357 and 6.5, under WLD, CFR-M, CCP-M, ZT-M, and CTR-M, respectively. GY under CTR-M were generally higher than the mean GY either in wet (12%) or in dry (18%) rainy season. At sowing and harvest time, moisture in the soil profiles under CTR-M treatment is more than WLD, CCP-M, ZT-M, and CTR-M treatments by 3–15% and 9–28%, respectively. Treatments CTR-M, CFR-M, and CCP-M showed adaptation potential to climate change in drought-prone dryland in eastern Sudan.

Keywords

Dryland farming · In-situ water harvesting · Conventional tillage · Conservation tillage · No-till · Residue mulching · Chiseling · Ridging

Introduction

Sudan is located in the eastern part of Africa, between latitudes 8.4° and 23.3°N and longitudes 21.5° and 39.0°E. It covers a total area (TA) of about 1.88 million km^2 and shares borders with seven countries: Egypt, Libya, Chad, Central Africa, Ethiopia, Eritrea, and South Sudan (Fig. 1). The total population of Sudan is estimated to be 39.0 million, growing at 2.7%, with more than 30 million people living in rural areas (CBS 2011).

Fig.1 Map of Sudan

Agriculture is the backbone of Sudan's economy, although the economic contribution of this sector in the country gross domestic product (GDP) is declining sharply due to the emergence of the oil industry in 2000 (UNEP 2007). Arable land constitutes about one-third of TA with about 21% of this land under cultivation. Over 35% of TA consists of pasture and rangelands (FAO 2012). The agriculture sector with its various sections (traditional, semi-mechanized, irrigated, and livestock) provides direct livelihood for about 70% of the population and 80% of poor people, especially in rural areas. It also contributes 40% of the country's GDP and more than 90% of the non-oil export earnings. More than 80% of labor is employed in agricultural and its related practices (Mustafa 2006). Agriculture supplies about 60% of the raw materials for agro-based industries (Elbashir et al. 2004). Currently, the sector faces several constraints including, but are not limited to, declining productivity, increasing poverty and food insecurity levels, massive land degradation, and climate change dilemma.

According to the average amount of annual rainfall (AR), four ecological zones can be identified: desert (AR $<$ 100 mm in 29% of TA); semi-desert (AR ranges between 100 and 300 mm in 19.6% of TA); low-rainfall savannah (AR ranges between 300 and 500 mm in 51.1% of TA); and high-rainfall savannah (AR $>$ 500 mm in 10.3% of TA). It has been estimated that the semiarid zone of the Sudan encompasses about 70% of the surface area of the country, and more than 70% of the Sudanese population lives in it, with herding and farming as the main sources of livelihood (Elhadary 2007).

Like many of its neighbors, Sudan is a country of inherently fragile ecosystems, frequent droughts, ongoing desertification, and as a result, the pressing challenges are to address the national priorities of food security, water supply, and public health (Anonymous 2018).

Climate Change

The Fifth Assessment Report (AR 5) of the Intergovernmental Panel on Climate Change (IPCC) by Niang et al. (2014) presents strong evidence that warming over land across Africa has increased over the last 50–100 years, and in particular, surface temperatures have already increased by 0.5–2 °C over the past 100 years. Data from 1950 onwards proved that climate change has changed the magnitude and frequency of some extreme weather events in Africa. The health, livelihoods, and food security of people in Africa have been affected by climate change (Cramer et al. 2014). African countries are the most vulnerable ones in respect to climate change due to the negative effects of extreme events of droughts, floods, and others, on: fragile infrastructure, vulnerable food production systems, and poor people with low adaptive capacity (Filho 2011).

Sudan's First National Communication (SFNC 2003) to the UNFCC, submitted in 2003, identified agriculture, water resources, and public health as most vulnerable sectors to climate change and climate variability. It provided an assessment of likely impacts of climate change on these sectors including decreasing annual rainfall, increasing both rainfall variability and average annual temperatures, all of them contribute to drought conditions in many areas. Changes in temperature and rainfall patterns represent a priority threat to food security in Sudan. Current increasing variability is a manifestation of long-term change of climatic conditions in the country, region, and globally. Changes in average temperature or precipitation often do not show strong signals, but the well-observed trends of decreasing annual rainfall and increased rainfall variability have contributed to drought conditions in many parts of Sudan (Hulme et al. 2001).

Relative to baseline expectations, the SFNC (2003) indicated an average warming range of 1–3 °C and average negative change in precipitation of −5.8% by 2030. Average annual rainfall has declined from about 425 $mm.year^{-1}$ during the 1941–1970 period to about 360 $mm.year^{-1}$in the 1970–200 period. This represents a decrease of annual rainfall of about 0.5% per year. It also revealed that there is a trend of greater rainfall variability, increasing at a rate of about 0.2% per year. Rainfall is already extremely erratic and varies widely over Sudan. The severity of drought experienced depends on the variability of rainfall both in amount, distribution, and frequency. Projections from IPCC (Niang et al. 2014) indicate a substantial warming trend across Sudan with projected daily temperature increases by 1.5–3 °C by the middle of the twenty-first century. In contrast, there is no clear projected trend in mean annual rainfall but large year-to-year variability exceeds any climate change signal by the mid-twenty-first century.

Agriculture is vulnerable to climate change in a number of dimensions. Higher temperatures eventually reduce yields and tend to encourage weed and pest proliferation. Greater variations in precipitation patterns increase the likelihood of short-run crop failures and long-term production declines. Although there might be gains in some crops in certain regions of the world, the overall impact of climate change on agriculture is expected to be negative, threatening regional food security in many parts of the developing world, that are still predominantly agrarian in nature, in particular and world over in general (Nedumaran et al. 2015).

Dryland Farming

Global population growth projections of 2–3 billion people over the next 40 years, combined with changing diets, result in a predicted increase in food demand of 70% by 2050 (UNESCO 2012). This challenge cannot be achieved without implementing more sustainable farming methods and responding to climate change in dry areas (Stewart and Thapa 2016). Drylands (DLs) are found primarily in arid, semiarid, and dry subhumid areas in which the ratio of annual precipitation to potential evapotranspiration falls within the range from 0.05 to 0.65 (UNESCO 2012). DLs cover more than 40% of the world's land area and are home to more than two billion people, about 90% of them are living in developing countries (Yazar and Ali 2016) and 50% are living in poverty (Stewart and Thapa 2016). Poverty, food insecurity, biodiversity loss, frequent droughts, and environmental degradation are widespread in these areas. This region is characterized by sporadic and highly unpredictable precipitation events accompanied by high temperatures that create the potential for plant water stress during the crop cycle (Peterson et al. 2012). Length of growing season in DLs is very short that hardly reaches 179 days (FAO 2000). The effects of climate change will lead to even more water scarcity and declining crop yields, leaving the people of these regions excessively vulnerable in the absence of appropriate technologies and risk management strategies (Rodriguez et al. 2016).

Although dryland and rainfed farming rely mainly on rainfall water to fully or partially compensate soil moisture deficit, they can be vastly different. Dryland farming (DLF) emphasizes water conservation, sustainable crop yields, limited fertilizer and other inputs, and wind and water erosion constraints. In contrast, rainfed farming often deals with disposal of excess water and water erosion constraints and strives for maximum crop yields using high levels of inputs (Stewart and Burnett 1987). Successful DLF requires the integrated management of soil, water, crops, and plant nutrients. In DLF, small-scale, resource-poor, usually subsistence-based farmers, widely referred to as small-holders, can be able to manage the multiple risks through diversification, flexibility, and adaptability (Mortimore and Adams 1999). The impacts of climate change in drylands are likely to lead to still more people and larger areas of land being affected by water scarcity and the risk of declining crop yields – with the peoples of drylands in developing countries least able to adapt due to poverty (Koohafkan and Stewart 2008).

Dryland Farming in Sudan

In Sudan, the two terms: dryland farming and rainfed farming are interchangeably used. Generally, there are two types of dryland farming in the country: traditional farming (TF) and semi-mechanized farming (SMF). A majority of Sudanese farmers (70%) rely on TF subsector for their sustenance which accounting for an area of around nine million hectares. It is generally a low input/low yield production system characterized by small farms ranging from 2 to 30 ha in size and relying on labor intensive cultivation with hand tools. Available estimates show that TF produces 90% of millet, 11% of sorghum, 48% of groundnuts, and 28% of sesame in the country (Abbadi and Ahmed 2006). Despite its importance, this subsector has suffered from low economic investment, resulting in negligible technical development. Recently, smallholder farmers in TF subsector have tended to use tractors and some agro-mechanical implements as a coping mechanism for managing risk associated with increasing uncertainty in the onset of rainy season and labor migration to urban areas.

In SMF, manual labors are partly replaced by tractors and harvesting machines. It was initiated by the British colonial power in Gadarif area in 1944 to meet the food needs of its army in East Africa (Elhadary 2010; UNEP 2007). Despite of its positive role in the economy, the sector has failed to fulfill its ambitions in securing food in the country and exporting some surplus. Not only that, SMF has been claimed to be responsible for the spread of environmental degradation by cutting of forests and resource use conflicts with herders (Elhadary 2015). Today, SMF occupies a large area estimated to be 6.5 million hectares of Sudan clay plains in the low- and high-rainfall savannah ecosystems. The principle crops cultivated are sorghum, sesame, groundnuts, and, to a lesser extent, cotton and sunflower (UNEP 2007). Annual crop yield depends on inputs that often vary from year to year, due to variability in: rainfall, formal credit, and the level of prices in the previous season. SMF subsector is the main producer of sorghum, producing 65% of the total production and thus plays a major role in determining whether the country enjoys a food surplus or suffers a food deficit in any one season (Elbashir and Ahmed 2006).

Sorghum Cultivation and Production in Gadarif State

In semiarid of Gadarif State, Eastern Sudan, DLF is practiced on about 2.6 Mha (Elhadary 2015). It is the basis of all subsistence farming and the dominating source of food. Crop production relies solely on unimodal pattern of rainfall extends from May to October ranging from 200 to 600 mm year^{-1}. Dryland Sorghum (*Sorghum bicolor* L. Moench) is widely grown as the main staple food grain in Sudan. Sorghum ranks second in importance after maize in Africa with cultivated area of about 24 million hectares and in low productivity level of 0.8 Mg.ha^{-1} (Maredia et al. 2000). Sorghum crop grows excellently on Black Cotton soils that are clayey due to higher fraction of montomorillonite clay (Vertic soils). Sorghum tolerates soil pH of 6.0–8.5 and mild levels of salinity and alkalinity. Sorghum crop is well adapted to tropical conditions and it withstands relatively higher temperatures during the growing period. It tolerates drought better compared to other cereals such as wheat or rice. Sorghum grows even in dry zones where annual rainfall is about 400 mm but thrives reasonably well in regions with annual rainfall between 400 and 600 mm. Long-term trials across different continents have shown that sorghum needs 500 mm moisture per season (Krishna 2014).

In Gadarif, sorghum is grown as a dual-purpose crop, the grain being used for human consumption and the straw for livestock feeding (Mohamed 2009). The predominant soils of the region are characterized by deep, cracking Vertisols. Such soil are often considered as problem soils mainly due to their stickiness and hardness natures when wet and dry, respectively which may hamper land preparation (Dudal 1980). However, much evidence has been produced that Vertisols can be transformed into productive cropland if their management addresses their inherent physical problems (Osman 2013). The most popular and conventional tillage system in DLF of Gadarif State is based on using a 75 HP tractor and trailed wide-level disk harrow (WLD) with mounted seed boxes. These disks have approximately 3.4–4.0 m working width, 1.8–2.0 ha.h^{-1} working rate, and 5–8 cm working depth. All crop residues are removed from the field prior a new crop is sown or planted in a fine, loose, and smooth soil. The rationale behind this practice is that: weeds are well controlled and the sowing or planting operation can be done effectively. By the onset of rainfall, first plowing is done to eradicate early weeds using WLD. A second disking is carried between mid-June and mid-July during which, planting is completed. Intensive tillage practices over bare soil could increase soil and water losses through increasing erosion, evaporation, and runoff processes (Elramlawi et al. 2009).

The productivity of sorghum under conventional management is very low and ranges between 285 and 642 $Kg.ha^{-1}$. This may attributed, among other many factors, to the adverse impacts of intensive use of WLD by constituting (forming) a subsurface plowpan (hardpan) and causing poor soil water retention, mono-cropping cultivation and absence of crop rotation, weeds and pests infestation, uneven rainfall distribution, and recurrent severe dry spell droughts due to increased climatic variability and tendency of Vertisols to form surface crusts which inducing runoff losses (IFAD 2010).

Problems Related to Conventional Tillage in Dryland Farming of Gadarif

Intense and repeated usage of WLD as conventional tillage could encourage soil compaction due to farm implement traffic and formation of plowpan (Elramlawi et al. 2009). These undesirable effects may greatly hinder water conservation by reducing water infiltration and, therefore, crop growth and production, unless corrected by use of an appropriate tillage method (Unger et al. 2012). Hardpan in the subsurface layer can be detrimental to root proliferation below the plough layers (Gill and Aulakh 1990). Also, hardpans have a high bulk density with a few macropores for roots to grow through (Rafiq 1990). The often form mechanical impedance that is great enough to reduce root growth markedly (Gerard et al. 1982; Hassan and Gregory 1999; Logsdon et al. 1987). A higher soil bulk density can decrease root length (Hill 1990) and trigger the formation of lateral roots (Ball-Coelho et al. 1998). Generally, favorable crop growth and yields, however, are obtained when such soils are loosened before or at planting with chisel plows or other types of subsoiling implements (Raper 2006). When adequately loosened, various types of mulch tillage implements, including no-tillage (ZT), can be successfully used. Disrupting the dense soil layers also enhances the potential for achieving greater soil water storage (Unger et al. 2012).

Soil and Water Management in Dryland Farming

Adaptation to climate change in DLF is one of the key solutions to persistent food insecurity in arid and semiarid areas. In sub-Saharan Africa (SSA), DLF is characterized by low productivity due to suboptimal performance related to management aspects rather than low physical potential (Rockström and Falkenmark 2000; SIWI 2001). Wani et al. (2012) state that evidence from water balance analyses in farmers' fields around the world shows that only a small fraction, less than 30% of precipitation, is used as plant transpiration to support plant growth while in arid areas about 90% of precipitation can be lost as evaporation and only 10% for transpiration. High water losses under dryland conditions accompanied by unevenly distributed and fluctuated rainfall pattern are the main limiting factors of crop production in DLF. Moreover, dry spells, as agricultural drought, which would be more frequent due to climate change (IPCC 2007), can adversely affect crop production especially during critical growth stages (Rockström et al. 2010). Agricultural droughts occur when the crop water demand cannot be met by either rainfall or soil water supply. Such drought effects can be enhanced by high-temperature events, which may give rise to large vapor pressure deficits that greatly enhance the evaporative demand beyond what can be sustained through soil water supply (Olesen 2014).

On the other hand, there are management-induced dry spells and droughts where the rainfall received is not stored and utilized in a productive way. These cases can be prevented by improved management practices than blaming on droughts (Rockström 2003). Within this context, increasing efficiency of water use by properly managing soil and water are crucial in securing food, alleviating poverty and achieving sustainable livelihood. Three aspects are engaged in efficient precipitation management which are: maximizing precipitation capture in the soil, minimizing evaporation of stored soil water, and maximizing plant water use efficiency (WUE). Water conservation is the main component of a dryland cropping system. It involves enhancing water infiltration into the soil (reducing runoff), suppressing soil water evaporation, and minimizing soil water use by non-crop plants (Unger 1990).

Soil tillage is one of the main primary tools, beside crop selection, available to manage precipitation storage in the soil. Tillage management, including type and timing, affect the amount of crop residue cover maintained on a soil surface and the soil pore size exposed to the atmosphere, which in turn control water capture and water retention (Peterson et al. 2012). Soil tillage has distinct effects on soil nutrient dynamics and moisture availability to crops (Krishna 2014). Tillage actually involves opening or loosening soil. It basically aims at obtaining better soil structure and tilth, so that seed germination, seedling establishment, and a good crop stand is achieved. Tillage systems can be grouped into: no or zero-tillage systems, stubble-mulch tillage systems, reduced tillage systems, and conventional tillage systems (Krishna 2014). In general, conventional tillage systems involve thorough plowing, turning, and exposing subsurface layers of soils to atmosphere. Intensive tillage has many adverse effects on the soil including losses in soil organic matter (Unger et al. 2006) and making soil vulnerable to surface erosion and loss of nutrients (Krishna 2014).

The use of conservation tillage (CT) measures, such as minimum till and no-till, under various water harvesting (WH) techniques, in some developing countries, was reported to conserve soil water (Rosegrant et al. 2002). In-situ WH (ISWH) and CT among other field management practices have been applied worldwide to improve water use and consequently crop production levels. In ISWH (often referred as water conservation), rainwater is collected where it falls. ISWH techniques include activities such as mulching, contour farming, and ridging. The purpose behind these methods is to ensure long holding of rainwater on the cropped area to ensure infiltration (Habitu and Mahoo 1999).

Ridge-furrow (RF) is the method of land preparation whereby the top soil is scraped and concentrated in a defined region to deliberately raise the seedbed above the natural terrain (Lal 1990). Crops usually grows on the ridges in rows, with one or more rows per ridge, even though in some cases, crops may be grown in the furrows to make advantage of the wetter condition of the soil under the furrow. It is an effective water management and erosion control practice when the system is on the contour (contour ridge-CR) and the slope of the land is less than 7% (Moldenhauer and Onstah 1977). RF is very effective in conserving water in the root zone in semiarid to subhumid regions, particularly when ridges have cross ties (CTR) in the furrows (knowing either as tied-ridging, furrow blocking, or basin tillage) (Gardner et al. 1999). In clay soil, tied-ridging (CTR) is likely to reduce surface runoff and increase retained water within the field if carefully designed across the slope or along the contour lines. However, runoff prevention does not ensure infiltration because some of the water temporarily stored on the surface may evaporate before it can infiltrate the soil (Unger 1990). Past and recent research works in Africa has shown that tied-ridging often leads to little or no runoff. In Zimbabwe, Piha (1993) and Vogel (1993); in Botswana, Carter and Miller (1991); and in Burkina Faso, Hulugalle and Malton (1990) have shown similar results of more retained water and less runoff.

Conservation tillage is defined as any tillage sequence, the object of which is to minimize or reduce loss of soil and water; operationally, a tillage or tillage and planting combination which leaves a 30% or greater cover of crop residue on the surface (SSSA 2018). With this definition, many practices have been considered as the measure of conservation tillage, including shallow surface tillage (reduced tillage RT), no-tillage, crop residue mulching, and subsoil mulching (Lampurlanés et al. 2002). CT can also be defined as "any tillage sequence that reduces loss of soil or water relative to conventional tillage; often a form of noninversion tillage that retains protective amounts of residue on the surface" (SCSA 1982). Compared to conventional tillage, there are several benefits from conservation tillage such as economic benefits to labor, cost and time saved, erosion protection, soil and water conservation (Glab and Kulig 2008) and increases of soil fertility or reduce nutrient loss (Wang and Gao 2004; Limousin and Tessier 2007). CT affects the chemical, biological, and physical properties of soil such as soil moisture, bulk density, and soil strength. When combined with prudent management of crop residues, crop rotations, and cover crops, CT is a useful technology for protecting soil and sustaining crop production (Blanco and Lal 2008).

One of the most successful soil management techniques in agricultural land is no-tillage management (ZT), and it is being applied worldwide (Barbera et al. 2012; Lieskovsky and Kenderessy 2014). Hence, during recent years, the trend is to adopt no-tillage or zero-tillage (ZT) systems. Practices such as residue burning and fallow period tillage have reduced SOC levels by as much as 60%. Current no-till dryland systems with intensified crop rotations have stabilized soil carbon and increased soil organic matter accretion (Hansen et al. 2016). No tillage systems are preferred because it protects soils from massive loss of soil organic carbon (SOC) as CO_2 and avoids erosion to a great extent (Duxbury 2005). In ZT land preparation systems adopted are equally important because they protect fields from erosion, reduce loss of soil moisture, and provide better crop stand. ZT is a method of crop production that involves no seedbed preparation other than opening the soil for placing seed at the desired depth (SCSA 1982). Adequate quantities of residues are often required to remain above the soil surface to provide cover and to protect the soil against erosion until the canopy of the next crop is well-developed (Gardner et al. 1999). Verhulst et al. (2011) found that ZT with residue retention had higher soil water content than management practices involving CT. In contrast, CT practices, and ZT in particular, do not always result in favorable soil conditions. Increased soil strength and decreased porosity and water infiltration have been observed by several authors under continuous no-tillage (López et al. 1996). Moreover, converting management practice from conventional to ZT methods may result in low crop yield during the few years after implementing ZT (VandenBygaart et al. 1999). Continuous ZT may result in developing shallow compaction and topsoil hindering after some years of application, which are manifested as increased bulk density (BD) and mechanical resistance, decreased macroporosity (Sasal et al. 2006), and low infiltration rate (Alvarez et al. 2009).

Chiseling with mulch along the contour (CCP-M) is also considered as CT practice (Lampurlanés et al. 2002). The primary objective of chisel plow is to loosen the compacted subsoil. It also improves grain yield by enhancing root growth and infiltrating more rainfall deeper in the soil profile particularly in soils with compacted low permeability sublayers (Salih et al. 1998).

Objectives of the Study

The objective of this study is to investigate the capability of conservation tillage (CT) and in-situ water harvesting (ISWH) systems to increase resilience of sorghum production system and attain sustainable and adaptable crop yield in dryland farming of Gadarif State, Sudan. The performances of the two systems along with conventional one are evaluated through different agronomic parameters such as sorghum grain yield (GY), water use efficiency (WUE), water use (ET), and soil water status at sowing and at harvest (SWS and SWH, respectively). The field treatments involve WLD, as a control treatment, no-till (ZT-M) and contour chisel plowing (CCP-M), both with residue mulching, as CT systems, and contour ridge-furrow (CRF-M) and contour tied-ridging (CTR-M CT), both with residue mulching, as ISWH systems.

Methodology

Site Description

The experiment was carried out at the Pilot Farm of the Faculty of Agricultural and Environmental Science, University of Gadarif, at Twawa area (long. 35.24° E, lat. 14.02° N), Gadarif State in Eastern Sudan, for three consecutive rainy seasons: 2011/12 (first season, FS), 2012/13 (second season, SS), and 2013/14 (third season, TS). The area has a mean altitude of 599 m above mean sea level and a semiarid climate with an average annual rainfall of 581 mm (Gadarif Meteorological Station 2014). Extensive use of WLD in the past led to the formation of a subsurface hardpan. The dominant cropping system is continuous summer sorghum, which accounts for over 90% of the total area for crop production. The study area is representative of a typical dryland farming region. The experimental site has a dark clay soil, soil pH was 7.9, electrical conductivity was 0.34 ds/m, and hydraulic conductivity was 3.2 mm/h Soil bulk density was 1.27 g/cm^3, gravimetric field capacity (F.C.), and permanent wilting point (P.W.P) were 42.3% and 30.8%, respectively. The averages of sodium (Na) content, calcium and magnesium (Ca + Mg) content, and sodium absorption ratio (SAR) were 1.54 mmol$^+$/l, 1.21 mmol$^+$/l, and 1.8 mmol$^+$/l, respectively.

Experimental Design and Methods

The experiment was set-up using a randomized complete block design (RCBD) with conventional tillage system (wide-level disk harrow with no residue mulching, WLD), two in-situ water harvesting techniques (contour ridge-furrow with residue mulching, CRF-M and contour tied-ridging with residue mulching, CTR-M) and two conservation tillage techniques (no-till with applying of herbicides and crop residue mulching, ZT-M and contour chisel plowing with residue mulching, CCP-M). The design was based on previous field studies in Twawa area including a comparison of different soil surface formation systems for maize production (Elramlawi et al. 2009). Plot size was 20 × 15 m and 3 m spacing between blocks and treatments with four replications. Details of the methods for tillage and crop residue in each treatment are described in Table 1. Locally recommended sorghum variety was used (i.e., Arfa'a Gadamak). The interrow and row spacing was 20 × 80 cm. Sorghum was seeded in each mid-July; the crop was harvested in each late-October.

Measurements, Calculations, and Statistical Data Analysis

Measurements included daily rainfall (DR), soil water at sowing (SWS), at early-season (SWE), at mid-season (SWM), at late-season (SWL) and soil water at harvest (SWH), water use (expressed as evapotranspiration, ET), grain yield of sorghum (GY), and water use efficiency (WUE). Rainfall during the experimental periods was measured using a rain gauge at the experimental site. Soil profile was sampled every 20 cm to ensure that soil water content was measured down to the maximum rooting depth (60 cm). Soil moisture content was determined gravimetrically and multiplied by BD and thickness of soil layers (200 mm) to be expressed in depth equivalent (mm).

Apparent water use during the growing season, expressed as evapotranspiration (ET, in mm) was calculated from seasonal rainfall (P) and soil water consumption during the growing season. Assuming no deep percolation or runoff to occur, the following simple equation was used to calculate ET:

Table 1 Agronomic treatments in Twawa, Gadarif State in Sudan (2011/12, 2012/13, and 2013/14)

Treatments	Description
WLD	Removing crop residues, first disking on bare soil (5–8 cm depth) for weed control using wide level disk harrow (WLD), second disking and seeding using WLD, weed control by hand 3 weeks after emergence
CRF-M	Ridging (15–20 cm height and 80 cm spacing) with mulch along the contour lines, manual seeding in holes (2 cm depth and 20 cm spacing) in furrow, weed control by hand 3 weeks after emergence
CTR-M	Ridging (15–20 cm height and 80 cm spacing) with mulch along the contour lines, constructing cross ties (8–12 cm and 1.5 m spacing), manual seeding in holes (2 cm depth and 20 cm spacing) on ridge, weed control by hand 3 weeks after emergence
ZT-M	Weed control using herbicides, mulching, manual seeding in holes (2 cm depth, 20 cm spacing between holes and 80 cm spacing between rows)
CCP-M	Plowing (20–25 cm depth) with mulch along the contour lines, manual seeding in holes (2 cm depth, 20 cm spacing between holes and 80 cm spacing between rows, weed control by hand 3 weeks after emergence

$$\mathrm{ET} = \mathrm{SWS} + \mathrm{P} - \mathrm{SWH} \qquad (1)$$

where

SWS and SWH: are the soil water (mm) at sowing and harvesting, respectively and P: is the seasonal rainfall (mm).

Sorghum grain yields were determined at harvest, and water use efficiency (WUE, in kg ha^{-1} mm^{-1}) was calculated from grain yield (GY) and ET according to:

$$\mathrm{WUE} = \frac{\mathrm{GY}}{\mathrm{ET}} \qquad (2)$$

Statistical analysis was conducted using IRRISTAT (2005). The data were subjected to an analysis of variance (ANOVA) using the F-test procedure. The mean separation was based on the LSD test at the 0.05 probability level (at $P \leq 0.05$).

Results

Variation in Annual Rainfall

During the 3-year experimental period (2011–2013), annual rainfall averaged 485 mm, ranging from 418 mm in the dry year 2013 to 542 mm in the wet year 2012, which were less than 581 $mm.year^{-1}$, the long-term average of Gadarif town (Gadarif Meteorological Station 2014). Growing seasonal rainfall (GSR) averaged 362 mm, ranging from 312 mm in 2013 to 407 mm in 2012. On average, rainfall during the growing season accounted for 74% of the annual rainfall, suggesting that

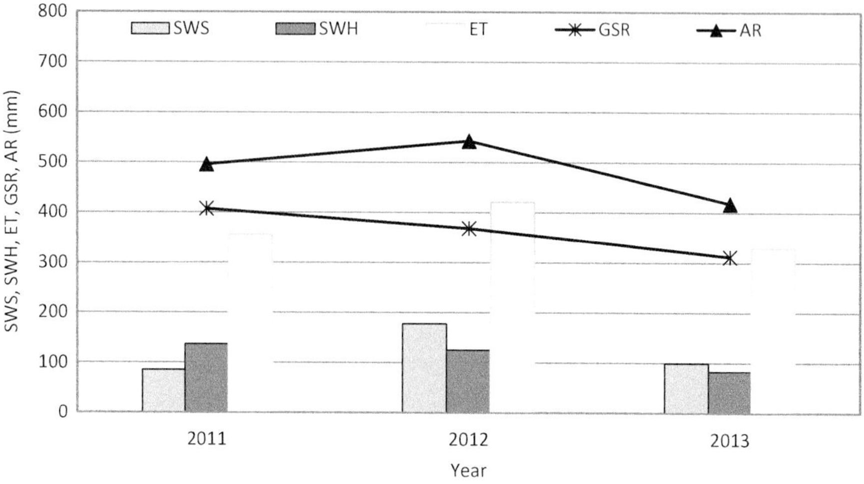

Fig. 2 Annual rainfall (AR), annual growing season rainfall (GSR), average amounts of soil water in the upper 0.6 m at sowing (SWS) and at harvest (SWH), and apparent water use (ET) in Twawa during the experimental period 2011–2013

the growing season for sorghum (July–September) is well synchronized to the rainy season (May–October). However, seasonal variations in rainfall were large (Fig. 2), and two to three dry spell events of more than 12 days often occurred during the growing season.

Grain Yields, WUE, Water Use, Soil Water in Responses to Treatments

Mean sorghum grain yields (GY), water use efficiency (WUE), water use (ET), soil water contents at sowing (SWS) and at harvest (SWH) in responses to ISWH and CT treatments are summarized in Table 2. Annual grain yields, WUE, ET, SWS, and SWH in responses to different treatments are shown in Fig. 2.

Grain Yields

Mean grain yields over the study period in responses to different treatments were 1894.7, 1944.8, 2198.4, 2183.6, and 2357.0 kg.ha^{-1} under WLD, ZT-M, CP-M, CFR-M, and CTR-M, respectively (Table 2). The grain yields under CTR-M were 7.2%, 7.9%, 21.2%, and 24.34% higher than CCP-M, CFR-M, ZT-M, and WLD, respectively ($P < 0.001$). CCP-M and CFR-M recorded higher grain yield over ZT-M by 13.03% and 12.28%, respectively and over WLD by 16.03% and 15.46%, respectively. Statistic analysis showed highly significant differences in grain yields between years ($P < 0.01$) and between treatments ($P < 0.01$). Annual variations in sorghum grain yield were large, ranging from 1489.2 kg ha^{-1} in the

Table 2 Effects of conventional tillage (WLD), conservation tillage (ZT-M and CCP-M) and in-situ water harvesting (CFR-M and CTR-M) on sorghum grain yield (GY), water use efficiency (WUE), water use (ET) and soil water contents at sowing (SWS) and at harvest (SWH), averaged over the whole experimental period (2011–2013) in Twawa, data for SWS, SWH, ET, and WUE from 0 to 60 cm soil depth

Treatment		GY (Kg. ha^{-1})		WUE (Kg. ha^{-1} mm^{-1})		ET (mm)		SWS (mm)		SWH (mm)	
WLD		1894.7	a	5.0	a	379.2	a	219.8	a	102.8	a, b
CFR-M		2183.6	b	6.2	b	356.1	a	239.2	b	132.0	c
CCP-M		2198.4	b	6.0	b	367.6	a	235.6	b, d	114.2	b, c, d
ZT-M		1944.8	a	5.3	a	371.8	a	203.5	c	104.4	a, c
CTR-M		2357	c	6.5	b	366.5	a	233.0	d	121.6	d
Source	DF	Pr>F		Pr>F		Pr>F		Pr>F		Pr>F	
Year	2	<0.01		<0.01		<0.01		<0.01		<0.01	
Treatment	4	<0.01		<0.01		0.184		<0.01		0.029	
LSD 5%		154.1		0.52				5.9		18.2	

Values with the same letter within a column are not significantly different at 5% level using IRRISTAT (2005)

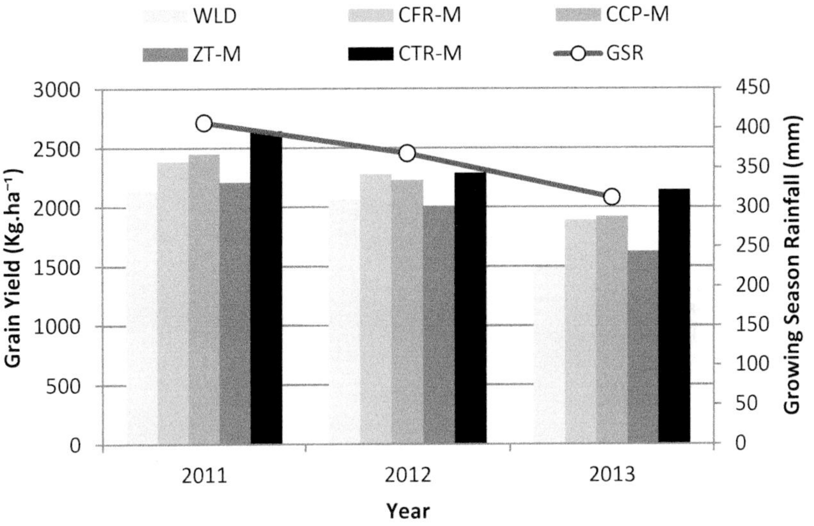

Fig. 3 Annual variations in sorghum grain yield in (GY) response to WLD, CFR-M, CCP-M, ZT-M, and CTR-M in Twawa during 2011–2013

lowest GSR 2013 to 2132.5 kg ha^{-1} in the highest GSR 2011, from 1888.3 to 2385 kg ha^{-1}, from 1917.8 to 2450 kg ha^{-1}, from 1621.8 to 2205 ha^{-1}, and from 2143.5 to 2640 ha^{-1} under treatment WLD, CFR-M, CP-M, ZT-M, and CTR-M, respectively (Fig. 3).

Water Use Efficiency

WUE averaged 5.0, 5.27, 6.03, 6.17, and 6.47 kg.ha^{-1} mm^{-1} under WLD, ZT-M, CP-M, CFR-m, and CTR-M, respectively (Table 2). Annual variations in WUE (kg ha^{-1} mm^{-1}) ranged from 4.5 (in the dry 2013) to 5.8 (in the wet 2012), from 5.7 to 7.1, from 6.1 to 6.7, from 4.9 to 6.1, and from 6.4 to 7.5, under treatment WLD, CFR-M, CCP-M, ZT-M, and CTR-M, respectively (Fig. 4). The WLD treatment usually had the lowest WUE while the CTR-M and CFR-M treatments had the highest WUE. No significant differences ($P > 0.05$) among CTR-M, CFR-M, and CCP-M treatments or between WLD and ZT-M had been received. The WUE was about 29%, 23%, and 21% higher under CTR-M, CFR-M, and CCP-M, respectively as compared to that under WLD. Moreover, CTR-M, CFR-M, and CCP-M increased WUE over ZT-M by about 23%, 17%, and 15%, respectively.

Water Use

Mean apparent water use (or ET) over the 3-year period was 379.2, 356.1, 367.6, 371.8, and 365.6 mm under WLD, CFR-M, CCP-M, ZT-M, and CTR-M, respectively (Table 2). No significant differences ($P > 0.05$) among treatments (WLD, CFR-M, CCP-M, ZT-M, and CTR-M) were observed. The ET under WLD was about 6%, 3%, 2%, and 3% higher than that under CFR-M, CCP-M, ZT-M, and CTR-M, respectively.

Annual variations in apparent water use (ET) by sorghum ranged from 330.2 mm in 2013 to 440.5 mm in 2012, from 330.8 to 402.8 mm, from 315.4 to 422.1 mm, from 334.2 to 420.4 mm, and from 333.4 to 416.2 mm, under treatment WLD,

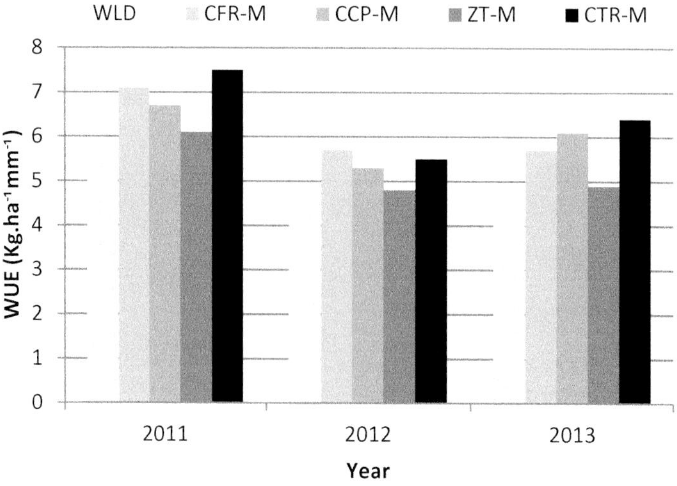

Fig. 4 Annual variations in water use efficiency (WUE) in response to WLD, CFR-M, CCP-M, ZT-M, and CTR-M in Twawa during 2011–2013

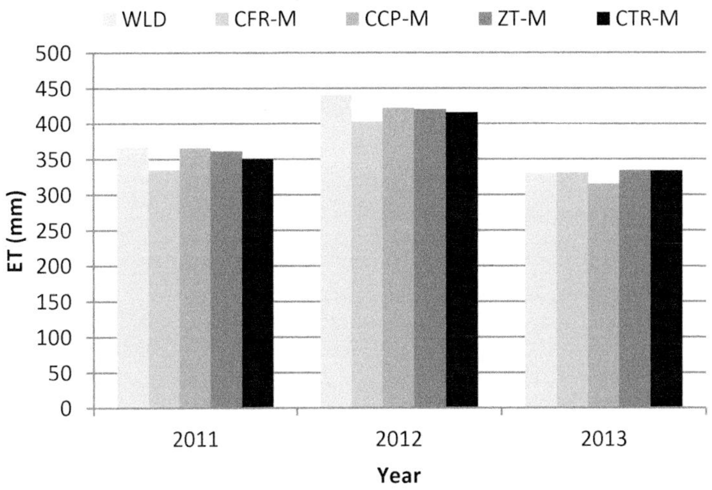

Fig. 5 Annual variations in water use (ET) in response to WLD, CFR-M, CCP-M, ZT-M, and CTR-M in Twawa during 2011–2013

CFR-M, CCP-M, ZT-M, and CTR-M, respectively (Fig. 5). The CFR-M treatment generally had the lowest water use, which were 8.6% lower than WLD treatments, except for 2013 when they were almost equal.

Soil Water Contents at Sowing and Harvest

Soil water contents at sowing and at harvest were significantly ($P < 0.01$) influenced by ISWH and CT treatments. Mean contents of SWS over the 3-year period were about 220, 239, 236, 204, and 233 mm and mean contents of SWH were about 103, 132, 114, 104, and 122 mm under WLD, CFR-M, CCP-M, ZT-M, and CTR-M, respectively (Table 2). At sowing time, the CFR-M treatment retained more water (3–15%) in the soil profile than the WLD, CCP-M, ZT-M, and CTR-M treatments. Moreover, at harvest time, the CFR-M treatment retained more water (9–28%) in the soil profile than the WLD, CCP-M, ZT-M, and CTR-M treatments.

Annual variations in soil water contents at sowing (SWS) ranged from 176 mm in 2013 to 281 mm in 2012, from 192 to 298 mm, from 189 to 297 mm, from 163 to 259 mm, and from 187 to 293 under treatment WLD, CFR-M, CCP-M, ZT-M, and CTR-M, respectively (Fig. 6).

Annual variations in SWH were associated with annual variation in GSR. Soil water contents at sowing and at harvest were also greatly influenced ($P < 0.01$) by ISWH and CT treatments. The CFR-M treatment had the highest SWS and SWH, which were 9–28%, 2–16%, 18–26%, and 3–9% more water in soil than the WLD, CCP-M, ZT-M, and CTR-M treatments, respectively (Figs. 6 and 7).

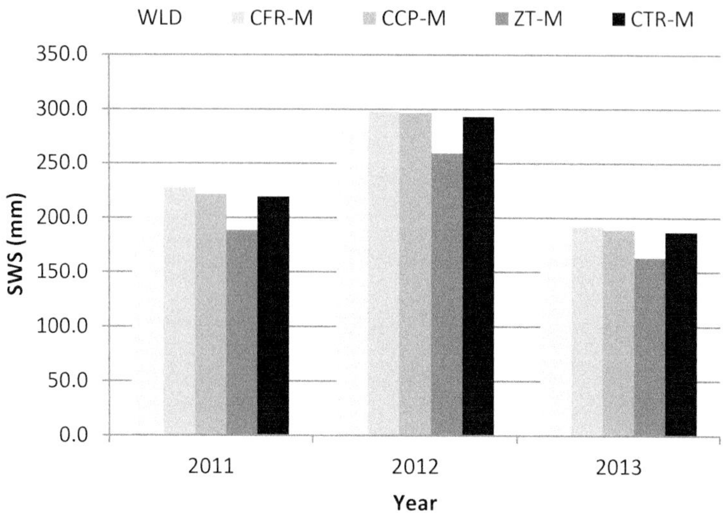

Fig. 6 Annual variations in soil water at sowing (SWS) in response to WLD, CFR-M, CCP-M, ZT-M, and CTR-M in Twawa during 2011–2013

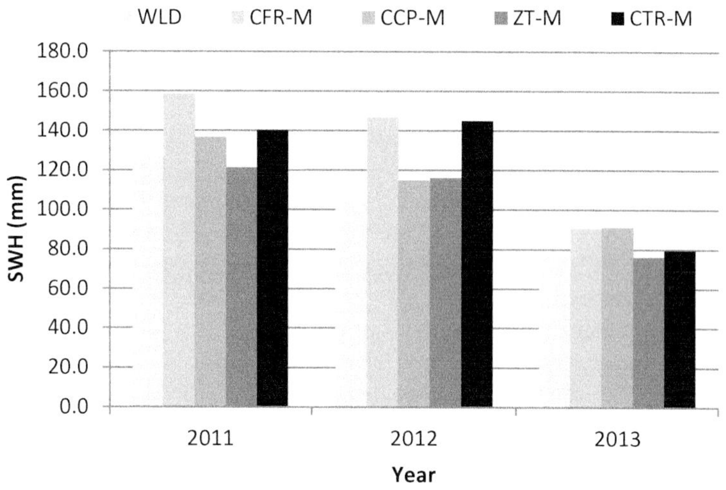

Fig. 7 Annual variations in soil water at harvest (SWH) in response to WLD, CFR-M, CCP-M, ZT-M, and CTR-M in Twawa during 2011–2013

Discussion

Rainfall distribution during the rainy season is the most determining factor to achieve high level of crop productivity, irrespective of the management system used. The highest GY under dryland farming system was attained in 2011 which ranked second, in terms of AR, after 2011, but first in terms of GSR. This is similar to the fact reported by IPCC (2007) and Peterson et al. (2012) that water losses accompanied by uneven-distributed and fluctuated rainfall pattern are the main limiting factors of crop production in DLF.

The dryland sorghum GY, WUE, and ET in Twawa showed stronger responses to ISWH treatments than CT Treatment rates during the experimental periods. The GY and WUE responses to ISWH and CT practices also varied greatly from year to year due to variable weather conditions (e.g., GSR and SWS). For instance, differences between years in mean sorghum GY about 30% (between 2011 and 2013) observed in this study, and these differences were mainly related to SWS and GSR. As Lampurlanés et al. (2002) reported, in terms of yields, the best tillage system is often affected by the weather experienced in that year. Giller et al. (2009) suggested that the variability in short-term crop responses to conservation agriculture (including minimum tillage or no-till, and mulch of crop residues) is principally the result of the interacting effects of crop requirements, soil characteristics, and climate.

The CTR-M treatment generally had the highest GY, about 7% and 21% higher than the CT treatments, ZT-M and CCP-M, respectively. GY under WLD and ZT-M ranked the lowest along the study period. Previous studies for dryland farming in Gadarif showed that yields under ZT were usually low and equivalent to those under WLD (Elramlawi et al. 2009; Mohamed 2009). At early season, the rapid entry of rainfall water through heavy (massive) and deep cracks in WLD and ZT-M treatments leads to low SWS in the root zone. Dry conditions at seeding impede seedling emergence and generally lead to low grain yield and nutrient uptake by maize (Cai et al. 1994 as cited by Wang et al. 2011). On heavy cracking soil under rainfed conditions in Sudan, Saeed and Eissa (2002) found a high average cracking percentage of about 56.7% under ZT system. Tillage is loosening the soil and improves its tilth, therefore, ZT resulting in a dense soil, and consequently reduces water infiltration and crop yields (Unger 1984). VandenBygaart et al. (1999) found that grain yield under ZT is low in the first few years after implementing ZT. They attributed such result to the declining in the number pores of 100–500 μm which may be crucial for the proper water storage, transmission, and root development.

Since large part of the previous crop (sorghum) was removed under conventional tillage, WLD treatment, during the dry season, soil crust formation was clearly observed. This resulted mainly from soil dispersion and reorientation of soil particles due to raindrop impact which may reduce soil infiltration. Under dryland conditions, especially where growing-season precipitation is limited, the amount of plant-available water stored in soil at planting time has a major effect on growth and yield of the crop (Unger 1990). Crop residue on the soil surface reduces runoff and increase infiltration by dissipating the energy of falling raindrops (Lal 1974), thereby reducing soil aggregate dispersion that results in surface sealing, and retarding the flow rate of water across the surface, thus providing more time for water infiltration (Unger 1990). Incorporating mulch with CCP-M, CTR-M, and CRF-M had showed advantage over WLD treatment in terms of improvement of water and crop yield. On the other side, ploughing in biomass poor hot tropical areas, as in WLD treatment, leads to accelerated oxidation of organic matter, thereby reducing water holding capacities of the soil (Johnson 1950), decreasing soil aggregate stability (Kemper and Koch 1966), and results in deterioration of soil quality (Johnston et al. 1943).

Plough land has, also, higher evaporation water losses, due to increased surface exposed to the atmosphere (Falkenmark et al. 2001). A compaction layer (15–20 cm below the plow layer) in soil of Gadarif was identified by Kevie and Buraymah (1976). They attributed such negative findings on agricultural production to the practice of monocropping (sorghum), the absence of proper crop rotation, and using of one kind of plow (wide level disk harrow). Regarding SWH, both ZT-M and WLD treatments failed to store enough rainwater in the root zone due to the existing of compacted layer. Such hard layer couldn't be loosened under ZT-M, and WLD and may hinder seed emergence, initial development, and root growth.

Higher SWS and SWH under CCP-M than ZT-M and WLD can be attributed to the effect of chiseling in mulch. This result is coincide with that of Burwell et al. (1966) who stated that cumulative infiltration before initial runoff was more closely related to surface roughness than to total pore space. Conservation tillage using CCP-M increases soil surface roughness, which may increase temporary surface water storage, and breaking underneath hard layer, while mulching prevents soil crusts formation and thus facilitate infiltration of water that would be lost as surface runoff. Moreover, ZT-M is expected to increase soil penetration resistance in the top layer (0–5 cm) as compared to CT (Alvarez et al. 2009). Residue mulching provides conditions for favorable plant growth as found by Cai et al. (2002), who observed a decreasing in surface soil temperature under mulch by 2–6.8 °C.

Sowing on the ridges, as CTR-M treatment, had the effect of raising the crop's rooting zone above the compacted layer and resulting in deep root growth. While sowing in the furrow, as in CRF-M treatment, had the advantage of the wetter conditions of the soil under the furrows, but still suffering from the impacts of plowpan. This resulted in better germination than WLD and ZT-M treatments. On the other hand, water use by plants may essentially depend on the amount of water stored in the soil. Water uptake during dry period is therefore greater and more uniform over time with CTR-M and CRF-M than with WLD and ZT-M.

The higher yields with CTR-M, CRF-M, and CCP-M treatments were due to significantly increased soil water storage in SWS. High SWS is beneficial from the viewpoint of better germination and early growth of crop (Cai et al. 1994 as cited by Wang et al. 2011), thus mitigating drought impacts. Water is the most important yield-limiting factor in dry years, while in wet years, yields may be limited by the availability of nutrients (Wang et al. 2011).

The changes in WUE had the same trend with the changes in grain yields over time. The CTR-M treatment usually had the highest WUE, while the WLD and ZT-M treatments had the lowest WUE. Annual variations in WUE were mainly influenced by annual GSR and tillage/mulch treatments, and their interactions.

Conclusion

The results of this study suggested that, over the short term, CTR-M, CCP-M, and CFR-M systems have the potentiality to enhance resilience of dryland farming in Gadarif State, Sudan. The application of CTR-M, CCP-M, and CFR-M significantly improved sorghum GY, WUE, and SWS comparing to ZT-M and WLD. Low crop yield is expected during the few years after shifting from WLD to ZT-M. Residue mulching has a positive impact in conserving soil and water and sustaining sorghum yield. These connoting responses to ISWH and CT suggest the need for local evaluation. Conservation tillage research in Sudan is relatively recent. Long-term results, necessary for an accurate evaluation of ISWH and CT, are not available yet.

References

Abbadi KA, Ahmed AE (2006) Brief overview of Sudan economy and future prospects for agricultural development. Expert opinion, Khartoum Food Aid Forum, Khartoum. http://nutritiontufts.edu/docs/sudan-economy and agricultural-development.pdf

Alvarez CR, Taboada MA, Gutierrez Boem FH, Bono A, Fernandez PL, Prystupa P (2009) Topsoil properties as affected by tillage systems in the Rolling Pampa region of Argentina. Soil Sci Soc Am J 73:1242–1250

Anonymous (2018) Annex_X_Part_E_Sudan. Ministry of Foreign Affairs of Denmark. http://um.dk/en/~/media/UM/English-site/Documents/Danida/Eval/978-87-7087-232-4/Annex_X_Part_E_Sudan.ashx. Retrieved 22 Mar 2018

Ball-Coelho BR, Roy RC, Swanton CJ (1998) Tillage alters corn root distribution in coarse-textured soil. Soil Tillage Res 45:237–249

Barbera V, Poma I, Gristina L, Novara A, Egli M (2012) Long term cropping systems and tillage management effects on soil organic carbon stock and steady state level of C sequestration rates in a semiarid environment. Land Degrad Dev 23:82–91

Blanco H, Lal R (2008) Principles of soil conservation and management. Springer, New York

Burwell RE, Allmaras RR, Sloneker LL (1966) Structural alteration of soil surface by tillage and rainfall. J Soil Water Conserv 21:61–63

Cai DX, Gao XK, Zhang ZT, Wang XB (1994) Effects of pre-sowing tillage measures on soil moisture retention and seedling emergence of spring crops. Soil Fert Sci China 2:10–13 (in Chinese)

Cai GX, Chen DL, Ding H, Pacholski A, Fan XH, Zhu ZL (2002) Nitrogen losses from fertilizers applied to maize, wheat and rice in the North China Plain. Nutr Cycl Agroecosyst 63:187–195

Carter DC, Miller S (1991) Three years experience with on-farm macrocatchment water harvesting system in Botswana. Agric Water Manag 19:191–203

CBS (Central Bureau of Statistics Sudan) (2011) Government of Sudan. http://www.cbs.gov.sd/. Retrieved 14 Oct 2017

Cramer W, Yohe GW, Auffhammer M, Huggel C, Molau U, da Silva Dias MAF, Solow A, Stone DA, Tibig L (2014) Detection and attribution of observed impacts. In: Field B, Barros VR, Dokken DJ, Mach KJ, Mastrandrea MD, Bilir TE, Chatterjee M, Ebi KL, Estrada YO, Genova RC, Girma B, Kissel ES, Levy AN, MacCracken S, Mastrandrea PR, White LL (eds) Climate change 2014: impacts, adaptation, and vulnerability. Part A: global and sectoral aspects. Contribution of working group II to the fifth assessment report of the intergovernmental panel on climate change. Cambridge University Press, Cambridge/New York, pp 979–1037. https://doi.org/10.1007/s11104-011-0728-8

Dudal R (1980) Soil-related constraints to agricultural development in the tropics. In: Proceedings symposium on properties for alleviating soil related constraints to food production in the tropics. IRRI-Cornell University, Los Baños

Duxbury JM (2005) Reducing greenhouse warming potential by carbon sequestration in soils: opportunities, limits and tradeoffs. In: Lal R, Stewart BA (eds) Climate change and global food security. Taylor and Francis, Boca Raton, pp 435–450

Elbashir A, Ahmed AE (2006) Food security policies in Sudan. Expert opinion, Khartoum Food Aid Forum, World Food Programme (WFP), Khartoum. https://reliefweb.int/sites/reliefweb.int/files/resources/99362D01BA1A3BE9C12571B200496287-wfp-sdn-08jun7.pdf

Elbashir A, Siddig FA, Ijaimi A, Nour HM (2004) Sudan poverty reduction and programs in agriculture. Food and Agriculture Organization (FAO), Khartoum

Elhadary YAE (2007) Pastoral adaptation and socio-economic transformations in the Butana Area, Al Gedarif State, Sudan. PhD thesis, University of Khartoum, Khartoum

Elhadary YAE (2010) Challenges facing land tenure system in relation to pastoral livelihood security in Gedarif State, Eastern Sudan. J Geogr Reg Plann 3:208–218

Elhadary YAE (2015) Mechanized rain-fed farming and its ecological impact on the drylands the case of Gedarif State, Sudan. In: Benkeblia N (ed) Agroecology, ecosystems, and sustainability. Advances in agroecology. CRC Press, Boca Raton, pp 121–137

Elramlawi HR, Saeed AB, Mohammed HI (2009) Effect of soil surface formation on yield and yield components of maize (*Zea mays* L.) in the north of Gadarif State. Sudan University of Science and Technology. J Sci Technol 10:82–97

Falkenmark M, Fox P, Persson G, Rockström J (2001) Water harvesting for upgrading of rainfed agriculture – problem analysis and research needs. Stockholm International Water Institute (SIWI), Stockholm

Filho WL (ed) (2011) Experiences of climate change adaptation in Africa, climate change management. Springer, Berlin, pp v–vi

Food and Agriculture Organization of the United Nations (FAO) (2000) Land resources potential and constraints at regional and country level, based on the work of Bot AJ, Nachtergaele FO, Young A. World soil resources report, 90. Land and Water Development Division, FAO, Rome

Food and Agriculture Organization of the United Nations (FAO) (2012) The land cover atlas of Sudan. FAO, Rome

Gadarif Meteorological Station (2014) Sudan Meteorological Corporation, Sudan

Gardner CMK, Laryea KB, Unger PW (1999) Soil physical constraints to plant growth and crop production. Land and Water Development Division, FAO, Rome

Gerard CJ, Sexton P, Shaw G (1982) Physical factors influencing soil strength and root growth. Agron J 74:875–879

Gill KS, Aulakh BS (1990) Wheat yield and bulk density response to some tillage systems on an Oxisol. Soil Tillage Res 18:37–45

Giller KE, Witter E, Corbeels M, Tittonell P (2009) Conservation agriculture and smallholder farming in Africa: the heretics' view. Field Crops Res 114:23–34

Glab T, Kulig B (2008) Effect of mulch and tillage system on soil porosity under wheat (*Triticum aestivum*). Soil Tillage Res 99:169–178

Habitu N, Mahoo H (1999) Rainwater harvesting technologies for agricultural production: a case for Dodoma, Tanzania. In: Kambutho PG, Simalenga TE (eds) Conservation tillage with animal traction. A resource book of Animal Traction Network for Eastern and Southern Africa (ATNESA), Harare, pp 161–171

Hansen NC, Allen BL, Anapalli S, Blackshaw RE, Lyon DJ, Machado S (2016) Dryland agriculture in North America. In: Farooq M, Siddique KHM (eds) Innovations in drylands agriculture. Springer International Publishing, Cham, pp 415–442

Hassan MM, Gregory PJ (1999) Water transmission properties as affected by cropping and tillage systems. Pak J Soil Sci 16:29–38

Hill RL (1990) Long-term conventional and no-tillage effects on selected soil physical properties. Soil Sci Soc Am J 54:161–166

Hulme K, Doherty R, Ngara T, New M, Lister D (2001) African climate change: 1900–2100. Clim Res 17:145–168

Hulugalle NR, Malton PJ (1990) Effects of rock bunds and tied ridges on soil water content and soil properties in the Sudan Savannah of Burkina Faso. Trop Agric 67:149–153

IPCC (2007) Climate change 2007, fourth assessment report. Cambridge University Press, Cambridge

IRRISTAT (2005) Biometrics and bioiformatics unit. International Rice Research Institute, Los Baños

Johnson WC (1950) Stubble mulch farming on wheatlands of the Southern High Plains. US Department of Agriculture/Circular 860. US Government Printing Office, Washington, DC

Johnston JR, Browning GM, Russell MB (1943) The effect of cropping practices on aggregation, organic matter control and loss of soil in the Marshall silt loam. Soil Sci Soc Am Proc 7:105–107

Kemper WD, Koch EJ (1966) Aggregate stability of soils from Western United States and Canada – measurement procedure, correlations with soil constituents. US Department Agriculture Technical Bulletin 1355. US Government Printing Office, Washington, DC

Koohafkan P, Stewart BA (2008) Water and cereals in drylands. FAO The Food and Agriculture Organization of the United Nations and Earthscan, London/Sterling

Krishna KR (2014) Agroecosystems soils, climate, crops, nutrient dynamics, and productivity. Apple Academic Press, Toronto

Lal R (1974) Soil erosion and shifting agriculture. In: Shifting cultivation and soil conservation in Africa. The Food and Agriculture Organization of the United Nations, FAO soils bulletin 24. FAO, Rome, pp 48–71

Lal R (1990) Ridge-tillage. Soil Tillage Res 18:107–111

Lampurlanés J, Angas P, Cantero-Martinez C (2002) Tillage effects on water storage during fallow, and on barley root growth and yield in two contrasting soils of the semi-arid Segarra region in Spain. Soil Tillage Res 65:207–220

Lieskovsky J, Kenderessy P (2014) Modelling the effect of vegetation cover and different tillage practices on soil erosion in vineyards: a case study in VRÁBLE (SLOVAKIA) using watem/sedem. Land Degrad Dev 25:288–296

Limousin G, Tessier D (2007) Effects of no-tillage on chemical gradients and topsoil acidification. Soil Tillage Res 92:167–174

Logsdon SD, Reneau RB, Parker JC (1987) Corn seedling root growth as influenced by soil physical properties. Agron J 79:221–224

López MV, Ardúe JL, Sánchez-Girón V (1996) A comparison between seasonal changes in soil water storage and penetration resistance under conventional and conservation tillage systems in Aragón. Soil Tillage Res 37:251–271

Maredia MK, Byerlee D, Pee P (2000) Impacts of food crop improvement research: evidence from sub-Saharan Africa. Food Policy 25:531–559

Mohamed NAM (2009) Effect of some in-situ rainwater harvesting and conservation tillage techniques on yield and yield components of sorghum (*sorghum bicolor-L*) in north of Gadarif State, Sudan. MSc thesis submitted to University of Gadarif, Gadarif

Moldenhauer WC, Onstah CA (1977) Engineering practices to control erosion. In: Greenland DJ, Lal R (eds) Soil conservation and management in the humid tropics. Wiley, Chichester, pp 87–92

Mortimore M, Adams WM (1999) Working the Sahel: environment and society in northern Nigeria. Routledge, London

Mustafa RH (2006) Risk management in the rain-fed sector of Sudan: case study, Gedaref area eastern Sudan. Institute of Agricultural and Food System Management, Faculty of Agricultural Sciences, Home Economics and Environmental Management, Justus Liebig University, Giessen

Nedumaran S, Jyosthnaa P, Singh NP, Bantilan C, Byjesh K (2015) Climate change and food security in Asia and Africa: agricultural futures. In: Singh NP, Bantilan C, Byjesh K, Nedumaran S (eds) Climate change challenges and adaptations at farm-level case studies from Asia and Africa. CABI climate change series. CABI, Wallingford, pp 86–114

Niang I, Ruppel OC, Abdrabo MA, Essel A, Lennard C, Padgham J, Urquhart P (2014) In: Barros VR, Field CB, Dokken DJ, Mastrandrea MD, Mach KJ, Bilir TE, Chatterjee M, Ebi KL, Estrada YO, Genova RC, Girma B, Kissel ES, Levy AN, MacCracken S, Mastrandrea PR, White LL (eds) Africa, climate change 2014: impacts, adaptation, and vulnerability. Part B: regional aspects. Contribution of working group II to the fifth assessment

report of the intergovernmental panel on climate change. Cambridge University Press, Cambridge/New York, pp 1199–1265
Olesen JE (2014) Rainfed intensive crop systems. In: Fuhrer J, Gregory PJ (eds) Climate change impact and adaptation in agricultural systems. CABI climate change series. CABI, Oxfordshire, pp 17–30
Osman KT (2013) Soils: principles, properties and management. Springer, Dordrecht
Peterson GA, Westfall DG, Hansen NC (2012) Enhancing precipitation use efficiency in the world's dryland agroecosystems. In: Lal R, Stewart BA (eds) Soil water and agronomic productivity. Advance in soil science. CRC Press, Boca Raton, pp 455–476
Piha MI (1993) Optimizing fertilizer use and practical rainfall capture in a semi-arid environment with variable rainfall. Exp Agric 29:405–415
Rafiq M (1990) Soil variability in agronomic research. Pak J Soil Sci 5:9–14
Raper RL (2006) In-row subsoiling southeastern soils to reduce compaction and improve crop yields In: Schwartz RC, Baumhardt RL, Bell JM (eds) Improving conservation technologies to compete for global resources and markets. Proceedings of 28th annual southern conservation tillage conference for sustainable agriculture, Amarillo. Rpt. no. 06-1. USDA-ARS Conservation and production research laboratory, Bushland, pp 85–94
Rockström J (2003) Water for food and nature in drought-prone tropics: vapour shift in rain-fed agriculture. R Soc Trans B Biol Sci 358:1997–2009
Rockström J, Falkenmark M (2000) Semiarid crop production from hydrological perspective: gap between potential and actual yields. Crit Rev Plant Sci 19:319–346
Rockström J, Karlberg L, Wani SP, Barron J, Hatibu N, Owei T, Bruggeman A, Farahani J, Qiang Z (2010) Managing water in rainfed agriculture – the need for a paradigm shift. Agric Water Manag 97:543–550
Rodriguez D, de Voil P, Power B (2016) Modelling dryland agriculture systems. In: Farooq M, Siddique KHM (eds) Innovations in drylands agriculture. Springer International Publishing, Cham, pp 239–256
Rosegrant M, Cai X, Cline S, Nakagawa N (2002) The role of rainfed agriculture in the future of global food production. EPTD discussion paper No. 90. Environmental and Production Technology Division, International Food Policy Research Institute, Washington, DC
Saeed AB, Eissa HE (2002) Influence of tillage on some properties of heavy cracking clay soils and sorghum yield in mechanized rainfed agriculture. University of Khartoum. J Agric Sci 10: 267–276
Salih AA, Babikir HM, Ali SAM (1998) Preliminary observations on effects of tillage systems on soil physical properties, cotton root growth, and yield in Gezira scheme, Sudan. Soil Tillage Res 46:187–191
Sasal MC, Andriulo AE, Taboada MA (2006) Soil porosity characteristics and water movement under zero tillage in silty soils in Argentinean Pampas. Soil Tillage Res 87:9–18
SFNC (2003) Sudan's first national communications under the United Nations Framework Convention on Climate, Higher council for environment and natural resources, Ministry of environmental & physical development, Republic of Sudan
Soil Conservation Society of America (SCSA) (1982) Resource conservation glossary. SCSA, Ankeny
SSSA (2018) https://www.soils.org/publications/soils-glossary#. Retrieved 29 Mar 2018
Stewart BA, Burnett E (1987) Water conservation technology in rainfed and dryland agriculture. In: Jordan WR (ed) Water and water policy in world food supplies. Texas A&M University Press, College Station, pp 355–359
Stewart BA, Thapa S (2016) Dryland farming: concept, origin and brief history. In: Farooq M, Siddique KHM (eds) Innovations in drylands agriculture. Springer International Publishing, Cham, pp 3–29
Stockholm International Water Institute (SIWI) (2001) Water harvesting for upgrading rainfed agriculture. Problem analysis and research needs. SIWI report no. 11. Stockholm

The International Fund for Agricultural Development (IFAD) (2010) Republic of the Sudan: supporting the small-scale traditional rainfed producers in Sinnar State (SUSTAIN). Design report: main report and annexes, Sudan
Unger PW (1984) Tillage systems for soil and water conservation. The Food and Agriculture Organization of the United Nations, FAO Soil bulletin 54. FAO, Rome
Unger PW (1990) Conservation tillage systems. In: Singh RP, Parr JF, Stewart BA (eds) Dryland agriculture, strategies for sustainability. Advances in soil science, vol 13. Springer, New York, pp 27–68
Unger PW, Payne WA, Peterson GA (2006) Water conservation and efficient use. In: Peterson GA, Unger PW, Payne WA (eds) Dryland agriculture. Agronomy monograph, vol 23, 2nd edn. ASA-CSSA-SSSA, Madison, pp 39–85
Unger PW, Baumhardt RL, Arriaga FJ (2012) Mulch tillage for conserving soil water. In: Lal R, Stewart BA (eds) Soil water and agronomic productivity. Advances in soil science. CRC Press, Boca Raton, pp 427–453
United Nation Education, Science and Cultural Organization (UNESCO) (2012) The United Nations world water development report 4: managing water under uncertainty and risk, vol 1. UNESCO, Paris
United Nations Environment Programme (UNEP) (2007) Sudan post-conflict environmental assessment. UNEP, Nairobi
van der Kevie, Buraymah IM (1976) Exploratory soil survey of Kassala Province. Soil Survey Administration, Wad Medani
VandenBygaart AJ, Protz R, Tomlin AD (1999) Changes in pore structure in a no-till chronosequence of silt loam soils, southern Ontario. Can J Soil Sci 79:149–160
Verhulst N, Nelissen V, Jespers N, Haven H, Sayre KD, Raes D, Deckers J, Govaerts B (2011) Soil water content, maize yield and its stability as affected by tillage and crop residue management in rainfed semi-arid highlands. Plant Soil 344:73–85
Vogel H (1993) Tillage effects on maize yield, rooting depth and soil water content on sandy soils in Zimbabwe. Field Crops Res 33:367–384
Wang ZC, Gao HW (2004) Conservation tillage and sustainable farming. Water Conserv Res 3:1–14. Agricultural Science and Technology Press, Beijing
Wang XB, Daia K, Zhang D, Zhang X, Wang Y, Zhaoa Q, Caia D, Hoogmoedc WB, Oenema O (2011) Dryland maize yields and water use efficiency in response to tillage/crop stubble and nutrient management practices in China. Field Crop Res 120:47–57
Wani AP, Garg KK, Singh AK, Rockström J (2012) Sustainable management of scarce water resources in tropical rainfed agriculture. In: Lal R, Stewart BA (eds) Soil water and agronomic productivity. Advances in soil science. CRC Press, Boca Raton, pp 347–408
Yazar A, Ali A (2016) Water harvesting in dry environments. In: Farooq M, Siddique KHM (eds) Innovations in drylands agriculture. Springer International Publishing, Cham, pp 49–98

Agriculture in the Face of Climate-Mediated Flooding in Tropical Africa: Technical Innovations of Fish Farmers in Southwestern Nigeria

125

Oyediran O. Oyebola, Jackson Efitre, Augustine E. Falaye, Taiwo M. Dada, and Funmilayo C. Idowu

Contents

Introduction 2576
Vulnerability of Agriculture to Climate Change in Tropical Africa 2576
Impacts of Climate Change–Mediated Flooding on Fish Farming and Food Security in Tropical Africa 2578
Need for Indigenous Technical Innovations for Sustainable Fish Farming in the Face of Climate Change-Mediated Flooding in Tropical Africa 2579
Case Scenario: Flooding, Fish Farming, and Indigenous Technical Innovations for Fish Farming despite Flooding in Southwestern Nigeria, West Africa 2580
Investigations on Technical Innovations for Fish Farming in the Face of Flooding in Flood–Prone States of Southwestern Nigeria, West Africa 2581
Methodology 2581
Results 2582
Conclusions and Recommendations 2590
References 2591

Abstract

Agriculture is highly vulnerable to climate change in tropical Africa. Climate change-mediated floods (CCMF) impact negatively on fish farming, with

O. O. Oyebola (✉) · A. E. Falaye · T. M. Dada
Department of Aquaculture and Fisheries Management, University of Ibadan, Ibadan, Nigeria
e-mail: olusegun.oyebola@yahoo.com; aefalaye@yahoo.com; teegurl19@gmail.com

J. Efitre
Department of Zoology, Entomology and Fisheries Sciences, Makerere University, Kampala, Uganda
e-mail: jefitre@gmail.com

F. C. Idowu
National Biotechnology Development Agency, Ogbomoso, Nigeria
e-mail: funmibukky03@gmail.com

W. Leal Filho (ed.), *Handbook of Climate Change Resilience*,
https://doi.org/10.1007/978-3-319-93336-8_163

consequences on food security. The need for technical innovations for sustainable fish farming despite CCMF necessitated this study. Three most famous but flood-prone fish farming states (Lagos, Ogun, Oyo) in Southwestern Nigeria were sampled for fish farmers with on-farm experience of CCMF but still farming. Respondents were sampled for practiced technical adaptive initiatives for sustenance of fish farming despite CCMF, using interview, group discussion, and field visits. All encountered respondents operated pond fish farming system. The respondents reported that CCMF was unpredictable and have increased from 2012 to 2016. Fourteen encountered adaptive technical innovations based on single/combined diverse techniques of pond fencing, pond embankment modifications, netting and caging, pond slope manipulations, and riverbank modification. Innovations were 35.7% state-shared and 64.3% state-specific.

Climate change-mediated flood threatens fish farming but can be technically mitigated using the identified innovations or their modifications in flood-prone fish farming communities in tropical African countries.

Keywords
Climate risks · Adaptive innovations · Flood · Fish farming · Food security

Introduction

This chapter utilizes the existing literature to discuss the vulnerability of agriculture sector to climate change in tropical Africa, the impacts of climate change-mediated flooding on fish farming and food security in tropical Africa, the need for technical innovations for sustainable fish farming in the face of climate change-mediated flooding in tropical Africa, and the scenario of fish farming and flooding in Southwestern Nigeria. As a follow-up to the existing literature, it reports the result of primary data gathering on types of technical innovations for fish farming in the face of flooding in selected flood-prone states in Southwestern Nigeria. It also gives suggestions on the needed precautions. It concludes on the needed developmental actions for improved uptake of the innovations across tropical African countries.

Vulnerability of Agriculture to Climate Change in Tropical Africa

There is increasing evidence that the Earth is warming with implications for present and future global climatic and environmental circumstances. Climate change impacts on all spheres of life, and African countries are among the most vulnerable (Idowu et al. 2011; Ipinjolu et al. 2014; Oladokun et al. 2015; New 2016; Freeman 2017). Sustainable response to mitigate climate change impacts would involve multi-sectoral approach to knowledge seeking for adaptation. This is especially necessary because sectors could vary in dimensions of vulnerability to climate change impacts. Over 80 percent of global agriculture is rain-fed (FAO

2003), or at least all agriculture subsectors have some degree of dependence on rainfall. According to Bates et al. (2008), production systems of subsectors of agriculture (crop), forestry, and fisheries depend greatly on the temporal and spatial distribution of precipitation and evaporation. Hence, the agriculture sector, with high dependence on climate variables, seems to be the most vulnerable to climate change impacts, especially in most tropical African countries. Meanwhile, this sector accounts for the biggest share of the economy of most of these developing countries.

High dependence of African agricultural productivity on climate-related factors has made the continent the most susceptible to climate change in the world (IPCC 2001). Africa is highly reliant on rain-fed agriculture whose importance varies within the region (Eriksen et al. 2008), while sub-Saharan Africa is the most important in this regard (Tagbo 2010). In West Africa, most farming activities are at subsistence level, with planting and cropping activities dependent on unpredictable climatic factors which subject the populace to high risk of hunger and poverty (Bevan 2007). Eriksen et al. (2008) claimed that climate change will be responsible for an expected 10% reduction in net productivity of major crops in Africa. Knowledge on sensitivity to changes in precipitation in the agriculture sector would be important in building resilience to climate change impacts for tropical agriculture. In this regard, building resilience would not only depend on monitoring the amount of precipitation but also on the time of the year that the precipitation occurs (Hély et al. 2006).

Fish production plays an essential role in livelihood, food security, and socio-economic benefits of millions of people around the world (FAO 2008; Olaifa 2015; FAO 2016). However, similar to other agricultural subsectors, the fish production sector in tropical Africa is susceptible to climate change and its impacts (Falaye 2013; Ipinjolu et al. 2014; FAO 2015). Climate change induces extreme weather events, sea level rise, and numerous other environmental changes (IPCC 2007), which will affect fish production with impacts on food security. Climate change has the capability to affect fishery production along many pathways. Ficke et al. (2007) assert that fish reproduction, growth, and migration patterns are all affected by temperature, rainfall, and hydrology. Fish can be produced through capturing of wild fisheries resources and by fish farming, otherwise known as aquaculture. However, there is a declining trend in global fish production from capture fisheries (Faturoti 1999), and climate change has propensity to increase the declining rate.

In West Africa, coastal states whose livelihood depends on aquaculture and fisheries are highly vulnerable to low productivity, as a result of saltwater intrusion arising from sea rise and increase in temperature (Mintzer 2010). Fish farming is on consistent growth (FAO 2016); it has potential for mitigating climate change, but it is also vulnerable to the impacts of climate-mediated flooding in tropical Africa (Oyebola and Fada 2017). The acknowledged impacts of climate change in the agriculture sector of tropical Africa would encourage people and communities to develop mitigation and adaptation strategies to lessen the potential impact and or avoid the adverse consequences.

Impacts of Climate Change–Mediated Flooding on Fish Farming and Food Security in Tropical Africa

The two most important climate risks linked to the fish farming subsector are flood and drought. Drought may be most strongly felt in evaporation of pond water, resulting in mortality and premature harvesting of pond fish. However, the impact of flood could be more grievous, as it results in loss of premature and mature fish, while motivating threat to indigenous fisheries owing to admixture of flooded life specimens (Oyebola and Fada 2017). This has effect on farmed fish health and compromises genetic integrity of stocked and wild fishes, with impacts on economic returns from farming investment, sustainability of fish farming, conservation of indigenous gene pool, and food security.

All the listed consequences have short- and long-term impacts and are very important in the livelihood. There is higher concern for food security because attainment of food security status is a global desire. Sasson (2012) observes that food insecurity is still a major global concern, as 1.0 billion people are suffering from starvation, undernutrition, and malnutrition. The key cause of food insecurity is inadequate food production. The African challenge is key to mitigating food insecurity in the world. The FAO (2011) claims that about 25% of the world population, representing an estimated 235 million people in sub-Saharan Africa, is currently food insecure. The Global Hunger Index in 2010 ranked sub-Saharan Africa as the second region with the highest hunger level in the world (Von et al. 2013). Climate change impacts negatively on food security in this region through reduction in fish production and great changes in the quality of water (Ayanwuyi et al. 2010).

Flood is one of the main disasters associated with climate change. Arid and humid areas could be flooded in tropical near-coastal regions, as flood generally emanates from cyclones that can drop a year's worth of rainfall in a day. Flood is the most prevalent disaster in North Africa; second most common in East, South, and Central Africa; and the third most common in West Africa (AWDR 2006). Fourteen countries, namely, Burkina Faso, Chad, Ethiopia, Ghana, Kenya, Liberia, Mali, Niger, Nigeria, Senegal, Sudan, Togo, Uganda, and Rwanda, are the worst hit by flood in the African continent (BBC News 2007). Flood is one of the major factors that prevent Africa's population from escaping poverty level (Action Aid 2006). Generally, flood disasters are usually traced to climate-mediated heavy downpour of rainfall. Climate-mediated flood emanates from unpredictable rainfall and rainfall pattern, and this has propensity to negatively impact on fish farming, with reflection on the decline in fish production. Flood will wash away pond-stocked fish, destroy production ponds by tearing down its embankments, and deposit toxic materials into ponds, leading to health challenge. All these will invariably lead to reduction in fish production (Adebayo et al. 2011; Falaye 2013). Extreme weather events have the potential to transmit disease across wide geographies (Dinesh et al. 2015).

According to Dinesh et al. (2015), extreme weather events such as flooding can distribute epizootic ulcerative syndrome (EUS) across fishes in kilometers of natural freshwater systems, with strong impacts on fish catches. The spread of fish diseases as a result of climate-mediated flooding will aggravate the negative impacts of climate change. The impact of climate change may vary across tropical African countries. Flood has affected Nigeria, with real devastating impacts on aquaculture sector (Falaye 2013; Ipinjolu et al. 2014). Impacts of flooding on fish farming have been notable in Nigeria, the second most important farmed-fish-producing African country.

Need for Indigenous Technical Innovations for Sustainable Fish Farming in the Face of Climate Change-Mediated Flooding in Tropical Africa

Fish farming is highly technical, and the socioeconomic status of farmers could influence technical adaptability. The socioeconomic status of a farmer would reflect the type of technical tool that such farmer would employ in adapting to environmental hazards, such as climate-mediated risks. Technical inefficiency would heighten the vulnerability of fish farming to climate-mediated hazards, such as flooding. Indigenous technical knowledge could show the efficiency of climate impact mitigation in poor and vulnerable communities of tropical African countries. Indigenous knowledge has been widely utilized in empowering resident communities against indigenous problems (Makhanu et al. 2007). Indigenous technological knowledge (ITK) is accumulated through cumulative and collective experiences, checked, validated, and revised through annual cycle of activities performed by local people (Shafie and Rahman 2009).

Application of indigenous knowledge in the face of a threat or hazard is often referred to as "coping strategy" (innovations by individuals/community to withstand stressor and thus survive it). Choice of skills and resources in adapting could vary according to the nature of hazard, available capacities to deal with specific threat, and priorities of variety of community in adapting to the threat/disaster (Shafie and Rahman 2009). However, indigenous technological knowledge (ITK) has been a key rescuer of fishers and common people dwelling in flood-prone areas, and this has resulted in innovative approaches in addressing miserable situations (Bordoloi and Muzaddadi 2014). Application of existing resource toward disaster management to coexist with the usual floods is necessary. Sustainable exploration of fisheries resources has been utilized in specific flood-prone communities. These include early warning on flood, construction of flood-safe fish ponds for flooded regions and post-flood situation, modified cropping methods, and fish preservation methods (Bordoloi and Muzaddadi 2014). These strategies were especially encountered among other ITK in Dhemaji District of Assam, India. There is the need to document the dimension of flood adaptation tools especially the technological approaches in fish farming across tropical Africa. There is a dearth of such scientific output in spite of the threat of the hazard in the region. This does not indicate that such technologies are not in existence; they may however be simply not yet adequately captured by scientific investigations and reported in tropical Africa.

Innovations toward sustainable development have received increasing global attention (Rennings 1999). Technological innovation is crucial to advancing sustainable development, achievable through its effects on the pillars of sustainability. It promotes economic vitality, environmental sustainability, and social progress (Constantinescu and Frone 2014). The European Institute of Innovation and Technology (2013) argues that radical technological changes can only occur if the economic, social, and environmental contexts are also changing in a sustainable way. Technological innovation can thus be considered as a key factor for sustainable development, assuming a catalytic role in economic, social, and environmental fields (Fokkema et al. 2005, Constantinescu & Frone 2014). Knowledge on indigenous technical innovations of fish farmers is indispensable in the development of coping strategies in order to ensure sustainable fish production, socioeconomic returns, environmental sustainability, and food security in the face of climate change in tropical Africa.

Case Scenario: Flooding, Fish Farming, and Indigenous Technical Innovations for Fish Farming despite Flooding in Southwestern Nigeria, West Africa

Floods are the most common and recurring disaster in Nigeria (Nigeria Post-Disaster Needs Assessment 2013). Floods are defined as extremely high flows of river, whereby water inundates terrains outside the water-confined major river channels (Tawari-Fufeyin et al. 2015). Runoff from inland drainage systems contributes greatly to flood condition. The degree of flood hazard is determined by occurrence of damaging consequences, known as flood risks, or by their impacts, conceived usually as loss of lives and material damage to society (Henry 2006). Flooding could cause destruction of homes, lives, and property and destruction of livelihoods, such as farming, poultry, animal husbandry, aquaculture, and fishing.

Nigeria ranks among top 20 countries exposed to flooding based on present and future projected scenarios till the 2070s, and coastal floods affect the low-lying areas in the southern part of the country, with increasing occurrences (Nkwunonwo 2016). Flood in Nigeria could take international dimension because incidence of floodwaters from Cameroun entering Nigeria through River Benue into River Niger, which caused lots of physical damages, such as destruction of farmlands and houses, has been recorded (Tawari-Fufeyin et al. 2015). According to Policy Brief (2012), climate change and the consequences of global warming, with the resultant impact of flooding, have been major concerns in Nigeria.

Historically, flooding in Nigeria dates back to the early 1950s, with coastal and fluvial floods affecting mainly coastal environments (Akintola 1994). These floods were influenced by seasonal interruption of major rivers and water, overtopping their natural and artificial defenses. The flood of year 2012, wherein heavy storms that lasted many days affected 30 out of the 36 states, with 24 considered severely affected, was notable as the worst flooding in more than 40 years. About 29,972 metric tons of fish (estimated to be worth of about 239.8 million naira) was lost due to flooding in the 14 Nigerian states considered in 2012 (NEMA 2013). The climate change-mediated, unpredicted, and persistent heavy rainfall, sometimes going on for hours and days, caused overflowing of riverbanks and drainage blockage, threatened dams which were completely submerged, and motivated excess water flowing out uncontrollably from reservoirs. The result of all these have been devastating flood disasters. Flooding is one of the most common environmental issues in the southern part of Nigeria, and flooding associated with heavy downpour of rainfall has resulted in the decline in fish production in Southwestern Nigeria (Adebo and Aiyelari 2011).

According to NEMA (2013), virtually all the southwestern states in Nigeria have reported flooding within the last 10 years. Adebayo et al. (2011) claim that washing away of key infrastructures, ponds, tanks, and stocked species, economic losses, diseases, and parasite infection through floodwater are rampant in Southwestern Nigeria owing to flooding. Fish of an average weight of less than 200 g to about 2 kg and above were reportedly washed away as a result of flooding in Southwestern Nigeria (Adebo and Ayelari 2011). Flood scenarios have mostly been reported in Ogun, Lagos, and Oyo States of Nigeria. Heavy rainfall resulting in partial collapse of a hydro dam occurred in Ogun State, and releases of excess water from Oyan Dam caused flood across Ogun and Lagos States in 2007, 2010, and 2012, usually coinciding with the rainy season. Falaye (2013) reported massive destruction of aquaculture facilities and loss of fish stocks in Oyo State during the incidence of flooding in 2012. Ibadan, the capital of Oyo State and the third largest city in Nigeria, is highly exposed to flooding owing to heavy downpour, resulting in damage to aquaculture facilities and fish stocks worth of over 300 million naira (US$1.9 million equivalent) in 2011 (IUFMP 2014). The reported damage could be an underestimate as a lot of fish farmers in remote areas might not have been captured in the data.

These flood-prone southwestern states contribute immensely to aquaculture fish production in Nigeria. The reported incidences are clear indication that flooding is a great threat to aquaculture development in Southwestern Nigeria. It would be ideal to investigate technical innovations of fish farmers in these flood-prone states of Nigeria, with the aim of understanding their readiness to cope with future flood occurrences, and to learn from their innovativeness for subsequent uptake in other tropical African countries.

Investigations on Technical Innovations for Fish Farming in the Face of Flooding in Flood–Prone States of Southwestern Nigeria, West Africa

Methodology

Study Area

Three most famous fish farming states with popular flood history (Lagos, Ogun, Oyo) in Southwestern Nigeria were sampled for fish farmers with on-farm experience of climate change-mediated flood (CCMF) but who were still farming. The map of Nigeria showing locations of the sampled states in Southwest region is presented in Fig. 1.

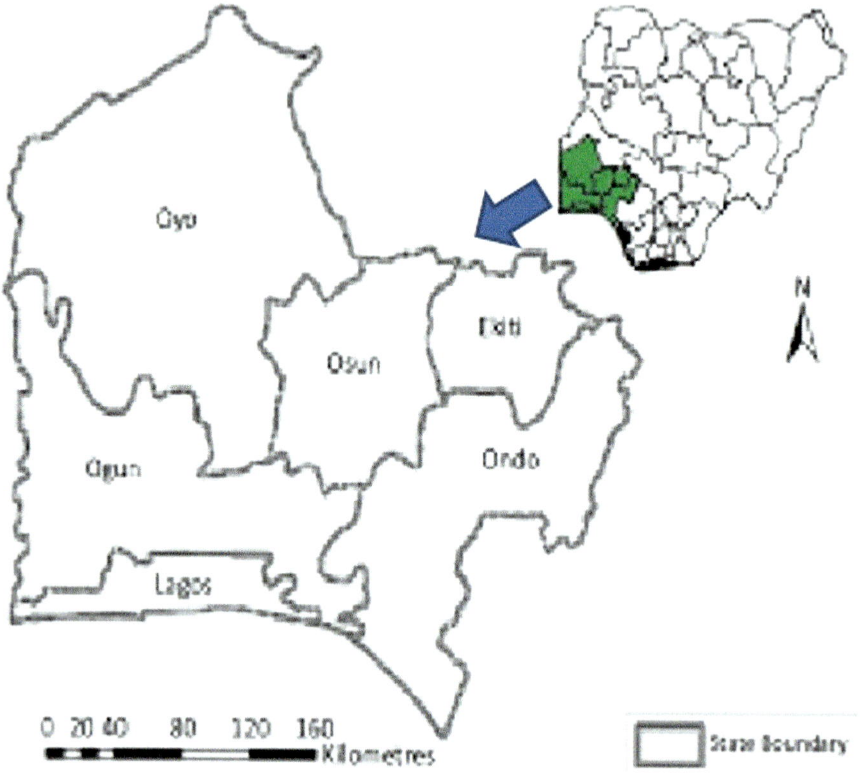

Fig. 1 Map of Southwestern Nigeria showing locations of the studied Oyo, Ogun, and Lagos States. (Adapted from Michael and Stephen 2017)

Data Sources, Processing, and Analysis

Data were sourced from fish farmers in the listed states in Southwestern Nigeria. The sampled fish farmers were selected after meeting and group discussion with fish farmers' association mobilized by fisheries extension officers from each of the states at one reconnaissance meeting held at each of the states. At the end of each meeting, 30 respondent fish farmers with flood experiences and who were still farming were selected per state. Each fish farmer was interviewed on experiences on climate-induced flooding and the technical adaptive innovations in place that could facilitate fish farming despite flooding. These innovations were documented during field visits to the farms. Responses on experiences on climate change-mediated flood were presented using descriptive tools of the SPSS software.

Results

Descriptive data on the observed farmer's experiences on the occurrence of flooding and encountered technical innovations across the states, including the documented

Table 1 Fish farmers' experiences on climate change-induced flooding in Southwestern Nigeria

	States		
Experiences	Lagos	Ogun	Oyo
Occurrence of climate change-induced flooding	100.0	100.0	100.0
Increased flooding (2012–2016)	100.0	100.0	100.0
Flood months in a year			
June	100.0	100.0	100.0
August	100.0	100.0	100.0
December/January	90.0	86.6	66.6

technical innovations, are discussed with reference to adaptability and necessary precautions for their uptake across tropical African countries.

Observed Experiences on Climate Change–Mediated Flooding among Fish Farmers in Southwestern Nigeria

Table 1 shows that all the sampled fish farmers reported that they had experienced climate change-induced flooding on their farms and that flooding rate increased in 2012–2016. This observation was in tandem with earlier scholars (discussed above) who reported 2012 as a significant period of outstanding flooding and increasing trend from 2012 onward. Although respondents observed that flood is unpredictable and it occurs at any month during these years, all of them admitted that June and August were notable flood months. However, contrary to expectation, 90.0, 86.6, and 66.6 percent of respondents, respectively, from Lagos, Ogun, and Oyo States reported that flood had been experienced in their farms during December/January periods. These are conventional dry months in this zone. The trend of experiences on period of flooding underlined the fact that flooding is unpredictable, as it occurred even at conventionally dry months of the year in Southwestern Nigeria. Therefore, resilience against this type of flood pattern would better be based on precautionary adaptive technical approaches, which should not be season biased. The proportion of respondents agreeing to have experienced an event could indicate popularity of such event. It could therefore be inferred that the situation would be relatively more serious in Lagos State and least in Oyo State, going by the relatively highest and least proportion of the respondents agreeing to the trend. Therefore, compared to Oyo and Ogun States, Lagos State should be on top priority in building resilience action against flooding in Southwestern Nigeria.

Observed Adaptive Technical Innovations of Fish Farmers for Uptake in Other African Countries

The flood-prone farmers were operating earthen pond system, and 14 adaptive technical innovations based on single/combined techniques were encountered across the selected flood-prone fish farms in Southwestern Nigeria (Table 2). The encountered innovations seemed not originally intended for climate change-mediated flood mitigation but were adaptable for the purpose. Some of the strategies were observed across the states, while some were state-specific. The divergence in state-based

Table 2 Encountered technical innovations for pond fish farming in the face of climate change-induced flooding in Southwestern Nigeria

Encountered adaptive innovations	States		
	Lagos	Ogun	Oyo
1. Pond perimeter fencing plus surface netting	+	+	−
2. Sandbag-fortified pond embankment	+	+	+
3. Riverbank barrage with sandbag	+	+	+
4. Pond surface netting only	+	+	+
5. Pond caging inside pond to overhead	+	−	−
6. Pond caging on dyke	+	+	+
7. Increased pond dyke with earthen materials	−	+	+
8. Surface netting cum sandbag-reinforced dykes	+	+	+
9. Increased dyke height with sandbag plus surface netting	−	+	+
10. Increased dyke height with sandbag minus surface netting	−	+	−
11. Dyke slope stepping plus surface netting	−	+	−
12. Dyke stepping plus pond caging on dyke	−	+	−
13. Dyke stepping plus surface netting plus pegged sandbagged ponds' slope base	−	+	−
14. Pond slope reinforcement with hard-core materials	−	−	+

35.7% states occurrence shared across states; 64.3% occurrence specific to state
Minus sign (−) indicates not encountered; plus sign (+) indicates encountered

innovations indicates the need for collation of the innovations that exist across the states, with subsequent creation of awareness for uptake of some of the innovations in states where they were not encountered and for their uptake at other flood-prone states in Nigeria. This approach would also be necessary for ensuring the spread of such innovation across other tropical African countries.

According to Bordoloi and Muzaddadi (2014), indigenous technical knowledge (ITK) for fish farming in flooded areas of Dhemaji District of Assam, India, included manipulation of pond dyke (construction of tall and wide dyke) for protection from flood, fencing of ponds (with low dyke height) with fine-meshed plastic nets to prevent fish from escaping during flood, strengthening of pond dyke (by turfing and plantation above the dyke), and the use of clay soil (from neighboring areas) for dyke construction, to protect dykes from erosion. The listed techniques were also observed at the study area; however, further dimensions were evident. The observed innovations at the study area were pond perimeter fencing combined with surface netting, pond embankments'/dykes' fortification, riverbank barrage with sandbag to prevent pond areas intrusion with floodwater from river, pond surface netting, pond caging from inside pond to overhead, pond cage placed on dyke to prevent fish escape during flood, increasing pond dyke with earthen materials, surface netting cum sandbag-reinforced dykes, increase in dyke height with sandbag plus surface netting or without surface netting, dyke slope stepping plus surface netting, dyke stepping plus pond caging placed on dyke, dyke stepping plus surface netting plus pegged sandbagged ponds' slope base, and pond slope reinforcement with hard-core materials. The complexity of the combinations of encountered approaches may be indicative of the myriads of challenges associated with flooding of the visited fish farms. Since human beings would normally innovate in order to cope in difficult conditions, the diversity of such innovation could be motivated by level of exposure.

Description of the Observed Technical Innovations

The observed technical innovations are discussed under five broad headings: ponds' fencing, pond embankment modifications, pond netting and caging, pond slope manipulations, and riverbank modification.

Pond Fencing

Pond fencing is one of the encountered adaptive innovations of fish farmers in the southwestern state of Nigeria. The technique is adaptable for mitigating climate change-induced flood in fish farming. In pond fencing, fish farmers made use of wire or iron mesh or mosquito net to enclose the pond area. Pond fencing is done in two ways: fencing of individual ponds and fencing of the total pond area (perimeter fencing). Pond fencing will prevent pond-stocked fish from escape during flooding. It also prevents mixing of fish of different sizes from different ponds within and across farm. Perimeter fencing would prevent entry of predators, such as frogs, toads, snakes, alligators, and invasive alien fish species carried by flood from entering ponds. However, fencing would not prevent entering of pond/farms by microbiota or pathogens carried by flood due to their microscopic nature.

Although fence design could vary in design, pond fencing is made with poles staked at an interval of about 2 m apart round the pond and then covered with galvanized wire mesh or mosquito net attached with nails to the staked poles. The nettings are normally buried into earth materials on dyke at the base part of the net. Hence, when flood occurs, erosion would not be able to lift the net, thus preventing the stocked fish from escape. Perimeter fencing of about 4 feet (1.2 m) high is the most common type of pond fencing (Fig. 2). However, pond fences must be checked

Fig. 2 A Typical flood-prone ponds supported with perimeter fencing in Southwestern Nigeria. Red arrows indicate pond fence; yellow arrow indicates flooded pond water

periodically for breakage and exposure of initially buried nets. Ideally, fences should be checked after each rain and corrected for any anomaly. A fenced pond can also be provided with pond surface netting either directly placed on pond surface on overhead (placing nets over the stakes) to prevent bird's invasions and ensure security from poachers. Pond netting is discussed fully under the section on netting and caging below.

Pond Netting and Caging

Netting of Ponds

This is the most common type of modification adopted by the fish farmers. Pond surface netting is not only done to adapt to climate-mediated flooding, but it is also utilized to safeguard ponds from predators. Sharp-edged pegs made from local wood materials, especially bamboo, are driven into pond dyke at regular intervals to hold the netting in place to cover the ponds surface. Twines are attached to the pegs such that several grids are formed, and then netting material is spread over the twine to cover the entire pond surface. Nets of observable varied mesh sizes, thickness, and tensile strengths were encountered. There is the need to characterize durability of the different netting materials. As stated under fencing, netting can be placed overhead following similar procedures but using longer stakes. Mesh sizes of the net may range from 1/2 inch to 2 inches, as these mesh sizes allow easy passage of feed pellets.

In situations of unpredicted climate-mediated flood, nettings are capable of restraining pond-stocked fish from escaping when the ponds overflow with water. However, this may be jeopardized if the earth materials holding stakes in place on the dyke are not eroded. Hence, it is advisable to drag stakes/pegs deep into the earth. Nettings also keep out the entering of various predatory invasive species, such as fish, birds, and reptiles from gaining entrance into the culture system. Pond netting works by entangling organism making attempt to enter the pond, being motivated by flood or by other factors. However, poorly constructed nets may also ghost-fish the stocked fish, if it is not lifted high enough above pond water level. Picture of typical netting is presented in Fig. 3a.

Caging of the Ponds

Caging of pond involves building of net cage enclosures over the pond. This is a specific innovation for preventing fish escape in flood scenario. The structure could be made of local materials, such as log metal or wooden materials, and it is usually built over the pond surfaces such that it looks like a pen or screen house. The wooden or metal material serves as the frame into which wire mesh net is usually attached. Chicken house mesh net grade is usually utilized for the upper part of the frame, while fine-meshed netting materials of smaller mesh size is usually utilized to surround the lower part of the frame structure. The height of the cage could vary, but the encountered specimens are usually between 2.0 m and 3.0 m. Some cages are erected on dyke of the pond, while others are erected to be rooted from inside the pond. Construction of the cages ensures that the stocked fish is retained in the pond

Fig. 3 (**a**) Pond netting and use of sandbags to reinforce dyke against flood; (**b**) caging of pond using local materials to prevent fish loss during flood; (**c**) overhead cage-like pond netting

whenever flooding occurs. It also prevents entry of predators and invasive fish species. However, it is advisable to utilize galvanized materials for the chicken mesh component in cages immersed in pond, as this would prevent metallic corrosion. Care should be taken when netting the cage such that the upper and lower nets are securely attached together so as to prevent escape of stocked fish at the point of joining the two netting materials.

Netting and caging would not prevent microbiota and pathogens from entering ponds during flood owing to their microscopic nature; and these techniques require periodic management practices which could attract additional running costs. Netting and caging are becoming more popular in the studied flood-prone states of

Southwestern Nigeria, and the respondents confirmed their relative efficiency compared to other techniques. This innovation could be adopted for use in other tropical African countries facing similar challenges. However, there is the need for further scientific inputs on details of their construction and management. Image of typical netting and caging innovations is presented in Fig. 3b. The image also shows the use of sandbags in reinforcement of the ponds' embankment. Stocking of fish inside net harpers and fish cage culture in rivers and reservoirs are also useful in adapting to flood risks.

Modification of Ponds' Embankments

Apart from increasing pond height using sandbags, ponds with pliable sandy soil, which is vulnerable to being easily washed away during floods, were observed at the study area to be constructed with sandbags at the base of the embankment to provide compact base. This was then topped with earth materials to achieve the desired height or to increase an existing height. This is becoming a common practice in flood-prone areas even when soil type is not sandy, as observed in Fig. 4a. In some instances, these sandbags at base of embankments (Fig. 4b) are pegged to improve stability. The pegs are usually made of bamboo and other local hardwood materials of about 60 cm height, made to have sharp end. The pegs are driven to pond base to provide brace for the sandbags. These manipulations make it difficult for the embankment to collapse during flooding and erosion. Sandbags are also usually utilized to increase the height of ponds in low-lying areas (Fig. 4c). This technical innovation is less expensive; it is effective in ensuring pond stability in the face of flooding and could be adaptable for similar flood scenarios across tropical Africa. However, it is important to use wood materials that will have minimal chemical reaction with pond water.

Pond Slope Manipulation

Slopes of most earthen ponds are prone to destruction during flooding, hence the need to manipulate pond slopes for stability. Two techniques of ensuring slope stability were encountered during the study. The first encountered method used hard-core stones to patch the pond slopes such that flooding would do little or minimal damage to the slope (Fig. 5). This innovation consolidates the slope and prevents washing away of the dyke or embankment. The method seems effective and has potential to last long; but it is expensive. It is important to allow the soldered hard-core materials to mature before stocking of fish. This is necessary because hard-core materials are soldered with cement materials, which would introduce toxic cement particles to stocked fish, if not properly cured before use. Reinforced concrete materials may also be utilized for this purpose with similar precautions but with relatively lower cost and longevity.

Apart from the normal grassing of dykes, stepping of pond slopes and lining of slopes with synthetic waterproof materials were also encountered. Although these innovations may not be flood-focused, slope stepping will especially have potential for reduction of rate of erosion of pond slope during flooding, thus reducing damage

Fig. 4 (**a**) Reinforcement of pond using sandbags; (**b**) filling up of eroded dyke with sandbags; (**c**) reinforcing pond base with sandbag

associated with climate-induced flooding. The steps observed were of varied lengths and numbers, but about 20.0 cm height was common.

River Embankment Modification

There is the need to fortify riverbanks against pond flooding. Most earthen pond fish farmers are located close to river; such that when flooding occurs, rivers discharge excess water to the adjoining ponds. Barraging of river and core trenching around fish farm area, to channel excess water away, were observed. The fish farmers diverted excess water from river away from their farms through the constructed trenches of about 1.0 m wide around their farms (Fig. 6a). Although river barrage is normally constructed to channel water from river to their pond, the role of this in flooding and flood management requires investigation. The use of sandbags and log for river barraging was encountered (Fig. 6b). However, dredging of river to increase the width of riverbank so as to facilitate movement of larger flood water volume was

Fig. 5 Stone patching of pond dyke (red arrow)

Fig. 6 (a) River barraging using logs of woods; (b) river barraging using sandbags

a common practice in flood management. The use of reinforced concrete and sluice gates for river channels to prevent destruction of the river embankment was also useful. However, barrages and trenches require periodic management for effectiveness.

Conclusions and Recommendations

Climate change-mediated flood threatens fish farming in tropical Africa, but it is being mitigated in Southwestern Nigeria using the described innovations. Developmental actions focusing on sharing and awareness creation on the innovations at flood-prone locations across tropical Africa are essential for sustainable fish farming in the face of climate change-mediated flooding in the region. This would improve uptake of the innovations. However, scientific endeavor to discover more of such innovations will be helpful for improvement, and this is recommended.

Acknowledgments Authors acknowledge the CIRCLE program of the Association of Commonwealth Universities and DFID, UK, for funding support.

References

ActionAid (2006) Climate change, urban flooding and the rights of the urban poor in African cities; key findings from six African cities. A Report by ActionAid international, October 2006. http://www.actionaid.org/sites/files/actionaid/climate_change_urban_flooding_and_the_rights_of_the_urban_poor_in_africa.pdf. Accessed 20 June 2018

Adebayo K, Dauda TO, Rikko LS et al. (2011) Emerging and Indigenous Technology for Climate Change Adaptation in Southwest Nigeria. ATPS Working paper series No. 56. African Technology Policy Studies Network. Nairobi, Kenya. https://searchworks.stanford.edu/view/10329439. Accessed 22 June 2018

Adebo GM, Ayelari TA (2011) Climate change and vulnerability of fish farmers in southwestern Nigeria. Afr J Agric Res 6(18):4230–4238

AWDR, African Water Development Report (2006) Freshwater resources in Africa. http://www.uneca.org/awich/AWDR%202006/Freshwater%20Resources%20in%20Africa.pdf. Accessed 20 April 2018

Akintola EO (1994) Flooding phenomenon in Ibadan region. Rex Charles Publication, Ibadan, pp 244–255

Ayanwuyi E, Kuponiyi E, Ogunlade FA et al (2010) Farmers perception of impact of climate changes on food crop production in Ogbomosho agricultural zone of Oyo state, Nigeria. Global J Hum Soc Sci 10(7):33–39

Bates BC, Kundzewicz ZW, Wu S et al (eds) (2008) Climate change and water. Technical paper of the intergovernmental panel on climate change. IPCC Secretariat, Geneva

Bevan J (2007). Between a rock and hard place: armed violence in African pastoral communities. www.undp.org/cpr/documents/armed_violence/AV_pastoral_communities.pdf. Retrieved on 6 Nov 2017

Bordoloi R, Muzaddadi AU (2014) Indigenous technical knowledge associated with disaster management and fisheries related activities in the highest flood affected district (Dhemaji) of Assam, India. Ind J Trad Knowl 14(3):407–415

BBC News. British Broadcasting Cooperation (2007). Rains threaten food hit Africa. Saturday, 15 September. Available at: http://news.bbc.co.uk/2/hi/6994995.stm#anchor. Accessed April 2018

Constantinescu A, Frone S (2014) The role of technological innovation in sustainable economic development. J Knowl Manag Econ Inf Technol 4:13–17

Dinesh D, Bett B, Boone R, et al. (2015). Impact of climate change on African agriculture: focus on pests and diseases. CGIAR Research Program on Climate Change, Agriculture and Food Security (CCAFS), Copenhagen, Denmark. www.ccafs.cgiar.org. Accessed 8 Mar 2018

Eriksen S, O'Brien K, Rosentrater L (2008) Climate change in Eastern and Southern Africa; impacts, vulnerability and adaptation. Global Environmental Change and Human Security Report 2008:2. http://www.bvsde.paho.org/bvsacd/cd68/climafrica.pdf. Accessed 2 June 2018

European Institute of Innovation and Technology, EIT (2013). Analysis of synergies fostered by the EIT in EU innovation landscape. www.eit.europa.eu. Accessed 24 Dec 2017

Falaye AE (2013) Status of aquaculture in Nigeria AU- IBAR Expert Planning Meeting on Aquaculture Development in Africa Mangochi, Malawi, 8–9 March 2013

Fokkema J, Jansen L, Mulder K et al (2005) Sustainability: necessity for a prosperous society. Int J Sustain 6:219–228

FAO, Food and Agriculture Organisation (2008) The state of the world fisheries and aquaculture-Sofia 2008. Fisheries and Aquaculture Department FAO, Rome. www.fao.org/docrep/fao. Accessed 14 Jan 2018

FAO, Food and Agriculture Organization (2011) Guidelines for measuring household and individual dietary diversity. FAO Rome. http://www.fao.org/fileadmin/user_upload/wa_workshop/docs/FAO-guidelines-dietary-diversity2011.pdf. Accessed 18 Dec 2017

FAO, Food and Agriculture Organization (2003) World Agriculture Towards 2015/2030. http://www.fao.org/documents/show_cdr.asp?url_file=/docrep/004/y3557e/y3557e00.htm Accessed 19 Nov 2017

FAO. Food and Agriculture Organization (2015) Assessing climate change vulnerability in fisheries and aquaculture: Available methodologies and their relevance for the sector, by Cecile Brugère and Cassandra De Young. FAO Fisheries and Aquaculture Technical Paper No. 597. Rome, Italy. www.fao.org/3/a-i5109e.pdf. Accessed 10 June 2018

FAO. Food and Agriculture Organization (2016) The State of World Fisheries and Aquaculture 2016. Contributing to food security and nutrition for all. Rome. www.fao.org/3/a-i5555e.pdf. Accessed 20 Dec 2017

Faturoti EO (1999) Fisheries potentials and investment opportunities in Nigeria. A paper presented at the Fisheries Society of Nigeria (FISON) Investors Workshop held at the Nigerian Institute for Oceanography and Marine Research (NIOMR), Lagos, Nigeria, 22 April 1999

Freeman OE (2017) Impact of climate change on aquaculture and fisheries in Nigeria; a review. Intl J Multidiscip Res Devpt 4(1):53–59

Ficke AD, Myrick CA, Hansen LJ (2007) Potential impacts of global climate change on freshwater fisheries. Rev Fish Biol Fis 17(4):581–613

Hély C, Bremond L, Alleaume S et al (2006) Sensitivity of African biomes to changes in the precipitation regime. Glob Ecol Biogeog 15:258–270

Henry P (2006) Levees and other raised ground. Am Sci 94(1):7–11

Ibadan Urban Flood Management Project, IUFMP (2014) Environmental and social management framework. The World Bank IBRD. IDA. http://projects.worldbank.org/P130840?lang=en. Accessed 14 Feb 2018

Idowu AA, Ayoola SO, Opele AI et al (2011) Impact of climate change in Nigeria. Iranica J of Ener Environ 2(2):145–152

Intergovernmental Panel on Climate Change, IPCC, (TAR) (2001) Third assessment report (TAR) of the intergovernmental panel on climate change. Parts 1, 2 and 3 synthesis report and policy makers summaries. Cambridge University Press, Cambridge, UK

IPCC, 2007: Climate Change 2007: Synthesis Report. Contribution of Working Groups I, II and III to the Fourth Assessment Report of the Intergovernmental Panel on Climate Change [Core Writing Team, Pachauri, R.K and Reisinger, A. (eds.)]. IPCC, Geneva, Switzerland

Ipinjolu JK, Magawata I, Shinkafi BA (2014) Potential impact of climate change on fisheries and aquaculture in Nigeria. J Fisher & Aquat Sci 9:338–344

Makhanu SK, Oteng'i SBB, China SS et al (2007) Indigenous Construction Technologies in Flood-Prone Areas of Western Kenya. Centre for Disaster Management and Humanitarian Assistance (CDMHA), Western University College of Science and Technology, Kenya. pp 1–11

Michael FO, Stephen OO (2017) ICT-based market information services utilization by small scale farmers in Southwest Nigeria. Rep & Opin 9(2):82–85

Mintzer R (2010) Destroying mangroves in West Africa detrimental to people, climate, and wildlife. Media Global, March 25, 2010

NEMA, National Emergency Management Agency (2013) Nigeria Post-Disaster Needs Assessment (PDNA) 2012 Floods: A report by The Federal Government of Nigeria. https://www.gfdrr.org/sites/gfdrr/files/NIGERIA_PDNA_PRINT_05_29_2013_WEB.pdf. Accessed 2 Jan 2017

New M (2016) Understanding, reducing and managing African climate risk. www.axa.com

Nkwunonwo UC (2016) A review of flooding and flood risk reduction in Nigeria. Glob J Hum-Soc Sc 16(2):22–42
Oladokun OJ, Adedara TM, Adedadamola JO (2015) Tourism and climate change: combating climate change effects on tourism participation in Nigeria. J Tour, Hospit Spo 5:1–6
Olaifa FE (2015). Food security: the perspective of aquaculture and fisheries. 2014/2015 Faculty lecture, Faculty of Agriculture & Forestry, University of Ibadan, delivered on 20 May, 2015. Ibadan University Press. 58pp
Oyebola OO, Fada PO (2017). Climate change: a potential risk to cage-culture and genetic structure of endemic fisheries in fish cage installed inland lake. African Great Lake Conference 2017, Entebbe, Uganda. 2–5 May, 2017
Policy Brief (2012). Floods in Nigeria; Re-visiting States Accountability on Emergency Response/Preparedness and Disaster Risk Reduction in Nigeria. http://www.actionaid.org/sites/files/actionaid/climate_change_urban_flooding_and_the_rights_of_the_urban_poor_in_africa.pdf. Accessed 20 June 2018
Rennings K (1999) Redefining innovation – eco-innovation research and the contribution from ecological economics. Ecol Econs 32:319–332
Sasson A (2012) Food security for Africa: an urgent global challenge. Agric & Food Sec 2012(1):2
Shafie H, Rahman S (2009) Traditional coping strategies of rural people living in flood-prone areas in north-West Bangladesh. Rangpur Dinajpur Rural Service (RDRS), Dhaka
Tagbo E (2010) Media coverage of climate change in Africa: a case study of Nigeria and South Africa. Reuters Institute for the Study of Journalism, University of Oxford. pp 2–3
Tawari-Fufeyin P, Megbuwe P, Adams OG (2015) Some aspects of historic flooding in Nigeria and its effects on some Niger-Delta communities. Am J Wat Res 3(1):7–16
Von GK, Headey D, Bene C et al (2013) Global hunger index: the challenge of hunger: building resilience to achieve food and nutrition security. Welthungerhife, International Food Policy Research Institute, and Concern Worldwide, Bonn/Washington, DC/Dublin

Investigation of Rainfall Characteristics in Sudano-Sahelian Region of Nigeria (1971–2006)

126

Aliyu Tambuwal Umar, Sani Abubakar Mashi, and Mansur Matazu Bako

Contents

Introduction: Some Facts About Sudano-Sahelian Region of Nigeria 2596
Rainfall and Length of Rainy Season in the Sudano-Sahelian Region of Nigeria 2597
Onset, Cessation, and Dry Spell Occurrences in the Sudano-Sahelian Region of Nigeria ... 2602
The Influence of the Onset, Cessation, and Dry Spells on Rainfall Characteristics in the Sudano-Sahelian Region of Nigeria 2615
Conclusion 2618
Cross-References 2619
References 2619

Abstract

Farmers especially in rural areas of Nigeria have for long relied on rain-fed farming for their livelihoods. Any perturbations in rainfall characteristics could adversely affect their livelihoods. It is against this background that this chapter examined the influence of variability in the dates of onset and cessation of the rains, dry spell occurrences on annual and seasonal rainfall amounts, and length of the rainy season in the Sudano-Sahelian region of Nigeria. Daily rainfall data for five selected meteorological stations in the region were collected from the Nigerian Meteorological Agency (NiMet), Abuja, Nigeria. The onset and cessation dates were computed using Walter's technique, while four (4) dry spell categories, namely, *very short* (DS1, 1–3 days), *short* (DS2, 4–7 days), *medium* (DS3, 8–14 days), and *long* (DS4, $\geq$15 days), were detected using multi-scale dry spell detection method adopted from Salack et al. Data were analyzed using

A. T. Umar (✉)
Department of Geography, Usmanu Danfodiyo University, Sokoto, Nigeria
e-mail: aliyuumar2@yahoo.com

S. A. Mashi · M. M. Bako
Nigerian Meteorological Agency (NiMet), Abuja, Nigeria
e-mail: saniamashi@nimet.gov.ng; b.matazu@nimet.gov.ng

W. Leal Filho (ed.), *Handbook of Climate Change Resilience*,
https://doi.org/10.1007/978-3-319-93336-8_165

stepwise multiple regression model. Results showed among other things that the onset and cessation dates of the rains, DS2, DS3, and DS4, had strong influence on rainfall characteristics examined in this study. It is recommended that farmers should adhere strictly to NiMet's seasonal rainfall predictions to prevent crop failures due to false onset of the rains and associated dry spells. This could enhance farmers' resilience to climate change and facilitates the attainment of food security in the region.

Keywords
Rainfall variability · Onset · Cessation · Dry spells · Sudano-Sahelian region

Introduction: Some Facts About Sudano-Sahelian Region of Nigeria

The Sudano-Sahelian region of Nigeria is located within the northern part of Nigeria between latitudes 10°N and 14°N (Fig. 1). A report on the population and funds committed to export crop farming by households of five selected states of Nigeria within the region is given in Table 1.

The northern Nigeria as a whole could be divided into three ecological zones, namely, the Guinea savanna, Sudan savanna, and Sahel savanna. The Sudano-Sahelian

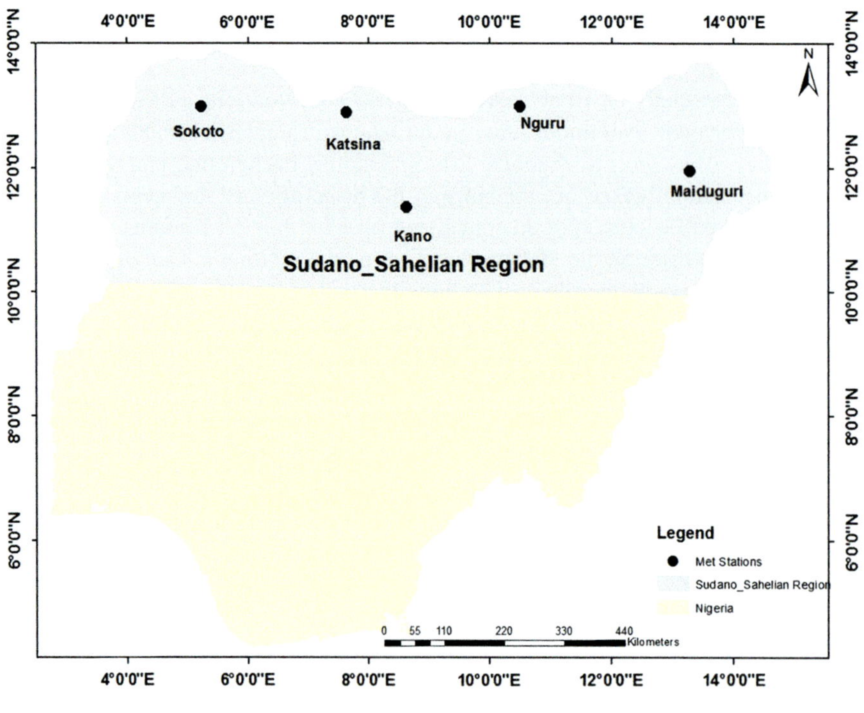

Fig. 1 Selected meteorological stations within the Sudano-Sahelian region. (Sources: Authors)

Table 1 Population and funds committed to export crop farming by households in five selected states of the Sudano-Sahelian region of Nigeria

Selected states	Population (millions)	Land mass (Km^2)	Population density (Km^2)	Funds committed to export crop farming by households in billions of naira(₦) as at 2004/2005
Sokoto	3,696,999	25,973.00	142	5.791
Kano	9,383,682	20,680.00	454	44.242
Katsina	5,792,578	26,785.00	216	18.443
Borno (Maiduguri)	4,151,193	71,130.00	58	2.897
Yobe (Nguru)	2,300,736	55,920	54	2.007

Sources: NPC (2006), CBN/NBS/MA&RD/FMC (2005)

region falls within the Sudan and Sahel savanna ecological zones. The major occupation in the area is rain-fed farming with few engaging in fishing, trading, weaving, making cottage, etc. The area enjoys tropical continental climate with two distinct seasons, wet and dry seasons. The wet season lasts for only 3–4 months (June to September), while the dry season lasts for about 8 months (October to May). Temperatures are generally high particularly during the dry season with a mean maximum temperature of 39 °C and mean minimum temperature of 19.5 °C. The mean annual rainfall in the region ranges from 600 mm in the Sudan savanna to only 400 mm in the extreme Sahel savanna zone (Aondover and Woo 1998; Ayoade 2011).

The country as a whole receives rainfall from the south westerlies which invade the country from the Gulf of Guinea coast, i.e., the tropical Atlantic. This moist airstream is overlain by the northeast trades which originate from above the Sahara and therefore are dry and dust laden. The zone of contact of the two air masses at the surface is a zone of moisture discontinuity, and it is known as the inter-tropical discontinuity (ITD). The ITD advances inland as far as 22–25 °N in August at the margin of the Sahara, i.e., considerably beyond Nigeria's northern border (Ayoade 2011), while it does not retreat equatorward beyond 4^0N latitude during the "Harmattan" dry season (Ayoade 2011). The amount, seasonal distribution, type of rainfall, and length of the rainy season as well as the general weather conditions experienced in the course of the year at any given location in the West African region depend primarily on its location relative to the position of the ITD and the associated weather zones (Ayoade 2011). The selected meteorological stations in the region include Sokoto, Kano, and Katsina which fall within the Sudan savanna and Maiduguri and Nguru which fall within the extreme Sahel savanna ecological zone.

Rainfall and Length of Rainy Season in the Sudano-Sahelian Region of Nigeria

Fluctuations and overall trends in annual and seasonal rainfall as well as the length of the rainy season/growing season are of paramount importance in rainfall climatological studies especially in the tropics where rainfall to a large extent determines the crop calendar of any given area. Fluctuations in annual rainfall at the selected meteorological stations and for the entire Sudano-Sahelian region are generated using 3-year running mean, which is presented in Fig. 2. The annual rainfall series for each year and for each selected station was derived from the daily rainfall records used for this chapter. It is interesting to note that the annual rainfall pattern at all the selected stations and Sudano-Sahelian region as a whole revealed a marked decline in rainfall in 1972–1973 and 1982–1984 periods which coincide with the droughts of the 1970s and 1980s that ravaged the West African Sahel as reported in the literature (Druyan 1989). The rainfall decline continued through the 1980s and 1990s at all the stations in Sudano-Sahelian region till about 1998 when rainfall began to appreciate in the region. In the late 1990s, annual rainfall pattern in the Sudano-Sahelian region showed an evidence of slight recovery in its annual rainfall. Several cases of floods had in the last 15 years or so been reported in the region (Umar et al. 2015).

Amount of rainfall that fell between the dates of onset and cessation of the rains were derived for each year and for each station from the daily rainfall records over the period 1971–2006. It is herein referred to as "seasonal rainfall amounts." A composite of such series (i.e., yearly averages of five selected stations) was also derived for the Sudano-Sahelian region. Interannual fluctuations in seasonal rainfall amounts at the selected Sudano-Sahelian stations and for the entire region were presented in Fig. 3. Unlike annual rainfall pattern obtained in Fig. 2, fluctuations in seasonal rainfall amounts showed a severe large-scale rainfall anomalies that occurred in the region from 1971 to about 1998. At Kano, for example, seasonal rainfall amounts showed a continued decline below the long-term mean from 1971 to about 1997. It is also interesting to note that all the stations and the region as a whole had also revealed a positive rainfall anomalies which began in the late 1990s.

Nicholson (2005) examined the question of the "recovery" of the rains in the West African Sahel using seasonal and August rainfall time series over the period 1998–2003 for 8 zones within latitudes 12 °N–20 °N and showed that, in the 3 zones to the south, seasonal totals exceeded the long-term means in all years and conditions were comparable to or wetter than those during the wet decade of the 1950s. The wettest years were 1998, 1999, and 2003. Using rainfall variability index, Oguntunde et al. (2011) examined the spatiotemporal patterns of rainfall in Nigeria over two climatic periods, 1931–1960 and 1961–1990, and showed that annual precipitation reduced by 7% between the two periods. The study further demonstrated that while more than 90% of Nigerian landscape showed no significant rainfall change during the 1931–1960 period, about 57% of Nigeria showed a significant decrease during the 1961–1990 period.

The length of the rainy season/growing period is the difference between the dates of onset and cessation of the rains at any given year. This had been generated for each year and for each of the selected stations used in this chapter. A composite of such series was also generated for the Sudano-Sahelian region using yearly averages of the five selected stations in the region. The length of the rainy season/growing period is an important rainfall characteristics since it is the period within which each crop is expected to grow. It is defined as the period within which rainfall spread for the crop utilization since crops have different water requirements at different stages

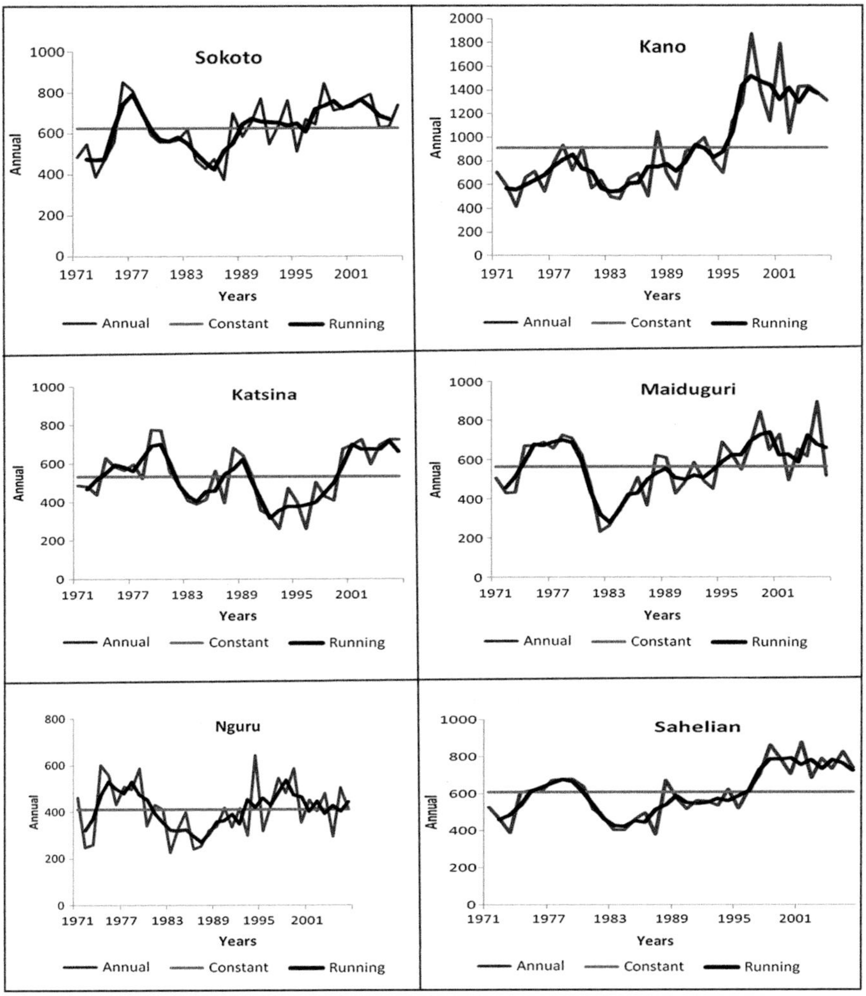

Fig. 2 Annual rainfall fluctuations in the Sudano-Sahelian region of Nigeria (1971–2006). (Sources: Authors)

of their growth (Odekunle 2004). Fluctuations in the length of the rainy season/ growing period at each of the selected stations and for the region as a whole were presented in Fig. 4. Length of the rainy season is characterized by oscillations at all the selected stations and the Sudano-Sahelian region as a whole. The drought years of the 1980s had shorter lengths of the rainy season at Maiduguri and Nguru stations with those of the 1970s having longer than normal lengths of the rainy season at Maiduguri only. Sokoto, Kano, Katsina, and the Sudano-Sahelian region as a whole showed oscillations in the length of the rainy season, an evidence of the high variability of this rainfall attribute in the region as previously examined by Umar (2010).

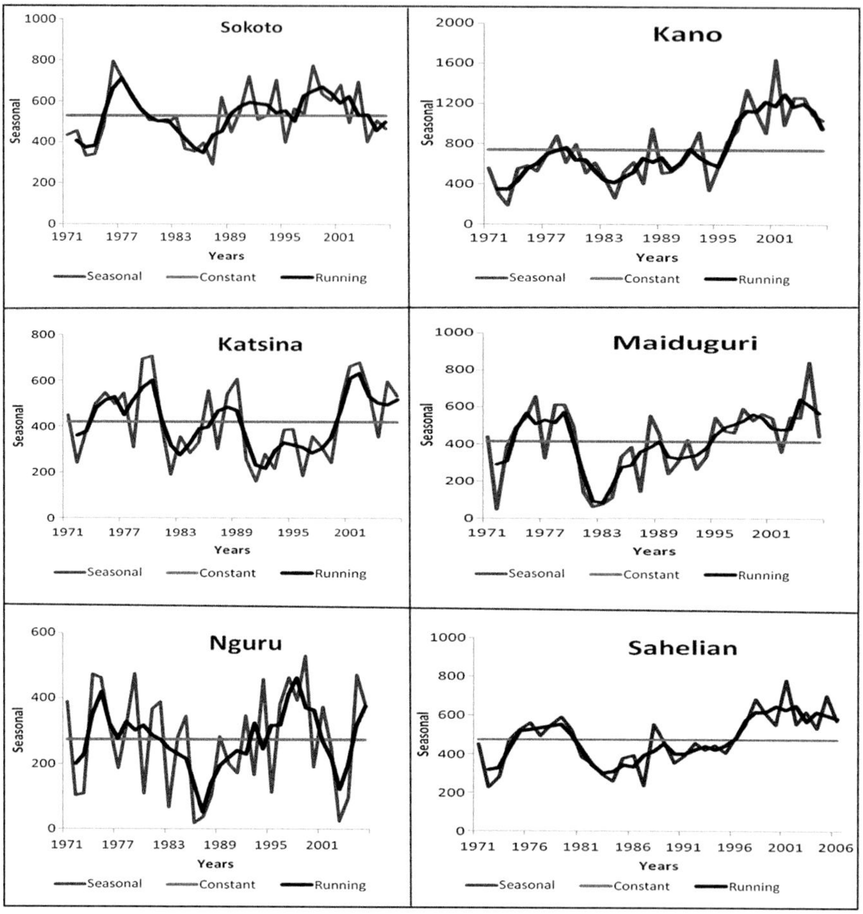

Fig. 3 Seasonal rainfall fluctuations in Sudano-Sahelian region of Nigeria (1971–2006). (Sources: Authors)

An overall trend in annual and growing season rainfall as well as the length of the rainy season at each of the selected stations and for the Sudano-Sahelian region as a whole are presented in Figs. 5, 6, and 7. Trend in the time series gives a general direction of the flow of such series. It could be observed in Figs. 5 and 6 that both annual and growing season rainfall series at all the selected stations and the region as a whole exhibited an upward trend despite the fact that in the 1970s and 1980s, droughts of different magnitudes struck the region. The general upward trend exhibited at all the stations and the region as a whole suggests the strong influence of the recent rainfall recovery in the Sahel which began about 1998 till date.

Sanogo et al. (2015) examined the spatiotemporal characteristics of the recent rainfall recovery in West Africa using daily(monthly) rainfall data from 167(254) stations across West Africa over the period 1980–2010 and the gridded African

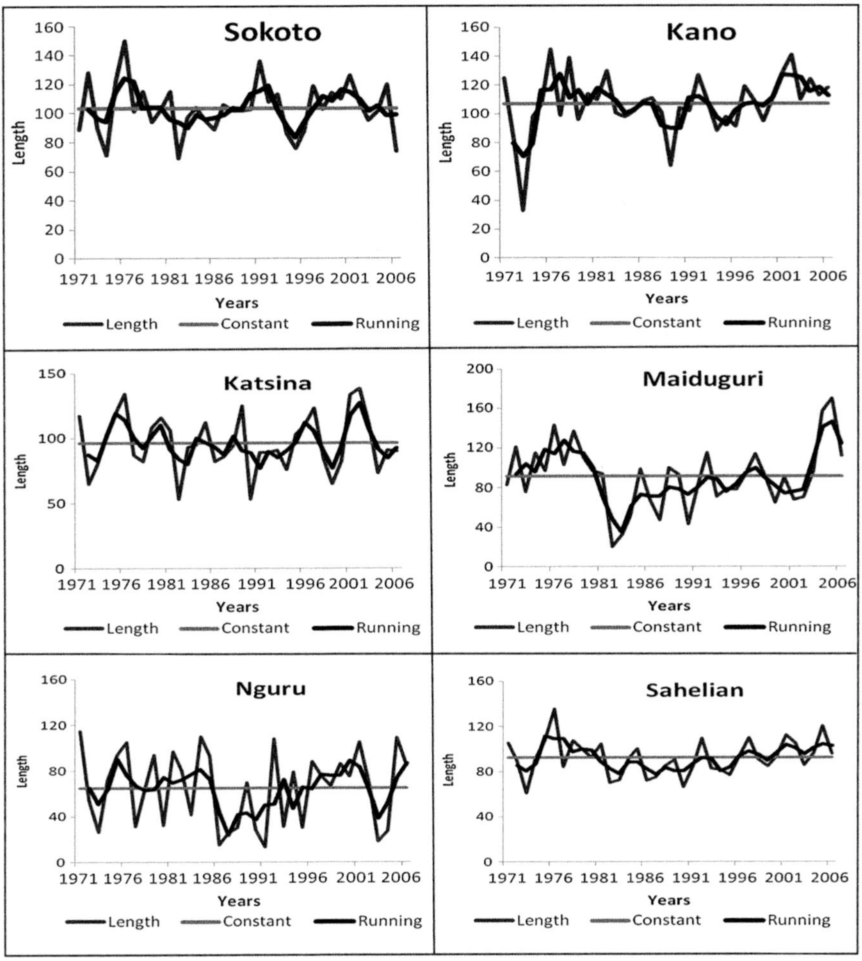

Fig. 4 Fluctuations in the length of the rainy season in the Sudano-Sahelian region of Nigeria (1971–2006). (Sources: Authors)

Rainfall Climatology Version 2 (ARC2) for the period 1983–2010. The study revealed that majority of the stations in the Sahel between the West Coast and 15°E shows a statistically significant positive rainfall trend for annual totals. The August to October period exhibited the largest rainfall recovery in the Sahel, and the date of the retreat of the rainy season significantly moved later into the year by 2 days decade^{-1}. Several scholars have attributed the dipole mode of rainfall anomalies in the Sahel to a number of dynamical factors ranging from north/south displacement of the tropical rain belt to factors affecting rainfall intensity in the region. Nicholson (2008) examined the intensity, location, and structure of the tropical rain belt over West Africa as factors in interannual variability over 2 years that were anomalously wet in the Sahel, 1955 and 1950, and 2 years, 1983 and 1984, that were anomalously dry and showed among

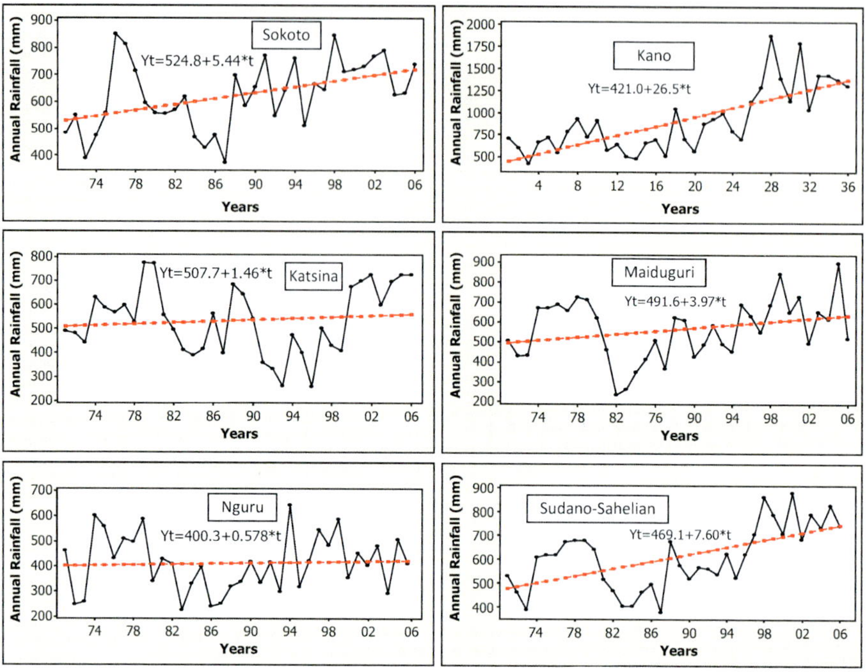

Fig. 5 Annual rainfall trends in Sudano-Sahelian region of Nigeria (1971–2006). (Sources: Authors)

other things that the contrast between wet and dry years is associated with a weakening and contraction of tropical rain belt.

The overall trends in the length of the rainy season for the selected stations and for the Sudano-Sahelian region are presented in Fig. 7. Kano, Maiduguri, and the Sudano-Sahelian region have shown a tendency for increasing length of the rainy season/growing period over the period 1971–2006. Sokoto, Katsina, and Nguru stations revealed a significant downward trend in the length of the rainy season. This could be attributed to effects of the 1970s and 1980s droughts in which most drought years had shorter lengths of the rainy season, resulting in the overall downward trend of this rainfall attribute over the period of the study. The implication of the reduced length of the rainy season is that crops that require longer period to mature and be ready for harvest may not do well under the condition of reduced length of the rainy season if not supported by irrigation.

Onset, Cessation, and Dry Spell Occurrences in the Sudano-Sahelian Region of Nigeria

The variability in the dates of occurrence of the onset and cessation of the rains as well as frequency and duration of dry spells annually is of great concern to farmers especially in semiarid environments characterized by high rainfall variability. A

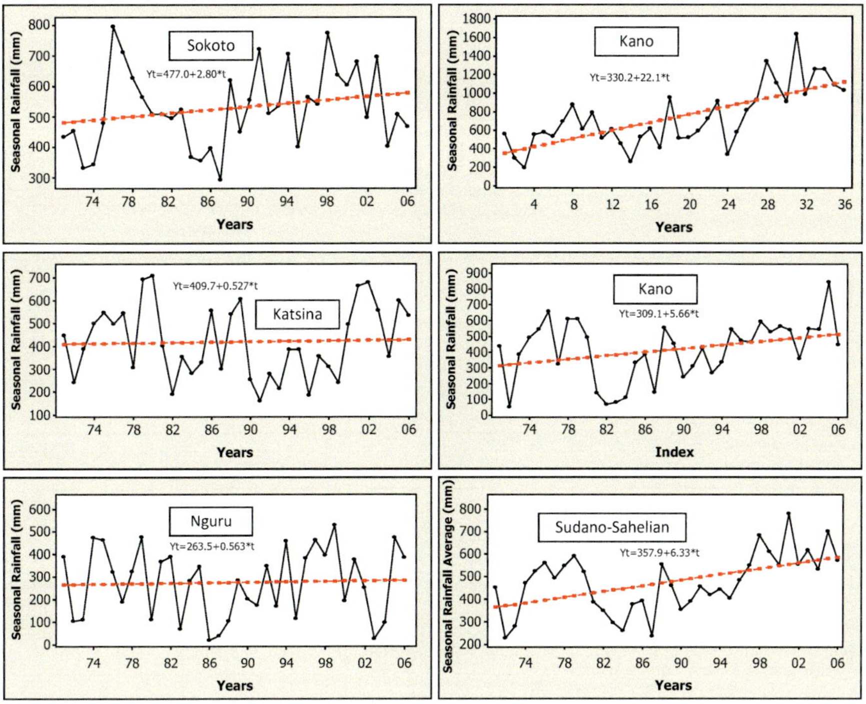

Fig. 6 Seasonal rainfall trends in the Sudano-Sahelian region of Nigeria (1971–2006). (Sources: Authors)

reliable prediction of rainfall onset and cessation times and thus the length of the growing season will greatly assist on-time preparation of farmlands, mobilization of seeds/crops, man power, and equipment and will also reduce the risk involved in planting too early or too late (Benoit 1977; Fasheun 1983; Olaniran 1983, 1984; Omotosho et al. 2000). The onset and cessation dates of the rains were defined by various authors depending on the threshold adopted, data requirement, and method for their derivation (Walter 1967; Ilesanmi 1972; Benoit 1977; Oduro and Adefolalu 1993; AGRHYMET 1996; Marengo et al. 2001; Odekunle et al. 2005; Usman and Abdulkadir 2013).

Walter (1967), for instance, defined the onset of the rains in Nigeria in terms of time of receiving an accumulated amount of rainfall in excess of 51 mm, while Benoit (1977) defined the onset of the rainy season in northern Nigeria as the date when accumulated rainfall exceeds and remains greater than one half of potential evapotranspiration for the remainder of the growing season, provided that no dry spell longer than 5 days occurs immediately after this date. Similarly, Oduro and Adefolalu (1993), AGRHYMET (1996), Marengo et al. (2001), and Odekunle et al. (2005) cited in Usman and Abdulkadir (2013) used cumulative rainfall and defined the onset of the rains as the end of a 30-day period within which one decade with at least 25 mm of rain is followed by two successive decades with at least 20 mm of rainfalls. The Oduro and Adefolalu's (1993) scheme which used the month of

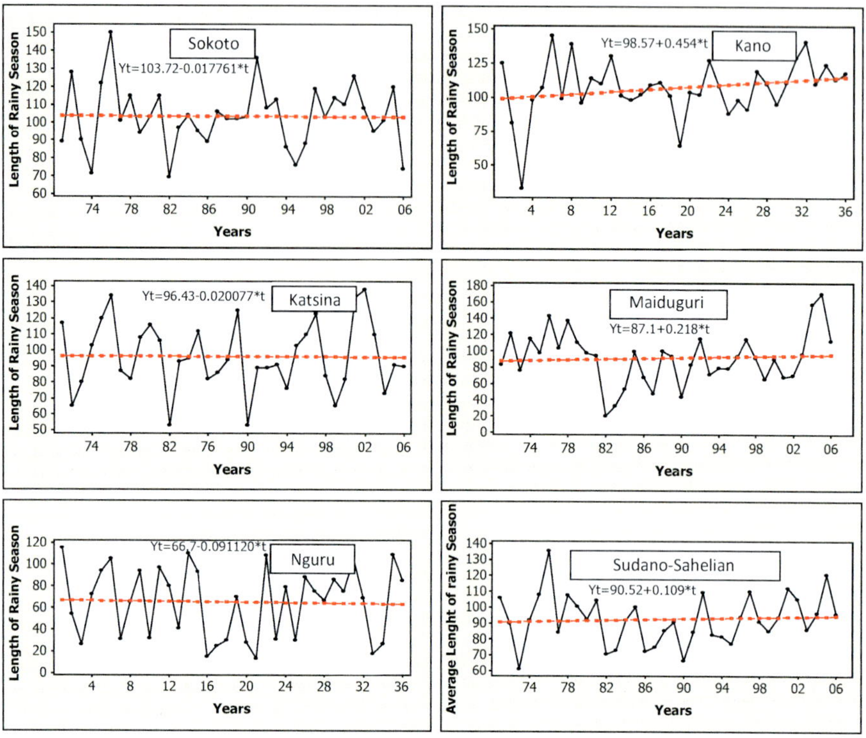

Fig. 7 Trend in the length of the rainy season in the Sudano-Sahelian region of Nigeria (1971–2006). (Sources: Authors)

positive departure from mean surface potential equivalent temperature (Qe) to determine the onset date of the rain which has been slightly modified by Omototosho et al. (Omotosho et al. 2000) is currently been used by the Nigerian Meteorological Agency (NiMet) in determining the onset of the rains over Nigeria (Usman and Abdulkadir 2013).

Generally speaking, the methods used for determining the dates of onset and cessation of the rains are divided into two broad categories: traditional methods (Ramadan fasting method) and scientific-based methods (Ati et al. 2002). The traditional (Ramadan) method used in Kano is named after the Moslem fasting period. This rule states that planting can begin with the first good rain after Ramadan provided that it is at least 7 months from the last effective rain of the previous season. If these dates are not close, a 7-month rule is applied and planting is advised to be 7 months after the last effective rain of the previous season. This method has been found to perform poorly (Ati et al. 2002). The scientific methods are further divided into three, namely, rainfall-based methods (Walter 1967; Ilesanmi 1972; Odekunle et al. 2005; Usman and Abdulkadir 2013), rainfall-potential evapotranspiration relation methods (Benoit 1977), and upper air-based data methods (Omotosho et al. 2000; NiMet 2018). None of these methods is all encompassing. The rainfall-based methods are however widely used since rainfall data are readily available unlike potential evapotranspiration (PE) records and upper air-based data which are scarce especially in developing countries.

The Walter's (1967) method is employed in this chapter for the computation of the onset and cessation dates of the rain for Sokoto, Kano, Katsina, Maiduguri, and Nguru and for the Sudano-Sahelian region as a whole over the period 1971–2006. The choice of Walter's technique is based on its inclination to statistics and uses only rainfall data which are readily available for the period 1971–2006 used in this chapter. The first essential step of this method is to derive the accumulated monthly rainfall of a given magnitude, which is used to define the onset and cessation of the rains. Walter (1967), for instance, defined the onset of the rains in Nigeria in terms of time of receiving an accumulated amount of rainfall in excess of 51 mm. The actual date of the onset of the rain according to Walter (1967) is as follows:

$$\text{Days in month} \times \frac{51 - \text{Accummulated rainfall in the previous month}}{\text{Total rainfall for the month}}$$

where the month under reference is that in which accumulated total rainfall is in excess of 51 mm of the rain. The end of the rain is the date after which no more than 51 mm of the rain is expected. Thus, the above formula is applied in the reverse order by accumulating the total backward from December to obtain the actual date of the cessation of the rain. The computed dates of onset and cessation of the rains for the selected stations and for the Sudano-Sahelian region are presented in Table 2.

It must be noted, however, that one of the major problems associated with rainfall studies especially in the developing countries is the issue of missing rainfall records especially if it requires a long rainfall records. The daily rainfall records are more prone to such cases of missing records due to several reasons ranging from technical, human, and political reasons. The daily rainfall records obtained from NiMet for the five selected stations used in this chapter had few cases of missing records at Maiduguri and Nguru meteorological stations only. At Maiduguri, for example, there were no daily rainfall records for the months of June to October, 1971. The Nguru station had no daily rainfall records for the months of July to October, 2003. To address this concern, monthly rainfall records for the missing months at Maiduguri and Nguru stations were sourced from the archives of the Nigeria Meteorological Agency (NiMet). The few missing daily rainfall records detected in the daily rainfall records file obtained from NiMet could certainly be attributed to operational problems since the monthly rainfall record's file was found to have complete rainfall records. Such operational problems could not be totally eliminated especially in developing countries like Nigeria. This does not however detract the quality of the data and analyses carried out herein using those records.

For the determination of the mean date of onset and cessation of the rains for each station and the region as a whole, the class interval method adopted by Bello (1996) was also employed in this chapter. The computed dates of onset and cessation of the rains were grouped into class intervals based on a 5-day time scale, which is the tolerable limit of dry spell duration which, if associated with the onset and cessation

Table 2 Onset and cessation dates of the rains in Sudano-Sahelian region of Nigeria (1971–2006)

Station				Maiduguri			Sokoto					
Year	Sokoto	Kano	Katsina	Onset	Nguru	SSRN	Cessation	Kano	Katsina	Maiduguri	Nguru	SSRN
1971	27th June	17th May	28th May	26th June	26th May	6th June	23rd Sept	8th Sept	21st Sept	16th Sept	17th Sept	11th Sept
1972	4th May	14th May	30th May	17th May	18th June	23rd May	8th Sept	1st Aug	2nd Aug	14th Sept	10th Aug	19th Aug
1973	25th June	2nd July	4th July	6th July	10th July	3rd July	22nd Sept	3rd Aug	21st Sept	19th Sept	4th Aug	1st Sept
1974	1st July	7th June	31st May	21st May	2nd July	12th June	9th Sept	12th Sept	10th Sept	12th Sept	11th Sept	11th Sept
1975	16th May	26th May	20th May	4th June	13th June	28th May	14th Sep	9th Sept	16th Sept	8th Sept	14th Sept	7th Sept
1976	23rd May	1st June	14th June	5th June	5th July	9th June	19th Oct	23rd Oct	25th Oct	25th Oct	18th Oct	21st Oct
1977	3rd June	7th June	23rd June	7th June	3rd July	15th June	11th Sept	13th Sep	17th Sept	17th Sept	2nd Aug	6th Sept
1978	11th June	7th May	16th June	28th April	9th June	27th May	23rd Sept	22ndSept	4th Sept	11th Sept	12th Aug	8th Sept
1979	10th June	5th June	4th June	3rd June	9th June	7th June	11th Sept	8th Sept	19th Sept	20th Sept	10th Sept	14th Sept
1980	23rd May	17th May	20th May	6th June	2nd July	1st June	2nd Sept	7th Sept	12th Sept	10th Sept	2nd Aug	31st Aug
1981	19th May	28th May	1st June	13th May	13th June	27th May	10th Sept	14th Sept	14th Sept	14th Aug	17th Sept	8th Sept
1982	1st July	15th May	11th June	12th July	26th June	19th June	7th Sept	21st Sept	2nd Aug	1st Aug	13th Sept	31st Aug
1983	1st June	15th June	15th June	2nd July	26th June	18th June	5th Sept	23rd Sept	15th Sept	2nd Aug	5th Aug	29th Aug

1984	7th June	15th June	13th June	15th June	9th June	6th June	18th Sept	2oth Sept	15th Sept	6th Aug	26th Sept	5th Sept
1985	7th June	10th June	2nd June	5th June	27th June	14th June	9th Sept	19th Sept	21 Sept	11th Sept	27th Sept	21st Sept
1986	22nd June	2nd June	5th June	5th July	18th July	16th June	18th Sept	18th Sept	24th Sept	9th Sept	1st Aug	7th Sept
1987	9th June	2nd June	25th June	20th June	12th July	20th June	24th Sept	20th Sept	18th Sept	5th Aug	6th Aug	2nd Sept
1988	13th June	2nd June	7th June	12th June	9th July	15th June	22nd Sept	10th Sept	8th Sept	19th Sept	7th Aug	7th Sept
1989	4th June	1st July	18th June	4th June	12th July	20th June	13th Sept	2nd Sept	20th Oct	4TH Sept	19th Sept	18th Sept
1990	30th May	6th June	11th June	21st June	7th July	12th June	9th Sept	17th Sept	2nd Aug	2nd Aug	3rd Aug	19th Aug
1991	4th May	23rd April	12th May	17th May	19th June	15th May	16th Sept	2nd Aug	8th Sept	7th Aug	1st Aug	20th Aug
1992	6th June	4th May	26th June	27th May	3rd June	1st June	21st Sept	7th Sept	22nd Sept	18th Sept	18th Sept	17th Sept
1993	26th May	1st June	2nd July	23rd May	8th July	12th June	15th Sept	16th Sept	30th Sept	1st Aug	7th Aug	1st Sept
1994	30th June	8th June	25th June	22nd June	18th June	21st June	23rd Sept	3rd Sept	8th Sept	8th Sept	4th Sept	9th Sept
1995	1st July	3rd June	5th June	23rd June	7th July	20th June	14th Sept	8th Sept	15th Sept	8th Sept	5th Aug	4th Sept
1996	1st June	3rd June	6th June	6th June	24th June	8th June	26th Aug	1st Sept	23rd Sept	6th Sept	19th Sept	9th Sept
1997	10th May	6th May	12th May	25th May	1st July	23rd May	5th Sept	1st Sept	11th Sept	15th Sept	13th Sept	3rd Sept
1998	16th June	16th May	20th June	12th June	6th July	14th June	26th Sep	2nd Sept	11th Sept	10th Sept	10th Sept	12th Sept

(*continued*)

Table 2 (continued)

Station				Maiduguri			Sokoto					
Year	Sokoto	Kano	Katsina	Onset	Nguru	SSRN	Cessation	Kano	Katsina	Maiduguri	Nguru	SSRN
1999	4th June	2nd June	5th July	1st July	1st July	21st June	25th Sept	4th Sept	7th Sept	3rd Sept	24th Sept	13th Sept
2000	12th June	16th May	14th June	12th June	27th June	4th June	29th Sept	3rd Sept	3rd Sept	9th Sept	9th Sept	5th Sept
2001	14th May	2nd May	13th May	3rd June	7th June	26th May	16th Sept	7th Sept	22nd Sept	7th Sept	29th Sept	10th Sept
2002	17th May	11th June	10th June	29th June	29th June	7th June	1st Sept	29th Oct	25th Oct	6th Sept	9th Sept	20th Sept
2003	10th June	19th May	28th May	11th June	16th June	30th May	12th Sept	5th Sept	14th Sept	14th Sept	17th Sept	7th Sept
2004	11th May	19th May	21st May	26th May	8th July	29th May	12th Aug	5th Sept	1st Aug	22nd Sept	3rd Aug	21st Aug
2005	15th May	13th May	6th June	11th May	4th June	24th May	11th Sept	2nd Sept	4th Sept	27th Oct	20th Sept	19th Sept
2006	23rd June	11th May	11th June	5th June	7th July	11th June	4th Sept	5th Sept	8th Sept	24th Sept	29th Sept	14th Sept

SSRN = Sudano-Sahelian Region of Nigeria
Sources: Authors

of the rains, could be regarded as tolerable (Bello 1996). For example, the earliest onset date at Sokoto over the period 1971–2006 is 4th May, and, therefore, 4th May to 8th May constituted a class, and it goes in that order. This was applied to all the selected stations and the Sudano-Sahelian region as a whole. The class with the highest frequency (modal class) was considered for each station, and the upper limit of such a class is taken as the mean onset date of the rain. In the case of cessation of the rains, the earliest cessation date of the rain at Sokoto from 1971 to 2000 is 12th August, and thus, 12th August to 16th August constituted a class. The same procedure was also applied but in a reverse order, and modal class was noted. In the case of cessation however, the lower limit of the modal class was taken as the mean date of the cessation of the rains.

Table 3 provides the mean dates of the onset and cessation of the rains in comparison to mean dates of the onset and cessation of the rains obtained by Bello (1996) and those predicted by the Nigerian Meteorological Agency (NiMet 2018). It is interesting to note that the mean dates of onset determined through Walter's technique as well as NiMet's predicted onset dates of the rain favored early onset of the rains when compared to mean dates of onset of the rains obtained by Bello (1996) using Benoit's technique. In the case of cessation of the rains, the mean dates of the cessation of the rains obtained through Walter's technique favored early cessation of the rains, while those obtained by Bello (1996) and those predicted by NiMet favored late cessation of the rain.

Both cases have a wide range of implications for farm operations in Sudano-Sahelian region of Nigeria. If the early onset of the rain in a given year is associated with early cessation of the rains, it could affect the length of the growing period within which crops are expected to utilize rainfall for their optimum growth. Similarly, if the late onset of the rain in a given year is associated with the early cessation of the rain, it could be very disastrous to crops because of the reduced length of the rainy season. The Walter's technique as well as NiMet's predictions which in both cases favored the early onset of the rains had high tendency of experiencing a dry spell of different magnitudes especially soon after the establishment of the onset of the rains. These dry spells, if of longer duration, could even lead to replanting at a great cost on the part of the farmers. It should however be noted that the late onset dates of the rains as favored by Benoit's method have very few chances of being associated with dry spells soon after the establishment of the onset of the rains as it frequently happens in the case of early onset dates of the rains.

The deviations of the computed dates of onset and cessation of the rains from the respective mean dates at each of the stations and for the Sudano-Sahelian region as a whole were generated and presented in Figs. 8 and 9. Adopting Bello's (1996) criteria of classifying the patterns of onset and cessation of the rains into early, normal, and late onset and cessation of the rains, it is obvious that both onset and cessation dates of the rains are highly variable in the Sudano-Sahelian region of Nigeria. At all the selected stations and the region as a whole, the number of years with early onset and cessation dates are almost equal to those with late onset and cessation dates of the rains as revealed in Figs. 8 and 9. The frequent deviations of the onset and cessation dates of the rains at all the selected stations and the region as

Table 3 Mean values of some of the rainfall characteristics in Sudano-Sahelian region of Nigeria

Station	Mean annual rainfall (mm)	Mean seasonal rainfall (mm)	Mean daily rainfall (mm)	Mean length of the rainy season (days)	Mean onset date computed using Walter's (1967) method	Mean onset date computed by Bello (1996) using Benoit's (1977) method	Predicted onset date by NiMet for 2018 rainy season	Mean cessation date computed using Walter's (1967) method	Mean cessation date computed by Bello (1996) using Benoit's (1977) method	Predicted cessation date by NiMet for 2018 rainy season	Predicted length of rainy season by NiMet for 2018 rainy season
Sokoto	625	529	1.75	103	4th June	17th June	4th June	13th Sept	20th Sept	10th Oct	128
Kano	912	739	2.58	107	27th May	15th June	5th June	9th Sept	22nd Sept	17th Oct	134
Katsina	535	419	1.46	96	6th June	–	11th June	10th Sept	–	7th Oct	118
Maiduguri	565	414	1.51	91	8th June	20th June	19th June	5th Sept	19th Sept	23rd Oct	126
Nguru	411	274	1.11	65	24th June	21st June	29th June	1st Sept	19th Sept	20th Oct	113
Sudano-Sahelian	610	475	1.68	92	8th June	–	–	8th Sept	–	–	–

Sources: Authors Bello (1996), NiMet (2018)

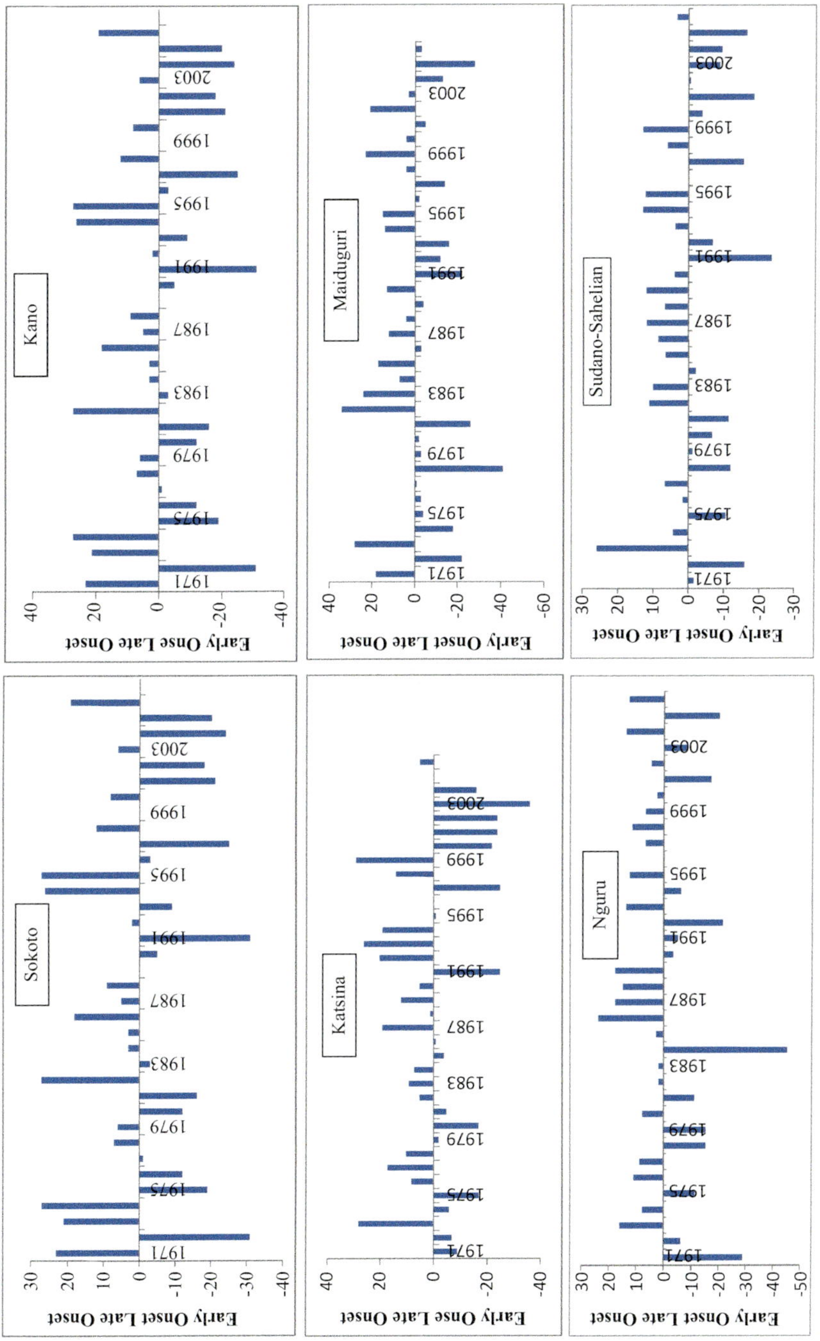

Fig. 8 Deviations from the mean onset date in Sudano-Sahelian region of Nigeria (1971–2006). (Sources: Authors)

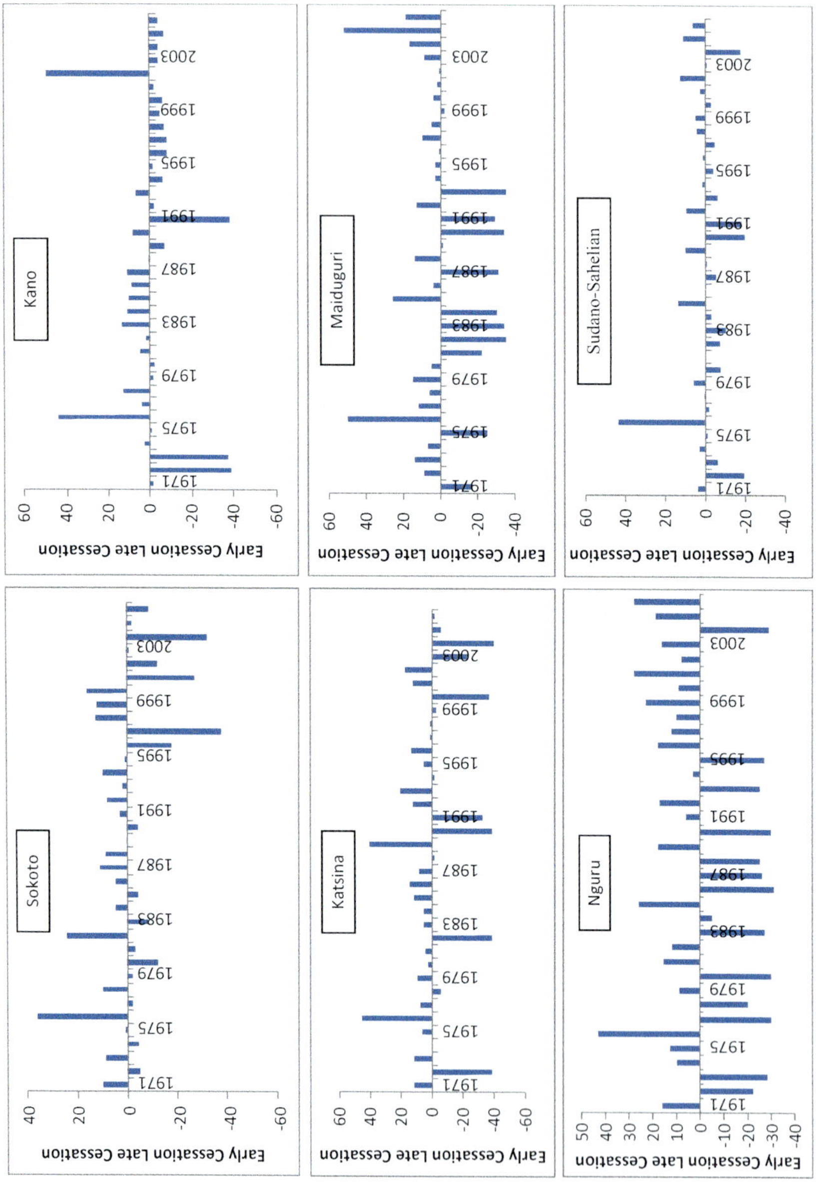

Fig. 9 Deviations from the mean cessation date in Sudano-Sahelian region of Nigeria (1971–2006). (Sources: Authors)

a whole from the respective mean dates of onset and cessation of the rains is a reflection of the dominant role of north-south annual course of the inter-tropical discontinuity (ITD) on rainfall dynamics in the region as widely discussed in the literature (Oladipo and Kyari 1993). These frequent deviations from the respective means could pose serious difficulties in ascertaining the accurate beginning of the rains and make planning of farm operations very difficult for the farmers even though the prospects for seasonal forecasting seem to be high in recent years as suggested by Palmer and Anderson (1994). This however calls for close monitoring of rain-producing systems and daily surface movements of the inter-tropical discontinuity (ITD) in the region which to a large extent determines the patterns of the variability in these rainfall attributes as revealed in this chapter.

Using daily rainfall records of 5 selected meteorological stations in Sudano-Sahelian region of Nigeria over the period 1971–2006, four (4) dry spell categories, namely, *very short* (DS1, 1–3 days), *short* (DS2, 4–7 days), *medium* (DS3, 8–14 days), and *long* (DS4, $\geq$15 days), were detected using multi-scale dry spell detection method adopted from Salack et al. (2011). A dry spell is defined as a sequence of days without rain preceded and followed by a wet day (Salack et al. 2011). Stern et al.'s (1982) threshold of at least 0.83 mm of rainfall to qualify as a rainy day is adopted for this chapter in deriving sequence of dry days between the onset and cessation dates of a rainy season. This was generated for each year and for each selected station. Salack et al.'s (2011) scheme was then adopted to classify the dry spells into four categories. Fig. 10 shows the total number of each dry spell category over the study period. It is obvious that the DS1 and DS2 categories have the highest frequency of occurrence at each of the selected stations. These dry spell categories do not however pose serious threat to crops because their effects could easily be reversed by a mere intervention of one good rain immediately after their occurrences.

A close observation of Fig. 10 shows that DS3 and DS4 categories had low frequency of occurrence compared to DS1 and DS2 at all the selected stations. However, these dry spell categories could easily pose great danger to crops especially if these occurred immediately after the onset date of the rain is established. It is pertinent to note that DS4 category is most dangerous of all the dry spells examined in this chapter. Kano recorded the highest of such dry spell incidences over the period of the study followed by Nguru. Some of the spectacular cases of DS4 category have been examined. In 1981, for example, soon after the onset date of the rains had been established as 19th May, a long dry spell of 26 days (28th May to 22nd June) occurred at Sokoto. Likewise in 1984 and 1987, a long dry spells of 16 days (18th August to 2nd September) and 15 days (4th August to 18th August), respectively, occurred at Sokoto. The cessation dates of the rain in both 1984 and 1987 at Sokoto are 18th September and 24th September, respectively. It is therefore not surprising since both 1984 and 1987 are well-known drought years in Sudano-Sahelian region of Nigeria as reported by Oladipo (1993).

Kano was not spared in this regard having recorded the highest number of such disastrous dry spells. For example, a long dry spell of 19 days (29th May to 16th June) occurred at Kano in 1981. In 1987, another long dry spell occurred at Kano

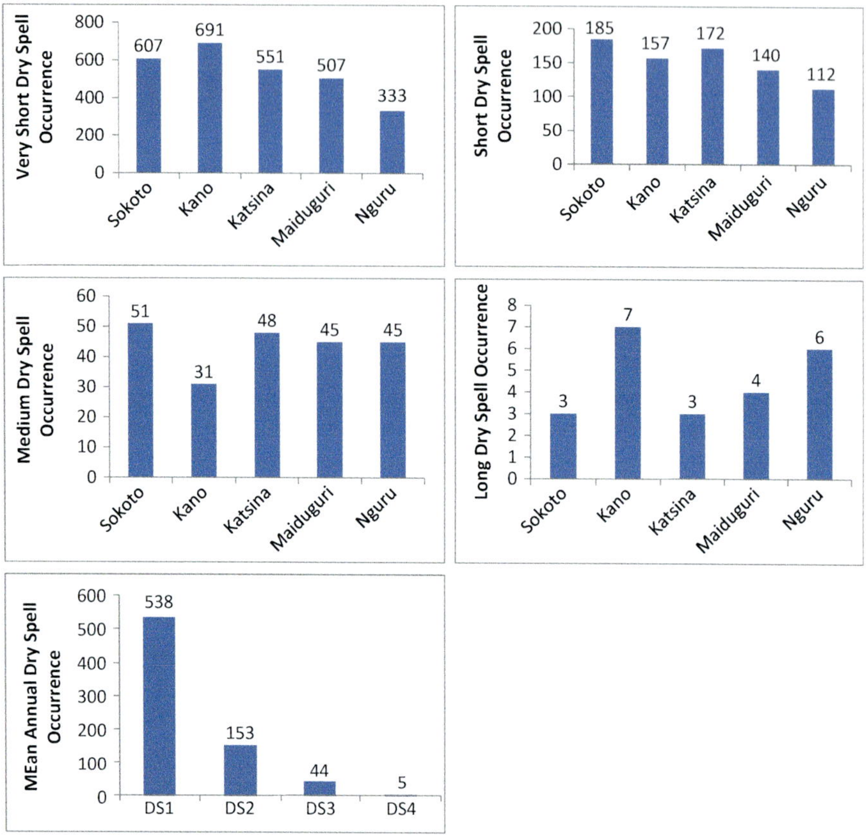

Fig. 10 Dry spell occurrence in the Sudano-Sahelian region of Nigeria (1971–2006). (Sources: Authors)

(5th June to 23rd June) just after the onset date of the rains had been established as 2nd June. This can certainly expose crops to serious water stress at their early stage of germination and can even lead to crop failure because of the associated heat stress, leading to high rate of evapotranspiration. Another catastrophic dry spell occurred at Kano in 1990 and lasted for 15 days (7th June to 21st June). It occurred soon after the onset date of the rain had been established as 6th June. In 2002, Kano witnessed another long dry spells of 23 days beginning from 7th October and lasted to the cessation date of the rain for that year (29th October). This shows that dry spells are part and parcel of the climatic characteristics of the Sudano-Sahelian region of Nigeria and can occur at any time of the rainy season. The year 2002 is regarded as part of the period during which the Sahelian region was experiencing gradual rainfall recovery.

At Maiduguri, a long dry spell occurred in the month of August, a rainiest month in the Sahel. This dry spell lasted between 1st August and 16th August, 1987. The occurrence of such dry spells in August suggests that the residential airtime of the rain-bearing air mass in the region was shorter than usual. In 2000 and 2002, respectively, Maiduguri also witnessed a long dry spell of 15 and 16 days, respectively. The former occurred between 24th June and 8th July, 2000, while the latter occurred from 30th June to 15th July, 2002. In the case of the former, it occurred during early phenological stage of crops, less than 2 weeks after the onset date of the rain had been established as 12th June. In the case of the latter, it occurred immediately after the onset date of the rain had been established as 29th June. This was quite injurious to crops at that early stage of their growth.

At Katsina, a long dry spell occurred between 29th May and 14th June, 1971. It occurred immediately after the onset date of the rain had been established as 28th May. It was followed again by a medium dry spell of 12 days (16th June to 27th June) just after an intervening rainfall of only 1.8 mm on 15th of June, 1971. It is not also surprising since the year 1971 was well acknowledged as part of the drought years in the Sahelian region of West Africa. At Nguru, a long dry spell of 28 days (29th May to 25th June) occurred in 1971, just 2 days after the onset date of the rains had been established as 26th May (see Table 4 for illustration). Another long dry spell occurred at Nguru in 1976. This lasted for 20 days (25th September to 14th October). The cessation date of the rain at Nguru for that year was determined as 18th October. The foregoing demonstrations underscore the need for NiMet to improve its capabilities to accurately forecast the occurrence, extent, and duration of the dry spells especially at the beginning and toward the end of the rainy season so that proper measures could be put in place to mitigate the effects of these dry spells on crops.

The Influence of the Onset, Cessation, and Dry Spells on Rainfall Characteristics in the Sudano-Sahelian Region of Nigeria

Several scholars had examined different rainfall characteristics such as trend and variability (Adefolalu 1986; Adeniyi 2014), onset and cessation of the rains (Oladipo and Kyari 1993; Bello 1996), and frequency and duration of dry spells (Sivakumar 1992; Sawa 2002). However, attempts to examine the contributions of different meteorological parameters on annual and growing season rainfall totals as well as the length of the rainy are scanty in the literature. The earliest works of Ayoade (1974) attributed the interannual variations in rainfall over Nigeria to factors such as continentality, altitude, and ITD movements. Anyadike (1981) investigated the relative contributions of atmospheric parameters to the rainfall of West Africa and concluded that the dominant contributor is always moisture transporting mechanisms – depths of the maritime winds (DMW) or meridional water vapor fluxes (MWVF) with secondary factor being precipitable water content (PWC).

Omotosho (1985) also examined the separate contributions of thunderstorms, disturbance lines (line squalls), and monsoon to total annual rainfall over Nigeria and reported that line squalls are the most important rain-producing systems, giving 47% of the mean total annual precipitation compared to 39% and 13% from

Table 4 Daily rainfall (mm) distribution at Nguru station in 1971

Day	Jan	Feb	Mar	Apr	May	Jun	Jul	Aug	Sep	Oct	Nov	Dec
1	0	0	0	0	0	0	0	5	0	0	0	0
2	0	0	0	0	0	0	1	0	19	0	0	0
3	0	0	0	0	0	0	0	1	0	0	0	0
4	0	0	0	0	0	0	0	0	7	0	0	0
5	0	0	0	0	0	0	9	6	0	0	0	0
6	0	0	0	0	0	0	0	0	0	0	0	0
7	0	0	0	0	0	0	28	0	23	0	0	0
8	0	0	0	0	0	0	33	4	0	0	0	0
9	0	0	0	0	21	0	9	0	0	0	0	0
10	0	0	0	0	0	0	0	14	0	0	0	0
11	0	0	0	0	0	0	0	0	0	0	0	0
12	0	0	0	0	0	0	0	11	13	0	0	0
13	0	0	0	0	26	0	0	0	0	0	0	0
14	0	0	0	0	0	0	11	0	0	0	0	0
15	0	0	0	0	0	0	21	32	0	0	0	0
16	0	0	0	0	0	0	0	0	0	0	0	0
17	0	0	0	0	0	0	0	0	0	0	0	0
18	0	0	0	0	0	0	0	0	25	0	0	0
19	0	0	0	0	0	0	0	2	0	0	0	0
20	0	0	0	0	0	0	2	0	0	0	0	0
21	0	0	0	0	0	0	0	33	0	0	0	0
22	0	0	0	0	0	0	0	19	0	0	0	0
23	0	0	0	0	0	0	0	10	0	0	0	0
24	0	0	0	0	0	0	3	15	0	0	0	0
25	0	0	0	0	0	0	0	0	0	0	0	0
26	0	0	0	0	0	2	0	0	0	0	0	0
27	0	0	0	0	0	17	0	0	0	0	0	0
28	0	0	0	0	14	0	0	0	0	0	0	0
29	0	–	0	0	**0**	0	0	2	0	0	0	0
30	0	–	0	0	0	0	22	0	0	0	0	0
31	0	–	0	–	0	–	0	0	–	0	–	0

Sources: Authors

thunderstorms and ordinary monsoon rain, respectively. The influence of inter-tropical discontinuity (ITD) on interannual rainfall characteristics in Nigeria was examined by Odekunle (2010) and reported among other things that although the ITD movement significantly accounts for interannual variations in rainfall, it does so in the northern part of the country only. Diatta and Fink (2014) also examined the statistical relation between remote climate indices and West African monsoon variability and showed among other things that both the Atlantic Multi-decadal Oscillation (AMO) and the Atlantic Meridional Mode (AMM) show a positive linear correlation to Sahel rainfall. Attempt is made in this chapter to examine the influence of the onset and cessation dates of the rain as well the dry spells on annual and

Table 5 Summary of the stepwise multiple regression model between the dependent and independent variables

Station	Annual rainfall predictor(s)	R^2	p-value	Seasonal rainfall predictor(s)	R^2	p-value	Length of rainy season predictor(s)	R^2	p-value
Sokoto	DS1	0.34	0.000	DS1	0.28	0.001	Onset	0.62	0.000
	DS4	0.43	0.041				Cessation	0.85	0.000
Kano	Onset	0.52	0.001	Onset	0.26	0.002	DS1	0.49	0.000
							DS2	0.69	0.000
							DS3	0.79	0.001
							DS4	0.85	0.001
Katsina	Onset	0.20	0.006	DS1	0.27	0.001	Cessation	0.45	0.000
							Onset	0.95	0.000
Maiduguri	Cessation	0.36	0.000	DS1	0.55	0.000	Cessation	0.61	0.000
	DS1	0.45	0.028	Cessation	0.67	0.001	Onset	0.87	0.000
	Onset	0.54	0.016						
Nguru	DS1	0.34	0.000	DS1	0.63	0.000	Cessation	0.59	0.000
	DS3	0.43	0.034	DS3	0.79	0.000	Onset	0.74	0.000
							DS2	0.78	0.022
							DS3	0.82	0.008
							DS1	0.91	0.000
							DS4	0.95	0.000
SSRN	DS1	0.31	0.000	DS1	0.46	0.000	Onset	0.49	0.000
	Onset	0.40	0.000	Onset	0.52	0.000	Cessation	0.88	0.000
	DS4	0.42	0.034	Cessation	0.53	0.024	DS1	89	0.000
							DS3	0.90	0.000
							DS2	0.91	0.000
							DS4	0.91	0.010

SSRN = Sudano-Sahelian region of Nigeria
Sources: Authors

growing season rainfall as well as the length of the rainy season/growing period in Sudano-Sahelian region of Nigeria using statistical methods – stepwise multiple regression models. The six set of predictors/independent variables include the following: the onset and cessation dates of the rain, very short dry spells (DS1), short dry spells (DS2), medium dry spells (DS3), and long dry spells (DS4), with annual rainfall series, seasonal rainfall series, and length of the rainy season each being a dependent variable. The summary of the regression models is presented in Table 5.

It is interesting to note that annual and seasonal rainfall series as well as the length of the rainy season were greatly influenced by the variability in the occurrence of the onset and cessation dates of the rains as well as the dry spells. At Sokoto, DS1 and DS4 jointly accounted for 43.0% of the explained variance in annual rainfall, while the onset and cessation dates of the rain accounted for 86.0% of the explained variance in the length of the rainy season. In the case of Kano, it was only onset date of the rain that influenced interannual variability in both the annual and seasonal rainfalls and accounted for 28.0% and 26.0% of the explained variance, respectively, while all the dry spell categories jointly accounted for 85.0% of the explained variance in the length of the rainy season.

At Katsina, onset and cessation dates of the rain jointly accounted for 95.0% of the explained variance in the length of the rainy season. At Maiduguri, the cessation, onset, and very short dry occurrence jointly accounted for 54.0% of the explained variance in annual rainfall variations, while cessation and onset dates of the rain accounted for 87.0% of the explained variance in the length of the rainy season. At Nguru, DS1 and DS3 accounted for 79.0% of the explained variance in seasonal rainfall variations, while all the independent variables considered for this chapter jointly accounted for 95.0% of the explained variance in the length of the rainy season.

In the case of the Sudano-Sahelian region as a whole, DS1, onset and cessation dates of the rain jointly accounted for 53.0% of the explained variance in seasonal rainfall variations while all the independent variables considered for this chapter jointly accounted for 91.0% of the explained variance in the length of the rainy season. It is expected that the onset and cessation dates of the rain will be much more significant in determining the interannual variations in the length of the rainy season since an early onset associated with late cessation of the rain will certainly ensure longer length of the rainy season while the reverse is the case. It is not also surprising that the dry spell occurrences showed a significant influence in annual and seasonal rainfall variations since the more the dry spells experienced especially medium and longer ones, the scantier the seasonal and annual rainfall will likely be, and therefore the highly variable the rainfall series will be at a given year as opined by Sivakumar (1992) and Sawa (2002).

The strong influence of the onset, cessation, and dry spell occurrences on annual and seasonal rainfall and the length of the rainy season in the Sudano-Sahelian region of Nigeria could be linked to the coupled air-sea interaction phenomenon of the El Niño/Southern Oscillation (ENSO). The warm phase of ENSO (i.e., El Niño) is associated with warmer-than-normal sea surface temperatures over the east-central tropical Pacific which reduces rainfall in Western and Central Africa, resulting in high frequency of occurrence of dry spells especially in the Sahelian region of West Africa. The cold phase of ENSO (i.e., La Niña) is associated with cooler-than-normal sea surface temperatures over the east-central tropical Pacific. This condition leads to the intensification of trades over the east-central tropical Pacific, resulting in anomalously high precipitation in different parts of the world including the West African subregion. Detailed discussions on the influence of El Niño and La Niña on rainfall in Africa are provided by Nicholson and Kim (1997) and Nicholson and Selato (2000). The southward displacement of the ITD is said to be responsible for the occurrence of early cessation of the rains and accounts for late cessation when it advances further inland (Ayoade 2011) and thus determines the length of the rainy season at any given year.

Conclusion

The pattern of the annual and seasonal rainfall and the length of the rainy season have been examined over the period 1971–2006. The pattern is characterized by oscillations and general downward trend in annual and seasonal rainfall as well as the length of the rainy season particularly during the 1970s and 1980s through 1990s. The Sudano-Sahelian region of Nigeria just like other parts of the West African Sahel began to experience above-normal rainfall similar to those of the 1950s from the year 1998 onward. The rainfall characteristics such as the onset and cessation of the rains as well as the dry spells have also been examined. While the dry spell occurrences had strong influence on the amounts of annual and seasonal rainfall amounts, the onset and cessation dates of the rainfall controlled the length of the rainy season in the region. This could have serious implications for crop production in the region, and thus farmers have been advised to adhere strictly to early warnings from the Nigerian Meteorological Agency (NiMet) on the likely occurrences of the dry spells and accurate dates of onset and cessation of the rains in the region. This could enhance farmers' resilience to climate change and helps in achieving food security in the region.

Cross-References

▶ Assessment of Climatic Variability Effect on Millet Production and Yield
▶ Impact of Climate Change on Growing Season in Nigeria: Seasonal Rainfall Prediction (SRP) as Assessment and Adaptation Tool
▶ Seasonal Climate Forecasts Among Farmers in Nigeria: Assessment of the Usability

References

Adefolalu DO (1986) Rainfall trends in Nigeria. Theor Appl Climatol 37:205–219
Adeniyi MO (2014) Variability of daily precipitation over Nigeria. Meteorol Atmos Phys 126(3–4):161–176. https://doi.org/10.1007/s00703-014-0340-6
AGRHYMET (1996) Methodologic de Suivi des zones a risque AGRHYMET Flash, Bulletin de la campague Agricoleou Sahel, centre regional AGRHYMET, vol 2, Niamey, pp 1–96
Anyadike RCN (1981) On the relative contributions of atmospheric parameters to the rainfall of West Africa. Arch Meteorol Geophys Bioklimatol A 30:87–98
Aondover T, Woo M (1998) Changes in rainfall characteristics in northern Nigeria. Int J Climatol 18:1261–1271
Ati OF, Stigter CJ, Oladipo EO (2002) A comparison of methods to determine the onset and cessation of the growing season in northern Nigeria. Int J Climatol 22:731–742
Ayoade JO (1974) A statistical analysis of rainfall over Nigeria. J Trop Geogr 29:11–23
Ayoade JO (2011) Introduction to climatology for the tropics. Spectrum Books Limited, Ibadan
Bello NJ (1996) Investigation of the spatial pattern of the characteristics of the onset and cessation of the rains in Nigeria. GeoJournal 43:113–123
Benoit P (1977) The start of the growing season in Nigeria. Agric Meteorol J 18:1232–1269
Central Bank of Nigeria (2005) Report on national survey of agricultural export commodities in Nigeria, 2004/2005
Diatta S, Fink AH (2014) Statistical relationship between remote climate indices and West African monsoon variability. Int J Climatol 34:3348–3367. https://doi.org/10.1002/joc.3912
Druyan LM (1989) Advances in the study of Sub-Saharan drought. Int J Climatol 9(1):77–90
Fasheun A (1983) Modelling of daily rainfall sequence far farm operating planning in Ibadan. Nig Meteorol J 1:102–109

Ilesanmi OO (1972) An empirical formulation of the onset, advance and retreat of rainfall in Nigeria. J Trop Geogr 34:17–24

Marengo JA, Liebmann B, Kousky VE, Filizola NP, Wainer L (2001) Onset and end of the rainy season in the Brazilian Amazon Basin. J Clim 1(14):833–852

National Population Commission (2006) Population estimates for Nigeria release by national population commission

Nicholson SE (2005) On the question of the "recovery" of the rains in the West African Sahel. J Arid Environ 63:615–641

Nicholson SE (2008) The intensity, location and structure of the tropical rainbelt over West Africa as factors in interannual variability. Int J Climatol 28:1775–1785. https://doi.org/10.1002/joc.1507

Nicholson SE, Kim J (1997) Relationship of the El-Nino/Southern Oscillation to African rainfall. Int J Climatol 17:117. https://doi.org/10.1002/(SICI)1097-0088

Nicholson S, Selato JC (2000) The influence of La- Nina on African rainfall. Int J Climatol 20(14):1761–1776. https://doi.org/10.1002/1097-008892000113020:143.0

Nigerian Meteorological Agency (NiMet) (2018) Seasonal rainfall prediction. ISSN:2346-7150

Odekunle TO (2004) Rainfall and length of the growing season in Nigeria. Int J Climatol 24:467–479

Odekunle TO (2010) An assessment of the influence of the inter-tropical discontinuity on inter-annual rainfall characteristics in Nigeria. Geogr Res 48(3):314–326

Odekunle TO, Balogun EE, Ogunkoya OO (2005) On the prediction of rainfall onset and retreat dates in Nigeria. Theor Appl Climatol 81:101–112. https://doi.org/10.1007/s00704-004-0108-x

Oduro AK, Adefolalu DO (1993) Instability indices for severe weather forecasting in West Africa. Atmos Res 30:51–68

Oguntunde PG, Babatunde JA, Leschied G (2011) Rainfall trends in Nigeria, 1901–2000. J Hydrol 411(3):207–218

Oladipo EO (1993) A comprehensive approach to droughts and desertification in northern Nigeria. Nat Hazards 8:235–261

Oladipo EO, Kyari JD (1993) Fluctuations in the onset, termination and length of the growing season in northern Nigeria. Theor Appl Climatol 47:241–250

Olaniran OJ (1983) The start and end of the growing season in the Niger River Basin Development Authority of Nigeria. Malays J Trop Geogr 9:69–78

Olaniran OJ (1984) The onset of the rains and the start of the growing season in Nigeria. Niger Geogr J 34:17–24

Omotosho JB (1985) The separate contributions of line squalls, thunderstorms and the monsoon to the total rainfall in Nigeria. Int J Climatol 5:545–552

Omotosho JB, Balogun AA, Ogunjobi K (2000) Predicting monthly and seasonal rainfall onset and cessation of the rainy season in West Africa, using only surface data. Int J Climatol 20:865–880

Palmer TN, Anderson DLT (1994) The prospects for seasonal forecasting – a review paper. Q J R Meteorol Soc 120:755–793

Salack S, Muller B, Gaye AT, Hourdin F (2010) Implications of low frequency dry spells in seasonal and interannual rainfall variability in Senegal, Poster presentation at the WCRP-UNESCO (GEWEX/CLIVAR/IHP) workshop on metrics and methodologies of estimation of extreme climate events. UNESCO Headquarters, Paris, 27–29 September, 2010. http://www.extremeworkshop.org

Salack S, Muller B, Gaye AT, Hourdin F (2011) Dry spell analysis: multi-scale detection, attribution of impacts and sources of uncertainty using an integrative approach. Paper presented at the Open Science Conference (OSC) of the World Climate Research Program (WCRP) in Denver, Colorado, USA, 25–28 October, 2011

Sanogo S, Andreas HF, Omotosho JB, Abdramane B, Robert R (2015) Spatio-temporal characteristics of the recent rainfall recovery in West Africa. Int J Climatol 11:177–204

Sawa BA (2002) Trends in the temporal variability of occurrence of wet and dry spells north of latitude 10°N in Nigeria. Zaria Geogr 15(1):34–41

Sivakumar MVK (1992) Empirical analysis of dry spells for agricultural applications in West Africa. J Clim 5:532–539
Stern RD, Dennett MD, Dale CI (1982) Analyzing daily rainfall measurements to give agronomically useful results. 1. Direct methods. Exp Agric 18:223–236
Umar AT (2010) Recent trends and variability in the length of the growing season in northern Nigeria. J Meteorol Clim Sci 8(1):40–52
Umar AT, Bello N, Dangulla M, Eniolorunda NB (2015) The changing frequency of extreme daily rainfall events and its implications for flood occurrence in Northern Nigeria. In: Iliya MA et al (eds) Climate change, environmental challenges and sustainable development conference, held at Usmanu Danfodiyo University, Sokoto, 23rd–26th November, 2015
Usman MT, Abdulkadir A (2013) On determining the 'real' onset date of seasonal rains in the semi-arid and sub-humid areas of West Africa. Nat Hazards 66:749–758
Walter WM (1967) Length of the rainy season in Nigeria. Niger Geogr J 10:123–128. 1.4

Climate Change Adaptation: Adoption of GMOs in Africa

127

Emmanuel Ejim-Eze

Contents

Introduction 2624
Definition of Terms 2626
Adaptation in Response to a Changing Climate 2628
Short-Term Adaptation 2629
Medium- and Long-Term Adaptation 2629
Climate Change, Agriculture, and Food Insecurity 2630
Agriculture Practices and Climate Change Adaptation 2632
Soil and Land Management 2632
Water Management 2633
Use of GE Crops in Agriculture and as Climate Change Adaptation Measures in Africa 2634
National Innovation Systems (NIS) and Agricultural Innovation Systems in Africa 2639
Africa's Biotechnology Capacity, Biosafety Status, and Adoption Impact 2639
Discussion and Conclusion 2641
Conclusion 2644
Policy Recommendations 2645
Cross-References 2646
References 2646

Abstract

Africa faces the risks of climate change due to variable climate as seen in low rainfalls and high temperature patterns. Drought affects crops and animal breeding, whereas warm temperature helps increase the incidence of insect pests and diseases that lower agricultural productivity. Beyond these are the issues of poor governance, political instability, and lack of capacity to increase productivity and foster agricultural developments.

E. Ejim-Eze (✉)
National Centre for Technology Management, Obafemi Awolowo University, Ile-Ife, Ile-Ife, Nigeria
e-mail: ezeigwe1949@gmail.com

W. Leal Filho (ed.), *Handbook of Climate Change Resilience*,
https://doi.org/10.1007/978-3-319-93336-8_168

Herdsmen in search of foliage and water engage in conflicts with farmers, and this has caused displacement of local farmers. This has further endangered food security in Africa and makes it hard for Africa to achieve zero hunger goal by 2030. Intergovernmental Panel on Climate Change (IPCC)'s fifth assessment report has also predicted that agricultural yields will decline by 2% per decade due to climate change, while demand for food is expected to increase by 14% per decade.

The aim of this paper is to identify adaptation strategies to manage climate change risks in Africa. Stress- and pest-resistant crops are required to support Africa's ability to adapt to variable rainfall and temperature patterns. This paper also reviewed the literature with the aim to find out if the adoption of genetically modified crops could be an adaptation strategy for Africa. It also looked at factors that could make or mar the adoption of GM crops as climate adaptation strategy in Africa.

Introduction

The cascading impacts of climate change are already experienced in different parts of the world, with a wide acceptance of the role of greenhouse gases contributing to global warming (Intergovernmental Panel on Climate Change (IPCC) 2007a; FAO 2016). The worst scenario could likely lead to global temperature rise of more than two degrees and related increase of extreme weather conditions. IPCC (2001) stated that changes in climate will affect socioeconomic sectors and natural resources such as water resources, agriculture, forestry, fisheries, ecological systems, and then human settlements and health conditions. Rising sea levels are overflooding lands and crops and contaminates water supplies with salt water while staple. Staple crops are close to their outer viable temperature ranges; irregular rainfall patterns and changing season's upset agricultural cycles and disrupt livelihoods. The IPCC's fifth assessment report confirmed findings of its previous reports on the evolution of the world's climate (FAO 2016). IPCC reaffirmed that the rise in global warming, rainfall variability, and extreme weather events will impact on global sea levels, ocean acidity, disappearance of glaciers, ecosys tem degradation, and rise in fire outbreaks and surges in pests.

Countries whose economies depend on climate-sensitive sectors as agriculture and natural resources are particularly vulnerable to climate change risks (Oxfam 2010). Majority of the world's poor depend on natural resources (water, forests, seas, soil, biodiversity, and so on) for food and livelihoods; but the ecosystem is fast losing its carrying capacity and its support quickly being affected by climate change. IPCC captured the proportion of human populations vulnerable to climate change which includes small holder farmers with inadequate resources, pastoralists, rural landless laborers, and urban poor (UN Millennium Project 2005; Oxfam 2010). Most of these human populations are in developing countries. It is also noted that countries with limited technological, institutional, human, and financial capacity are more vulnerable to climate change risks. Developing countries fall into the category of these countries. They have less capacity to mitigate or adapt to climate change risks, just as they are to other risks. Poverty seems to be a major factor (Oxfam 2010), which increases vulnerability to climate change risks. Poverty, more than any other factor, determines vulnerability to climate change and limits adaptive capacity. Limited access to arable and fertile land, finance, credit, information, health care, personal mobility, education, and lack of empowerment opportunities all limit adaptive capabilities. Gender inequalities coupled with poverty magnify women's vulnerability to climate change and limit their ability to adapt. Poor agricultural communities in developing countries are more vulnerable to climatic change risks (Maskrey et al. 2007).

Vulnerability includes all characteristics and circumstances of a community, system, or asset that makes it susceptible to the damaging effects of climate change and other hazards. From the ecological perspective, people living on arid or semiarid lands, in low-lying coastal areas, and in dry lands, waterlogged areas, flood-prone areas, or small islands are particularly vulnerable to climate change (IPCC 1996). Africa is endowed with regions which have the above ecological features. Africa is known as one of the most vulnerable continent to climate change and climate variability (IPCC 2007b). Sub-Saharan Africa (SSA) seems to be more vulnerable because as it is located near an encroaching large desert. The region is susceptible to drought and floods and high incidence of diseases (WDR 2008). Economic growth in the region has being largely driven by natural resources (affected frequent burst and boom of the international market price), with limited infrastructures and technologies.

Changes in climate conditions as seen in rising temperatures, flooding, drought, and desertification constitute serious threats to agriculture, as seen in seasonal yield losses and food insecurity especially in the developing countries (IPCC 2009). The impact of variable rainfall patterns, flooding, and drought coupled with high temperatures predisposes the region to insect, pest, and crop diseases and increases the chances of other disease outbreaks (Gregory et al. 2009; Heisey and Rubenstein 2015).

Climate change and its risks factors affect lower agricultural productivity in Africa. Agriculture is said to contribute about 35% of the continent's GDP and making up 70% of her labor force (Falck-Zepeda et al. 2013). Yet it has remained a great challenge to end hunger and multidimensional poverty in Africa. Drought and desertification have led herdsmen to search for more feed and water. Fulani herdsmen destroyed farmlands and engaged in conflicts with farmers, and these have caused displacement of local farmers. The worst cases are seen in Nigeria with the incessant killing of local farmers and destruction of their villages by these herdsmen. This has further endangered food security in Africa and makes it hard for Africa to achieve zero hunger goal by 2030. Intergovernmental Panel on Climate Change (IPCC)'s fifth assessment report has also predicted that agricultural yields will decline by 2% per decade due to climate change, while demand for food is expected to increase by 14% per decade.

There is an urgent need for climate change adaptation in Africa to avert future human disasters. Climate change adaptation is described as actions that people and institutions make in anticipation of or in response to a changing climate (Oxfam 2010). Whereas there has been attention on mitigation measures, adaptation should be integrated into climate change strategy.

Mitigation efforts channeled to reduction of greenhouse gas emissions may not reduce the adverse impacts of climate change already being felt (Global Climate Action 2009) with adaptation strategies. Adaptive capacity is the potential of individuals, communities, and societies to be actively involved in the processes of change, in order to minimize negative impacts and maximize any benefits from changes in the climate. One of the adaptation strategies as suggested by some is the adoption of genetically modified organisms (GMOs)/crops in agricultural production in Africa. Studies on climate change and agriculture have recommended adoption of stress-resistant crop varieties as a feasible strategy (FAO 2009). GM crops have been adopted by a number of farmers in other developing countries including those in Argentina, Brazil, China, and India (Falck-Zepeda et al. 2013).

The potentials of GM crops to address Africa's challenges have been acknowledged (Juman and Serageldin 2007). Yet only Burkina Faso, Egypt, and South Africa have approved commercial use of genetically engineered (GE) crops, whereas confined field testing is performed in Kenya, Nigeria, Uganda, and other countries (Falck-Zepeda et al. 2013). Zambia completely banned all GM crop imports, whereas Angola, Lesotho, Malawi, and Zimbabwe also banned imports unless processed or milled. Despite the potential benefits of higher productivity, better pest, and weed management and introduction of stress (drought and salinity)-resistant varieties, policymakers and farmers in Africa are at crossroads on the adoption and use of GM crops. Are there other climate change adaptation strategies that are available to Africa while we wait for more knowledge about benefits, costs, and safety of GMO in the African context? This chapter will help provide more information to assist African policymakers, farmers, and other stakeholders to make informed decisions on the adoption of GM crops as an adaptation strategy to climate change.

African governments and those of other developing countries are even asked to regulate GM crops in their respective countries. Biosafety regulations and regimes are quickly drafted and modeled for them. There are several unconfirmed reports on how these biosafety regulations were developed and legitimatized in African countries. There are more questions than answers. Can you regulate what you do not have knowledge about? Have these countries without technological capabilities regulate any technology use or trade before then without the required capability? What risks management capabilities are there to manage any possible negative risk occurrence of GM crop commercialization in Africa?

Definition of Terms

Climate Change

IPCC (2001) defined climate change a change in climate over time, whether due to natural variability or as a result of human activity. In other words it encompasses both natural variability and anthropogenic changes and hence refers to the observed and projected increase in average global temperature and the associated impacts, including an increase in extreme weather events; melting of icebergs, glaciers, and permafrost; sea-level rise; and changes in the timing and amount of rainfall.

Climate Change Risks

Risk is the probability that a situation will produce harm under specified conditions. It is a combination of two factors: the probability that an adverse event will occur and the consequences of the adverse event. Risk encompasses impacts on human and natural systems and arises from exposure and hazard (Australian Greenhouse Office 2003).

Vulnerability to Climate Change

Vulnerability to climate change has been defined as the degree to which a system is susceptible to, or unable to cope with, adverse effects of climate change, including climate variability and extremes. Vulnerability is a function of the character, magnitude, and rate of climate variation to which a system is exposed, its sensitivity, and its adaptive capacity. Vulnerability is defined by the IPCC as the extent to which a natural or social system is susceptible to sustaining damage from climate change impacts and is a function of exposure, sensitivity, and adaptive capacity (IPCC 2001).

Climate Change Adaptation

Climate change adaptation is described as actions that people and institutions make in anticipation of or in response to a changing climate.

Maladaptation

Maladaptation is any changes in natural or human systems that inadvertently increase vulnerability to climatic stimuli; an adaptation that does not succeed in reducing vulnerability but increases it instead (IPCC TAR 2001).

Genetically Modified Organisms (GMOs)

According to the European Union's definition (Directive 2001/18/EC), "genetically modified organism (GMO)" means an organism, with the exception of human beings, in which the genetic material has been altered in a way that does not occur naturally by mating and/or natural recombination. They are those organisms that have been modified by the application of recombinant DNA technology or genetic engineering, a technique used for altering a living organism's genetic material (FAO 2016). GMOs have alternative names such as genetically engineered (GE). One will often see genetically modified (GM) or genetically engineered crops, plants, or animals used. They can be interchanged.

Insect-Resistant Trait

Insect resistant trait conferred on crops, allowing them to synthesize crystalline proteins that are lethal to specific insects (originates from the bacteria *Bacillus thuringiensis*).

Adaptation in Response to a Changing Climate

The need for climate change adaptation cannot be overemphasized. The impacts of climate change are already occurring, and there are several cases with varying levels of human disasters across the globe. The impacts are projected to become even more severe in the future. The most intriguing aspect of climate change is on the possibility of pushing the poorest and most vulnerable beyond their capacity to respond. Adaptation to the impacts of climate change is a necessity for all nations, including developing and developed nations.

Climate change adaptation is about actions that people and institutions make in anticipation of, or in response to, a changing climate (Oxfam 2010). This includes changes to the things they do and/or the way they do them. Adaptation prevents or manages risks or exploits opportunities, by making changes in natural or human systems against climate change impacts.

Adaptation processes can be autonomous or planned. Every economic sector (water, agriculture, infrastructure/settlement, human health, tourism, transport, and energy) has different options to adapt to climate change (IPCC 2007a).

According to the Committee on Approaches to Climate Change Adaptation (2010), adaptation measures are designed based on the following concepts.

(i) Risk avoidance: This includes preventive measures against the occurrence of estimated climate change impacts. Examples are disaster prevention facilities and regulation of development in vulnerable areas.
(ii) Reduction of negative impacts: Measures for reduction of damage caused by impacts. Examples include measures to reduce the damage from disasters, recovery assistance, etc.
(iii) Risk sharing: Measures to transfer the risks of possible impacts by spreading to a third party or other stakeholders. This includes introduction of climate change disaster insurance as seen in agricultural sector.
(iv) Risk acceptance: Accepting impacts with low probability of occurrence, by not taking any given measures today (maintaining the status quo) or by delaying the implementation of measures while monitoring the situation.
(v) Exploitation of opportunities: Climate change risk with positive impacts can bring in new business and other opportunities depending on the sector and region. Government and other stakeholders are to be proactive to utilize new opportunities. This should be based on projections of future climate, impacts, and societal trends. These projections involve uncertainty, and the challenge lies on formulate adaptation planning under such uncertainties.

Adaptation to climate change impacts may involve short-, medium-, and long-term measures. Short-term adaptation measures are required urgently in order to prevent and moderate impacts that are already experienced, while medium- and long-term adaptation measures are to prevent and mitigate projected future impacts. These different measures as well are described by the Committee on Approaches to Climate Change Adaptation.

Short-Term Adaptation

Short-term adaptations are measures needed for instant prevention and moderation of impacts being experienced at the present (on the scale of months) months but normally measured on seasons to annual bases. These are immediate climate change adaptation measures for impacts with high probability of occurrence despite mitigation efforts. Adaptation measures that ameliorate the impacts of present-day climate variability include sea defenses, institutional adaptations, plant breeding, and adoption of new technologies in agriculture (IPCC 2001). However, measures, such as provision of subsidies, changing planting and harvesting dates, and mixing less productive, drought-resistant varieties and high-yield water-sensitive crops (Bradshaw et al. 2004), are more of short-term measures than the one stated on by IPCC authors. Other examples include the introduction of heat-resistant crop varieties and promotion of appropriate cultivation methods, to address the declining crop quality. Mulching and enhancement of field humus content are other measures used traditionally in Africa.

It is important to consider the latest climate change projections and their levels of uncertainty. One has to take cognizance of future projections and the cost and time necessary for implementation of adaptation measures. Other options worth taking into consideration include (1) implementing measures as urgent as possible to prevent the future predicted impacts, whatever budget cost; (2) implementing measures based on developments, experience, and new knowledge about future projections; and (3) accepting a certain amount of minor risk of damage while enhancing measures wherever possible.

Medium- and Long-Term Adaptation

These are measures to prevent and mitigate projected future impacts. This are planned responses to climate change (and its associated impacts) in the medium and long term (on the scale of 10–100 years). These measures are planned based on sets of scenarios using climate change projection models, continuous improvement of adaptive capacity in the society, and different sectors based on risk assessments. FAO (2007) also described long-term adaptations as major structural changes to overcome adversity such as changes in land use to maximize yield under new conditions and application of new technologies, new land management techniques, and water-use efficiency-related techniques. Measures are needed to develop adaptive capacity to forestall possible impacts, by weighing the risks of impacts that may occur in the medium and long term, by managing the impacts, reducing vulnerability, and building resilience. Examples are the building of watershed dams and development of agricultural infrastructure. Other measures are to protect against the loss of forest cover, bleaching of coral, etc.

Climate Change, Agriculture, and Food Insecurity

Agricultural production as a human endeavor has been known to be highly affected by climate, and unprecedented variability of the climate accounts for recent trends in food security (Hansen 2002; Schmidhuber and Tubiello 2007 as cited in Sultan and Gaetani 2016). Rising temperature has remained as a global problem and will continue to be with uncontrolled increase in global greenhouse gas emissions. Temperatures rise by 2 °C (IPCC 2007) will upset global climate conditions. Increase in temperature could increase or decrease precipitation leading to increase in frequency and intensity of droughts and floods in different regions. High temperatures compound the effects of water shortage and low rainfalls and may make less difference with increased downpours (Thornton et al. 2009). Table 1 below shows changes in temperatures and precipitation patterns and possible impacts on agriculture.

Flooding, drought, and desertification are serious treats to agriculture leading to low productivity and food insecurity especially in the developing countries (IPCC 2009; Byamukama et al. 2011). The effects of drought are crosscutting, with several direct impacts on agriculture, water resources, and natural vegetation as well as indirect effects on health and the economy. The impacts of drought are confounded by environmental degradation, including soil erosion, water pollution, and deforestation. The effect on agriculturalists and pastoralists depends on how much surplus they produce and the relative terms of trade (IPCC 2001).

Africa is notable for a high-yield loss due to drought (Li et al. 2009). Some vulnerable hot spots identified in Africa include arid and semiarid systems in the Sahel, arid and semiarid rangeland systems in parts of eastern Africa, the coastal

Table 1 Climate change impacts on agriculture

Climate event	Possible impact on agriculture
Warmer and fewer cold days and nights	Increased yields in colder environments
Warmer and more frequent hot days and nights over most land areas (virtually certain)	Decreased yields in warmer environments Increased insect pest outbreaks
Warm spells and heat waves increasing in frequency over most land areas (very likely)	Reduced yields in warmer regions due to heat stress Increased crop damage due to wildfire
Heavy precipitation events increasing in frequency over most areas (very likely)	Damage to crops, soil erosion, inability to cultivate land due to waterlogging of soils
Drought-affected area increases (likely)	Increase in land degradation and soil erosion Lower yields from crop damage and failure Increased risk of wildfire, loss of arable land Salinization by irrigation/irrigation water
Intense tropical cyclone activity increases (likely)	Damage to crops, storages, and agricultural infrastructure
Extremely high sea levels increase in incidence (excludes tsunamis) (likely)	Salinization of irrigation water, loss of arable land, and increase in migration

Source: FAO (2008)

regions of eastern Africa, and dry lands in southern Africa (Thornton et al. 2009 as cited in Zawedde 2011). West Africa is another hot spot in sub-Saharan Africa and among the most persistent and early emerging prominent hot spots with observed and forecasts of increased mean temperature and extreme hot-season occurrence (Turco et al. 2015).

Zawedde (2011) noted that Africa will lose biodiversity resources due to drought affecting foods and plant-based medicines. Rain-fed agriculture will be affected by precipitation pattern variability, and crop yields will dwindle. Yields from rain-fed agriculture in some countries could be reduced by up to 50% (Müller et al. 2011). Cereal production will decline by 22% for wheat, 14% for rice, and 5% for maize by 2050 due to climate change. The impact of flooding and drought will increase the potential of disease transmissions; and temperature increase enhances the prevalence of diseases in plants. Rising temperatures affect the diversity of insect, pest, and crop diseases and double the occurrence of disease outbreaks (Gregory et al. 2009).

Intergovernmental Panel on Climate Change (IPCC)'s fifth assessment report has predicted that agricultural yields will decline by 2% per decade due to climate change, while demand for food is expected to increase by 14% per decade (IPCC 2014; Deressa et al. 2014). Agricultural productivity would be threatened seriously by climate change as the latter threatens African production systems, infrastructures, markets, and livelihoods (Müller et al. 2011). The continent is probably the most vulnerable to climate change because of multiple crises as seen in poor governance and economic policies, continued dependence on natural resources, and weak adaptive capacity. According to the IPCC's fourth assessment report (2007a), between 75 and 250 million people may be exposed to increased water stress due to climate change by 2020 in Africa, and this will adversely affect livelihoods in the region. According Müller et al. (2011), areas suitable for agriculture, growing season durations, and yield potentials will be affected by climate change. With the potential reduction of yields from rain-fed agriculture by up to 50% in some countries, climate change may have particularly serious consequences in Africa, where some 800 million people suffer malnutrition.

Many SSA countries have adopted a national adaptation programme of action (NAPA) in compliance with international policies such as the United Nations Framework Convention on Climate Change (UNFCCC 2007). Some countries have developed a number of national policies related to adaptation and mitigation impacts of climate change (WDR 2008). In relation to agriculture, such policies advocate for adaptation to climate change through adoption of stress-resistant crop varieties, improved water and soil management, land management systems, and market and extension services as well as enhance communication (Nelson et al. 2009).

It is also important to state that agriculture is not only affected by climate change; it also contributes to climate change directly and indirectly through the emissions of the three prime greenhouse gases (GHG), namely, carbon dioxide, methane, and nitrous oxide, to the environment (FAO 2016). These gases according to IPCC originating from agriculture, forestry, and other land use (AFOLU) are caused mainly by deforestation, livestock production, and soil and nutrient management. They contributed to about 21% of total global GHG in 2010 (FAO 2016).

Agriculture Practices and Climate Change Adaptation

Dinesh and Vermeulen (2016) have stated that portfolios of agricultural practices and technologies indigenous to some agroecological zones have already proven benefits successful in ensuring food security, resilience, and productivity. Climate change impacts, vulnerability, adaptive capacity, and barriers to adaptation are location-specific and change over time. Adaptations sometimes occur in response to prevailing local climate conditions even without having formal scientific knowledge on the basis of the responses. Proven approaches build on existing indigenous practices and knowledge to maximize benefits to climate change adaptation in farming, livestock, and fisheries.

According to Oxfam (2010), strengthening the carrying capacity of natural resources or the ecosystem has proved to be a good adaptive measure. Measures such as soil, land, and water management have helped farmers to adapt to changes in climate. Let us look below and see how the management of these resources helps in building climate change adaptation.

Soil and Land Management

Soil and land management focuses on improving the agricultural productivity of the soil. Indigenous knowledge and other sustainable techniques are keys to maximize benefits from degrading arable land due to climate change adaptation. Indigenous practices in the West African Sahel such as contour stone bunds, half-moons, and zaï help control the impacts of rainfall runoff and erosion while increasing soil organic matter in the area (Dinesh and Vermeulen 2016). Trench construction has helped to manage erosion, increase groundwater recharge, enhance soil fertility, and improve vegetation.

Watershed restoration can also help local communities to mitigate the degrading impacts of droughts and human activities on the surrounding land (Oxfam 2010). Soil management involves mulching, which improves the soil's organic content and enhances water retention and drainage. This will help crops grow in areas with heavy downpours. Reforestation of forest reserves helps in the management of local high temperatures; provides additional income; protects against soil erosion, landslides, and local flooding; and protects biodiversity and increases foliage for fodder in times of scarcity. Integrated soil fertility management (ISFM) involves use of appropriate quantity of organic and inorganic fertilizers together with disease- and pest-resistant germplasm and good agronomic practices. ISFM is widely promoted across Africa. Malawi has trained about 30,000 farmers and several hundred farmer associations and agricultural extension workers in ISFM.

Water Management

Agriculture is said to be high consumer of about 70% of the world's freshwater resources (Dinesh and Vermeulen 2016). Improved water management involves water capture and retention measure from rainfall or runoff and innovative irrigation practices. Studies also have shown that irrigation is an important adaptation measure that will significantly reduce the negative impacts associated with changes in climate (Mendelsohn and Dinar 2003; Kurukulasuriya and Mendelsohn 2008a, b; Mendelsohn and Seo 2007). Alternate wetting and drying have a potential of 30% and 48% reductions in water use and methane emissions, respectively, in rice production. Farmer groups of about 3,000 at 200 sites in Benin and Togo have designed and developed low-cost irrigation systems using bunds and drainage canals (Dinesh and Vermeulen 2016). This has improved rice yields to 3.5–4 tonnes per hectare. Other emerging water harvesting models include mechanized water harvesting systems in Syrian rangelands and small basins and semicircular bunds in Jordan and Egypt. Morocco has a system made up of trees and shrubs in combination with contour ridges efficient in collecting runoff in areas with rainfall of 100–200 mm. Soil fertility and crop management innovations also improve water-use efficiency.

Beyond the soil, land, and water management practices, there are other approaches that have helped farmers to adapt to the changing climate. According to Dinesh and Vermeulen (2016), conservation agriculture is used in about 157 million hectares, covering up to 11% of the world's arable land area. Brazil, Argentina, and Paraguay have adopted conservation agriculture. This agriculture model combines three practices: no-till, retention of crop residues in fields, and regular fallow periods. It is useful in ecological climes with dry and extreme rainfall predictions.

Crop farming management aims to improve crop production under climate change in compliment to soil management, agroforestry, and water management. Crop-specific innovations include breeding of more resilient crop varieties, diversification, and intensification. Burundi, Rwanda, and Uganda engage in coffee-banana intercropping in response to increasing temperatures which affect coffee production. Growing the two crops together can increase incomes compared to mono-cropping either crop alone.

In response to changing rainfall patterns, farmers now set up a community nursery to grow seedlings to replace crops lost to floods and drought. The nursery beds also serve as field trial platforms for new adapted seedlings or entirely new crops more suited to the changing conditions. Farmers also select to plant late or early millet varieties or to adapt by planting late or early in response to the varying onset of planting seasons (Dingkuhn et al. 2006). Adjustment in sowing time (22% households), use of drought-tolerant varieties (15%), and shifting to new crops (25%) were the three major adaptation practices used by farmers in the study area.

Mixed farming has also become a normal practice in response to climate change. Mendelsohn observed that farmers in Africa make various adaptations in response to climate by choosing whether to grow crops or livestock (Mendelsohn and Seo 2007; Seo and Mendelsohn 2008). Farmers also alternate and switch crops and livestock in response to climate change (Kuruklasuriya and Mendelsohn 2008c; Seo and Mendelsohn 2007, 2008). Farmers in warmer climes tend to choose goats and sheep as opposed to cattle and chicken. In Nyando, Kenya, heat-, drought-, and disease- resistant crossbreeds of native sheep and goats are being introduced to increase resilience and productivity of farming households.

Oxfam (2010) noted that access to weather and climate information, coupled with engagement of other farming communities to share experience, is the adaptation measure that has been adopted in response to climate change. Climate information services help farmers cope with climate variability. Weather forecasts and advisory services designed for farmers are made available to about 7.4 million rural people in Senegal (Dinesh and Vermeulen 2016). Similar climate information service has being deployed in four regions in Colombia. The Colombian model uses both scientific and indigenous knowledge to provide agroclimatic information to about 1,500 farmers. They have access to farm planning information on varieties choices, water harvesting, optimization of fertilizer application, and flexible planting dates. Table 2 below gives a series of agricultural practices adopted by farmers in three West African countries.

Use of GE Crops in Agriculture and as Climate Change Adaptation Measures in Africa

Proponents of GE crops present these technologies as a distinct option to promote food security and sustainable agriculture for developing countries (Falck-Zepeda et al. 2013). GEs have the potential of addressing both biotic and abiotic constraints to food production systems in Africa. The adoption of GM crops may enhance productivity, increase nutritional values of staple crops, improve management of pests and weed, and reduce the impacts of drought and salinity (Zawedde 2011; Falck-Zepeda et al. 2013).

Despite the documented benefits to farmers in developing countries, including those in Africa, many policymakers and farmers in sub-Sahara remain hesitant about the use of GM crops, and they need more information about their potential, benefits, costs, and safety in the African context. Limited progress has been made from several stakeholder discussions facilitated by the African Union (Falck-Zepeda et al. 2013). Such discussions have recommended more policy research on economic, social, ethical, environmental, intellectual property, and trade issues relevant to Africa. We are yet to see the results of the policy studies and their impact on GM crop adoption in Africa. The relatively limited research on the impact of GE crops implies that its contribution to the policy debate on GE crops and biotechnology in the continent has also been limited.

South Africa is the leading country in Africa for commercial production of GE crops. She was the first in Africa to commercially produce GM crops with the release of insect-resistant (Bt) cotton in 1997–1998 (Falck-Zepeda et al. 2013). Herbicide-tolerant (HT) cotton was made available for commercial production in the 2001–2002 season along with HT soybeans. Commercialization of HT maize

Table 2 Agricultural practices and climate adaptation measures in three West African countries

Climate adaptation measures as used by farmers	Nigeria	Sierra Leone	Liberia
Ground water harvesting	11.9	37.6	18.1
Mulching/use of cover crops	74.1	5.4	5.0
Cultivating in wetlands/river valleys (e.g., Fadama)	17.8	45.5	14.9
Construction of drainage systems	26.3	12.9	8.1
Planting of trees	23.3	4.9	3.7
Use of resistant varieties	38.7	5.0	5.0
Processing crops to minimize postharvest losses	51.8	15.7	4.3
Expansion of cultivated land area	26.3	44.3	27.6
Increased use of fertilizers, seeds	57.0	60.0	11.6
Intensive manure application	53.0	20.0	5.0
Increased weeding	63.9	62.2	74.4
Use of chemicals, herbicides, pesticides, etc.	55.3	10.0	11.3
Moved to a different site	19.1	51.8	47.8
Change in the timing of land preparation activities	30.1	10.3	18.7
Changes in planting dates	38.0	40.4	21.8
Practicing zero/minimum tillage	38.5	4.5	3.4
Changes in harvesting dates	30.0	36.4	15.8
Multiple cropping (planting of many crops in the same piece of land)	42.9	49.1	37.2
Mixed farming (crop and animal production)	29.4	9.7	7.7
Relay cropping-planting and harvesting in succession	14.2	39.8	2.9
Intercropping – main crops planted with subsidiaries at low densities	19.0	42.4	14.7
Decreasing animal stock	11.9	0.7	1.9
Change from animal production to crop production	8.4	0.2	1.7
Prayers for God's intervention	58.1	70.0	60.6

Source: ATPS (2011)

seeds followed in 2003–2004. GM cotton containing the combined or "stacked" trait (Bt and HT) was released for the 2005–2006 season, and Bt/HT maize was released for the 2007–2008 production season.

The results on the performance and socioeconomic impact of the introduction of Bt cotton (with a focus on smallholder farmers) showed that yield increased with the use of Bt cotton compared to conventional varieties (IRPRI 2013). The differences between the Bt yield advantages of small- and large-scale farmers were down to the financial and human capital constraints that cause smallholders to invest in chemical pest control. The average insecticide application level of smallholder farmers on the Makhathini Flats was captured to be lower than 50% of the optimal level.

Also the yield increase with Bt cotton when compared to conventional cotton depended on the level of bollworm infestation for that planting season. The yield advantage differed across farmers, farms, regions, and seasons (Fok et al. 2007 as cited in Falck-Zepeda et al. 2013). The Bt cotton still has to battle with other insects apart from bollworms and requires some chemical spraying to prevent attack. The manner of chemical application will also affect the plant performance and yield.

Both large-scale and smallholder farmers were reported to record significant savings on insecticides. However farmers had concerns on the Bt cotton; see technology fee which was later reduced after the introduction season (Gouse 2009 as cited in Falck-Zepeda et al. 2013).

The case of Bt maize in South Africa was not quite different except that the interest of the farmers differs based on the fact that cotton is more of a cash crop. It was found out that the average yield increases (against stem borer) of 10–11% and 0–32%, respectively, for commercial and smallholder farmers that adopted Bt maize from 2001–2002 to 2007–2008 (Gouse et al. 2010).

The results were different for seasons with low stem borer infestation. There were no significant difference in yields between farmers who planted the Bt maize and the conventional hybrid maize. However the farmers that adopted Bt maize were worse with the technology fee for seasons without insect infestations.

Annecke and Moran (1982) noted that it is a difficult task to predict seasonal stalk borer infestation on maize due to climate changes with variable rainfall patterns. Whereas Bt maize was a measure of insurance against unforeseeable stalk borer outbreaks for commercial farmers, technology fees for Bt maize were threats to smallholder farmers. Bt maize seeds were 23–30% costly than conventional seeds. Expenditures on insecticides decreased for commercial farmers but not for the small-scale ones, who do not necessarily use insecticides. The introduction of HT maize gave rise to many farmers in South Africa preferring HT maize seed, as its laborsaving benefit is valued higher than the higher yield benefit of Bt seeds. Some smallholders also showed preference for the stacked Bt/HT maize, whereas others saw the stacked Bt/HT maize to be very expensive and opted for HT.

However, the adoption of GM crops in Africa suffered setbacks due to the incidence in Burkina Faso. Bt cotton with inherent resistance to the bollworm pests (that can destroy up to 90% of a farmer's yield) was introduced in Burkina Faso in 2008 (Gakpo 2018). It was reported that cultivated acres of insect-protected cotton (Bt) which grew till about three quarters of the cotton output in the country were of the Bt strains. The new cultivar helped control pests on cotton farms, reducing the use of pesticides by up to 70%. A national study showed the introduction of Bt cultivars in Burkina Faso led to a 22% increase in yield over conventional cultivars, and households recorded an average profit gain of 51%. Bt cotton was grown in Burkina Faso from 2000 to 2014. The cultivation of the GE (Bt) cotton was halted in 2015 due to production of low-quality cotton with shorter fiber lengths. The withdrawal of Bt cotton has not gone without controversy, with farmers claiming yield losses even with more application of pesticides. The government of Burkina Faso is already searching for alternatives to avoid the ugly trend in India where campaigners blame the death of more than 270,000 cotton farmers to the adoption of GMOGM cotton cultivation. The Guardian (2018) linked the deaths to piles of pressure on poorly paid growers with burdens of cycles of unmanageable deaths from high-priced GM seeds.

The above cases of South Africa and Burkina Faso are among the few examples of GM crop impact assessment in Africa. Smale et al. (2009) and Qaim (2001) (as cited in Falck-Zepeda et al. 2013) reviewed such impact assessment in African and other developing countries' agriculture as it relates to farmers, households, communities, and trade and the institutional context in which these technologies may be deployed for potential use by farmers. There are lessons drawn from GM crop impact assessment in Africa (Falck-Zepeda et al. 2013). Averagely, adoption of GM crops has been profitable to farmers, but this masks variability in agroclimates, host cultivars, and farming practices. There are too few traits examined in few case studies and generalizations which cannot be drawn. The need to address crosscutting issues such as impacts on poverty, gender, and public health is vital. The other issues are that of the environmental and health risks associated with GMOs demand national risk assessment frameworks with specificities for each GM crop before adoption and commercial release to the public (Paarlberg 2008). More so, the issue of accessibility to poor farmers should be handled by development of local GM crop varieties from indigenous plants (Cohen 2005), addressing native production constraints. Such development needs research and development capabilities that will take the local farmers' need and local conditions into consideration.

The Makhathini smallholder experience as cited above can always be a reference when discussing GM crop adoption in Africa. Technological solutions (like the adoption of Bt cotton) alone cannot resolve the fundamental management challenges of smallholders (such as lack of knowledge regarding insects and pest control, limited access to inputs, or evolution in pest pressure) and agriculture in Africa. The particular case of the Makhathini Flats and the wider story of cotton in South Africa emphasize that although all agricultural systems require adequate investment and appropriate technologies, their viability is determined by the policies and management that facilitate sustainable and profitable production. More so the international commodity price with its burst and boom cycles affects agricultural crop sales from Africa. Despite the Bt cotton and more of stacked (Bt/HT) varieties still the varieties of choice for smallholder producers in South Africa, production levels are affected by limited relatively low price of cotton in the international market. The introduction of Bt cotton and stacked Bt/HT varieties may affect the supply of cotton but did not really changed the demand on the long run. Whereas many smallholders continued producing, commercial farmers left the sector for greener pastures. The agricultural sector may not flourish just with the diffusion of the innovation (adoption of GM crops).

Despite the potential economic benefits to accrue from the adoption of GM crops in Africa, opponents of GM crops have raised a wide range of concerns with respect to food safety, environmental effects, and socioeconomic issues (FAO 2016). Concerns on the food safety and health relate to possible toxicity and allergenicity of GM foods and its products. Environmental risks include the contamination of traditional varieties due to gene flow, uncontrolled gene dispersion and introgression of the transgenes into the local ecosystem, effect on nontarget organisms, and evolution of pest resistance and loss of biodiversity. Opponents also emphasized that GMs are largely developed by the capitalists in developed countries to be used in large-scale agriculture. They imply that GMOs are capital intensive and not designed for traditional agriculture in Africa and other developing countries and will also empower these private firms to monopolize agricultural market (Moschini and Lapan 1997; Falck-Zepeda et al. 2000, as cited in Falck-Zepeda et al. 2013). Private sector-led agricultural R&D investments are in a different pathway compared to previous agricultural innovation processes mainly driven by the public sector. GM crop adoption will perform better in such private-led agricultural sector.

Moreover GM crops will impact on socioeconomic status of local communities by disrupting traditional communities and livelihoods, dependency on private sector, production risk increasing due to the rise of monocultures, and the decline of smallholder crop diversification. It was bound to affect export markets to Europe, which is sharply divided on importation of GM crops. Other social and ethical concerns relate to restriction of access to genetic resources and new technologies, loss of traditions (saving seeds for next planting season), and private sector monopoly leading to loss of income of resource-poor farmers.

According to Falck-Zepeda et al. (2013), these concerns may or may not be unique to GE crops. They may also belong to a larger set of general concerns over the role of science and technology in contributing to poverty alleviation and development. Furthermore, it is not always clear whether these GE crops may be particularly important if they help solve specific crop productivity constraints in Africa. This is true especially of those productivity constraints that have not been resolved by conventional means, including conventional plant breeding and integrated pest management, and in those situations where other control/productivity enhancement approaches may not be accessible to farmers.

Recently, some organizations have surfaced and they look like they were designed for the mission of tackling issues of adoption of GM crops (Falck-Zepeda et al. 2013). They include social justice, biodiversity protection, small farmer livelihood assurance, and food security (such as the Third World Network, Via Campesina, Greenpeace, Oxfam, or GRAIN, Friends of the Earth International) have argued and conducted campaigns for and against the deployment of GE technologies in developing countries, including those in Africa.

A more balanced position would consider GE crops neither as Frankenfoods nor as miracle crops per se but rather as a set of technologies with unique attributes, different from past innovations, such as the green revolution's maize and wheat varieties. Many of these technologies have been produced using research and development (R&D) inputs with intellectual property owned by the private sector. These attributes need to be characterized, discussed, and in some cases addressed to ensure their compatibility with and support of poverty alleviation efforts, especially in the agricultural context of SSA.

In the current state of the art, GE may be still expensive for majority of farmers in Africa who are primarily subsistent farmers, but the greater consequence lies on its monocultural cropping systems. It is designed for mainly large-scale farming, and this has to adopt integrated pest and weed management system to avoid pesticide resistance in planting of Bt and Ht varieties. In the face of climate change risks, Africa will be neophytes in managing risks from pesticide resistance. As in the case of Bt cotton in Burkina Faso, it would look like opening Pandora's box for Africa if she wholly adopt GM crops without the necessary capabilities and infrastructure.

National Innovation Systems (NIS) and Agricultural Innovation Systems in Africa

Muller et al. (2011) used the IPCC confidence rating and got a very high confidence that climate change will negatively affect at least parts of African agriculture (14 of 14 studies). There was also high confidence rating that African agriculture will be partly affected positively by climate change on a 12 of 14. Beyond the effect of climate change, agricultural production system in Africa is heavily affected by geometric population growth rates. Africa then has huge potentials to increase her agricultural productivity, with yield gaps of 10% in Egypt and about 90% in Angola (Neumann et al. 2010 as cited in Mueller et al. 2011). The inefficiency in African agriculture production system can be explained by limited access to information (climate information system), lack of access to knowledge support, access to market, farm inputs (fertilizer, improved seedlings, pesticides, and others), access to finance, etc. (Markelova 2009; Wichelns 2003; Muller et al. 2011; IFIPRI 2013). The lack of access to knowledge support and useful information shows there is a need to build an active national innovation system (NIS) that can foster a working agricultural innovation system. Some challenges are common to the dissemination of all new technologies in SSA. Challenges unique to GE crops may require innovative approaches for addressing the constraints in a meaningful way. However, credible objective information on the impacts of GM crops based on their cost, risks, and benefits are not always easy to find and to digest, given the complexity of some of the issues at stake. GM crop diffusion in SSA will need an active NIS; but these systems are weak.

Linkages, interactions, and learning mechanisms among the component actors are notably weak and/or often nonexistent. Empirical evidence revealed several linkage gaps and missing links among and between the actors in the systems (Agbamu 2000; Egyir 2009 as cited in ATPS 2011). These gaps must be strengthened if Africa must make meaningful progress on the issue of adoption of GM crops.

Africa's Biotechnology Capacity, Biosafety Status, and Adoption Impact

Africa lacks the capacity to adapt, innovate, and transform its agriculture sector especially when it comes to using biotechnology tools and applications (Chambers et al. 2014). R&D intensity and its level of investments in SSA translate to a relatively poor biotechnology innovative capacity. This in turn affects the capacity for conducting GM-related R&D and biosafety programs in Africa. According to Chambers et al. (2014), the increase in spending on public agricultural R&D in SSA has been seen in a few countries (Kenya, Nigeria, Tanzania, and Uganda); however the increase in the numbers of researchers has been coupled with a decrease in the share of PhD holders. These trends are inconsistent with the goal of a technologically adept and sophisticated agriculture sector.

GM crops unlike others have to comply with biosafety regulations and risk assessments. Lack of biosafety regulatory frameworks and of GMO technologies contributes to low level of GM crop adoption in Africa (IFIPRI 2013). Currently, there are no indigenous GM crops developed in Africa except in Egypt. Most GM crops under development for SSA use genetic constructs or transformation procedures developed elsewhere, and plant germplasm materials are usually selected and developed internally in the country of interest.

Biotechnology capacity that can develop those GMO traits of interest to national priorities has to be weighed against the possibility of accessing such technologies developed elsewhere. From a science and technology perspective, African decision makers can opt for different innovative capacity systems to deliver products to farmers. Some countries in Africa, though, lack even the minimal investments in R&D capacity necessary to conduct adaptive R&D in their own national research systems.

Furthermore, some countries in the region have expressed a concern over the potential risks of these technologies to their farmers. Such concerns are not insurmountable hurdles, as there are practical and feasible approaches that can empower countries to ensure an acceptable level of safety after a scientific assessment while ensuring innovation and technology transfer to their farmers.

African seed systems and germplasm delivery mechanisms face a number of constraints that have been described in the literature. Technology delivery and dissemination are limited by institutional weaknesses and the insufficient development of national seed systems in Africa. These constraints may affect GM crop seeds. Yet they can be magnified as the need arises to ensure a working system that facilitates knowledge and information exchange about the use of the technology and market signal transmissions, including price premiums or value-added paid for competing commodities in markets.

Farmers who will consider the adoption of GM crops in Africa will face many institutional challenges (Chambers et al. 2014). Access to credit and complementary inputs, their ability to manage production risk, and other binding institutional policy constraints play determinant roles in their decisions. There are also institutional issues such as intellectual capital exploitation from developed GM seeds using licenses. The collection of license fees on GM seeds from traditional farmers, who are used to reusing seedling from one season to another without paying to anyone, is a big challenge in Africa. A very important issue identified in a growing number of studies is farmers' access to knowledge and information about the use of the technology and its market potential. Many of the GM crops available for adoption in and outside Africa have demonstrated their technical capability to provide benefits to farmers. However, a farmer's ability to tap into those potential benefits can be limited by institutional issues. Policymakers will have to consider support structures and institutional environment to maximize the benefits and minimize the risks of GM crop adoption in Africa. In this regard, the policy environment will strive to avoid potential cases of "technological triumphs but institutional failures" observed (Gouse et al. 2005). This buttresses the role of an active national innovation system.

Discussion and Conclusion

The real management of climate change risks is by tackling the real and actual risks. This has been highlighted by the IPCC as a key adaptation action (FAO 2016). Sometimes climate change risks are wrongly perceived, and interventions may result to maladaptation. Adaptation to climate change is highly local, but not without supports from foreign actors to identify and reduce vulnerability in the adaptation programs. Adaptation program (planning and implementation) can be enhanced through complementary actions from all actors or stakeholders (IPCC 2014). However, as expected in the coordination of sustainable development goals, national governments can coordinate adaptation efforts of local and subnational governments, for instance, by protection of vulnerable groups, facilitating diversification of economy, providing useful information, and crafting policy and legal frameworks and financial support. This requires appropriate institutions and policies, which are mostly sector- and/or risk- specific. In this regard, the need for an active national innovation system cannot be overemphasized. Local solutions and appropriate technologies must be adapted to solve local challenges. Weather stations, weather and climate projection tools, yield response models, environmental monitoring tools, and vulnerability assessments can help to predict how local climate conditions will change in the future and to estimate their impact on agricultural production. IPCC findings also suggest that adaptation is shown to be successful and sustainable when linked to effective governance systems, civil and political rights, and literacy. Some countries have set up units or institutions responsible for adaptation planning and coordination across ministries and processes to ensure alignment with development priorities (Oxfam 2010).

Underestimating of peculiarities of adaptation as a social process can lead to maladaptation and achievement of unintended adaptation outcomes IPCC (2014). Adaptation can reduce the risks of climate change impacts, but constraints can interact to impede interventions. Common constraints on implementation are limited financial and human resources, limited integration or coordination of governance, uncertainties about projected impacts, different perceptions of risks, competing values, absence of key adaptation leaders and advocates, and limited tools to monitor adaptation effectiveness and insufficient research. There is some measure of uncertainty that the adoption of GM crops will serve as an effective adaptation strategy amidst the numerous constraints and systemic failure in Africa.

More so climate change impacts are sometimes with uncertainty. For example, climate models for the Sahel conflicts with predictions of increasing and decreasing rainfall could result to maladaptation by adapting to either condition (Oxfam 2010). In order to avoid maladaptation, mechanisms should be introduced to validate adaptation options.

The right policy mix and integrated approach can be utilized in climate change adaptation. Biotechnological solution using GM crops could be a one strategy, and there is no-one-size-fits-all solution to adaptation, which is a social process. A longer-term approach, using sustainable development concepts, will increase the possibility that more immediate adaptations will also enhance future options and preparedness.

However, one has to state the obvious that development is path dependent, and Africa failed to build capabilities to exploit the comparative advantages that she had on her abundant natural resources. Agricultural and other exports from natural resources had been the mainstay of most African economies. The cascading impacts of climate change have threatened Africa's further dependence on agriculture. The natural environment is stressed and under pressure to sustain traditional agriculture in Africa. Huge population growths also put pressure on the carrying capacity of the environment to support agriculture demands and its productivity in SSA. Climate change may not be new today, but the difference today is based on the enormous contribution and pressure from human activities, which destabilizes the climate system and supporting roles of the ecosystem. Men have used portfolios of agricultural practices and technologies indigenous to their agroecological zones. These technologies have already proven benefits successful in ensuring food security, resilience, and productivity. Yet Africa did not improve or update those adaptive practices to face the current climate change risks. Africa did not build on knowledge systems that helped develop adaptive capacity to the changing climate. Studies cited above showed that some farmers adapted in response to prevailing local climate conditions even without having scientific knowledge on the basis of their actions or practices. Measures such as soil, land, and water management have helped farmers to adapt to changes in climate. While it is commendable that some farmers from Morocco to Malawi are now trained on how to engage in integrated soil management, it is still surprising how Africa has not prepared to exploit even the positive risks and opportunities created by change in climate. Lack of adaptation in this regard portends a bleak future for Africa. Lack of these adaptive capabilities or basic knowledge system will affect the transfer of technologies such as GM crops.

According to NASEM (2016), the available evidence indicates that GE soybean, cotton, and maize have generally had favorable economic outcomes for farmers who adopted them. However, outcomes have been heterogeneous depending on pest abundance, farming practices, and agricultural infrastructure. The Bt crops generally decreased yield losses and insecticide use on both small and large farms when compared to conventional varieties. The South African cases cited above before the discussion section show that scale affects the benefits derived from planting of Bt crops. More so, change in climate may affect predictions on performance and outcomes of Bt crops in Africa and elsewhere.

Bt crops have also led to higher insect biodiversity on farms than planting conventional varieties which were treated with synthetic insecticides (NASEM 2016). Locations without resistance-management strategies experienced damaging levels of resistance with some insect pest evolving. Herbicide-resistant (HR) crops sprayed with the herbicide glyphosate often had small increases in yield in comparison with non-HR counterparts. In areas where planting of HR crops led to heavy reliance on glyphosate, some weeds evolved resistance and present a major agronomic problem.

Africans cannot afford to suffer pesticide resistance in the face of climate change. It will be a twin challenge to manage microbial drug resistance to diseases in Africa and now pesticide resistance. GMO crops may end up creating more challenges than solutions in African agricultural sector.

It is quite obvious that the proponents of GM crop adoption failed to acknowledge the facts that the lack of capabilities to build an agricultural system with high productivity will also affect the capability to manage benefits from the adoption and production of GM crops in Africa. These capabilities are not far from poor financial resources, technological capabilities, and low critical mass of scientific personnel. Strengthening the carrying capacity of natural environment or the ecosystem can serve as great adaptive measures to improve the productivity of agricultural sector in Africa. However, lack of infrastructure, technological capability, and critical mass of skilled human capital affects the capability of SSA to have active production systems. These in turn affect sectors such as agriculture, which depends on climate.

The problem with Africa is systemic and should be addressed as such. Most African countries have weak national innovation systems. The agricultural production system does not benefit from existing poor research and development structures. The educational system, science and technology spheres, public structures such as environment ministries, trade and commerce, etc. are not linked to tackle national priorities. The drive of sustainable development is to make economies adapt to changes in climate by changing the way all other sectors of the economy function. This is the front line of national climate adaptation. Is climate change adaptation a national priority for African countries? Until then, it will be a high hill task for one measure such as adoption of GM crops to solve adaptation challenges.

Africa is not ready to walk the talk by going through a learning curve that involves development and deployment of appropriate technologies. Technology transfers have failed in other sectors. Biotechnological innovations involve the issue of human lives and have environmental concerns.

According to the US committee from the National Academies of Sciences, Engineering, and Medicine (NASEM 2016), the complex nature of assessing long-term environmental changes often makes it difficult to reach definitive conclusions. That was illustrated by the case of the decline in overwintering monarch butterfly populations. Conclusions are yet to be made on the cause of the decline. However, the impending ban on glycophosphate herbicides in Europe and in other countries shows that even with lack of scientific evidence of danger and possible risks, evidence could emerge but much later with its consequences.

More so, it is important to also consider socioeconomic impacts of growing GM crops, which is more adapted to large-scale farming of a single plant or "monoculture." This may not be the most appropriate to most of the African farmers, who are mostly small-scale farmers with less than two hectares of land. Mixed cropping system is dominant in African traditional agricultural setting. Adoption of GM crop will tend to achieve economy of scale, and this will work adversely to the synergistic effects of mixed cropping system in Africa.

Africa is rather cautious. She is right to do so in view of several uncertainties as discussed earlier. The quick fix of developing countries' biosafety laws and hastening of the diffusion of GM crops in Africa will fail if not properly handled. The case of withdrawal of GM cotton in Burkina Faso as cited in this chapter is a good example. Africa is a technology laggard, and she has to learn what it takes to adopt and adapt technologies for her own use. On GMO technologies, Africa should be allowed to learn, or else it will go the way of industrialization programs in Africa. Africa did not industrialize, and now she instead shows interests in leapfrogging without technological learning. The failure of industrialization in Africa is affecting her development path today. Development is said to be path dependent. Africa is yet to embrace sustainable development. Climate change adaptation will put Africa in the right track. However, there is no one-size-fits-all approach to it. Africa has to build national innovation systems to rediscover herself.

Conclusion

The concerns raised by opponents of GM crop adoption in Africa are genuine even though some counter those views base on lack of scientific evidence. However, lack of evidence on glycophosphate herbicide yesterday does not mean it was safe. The concerns raised by African countries on the safety of GM crops are easily waived outside by the calls for Africa to regulate GM crop adoption and diffusion, it is obvious that Africa lacks the capacity to regulate GM crop diffusion in her territory. The levels of research and development, science and technology budgets, human capacity building in biotechnology safety, policy analysis, and learning are too low to enable Africa to reach logical conclusions and make the right decisions on GM crops. Moreover, the proponents of adoption of GM crops as a climate change adaptation measure in Africa are yet to convince us on how GM crops will serve that purpose with weak capabilities and poor institutions and production systems. Bt/HT seed technologies are farm inputs just like fertilizers, herbicides, or irrigation technologies. Contrary to the technologies of the green revolution, they might be able to improve the yields of farmers with limited ability or means to control insects and herbs. Staggering results could easily be seen for small-scale/subsistence farmers that make up the large population of farmers in Africa. Few socioeconomic impact assessments on GM crops as even discussed in this chapter show that improved productivity attributed to the initial adoption of GM crops can be easily changed with burst in international commodity market prices. Extreme climate changes (as often witnessed in some parts of Africa) can also eclipse the marginal profits that accrue to farmers that adopt GM crop technologies.

GM crops may not be particularly important if crop productivity constraints are effectively managed in Africa. Results of impact assessment of GM crops in countries such as the United States, Germany, Spain, and others show little or no significant differences on yields on the use of GE crops compared to non-GE crops use by farmers. The committee of the US National Academies of Science, Engineering, and Medicine (2016) stated that there is no evidence from the US Department of Agriculture (USDA) data that they have substantially increased the rate at which US agriculture is increasing yields. GM crops may not increase agricultural productivity in Africa at the long run. The adoption of GM crops in Africa cannot by itself overcome management failure and governance challenges that seem to be endemic in African agriculture and that were also the limiting factors in the green revolution. The experience with Bt cotton on the Makhathini Flats emphasizes that technology-induced advances might be short lived in the absence of the correct management structures, regulations, cooperation, and competition.

The claims by opponents to GM crop adoption in Africa that GM crops are designed for large-scale farming may sound true when poor farmers are faced with rising cost of GM seeds. GM cropping system seems to support more of monoculture cropping rather than the mixed cropping system prevalent in the African traditional subsistence farming system. This often leads to the emergency of pesticide resistance and evolution of crop pests with no immediate remedy.

Development is path dependent, and the general concerns over the failure of previous transfers of technology to narrow poverty and inequality will also affect the adoption of biotechnology, particularly GM crops in Africa. GM crops are already perceived as tools of creating new monopoly from companies in the developed economies. There is already a great doubt on how African farmers can take it as an adaptive measure to climate change. Adaptation is a social process, and the adoption of GM crops as an adaptive measure must take cognizance of the social context in Africa.

Policy Recommendations

The relatively limited research on the impacts of GMO crops in Africa shows why policy debates on GM crops and biotechnology in the continent has also been limited. The gaps in capabilities demand that African must build the necessary capacity in platform technologies such as nanotechnology, ICTs, and biotechnology. Africa can regulate biotechnology to the extent that she has developed capabilities in biotechnology, technology management, risk management and other related areas.

The author therefore makes the following recommendations:

- There should be policy researches on economic, social, ethical, environmental, intellectual property, and trade issues on GM crops and its impact as relevant to Africa.
- Science, technology, and innovation policies should be crafted to deal with the biotechnology and issues of GM crops. Strengthening of national innovation systems will help address systemic failures in the development and adaptation of appropriate technologies needed for economic growth and development.
- Countries in Africa should mainstream climate change adaptation as a national priority. This adaptation will help Africa to benefit from exploitation of positive risks of climate change.

Cross-References

- ▶ Building Pastoral and Agro-pastoral Community Resilience Against Drought in the Context of the Paris Agreement: The Case of Isiolo County, Kenya
- ▶ Climate Change and Variability in the Mixed Crop/Livestock Production Systems of Central Ethiopian Highland
- ▶ Drip Watering Regimes on Growth Performance, Yield, and Water Use Efficiency of Sorghum in Semi-arid Environment of Tanzania: Effects
- ▶ Enhancing the Capacity of Vulnerable Community to Climate Change: Role of Quality Declared Seed Production Model in Semi-Arid Areas of Central Tanzania
- ▶ Fertilization Strategies Based on Climate Information to Enhance Food Security Through Improved Dryland Cereals Production
- ▶ Impact of Smart Crop-Livestock Diversification as Climate Change Adaptation Strategies on Farmers' Living Conditions, Tahoua State, Niger Republic
- ▶ Intensifying Maize Production Under Climate Change Scenarios in Central West Burkina Faso
- ▶ Managing Vulnerability to Drought and Enhancing Smallholder Farmers Resilience to Climate Change Risks in Zimbabwe

References

African Technology Policy Studies Network (ATPS) (2011) Agricultural innovations for climate change adaptation and food security in West Africa: the case of Nigeria, Sierra Leone and Liberia. Working paper series no 61 of ATPS

Agbamu JU (2000) Agricultural research extension systems: an international perspective. Agricultural Research and Extension Network Paper No. 106. ODI London, 7

Annecke DP, Moran VC (1982) Insects and mites of cultivated plants in South Africa. Butterworths, Durban

Australian Greenhouse Office (2003) Climate change – an Australian guide to the science and potential impacts, edited by Barrie Pittock

Bradshaw B, Dolan H, Smit B (2004) Farm-level adaptation to climatic variability and change: crop diversification in the Canadian Prairies. Clim Chang 67:119–141

Byamukama B, Carey C, Cole M, Dyszynski J, Warnest M (2011) National strategy on climate change and low carbon development for Rwanda baseline report. Smith School of Enterprise and Environment, University of Oxford, Oxford, UK. Retrieved from http://www.smithschool.ox.ac.uk/wp-content/uploads/2011/03/Rwanda-Baseline.pdf

Chambers JA, Zambrano P, Dalck-Zepeda J, Gruere G, Sengupta D, Hokanson K (2014) GM agricultural technologies for Africa. A state of affairs. International Food Policy Research Institute and African Development Bank, Washington, DC

Cohen JI (2005) Poorer nations turn to publicly developed GM crops. Nature 23(1):27–33

Committee on Approaches to Climate Change Adaptation (2010) Approaches to climate changeadaptation. Retrieved from https://www.env.go.jp/en/earth/cc/adapt_guide/.../approaches_to_adaptation. Accessed 30 Jan 2018

Deressa T, Pugliese J, Kimenyi MS (2014) Genetically engineered crops: key to climate adaptation and food security in Africa? Retrieved from https://www.brookings.edu/blog/africa-in-focus/2014/09/04/genetically-engineered-crops-key-to-climate-adaptation-and-food-security-in-africa/. Accessed 30 Jan 2018

Dinesh D, Vermeulen SJ (2016) Agricultural practices and technologies to enhance food security, resilience and productivity in a sustainable manner: messages to SBSTA 44 agriculture workshops. CCAFS working paper no. 146. CGIAR Research Program on Climate Change, Agriculture and Food Security (CCAFS), Copenhagen. Retrieved from http://hdl.handle.net/10568/71050. Accessed on 2 May 2015

Dingkuhn M, Singh BB, Clerget B, Chantereau J, Sultan B (2006) Past, present and future criteria to breed crops for water-limited environments in West Africa. Agric Water Manag 80:241–261. https://doi.org/10.1016/j.agwat.2005.07.016

Egyir IS (2009) The plantain ASTI system in Ghana. Paper presented at the CTA Training of Trainers (ToT) Regional Workshop on Agricultural Science, Technology and Innovation (ASTI) Systems held at Sheraton Hotel and Towers, Abuja, 24–28 August 2009

Falck-Zepeda JB, Traxler G, Nelson RG (2000) Rent creation and distribution from biotechnology innovations: the case of Bt cotton and herbicide-tolerant soybeans in 1997. Agribusiness 16:21–32

Falck-Zepeda JB, GruÃre, GP, Sithole-Niang I (eds) (2013) Genetically modified crops in Africa: economic and policy lessons from countries south of the Sahara. IFPRI books, International Food Policy Research Institute (IFPRI), number 978-0-89629-795-1, June 2013

Food and Agricultural Organization-FAO (2007) Adaptation to climate change in agriculture, forestry and fisheries: perspective, framework and priorities. Retrieved from http://www.fao.org/nr/climpag/pub/adaptation_to_climate_change_2007. Accessed 20 Oct 2018

FAO (2008) Climate change adaptation and mitigation in the food and agriculture sector. Food and Agriculture Organization of the United Nations, Italy

FAO (2009) One sixth of humanity undernourished – more than ever before. Food and Agriculture Organization of the United Nations, Italy. http://www.fao.org/news/story/en/item/20568/icode/. Accessed 28 Nov 2009

Fok M, Hofs JL, Gouse M, Kirsten JF (2007) Contextual appraisal of GM cotton diffusion in South Africa. Life Sci Int J 1(4):468–482

Food and Agricultural Organization-FAO (2016) The state of food and agriculture, climate change, agriculture and food security. www.fao.org/publications. Accessed 13 May 2018

Gakpo JO (2018) Burkina Faso is moving forward with GMO research. Retrieved from https://allianceforscience.cornell.edu/blog/2018/01/burkina-faso-is-moving-forward-with-gmo-research/. Accessed 10 May 2018

Gouse M, Kirsten J, Shankar B, Thirtle C (2005) Bt cotton in Kwazulu Natal: technological triumph but institutional failure. AgBiotechNet 7(134):1–7

Gouse M (2009) Ten years of Bt cotton in South Africa: putting the smallholder experience into context. In: Tripp R (ed) Biotechnology and agricultural development: transgenic cotton, rural institutions and resource-poor farmers. Routledge, New York, pp 200–224

Gouse M, Kirsten JF, Piesse J, Thirtle C, Poulton C (2010) Insect resistant and herbicide tolerant maize adoption by South African smallholder farmers – making sense of seven years of research. Presented at the 14th international consortium on applied bioeconomy research conference, bioeconomy governance: policy, environmental and health regulation and public investments in research, 16–18 June, Ravello

Gregory PJ, Johnson SN, Newton AC, Ingram JSI (2009) Integrating pests and pathogens into the climate change/food security debate. J Exp Bot 60:2827–2838

Hansen JW (2002) Realizing the potential benefits of climate prediction to agriculture: issues, approaches, challenges. Agric Syst 74:309–330. https://doi.org/10.1016/S0308-521X(02)00043-4

Heisey PW, Rubenstein KD (2015) Using crop genetic resources to help agriculture adapt to climate change: economics and policy. Economic information bulletin number 139

Intergovernmental Panel on Climate Change (IPCC) (1996) In: Watson RT, Zinyowera MC, Moss RH (eds) Climate change 1995. Impacts, adaptations and mitigation of climate change: scientific-technical analysis. Cambridge University Press, Cambridge, UK/New York, pp 289–324

Intergovernmental Panel on Climate Change (IPCC) (2001) Climate change 2001: impacts, adaptation, and vulnerability. IPCC, Cambridge University Press, Cambridge, UK

Intergovernmental Panel on Climate Change (IPCC) (2005) Guidance notes for lead authors of the IPCC fourth assessment report on addressing uncertainties. IPCC, Geneva

Intergovernmental Panel on Climate Change (IPCC) (2007a) In: Parry ML, Canziani OF, Palutikof JP, van der Linden PJ, Hanson CE (eds) Climate change 2007: impacts, adaptation and vulnerability: contribution of Working Group II to the Fourth Assessment Report of the Intergovernmental Panel on Climate Change. Cambridge University Press, Cambridge, UK/New York

Intergovernmental Panel on Climate Change (IPCC) (2007b) Climate change impacts, adaptation and vulnerability-working group II contribution to the Intergovernmental Panel on Climate Change: summary for policymakers. IPCC Secretariat, Geneva

Intergovernmental Panel on Climate Change (IPCC) (2009) The Intergovernmental Panel on Climate Change. Retrieved from www.ipcc.ch. Accessed 22 Sept 2009

Intergovernmental Panel on Climate Change (IPCC) (2014) In: Core Writing Team, Pachauri RK, Meyer LA (eds) Climate change 2014: synthesis report. Contribution of Working Groups I, II and III to the Fifth Assessment Report of the Intergovernmental Panel on Climate Change. IPCC, Geneva, 151 pp

Juman C, Serageldin I (eds) (2007) Freedom to innovate: biotechnology in Africa's development. A report of the high-level African panel on modern biotechnology. African Union/New Partnership for Africa's Development, Addis Ababa/Pretoria

Kurukulasuriya P, Mendelsohn R (2008a) A Ricardian analysis of the impact of climate change on African cropland. Afr J Agric Resour Econ 2:1–23

Kurukulasuriya P, Mendelsohn R (2008b) Crop switching as an adaptation strategy to climate change. Afr J Agric Resour Econ 2:105–126

Li YP, Ye W, Wang M, Yan XD (2009) Climate change and drought: a risk assessment of crop-yield impacts. Clim Res 39:31–46

Markelova H, Meinzen-Dick R, Hellin J, Dohrn S (2009) Collective action for smallholdermarket access. Food Policy 34:1–7

Maskrey A, Buescher G, Peduzzi P, Schaerpf C (2007) Disaster risk reduction: 2007 global review. Consultation edition. Prepared for the Global Platform for Disaster Risk Reduction first session, Geneva, pp 5–7

Mendelsohn R, Dinar A (2003) Climate, water, and agriculture. Land Econ 79:328–341

Mendelsohn R, Seo N (2007) An integrated farm model of crops and livestock: modeling Latin American agricultural impacts and adaptation to climate change. World Bank policy research working paper 4161

Moschini G, Lapan H (1997) Intellectual property rights and the welfare effects of agricultural R&D. Am J Agric Econ 79:1229–1242

Müller C, Cramera W, William L, Harea B, Lotze-Campena H (2011) Climate change risks for African agriculture. PNAS 108(11):4313–4315

National Academies of Sciences, Engineering, and Medicine-NASEM (2016) Genetically engineered crops: experiences and prospects. The National Academies Press, Washington, DC. https://doi.org/10.17226/23395

Nelson GC, Rosegrant MW, Koo J, Robertson R, Sulser T, Zhu T, Ringler C, Msangi S, Palazzo A, Batka M, Magalhaes M, Valmonte-Santos R, Ewing M, Lee D (2009) Climate change: impact on agriculture and costs of adaptation. Food policy report, no 21. International Food Policy Research Institute, Washington DC, p 19

Neumann K, Verburg P, Stehfest E, Müller C (2010) A global analysis of the intensification potential for grain production. Agric Syst 103:316–326

Oxfam (2010) Climate change adaptation, Oxfam International Research report. http://www.oxfam.org.uk/resources/policy/climate_change/climatechange-adaptation.html. Accessed 30 Apr 2018

Paarlberg R (2008) Starved for science: how biotechnology is being kept out of Africa. Harvard University Press, Cambridge, MA/London

Qaim M (2001) A prospective evaluation of biotechnology in semi-subsistence agriculture. Agric Econ 25:165–175

Schmidhuber J, Tubiello FN (2007) Global food security under climate change. Proc Natl Acad Sci USA 104:19703–19708. https://doi.org/10.1073/pnas.0701976104

Seo N, Mendelsohn R (2007) An analysis of livestock choice: adapting to climate change in Latin American farms. World Bank policy research working paper 4164. World Bank, Washington, DC

Seo N, Mendelsohn R (2008) A structural Ricardian analysis of climate change impacts on African farmers. World Bank policy research working paper 4603. World Bank, Washington, DC

Smale M, Zambrano P, Gruère G, Falck-Zepeda J, Matuschke I, Horna D, Nagarajan L, Yerramareddy I, Jones H (2009) Measuring the economic impacts of transgenic crops in developing agriculture during the first decade: approaches, findings, and future directions. IFPRI food policy review 10. International Food Policy Research Institute, Washington, DC

Sultan B, Gaetani M (2016) Agriculture in West Africa in the twenty-first century: climate change and impacts scenarios, and potential for adaptation. Front Plant Sci 7:1262. https://doi.org/10.3389/fpls.2016.01262

The Guardian (2018) India's farmer suicides: are deaths linked to GM cotton? – in pictures. https://www.theguardian.com/global-development/gallery/2014/may/05/india-cotton-suicides-farmer-deaths-gm-seeds. Accessed 14 May 2018

Thornton PK, Jones PG, Alagarswamy G, Andresen J (2009) Spatial variation of crop yield response to climate change in East Africa. Glob Environ Chang 19:54–65

Turco M, Palazzi EJ, von Hardenberg, Provenzale A (2015) Observed climate change hotspots. Geophys Res Lett 42:3521–3528. https://doi.org/10.1002/2015GL063891

UNFCCC (2007) Report on the African RegionalWorkshop on Adaptation. FCCC/SBI/2007/2. UN Office at Geneva. http://unfccc.int/resource/docs/2007/sbi/eng/02.pdf/. Accessed 28 Feb 2018

UN Millennium Project (2005) Halving hunger: it can be done: report of the task force on hunger. The Earth Institute at Columbia University, New York. Retrieved from www.unmillenniumproject.org/documents/mainreportcomplete-lowres

Wichelns D (2003) Policy recommendations to enhance farm-level use of fertilizer and irrigation water in sub-Saharan Africa. J Sustain Agric 23:53–77

World Development Report (WDR) (2008) Agriculture for development. The World Bank, Washington, DC

Zawedde BM (2011) Potential role of GMOs in adapting agriculture to climate change in Sub-Saharan Africa. In: Proceedings of the international conference on agro-biotechnology, biosafety and seed systems in developing countries, pp 63–70

Climate Smart Adaptations in the African Tropics: Scaling Weather Information for Decision Support Outcomes in Nigeria Savannahs

128

David Olufemi Awolala

Contents

Introduction 2652
Weather Information and Agricultural Risk Management Decisions 2655
Weather Information as Pathway Towards Climate Resilience 2657
Data and Methodology 2658
  Study Area, Sampling Technique, and Data Collection 2658
  Data Analysis 2660
Results and Discussions 2662
  Socioeconomic Characteristics of the Respondents 2662
  Weather Information in Adaptation Decision Process 2665
  Understanding Farmers Need for Weather Information 2665
  Farm-Level Climate Resilient Strategies 2667
  Weather Information Required for Adopting Climate Resilient Strategies 2668
Modelling Drivers of Climate Resilient Decisions 2670
  Resilient Decision-Making in Sudan Savannah: Heckman Outcome Model 2670
  Access to Weather Information in Sudan Savannah: Selection Model 2672
Predicted Impact of Climate Resilient Practices: Sudan Savannah 2673
  Resilient Decision-Making in Guinea Savannah: Heckman Outcome Model 2674
  Access to Weather Information in Guinea Savannah: Selection Model 2676
Predicted Impact of Climate Resilient Practices: Guinea Savannah 2677
Conclusions and Policy Issues 2678
References 2680

D. O. Awolala (✉)
Department of Agricultural and Resource Economics, Federal University of Technology Akure, Akure, Nigeria

Research Fellow, Adaptation Finance Fellowship Programme of the Thailand Research Development Institute Bangkok, Thailand and Frankfurt School of Finance and Management, Frankfurt, Germany
e-mail: ddawolala@gmail.com

W. Leal Filho (ed.), *Handbook of Climate Change Resilience*,
https://doi.org/10.1007/978-3-319-93336-8_169

Abstract
Extreme weather events and climatic variability exert tremendous influence on the livelihoods and decision-making processes of rural households in managing loss and damage in the dry savannah areas of Nigeria. Smallholders have long relied on traditional weather monitoring but unfortunately, increasing uncertainties in weather events have rendered indigenous knowledge obsolete. Renewed interest climate resilient development planning has raised the prospect of improved weather information as a decision support tool for future risk preparedness and early action. This study explored weather and climate forecast information that should be upscale in facilitating climate resilient adaptation decisions in sudan and guinea savannah areas of Nigeria. Rainfall probabilities are the most highly demanded weather forecast element by farmer users in the savannah areas. Onset date is the most profiled weather information required in adapting with delayed planting and drought resilient varieties, dry spell distribution will facilitate adapting with minimum tillage and cover cropping practices, while length of growing season is needed to take decision on early maturing varieties. Heckman probit estimates with Wald statistic of -0.614 and p-value of 0.0002 reveals gender, farming experience, annual farm income, and persistent occurrences of erratic rainfall as the significant factors that will catalyze rural decisions to adapt with resilient practices in the sudan savannah while farmers' distance to weather station in addition will foster decisions on resilient practices in the guinea savannah of Nigeria.

Keywords
Weather variability · Forecast information · Adaptation decisions · Resilience · Dry savannahs · Nigeria

Introduction

The Fifth Assessment Report (AR5) expressed climate impacts as jeopardizing economic potentials and threatening food basket of economies that heavily rely on rain-fed agriculture, especially in the arid region of tropical Africa (IPCC 2015). A central concern is the huge possibility of trapping farming households into poverty, smallholders who need financial cover, driven by risk assessment and local needs to strengthen their resilience (IFAD 2016). Africa will be disproportionately affected by increasing unpredictability of weather patterns and extreme weather events due to their dependence on rain-fed agriculture and poor adaptive infrastructure. Swiss Re (2010) confirmed that extreme weather and climate-induced shocks have continue exposing local producers to the challenges and costs of protecting livelihoods against widening food gap and economic losses, thereby making income security unpredictable from year to year. Over half of the world's population lives in regions highly vulnerable to extreme events and scientific predictions have provided ideas about how the magnitude of these changes will undo decades of development efforts.

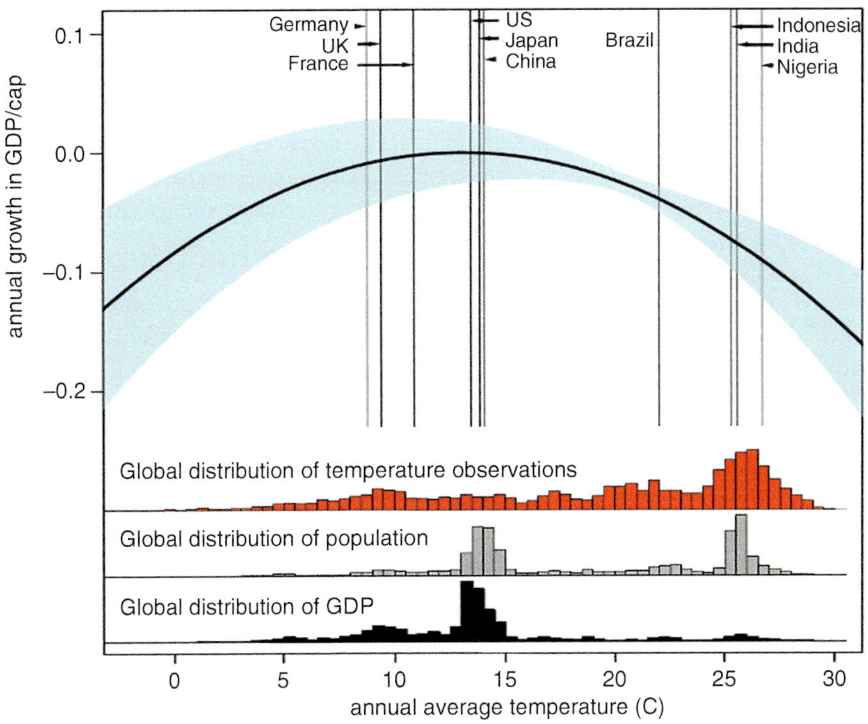

Fig. 1 Annual average temperature and changes in annual GDP growth rates. (Source: Burke et al. in *Nature* (2015))

While climate change is a global phenomenon, its negative impacts are severely felt by poor people and developing countries due to their high dependence on natural resources and limited capacity to cope (UNFCCC 2002; World Bank 2010).

Impact of climate shocks to the economy, vulnerable populations, ecosystems and infrastructure threatens achieving Sustainable Development Goals (SGDs) such as poverty alleviation strategies and food security agenda (UNDP 2015). The most recent scientific findings reveal that climate change will worsen global economy by pushing many countries beyond the optimal temperature for economic activities, mostly poorer countries in the tropics but a smaller positive impact on richer countries in the North.

Burke et al. (2015) expressed that if future adaptation mimics past adaptation, unmitigated climate change is forecasted to reshape the global economy such that average global per capita incomes will decrease by 23% in 77% of countries in the world relative to the current levels by 2100. The average income in the poorest 40 countries is predicted to fall by 75%, signaling a potential to widen the global inequality gap, relative to scenarios without climate change. Figure 1 shows that if annual average temperatures cross a critical heat mark of 13 °C, wealthy economies such as the United States, Japan, and China will experience a drop-off in economic output, while emerging economies such as Indonesia, India, and Nigeria will experience much larger drop-off in economic growth. Poorer countries will suffer even steeper losses because they have already past the temperature threshold. The likelihood of a global economic contraction under future warming scenarios and these results provide the first evidence in both rich and poor countries that economic activity in all regions is connected to the global climate and establish a new empirical foundation for modelling economic loss in response to climate change (Revesz et al. 2014; Pindyck 2013).

A glimpse at the current climate consequences have seen Africa, for instance, battered by heat and drought. For instance, in the past three decades, extreme climate events such as droughts have cost Mali 9% loss in GDP, Niger has lost 18% of its GDP), Ethiopia has lost 7% of its GDP), Zambia has lost $13.8 billion GDP (0.4% loss), and Zimbabwe had lost 12% in GDP (Mulobela 2016).

Nigerian agriculture characterized by strong dependence on rain-fed agriculture and low-adaptive capacity makes it one of the most vulnerable countries to climate change among developing African economies. It is highly susceptible to increased frequency and intensity of drought risk event because almost 65% of its land covers prone to drought and desertification (NEST 2004; Anuforom 2010). The impacts of long droughts and deficit rainfall will be mostly felt by the rural population who represent over 80% of the Nigerian population who have less means to adapt and/or diversify their livelihoods (World Bank 2013). Nigeria may lose about 11% of GDP ($460 billion) to drought risk event by 2020 and about 25% by 2050 (Ozor 2009). Policy-makers' have thus considered several alternatives for enhancing farmers' adaptive capacity in the face of changing climatic parameters (Wheeler and von Braun 2013). A national strategy towards climate resilience therefore is to strengthen local resilience and adaptive capacity of vulnerable smallholders to facilitate their investments in Climate Smart Agriculture National Adaptation "Action" Plan towards a resilient agricultural sector (CCAFS 2015; IFAD 2016).

One of the strongly identified pathways to meet this Action Plan is the utilization of weather and climate information to facilitate smallholder decisions for adapting to changes in weather parameters. Access to weather and climate forecasts is of tremendous significance in risk management decision-making literature (UNFCCC 2015). Such studies provide basic socioeconomic contexts within which rural households could be reached with necessary climate information. It is a form of risk management, because it utilizes knowledge and information on the past climatic trends and events to make some reliable forecast decisions (Kiem and Verdon-Kidd 2011). There is a wide range of weather and climate observations that can considerably assist to control and reduce the impact of extreme climate events. These observations and forecasts are crucial to the socioeconomic and environmental impact in different regions of the world, especially in the Sub-Sahara Africa and must be disseminated to enable users establish policies and strategies for planning their livelihood activities (ADF 2010). Early warnings from seasonal predictions could facilitate adaptation and manage weather shocks by farming households.

In line with the Nigeria plan on "Climate Smart Agriculture" under the Nigeria's Intended Nationally Determined Contributions (INDCs) of COP 21 Paris

Agreement, climate information service is foremost as safety net to enable smallholder population take adaptation decisions. Mitigating impacts through access to reliable and timely weather forecasts and other adaptive mechanisms are foremost in Africa's policy dialogues and socioeconomic development agendas. Nevertheless, Nigeria faces considerable challenge in adopting climate resilient practices due to large underutilization of weather forecast information by smallholders in the agricultural sector. It is therefore an important step to consider empirical investigation on whether farmers' needs for agro-weather and climate information are usually taken into account when supplying weather information by providers in Nigeria.

Weather Information and Agricultural Risk Management Decisions

Over past centuries, smallholders have relied on traditional weather knowledge methods to understand when and where to carry out their agricultural activities. Unfortunately, climate change and extreme weather events have rendered this knowledge obsolete by affecting rainfall patterns, length of growing periods, temperatures, among others (IFAD 2011). This illustrates the need to identify scientific forecasting processes, methods, and tools which may assist African smallholders adapt on local scale (IPCC 2014).

Jones and Carabine (2013) observed that the need for medium- and long-term climate information to support climate-resilient development planning in Africa is clear to generate long-term strategies for loss and damage, adaptation finance, climate resilient agricultural practices, and other variants such as climate-compatible development. Access to seasonal climate forecasts can benefit farmers by allowing them to make more informed decisions about their farming practices (Gunda et al. 2017). Access to weather information is often positively correlated to changes in farming practices (Wood et al. 2014) with seasonal forecasts, in particular, having considerable potential to improve livelihoods in regions with high interannual rainfall variability (Hansen et al. 2011). Such forecasts can be used by farmers, for example, to inform their crop diversification strategies by helping them decide which crops to plant (Crane et al. 2010).

Successful climate change adaptation outcomes are supported by decision-making that is informed by the best available climate science (Meinke et al. 2010). Successful climate change adaptation therefore needs to be supported by realistic and practically useful information on climate change impact such that robust and effective management decisions can be made. This difference in approaches leads to a fundamental disconnect between scientists and end users (Kiem and Verdon-Kidd 2011; Meinke et al. 2010). This gap often prevents the effective exchange of information between climate change information providers and end users across a range of sectors (Li et al. 2012; Ekstrom et al. 2011).

Climate science and weather is capable of supporting planners in making informed decisions on future investments aimed at optimizing the use of scarce resources available. In 2009, at the Third World Climate Conference (WCC-3) in Geneva, the Conference declaration established a Global Framework for Climate Services (GFCS) to strengthen the production, availability, delivery, and application of science-based climate prediction and services. Climate services are described as the provision of climate products to assist user's decision-making and planning in climate sensitive livelihoods, hence a climate product is a process of synthesizing climate data and information (Kadi et al. 2011). Given that climate services depend critically on predictions of time-evolving regional climate from seasonal-to-interannual, multidecadal, and beyond, it is necessary that climate prediction science must be accorded an important part in organizing beneficial climate services. Effective implementation of the GFCS is expected to contribute to social, economic, and environmental benefits through more effective climate risk management and increased capacities for adaptation to climate variability and weather seasonality (African Development Forum [ADF] 2010).

Over one-third of the World Meteorological Organization (WMO) Membership already had the capability to provide some form of operational Seasonal to Inter-annual Prediction (SIP) (Kimura 2001). In Africa alone, following an initiative from the African Centre for Meteorological Applications for Development (ACMAD), most sub-Saharan Meteorological Services possess an in-house single season empirical rainfall prediction model. Despite the advanced technology behind seasonal prediction, there is a lack of evidence for – and detailed understanding of – gaps in the uptake of weather forecast information for long-term strategies of climate resilience, particularly for sub-Saharan Africa (African Climate Policy Centre [ACPC] 2011).

The United Nations Development Programme [UNDP] (2011) confirms that Africa faces considerable challenges in adapting to the long-term impacts of climate change and extreme climate events (World Bank 2010). Policy-makers have to contend with the high vulnerability due to poor weather and climate communication across many communities in developing countries. Ensuring that decision-makers and end users, especially farming households are able to respond to the medium- and long-term implications of climate change is a critical factor in promoting climate-resilient development (Ziervogel and Zermoglio 2009).

In West Africa, weather information products and services including historical data outputs have been used to inform agriculture policies, plans, and practices. Some West African countries have developed regional and national food security outlooks and famine early warning systems based on climate products and services to manage increasing climate risks (Kadi et al. 2011). While improvements in the ability to model African climate are being made, the ability to communicate agro-weather information remains poor while messages are often contradictory and unsuitable to inform farmers' adaptation. There is a poor connection between weather information services by forecast providers and information requirements at farmers' level.

Methods and channels used in communication also determine whether or not climate information will be available in time for users to make timely decisions. At present, Africa is still the only continent with the fewest information and telecommunication systems. A whole range of up-to-date media resources are widely available across major cities, yet it cannot be denied that the rural population, which dominate the largest percentage of the continent, do not appear to benefit from these information technology sources for transmitting climate and weather forecasting information. In rural communities, information is often difficult to circulate extensively due to their isolation and limited telecom services. Only a few tools and portals provide climate change information that is tailored and communicated specifically to decision-makers (Ziervogel and Zermoglio 2009).

Weather Information as Pathway Towards Climate Resilience

Weather forecasting is a term used to describe prediction of atmospheric conditions of a given location, supplemented by a variety of statistical and empirical techniques and by technology (Cahir 2013). Weather forecasts are issued to save crops and signal what to expect in our atmospheric environment. Modern-day skillful weather forecasts are based on the numerical weather prediction approach, involving computer-based modeling from a set of initial atmospheric and environmental conditions. Weather and climate information are data-rich products from many government organizations that provide fundamental ongoing information for farmers, farmer organizations, agricultural consultants, industry, and other agriculturally focused government organizations (NIMET 2013).

Seasonal weather and climate variations have a significant impact more directly on agricultural and food security for which the variations can significantly translate into diminishing agricultural outputs. Over the years, farmers have devised different measures to limit their losses to unpredictability in weather factors but the increasing scale, frequency, and magnitude of extreme weather events and climate change have undermined farming households' ability to adjust with huge impacts (Sulewski and Kloczko-Gajewska 2014; Traore et al. 2014). Weather variability has caused changes in agricultural production cycle such that by depending only on indigenous knowledge, farmers can no longer predict arrival of rainy season because of change in wind pattern, variability in start and duration of rainfall seasons, planting dates, harvesting, or even the right crops to cultivate at a particular time (Madzwamuse 2010).

Severe financial losses over the years has increased the need for farmers to depend on weather forecast information to complement and enhance their knowledge and enabling predictions beyond the timeline of traditional and indigenous knowledge. Weather forecast information has shown potential for improving resilience of agriculture to climate shocks and uncertainty (George 2013; Oyekale 2015a). The application of such information in farm management decision will be of immense benefit to farmers. Access to weather and climate information forms the fundamental basis upon which many agricultural decisions are made and very useful in diversification of livelihoods and managing risks and mitigating adverse effects of climate change (FAO 2013, 2015).

While adequate and normal rainfall pattern is desirable, excess or deficiency present serious hazard and must be managed in order to mitigate their negative impact in production activities. Prior information of expected rainfall pattern is therefore valuable for planning and decision-making (NIMET 2012). Weather information is very important before carrying out major farming operations such as land clearing and preparation, planting, fertilizer application, disease/pest control, spraying of chemicals, irrigation, among others.

Most often times, rural communities usually use and depend on traditional weather predictions in their production decisions. This can be attributed to poor access of scientific forecast, small farm size, low literacy level, sociological and cultural factors (Mugabe et al. 2010). Farmers have developed a variety of traditional indicators for predicting weather situation to decide on farm management decisions. Consequently, some traditional weather and climate indicators suffer from climate change, as they are no longer based on existing climate behavior (Sanni et al. 2012). However, commercial farmers do rely on the scientific weather forecast for their daily farm activities and critical farm decisions. Therefore, the reliability and use of weather forecast/predictions depends on a number of factors, which range from socioeconomic, scientific, environmental, cultural, and agro-meteorological factors (Oyekale 2015b; Pal and Laha 2014).

To build resilience against adverse impacts of extreme climate events and climate variability, farmers have to adjust to harsh weather conditions by relying on weather forecast services for their farm operations, which would help moderate potential damage and adverse consequences (Frisvold and Murugesan 2013; Oyekale 2015a). The accessibility and usefulness of weather information are therefore critical factors that will affect farmer's ability to adapt to climate change (Alston 2013; Bryan et al. 2012; Serna 2011). Weather forecast is therefore a considerable potential for weather risk management and enhanced farmers' resilience by adapting with climate smart adaptation practices (Goddard et al. 2010).

Detail empirical study is required to capture the interface between agro-weather information services and actual needs of farming households for decision-making in the savannah agroecological zones of Nigeria which dominates over 60% of its cultivable land surface. To enhance delivery of weather and climate information products and services that are specific to meet needs of rural households, this study will analyze weather information that will be scale-up in support of climate resilient adaptation decisions of farming households in the savannah areas of Nigeria.

Data and Methodology

Study Area, Sampling Technique, and Data Collection

Nigerian case in the African tropics was studied in the middle savannah belt region of Nigeria. The savannah Agro-Ecological Zone (AEZ) which covers an area of about 32,500 sq. km is located in Kwara State. The state contains two distinct agroecologies namely the Sudan savannah grasslands in the north and Guinea savannah in the south. Kwara State enjoys both dry and wet seasons; however, the study site characterized by declining rainfall with 30% of the total distribution from

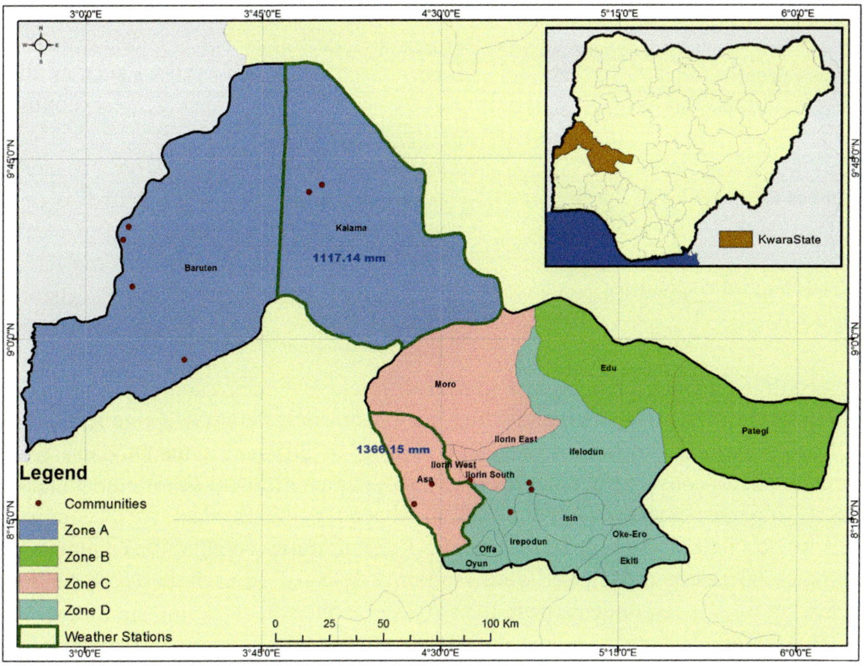

Fig. 2 Surveyed Villages in the Sudan and Guinea Savannahs of Kwara State, Nigeria

about 1150 mm annual rainfall (Oguntunde et al. 2011). The middle belt is the most extensive AEZ with the capacity to produce all major food crops grown in the country. It is the most luxuriant region of the savannah vegetation belts, as the food basin of Nigeria (Sowunmi and Akintola 2010). The sudan and guinea savannah zones are prominently known for commercial agricultural productions of cereals such as maize, sorghum, and millets as major source of livelihood but subsistent roots and tubers for households consumption. Rain-fed agriculture dominates food crop production because irrigation accounts for less than 10% of its total cultivated area. The Savannah zones are responsible for the largest quantities needed for food processing industries which included the north and south agricultural zones of Kwara State. Guinea savannah is the ecology with the highest maize yield potential in the West and Central Africa (Bamire et al. 2010; Badu-Apraku et al. 2014). In Kwara State, maize dominate food crop production by contributing over 60% of the total food production, hence strategic socioeconomic stability in dry savannah zones and for the Food Security Agenda of Nigeria (Onuk et al. 2010) (Fig. 2).

Multistaged sampling technique was used to select sample respondents. Kwara State was purposively selected because of its increasing aridity and characterized by two distinct agroecological zones (Sudan and Guinea savannahs) with very high maize production concentration in the middle belt of Nigeria. Four local government areas (LGAs) (two from each AEZ) were selected purposively to represent LGAs with or without weather station as presented in Table 1. Smallholders within each

Table 1 Distribution of study sites and respondents by the agroecological zone

Agroecology (and agricultural zone of Kwara State)	Local government area (LGA)	Number of sampled respondents
Sudan savannah (agricultural zone A)	Baruten	78
	Kaiama *(With Weather Station)*	95
Guinea savannah (agricultural zone D)	Ifelodun	87
	Asa *(With Weather Station)*	53
Total sampled respondents		313

Field data, 2017

agroecological system are assumed to have differences in their traditional knowledge and weather information demand which may result in different adaptive responses. As climate change may have different impacts in different agroecological zone, farmers in respective agroecologies may also practice different smart climate adaptation strategies (Lema and Majule 2009), shaped by biophysical, socioeconomic, and socio-cultural context of the areas. Four farming communities known for commercial maize cultivation were randomly selected in each LGA. A random sampling of 313 respondents from the two savannahs was made for this study.

Household data were collected through a farm-level survey using structured questionnaire. During the pretest of the questionnaire, households from nonsampled communities were identified and interviewed prior to the actual interview of the target sample households. Based on the limitations identified in the pretest, the questionnaire was amended and enriched for the actual interview. In addition, household level data were collected through the focus group discussions (FGDs) held with heads of male and female farmers' groups in each village comprises 10–15 individuals. These exercises were moderated using a checklist which comprises climate change parameters in the area, resultant impacts, farmers' response, and what factors influenced their adaptation decisions to provide clear understanding on their extent of dependence on weather forecast in taking decisions. Key informant interviews (KII) were held with knowledgeable persons from the community such as the farmers' cooperative' leaders and agricultural staff of the Agricultural Development Project (ADP) Field Offices in Kwara State, individuals who have access to information on weather forecasts, climate change impacts, and constraints to adapting to climate change. Personal interviews with some farmers who have specifically practiced climate-smart adaptation strategies will be conducted to provide clear information on their extent of dependence on weather forecast in taking decisions.

Data Analysis

Descriptive Statistics

Frequency distributions, means, percentages, minimum, maximum, coefficient of variations (CV) were used to describe socioeconomic characteristics of the respondents, the most valued weather forecast information, existing climate-smart adaptation practices and reliance on agro-weather forecast information in the drylands.

Weighted Average Index

The importance of weather forecast information was assessed using Weighted Average Index (WAI) analysis. The WAI was also used to rank importance of weather data forecast elements. Respondents were asked to score the weather synopses reports and data forecast elements based on a 0–2 Likert scale, that is, "high," "moderate" and "low." The WAI was estimated using the relationship:

$$\text{WAI} = \frac{F_2W_2 + F_1W_1 + F_0W_0}{F_2 + F_1 + F_0} \tag{1}$$

Given that:

$$\text{WAI} = \frac{\sum F_iW_i}{\sum F_i} \tag{2}$$

where F = frequency, W = weight of each scale, and i = weight (2 = high importance; 1 = moderate importance and 0 = low importance). The WAI computation was used to rank importance of weather information: synopses reports and forecast elements to examine the required and valued placed on agro-weather information for adaptation decisions by rural farmers.

Heckman Probit Selection Model

When a decision process by farmers to adopt an innovative strategy requires more than one step, models with two-step regressions are usually used to correct for the selection bias generated during the decision-making processes which implies the use of Heckman's sample selectivity probit model (Heckman 1976; Maddison 2006). Two probits were estimated as the access model and decision model. Heckman's two-step procedure was used to determine effect of farmers' application of weather information in taking climate-smart adaptation decision.

The model specification started with first step of analyzing determinants of access to weather information (selection model) and the second step was adoption of climate-smart strategies, conditional on the first stage of farmers' access to weather information (outcome model). The probit model for sample selection assumes that there exists an underlying relationship. The latent equation given by:

$$y_j^* = x_j\beta + u_{1j} \tag{3}$$

Such that the binary outcome is only observe given by the probit model as:

$$y_j^{probit} = \left(y_j^* > 0\right) \tag{4}$$

The dependent variable is observed only if the observation j is observed if the selection equation:

$$y_j^{select} = \left(z_j\delta + u_{2j} > 0\right) \tag{5}$$

$$\begin{aligned} &u_1 \sim N(0,1) \\ &u_2 \sim N(0,1) \\ &corr(u_1, u_2) = \rho \end{aligned}$$

where x is a k- vector of regressors, z is an m-vector of repressors; u_1 and u_2 are error terms jointly normally distributed, independently of x and z with zero expectations. When $\rho \neq 0$, standard probit techniques applied to Eq. (7) gives biased results. Hence, the Heckman probit provides consistent, asymptotically efficient estimates for all parameters in such models (StataCorp 2003).

Marginal analysis was performed to determine effect of farmers' use of weather information for climate-smart adaptation decisions. Marginal effect of a unit change in an independent variable on the probability $P(Z = 1| X = x)$ was obtained given that all other variables are held constant and mathematically expressed as:

$$\frac{\delta P(Z_i = 1|x_i)}{\delta x_i} = \frac{\delta E(Z_i = 1|x_i)}{\delta x_i} = \varphi\left(x_i^{'}\beta\right) \tag{6}$$

Results and Discussions

Socioeconomic Characteristics of the Respondents

The descriptive statistics shows that 83.1% are male respondents while the rest are female counterparts in the study. This explains the dominance of male farmers in farming than females. This could be attributed perhaps some socioeconomic factors which might have led to capital limitation and inequality differences which have limited female investments in agricultural production. During focus group discussions (FGDs), it was observed that religion may have also prevented large part of the female population to participate in the study in reverence to Muslim veneration which prevents women from public interactions.

An average farmer is 45 years old which indicates that most of the respondents are in their productive working age. The oldest is 87 years while the youngest is 17 years. It implies that most respondents are within economically active age when their ability to take risky decisions such as use of weather forecasts as basis for climate risk adaptation planning is high. An average person has completed junior secondary education (mean = 8.91 years) which is likely to positively influence their decisions to use weather forecast information as basis of early action (adaptations) against weather and climate risks in the study area.

Experience plays a key role in risk management decisions. Farmers' experience distribution indicates that an average respondent has spent average of about 24 years in maize production. The farmer with the longest farming experience has spent 76 years while the youngest has spent 2 years. These respondents are expected to have good knowledge of different adaptation options fitting to the agroecology of the area. It is of the opinion that experience will help facilitate use of weather forecast for adopting climate smart agricultural practices. Average household size consists of 11 members. The largest family consists of 62 persons. This result provides a quick check that majority of the respondents enjoy social capital relationships. This is expected to influence use of weather services through interactions compared with small family members who are constrained by joint household decisions in enhancing adaptation decisions. Size of farm cultivated with maize ranges between 0.5 and 50 hectares.

The distribution on Table 2 shows that there are 75% medium holder farmers, 16% are smallholders, while 9% are large-holder famers with farm size above 10 hectares in the sample. This study support previous studies that majority are medium-scale farmers in the savannah middle belt. Average farm size of 5.4 hectares cultivated with maize agrees with the Federal Ministry of Agriculture and Water Resources (FMA&WR) and Food and Agriculture Organization (FAO) that the savannah middle belt accounts for over 60% maize production in Nigeria. Given this result, medium-scale farmers could consider reliance on weather forecast information in protecting their investments against erratic rainfall, false start rain, late onset, and dry spells common to the study area.

Interestingly, 60.4% of the respondents take food crop production as their primary livelihood while 34.2% engage in other employment alternative for nonfarm incomes. This result encourages use of weather information in taking decisions on smart climate practices as important step to secure their household means of livelihood from losses which can be associated with increasing weather variability. Further observations reveal that 65.8% associated with a farmers' group while 34.2% do not. Farmers' group membership is expected to contribute strongly to information-sharing that will allow for early action by respondents, especially access and use of scientist weather forecast information in adopting climate smart practices decisions.

Informal loan access was dominated credit financing profile of the respondents. Only about 36% have accessed informal loan while 25% have accessed formal loans of banks. Farmers' lack of access to formal financial resources might be discouraging to farmers pursuing access and use of weather information which might, in turn, prevent early action and preparedness that could lead to adapting with smart climate practices.

Access to scientific weather information, as shown in Table 2 explains that 68% have no access to early warnings compared with 32% with access. Greater number of the population were cut-off from access to scientific weather forecast services with negative effect on adaptation decisions based on early warnings to prevent high economic losses. Serious attention is needed on the 68% with no access to weather information services. If not adequately addressed, this present situation could pose

Table 2 Socioeconomic characteristics of respondents

Farmers' characteristics	Obs (%)	Mean	Std. Dev.	Std. error of mean	Min.	Max.
Sex						
Male	83.07	0.83	0.375	0.021	0	1
Female	16.93					
Age (years)	100	45.66	14.09	0.797	16	87
Level of education (years)	100	8.91	6.09	0.341	1	16
Farmers experience (years)	100	23.98	13.44	0.760	2	76
Household size (number)	100	11.02	7.76	0.439	1	62
Farm size (maize) (ha)						
< 2.00	16	5.41	7.28	0.412	0.5	50
2.00–10.00	75					
> 10.00	9					
Primary livelihood (farming)						
Yes	60.4	0.60	0.490	0.028	0	1
No	39.6					
Farmers' group membership						
Yes	65.8	0.66	0.475	0.027	0	1
No	34.2					
Bank loan						
Accessed	24.9	0.25	0.433	0.024	0	1
No access	75.1					
Informal loan						
Accessed	36.1	0.36	0.481	0.027	0	1
No access	63.9					
Agric. extension service						
Accessed	55.3	0.55	0.498	0.028	0	1
No access	44.7					
Market linkage						
Yes	70.6	0.71	0.456	0.026	0	1
No	29.4					
Distance to weather station	100	1.16	5.102	0.288	1	60
Scientific weather information						
Accessed	31.9	0.32	0.467	0.026	0	1
No access	68.1					
Rainfall forecast						
Used	35.5	0.35	0.479	0.027	0	1
Never used	64.5					
Climate resilient practice						
Adopted	44.7	0.45	0.498	0.028	0	1
No adoption	55.3					

(continued)

Table 2 (continued)

Farmers' characteristics	Obs (%)	Mean	Std. Dev.	Std. error of mean	Min.	Max.
Farmers income (₦) (annual)						
<₦216,000 (annual minimum wage)	31.4	690,936	1,043,907	59,005	8,000	8,000,000
₦240,000 – ₦480,000	20.8					
> ₦480,000 – ₦720,000	47.8					

Source: Field data, 2017. 1$ equivalent to ₦305 during this study

serious problem of vulnerability to drought risk and pest infestation in maize production as well as low adoption of climate smart practices as mentioned during focus group discussions with men and women groups since maize crop cultivation is under serious threat of unpredictable rainfall conditions.

Average income was ₦690,936.00 per annum translated into a ₦172,734.44 seasonal income from maize crop. In terms of percentile distributions, ₦216,000.00 obtained as the 25th percentile annum income while ₦480, 000.00 represents the 75th percentile annual income. The maximum income was ₦8million per annum while ₦8000.00 was the minimum income in the savannah area.

Weather Information in Adaptation Decision Process

Rainfall Changes with Impact on Agricultural Livelihoods

Figure 3 presents the distribution of the most impactful rainfall changes in the past 5 years as experienced by the sampled farmers. Erratic rainfall is the most perceived impactful rainfall changes on agricultural production and livelihood activities by 75.4% of sampled respondents, delayed rainfall is the next impactful to farmers by 72.2%, while late onset is considered the third most impactful rainfall changes to the respondents in the past 5 years expressed in the distribution. Furthermore, early cessation, dry spell experience and short length of growing season are other rainfall changes elements which are critical to food security in the savannah belt of Nigeria.

Understanding Farmers Need for Weather Information

An important insight into the most important weather forecast elements which farmers most demanded in the savannah middle belt is shown in Table 3. Weighted importance index indicated that rainfall probabilities (weight = 1.44) was the most important weather forecast element to the respondents while relative humidity (weight = 1.12) was second most important weather elements while air temperature (weight = 1.09) was ranked third in the order of importance by respondents. The dry ecology nature of the savannah region confirms reason for which farmers highly develop interest in weather information which has to do with rainfall probability due to its dominant characteristics such as occurrence, spatial and statistical distributions

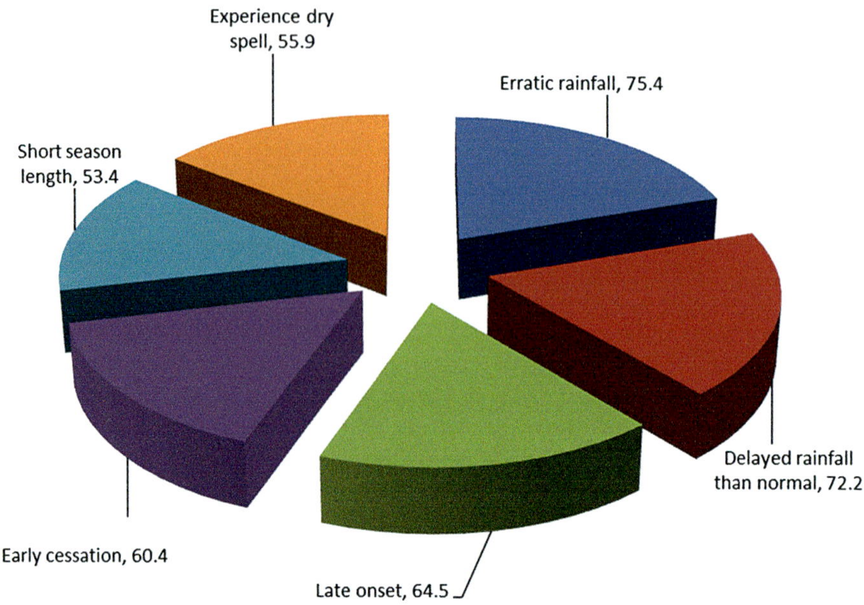

Fig. 3 Distribution of most impactful rainfall changes in the past 5 years

Table 3 Ranking distribution of importance of weather forecast elements

Weather forecast elements	Weighted importance index	Rank
Rainfall probabilities (n = 309)	1.44	1
Air temperature (n = 309)	1.09	3
Wind speed and direction (n = 310)	0.99	5
Relative humidity (n = 310)	1.123	2
Sunshine hours (n = 311)	1.009	4
Growing degree days (n = 313)	0.843	6

Source: Field data, 2017

which greatly affect food crop production in the study area. It is upon rainfall probabilities information that farmers' agricultural planning investment decisions are dependent in the arid and semi-arid agro-climatic region. Vital associated information required from the rainfall probabilities has to do with details of false start rain, onset data, and cessation date as indicated during focus group discussions, hence of serious interests to farmers in the study area.

Further results on Fig. 4 reveals that rainfall probabilities is the most highly used weather forecast element as expressed by 33.9% of the respondents, sunshine hour duration are being used by 26.5% while growing degree days as parameter for measuring length of crop developments are been used by 26.25% in taking on-farm planning decisions in the study area. However, 66.1% of the respondents never used information on rainfall probabilities while 73.5% have never used

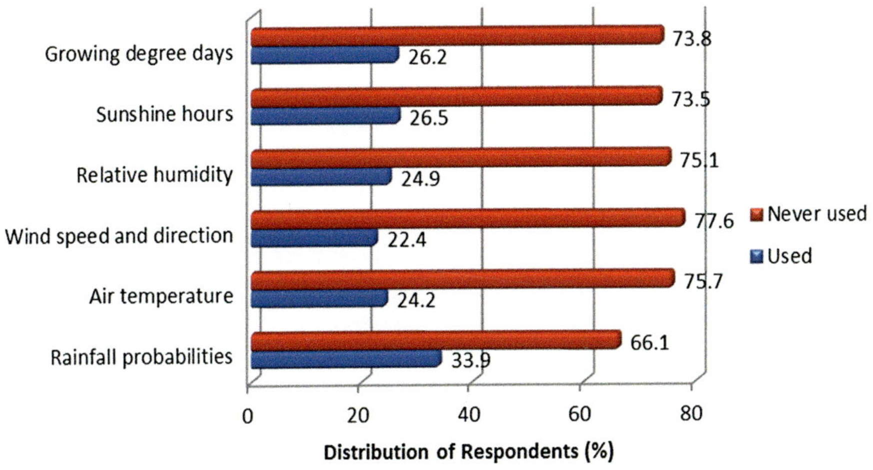

Fig. 4 Distribution of weather elements by importance

Table 4 Distribution of adopted climate adaptation strategies

	North (Sudan savannah)	South (Guinea savannah)
Basic infrastructure	Response rate (Yes = 1) (%)	Response rate (Yes = 1) (%)
Delayed planting date	47 (27.16%)	49 (35.00%)
Minimum tillage	51 (29.47%)	45 (32.14%)
Early maturing variety	57 (32.94%)	65 (46.42%)
Drought resilient variety	46 (26.58%)	62 (44.28%)

Source: Field data, 2017

sunshine hour information in planning their agricultural investments decisions. The implication of this result could be that there should be improvement in farmers' awareness of the importance of considering weather forecast information farm planning and also to enhance scaling up smart climate agricultural practices in the study area.

Farm-Level Climate Resilient Strategies

The distribution of respondents in Table 4 presents the present extent of adaptation to declining rainfall, dry spell, and drought events by practicing innovative climate smart agricultural strategies. In the sudan savannah area (north agricultural zone), 32.94% adapted with early maturing crop variety while 29.47% of the respondents adapted with minimum tillage during 2017 farming season. In the guinea savannah area (south agricultural zone), 46.42% adapted with early maturing crop variety while 44.28% adapted with drought resistant varieties in the same farming season. Further responses show the adoption of other strategies which farmers have used in the middle belt.

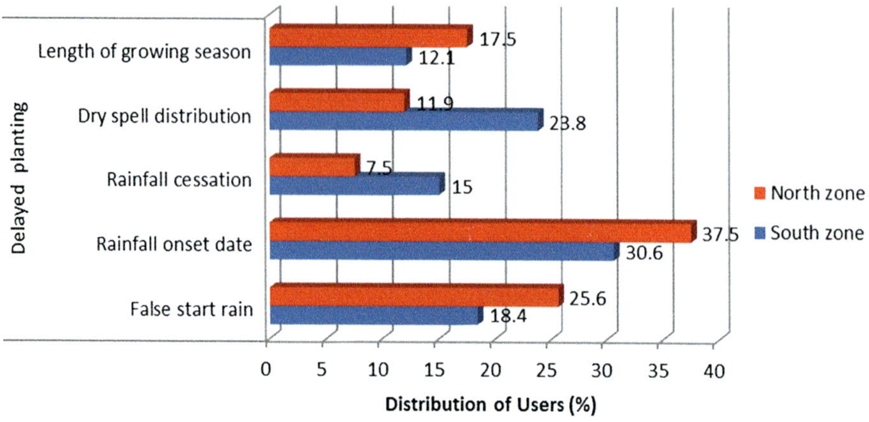

Fig. 5 Weather information required for adopting delayed planting

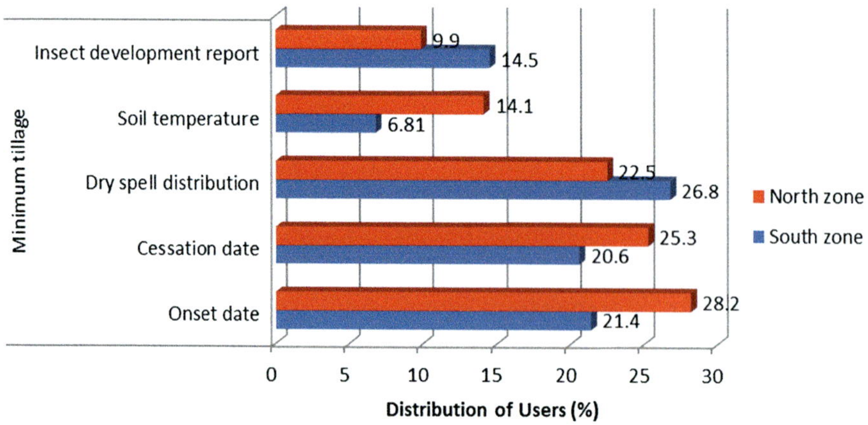

Fig. 6 Weather information required for minimum tillage practise

Weather Information Required for Adopting Climate Resilient Strategies

Rainfall onset date is the major weather information required by farmers in the sudan (north) and guinea savannah (south) zones to facilitate their decisions to adapt with delayed planting of crops as revealed in Fig. 5. Rainfall onset (sudan savannah) and dry spell distribution (guinea savannah) are the most required by smallholders to decide on adapting with minimum tillage as shown in Fig. 6. To adapt with early maturing varieties, farmers would adapt if they have access to information about rainfall onset date (sudan savannah) in the north zone and length of growing season (guinea savannah) in the south zone as expressed in Fig. 7.

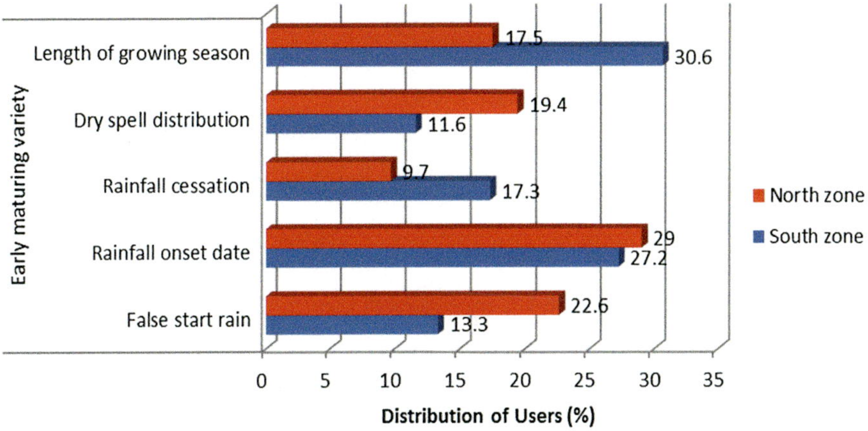

Fig. 7 Weather information required for adopting early maturing variety

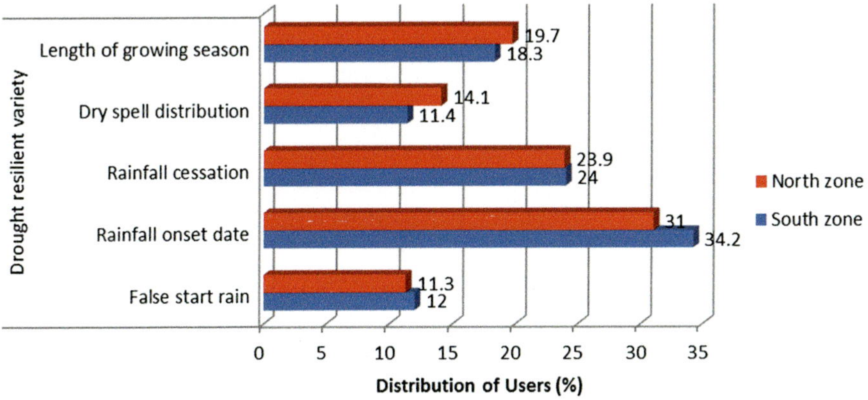

Fig. 8 Weather information required for adapting with drought resilient varieties

Further results presented in Fig. 8 show that in both sudan and guinea savannahs, rainfall onset date is needed to adapt with drought resilient variety by farmers in both agricultural zones, while dry spell distribution in both sudan and guinea savannahs are the most required by farmers to take adaptation decisions with cover cropping as presented in Fig. 9.

It is imperative that Nigerian Meterorological Agency (NIMET) should give prompt attention to these weather forecast information which would trigger early adapation decisions of smallholders. Information demanded from farmers should be given consideration in weather forecast analysis and interpretations for agricultural risk management decisions should form the basis of seasonal or interannual weather monitoring exercises.

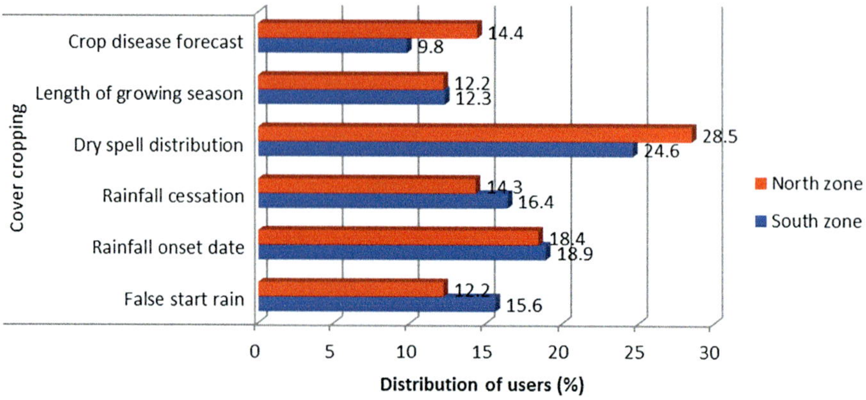

Fig. 9 Weather information required for adapting with cover cropping

Modelling Drivers of Climate Resilient Decisions

Resilient Decision-Making in Sudan Savannah: Heckman Outcome Model

The model was tested for its appropriateness by comparing the dependence of the error terms in the outcome and selection equations. Heckman model results from the sudan savannah zone (north agricultural zone) of Kwara State, Nigeria provide evidence of a sample selection problem since rho was significantly different from zero (Wald test for independent equations = 50.88, with P = 0.0002). Therefore, it was appropriate to use the Heckman probit model. The likelihood function (Lf) of the Heckman was significant (Wald $\chi^2 = -0.614$, with P = 0.0002) meaning that the model had a strong explanatory power. Table 5 presents result from the ML estimation together with the marginal effects, which is the expected change in the probability of adapting and/or not adapting to with climate smart strategies given a unit change in an independent variable from the mean value, ceteris paribus.

Gender, farming experience, annual farm income, and persistent experience of erratic rainfall significantly increases the likelihood of farmers' decision to adapt with resilient practices in the Sudan savannah on the condition that they had access to weather information as indicated in the sample model. The positive coefficients of gender, years of farming experience, annual farm income, and persistent experience of erratic rainfall significantly increase the likelihood of farmers' decision to adapt with climate resilient practices given their access to weather forecast information. Male farmers are therefore more likely to adapt with climate resilient practices than female farmers. This result is in line with the argument that male-headed households are often considered to be more likely to get information about new technologies than female-headed households. This might be associated with some sociocultural factors such as religion and cultural background which restrict women from taking

Table 5 Heckman results of adaptation decision and access to weather information

Variables	Climate smart adaptation (outcome model) Coefficient	Std. Err.	Access to weather forecast (selection model) Coefficient	Std. Err.
Sex	0.233**	0.115	−0.238	0.414
Age	−0.002	0.004	−0.010	0.015
Education	−0.012	0.017	0.192***	0.029
Farmer's experience	0.011***	0.004	0.004	0.017
Household size	0.002	0.006		
Farm size	0.003	0.004		
Farmer's group membership	−0.160**	0.081	0.137	0.322
Bank credit	0.179*	0.112	−0.029	0.319
Informal loans	−0.024	0.081	−0.938	0.307
Agric. Extension service	−0.031	0.098	0.331***	0.377
Market linkage	0.004	0.095	−0.021	0.020
Farm income	7.46e-08**	3.88e-08	2.39e-07	2.33e-07
Erratic rainfall	0.217**	0.093	0.256	0.360
Number of dry spells			0.022	0.031
Weather station distance	−0.009**	0.004		
Weather advisory services	0.082	0.187	−0.323	0.598
Weather local interpretation	0.160	0.182	0.581	0.535
Repeated operational cost			−1.63e-07	8.85e-07
Number of times listening to weather	−0.001	0.001	0.001	0.001
Radio-battery expenses	−3.42e-07	1.52e-06	−6.66e-06*	5.06e-06
Radio			−0.012	0.373
Simple mobile handset	−0.034	0.088	0.568*	0.353
Smartphone	−0.060	0.086	0.240	0.428
Constant	0.466	0.350	−0.964	0.806
Total observations	173			
Censored obs	53		Uncensored obs	120
Rho	−0.614		Wald chi2 (20)	50.88
Prob > chi2	0.000			
Sigma	0.375			

*Statistical significant at 0.1
**Statistical significant at 0.05
***Statistical significant at 0.01
Source: Field data, 2017

household decisions. Men are responsible for most of all decision-making processes in the savannah belt.

The longer farmers spent years in farming, the more likely that they would adapt with climate resilient strategy. This result is well-connected to previous experiences which farmers might have had from different indigenous strategies. The current speed of climate change arising from erratic rainfall and dry spells has modified known variability patterns such that farmers have been confronted with economic losses which they are ill-equipped to handle, despite their farming experience. This explains the need for anticipatory and planned adaptation as preparedness to prevent losses at a local scale.

Improvements in farmers' income by an infinitesimal change would lead to more likelihood of taking resilient decision actions. This is likely attributed to marginal productivity of majority of the farmers since they are medium-scale farmers. Farmers with high income are more likely to give trial to different social and climate protections as safety nets to prevent or reduce future losses. Increasing occurrences of erratic rainfall observed by farmers increases farmers' likelihood to take up resilient practices in the dry savannah zone. Increasing magnitude of unpredictability in climate variability increases farmers' propensity to respond by taking early action towards protection and resilience.

The inverse coefficient of membership of farmers' group explains that as farmers become member to group associations, the less likely they adopt climate resilient technologies despite access to weather information. Farmer groups are known for their strong attachment hence they trust decision-making processes existing within their social networks. They are more likely to adapt with indigenous safety nets during weather shock conditions which explain reason for significant decrease in the probability of adopting climate smart practices.

Distance to weather station prevents climate resilient adoption by farmers. This shows that the farther farmers are away from weather station location, the less likely they will adopt climate smart strategy. Farmers trusted weather information based on the closest meteorological station to them hence, they are more likely to make use of weather forecasts that are specific to their geographies and matches with their location weather or climatic conditions, rather than general forecast being communicated. In locations where weather stations are cited, average distance is too far from the farmers which explain reason why it would significantly reduce probability of taking climate smart decisions.

Access to Weather Information in Sudan Savannah: Selection Model

The selection model on Table 5 indicated that education, agricultural extension services, and ownership of simple mobile handset showed positive significant effects which increase the likelihood of farmers' access to weather information. However, expenses on radio battery showed negative significant effect which decreases the likelihood of farmers' access to weather information.

Education is an important factor which determines access to weather information. Farmers' education has a positive significant effect on farmers' access to weather information. As farmers become more educated, it is more likely that they will have more access to weather forecast information. The competency to interpret weather predictions and adapt their agricultural production management decisions are more enhanced based on better understanding of prevailing and future weather conditions which guides farmers in their decision-making processes. Likewise, the more farmers have access to agricultural extension services, the more likelihood that they will have access to weather information. Periodic extension education and training services serve as opportunities to share current information on weather fluctuations with farmers and advisory services on agro-weather related management technologies and techniques. Therefore, functional extension service system has great potential in facilitating early action on disaster risk management through adaptation with climate smart practices.

Ownership of simple mobile handset device also has a positive significant effect on farmers' access to weather information. Increase in ownership of simple mobile telephones increases the likelihood that farmers will have access to weather information. Weather services such as rainfall forecasts, advisory planting calendars, and warning alerts are communicated via text messages; wider spatial scale are covered within short time and farmers can read text messages over and over again to enhance their decision-making. These are important drivers that will ensure rural access to weather services to facilitate adopting climate smart adaptation technologies in the dry savannah agroecological zones of Nigeria.

Predicted Impact of Climate Resilient Practices: Sudan Savannah

The marginal impact analysis of the various factors presented in Table 6 revealed marked differences in the ability to adapt with climate smart practices by respondents. Male farmers have more probability of adapting with climate smart strategy given that a unit change from being a female to male farmer increases the probability of adapting with climate smart technologies by 23.3%.

The marginal effect of an additional unit increase in farmers' experience shows that there is 11.73% increase in probability that farmers would adapt with climate smart technologies on average, ceteris paribus. Good and bad previous experiences and increasing vulnerability to weather variability will influence continuous practice. Likewise, the marginal effect of farm income suggests that by an infinitesimal increase in income, there is more likelihood that farmers would adapt with climate smart strategy on the average, ceteris paribus. It is an indication that farmers would have capacity to sustain the cost of adaptation finance. Effect of erratic rainfall show that if persistent erratic rainfall increases by a unit, the probability that farmers will adapt with climate smart adaptations increase by 21.7%.

The negative marginal effect of membership of a farmers group shows that as group social interaction increases by a unit, the probability that farmers would adapt with climate smart practices will decline by 16.0% on average, ceteris paribus. The provision of safety nets and local interest-free credit arrangement might have been responsible given that the purpose of their establishment could be remodified to enable farmers benefit from modern scientific climate smart practices. Likewise, on the average, marginal effect of an increasing unit distance of weather station will

Table 6 Marginal impact of adapting with climate smart strategy in sudan savannah

Variable	$\delta y/\delta x^a$	Std. Err.	Z value	P > |z|
Sex	0.233**	0.115	2.02	0.043
Age	−0.002	0.004	−0.53	0.579
Education	−0.012	0.017	−0.73	0.463
Farmer's experience	0.117**	0.004	2.39	0.017
Household size	0.002	0.006	0.45	0.655
Farm size	0.003	0.004	0.83	0.408
Farmer's group membership	−0.160**	0.081	−1.98	0.048
Bank credit	0.179*	0.112	1.59	0.111
Informal loans	−0.024	0.081	−0.30	0.764
Agric. Extension service	−0.031	0.098	−0.32	0.750
Market linkage	0.004	0.095	0.04	0.964
Farm income	7.46e-08**	0.000	1.92	0.055
Erratic rainfall	0.217**	0.093	2.32	0.020
Weather station distance	−0.009**	0.004	−2.14	0.033
Weather advisory services	0.082	0.187	0.44	0.661
Weather local interpretation	0.160	0.182	0.88	0.378
Number of times listening to weather	−0.001	0.000	−0.54	0.589
Radio-battery expenses	−3.42e-07	0.000	−0.23	0.822
Simple mobile handset	−0.034	0.088	−0.39	0.694
Smartphone	−0.061	0.086	−0.70	0.481
Number of dry spells	0	0	0	0
Repeated operational cost	0	0	0	0
Radio	0	0	0	0

[a]y = linear prediction (predict) = 0.8680
*Statistical significant at 0.1
**Statistical significant at 0.05
Source: Field data, 2017

result in 0.9% decrease in the probability that farmers would adapt with climate smart technologies.

Resilient Decision-Making in Guinea Savannah: Heckman Outcome Model

Heckman probit outcome model in Table 7 present results in the guinea savannah zone (south agricultural zone) of Kwara State, Nigeria. Persistent erratic rainfall and distance to weather station significantly increase the likelihood of farmers' decision to adapt with climate resilient practices. In the other way round, market linkage and number of times listening to weather forecast on radio are significant negative factors affecting farmers' decision to take up climate smart strategies. Institutional factors namely access to bank credit, access to informal credit, and extension service are positively correlated with climate smart absorption although they do not significantly influence farmers' decision-making processes in the guinea savannah area.

Table 7 Result of adaptation decision and access to weather information in guinea savannah

	Smart climate adaptation (outcome model)		Access to weather forecast (selection model)	
Variables	Coefficient	Std. Err.	Coefficient	Std. Err.
Sex	−0.409	0.293	1.084**	0.547
Age	0.001	0.010	−0.025	0.017
Education	−0.005	0.014	0.111	0.033
Farmer's experience	0.011	0.010	0.039**	0.016
Household size	0.024	0.024	−0.016	0.033
Farm size	0.016	0.050		
Farmer's group membership	0.214	0.235	−0.027	0.381
Bank credit	0.228	0.228		
Informal loans	0.123	0.161	0.577	0.357
Agric. extension service	0.194	0.242	0.331	0.365
Market linkage	−0.403*	0.229	0.766*	0.422
Farm income	−3.11e-07	2.95e-07	3.02e-08	5.11e-07
Erratic rainfall	0.359*	0.220	−0.037	0.504
Number of dry spells			0.092*	0.051
Weather station distance	0.116*	0.067		
Weather advisory services	−0.114	0.247	1.731***	0.407
Weather local interpretation	−0.390*	0.289	1.429***	0.393
Repeated operational cost			−3.50e-06	3.33e-07
Number of times listening to weather	−0.001	0.001	0.001	0.001
Radio-battery expenses	4.05e-06	3.43e-06	8.26e-07	6.84e-06
Radio			0.536	0.497
Simple mobile handset	−0.168	0.164	−0.542	0.468
Smartphone	−0.145	0.169	−0.117	0.392
Constant	1.506	0.815	−3.219	1.093
Total observations	139			
Censored obs	85		Uncensored obs	54
Rho	−1.000		Wald chi2 (20)	25.70
Prob > chi2	0.175			
Sigma	0.526			

*Statistical significant at 0.1
**Statistical significant at 0.05
***Statistical significant at 0.01
Source: Field data, 2017

Increasing frequency of erratic rainfall increases farmers' likelihood to take up climate smart strategy in maize production. As the magnitude of rainfall variability increases, so also will the propensity of farmers to respond to climate smart strategies increases as resilient action taken strategy given their access to weather information. The positive coefficient of distance to weather station reveals that the more farmers are far away from weather station, the more likely they will adapt with climate smart strategy given their access to weather information. In view of the increasing variability in weather elements, farmers are interested in future understanding of weather patterns such as rainfall onset, dry spell distribution and length of growing season and their overall implications on their agricultural livelihoods. So, to ensure that they take early cautions that prevent income losses given that they have access to weather information perhaps through agricultural extension services, mobile handset, and radio services, they will be more likely to adopt climate smart strategy as form of early action and drought risk preparedness against weather shocks conditions.

Access to Weather Information in Guinea Savannah: Selection Model

The selection model on Table 7 explained that sex, length of farmers experience, access to extension service, market linkage, dry spells experience, advisory weather information, and local interpretation of weather information showed positive significant effects that increase the likelihood of farmers' access to weather information. However, age showed negative significant effect which decreases the likelihood of farmers' access to weather information.

Male farmers are more likely to have access to weather information than females. Usually, decision-making processes within household units are traditionally considered as the obligations of male gender. Monitoring of weather information is commonly associated with men than women, situation which perhaps is strongly influenced by sociocultural influences which empowers men but restricts women in household decision-making participation. As farmers' length of farming experiences keep increasing, so also the likelihood of farmers' access to weather information. Previous experiences in farm management decisions might also responsible, hence reliance on previous knowledge on traditional weather risk monitoring might no longer be practical in view of the frequent cases of erratic rainfall, late onset, and dry spells which are affecting food crops yield and households farm incomes. Current weather information and advisory services are therefore important uptake of climate resilient agricultural practices in the guinea savannah.

Agricultural extension service shows a positive significant effect on farmers' access to weather information. The more farmers have contact with agricultural extension services, the more likelihood that they will have access to weather information. Effective extension service delivery system has great potential in facilitating early action on disaster risk management through agricultural adaptation with climate smart strategies. The more farmers enjoy market linkage and interactions, the more the likelihood that they will have access to weather information. This perhaps can be attributed to communication processes among agricultural stakeholders such as fellow farmers, input suppliers, output marketers, processors, and others who will provide certain information that will facilitate farmers' use of weather information.

Increasing frequency of dry spells experienced by farmers will increase farmers' likelihood of access to weather information. Agricultural investment is the dominant means of livelihood in the study area and hence increasing frequencies of dry spells will increase farmers' readiness to access weather information services. Advisory weather information likewise has a positive significant effect on farmers' access to weather information. When weather information are robustly provided with advisory services on when to start land preparations, time to plant, time to apply soil nutrients and pesticides, time to harvest, among others, farmers will be more inclined to access weather information services and not just limited to statistics, patterns or distributions but including advisory services.

Similarly, local interpretation of weather information to farmers who are end users significantly influence farmers' access to weather information. Interpretation of weather forecast in addressing agricultural risk management approaches will facilitate farmers' continuous application of weather information. It is important that farmers will value forecasts with rich meanings on local agricultural impacts and follow-up actions that are specific to geographical location and/or farm sites, thus they will be more inclined to access weather information services.

Predicted Impact of Climate Resilient Practices: Guinea Savannah

The positive marginal effect of erratic rainfall perceived by farmers in Table 8 expressed that if erratic rainfall increases by a unit, the probability that farmers will adapt with climate smart adaptations increases by 35.9%, ceteris paribus. The agro-ecological zone is highly vulnerable to fluctuations in weather parameters that could trigger loss and damage in agricultural investment if appropriate early warnings and alerts are not followed through preparedness and early action. This rationale is responsible for the increase in the probability that farmers will adapt with climate smart adaptations.

An increasing unit in distance of weather station suggests that there will be 11.6% increase in the probability that farmers would adapt with climate smart technologies on average, consequent that they have access to weather information ceteris paribus. This result is very important for facilitating climate resilient food crop production and enhancing early warning alerts against weather shocks ahead of time prior to their occurrences. Nigeria lags behind other developed countries in terms of the required number of weather stations recommended by the World Meteorological Organization (WMO). The situation necessitate calls for adaptation finance as a way out to meet up national plans on real time weather tracking infrastructure to enhance risk preparedness and early adaptation actions at farm level.

Nevertheless, the negative marginal effect of farmers' market linkage shows that as network of community interactions increases by one unit, the probability that farmers would adapt with climate smart practices will decline by 40.30% on average, ceteris paribus. This signals serious attention to policy-makers in quickly addressing the operations of all actors in the market linkage chain for a positive behavioral change towards climate smart strategies in terms of its cost, input requirement, and the associated benefits. Likewise, a unit increase in local interpretation of weather information reveals that there is 39% decrease in probability that farmers would adapt with climate smart technologies. Incompetency in understanding information

Table 8 Marginal impact of adapting with climate smart strategy in guinea savannah

Variable	δy/δx[a]	Std. Err.	Z value	P > |z|
Sex	−0.409	0.293	−1.40	0.136
Age	0.001	0.010	0.04	0.968
Education	−0.005	0.014	−0.40	0.686
Farmer's experience	0.011	0.010	−1.08	0.280
Household size	0.024	0.024	0.98	0.325
Farm size	0.016	0.050	0.32	0.751
Farmer's group membership	0.214	0.235	0.91	0.363
Bank credit	0.228	0.228	1.00	0.318
Informal loans	0.123	0.161	0.76	0.446
Agric. Extension service	0.194	0.242	0.80	0.422
Market linkage	−0.403*	0.229	−1.76	0.079
Farm income	−3.11e-07	2.95e-07	−1.06	0.291
Erratic rainfall	0.359*	0.220	1.63	0.103
Weather station distance	0.116*	0.067	1.72	0.085
Weather advisory services	−0.114	0.247	−0.46	0.643
Weather local interpretation	−0.390*	0.289	−1.35	0.178
Number of times listening to weather	−0.001	0.001	−1.79	0.073
Radio-battery expenses	4.05e-06	3.43e-06	1.18	0.238
Simple mobile handset	−0.168	0.164	−1.02	0.306
Smartphone	−0.145	0.169	−0.86	0.389
Number of dry spells	0	0	0	0
Repeated operational cost	0	0	0	0
Radio	0	0	0	0

[a]y = Linear prediction (predict) = 1.102
*Statistical significant at 0.1
Source: Field data, 2017

on weather variables would lead to inaccurate and wrongly interpreted weather forecast information. This in turn may result in serious crop and income losses to farmers. This emphasizes that location specific weather information and advisory services, not just generalized weather statistical summaries, will assist in decision-making processes, hence will increase likelihood to adopt with climate resilient practices in the guinea savannah area.

Conclusions and Policy Issues

This study identifies rainfall probabilities as the most highly useful weather forecast element by farmer users in the savannah areas. Onset date is the most required weather information needed in adapting with delayed planting and drought resilient varieties, dry spell distribution are needed to facilitate adapting with minimum tillage and cover cropping practices while length of growing season is needed to take decision on early maturing varieties.

There is more male access to weather information than female. Given that decision-making processes is considered the responsibility of men. Farmers' education is critical to access to access of weather predictions for agricultural risk adaptation decisions. Extension education and training services also serve as opportunities to share robust information with farmers and to provide advisory services on agro-weather related management improvements. Market linkage and interactions were found to increase access to weather information. Local interpretations of weather information to farmers as principal end users in the agricultural sector significantly influence their access to weather information. Proximity of weather stations to farmers is also a determinant. Farmers are more inclined to use weather forecasts which are specific to their geographies, matches with their location conditions rather than general overall forecast which are not useful for taking specific decision. Institutional factors including access to bank credit and informal credit are positively correlated in facilitating with climate resilient decision in the savannah dryland regions. In the other way round, market linkage and number of times listening to weather forecast on radio are significant negative factors affecting farmers' decision to take up climate smart strategies.

The outcomes from this study have led to some recommendations for future upscaling of weather information as decision support for climate smart adaptations in the Nigeria Savannahs.

The pathway to strengthen farmers' decision support for risk preparedness and early action is to facilitate their involvement in climate resilient agricultural practices through future risk preparedness and enhanced capacity to take early adaptation decisions. There is need to develop initiative that will harness male farmers as effective channel of communicating agro-weather information in dry savannah areas. This is a first necessary step in view of sociocultural influences which empowers men but restricts women in household decision-making processes. Improvements in farm level education program and services will assist low and uneducated farmers in understanding agricultural impacts of weather forecast results when they are utilized or not. There should be a professional partnership and good coordination between Nigerian Meteorological Agency (NIMET) and the Agricultural Extension Department of Federal Ministry of Agriculture who take responsibility for farmers' education and training services. This partnership will give opportunities to provide weather services that match exactly farmers' needs in taking climate resilient investment decisions and enhance robust agro-weather communication with farmers to prevent weather shocks in the dry savannah areas. Market linkages should be strengthened as communication feedback mechanisms in reaching farming households with up-to-date weather services. This may include but not limited to fellow farmers, input suppliers, output marketers, processors, and others in the input-output supply chain.

In ensuring early cautions, mobile handset and radio services should be used as major channel of weather information communications with farmers. This will ensure a wider coverage and quickly facilitate adoption of climate smart strategies when such information is considered useful when they are found specific to their locations and impacts interpreted locally, not just generalized weather statistical summaries.

It becomes imperative that the Nigerian Meterorological Agency (NIMET) should give prompt attention to these weather forecast information which would trigger early adapation decisions of smallholders. Farmers information demands should be considered first in developing weather forecast analysis and interpretations for agricultural risk management decisions should form the basis of seasonal or interannual weather monitoring exercises. Location-specific weather information and advisory services will also enhance decision-making.

Nigeria lack sufficient weather tracking infrastructure in terms of the required number of installed automatic weather stations (AWS) recommended by the World Meteorological Organization (WMO). This situation calls for adaptation finance in investing into more installations of more automatic weather stations to meet up national plans of promoting the use of weather information by millions of end users in Nigerian agricultural sector. The competency of weather service providers should receive urgent attention. Incompetency in understanding and interpreting weather variables would lead to inaccurate decisions that could cost farmers so much loss and damages. A good coordination between the meteorologists and agricultural institutions such as universities and research institutes could help bridge knowledge gaps in communicating weather product services and agricultural impacts interpretations. These socioeconomic drivers will enhance farming households' adaptation with climate resilient practices in the savannah areas of Nigeria.

Acknowledgments I acknowledge the generous funding for this study by the International Development Research Centre (IDRC) under the Adaptation Finance Fellowship Programme, the Thailand Development Research Institute, Bangkok, and the Frankfurt School of Finance and Management, Frankfurt, which manage the partnership program and my participation.

References

African Climate Policy Center (2011) Assessment of Africa's climatic records and recording networks including strategic for rescuing of climatic data. Working paper 3 of the African Climate Policy Center (ACPC) of United Nations Economic Commission for Africa (UNECA) Under the Climate for Development in Africa (Climdev Africa) Programme

African Development Forum (2010) Acting on climate change for sustainable development in Africa. Climate risk management: monitoring, assessment, early warning and response. Issues paper 4. Seventh African Development Forum, 10–15 October 2010, Addis Ababa

Alston M (2013) Gender mainstreaming and climate change. Women's Studies International Forum, pp 1–8

Anuforom AC (2010) Demonstration and assessment of climate change in Nigeria and development of adaptation strategies in the key Socio-economic sectors: meteorological approach. A paper presented to the National Stakeholders' workshop on "developing national adaptation strategy and plan of action for Nigeria". NiMet, Abuja

Badu-Apraku B, Asuboah RA, Fakorede B, Asafo-Adjei B (2014) Strategies for sustainable maize seed production in West and Central Africa. IITA, Nigeria, p 140

Bamire AS, Abdoulaye T, Sanogo D, Langyintuo A (2010) Characterization of maize producing households in the dry savanna of Nigeria. Country Report – Nigeria, IITA, Ibadan, 46 pp

BNRCC (2011) Reports on research projects on impacts and adaptation– climate change in Nigeria: a compendium of reports commissioned by building Nigeria's Response to Climate Change

(BNRCC) Project implemented with the financial support of the Canadian International Development Agency (CIDA) and executed by CUSO-VSO and ICF Marbek, both of Ottawa, Ontario, Canada
Bryan E, et al. (2012) Adapting agriculture to climate change in Kenya: household strategies and determinants. J Environ Manage 114
Burke M, Hsiang SM, Miguel E (2015) Global non-linear effect of temperature on economic production. Nature Research Letter. https://doi.org/10.1038/nature15725
Cahir JJ (2013) Weather forcasting encyclopedia Britannica. http://www.britannica.com/EBchecked/topic/638321/weather-forecasting
Climate Change, Agriculture and Food Security (2015) Implementing climate-smart agriculture for enhanced food security and resilience. Progress report led by the International Center for Tropical Agriculture (CIAT) and other Research Centres
Crane TA, Roncoli C, Paz J, Breuer N, Broad K, Ingram KT, Hoogenboom G (2010) Forecast skill and farmers' skills: seasonal climate forecasts and agricultural risk management in the southeastern United States. Weather Clim Soc 2(1):44–59. https://doi.org/10.1175/2009WCAS1006.1
Ekstrom JA, Moser SC, Torn M (2011) Barriers to climate change adaptation: a diagnostic framework. California Energy Commission. www.energy.ca.gov/2011publications/CEC-500-2011-004/CEC-500-2011-004.pdf
FAO (2013) Climate smart agriculture source book. United Nations. Food and Agriculture Organization, Rome
FAO (2015) Barriers, incentives and benefits in the adoption of climate smart agriculture: enhancing agricultural mitigation within East Africa Dairy Development project (EADD) in Kenya. FAO, Rome
Frisvold GB, Murugesan AM (2013) Use of weather information for agricultural decision making. Am Meteorol Soc. https://doi.org/10.1175/WCAS-D-12-00022.1
George B (2013) Use of weather information for agricultural decision making. J Weather Clim Soc 5
Goddard L, Aitchellouche Y, Baethgen W, Dettinger M, Graham R, Hayman P, Meinke H (2010) Providing seasonal to interannual climate information for risk management and decision-making. Proced Environ Sci 1:81–101. https://doi.org/10.1016/j.proenv.2010.09.007
Gunda T, Bazuin JT, Nay J, Yeung KL (2017) Impact of seasonal forecast use on agricultural income in a system with varying crop costs and returns: an empirically-grounded simulation. Environ Res Lett (12):3
Hansen JW, Mason SJ, Sun L, Tall A (2011) Review of seasonal climate forecasting for agriculture in Sub-Saharan Africa. Exp Agric 47:205–240. https://doi.org/10.1017/S0014479710000876
Heckman JJ (1976) The common structure of statistical models of truncation, sample selection and limited dependent variables and a simple estimator for such models. Ann Econ Soc Meas 5:475–492
IFAD (2011) Rural Poverty Report 2011. International Fund for Agricultural Development, Rome
IFAD (2016) IFAD strategic framework 2016–2025: enabling inclusive and sustainable rural transformation, Rome, Italy
IPCC (2014) Climate change 2014: impacts, adaptation, and vulnerability. Contribution of Working Group II to the Fifth Assessment Report of the Intergovernmental Panel on Climate Change
IPCC (2015) Climate change 2014 synthesis report. Fifth Assessment Report
Jones L, Carabine E (2013) Exploring political and socio-economic drivers of transformational climate policy: early insights from the design of Ethiopia's Climate Resilient Green Economy strategy. Research reports and studies. Overseas Development Institute, London
Kadi M, Njau LN, Mwikya J, Kamga A (2011) The state of climate information services for agriculture and food security in East African countries. CCAFS Working paper no 5. Copenhagen
Kiem AS, Verdon-Kidd DC (2011) Steps towards 'useful' hydroclimatic scenarios for water resource management in the Murray-Darling Basin. Water Resour Res 47:W00G06. https://doi.org/10.1029/2010WR009803
Kimura Y (2001) A survey on the present status of climate forecasting. Available from the WMO Secretariat on CD-ROM: World Meteorological Organisation Commission for Climatology, Thirteenth Session, Geneva

Lema AM, Majule AE (2009) Impacts of climate change, variability and adaptation strategies on agriculture in semi-arid areas of Tanzania: the case of Manyoni District in Singida Region, Tanzania. Afr J Environ Sci Technol 3(8):206–218

Li ZH, Yang J, Shi CE (2012) Urbanization effects on fog in China: field research and modeling. Pure Appl Geophys 169(5–6):927–939

Maddison, D. (2006) The perception of and adaptation to climate change in Africa. CEEPA. Discussion paper no 10. Centre for Environmental Economics and Policy in Africa. University of Pretoria, Pretoria

Madzwamuse M (2010) Climate governance in Africa: adaptation strategies and institutions. Heinrich Böll Stiftung (HBS)

Meinke H, Howden M, Struik PC, Nelson R, Rodriguez D, Chapman SC (2010) Adaptation science for agriculture and natural resource management: urgency and theoretical basis. Curr Opin Environ Sustain 1:69–76

Mugabe FT, Mubaya CP, Nanja DH, Munodawafa A, Gondwe P, Mutswangwa EA, Changonda C, Masere P, Makuvaro V, Dimes J, Murewi C (2010) Using indigenous knowledge for climate adaptation in Southern Zambia and South-Western Zimbabwe. Zimb J Sustain Dev Afr 11 (2):237–256

Mulobela, W. (2016) Early warning can cut climate disaster losses in Africa. Reports from the Thomson Reuters Foundation in Benin Republic, South Colonnade. http://news.trust.org/

Nigerian Environmental Study/Action Team (2004) Climate change in Nigeria: a communication guide for reporters and educators. NEST, Ibadan, pp 5–16

Nigerian Meteorological Agency (2012) Seasonal rainfall prediction: socio-economic implications for Nigeria. The 2013 annual seasonal rainfall prediction, Nigerian Meteorological Agency, Abuja

NiMet (2013) NIMET (Establishment) Act, 2003. http://nimet.gov.ng/nimet-establishment-act-2003

Oguntunde PG, Abiodun BJ, Lischeid G (2011) Rainfall trends in Nigeria, 1901–2000. J Hydrol 411(3):207–218

Onuk EG, Ogara IM, Yahaya H, Nannim N (2010) Economic analysis of maize production in Mangu local government area of Plateau State, Nigeria. PAT 6(1):1–11. ISSN: 0794-5213. Online copy: www.patnsukjournal.net/currentissue

Oyekale AS (2015a) Access to risk mitigating weather forecasts and changes in farming operations in East and West Africa: evidence from a baseline survey. Sustainability 7:14599–14617. https://doi.org/10.3390/su71114599

Oyekale AS (2015b) Factors explaining farm households' access to and utilization of extreme climate forecasts in Sub-Saharan Africa (SSA). Environ Econ 6:91–103

Ozor N (2009) Implications of climate change for National development: the way forward. Debating policy options for national development. Enugu Forum Policy

Pal D, Laha AK (2014) Credit off-take from formal financial institutions in rural India: quantile regression results. Agric Food Econ (2):9. http://www.agrifoodecon.com/content/2/1/9

Pindyck RS (2013) Climate change policy: what do the models tell us? J Econ Lit 51:860–872

Revesz RL, Howard PH, Arrow K, Goulder LH, Kopp RE, Livermore MA, Oppenheimer M, Sterner T (2014) Global warming: improve economic models of climate change. Nature 508 (7495):8–10

Sanni SA et al (2012) Traditional capacity for weather prediction, variability and coping strategies in the front line states of Nigeria

Serna JM (2011) Drought assessment Northern Eastern Kenya (Wajir East, South and Mandera). Northern Eastern Kenya Drought Assessment Report

Sowunmi FA, Akintola JO (2010) Effect of climatic variability on maize production in Nigeria. Res J Environ Earth Sci 2(1):19–30. ISSN 2041-0492

StataCorp (2003) Stata base reference manual, G – M, release 8, vol 4. Stata Corporation, College Station

Sulewski P, Kloczko-Gajewska A (2014) Farmers' risk perception, risk aversion and strategies to cope with production risk: an empirical study from Poland. Stud Agric Econ 116 (2014):140–114. https://doi.org/10.7896/j.1414

Swiss Re (2010) Innovative insurance solutions are key to helping local communities finance the rising costs of climate risks. Publication of the Swiss Reinsurance Company, New York

Traore B, Van-Wijk MT, Descheemaeker K, Corbeels M, Rufino MC, Giller KE (2014) Evaluation of climate adaptation options for Sudano-Sahelian cropping systems. Field Crop Res 156:63–75. https://doi.org/10.1016/j.fcr.2013.10.014

UNDP (2011) Human development report 2011. Sustainability and equity: a better future for all. Palgrave Macmillan Publishers, New York

UNFCCC (2002) Report of the conference of the parties on its seventh session, held at Marrakesh from 29 October to 10 November 2001. Addendum 4 part two: action taken by the conference of the parties. FCCC/CP/2001/13/Add. 4. UNFCCC Secretariat

UNFCCC (2007) Climate change: impacts, vulnerabilities and adaptation in developing countries, report issued by Climate Change Secretariat (UNFCCC). Bonn, Germany

UNFCCC (2015) Adoption of the Paris agreement. Proposal by the President, Conference of the Parties Twenty-first session. Draft decision -/CP.21

United Nations Development Programme [UNDP] (2015) Climate change adaptation. Programme on climate information for resilient development in Africa (CIRDA)

Verdon-Kidd DC, Kiem AS, Austin EK (2012) Decision making under uncertainty – bridging the gap between end user needs and climate science capability. National climate change adaptation research facility, Gold Coast, p 116

Wheeler T, von Braun J (2013) Climate change impacts on global food security. Science 341 (6145):508–513. https://doi.org/10.1126/science.1239402

Wood SA, Jina AS, Jain M, Kristjanson P, DeFries RS (2014) Smallholder farmer cropping decisions related to climate variability across multiple regions. Glob Environ Chang 25:163–172. https://doi.org/10.1016/j.gloenvcha.2013.12.011

World Bank (2010) Africa infrastructure: a time for transformation. World Bank, Washington, DC

World Bank (2013) Weather and climate resilience: effective preparedness through national meteorological and hydrological services (NMHSs) in the World Bank Development report of Rogers DP, Tsirkunov VV (2013) Directions in development – environment and sustainable development. World Bank, Washington, DC

Ziervogel G, Opere A (eds) (2010) Integrating meteorological and indigenous knowledge-based seasonal climate forecasts for the agricultural sector: lessons from participatory action research in sub-Saharan Africa. International Development Research Centre, Ottawa

Ziervogel G, Zermoglio F (2009) Climate change scenarios and the development of adaptation strategies in Africa: challenges and opportunities. Clim Res 40(2–3):133–146. https://doi.org/10.3354/cr00804

Climate Change Education for Sustainable Development: Lessons for Nigeria

129

Daniel Mailumo, Simeon Igbe, and Pius Mailumo

Contents

Introduction 2686
Literature Review 2687
Nigeria's Vulnerability to Climate Change 2689
Importance of Climate Change Education for Sustainable Development 2692
Lessons of Nigeria 2693
Conclusion 2695
References 2696

D. Mailumo (✉)
Department of Government and Public Administration, College of Advanced and Professional Studies, Makurdi, Nigeria

Association of Entrepreneurs and Technology Managers of Nigeria (AETMAN), Eco-system Based Adaptation For Food Security Assembly (EBAFOSA), Makurdi, Nigeria
e-mail: danaondofa@gmail.com

S. Igbe
Association of Entrepreneurs and Technology Managers of Nigeria (AETMAN), Eco-system Based Adaptation For Food Security Assembly (EBAFOSA), Makurdi, Nigeria
e-mail: aetmaninc@yahoo.com

P. Mailumo
Association of Entrepreneurs and Technology Managers of Nigeria (AETMAN), Eco-system Based Adaptation For Food Security Assembly (EBAFOSA), Makurdi, Nigeria

University of Calabar, College of Education, Katsina-Ala, Nigeria
e-mail: Mailumopius50@gmail.com

W. Leal Filho (ed.), *Handbook of Climate Change Resilience*,
https://doi.org/10.1007/978-3-319-93336-8_170

Abstract

Climate change is caused by human activity through the enhancing of the greenhouse gas emission and has grave consequences on natural and human systems on all continents and across the oceans. The warming trend of the Earth's atmosphere started with the industrial revolution and has now become a serious concern and complex global problem especially in developing countries that are the least responsible for climate change, yet the most at risk because of their inability to tackle the scourge. The world needs to make progress on many fronts including development and transfer of green technologies, greenhouse gas emissions reductions, and the establishment of effective policies. However, education, training, and awareness creation are equally of importance to creating an informed and knowledgeable workforce. This chapter scrutinizes climate change education and awareness creation as important tools toward achieving sustainable development and lessons for Nigeria as a developing country that is vulnerable to the effects of climate change. The chapter revealed that Nigeria lags behind in this sphere of development as efforts so far laid have yet to gain significant response. In the midst of other countries progress has been made toward helping the general public especially the next generation to understand and relate to the issues, make lifestyle changes to reduce greenhouse gas emissions, and adapt to the changing local conditions. The chapter recommends that innovative educational approaches should be provided if we aim to have an audience that will understand, address, mitigate, and adapt to the impacts of climate change, which will put our world on a more sustainable development path and will build a generation that is climate aware thereby strengthening capacity for appropriate response strategy.

Keywords

Climate change · Sustainable development · Education · Environment · Emissions

Introduction

Before the industrial revolution of the sixteenth century, man's economic activities were simple and with little or no devastating effect on the environment. Today's need to accumulate wealth by man necessitated a change in these economic activities to a more sophisticated system that would produce man's need in abundance and satisfy the growing demand. Since the Industrial Revolution, man has been able to take better control of his natural habitat and exploit it for his natural benefit and advancement. As a result, industries and their growth all over the world are now considered an important index for measuring economic development. While significant attention was paid to the development of cottage industries in Europe in the sixteenth and seventeenth centuries to process the primary goods produced in the colonies, heavy industries became the vogue of the nineteenth century with the discovery of coal, iron ore, and other heavy metals. The discovery of oil added impetus to this growth and exacerbated the environmental damage that was occasioned by light or cottage industries of the olden days (Edenhofer et al. 2014).

While man profited from the automation of the means of production and the consequent large-scale production that it facilitated, there was no corresponding concern regarding the environmental degradation it occasioned, let alone evolving ways of controlling it. This was largely the position until the tail end of the nineteenth century, when man woke up from his slumber only to be confronted with the devastating effects of industry whose effects was climate change. It, therefore, became apparent that some significant measures must be taken to sustain the change in climate and that the harmful effect of industry could not be ignored.

This new wave of climate consciousness spread rapidly throughout the developed nations bringing in its wake a new realization that development need not be accompanied by environmental degradation. It is the realization of global nature of pollution caused by industry and its effects across national boundaries that led to the convening of the United Nations Conference on the Human Environment in 1972 (Papp 1997; Rengger 1990), the United Nations Environment Programme (UNEP), the United Nations Framework Convention on Climate Change (UNFCCC), and the United Nations Conference on Sustainable Development (UNCSD). These conferences brought together the industrialized and the developing nations of the world to address the common concern for healthy and productive environment. This came from the realization that the problem of climate change was far more complicated as it is a trans-boundary and transnational phenomenon and not one that would be handled by a single country.

While efforts made at national, regional, and international levels toward tackling the climate challenge moved steadily in relation to the zeal of actors involved, it has not been quite the same for Nigeria as the country experienced a number of challenges including high level of illiteracy owing to inadequate access to education and training as well as poor quality education. This chapter which draws on the discussion to be held at a symposium on climate change adaptation, University of Ibadan, Nigeria, outlines the main issues around the question of climate change education for sustainable development – lessons for Nigeria. A number of key issues are central to the debate, they include the introduction, literature review, Nigeria's vulnerability to climate change, importance of climate change education for sustainable development, lessons for Nigeria, and conclusions.

Literature Review

Climate change is an undeniable environmental threat of the twenty-first century (Drexhage et al. 2007; The World Bank 2007). Climate change has the dubious accolade of being the most complex and difficult environmental problem which humankind has ever faced and if unchecked has the potential to undermine the support systems upon which the vast majority of the world's population depends for subsistence (Mabey et al. 1997). Environmental scientists are increasingly concerned that the rising levels of carbon dioxide may change the Earth's climate (Raven et al. 1998). Prominent leaders, including Barrack Obama, Al Gore, Nicolas Sarkozy, and Ban Ki-Moon, have all cited climate change as an international security threat (Ki-Moon 2014; Schoch 2011). Climate change affects the whole world through the poorest people who contribute least to the change are the ones who suffer the most (World Economic Forum 2016; Ebele and Emodi 2016). At the same time, provision of climate change education and training services are equally important for sustainable development since education is the catalytic factor which leads to human resource development for improved socioeconomic opportunities.

According to O'Neill, 2017, the earth's climate has never been static. For the past 2.7 million years, it has shown a pattern of alternating long ice ages, governed by cycles in the earth's orbit around the sun. Man has from time immemorial relied on the environment for survival. Since the importance of the environment cannot be overemphasized, the need for its sustainability is even more compelling because the environment can transform people's lives for the better. Energy and environment have the potential to free millions of women and children from the daily grind of water and fuelwood collection, and the provision of artificial lighting can extend the working day, providing also the invaluable capability to invest time in education, health, and the community. They open a window to the world through the radio, television, and telephone. In the aggregate, they are powerful engines of economic and social opportunity (World Bank 2000).

It is in the light of this that education for sustainable development gained currency when the United Nations General Assembly declared 2005–2014 as the UN Decade of Education for Sustainable Development (McKeown 2016). This spurred efforts toward Global Action Programme (GAP) on Education for Sustainable Development (ESD) where many countries are poised to raise awareness, move to capacity building, and implement of good practice (Laurie et al. 2016). Longhurst (2014) defined Education for sustainable development as the process of equipping students with the knowledge and understanding, skills, and attributes needed to work and live in a way that safeguards environmental, social, and economic well-being both in the present and for future generations (National Academy of Science 2003). This is in consideration of the concepts of global citizenship, environmental stewardship, social justice, ethics, well-being, and sustainable futures.

Literature reviewed showed that climate change education has recently received increasing international attention, in particular under the United Nations Framework Convention on Climate Change (UNFCCC) and other conventions which aimed at achieving sustainable development. Article 6 of the UNFCCC on Education, Training and Public Awareness adopted at the "Rio Earth Summit" in 1992 calls on governments to develop and implement education and training programs, including the strengthening of national institutions; training of scientific, technological, and managerial personnel; as well as implementing public awareness programs on climate change and its effects (UNITAR 2013).

At the continental level, the Arusha Declaration on Africa's post Rio+20 Strategy for Sustainable Development was signed in September 2012 by Africa's

Environmental Ministers at the fourteenth session of the African Ministerial Conference of the Environment (AMCEM) which recognized the ongoing United Nations Decade of Education for Sustainable Development (UNDESD) 2005–2014 and the Tbilisi Communique – Education Today for a Sustainable Future – adopted by the Tbilisi Intergovernmental Conferences on Environmental Education for Sustainable Development (Tbilisi+35), held on 6 and 7 September 2012. It furthermore resolved, under clause 17 (AEETAP 2015):

> To agree to strengthen environmental education and training and develop an action plan for Africa, covering formal and non-formal education, capacity building and information networking components, among others, and to explicitly include a focus on technology enhanced learning in this action plans.

Previously, no formal education program on climate change exists in many countries including Nigeria, but after the recognition by the UNFCCC, many countries like Benin, Malawi, Indonesia, and others found reason to develop climate change learning strategy in order to process and strengthen their national capacities. A key objective of the National Climate Change Learning Strategy is to strengthen national education and training institutions and to foster linkages between them and the climate change community (UNITAR 2013).

Toward this end, the Government of Malawi launched in 2000 a vision for the country which articulates the long-term development perspective for the country up to 2020. It is a framework for national development goals, policies, and strategies (UNITAR 2013). This initiative urges the Government of Malawi to ensure that greenhouse gas emission are monitored and reduced. In the same vein, the Indonesia National Climate Change Learning Strategy was launched in August 2013 with the aim of building capacity and integrating climate change learning in the national education system.

Nigeria's Vulnerability to Climate Change

Nigeria is not only part of the global environment from which, by that very fact, it cannot insulate itself, but it also belongs in the class of the developing countries of the Global South, which are at the receiving end of the climate crisis. Many reasons account for this. Developing countries do not have the capacity to manage a green environment; industrialization level is still recrudescent and cannot match the green manner the developed North is already moving toward; the South remains the dumping ground for all sorts of toxic wastes, crude technology, and environment-endangering activities of both the multinationals and locals. This implies that Nigeria and African countries are particularly vulnerable to the effects because of their overdependence on rainfed agriculture, compounded by factors such as widespread poverty and weak capacity (Folarin 2012). For instance, over 102 lives were lost in August 2011 Ibadan flood while Lagos remained completely still for a few days in July 2011 as a result of continuous high intensity rainfall over a period of 24 h.

Villages have been taken over by flood in Ilushi, Urho, Urhowa and Inyelen in the Niger Delta region in Nigeria (FGON 2012).

The developing countries felt that countries of the North should take responsibility and pay for solving the problem. It is established that the industrial countries cause the greatest problem. Data from the Pew 2007 Global Attitudes Project show that in 34 out of 37 countries surveyed, the United States is named by a majority as the country that is "hurting the world's environment the most" (Campbell et al. 2007) while developing countries like Nigeria are also more vulnerable to the impacts of climate change. Their economies are more dependent on climate-sensitive sectors such as agriculture and forestry, and they lack the infrastructure or resources to respond to the results of change in climate (World Bank Group 2015; The World Bank 2007).

The results of Nigeria's vulnerability analysis with climate change show that more than 13 million people are presently at risk and may be relocated due to climatic variations and sea level changes. If sea level should rise by about 1.0 million with the projected climate change, the number of people that may be relocated assuming there is no development would be more than 48 million (Mailumo 2018; IPCC 2015). With development and with the projected climate change, the number of people that may be relocated would be more than 92 million. Thus, dangerous climate change has the potential to deliver powerful systemic shocks to human development.

Oladipo (2010) has highlighted some of the socioeconomic and sociocultural sectors that can be adversely affected by climate change in Nigeria. The Niger Delta of Nigeria is estimated with an accelerated sea level rise (ASLR) of about 0.5 m, and about 5% of the delta could be lost (UNEP 2016). Also, with nearly 70% of the country's population depending on agriculture for the country's GDP, the country is highly vulnerable to climate variability and long-term climate change which could result in higher food prices and lower domestic revenues (Oladipo 2010).

Again, in Nigeria, for example, the climate change phenomenon can be seen in persistent drought in the North, desert encroachment, water stress/scarcity, rising sea level, and inundation of coastal lands by sea water in the South, while above average daily minimum and maximum temperatures are common. Also, increasing floods and erosion are being experienced all over the country. In addition, the country is experiencing the loss of biodiversity and threats to endangered species, oil, and gas pollution as well as severe ecological degradation. The country is also experiencing changing rainfall patterns. This has the potential to devastate rainfed agriculture on which so much of the population of Nigeria depend, while the increased occurrence of drought will eventually lead to a decline in agricultural yields and diminished food security (Federal Ministry of Environment 2015). The desert encroachment in the Northern part of the country exerts pressure on food production, demographic balance, and ecological disasters. The onslaught of dry land and sand attack has continued to displace thousands of farming communities in the North thereby creating the stimulus for social tensions in the Middle Belt.

In the same way, the Niger Delta region has been massively polluted by the activities of multinational oil companies to flare associated gas from oil exploration

(Mailumo 2018). The Niger Delta has put Nigeria on the map as the eight largest oil exporters in the world but is being degraded by unsustainable practices and a legacy of pollution and oil spills (Elenwo and Akankali 2014). The Federal Government is determined to ensure effective restoration and management of Nigeria's ecosystems through the creation of appropriate mechanisms and incentives such as improved policy planning and awareness creation on climate change adaptations; he insisted that "during the course of this century the resilience of many ecosystems is likely to be exceeded by an unprecedented combination of change in climate and other drivers, if greenhouse gas emissions and other changes continue at or above current rates."

In view of the foregoing, therefore, Nigeria has taken the challenge of climate change seriously. The first national communication was produced in November 2003. A stakeholders' initiation workshop on the Second National Communication (SNC) took place in December 2006. Nigeria now has a Department of Climate Change (DCC) in the Federal Ministry of Environment, Abuja, Nigeria. DCC is created to implement the convention and the protocol activities. It also coordinates the activities of the interministerial committee on climate change whose membership is drawn from the ministries of finance, agriculture, education and water resources, energy commission, Nigeria National Petroleum Corporation (NNPC), Foreign Affairs, Nigerian Meteorological Agency (NIMET), Industry, NGOs (Nigerian Environment Study/Action Team) and the academia (Federal Ministry of Environment 2012).

The country has provided leadership in some areas and has continued to partner with the relevant bodies like UNESCO in the global efforts to combat climate change, the country has maintained its commitments and obligations to the environment conventions, treaties, protocols, and agreements to which it is signatory (FGN 2012). It is in view of this that, His Excellency, President Mohammadu Buhari stated in his inaugural speech (May 2015) that Nigeria is committed to tackling climate change. Nigeria's INDC demonstrate its determination to contribute to the success of the Paris climate summit in December 2015. The President of Nigeria on 26 November 2015 approved the Nigeria Intended Nationally Determined Contributions (INDC) (Federal Ministry of Environment 2015).

It is also in furtherance with the 1999 Constitution of Nigeria reinforcing the policy and legal basis of sustainable development as stated therein that Nigeria has over the years boosted its leadership credentials in Africa and in the international bodies over climate change education. The Nigeria's National Assembly has also taken the issue of climate change seriously and has created special committees for environment and climate change in the two chambers where all legislative issues concerning environment, climate change and education are thoroughly deliberated before laws are made (Eleri 2011). We hope that in no distant future, all matters in this area shall have legal backing. As such, the country will be better positioned for the healthier implementation of policies toward climate change education for sustainable development.

However, as at today, there has been a global transformation that constitutes ecological and socioeconomic challenges with an urgent need for solution from all countries of the world. Although Nigeria does not contribute much to global greenhouse gas (GHG) emissions, the country is likely to suffer from the projected climate change because of significant dependence of the population and economy on such climate – sensitive sectors such as agriculture, forestry and biodiversity (World Bank Group 2015). Therefore, the best way to address climate change may be to orient national development action on education toward sustainable development pathway (Singh 2011).

Importance of Climate Change Education for Sustainable Development

The Ministers and Heads of Delegation attending the twentieth session of the Conference of Parties and the tenth session of the Conference of Parties serving as the meeting of the parties to the Kyoto Protocol, held in Lima, Peru, from 1 to 12 December, 2014 had reaffirmed the importance of Article 6 of the Convention and Article 10 (e) of the Kyoto Protocol; they recognized that education, including formal, nonformal, and informal education, and public awareness programs should promote the attitudes and behavior needed to prepare our societies to adapt to the impacts of climate change (UNFCCC 2014). This was further reaffirmed at the 3rd Dialogue on Article 6 of the convention which took place on the 2 and 3 June, 2015 where more than 130 representative of Parties, intergovernmental organizations (IGOs) and nongovernmental organizations (NGOs) as well as experts, practitioners, and stakeholders shared experiences and exchanged ideas, best practices, and lessons learned regarding climate change education and training (UNFCCC 2015; UNSDGs 2015).

In carrying out their Global Action Programmes (GAP) as commitments under Article 6 of the UN Convention, the parties agreed to promote and facilitate appropriate national laws and regulations within their respective capacities, develop and implement educational and public awareness programs on climate change and its effects, provide adequate response in addressing climate change and its effects, and provide public access to information on climate change and its effects as well as training of scientific and managerial personnel (UNFCCC 2015). It is important to emphasize that education for sustainable development's crucial importance for all the other 16 SDGs with its overall aim to develop crosscutting sustainability competencies in learners is an essential contribution to all efforts to achieve the SDGs, enabling individuals to contribute to sustainable development by promoting societal, economic, and political change as well as by transforming their behavior (UNESCO 2017).

Education for sustainable development encourages different disciplines to enter into dialogue, make connections, share knowledge, and work together on emergent areas. It aims to develop students' ability to understand and evaluate connections between big issues, such as inequality, public health, global consumption, biodiversity loss and the limits of natural systems.

Education for sustainable development is "future-facing" in the sense that students are encouraged to think about current and emergent and future situations, relevant to their studies, and in so doing gain a wider socioeconomic and environmental perspective on the relevance of their work.

Education for sustainable development encourages students to develop critical thinking and to take a wide-ranging, systemic and self-reflective approach to novel situations that can arise from complexity. An ability to anticipate and prepare for predictable outcome and be ready to adapt to unexpected ones is an important goal (Longhurst 2014).

Lessons of Nigeria

John Stuart Mill says, "Not only does education include whatever we do for ourselves and whatever is done for us by others, for the express purpose of bringing us somewhat nearer to the perfection of our nature (Aggarwai 2010). We believe that the current level of environmental destruction is unique in the history of humanity and that if it is left to continue at the current rate, we will, within our generation, experience the initial phases of a trend that seriously risks destroying the very fabric of our lives. This spurred the works of environmental education researchers to bring to the fore environmentally sound and sustainable policies and have shifted from the personal to the economic realities. Most prominently, climate change education for sustainable development encourages different disciplines to enter into dialogue, make connections, share knowledge, and work together on emergent areas.

Education for sustainable development is also future-facing as it encourages thinking about current, emergent, and future structures. This helps in gaining a wider socioeconomic and environmental perspective on its relevance. This has been confirmed because climate change is one of the most pressing challenges the world faces today. It affects all aspects of the development agenda including education, poverty eradication, health care, economic growth, and disaster risk reduction (World Economic Forum 2016). These pushed for the Rio+20 Conference on Sustainable Development in 2012 initiated by the United Nations as a process to shape international and national development priorities for all countries until at least 2030 (Ansuategi et al. 2015). It is indeed pertinent to point out here that despite Nigeria's accession to the various agreements and protocols regarding education for sustainable development, it failed to include it in the handbook "Climate change and sustainable development" published by the Federal Government of Nigeria in 2012, a somewhat contradiction of a forward looking effort at this critical time of capacity building toward providing meaningful solution to the challenge at hand.

In Kenya, government has responded positively to the Rio+20 outcomes and recent adoption of post 2015 Development Agenda, while higher education institutions in Kenya have stepped up their efforts on sustainable development issues, universities are also actively adopting low-carbon strategies, mainstreaming environmental sustainability across the curriculum, and engaging with communities and stakeholders at all levels for sustainable solutions. Currently, there are over

62 universities in Kenya that have very meaningful and positive impacts in achieving sustainable development in Kenya (UNEP 2011). In Nigeria, only 10 out of over 200 universities have enlisted in the Global Universities Partnership on Environment and Sustainability (GUPES) without commensurate policy and implementation strategies for the program (UNEP 2016). This is at a time the United Nations Environment Programme (UNEP) has registered over 430 universities across the globe on GUPES which is the result of a consultative forum organized by UNEP and its partners to increase successful engagement with universities (UNEP 2016). Given the importance of education to sustainable development, it is essential for Nigeria to ensure that sufficient expertise is readily available to support capacity needs.

In the United States of America, kindergarten through 12th grade curricula include environmental/climate change education modules, and environmental educators who work in the private sector provides highly sought after extra curricula activities to school districts unable to develop their own programs and to parents who want their children to have more positive environmental experiences (Sortore 2010). This is in furtherance to the guidance provided by the international environmental education goals and strategies adopted by UNESCO; it is expected that Nigeria should key into these initiatives.

For Nigeria, it is high time to link education and sustainable development process as they both require the promotion and implementation of a more ambitious climate agreement which will help to achieve SDGs. Contemporary perspective proves that humans collectively produce and reproduce their environment through their actions and are therefore capable of acting responsibly for a sustainable future. In order to design climate change education that will effectively contribute to a sustainable future, we argue for a learning framework that is consistent with contemporary climatic or environmental circumstances.

Basically, education within the sustainable development agenda is founded on principles drawn from a rich history of international instruments and agreements. These principles state that education is both a fundamental human right and an enabling right, that is, it enables other human rights; it is a public good and a shared societal endeavor, which implies an inclusive process of public policy formulation and implementation (UNESCO 2016). Thus, education and sustainable development is interlinked because education and development are based on human rights and dignity, justice and shared responsibility.

This is in furtherance to the fact that, at present, all SDG targets are universal, and each of the 17 goals are involved in learning, training, or at least in raising awareness. Therefore, education can be said to be a critical factor in addressing environmental and sustainability issues and ensuring human well-being. This is to say also that, increased educational attainment helps transform lives by reducing poverty, improved health, advancing technology, and increasing social cohesion (UNESCO 2016). Furthermore, climate-literate citizens will bring about change toward sustainable development; by this, people need to learn to understand issues and their causes, and they need to develop the values and necessary competences that will help them respond to solutions and climate practices. All of these are presented in the 17 Sustainable Development Goals (SDGs) below.

Goal 1	Education is critical to lifting people out of poverty
Goal 2	Education plays a key role in helping people move towards more sustainable farming methods, and in understanding nutrition
Goal 3	Education can make a critical difference to a range of health issues, including early mortality, reproductive health, spread of disease, healthy life styles and well-being
Goal 4	Education first initiative to accelerate progress
Goal 5	Education for women and girls is particularly important to achieve basic literacy, improve participative skills and abilities, and improve life chances
Goal 6	Education and training increase skills and the capacity to use natural resources more sustainably and can promote hygiene
Goal 7	Educational programmes, particularly non-formal and informal, can promote better energy conservation and update of renewable energy sources
Goal 8	There is a direct link among such areas as economic vitality, entrepreneurship, job market, skills and levels of education
Goal 9	Education is necessary to develop the skills required to build more resilient infrastructure and more sustainable industrialization
Goal 10	Where equally accessible, education makes a proven difference to social and economic inequality
Goal 11	Education can give people the skills to participate in shaping and maintaining more sustainable cities, and to achieve resilience in disaster situations
Goal 12	Education can make a critical difference to production patterns (e.g. with regard to the circular economy) and to consumer understanding of more sustainably produced goods and prevention of waste
Goal 13	Education is key to mass understanding of the impact of climate change and to adaptation and mitigation, particularly at the local level
Goal 14	Education is important in developing awareness of the marine environment and building proactive consensus regarding wise and sustainable use
Goal 15	Education and training increase skills and capacity to underpin sustainable livelihoods and to conserve natural resources and biodiversity, particularly in threatened environments
Goal 16	Social learning is vital to facilitate and ensure participative, inclusive and just societies, as well as social coherence
Goal 17	Lifelong learning builds capacity to understand and promote sustainable development policies and practices

Source: UNESCO 2016

Conclusion

The submission of this chapter is that climate change education for sustainable development is now understood to broaden our horizons, deepen our insight, refine our reactions and stimulate our feelings as we move forward toward tackling the greatest challenge of the twenty-first century. The geopolitics of climate change decision-making is rapidly changing with increase in scale, scope, visibility, and complexity. It therefore behooves the society to provide responses for a broader journey toward sustainable development. Again, climate change education is now critical for the acquisition of knowledge, skills, values, it provides action competencies and also informed citizens in the development of an ecological sustainable and socially just society. Climate change education is also vital for poverty reduction, and it ensures human well-being.

Although Nigeria participated in international efforts made at the outcome document of the United Nations Conference on Sustainable Development (Rio+20) in Rio de Janeiro, Brazil, where member states agreed to "promote education for sustainable development and to integrate sustainable development into education beyond the United Nations Decade of Education for Sustainable Development," in Nigeria, education hardly appears in national climate policies even as the country is party to the United Nations Conventions as has the obligation to do so as prescribed by Article 6. There is now the need for Nigeria to learn from experience adopted by other countries in addressing the global strategic threat through education for sustainable development. To this end, the chapter recommends that Nigeria should work toward a creative and innovative ways of addressing climate change education for sustainability issues, this can be done through the development of programs that enhance networking and partnerships. Another important recommendation is that Nigeria must work with the mantra "think global and act local," this will help for national awareness policies, activities, and initiatives that have a bearing on sustainability.

References

AEETAP (2015) Strengthening sustainable development in Africa – 2015–2024. Arusha: Africa Environmental Education and Training Plan – African Ministerial Conference of the Environment.

Aggarwai JC (2010) Teacher and education in a developing society, 5th edn. Vicas Publishing House PVT Ltd, New Delhi

Ansuategi A, Greno P, Houlden V, Markandya A, Onofri L, Picot H, Tsarouchi G (2015) The impact of climate change on the achievement of the post-2015 sustainable development goals. London: CDKN-HR: Wallingford and Metroeconomica

Campbell KM, Fuerth L, Gulledge J, Lennon ATJ, McNeill JR, Mix D, Ogden P (2007) The age of consequences: the foreign policy and national security implications of global climate change. Center for New American Security, Washington, DC

Drexhage T, Murphy D, Brown O, Cosbey A, Dickey P, Parry J, Ham JV (2007) Climate change and foreign policy – an exploration of options for greater integration. International Institute for Sustainable Development, Winnipeg

Ebele N, Emodi N (2016) Climate change and its impact on Nigerian economy. J Sci Res Rep 10(6):1–13, 2016

Edenhofer O, Pichs-Madruga R, Sokona Y, Farahan F, Kadner S, Seyboth K, Adler A (2014) Climate change 2014: mitigation of climate change. Contribution of working group to the fifth assessment report of the Inter-governmental Panel on Climate Change. Cambridge University Press, New York

Elenwo E, Akankali J (2014) Environmental policies and strategies in Nigeria oil and gas industry: gains, challenges and prospects. Retrieved on 28 Dec 2016 from www.scirp.org/journal/nr

Eleri E (2011) Nigeria – empty handed at Durban. Retrieved on 6 Dec 2017 from www.econigeria.com/wp/-content/uploads/2011/11/cop

Federal Government of Nigeria (2012a) Nigeria's path to sustainable development through green economy. Country report to the Rio+20 summit

Federal Government of Nigeria (2012b) Climate change and sustainable development in Nigeria – National Report for cop 18/Cmp8

Federal Ministry of Environment (2015) Nigeria's intended nationally determined contributions. Federal Ministry of Environment, Abuja

Folarin S (2012) Interrogating Nigeria's international role conceptions in an age of global climate change. In: Okpaga A, Bako R (eds) Global economic crisis, climate change and environment: the challenges for Nigeria. Fulbright Alumni Association of Nigeria, Makurdi

IPCC (2015) Climate change 2014 – synthesis report. Contribution of working groups I, II and III to the fifth assessment report of the intergovernmental panel on climate change. World Meteorological Organization (WMO), Geneva

Ki-Moon B (2014) UNEA – reconciling the needs of people and planet. In 'Our planet' – the magazine of the United Nations Environment Programme. Retrieved on 3 Dec 2014 from www.unep.org/ourplanet

Laurie R, Nonoyama-Jarumi Y, McKeown R, Hopkins C (2016) Contributions of education for sustainable development (ESD) to quality education: a synthesis of research. Centre for Environmental Education, London

Longhurst J (2014) Education for sustainable development – guidance for UK higher education providers. Higher Education Academy, London

Mabey N, Hall S, Smith C, Gupta S (1997) Argument in the greenhouse. The international economics of controlling global warming. Routledge, London

Mailumo AD (2018) Climate change and its impact on Nigeria's foreign policy: 200–2015. Unpublished thesis presented for the award of PhD in Political Science, Benue State University, Makurdi

McKeown R (2016) Education for sustainable development: the Oregon context. Portland University, Portland

National Academy of Sciences (2003) Our common journey – a transition toward sustainability. National Academy Press, Washington, DC

Oladipo E (2010) The global framework for mitigating climate change. In: Eze O, Oche O (eds) Climate change and human security in Nigeria. Nigerian Institute of International Affairs, Lagos

Papp D (1997) Contemporary international relations – frameworks for understanding, 5th edn. Allyn and Bacon, London

Raven P, Berg L, Johnson G (1998) Environment, 2nd edn. Saunders College Publishing, Orlando

Rengger, N (1990) Treaties and alliances of the world. 5th edn. United Kingdom: Longman

Schoch C (2011) Rethinking climate change as a security threat. Retrieved on 12 Dec 2012 from http://www.iied.org/Pubs/display.Php. 1710 IIED

Singh MP (2011) Global warming and forest. Daya Publishing House, New Delhi

Sortore E (2010) Environmental science. Apple Academic Press, Oakville

The World Bank Report (2000) Fuel for thought – an environmental strategy for the energy sector. International Bank for Reconstruction and Development/The World Bank, Washington, DC

UNCSD (2012) The future we want. Retrieved on 5 Mar 2014 from www.uncsd.org

UNEP (2016) UNEP's environmental assessment of Ogoniland. Retrieved on 23 Dec 2016 from http://www.unep.org/nigeria.retrieved

UNESCO (2017) Education for sustainable development goals – learning objectives. UNESCO, Paris

UNFCCC (2015) Adoption of the Paris Agreement. United Nations conference on environment and development (1992). The Regency Press Corporation, London

UNITAR (2013) Guidance note for developing a national climate learning strategy – strengthening human resources and skills to advance green, low emission and climate resilient development. United Nations Institute for Training and Research, Geneva

United Nations Environment Programme (2011) Environmental assessment of Ogoniland. United Nations Environment Programme, Nairobi

United Nations General Assembly (2016) Review of activities and resources, devoted to address climate change in the United Nations system organizations. United Nations General Assembly, Washington, DC

UNSDGS (2015) 17 sustainable development goals, 17 partnerships. Retrieved on 17 Nov 2015 from https://sustainabledevelopment.un.org/partnerships

World Bank Group (2015) Global economic prospects – having fiscal space and using it. IBRD/World Bank, Washington, DC

World Economic Forum (2016) The global risks report 2016, 11th edn. World Economic Forum, Geneva

Enhancing the Capacity of Vulnerable Community to Climate Change: Role of Quality Declared Seed Production Model in Semi-arid Areas of Central Tanzania

130

E. Y. Swai, F. Njau, and M. Farrely

Contents

Introduction 2700
The Overall Objective and Specific Objectives 2702
Overall Objective 2702
Specific Objectives 2702
Rationale of Introducing Quality Declared Seeds in Semi-Arid Areas of Central Tanzania 2702
Methodology 2703
Selection of Study Sites 2704
Criteria Used for Selection of Case Study Sites 2704
Description of the Study Area 2705
Situation Analysis in Kikombo and Idifu Wards 2705
Willingness of Vulnerable Smallholder Farmers to Engage in QDS Production 2706
Willingness to Buy Improved Seeds 2707
Data Analysis 2708
Results and Discussions 2708
Characteristics of Smallholder Farmers at Kikombo and Idifu Wards 2708
The Source of Income and Occupation for Managing Risks 2709
Farmers' Perceptions on Weather Pattern and Climate Changes and Adaptation Measures in Kikombo and Idifu Wards 2710
Climate Vulnerability and Capacity Analysis 2710
The Perception on Weather Pattern Changes and Adaptation Measures 2711

E. Y. Swai (✉)
Agricultural Research Institute (ARI) Hombolo, Dodoma, Tanzania
e-mail: eyswai29@gmail.com; eyswai@yahoo.com

F. Njau
Institute of Rural Development Planning (IRDP), Dodoma, Tanzania
e-mail: frabe59@gmail.com

M. Farrely
Tanzania Organic Movement for Agriculture, Dar es Salaam, Tanzania
e-mail: mrfarrelly@gmail.com

W. Leal Filho (ed.), *Handbook of Climate Change Resilience*,
https://doi.org/10.1007/978-3-319-93336-8_176

Knowledge on Crop Yields, Food Situation, and Food Shortage Coping Strategies at Kikombo and Idifu Wards 2712
Farmers' Perceptions on the Main Sources of Crop Seeds 2715
Farmers Perception on Access to Improved Seeds and Their Limitation 2715
Farmers' Perceptions on the Willingness to Produce Quality Declared Seed 2717
Willingness to Buy Improved Seeds in Kikombo and Idifu Wards 2718
Conclusion 2719
References 2720

Abstract

In sub-Saharan Africa the rainfed agricultural sector is important and relies heavily on weather-sensitive agriculture. In the semi-arid areas of Tanzania, the overall quantity of rainfall has been declining over time leading to insecure harvests. The study was undertaken in semi-arid areas of Central Tanzania to explore the farmers' perceptions on the main sources of food crop seeds, access to improved seeds and their limitations, willingness to engage in the production of Quality Declared Seed, and willingness to buy improved seeds. The study employed qualitative and quantitative tools to gather information on the crop production. Findings revealed that 76.4% (n = 400) of households are using poor quality recycled seeds mainly attributed to the lack of improved seeds (44.1%) and high price of improved seeds (41.8%) inter alia. As a result yields of major food crop were low (<1 t/ha) resulting to only 5% of household harvest crop produce that can last for more than 12 months and the rest of household harvest crop produce that can last for 3–9 months. Therefore in order to unravel the problem of food insecurity in uncertain climate of rural Tanzania, production of Quality Declared Seed tolerant to drought was introduced and has gained momentum.

Keywords

Semi-arid · Central Tanzania · Quality Declared Seed · Vulnerable farmers

Introduction

A number of countries in Africa already face semi-arid conditions that make agriculture challenging, and climate change will be likely to reduce the length of growing season as well as force large regions of marginal agriculture out of production. This would adversely affect food security in the continent (Boko et al. 2007). Predictions on the impact of climate and variability in Tanzania revealed that the interior regions including central regions of Tanzania, namely, Dodoma and Singida, are expected to experience a reduction in precipitation up to 20%, prolonging the dry season and increasing the risk of drought (Hulme et al. 2001; Paavola 2008). Similarly, in Tanzania, temperatures are predicted to rise 2–4 °C by 2100, warming more during the dry season. Worse still, the drier areas of Dodoma, Singida, and Shinyanga are top in the list of drought-stricken areas (NAPA 2007). Extreme daily temperatures above a certain threshold may have damaging consequences on crop yields (Schlenker and Lobell 2010). Furthermore, prediction shows that areas with unimodal rainfall in Tanzania will experience decreased rainfall of about 5–15%, and as a result maize production in the drier regions will decrease by 84% (NAPA 2007).

Considering the effects of climatic change on agriculture subsector which is already being felt by the majority of the population in semi-arid areas of Central Tanzania, measures focusing in preventing the destructive impact of climate change on food insecurity are of critical importance. To this end, a project titled "Ecovillage Adaptation to Climate Change in Central Tanzania (ECO-ACT)" funded by the European Union (2015/2016 to 2018/2019) launched some strategies for lifting vulnerable farming communities out of hunger and poverty in four villages of Chamwino and Dodoma districts. Some of the technologies rolled out in study villages related to agriculture to address a chronic problem of food insecurity encompass Quality Declared Seed (QDS) production, introduction of drought-tolerant early-maturing seeds, sustainable soil fertility management options, soil water management technologies, and control of soil erosion, among others. Of these initiatives, introduction of community seed production famous as "Quality Declared Seed" laid down a strong foundation for addressing challenges on access to improved seeds for building resilience on household food security. Estimates of 76.4% of households are using poor quality recycled seeds in Kikombo and Idifu wards despite the existence of a well-defined seed system in Tanzania. Among others, limited access to appropriate improved seed for majority of vulnerable farmers in Central Tanzania is contributing to insecure harvest.

Improved seed is an important strategic input for building resilience of crop production in drought-prone areas of Central Tanzania. Tanzania requires 100,000 t of seed annually. However, only 13,000 t of improved seeds are marketed annually out of which 5000 t are of maize (Barnett et al. 2011). Implying that, most of the commercial seed companies are mainly engaged in production of high-value crops such as maize and horticultural crops which are more suited in subhumid zones. In the contrast, there has been very little investment on production of seeds for crops such as pearl millet and sorghum which are more favored in drought-prone areas of semi-arid zones of Central Tanzania. Adding to the problem, most of the seed companies are mainly concentrated in potential areas of Southern Highlands and Northern part of Tanzania, being above 450 km far away from central regions of Tanzanian. Implying that, transportation cost will certainly add cost which is unbearable by limited constrained resource farmers. This calls for alternative gender-oriented and affordable innovations such as introduction of Quality Declared Seed. According to the Tanzanian Seed Act of 2003, Quality Declared Seed means seed produced by a registered smallholder farmer which conforms to the specified standards (http://www.saflii.org/tz/legis/num_act/sa200368.pdf).

The QDS is generally considered to be cost-effective, and indeed it is an entry point for resource constrained vulnerable farming communities to access improved seeds. The advantages of improved drought-tolerant varieties in adapting to the adverse effects of climate change and variability cannot be overemphasized. They have got key important attributes in most cases high yielding and intended to overcome major production constraints including increased incidences of drought episodes, low yield, diseases, and insect pest infestations, among others. The introduction and popularization of QDS is not intended to compete or replace the existing schemes of seed quality control mechanisms in target area of Central Tanzania. It is compelled by the fact that most of the seed companies are not economically attracted to the production of open pollinated varieties (OPV) seeds. The tendency of limited resource farmers to recycle seed regularly without replacing it with a new stock is very common in the rural setting of Central Tanzania. The QDS model is therefore an approach which is designed to support vulnerable farming communities to access drought-tolerant quality seeds within their vicinity and at affordable price. To this end, the study is built up on the Ecovillage Approach that aims at creating a model of good practice in climate change adaptation to strengthen the capacity of vulnerable farming communities to address the problem of food insecurity in semi-arid zones of Central Tanzania through introduction of most affordable community seed production employed in the production of Quality Declared Seeds.

The Overall Objective and Specific Objectives

Overall Objective

The study is built up on the Ecovillage Approach that aims at creating a model of good practice in climate change adaptation to strengthen the capacity of vulnerable farming communities to address the problem of food insecurity in semi-arid zones of Central Tanzania through introduction of most affordable community seed production employed in the production of Quality Declared Seeds.

Specific Objectives

The study was undertaken in semi-arid areas of Central Tanzania to achieve the following:

(i) To explore the farmers' perceptions on the main sources of food crop seeds
(ii) To explore farmers' perceptions on access to improved seeds and their limitations
(iii) To examine the willingness of vulnerable farming communities to engage in the production of Quality declared seeds that can endure in harsh and uncertain climate
(iv) To assess the farming communities on willingness to buy improved seeds in the study villages.

Rationale of Introducing Quality Declared Seeds in Semi-Arid Areas of Central Tanzania

In the semi-arid area of Tanzania, both formal and informal seed systems are generally weak, thus making it difficult for vulnerable farming communities to access high-yielding varieties that can endure the harsh and adverse effects of climate change. In general there is a limited use of improved high-yielding drought-tolerant varieties by majority of rural vulnerable farming communities in semi-arid zones of Central Tanzania such as sorghum, pearl millet, sunflower, and groundnuts, among others, mainly attributed to the lack of awareness of the tangible benefits of certified seed, high seed prices, and fake seeds obtained from formal seed system inter alia. Notwithstanding, the gap for the use of improved certified seed that can withstand adverse effects of climate change and variability cannot be solved by the current existing seed systems. Worse still, there are few improved varieties for sorghum, pearl millet, and groundnuts that are available outside of the existing maize-based commercial seed subsector in Tanzania. For example, some of the improved seeds suited in harsh agroecology of Central Tanzania that cannot be accessed from seed companies and seed stockists include sorghum, pearl millet, and groundnut despite the fact that they were released long time ago. Majority of vulnerable rural farming communities in semi-arid areas of Central Tanzania have tendencies of recycling the seeds. As a result, most seed companies are not economically attracted to the open pollinated varieties because they fear that farmers will not regularly renew their seed stock.

Therefore, improving the seed system in semi-arid areas of Central Tanzania which would ensure that high-yielding resilient varieties are available for majority of vulnerable farming communities when needed is a key to addressing challenges of food insecurity and poverty. This will require introduction of sustainable seed systems such as production of "Quality Declared Seed" that will not only ensure availability of seed at affordable price but would stimulate the involvement of vulnerable farming communities to engage in farming. In a situation where seed cannot be accessed from known common sources, concerted efforts for engaging Agricultural Research Institutes in the country mandated for specific cereal and legume crops are of paramount importance. To this end, the introduction and popularization of QDS Production Model is crucial since this will ensure that improved seeds via this model are multiplied and distributed to members of the community at a relatively cheap price and on time.

Methodology

The procedure for piloting QDS model in case study sites followed the Ecovillage Approach, and it was a rigorous process which consisted of four main stages. The first three stages were compulsory for all interventions being implemented in Ecovillage Adaptation to Climate Change in Central Tanzania Project. The fourth stage was specific to an intervention itself and aims at addressing a specific constraint that emerged under major thematic areas notably agriculture, livestock, natural resource, and energy.

In view of the above, key stages adopted in Ecovillage in the introduction of QDS model in four villages of Dodoma and Chamwino districts include the following:

(i) Ecovillage approach used for selection of four most vulnerable villages using climate vulnerability and capacity analysis tool.
(ii) Situation analysis of four selected villages (i.e., Chololo and Kikombo in Dodoma Municipal Council; Idifu and Miganga in Chamwino District Council) on the following:
   - Assessment on sources of crop seeds and access to an improved seed industry in semi-arid areas of Central Tanzania. Central factors which influence the uptake of technology, namely, education level, main occupation, and source of household income.
   - Farmers' perceptions on the factors which have direct influence on crop production and household food situation such as element of weather and impact of climate change.
   - Factors which may influence production of QDS notably weather, climate change, and climate variability.
   - Measures undertaken by vulnerable smallholder farmers to address problem of food shortage.
(iii) Approaches for introduction and popularization of Quality Declared Seeds model in Kikombo and Idifu wards.

Selection of Study Sites

Criteria Used for Selection of Case Study Sites

The procedure used for selection of study sites consisted of two steps which comprised consultation of agriculture extension officer in charge and the use of Climate Vulnerability and Capacity Analysis tools on selected villages.

Consultation of District Extension Officers

The District Agriculture, Irrigation and Cooperative officers in four districts of Chamwino, Bahi, and Kongwa in Dodoma Region were consulted in the identification of villages. The criteria for selection were based on the wards having villages which are more prone to the adverse impact of climate change and variability.

Climate Vulnerability and Capacity Analysis (CVCA)

In order to map strategies to facilitate a community-driven climate change adaptation initiatives, a combination of participatory tools from the Climate Vulnerability and Capacity Analysis (CVCA) toolkit as outlined in *Climate Vulnerability and Capacity Analysis Handbook* – CARE 2000 (http://www.careclimatechange.org) – was used.

Characteristics of Selected Study Sites

Dodoma Municipal District and Chamwino District Council where the two wards are situated were selected on the basis that the areas are frequently affected by the impacts of climate change. Also, both districts fall within the semi-arid areas of

Tanzania where there are frequent droughts and food shortage predominantly affecting the livelihood of the local community.

Description of the Study Area

The study was conducted in two wards, namely, Kikombo (i.e., outskirts of city) in Dodoma City, and Idifu (typical rural area) in Chamwino District Council of Dodoma Region, and four target villages are depicted in Fig. 1.

Situation Analysis in Kikombo and Idifu Wards

Situation analysis was undertaken in two wards of Dodoma and Chamwino districts which were identified to be more prone to the impact of climate change using CVCA tool. The primary data were collected from four villages (i.e., Chololo and Kikombo in Dodoma District; Idifu and Miganga in Chamwino District) on the number of variables which encompasses level of education, main occupation, source of household income, crop production, household food, sources of crop seeds, access to improved seeds and their limitation, farmers' perceptions such as element of weather and impact of climate change, and existing smallholder farmers' initiatives to respond to the problem of food shortage.

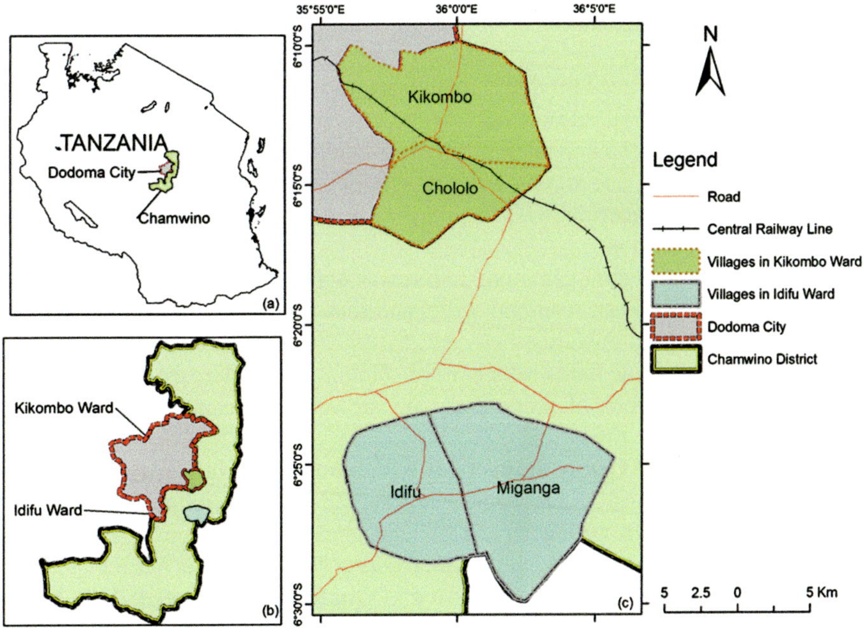

Fig. 1 Location of the action sites in semi-arid areas of Central Tanzania

The study involved the use of multiple approaches including interviews, questionnaires, and focus group discussions (FGD). A total of at least 200 respondents from each ward, namely, Kikombo and Idifu, gave a total of 400 participants. These participants involved in the interviews were almost 13.6% of 1,404 total number households in the Kikombo ward and 8.3% of 2393 total household for Idifu ward. The field survey entailed the administration of questionnaire instruments to heads of farming households or their representatives. The administration of questionnaires at the household level was complemented by focus group discussions.

Willingness of Vulnerable Smallholder Farmers to Engage in QDS Production

The willingness to engage in QDS production consisted of several steps including the following:

(i) **Readiness to engage in production of Quality declared seeds:** Farmers' assessment on willingness to commit piece of their land and engage in production of quality declared seeds. This information was collected during situation analysis which was undertaken at Kikombo and Idifu wards.

(ii) **Identification of seed producers:** Broadly, identification of seed producers had several steps which include the following:
   - Awareness creation at sub-village level by extension officers and researchers.
   - Preliminary registration at sub-village leader and extension officers in target villages.
   - Screening for seed producer candidates was carried out, and it involved all farmers who had registered.
   - Criteria for selection of farmers who qualify for QDS productions were shared which encompass farmers who are willing to adhere to the strict regulation concerning QDS and that tally with Tanzania Seed Act of 2003 without coercion.
   - Level of education came out, and it was noted that seed producers require someone who can read and write, and thus minimum cutoff point will be standard seven onward.
   - Farmers having fields which have minimum isolation distance for each specific category of seeds and who are ready to commit land for seed production.

(iii) **Registrations of QDS producers at Kikombo and Idifu wards:** Each QDS seed producer must be registered with the Tanzania official seed certification institute (TOSCI). Ordinarily, TOSCI personnel keeps records of seed producers, and this is an agreement with existing Tanzania Seed Act regulations.

(iv) **Capacity building on the production of Quality declared seed:** The experts from TOSCI and national research system were the main facilitators involved in imparting knowledge and providing tailor-made training materials on QDS production. Broadly the main topics covered during training exercise include seed regulations in Tanzania as follows:

- Category of seeds, and specifically seed producers were introduced on seed classes (basic and certified seeds) which could be sourced for production of QDS.
- Recommended isolation distance for different crops.
- Field hygiene.
- Agronomic practices.
- Procedures employed during field inspection by the Tanzania Official Seed institute.
- Harvesting, drying, and sample collection ready for submission to TOSCI in Morogoro, Tanzania, for quality assessment.

Willingness to Buy Improved Seeds

The willingness of vulnerable smallholder farmers to buy improved seed consisted of several fundamental steps which include the following:

(i) **Conducted meeting:** Participatory meeting which engaged both project implementers and village leaders to agree on modalities on why seeds should be sold and identification of influential trusted vendors across pilot area.

(ii) **Price for QDS seeds**: Consensus on price for selling improved seeds, and this engaged ECO-ACT project team members, lead farmers, and village government.

(iii) **Sensitization meeting:** Awareness creation was carried out 2 months prior onset of rainy season (i.e., October–November 2017) to inform the communities about the availability of improved seeds and price for each crop type. The dissemination of information employed the use of loud speaker, publicity on village notice board, village assembly meeting, and engagement of hamlet/sub-village leaders.

(iv) **Seed supply:** Supply of improved seeds to vendors identified in each participating four villages of Kikombo and Idifu wards spans from October 2017 to end of January 2018. The amount of seed supplied varied substantially depending on demand.

(v) **Record keeping:** Record keeping of seed buyers for quantity bought was kept in each target village for easy tracking of households' bought seeds and quantity of seeds purchased.

(vi) **Monitoring on seed stock:** This was done regularly to ensure that vendors have adequate seed stock to sell during the period from October 2017 to the end of January 2018.

(vii) **Evaluation:** Evaluation carried out to gauge as to what extent vulnerable smallholder farmers are able to access improved seeds from village vendors and finally establish actual number of farmers who bought seeds during 2017/2018 cropping season.

(viii) **Field sampling**: Thirty fields in each participating village were sampled during 10–15th of February 2018 to assess if farmers obtained seeds from vendors planted in their fields.

Data Analysis

Primary data from interview were edited, coded, entered, and analyzed using computer program Statistical Package for the Social Sciences (SPSS) software version 21. Descriptive statistics to give frequencies and cross tabulation was undertaken. Tables, graphs, and bar charts were used to present different variables.

Results and Discussions

Characteristics of Smallholder Farmers at Kikombo and Idifu Wards

Results indicated that female respondents in Kikombo and Idifu wards were 58.5%, while male respondents were 41.59% (Table 1). Similarly, the study sought to find out formal educational levels of respondents in the four target villages of Kikombo and Idifu wards. Results indicated that 262 (65.5%) of respondents have primary school education (class 1 to 7). Other respondents have secondary education (i.e., from 1 to 4) 9 (2.2%), college education 12 (3.0%), and adult education 5 (1.2%). The remaining 112 (28.0%) have not undergone formal education. Based on the findings, most of the respondents had low level of education. This finding suggests that the introduction of Quality Declared Seed production intervention in the study

Table 1 Demographic characteristics of respondents (n = 400) at Kikombo and Idifu wards in Chamwino and Idifu districts

Variable	Characteristic	Ward				Total	
		Kikombo		Idifu			
		Frequency	%	Frequency	%	Frequency	%
Sex	Male	62	15.5	104	26	166	41.5
	Female	129	32.2	105	26.2	234	58.5
Education level	Primary	116	29	146	36.5	262	65.5
	Secondary	3	0.8	6	1.5	9	2.2
	Collage	3	0.8	9	2.2	12	3
	Adult education	3	0.8	2	0.5	5	1.2
	No formal education	66	16.5	46	11.5	112	28

Table 2 Source of household income at Kikombo and Idifu wards (n = 400)

	Wards of residence					
	Kikombo		Idifu		Total	
Source of income[a]	Frequency	%	Frequency	%	Frequency	%
Livestock sales	81	20.2	134	33.5	215	53.8
Crop sales	147	36.8	176	44	323	80.8
Casual labor	35	8.8	79	19.8	114	28.5
Honey selling	2	0.5	3	0.8	5	1.2
Hire of oxen	7	1.8	14	3.5	21	5.2
Sales of hides/skin	10	2.5	1	0.2	11	2.8
Others	99	24.8	58	14.5	157	39.5

[a]Dataset based on multiple responses

villages should engage farmers who can read and write. Teaching methodology for QDS producers must be well designed to impart knowledge to farmers with low level of education. This would require the use of varieties of teaching methods for enhancing learning process. Training should be complemented by actions, and seed producers should be fully engaged in the entire process of seed production to broaden their knowledge. To achieve this, the researchers and district crop subject matter specialist regularly visited seed producers for technical backstopping.

The Source of Income and Occupation for Managing Risks

Sources of Income

The study sought to examine main sources of income for target villages at Kikombo and Idifu wards before embarking on QDS production to understand on how seed producers will continue to be unaffected due to extreme events impacted by weather, climate change, and climate variability. The results in Table 2 indicated that the main sources of income at household are crop sales 323 (80.8%), sale of livestock 215 (53.8%), casual labor 114 (28.5), hire of oxen 21 (5.2%), sales of hides and skin 11 (2.8), and honey selling 5 (1.2). In addition, 157 (39.5%) get their income through other sources such as selling of vegetables, remittances, selling local brews, making local brooms, and selling livestock products among others. The present study confirmed that it is worthy investing on the production of QDS since majority of rural dwellers in the target villages derived their livelihood from different sources (Table 2). For instance, in case there is a crop failure impacted by drought, the QDS producers who are keeping livestock can easily recover the loss.

Occupation of Respondents at Idifu and Kikombo Wards

Findings showed that most respondents in the area are involved in both crop farming and livestock keeping 237 (59.2%), crop farming 151 (37.8%), and livestock keeping 13 (3.2%) (Table 3). Findings revealed that concerted efforts undertaken

Table 3 Main occupation at Kikombo and Idifu wards

Main occupation[a]	Wards of residence: Kikombo Frequency	%	Idifu Frequency	%	Total Frequency	%
Crop farming	90	22.5	61	15.2	151	37.8
Livestock keeping	13	3.2	0	0	13	3.2
Crop farming and livestock keeping	92	23	145	36.2	237	59.2
Formal employment	2	0.5	0	0	2	0.5
Petty business	55	13.8	31	7.8	86	21.5
Others	22	5.5	11	2.8	33	8.2

[a]Dataset based on multiple responses

to unravel constraints affecting this crop farming enterprise are of significant importance. The introduction of QDS which aims at building resilience on crop production envisioned to overcome the problem of food insecurity and poverty among smallholder farmers. All the same, the in-depth understanding of the main the occupation of target villages provide clues on the level of investment required to carter for the projected demand of improved seeds.

Farmers' Perceptions on Weather Pattern and Climate Changes and Adaptation Measures in Kikombo and Idifu Wards

The seed industry plays an important role in increasing food production; however, target villages of Kikombo and Idifu wards rely entirely on rainfed agriculture. The implication is that crops grown in the area are more prone to the adverse impact of weather and climate change.

Climate Vulnerability and Capacity Analysis

Findings revealed that drought was the second most important hazard that significantly affects livelihood resources across study area with a significant effect on water availability, agriculture, and the human population. In general, the existence of drought and insufficient rains has high impacts on crop production. The reported coping strategies by farmers include labor selling and food borrowing at high interest rates (returning double or more of borrowed amount), and this phenomenon in local language is known as "songoleda." These coping strategies suggested that the problem of drought and frequent famine in the study area would require the introduction of drought-tolerant/early-maturing high-yielding varieties. It is against this background that an intervention addressing the challenges of drought such as the use of improved drought-tolerant varieties was introduced and rolled out across study area.

Table 4 The perception of local community on weather pattern changes noticed at Kikombo and Idifu wards

Weather pattern changes noticed[a]	Ward of residence				Total	
	Kikombo		Idifu			
	Frequency	%	Frequency	%	Frequency	%
Reduced amount of rainfall	148	37.5	172	43.5	320	81
Increase in drought	143	36.2	166	42	309	78.2
Rainfall unpredictability	142	35.9	140	35.4	282	71.4
Increase in temperature	123	31.1	106	26.8	229	58
Early cessation of rainfall	93	23.5	115	29.1	208	52.7
Late onset of rainfall	75	19	87	22	162	41
Early onset of rainfall	59	14.9	67	17	126	31.9
More wind	57	14.4	38	9.6	95	24.1
Late cessation of rainfall	32	8.1	39	9.9	71	18
More floods	35	8.6	35	8.9	69	17.5
Others	0	0	1	0.3	1	0.3

[a]Dataset based on multiple responses

The Perception on Weather Pattern Changes and Adaptation Measures

The perception of the local community on noticing changes in weather patterns at Kikombo and Idifu wards is given in Table 4. The changes noticed include reduced amount of rainfall (81.0%), increase in drought (78.2%), rainfall unpredictability (71.4%), increase in temperature (58.0%), and early cessation of rainfall (52.7%). Similarly, late onset of rainfall (41.0%), early onset of rainfall (31.9%), more wind (24.1%), and more floods (Table 4) are noticed. In order to improve decision-making of QDS seed producers of what type of seeds to produce and when to plant, detailed information on weather pattern is crucial, findings provided a signal on the types of seeds to be grown in the target villages which can endure in harsh environment. Deliberate efforts were geared on the production of high-yielding drought-tolerant seeds in order to cope with the negative effects of the changes in weather patterns. Knowledge on the use of soil water management interventions such as deep plowing and the use of tied ridges and ripping was also imparted for building resilience on the production of QDS as well as crop production at large.

Farmers' Perceptions on Impacts of Climate Change and Adaptation Measures

The impacts of climate change to the local community and livelihoods in Kikombo and Idifu wards varied across locations. Impacts identified by the local community were crop failure 373 (94.4%), water shortage 291 (73.7%), and shortage of pasture 250 (63.3%). Other climate change impacts identified in the area as indicated in Table 5 were outbreak of livestock diseases 193 (48.9%), outbreak of human diseases 190 (48.1%), outbreak of crop diseases 158 (40.0%), livestock deaths

Table 5 The perception of local community on impacts of climate change. (N = 395) in Kikombo and Idifu wards

Impacts of climate change[a]	Ward of residence				Total	
	Kikombo		Idifu			
Crop failure	178	45.1	195	49.4	373	94.4
Water shortage	144	36.5	147	37.2	291	73.7
Pasture shortage	118	29.9	132	33.4	250	63.3
Outbreak of livestock diseases	96	24.3	97	24.6	193	48.9
Outbreak of human diseases	83	21	107	27.1	190	48.1
Outbreak of crop diseases	70	17.7	88	22.3	158	40
Livestock deaths	65	16.5	78	19.7	143	36.2
Destruction of infrastructures	24	6.1	30	7.6	54	13.7
Others	4	45	3	0.8	7	1.8

[a]Dataset based on multiple responses

143 (36.2%), and destruction of infrastructures 54 (13.7%). Findings showed that crop failure is the most vulnerable compared to the other livelihoods examined in the target villages. While, it could be a major reason contributing to insecure harvests in semi-arid areas of Central Tanzania.

Further analysis noted that crop varieties grown by majority of vulnerable smallholder farmers can no longer perform well under the changing climate. Prior knowledge on the consequence of climate change on the production of QDS is required in order to plan well in advance on how risk associated with crop failure could be addressed. These findings underscore the need to introduce drought-tolerant seeds which have recently been released to replace the varieties which succumb to drought. For example, NACO Mtama 1 (improved sorghum variety) was introduced and popularized in the study villages. While other options quite often are overlooked by smallholder farmers such as the use of fertilizers and appropriate plant density, weeding regime must be a part for enhancing the resilience on QDS production.

Knowledge on Crop Yields, Food Situation, and Food Shortage Coping Strategies at Kikombo and Idifu Wards

Crop Yields of Major Crops Grown in Dodoma and Chamwino Districts

The average crop production across Kikombo and Idifu wards is given in Table 6. The findings indicate that the average crop yields per unit area of local crop varieties were generally low ranging between 218.3 kg/ha for sesame and 719.5 kg/ha for paddy. Findings revealed that yields for all major crops grown in the target areas were less than a ton/ha, which partly may account for food insecurity in the target villages. The knowledge on crop yields provide highlights on the selection of crops varieties for QDS production. Findings clearly demonstrated that there is huge potential of increasing crop yields per unit areas as opposed to a current a practice of increasing area

Table 6 Average crop production (Kg/ha) of local cultivars in Kikombo and Idifu wards

Location of study Area	Crop type	Grain yield (Kg/ha)
Kikombo ward	1. Pearl millet	567.0
	2. Sorghum	457.0
	3. Maize	480.5
	4. Sunflower	557.5
	5. Cow peas	163.8
	6. Groundnuts	330.0
	7. Sesame	218.3
	8. Paddy	719.5
Idifu	1. Pearl millet	558.7
	2. Sorghum	536.0
	3. Maize	466.8
	4. Sunflower	246.3
	5. Cow peas	515.5
	6. Groundnuts	276.8
	7. Sesame	754.8

Table 7 Number of months the harvested food last revisit original report

	Ward of respondent		
Months	Kikombo	Idifu	Total
Less than 3 months	19 (4.9)	15 (3.9)	34 (8.8)
3–6 months	61 (15.8)	55 (14.2)	116 (30)
6–9 months	48 (12.4)	39 (10.1)	87 (22.5)
9–12 months	54 (14)	76 (19.6)	130 (33.6)
More than 12 months	3 (0.8)	17 (4.4)	20 (5.2)

Figures in brackets are percentages

under cultivation. Therefore decision on which QDs seeds to produce should also take into account other compounding factors such as limited access to chemical fertilizers among others. Attention should be paid on varieties that can couple with upgrading strategies of limited resources of endowment farmers.

Household Food Situation in Kikombo and Idifu Wards

It is known that the level of household food security or insecurity can be measured by different ways depending on the purpose of a study. For instance, the level of food insecurity can be measured by looking on the average amount of available grain per household per year, time that the grain available lasts, daily caloric intake, and income level of the household. Food shortage in Kikombo and Idifu wards at households is regular as indicated in Table 7. Only 32.5% of the respondents stays with the harvested food for 12 months, 29% are food secured for 3–6 months, 21.8% for 6–9 months, 8.5% for less than 3 months, and only 5% for more than 12 months.

Table 8 Mean value for months the harvested food lasts

Variables	Mean (M) + SD		t-value	Df	Sig. (two-tailed)	Mean difference	95% CI of the difference	
	Kikombo	Idifu					Lower	Upper
Months harvested food lasts	7.67 + 3.52	8.82 + 3.98	−3.048	384.58	0.002	−1.16187	−1.91123	−41,250

An independent samples t-test (Table 8) was computed to compare the month's food lasts after harvest for Kikombo and Idifu wards. There was significant difference in scores for Idifu ($M = 8.82$, $SD = 3.52$) and Kikombo ($M = 7.67$, $SD = 3.98$; $t\ (385) = -3.032$, $p = 0.002$, two-tailed). This means that Idifu ward is more food secured compared to Kikombo ward as it was verified by t-test, whereby in average, in Idifu the harvested food lasts for about 9 months, while in Kikombo the harvested food lasts for 8 months. However, the situation is not good because more than half (59.25%) of the respondents' food does not cater for the annual household food requirements.

In view of these, concerted efforts geared on the use of improved early-maturing and drought-tolerant crop varieties coupled with improved agronomic practices are required. Leach and Kilama (2009) argue that hunger is quite widely experienced by Tanzanian families whereby two in five rural children and more than one in five urban children were stunted mid-decade (stunting indicates long-term malnutrition) and one in five overall was underweight (a composite indicator of long- and short-term malnutrition). In this case, reducing the number of meals per day and low quality of meals consumed are widespread response to hardship and food shortage of many rural households (Shepherd et al. 2011). As in many sub-Saharan African countries, even in areas with surplus production, farm households sell their surplus product immediately after harvest.

Food Shortage Coping Strategies at Kikombo and Idifu Wards

The strategies found at study villages in addressing problem of food shortage are mainly short-term solutions (Table 9). Results showed that buying is the most used method (82.3%) followed by casual labor (54%), remittances (51.1%), food aid (24.8%), borrowing (16.5%), livestock selling (16.5%), and other ways like pot making and small business (29.3%). Indeed, food insecurity in Africa is considered as a challenge across the region and that its causes are complex, attributed to multiple and often intertwined factors (Pauw and Thurlow 2011). The main concerns are the impacts of climate change, increase in food prices, and loss of subsistence and traditional food crops and cash crop (FAO and WOCAN 2010). During focused group discussion, participants also revealed that the use of drought-tolerant seeds is an important method in controlling food shortage in their villages. The measures which aim at improving household food security were initiated in 2015/2016 cropping season and involved the use of improved seeds and production of QDS, coupled with good agronomic practices and soil water management among others.

Table 9 Food shortage coping strategies at Kikombo and Idifu wards during bad seasons

Control method[a]	Frequency	Percentage
Buying	329	82.3
Begging	27	6.8
Borrowing	66	16.5
Selling of livestock	66	16.5
Casual labor	216	54
Food aid	99	24.8
Remittances	206	51.5
Others	117	29.3

[a]Dataset based on multiple responses

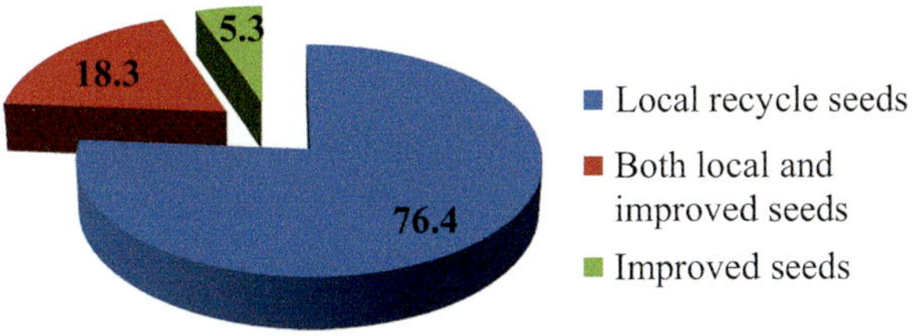

Fig. 2 Farmers' perceptions on main sources of crop seeds at Kikombo and Idifu wards during 2015/2016 cropping season

Farmers' Perceptions on the Main Sources of Crop Seeds

Farmers' perceptions on the main sources of crop seeds are depicted in Fig. 2. Before the introduction of QDS at Kikombo and Idifu wards, the study was conducted to explore farmers' perceptions on the main sources of crop seeds during 2015/2016 cropping season. The findings revealed that 76.4% of the respondents use local seeds, 18.3% use both local and improved seeds, and only 5.3% of respondents use improved seeds.

Farmers Perception on Access to Improved Seeds and Their Limitation

Farmers Perception on Access to Improved Seeds

Farmer's perceptions on access to improved seeds are given in Table 10. Interestingly findings revealed that among the improved seed users in Kikombo and Idifu wards, 69.4% grow pearl millet (Shibe and Okoa varieties), 41.1% cultivate maize (TMVI, Staha Seedco, Catuman), 37.6% cultivate sunflower (Record, Zebra), 7.1% grow groundnuts (Nyota), 10.6% grow simsim (short and Naliendele), and 9.4% grow sorghum (Pato, Macia), and paddy and cow pea hold 1.2% each (Table 10). However, 3.5% of the respondents using improved seeds do not know the names of such seeds. Further analysis noted that most of the improved seeds grown in Kikombo and Idifu wards were released in the period between 1980 and 1995. For

Table 10 Types of improved seeds used during the past growing seasons (n = 400)

	Ward of respondent[a]					
	Kikombo		Idifu		Both	
Crop type	Frequency	%	Frequency	%	Frequency	%
Pearl millet (Shibe[b])	23	27.1	18	21.2	41	48.2
Sunflower (Record)	12	14.1	12	14.1	24	28.2
Maize (TMVI)	3	3.5	0	0	3	3.5
Sunflower (Zebra)	5	5.9	3	3.5	8	9.4
Maize (Staha)	16	18.8	12	14.1	28	32.9
Pearl millet (Okoa)	7	8.2	11	12.9	18	21.2
Sorghum (Pato)	5	5.9	3	3.5	8	9.4
Groundnuts (Nyota)	3	3.5	3	3.5	6	7.1
Simsim (Short)	1	1.2	5	5.9	6	7.1
Maize (Seedco)	0	0	3	3.5	3	3.5
Simsim (Naliendele)	0	0	3	3.5	3	3.5
Paddy (Fire jike)	0	0	1	1.2	1	1.2
Cow peas	0	0	1	1.2	1	1.2
Maize (Katumani)	0	0	1	1.2	1	1.2
Improved pepper seeds	1	1.2	0	0	1	1.2

[a]Dataset based on multiple responses
[b]Words in brackets are names of improved variety

instance, improved pearl millet varieties, namely, Shibe and Okoa, cannot be found in agro-dealer shops except at Hombolo Agricultural Research Institute mandated for semi-arid technologies.

However, field visit conducted in study villages during 2015/2016 cropping season noted that improved varieties claimed to be grown were not vividly seen in their fields. It is an indication that smallholder farmers have been recycling the same seeds probably obtained quite a long time ago. About 95% of improved varieties claimed to be grown by respondents were released in the 1980s and 1990s. Thus, this suggests that types of seeds grown by majority of farmers in true sense are not improved seeds which have a proven record of high productivity and drought-tolerant attributes. The conclusion is that majority of vulnerable smallholder farmers are not aware of the benefits that could be realized when using improved seeds. Farmers' decision to adopt or not adopt new technology is based on the profitability and risks associated with the new technology (Kaliba et al. 2000).

Farmers' Perception on the Limitation of Improved Seeds

The vulnerable smallholder farmers at Kikombo and Idifu wards pointed out reasons for not using improved seeds (Table 11). These include unavailability of improved seeds in the village (44.1%), high cost of improved seeds (41.8%), and lack of knowledge on the use of improved seeds (11.2%). Of all the reasons mentioned above, unavailability and prices of improved seeds were

Table 11 Farmers' perceptions on the limitation of improved seeds (n = 340)

	Ward of respondent					
	Kikombo		Idifu		Total	
Reasons[a]	Frequency	%	Frequency	%	Frequency	%
Because of sand soil	3	0	0	0	3	1
Low productivity	2	0.7	1	0.3	3	1
Not available in the village	41	13.5	93	30.6	134	441
High cost/price	86	28.3	41	13.5	127	41.8
Lack of knowledge	9	3	25	8.2	34	11.2
Used to local seeds	7	2.3	10	3.3	17	5.6
Does not resist drought	2	0.7	2	0.7	4	1.3
Seeds are bought at very late time	2	0.7	5	1.6	7	2.3
They need chemical treatment	0	0	1	0.3	3	1
Do not know	0	0	8	2.6	8	2.6

[a]Dataset based on multiple responses

the major barriers on the uptake of improved seed interventions by vulnerable farming communities. Earlier findings indicated that Tanzania requires 100,000 t of seed annually. However, due to inefficiencies in the seed sector, only 13,000 t of improved seed are marketed annually, out of which 5000 t are of maize (Barnett et al. 2011). Therefore, in order to unravel a problem on the access to improved seeds, the use of community seed production approach notably QDS model was rolled out in Kikombo and Idifu wards during 2015/2016 to 2017/2018.

Farmers' Perceptions on the Willingness to Produce Quality Declared Seed

The study was conducted to examine the willingness of vulnerable farming communities at Kikombo and Idifu wards to engage in the production of Quality Declared Seeds that can endure in harsh and uncertain climate. The findings indicated that 87.7% of the respondents were ready to provide plots and produce seeds at community level. Seed is the most important strategic inputs in agricultural production that determines the quantity and quality of output. Study has shown that the use of quality seed has potential to double or even triple the yields of most crops, producing higher quality of agricultural produces and satisfying consumer demands (CTA 2014).

The QDS producer groups were established in each four villages of Kikombo and Idifu wards. Seed producers developed operational structures for leadership. Each seed producer group in each participating village had a chairperson, secretary, and treasury. The chairpersons provide group leadership, secretary to record all group proceedings include data coming from farmers and the farmer to farmer

Table 12 Number of farmers trained on the QDS production in Kikombo and Idifu wards during 2015/2016 cropping season

District	Ward	Village	Gender		Total
			Male	Female	
Chamwino	Idifu	Idifu	15	10	25
		Miganga	22	19	41
Dodoma	Kikombo	Chololo	9	2	11
		Kikombo	22	16	38
	Total		**68 (59.1)**	**47(40.9)**	**115**

Figure in brackets is percentage

as a link person between farmers and extension. In 2015/2016 a total of 32 farmers (male, 22; female, 10) and 2 institutions across study villages were registered officially by TOSCI after qualifying the minimum standard stipulated in Seed Act of 2003. Capacity building of QDS producers and lead farmers was conducted before inception of 2015/2016 cropping season. A total of 115 farmers including non-seed producers across target villages received training on the production of quality seed (Table 12). The number of men trained was 68 (59.1%) and women 47 (40.9%) less by 10% to achieve 50%.

Willingness to Buy Improved Seeds in Kikombo and Idifu Wards

Previous study has shown that the use of quality seed has potential to double or even triple the yields of most crops, producing higher quality of agricultural produces and satisfying consumer demands (CTA 2014). It is against this background that the Ecovillage Adaptation to Climate Change in Central Tanzania Project envisioned to transform the vulnerable smallholders farmers from seed dependence syndrome to self-reliance. An approach employed to catalyze smallholder farmers to buy improved seeds involved popularization and evaluation of drought-tolerant varieties, namely, sorghum (NACO Mtama 1 variety), sunflower (Record variety), and pigeon pea (Ilonga medium duration variety), among others. In the first two cropping seasons, i.e., 2015/2016 and 2016/2017, smallholder farmers received seeds for testing. This approach is built up on the assumption that majority of limited resources farmers cannot dare to risk on new technology without a prior knowledge on the performance of seeds.

During 2017/2018 no free seeds were provided; instead modalities for selling QDS seeds produced by their fellow farmers were initiated as stipulated in section "Willingness to Buy Improved Seeds." above. Interestingly, during the period between October 2017 and January 2018, a total of 1.2 t of improved sorghum variety (NACO Mtama 1) was bought by smallholder farmers across study villages. By end of January 2018, the number of farmers who bought improved seeds was 453 (M = 185; F = 268). Findings showed that percentage of female who bought improved seeds was 59.2% and men 40.8%.

Interestingly, during 2017/2018 cropping season, improved seeds for sorghum couldn't be available from any known source in Dodoma Region. As a result, stockist took advantage of buying the QDS produced at Kikombo ward; QDS producers sold stock of improved sorghum (NACO Mtama 1) they had produced in 2016/2017. Above all, individual farmers bought sorghum seeds from the ECO-ACT seed producers. These initiatives pave a way for the majority of vulnerable smallholder farmers in semi-arid zone of Tanzania to access drought-tolerant improved seeds with proven record.

Conclusion

In Tanzania, the QDS model officially became operational in 2003 (Seed Act 2003). Despite its good intention to unleash problem on access to improved seeds for the majority of limited resource-endowed farmers, the vast areas in Central Tanzania still have limited access to improved seeds. Findings revealed that to date majority of farming communities have continued to use inferior recycled seeds, thus resulting in insecure harvest. In view of the above, "Ecovillage Adaptation to Climate Change in Central Tanzania Project" currently implemented in semi-arid areas of Central Tanzania explored the innovative approaches required in introducing and sustaining the QDS model for building resilience on crop production. Indeed, the QDS is considered to be the most affordable model required for addressing the challenge of food insecurity for limited resource farmers in semi-arid zones. However, customizing QDS model in semi-arid zone such as Central Tanzania relies on a number of factors which encompass the following:

(i) Information on common food crops grown in the area, food situation, responses to food shortage, and willingness of community member to engage in production of QDS is imperative.
(ii) In-depth understanding of perceptions of farming communities on the impact of climate change and variability on crop production for operationalizing QDS model.
(iii) Farmers who can read and write, in the case of Tanzania, a minimum education level would be a standard seven for in-depth understanding of seed regulations.
(iv) In the case of new variety, the popularization is inevitably important, and this would require a setup of demonstration plots prior to engaging in QDS production.
(v) Formation of seed producer group should engage well-motivated farmers already farming with good track of records on crop production and applying good agronomic practices in their fields.
(vi) Formation of seed producer group either at the village or ward level is very crucial in order to lower the costs associated with inspection and registration

by authorized institute designated for quality control, and Tanzania official seed certification institute (TOSCI) is mandated for seed quality control.

(vii) Assessment on willingness to buy improved seeds is very important in mapping of farmers' preference for building confidence of seed producers to sustain in their business.

(viii) The study recommends producing small quantity of QDS in the first place while exploring trend on seed amount bought.

(ix) The study recommends seed producers to focus primarily on open pollinated varieties as opposed to hybrid seeds which can be recycled for at least two seasons due to limited purchasing power of smallholder farmers.

Acknowledgment The authors wish to acknowledge the European Union through Global Climate Change Alliance, Tanzania, for financial assistance and technical support during the implementation of the project in semi-arid areas of Central Tanzania.

References

Kaliba AR, Verkuijl H, Mwangi W (2000) Factors affecting adoption of improved maize seeds and use of inorganic fertilizer for maize production in the intermediate and lowland zones of Tanzania. J Agric Appl Econ 32(1):35–47

Barnett C, Munhamo C, Pinto Y (2011) East Africa synthesis report (Kenya, Tanzania, Uganda). Country case studies on the pass value chain strategy/approach and its impact/ effect on smallholder farmer yields in Africa

Boko M, Niang I, Nyong A, Vogel C, Githeko A, Medany M, Osman-Elasha B, Tabo R, Yanda P (2007) Africa. In: Parry ML, Canziani OF, Palutikof JP, van der Linden PJ, Hanson CE (eds) Climate change 2007: impacts, adaptation and vulnerability. Contribution of working group II to the fourth assessment report of the intergovernmental panel on climate change. Cambridge University Press, Cambridge, UK, pp 433–467

CTA (2014) Seed systems, science and policy in east and Central Africa. CTA, Wageningen

FAO and WOCAN (2010) Supporting women producers to respond to the challenges of food insecurity. a paper presented in women meeting by FAO, WOCAN, November 2009 Rome, Italy. 8pp

Hulme M, Doherty R, Ngara T, New M, Lister D (2001) African climate change: 1900–2100. Clim Res 17:145–168

Leach V, Kilama B (2009) Institutional analysis of nutrition in Tanzania. Repoa special paper 09.31. REPOA, Dar es Salaam, TZ

National Adaptation Programme of Action (2007) United Republic of Tanzania, Vice President's Office, Division of Environment, Dar es Salaam

Paavola J (2008) Livelihoods, vulnerability and adaptation to climate change in Morogoro, Tanzania. Environ Sci Pol 11:642–654

Pauw K, Thurlow J (2011) Agricultural growth, poverty and nutrition in Tanzania. Food Policy 36:795–804

Schlenker W, Lobell DB (2010) Robust negative impacts of climate change on african agriculture. Environ Res Lett 5:014010

Shepherd A, Kayunze K, Vendelin S, Darko E, Evans A (2011) Hidden hunger in rural Tanzania: what can qualitative research tell us about what to do about chronic food insecurity? Working paper 206. Chronic Poverty Research Centre. ISBN 978-1-908536-02-0

Assessment of Information on Successful Climate-Smart Agricultural Practices/Innovations in Tanzania

131

Paschal Arsein Mugabe

Contents

Introduction Background 2722
Materials and Methods 2725
Case Study 2725
Data Collection 2727
Literature Review 2728
Archival 2729
Household Survey 2729
Online Survey 2729
Results and Discussion 2730
Conclusion 2735
Recommendations 2738
References 2740

Abstract

This chapter examines some of the key practice/innovations achieved by agricultural communities to achieve the transformation to climate-smart agriculture (CSA). Data were collected using household survey, online survey, literature review, and archives. The results reveal that most of the CSA innovations and practices are geared at improving both crops and livestock. The potential adaptation includes promoting climate change resilient traditional and modern knowledge on sustainable pasture and range management systems, promoting development and implementation of land use plans countrywide, and enhancing development of livestock infrastructure and services. These CSA practices and policies are having a widespread impact on food security, are helping farmers to

P. A. Mugabe (✉)
College of Engineering and Technology (COET), University of Dar es Salaam, Dar es Salaam, Tanzania
e-mail: pmugabe@udsm.ac.tz; paschal.mugabe@gmail.com

W. Leal Filho (ed.), *Handbook of Climate Change Resilience*,
https://doi.org/10.1007/978-3-319-93336-8_180

adapt to climate change, and are contributing to climate change mitigation. These practices show the potential of agriculture to adapt to a changing climate; to be more resilient and protect farmers against future changes in weather patterns, pests, and diseases; and to slow the rate of climate change. However, this chapter also found that while a number of CSA-related programs and projects are ongoing, and a number of different CSA practices are being used in various parts of the country, these CSA systems are not widely spread across the country and stand as limiting factors, partly because of limited financial resources, capacity, knowledge, and policy support. However, the potential and the opportunity are that various CSA interventions have been implemented and continue to be implemented, with variable success, using different approaches by individual farmers as well as national, regional, and international organizations in Tanzania.

Keywords
Climate-smart agriculture · Mitigation · Adaptation · Food security

Introduction Background

Climate change and related global concerns dominate the current environmental agenda, as evidenced by the recent wave of articles, symposia, workshops, and other scientific and lay forums dealing with this issue. Africa's climate is changing. Across the continent rainfall patterns are set to alter. In many areas droughts will become more frequent, more intense, and last longer. In others, new patterns of rainfall will cause flooding and soil erosion. And it has been doing so. Climate change is emerging as one of the major threats to development across the continent. Africa's agriculture must undergo a significant transformation to meet the simultaneous challenges of climate change, food insecurity, poverty, and environmental degradation (Nyasimi et al. 2014). In sub-Saharan Africa, heavy reliance on rain-fed agriculture renders farming communities more vulnerable to effects of climate change and variability resulting in widespread food insecurity and poverty. Climate change and variability present new challenges particularly for smallholder farmers whose mainstay is agriculture. Providing food for the future will require a holistic approach to produce on less and by using appropriate methods. Farming communities and researchers are obliged to reevaluate mainstreaming of farming practices and techniques and look for ways of securing food through adoption of conservation agriculture, integrated livestock keeping, fish farming, and agroforestry (URT 2015). Current debates among many prominent development actors feature climate-smart agriculture (CSA) as the solution to the global challenge of achieving food security and economic development under a changing climate. The FAO, member and host of the Global Alliance for Climate-Smart Agriculture, has defined CSA as agricultural production systems that "sustainably increase productivity, resilience (adaptation), reduce/remove greenhouse gases (mitigation) and enhance achievement of national food security and development goals" (URT 2015). Global investment in agricultural research and development is dominated by commercial and developed country interests, which can result in the design and promotion of technologies that fail to incorporate the knowledge and needs of smallholder farmers in developing countries or that degrade the natural resources on which they depend. Agriculture in sub-Saharan Africa is highly vulnerable to climate change, and urgent actions are needed to combat its impacts and maintain or improve food security and livelihoods. Eastern Africa is among the regions in the world most affected by climate change and is also among the most vulnerable regions. In the subregion approximately 75.5 million people are economically involved in agriculture, either in full-time employment or as a main household livelihood. Additionally, approximately 95% of the food in the subregion is grown under rain-fed agriculture and is highly vulnerable to adverse weather conditions such as droughts, dry spells, and uneven or erratic rainfall. Smallholder farmers in East Africa need information and knowledge on appropriate climate-smart agriculture (CSA) practices, technologies, and institutional innovations in order to effectively adapt to changing climatic conditions and cope with climate variability. The climate-smart agriculture (CSA) approach offers an opportunity of achieving triple wins of food security, adaptation, and mitigation. This research intended to examine some of the key technical responses required to achieve the transformation to CSA. Building on case study of Tanzania, the research outlines a range of practices, approaches, and tools aimed CSA practices. Tanzania has been referred to as one of Africa's sleeping agricultural giants (Binswanger-Mkhize and Gautam 2010). Endowed with natural resources, an abundance of arable land, and plenty of available water resources, Tanzania has tremendous agricultural potential. Direct access to the Indian Ocean and with six neighboring countries growing rapidly, Tanzania is provided with ample market opportunities to use that agricultural potential for its economic transformation (ECDPM & ESRF 2015). Tanzania depends largely on agriculture (as indicated in Table 1), which is strongly dependent on natural resources such as land, forests, air, and water. Sustainable utilization of these resources is vital for the growth and sustainability of the sector. However, agriculture is vulnerable to the effects of climate change associated with global warming. Smallholder farmers dominate the agricultural sector with average farm sizes of between 0.2 and 2.0 ha, depending on

Table 1 Share of agriculture in Tanzania's national gross domestic product: 2005–2013

Year	Agriculture
2005	26.9
2006	25.3
2007	25.0
2008	24.9
2009	23.8
2010	24.1
2011	24.7
2012	24.8
2013	24.7

Source: Modified from URT (2013)

the location. Yields have been mostly stagnant, and agricultural productivity gains have been based more on the expansion of cultivated land, which is one of the major drivers of deforestation and land degradation in the country. Studies by the Tanzania Meteorological Agency (TMA) have shown that some of the previously known highly productive areas such as the Southern and Northern Highlands will continue to be affected by declining rainfall, frequent droughts, and significant increase in spatial and temporal variability of rainfall with long-term implications in the agricultural sector planning and resource allocation, such as improved seed pesticides and even the shifts in types of agricultural produce (URT 2009).

To deal with climate change problem that affects agricultural prosperity of Tanzania, there have been many attempts to solve the problem. CSA emerged in Africa around early 2010s, several CSA-related programs and projects are ongoing, and a few different CSA practices are being used in various parts of the country; these CSA systems are not widely spread across the subregion, partly because of limited financial resources, capacity, knowledge, and policy support. However, various CSA interventions have been implemented and continue to be implemented, with variable success, using different approaches by individual farmers as well as national, regional, and international organizations in the Eastern Africa subregion. There is no single guiding policy, and there is a lack of clear understanding about the concept and how to be implemented as well as research gap in this approach. This chapter aims to clear this gap by generating the knowledge to design climate-smart agricultural systems for agricultural communities. Drought spells and extreme precipitation events are likely to increase in changing climate and endanger the stability of agricultural production. Drainage fosters mineralization of organic soils and deteriorates the soil and water quality, inducing a high risk of crop failure. Due to subsidence, the remaining lifetime of such systems is limited by the (rising) sea water level and underlying low-quality soil layers. Organic soils are the hotspots of vulnerability, adaptation needs, and GHG emissions in temperate and boreal. The proposition is that wet organic soils could be used as risk insurance in dry periods on farm or regional level, while water and soil management could prevent yield losses due to traffic ability problems in wet conditions. All these problems call for more research on CSA. This chapter provides and distributes evidence that climate-smart management improves yield stability and quality as well resilience to climate change while providing strong GHG mitigation and improved soil and water quality. Since agriculture accommodates high share in Tanzanian gross domestic product (as indicated in Table 1), there is a need to enhance CSA as a guiding framework in improving agricultural sector. This chapter expects that the strong potential for adaptation to increased climatic change on farmed organic soil will facilitate mitigation of the largest GHG source from agriculture in Tanzania. Study country represents typical land use systems and unique challenges with a small share of organic soils drained for agriculture. Stakeholders are involved in bidirectional manner as international, national, regional, and district partners, by questionnaires, and these results will be disseminated widely to the scientific community, practitioners, decision-makers, media, and the public.

Despite the relevance of CSA and its potential use as an approach/tool toward climate change adaptation, mitigation, and increased food productivity, there is a paucity of scientific works in this field. This research contributes toward filling this gap. It describes what CSA is, discusses a variety of ways of pursuing it, and outlines – based on case studies from Tanzania – the various means via which it can be achieved. Finally, it outlines some new opportunities in the fields of climate change and agriculture which exist and which may help to improve the livelihoods of Tanzanian smallholder farmers and pastoralists. The aim is to assess different innovations, opportunities, and factors for scaling up and out of climate-smart agriculture (CSA) in Tanzania. Building on a methodology developed, the research embarked on this chapter with the aim to identify and assess the innovations and practices of CSA in Tanzania.

Materials and Methods

Case Study

As a base of the analysis, Tanzania was used as a case study. The review was based on different CSA practices operating in different parts of the country in the form of projects and programs. Tanzania is located south of the equator, lying mostly between latitudes 1° and 12°S and longitudes 29° and 41°E. It is constituted by Tanzania mainland and Zanzibar with a total area of 945,087 km^2 comprised of land area of 883,749 km^2 (881,289 km^2 mainland and 2,460 km^2 Zanzibar) and 59,050 km^2 of inland water bodies and part of the Indian Ocean. Tanzania mainland encompasses island of Mafia (518 km^2), while Zanzibar comprises islands of Unguja (1,666 km^2) and Pemba (795 km^2) as major islands. There are numerous smaller islands both in the mainland and Zanzibar. Tanzania shares borders with Kenya and Uganda to the north; Rwanda, Burundi, and Democratic Republic of the Congo to the west; Zambia and Malawi to the southwest; Mozambique to the south; and the Indian Ocean to the east (URT 2013). The climate in the country is diverse because of proximity of the ocean and inland lakes; wide altitudinal range, which governs temperature; and latitude. It is characterized by two main rain seasons, namely, the long rains and the short rains which are associated with the southward and northward movement of the Intertropical Convergence Zone (ITCZ). The long rains (*masika*) begin in the mid of March and end at the end May, while the short rains (*vuli*) begin in the middle of October and continue to early December (as indicated in Figs. 1 and 2). The northern part of the country including areas around Lake Victoria Basin, Northeastern Highlands, and the Northern Coast experiences a bimodal rainfall regime. Central, south, and western areas have a prolonged unimodal rainfall regime starting from November, continuing to the end of April. In general, annual rainfall varies from 550 mm in the central part of the country up to 3,690 mm in some parts of southwestern highlands. Most of the country receives less than 1,000 mm, except the highlands and parts of the extreme south and west where 1,400–2,000 mm can be expected. Average rainfall in the central regions is around 600 mm. Modeling of the

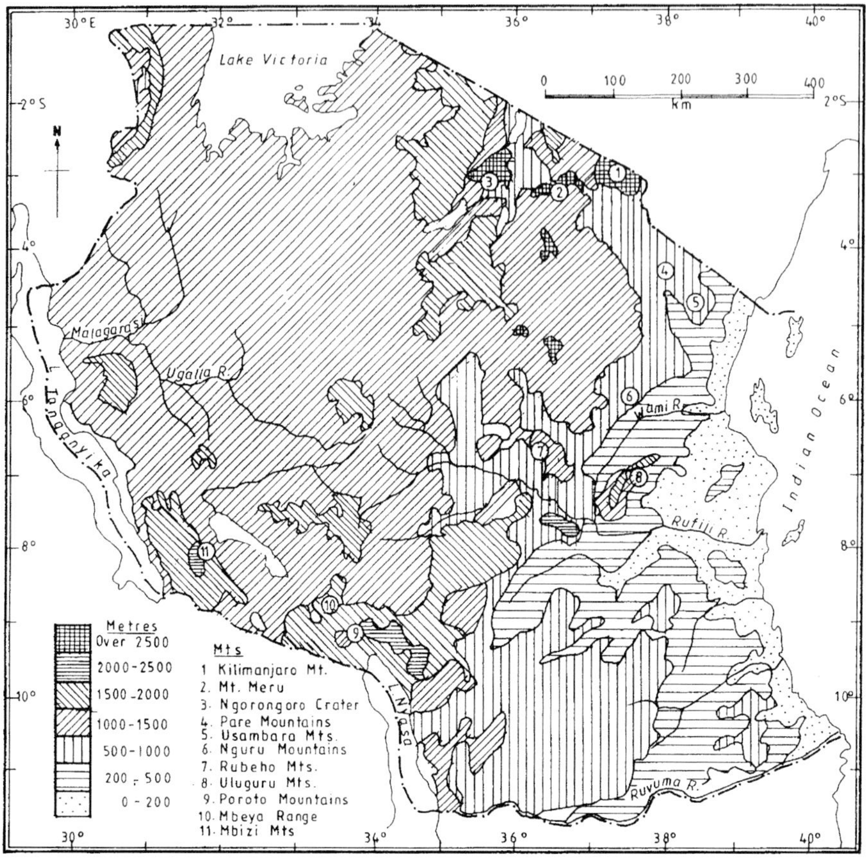

Fig. 1 Topographic map of Tanzania. (Source: Modified from Basalirwa et al. 1999)

future climate of Tanzania climate projections shows that the mean temperatures will increase throughout the country particularly during the cool months by 3.5 °C, while annual temperatures will increase between 2.1 °C in the northeastern parts to 4 °C in the central and western parts of the country. Predictions show that the mean daily temperature will rise by 30–50 °C throughout the country and the mean annual temperature by 20–40 °C. There will also be an increase in rainfall in some parts, while other parts will experience decreased rainfall. Predictions further show that areas with bimodal rainfall pattern will experience increased rainfall of 5–45% and those with unimodal rainfall pattern will experience decreased rainfall of 5–15% (URT 2007). The climate of Tanzania (Fig. 2) is characterized by two main rain seasons, namely, the long rains and the short rains which are associated with the southward and northward movement of the ITCZ (URT 2011). The long rains (masika) begin in the mid of March and end at the end May, while the short rains (vuli) begin in the middle of October and continue to early December. The

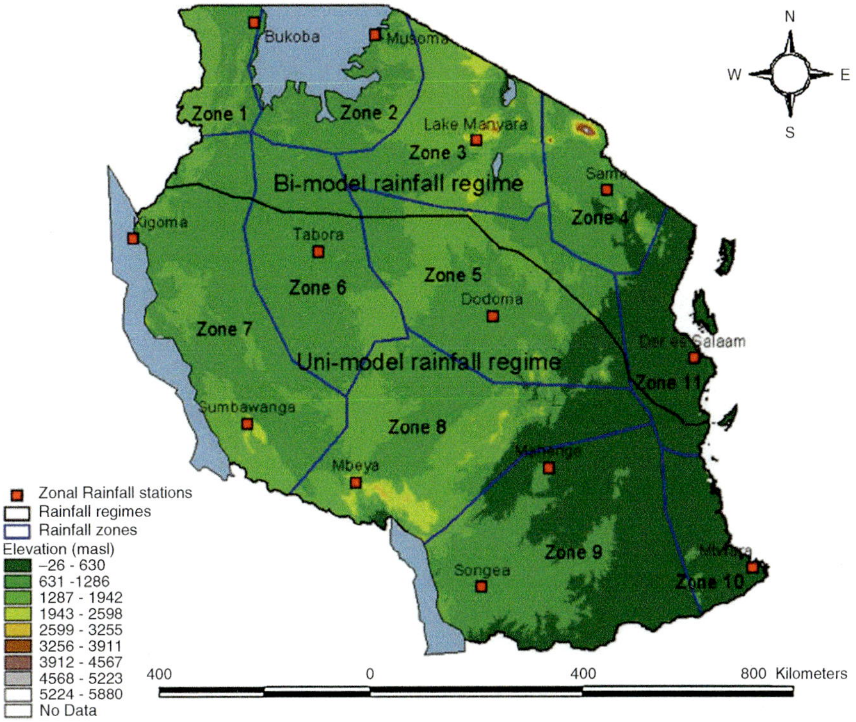

Fig. 2 Tanzania rainfall zone. (Source: Wambura et al. 2014)

northern part of the country including areas around Lake Victoria Basin, North-eastern Highland, and the Northern Coast experiences bimodal rainfall regime, whereby the first maximum occurs in the period of March, April, and May (MAM) while the second maximum in the period of October, November, and December (OND). Central, south, and western areas have a prolonged unimodal rainfall regime starting from November continuing to the end of April. Annual rainfall varies from 550 mm in the central part of the country up to 3,690 mm in some parts of southwestern highlands. There are about 12 rainfall zones falling in the 2 regimes (Fig. 2), with each rainfall zone having homogeneous rainfall within its zone. Table 2 shows the summary of rainfall amount in the rainfall zones taken from the zone rainfall stations.

Data Collection

The phases were conceptualized in a systemic design using a mixed-methodology approach including both qualitative and quantitative measurements to contextually collect, analyze, and integrate data (Table 3).

Table 2 Summary of monthly rainfall amount (mm), 1980–2010

Station	Minimum	Average	Maximum
Bukoba	47.5	168.3	333.3
Musoma	15.1	74.5	174.4
Same	3.6	44.5	99.7
Tabora	0.8	79.1	201.3
Lake Manyara	16.2	82.2	213.5
Dodoma	0.0	49.6	136.7
Sumbawanga	2.0	75.3	184.7
Mahenge	9.7	195.3	485.0
Dar es Salaam	0.0	81.3	238.1
Mbeya	0.2	99.3	230.1
Songea	0.8	89.1	263.4
Mtwara	9.4	88.8	220.1

Source: Modified from Wambura et al. (2014)

Table 3 Summary of main approaches and techniques of the study

		Objective			
Method	Techniques	1	2	3	4
Literature review	Content/meta-analysis	+	+	+	+
Archival	Content analysis		+		
Online survey	Content analysis			+	+
Household survey	Available data sets	+			

Literature Review

The review was based on current knowledge of practices and approaches and opportunities for CSA in East African level with a special focus on Tanzania. The review used interdisciplinary findings as well as theoretical and methodological contributions to institutional arrangement on CSA and climatic effects on food security, specifically the review of literature on the socioeconomic characteristics of farmers, food production systems, climate change adaptation and mitigation, policies, research reports, books, and journal articles. Data and information collected included adaptation and mitigation measures in use, case outlines of successful climate-smart agriculture, indicators of development and governance, and national policies and strategies of CSA. Available local and international literature on climate-smart agriculture required a desk review and gathering of information from available sources which included reports and documents on CSA in the subregion, CSA meeting and conference proceedings, annual progress reports for some CSA projects, and organizational databases and the Internet. These documents were reviewed over the entire period of the study, and key issues were identified for further questioning and consideration. Information from the desk review was also used to inform the design of the data collection tool for online survey. Data and information collected from these sources were collated and analyzed and conclusions made.

Archival

Data from the secondary sources on institutional arrangements on CSA, adaptation, mitigation (CO_2 emissions), crop production, land, climate change, and agriculture was gathered. Data and information collected included adaptation and mitigation measures in use, case outlines of successful CSA, observed temperature and rainfall, vulnerability to climate change and impacts, socioeconomic and demographic characteristics of farmers, crop yield, indicators of development and governance, and national policies and strategies. At its most recent, this research will collect data with reference to the 2017 agricultural production season which is considered a baseline year for this report except as otherwise specified.

Household Survey

The data sets from household survey of 2014 conducted for author's PhD thesis on issues related to adaptation, mitigation, and increased productivity were analyzed and used in this chapter due to its relevance. About 300 household survey questionnaires were used for this purpose. Additionally, some research reports and publications used household survey which was also applied for this chapter.

Online Survey

The survey targeted a wide variety of professionals, academia, and students from Symposium on Climate Change and Droughts Resilience in Nairobi from 7th to 9th of November 2017 to sample respondents for the chapter. Those researchers from academic field were categorized as academicians. Also, the participants of CSA workshop in January 2017 are organized by the African Climate Smart Agriculture Alliance (ACSAA), together with the Tanzanian Ministry of Agriculture Livestock and Fisheries (MALF) in Dar es Salaam, Tanzania. Additionally, this sampling approach was expected to make the results more useful to the individual organizations and policy-makers. The online survey was designed with input from the students and staff of Research and Transfer Centre, Sustainability and Climate Change Management, HAW. This helped to ensure that the questions and response options were easily understood and the questions were appropriate for the audience. The survey took about 10–15 min for most respondents (see the professional background of the respondents in Fig. 3). "Climate Change Information" explored how respondents treat the topic of CSA and climate change in their community. About 56 responses were used for this analysis. The age structure was 18–24 (1.79%),

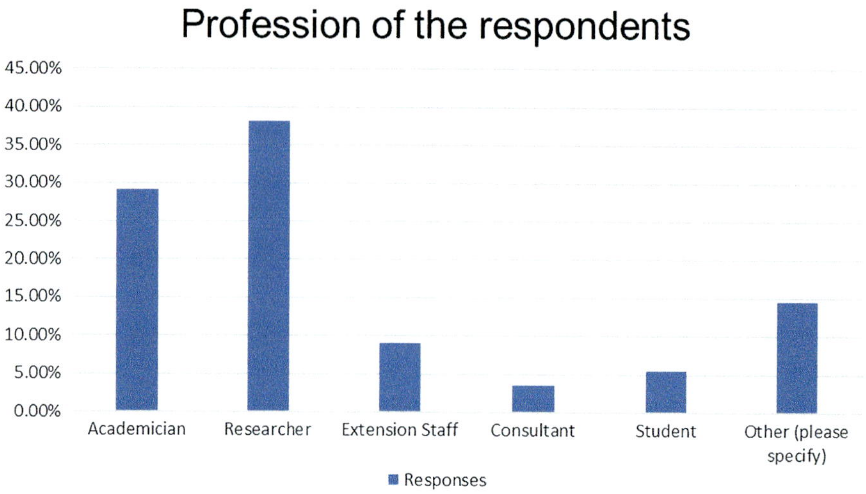

Fig. 3 Professional background of respondents of the online survey. (Source: Author, 2017)

25–34 (33.93%), 35–44 (26.79%), 45–64 (32.14%), and above 65 (5.36%) with male 65.45% and female 34.55%.

Results and Discussion

Adaptation. The results reveal that most of the CSA innovations and practices are geared at improving both crops and livestock. In Tanzania, the potential adaptation in the livestock sector includes promoting climate change resilient traditional and modern knowledge on sustainable pasture and range management systems, promoting development and implementation of land use plans countrywide, and enhancing development of livestock infrastructure and services. Other adaptation options include promoting development of livestock insurance strategies, strengthening weather forecast information, promoting livelihood diversification of livestock keepers, and promoting improved traditional livestock keeping systems (URT 2017). In fisheries the strategic intervention includes those for enhancing monitoring of fisheries' habitat and species, facilitating enhancement and/or development of integrated data management systems in the fisheries sector, protecting aquaculture and enhancing protection of aquatic systems, and supporting alternative livelihood initiatives for fisheries community.

Mitigation. According to URT (2012), despite its negligible GHG emissions, Tanzania can participate in mitigation activities to contribute to its sustainable national development (URT 2012). In agriculture, potential mitigation actions were identified to be promoting agroforestry systems, practices, and technologies. Potential mitigation interventions in livestock include promoting manure management practices, promoting appropriate technology for animal feed production, promoting waste management in abattoirs, and improving rangeland's productivity and complementary activities. Farmers who were not using scientific weather information cited unreliability and inaccessibility of weather information as the main reason. Women particularly reported a lack of adequate weather information to plan for their farming activities. Moreover, weather information is usually passed on to the women through their husbands or village chiefs, and, in most cases, they do not know how to use the information. While there are multiple initiatives in East Africa that are aimed at producing and delivering climate information services for farmers (Conway et al. 2010; Lu 2009), many challenges still remain in terms of accurate and timely weather and climate forecasts to support farmers' efforts to adapt to a changing climate and increasing climate risks.

The use of furrows. They are used to trap rainwater and are tied at the end to prevent water flow out of the furrow. Furrows are suitable for inter cropping especially cereal and beans. However, the peculiar thing is that similar methods have been useful across the country in areas with similar geophysical characteristics. Generally, contour farming is more effective in Tanzania in areas with slope of 4–6%, and all farm operations are done along the contour (Mati 2007). The wide variation of agricultural ecological zone (AEZ) has been key to determine the wide range of CSA practiced in Tanzania. At some point as CSA practices work better in a combination, climate-smart villages have been a potential way to practice and scale out best bet CSA practices.

Agroforestry. Outstanding traditional technologies/innovations and practices, which follow the concept of CSA as documented by the Ministry of Agriculture, Livestock, and Fisheries (MALF) in Tanzania, include the Chagga (*Kihamba*) agroforestry system, the Haya (*Kibanja*) agroforestry system, the *Luguru* ladder terraces, and the *Matengo Ngoro* pits (soil and water conservation). The Chagga agroforestry system is among the oldest and most sustainable forms of upland farming that has been able to support one of the highest rural population densities in Africa (FAO 2013). All these systems are among the essential agricultural technologies that require complementarities with other resources such as land and water. However, in some communities such as coastal zones, the practice of new technological innovations among the poor communities is so low. For example,

Table 4 Percentages of usage of agricultural inputs in Western Bagamoyo by March 2014

Input	% who used
Hand hoe	54
Tractor	1
Plough	1
Improved seeds	19
Chemical fertilizer	11
Organic manure	16
Pesticide for cultivating	46
Pesticide for storing	25

Source: Mugabe (2016)

Table 4 shows household survey conducted by author in Western Bagamoyo which revealed little knowledge about the use of agricultural inputs. Some of the respondents commented that they do not need to use additional fertilizer because they believed that their land was fertile enough.

Sustainable land use. The result also shows that attempts to mobilize small-holder farmers to invest in sustainable land use often fail when land conflicts are not sorted out well and when there is no feeling of land security. Crop producers are often more willing to invest in their land for higher and sustainable production when they are sure to use it for a long period and hence benefit from their investment (Kami et al. 2016). Tanzania's land base is made of 70% village lands, 28% protected areas including national parks and wildlife reserves, and 2% general land (urban or private) (Kami et al. 2016). Climate-smart villages have been a method to practice and demonstrate climate-smart practices. Chololo Ecovillage, an initiative funded by the EU to improve livelihood of the poor in an arid land of Central Tanzania, has been showing good results in mitigation to climate change (https://chololoecovillage.wordpress.com; Kalumanga et al. 2014).

Crossbreeding. As for livestock production, the existing adaptation methods are strengthening crossbreeding for resistant breeds, strengthening tick and tsetse control programs, strengthening livestock extension services, improving livestock marketing infrastructure, enhancing research and development, and promoting zero grazing. The potential adaptation methods are changing land use patterns, tsetse fly control, integrated pest and disease control, sustainable range management, infrastructure development, research and development, education of farmers/livestock keepers, advocating zero grazing, and controlling movement of livestock. Another best bet innovation/practice of CSA evidence in Tanzania is the project *Strengthening the Capacity for Climate Change Adaptation through Integrated Land and Water Management using the CSA approach in Kiroka Village, Morogoro* (Mahoo 2014). System of Rice Intensification (SRI)) principle is transplanting single seedlings, transplanting younger seedlings at 8–12 days old, wide plant spacing of 25 × 25 cm minimum water applications during vegetative growth period, keeping soils moist but well-drained and aerated, frequent weeding with a simple mechanical hand weedier, and applying of organic matter, in preference to chemical fertilizer and seed selection. The combined increase in demand for food and scarcity of water in Tanzania and worldwide necessitates saving water and increasing its productivity of which SRI fits in as demonstrated in Figs. 4, 5, and 6.

In the subhumid-humid zones, the main CSA approaches in use include swampland farming, short duration and drought-tolerant crops, flexible farming calendars, dry season cropping, crop diversification, and multilevel tree crop farming. Apart from agronomic practices, adaptive capacity of farmers is being built by field schools and climate-smart villages, improving land tenure and access, conflict management, as well as early information and warning services by meteorological departments (Msaki et al. 2015). This was reported in Tanzania by FARA, whereby the adaptive capacity of farmers is being built by field schools and climate-smart villages. Example of such is participatory soil management by the African Highland Initiative (AHI) in Tanzania which increases N use efficiency through maximizing N uptake by crop. Other

Fig. 4 SRI: 25 × 25 cm – 60 days after transplanting. (Source: Mahoo 2014)

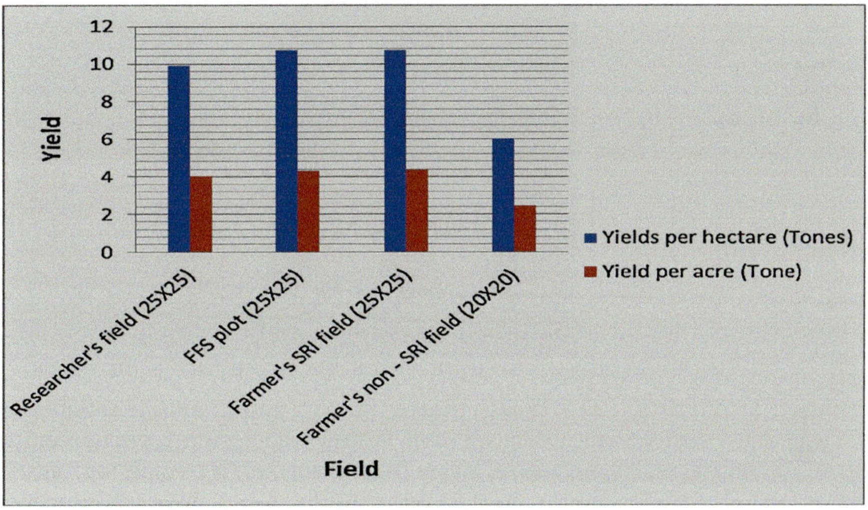

Fig. 5 Grain yield of rice from different treatments. (Source: Mahoo 2014)

adaptive measures are improving land tenure and access, conflict management, as well as early information and warning services by meteorological departments.

Several factors influence farmers' ability to adopt CSA practices. Among the key factors include (a) availability and access to resources needed to use the practices such as land, labor, and financial capital, (b) potential benefits to be accrued vis-à-vis other practices, (c) whether they have the required skills and information to use it, (d) ability to cope with challenges that might arise during or after using the practices, and (e) compatibility with local social and cultural practices. For farmers in Lesotho District, Tanga, region in Tanzania, the ability to adopt CSA practices is influenced by the above factors, albeit for different technologies and practices. Improved crop

Fig. 6 Rooting under SRI (top) and conventional (down) at harvesting. (Source: Mahoo 2014)

varieties, agroforestry, and scientific weather forecast information were the main CSA practices farmers were using, with similar patterns.

While most farmers are aware of many CSA practices and innovations, only a small number of the farmers are adopting the practices. Most farmers indicated their willingness to use CSA practices, but are constrained by several factors including cultural practices, such as tenure and ownership rights, labor requirements, high investment costs, and lack of skills and knowledge on how to use the practices. Irrigation and Savings and Credit Co-Operative Society (SACCOs), for example, require high initial investments. These two practices had the largest discrepancies between awareness and use. Studies have shown that belonging to a rural social or marketing group enhances social and financial capital allowing trust, idea, and information exchange. Thus, membership to a group (e.g., a SACCO) can increase the uptake of a CSA practice. However, SACCO membership will require subscription fee that most farmers cannot afford, even if they would like to belong to the SACCO. On-farm diversification through different farming enterprises such as different varieties of crops and different types of livestock (including beekeeping, poultry) and fish farming is an important risk management strategy and can cushion farmers and their families during bad seasons or years. In addition, farmers can engage in non-farming activities such as value addition enterprises. Livelihood diversification, on-farm and off-farm, is a key risk management strategy and can also increase farmers' adaptive capacity.

Most of the farmers (over 98%) reported that they receive agricultural information orally. This is applicable for crop and livestock production as well as tree planting activities. For crop production, farmers need information at the time of planting, use of proper seed, proper seed spacing, crop rotation, use of traditional and scientific weather information, and appropriate land preparation practices, such as no-tillage. For livestock production, farmers would like information on proper feeds and fodder production, management, vaccinations, deworming, reducing the number of herd, and zero-grazing system. Government extension services are the main source of oral information (75%). Other sources include a farmer's own experience (26%), traditional knowledge (11%), researchers (7%), neighbors (6%), and agri-service providers and seed companies. Farmers reported that the information they receive can enable them to start preparing for the changing climate up to 20 years into the future. This was reported by more than 99% of the respondents. Improving agricultural productivity was the major motivation for seeking and using information and technologies on crop husbandry, new varieties and techniques, and improvement of soil fertility. More than half (55%) of the respondents indicated that the men make decisions on how information is used within the household, while 23% said that it is the woman. Thirteen percent indicated that both women and men are involved in the decision-making process. In terms of use of new technologies, 45% of respondents reported that both men and women are involved. It is, therefore, critical to ensure that information is shared within the entire household, including both women and men. The information also needs to be packaged to suit the different audiences.

Knowledge of farmer and farm characteristics facilitates the appropriate targeting of agricultural technologies to households and locations where they are most suited and therefore stand good chances of adoption. It also helps to assess the willingness to take risks (Charness and Viceiza 2011). As climate change to be taken seriously, communities have been surveyed to declare if they know that climate change has been happening and if after all it has an impact on their livelihoods. The surveys conducted in March 2014 in Tanzania show that the communities in the surveyed areas are aware that climate change is real. However, while communities could directly mention that there is climate change, 80% of respondents identified the indicators of climate change such as unpredictable rains but without being able to link unpredictable rains to climate change. Farmers acknowledging that climate change is happening should be a first step to adapt CSA.

Conclusion

In this chapter, key stakeholders involved in promoting CSA can be grouped according to four broad categories: government, development agencies (including international and local organizations, NGOs, and CBOs), research and academic institutions, and the private sector. There are several barriers that prevent smallholder farmers in Tanzania from adopting CSA practices and technologies, and so far existing policies and actions to remove these barriers remain inadequate. A good understanding of what these barriers are and how they impinge on adoption of CSA practices is essential. Equally essential are actions that favor the removal of these barriers while at the same time promote adoption of CSA practices. These barriers can be classified under two broad categories. The first relates to the physical means or resources required to practice CSA. These can be considered as the hardware barriers and include physical inputs such as land, human resources, equipment, infrastructure, and finances. The second, referred to as the nonphysical or software barriers, relates to the institutional, cultural, policy, and regulatory environment; information, knowledge, and skills; technologies and innovations; and governance, among others.

Effective implementation of CSA often requires an integrated landscape management strategy that synergistically combines a diversity of land uses and economic activities. The Usambara landscape was a useful case for considering how local stakeholders could play an active role in assessing and defining a holistic program of CSA investment within a priority landscape. Research and planning tools, such as the one developed for this chapter, provide a framework for landscape stakeholders to identify the concrete steps that can move them toward a climate-smart landscape. Such tools should continue to be refined, and, as more landscapes begin to plan and implement climate-smart activities, experiences from using them should be documented systematically and shared widely to support the mainstream practice of climate-smart agriculture.

In spite of the development of several climate-smart agriculture (CSA) technologies and the positive gains arising from those technologies, wide-scale adoption of CSA practices remains problematic in Tanzania. There are several barriers that prevent smallholder farmers from adopting CSA technologies, and so far existing policies and actions to remove these barriers remain inadequate. A good understanding of what these barriers are and how they impinge on adoption of CSA practices is essential.

As in many other developing countries, the majority of the poorest groups are composed of women, the disabled, and those living with HIV/AIDS. In many cases CSA programs have targeted these groups and aimed to mainstream gender into their activities. As a result, CSA activities implemented have had an overall positive impact on social equity and gender equality among the target group/s as well as broader communities in general. Many of the activities have had a direct positive impact on gender equality in the form of enhanced food resources and incomes, especially for women who constitute approximately 75% of the rural agricultural labor force in the country.

Other practices that may especially support CSA include integrated pest management, construction of lagoons and wetlands on farms to increase water storage, conservation of riparian buffers, and improved soil fertility management. These practices are encouraged by voluntary certification programs and government recommendations alike (Milder et al. 2015). In the subhumid-humid zones, the main

CSA approaches in use include swampland farming, short duration and drought-tolerant crops, flexible farming calendars, dry season cropping, and crop diversification and multilevel tree crop farming. Conservation agriculture is an approach to farming that can sustainably increase yields from cereal, legume, fodder, and cash crops. The various practices that constitute this approach follow key principles that aim to conserve the soil, rainwater, and soil nutrients and stabilize land production while reducing production costs.

Improving the use and access of weather information is a central area of adaptation and a promising area of development for CSA. Through various projects implemented, institutional actors have worked to develop strategies for providing locally specific weather information that can improve farmers' and pastoralists' ability to make timely farm-level decisions and to ensure that information is accessible in local languages via efficient and appropriate channels. The efforts also aimed to complement weather information with additional advice on the best agricultural practices relevant for the available weather data and to facilitate a support system that enables farmers to take informed action including the support of input market, credit mechanisms.

Current practices and their alternatives formed the subject of several other studies that looked at intercropping, crop diversification, improved soil preparation, and water conservation practices and surveyed their appeal and adoption by farmers. Despite the popularity of intercropping systems, information on the effects of intercropped populations and row orientation in sorghum-cowpea intercropping is limited. Crop competition/interaction can have significant impacts on growth factors, such as soil moisture and solar radiation interception, which limit yields. A better understanding of these systems is important to assist efforts to improve crop management and increase food grain production.

The landscape approach, the Rangeland Rehabilitation and Management Programme, focuses on capacity building and makes use of two well-developed tools: planned grazing and vision setting. Planned grazing is a technical solution to land degradation, while vision setting provides the human or managerial context for improved practice. Technical demonstrations proved the value of properly managed animal herds as plant regenerators and led to community planning and the successful implementation of a grazing plan in their dry season reserve. The plan's most important elements involve dividing a controlled area into blocks, calculating the number of grazing days for the herds in each block based on community-informed forage assessment, and combining animals into as few herds as possible, which move through the blocks by a predetermined sequence according to water availability, grazing competition, distance, and other factors. Most importantly, animals are gathered into tight herds as they graze to maximize soil disturbance and graze a different section of the block in use each day to eliminate overgrazing. The immediate results included improved land health, livestock survival and productivity, youth involvement, and community unity. The successful land restoration can be attributed to the focus on enhancing the four ecosystem processes that together determine ecosystem health and productivity: water cycle, mineral cycle, energy flow, and plant and living communities. Much of the training focuses on "eco-literacy," combined with techniques for low-stress handling of large herds. This has restored the traditional value of herding and herders in the eyes of the community. Not only does this process reverse land degradation, it is also reversing a long-term trend of carbon release from soil to atmosphere.

Recommendations

Relevant information gathered to ensure effective flow of CSA information in Tanzania listed by respondents in the survey were technology transfer, training, mass media and advocacy, research, foreign aid, foreign direct investment in agriculture and financing, national and international policies, and measuring, reporting, and verification (MRV). The most important institutions that should work together in order to improve the livelihood situation of people were ranked as government, NGOs, researchers, individuals, private, companies and farmers.

In order to increase the uptake of climate-smart agriculture and reach the global target of 25 million smallholder households practicing CSA by 2025 across Africa, it is important that governments, investors, scientists, policy-makers, and farmers in general understand more about which possible climate change scenarios are expected for particular areas. It is acknowledged that the global community does not know enough about all the effects of climate change to predict precise impacts, nor in time or local scale. Agroecological zones (AEZs) have however provided a useful way to categorize the East African in order to better understand some of the likely climatic changes to subregional environments.

The current climate change discourse is very much promoted by international NGOs and some civil society organizations with little contribution from local farmers and communities. An indigenous (African) critical consciousness to climate change is still lacking. It is therefore important that this consciousness is cultivated and raised at all levels in order to change perceptions of climate change for Tanzania to take responsibility for addressing the challenges it present. Most of the challenges can be addressed through adoption of CSA. Whereas resource constraints may limit the practice of CSA, increased consciousness about climate change can enable farmers and farm communities to generate the resources to enable them to practice CSA (Barnard et al. 2015).

Commercial banks and microfinance institutions are generally absent in most rural areas of Tanzania. However, a variety of community savings and lending models have been implemented successfully throughout Tanzania. These include rotating savings and credit associations and village savings and loan groups. These groups pool household resources and agricultural profits to create a structured collective mechanism for savings and short-term lending and investment in household economic activity. With or without these various forms of financial support, farmers can hedge financial risks by adopting CSA piecemeal or on a fraction of their farm. Doing so reduces short-term expenditures and allows the farmer to guard against the possibility that yields will drop before they rise. Piecemeal adoption is a logical strategy at the household level, but makes it difficult to institutionalize CSA rapidly over large areas.

Smallholder farmers avoid taking risk in their farming operations. There should be enough financial guarantees to secure their family livelihoods. Doing so reduces short-term expenditures and allows the farmer to guard against the possibility that yields will drop before they rise. Piecemeal adoption is a logical strategy at the household level but makes it difficult to institutionalize CSA rapidly over large areas.

Prioritize research and development gaps in crop and livestock such as improving the productivity of breeds of sheep and goats that can cope with harsh environmental conditions; identify types of CSA support needed most by stakeholders incorporating scientific expertise with indigenous knowledge. Strengthen the capacity of national institutions working with community-/farmer-based organizations to innovate and develop community action plans, preferably on a landscape (micro-catchment) basis.

CSA represents a long-term investment in the fertility and sustainability of a farm. Hence, secure land tenure is a key prerequisite for investments in climate change adaptations related to land and water management. This is because clear, long-term individual (or household) rights to land can encourage such long-term investments, while the absence of such rights may discourage it. Issuance of authoritative individual-, household-, or community-level certificates of land ownership by recognized central and/or local government authorities would secure farmers' rights to land. In communities where certificates of land ownership may not be feasible, for example, where land is communally owned and rights to the land are defined by written or unwritten customs and governed by decision-making authorities at the extended family, clan, chief, or village levels, adoption of perennial systems including agroforestry may help individuals and/or households establish, claim, and exercise tenure rights that can encourage them to adopt CSA.

Changing peoples' mind-sets on climate change and CSA is a challenge. Most, if not all, smallholder farmers in East Africa are already used to and conversant with conventional farming methods using traditional hand tools. Departing from these methods is sometimes a challenge, especially if the new methods are not highly mechanized. CSA differs from conventional farming knowledge and farming traditions in terms of methods for soil preparation, seeding, fertilizer application, soil erosion control, and residue and water management. In most cases, ploughing is not required for successful planting under CSA. Adopting CSA therefore requires substantial changes not only from conventional farming knowledge and practices but also mind-sets.

Countries are at different levels of promotion and implementation of CSA. At the same time, while Tanzania could have similarities with any other country, there are also many significant differences in terms of soils, climate, culture, and finances, among other factors. Countries need to be supported based on their current status and needs as there is no blanket approach that can be used to support CSA across the subregion.

A number of projects and programs have been conducted focusing on the promotion of CSA. There is now a need to focus more on the results of all this work with regard to adoption by farmers and long-term impact at grassroots level regarding the three CSA pillars among those who have been sensitized, trained, and supported to adopt CSA practices. This can be supported by the development of, and capacity building in, improved monitoring and evaluation systems for CSA adoption, as well as support to countries to conduct CSA baseline studies, develop CSA databases, and integrate CSA data collection into national agricultural surveys or censuses.

CSA needs to be mainstreamed within national and regional CAADP Compacts and National Agricultural Investment Plans (NAIPs). This can be done through the development of CSA investment frameworks, the integration of CSA into new NAIPs, and the screening of current NAIPS for CSA opportunities. The exercise has already commenced in Eastern Africa, with Uganda having been supported on this; however, as per the desire of the regional economic communities (EAC in particular), the wish is for all countries to receive the same support and rise to the same levels.

Overall many strides are being made in Tanzania in terms of CSA promotion, and the chapter should be commended for the efforts being undertaken. However, a lot still needs to be done, and support from all stakeholders as well as international, regional, and national partners is called upon to transform the agricultural systems in Tanzania into climate-smart agricultural systems that contribute to national food security by sustainably increasing productivity and incomes, building resilience and adapting to climate change, and developing opportunities for reducing greenhouse gas emissions compared with expected trends.

References

Barnard J, Manyire H, Tambi E, Bangali S (2015) Barriers to scaling up/out climate smart agriculture and strategies to enhance adoption in Africa. FARA. Forum for Agricultural Research in Africa, Accra

Basalirwa CPK, Odiyo JO, Mngodo RJ, Mpeta EJ (1999) The climatological regions of Tanzania based on rainfall characteristics. Int J Climatol 19:69–80

Binswanger-Mkhize H, Gautam M (2010) Towards an internationally competitive Tanzanian agriculture. Draft World Bank paper. World Bank, Dar es Salaam

Charness G, Viceiza A (2011) Comprehension and risk elicitation in the field: evidence from rural Senegal. IFPRI Discussion Paper 01135

Conway D, Lisa E, Schipper F (2010) Adaptation to climate change in Africa: challenges and opportunities identified from Ethiopia. Glob Environ Change 21:227–237

ECDPM & ESRF (2015). Assessment of policy coherence for development for food security in Tanzania: Tanzania Food Security Profile

FAO (2013) FAO report: global agenda for sustainable livestock. FAO. http://www.livestock dialogue.org/

Kalumanga V, Msaki MM, Bwagalilo F (2014) Climate change adaptation in semi-arid areas: a gender perspective. Int J Ecosyst 4(2):53–59. https://doi.org/10.5923/j.ije.20140402.02

Kami J, Kalenzi D, Mwita V, Flintan F, Liversage H (2016) Making village land use planning work in rangelands: the experience of the sustainable rangeland management project Tanzania. Paper

prepared for presentation at the "2016 World Bank conference on land and poverty", Washington, DC, 14–18 March 2016
Lu X (2009) Applying climate information for adaptation decision-making: a guidance and resource document. National Communications Support Programme (NCSP) UNDP-UNEP-GEF, New York
Mahoo H (2014) Strengthening the capacity for climate change adaptation through integrated land and water management using the CSA approach in Kiroka Village, Morogoro. Presented at Tanzania National Climate Change and Agriculture workshop: sharing evidence and experience on climate-smart agriculture, New Africa Hotel, Dar es Salaam, 15–16th October 2014
Mati B (2007) 100 ways of managing water for smallholder agriculture in Eastern and Southern Africa. A Compendium of Technologies and Practices. SWMnet Working Paper, pp 13–52
Milder JC, Arbuthnot M, Blackman A, Brooks SE, Giovannucci D, Gross L, Kennedy ET, Komives K, Lambin EF, Lee A, Meyer D, Newton P, Phalan B, Schroth G, Semroc B, Van Rikxoort H, Zrus M (2015) An agenda for assessing and improving conservation impacts of sustainability standards in tropical agriculture. Conserv Biol 29(2):309–320
Msaki MM, Emmanuel T, Solomon B (FARA) (2015) State of knowledge on csa in Africa, case studies from Rwanda, Tanzania and Zambia Forum for Agricultural Research in Africa, Accra, Ghana
Mugabe PA (2016) Climate variability and food security in Tanzania: the case of Western Bagamoyo. PhD thesis
Nyasimi M, Amwata D, Hove L, Kinyangi J, Wamukoya G (2014) Evidence of impact: climate-smart agriculture in Africa. CCAFS working paper no. 86. CGIAR Research Program on Climate Change, Agriculture and Food Security (CCAFS), Copenhagen. Available at: http://ccafs.cgiar.org/publications/evidence-impact-climate-smart-agriculture-africa
United Republic of Tanzania (2017) Climate smart agriculture guideline. Ministry of Agriculture, Livestock and Fisheries, Dar es Salaam
United Republic of Tanzania (URT) (2007) National Adaptation Plan of Action (NAPA). Vice President's Office, Division of Environment, Dar es Salaam
URT (2009) The Fertilizers Act. Government Printer, Dar es Salaam
URT (2011) National Climate Change Strategy and Action Plan, 2011. Division of Environment, Vice President's Office, Dar es Salaam
URT (2012) National Climate Change Strategy. Vice President's Office, Division of Environment, Dar es Salaam
URT (2013) The economic survey. Ministry of Finance, Dar es Salaam
URT (2015) Ministry of Agriculture, Food Security and Cooperatives and Vice President's Office, Dar es Salaam
Wambura F, Tumbo S, Ngongolo H, Mlonganile P, Sangalugembe C (2014) Tanzania CMIP5 climate change projections. Proceedings of the International Conference on Reducing Climate Change Challenges through Forestry and Other Land Use Practices, Dar es salaam.

Impact of Climate Change on Growing Season in Nigeria: Seasonal Rainfall Prediction (SRP) as Assessment and Adaptation Tool

132

Paul Akeh Ugbah, Olumide Olaniyan, Sabastine Dekaa Francis, and Adamu James

Contents

Introduction 2744
  Background 2744
  Study Area Description 2747
Literature Review 2748
Methodology 2750
  Materials and Method 2750
Results and Discussion 2752
  Evidence of Climate Change in Nigeria 2752
  Comparison of the Spread/Pattern of Predicted and Observed Rainfall Parameters 2757
  Evaluation of the Skill of the NiMet SRP Model Forecast: Standard Deviation, Centered Root Mean Square Error (RMSE), and Correlation Between Forecast and Observed 2763
  Performance of the NiMet SRP 2763
Conclusion 2767
References 2768

Abstract

Information on climate over many years shows signals of a changing climate in Nigeria and generally on a global scale caused by temperature increase. Climate change effects are increasing, making it necessary to appropriate actions informed by sound climate knowledge to mitigate the effect of these changes. This is achievable by integrating knowledge of climate change into local, national, and global policy processes. This chapter presents an analysis of temperature and rainfall data collected from 41 synoptic stations of the Nigerian Meteorological Agency (NiMet) spread across Nigeria between 1981 and 2017. It also looks at

P. A. Ugbah · O. Olaniyan (✉) · S. D. Francis · A. James
National Weather Forecasting and Climate Research Centre, Nigerian Meteorological Agency, Abuja, Nigeria
e-mail: Olumydas2016@gmail.com

W. Leal Filho (ed.), *Handbook of Climate Change Resilience*,
https://doi.org/10.1007/978-3-319-93336-8_183

the effectiveness of the Nigerian Meteorological Agency's Seasonal Rainfall prediction (SRP) as a climate Smart Agricultural tool to mitigate and adapt to the effects of climate change in Nigeria. The analysis showed that there is an increasing trend in annual mean maximum and minimum temperature anomalies. There has been a persistent increase in the maximum temperature anomalies especially in the past five (5) consecutive years (2013–2017). Rainfall analysis of recent years depicts positive standardized anomalies of above 0.5. An assessment of the SRP tool showed improvement and increase in skill of performance of the model in predicting seasonal rainfall onset, cessation, length of season, and annual rainfall amount across the country from 2012 to 2015, a period for which data was available. The skill was determined by calculating the percentage of forecast accuracy at 95% confidence level for 41 stations across the country. The analysis also shows that NiMet SRP could predict onset and cessation dates of rainfall with a skill above 70%, followed by length of season at a skill of 60% and the lowest skill of 51% in prediction of rainfall amount. The correlation between forecast and observed is 90% for all the rainfall parameters. The good skills suggest that NiMet SRP could serve as a useful adaptation tool to mitigate the effect of climate change on Agriculture over Nigeria and therefore recommended for use by stakeholders.

Keywords
Climate change · Climate smart Agriculture · NiMet SRP · Onset · Cessation · Length of season · Rainfall amount

Introduction

Background

Climate change is real in Nigeria as evidenced in the black flood which submerged more than half of the country during the peak of the rainy season in 2012. The flood lasted several days and caused great loss of lives, properties, and agricultural produce. It also affected transportation and general livelihoods. The continuous variability, and unpredictable nature of the present climate combined with the threat it poses to mankind and his environment has led to the quest for climate smart tools that can effectively predict seasonal rainfall particularly in Nigeria where farmers majorly practice rain-fed agriculture (Kerandi and Omotosho 2008; Omotosho et al. 2000; Mugalavai et al. 2008).

Climate change affects different regions, ecosystems, and sectors of the economy in many ways. These effects depend not only on the sensitivity of those systems to climate change but also on their ability to adapt to risks and changing conditions. Climate change affects agriculture in many ways with a direct impacts on food production, poor and unpredictable yield. Farmers, particularly in Africa are more vulnerable due to low levels of adaptation and mitigation (IPCC 2015). Therefore, adaptation becomes necessary. Experience and studies have shown that, in general, local knowledge of current climate risks is substantial, and the perception of local changes in weather patterns (climate variability) is evident. These changes are attributable to anthropogenic influences such as burning of fossil fuels, cutting down forests, and developing land for farms, cities, and roads. Greenhouse gases are released into the atmosphere through some of these human activities leading to shifts in natural climate cycle (Buba 2004; Odjugo 2007). To effectively adapt to climate change, farmers, policy makers, and other stakeholders must understand its potential effects on crop production. A shortened growing season caused by reduced precipitation and increased temperature can impact crop productivity, particularly in areas where growing seasons are already short. (Farauta et al. 2011). This could push some of these areas to a point where cropping may fail with a disastrous consequences on livelihood of farmers. Increase in length of growing season however may present more growing opportunities for farmers due to extended period of rainfall. Therefore, having understanding of variability of growing season will enable farmers to be able to identify appropriate tools of adaptation to these variabilities. One of such tools is climate smart agriculture which will enable them to adapt to shorter growing seasons by planting drought and heat tolerant crops or varieties that mature more quickly.

Climate smart agriculture is an agricultural practice that is aimed at improving agricultural production, increase output and food security, as well as sustain development through use of science based information (Lipper et al. 2014). The practice can also reduce risk of losses and damages arising from crop failure (Omotosho et al. 2000). It can enhance resilience of agriculture to climate stress (FAO 2010, 2013; Neufeldt et al. 2011). NiMet produces the Seasonal Rainfall Prediction (SRP) as a climate smart tool that can be used to reduce the negative impacts of the effect of climate change on crop production. It also provides vital information for informed decision-making, policy formulation and planning in agriculture, water resources, and environment management in Nigeria. Timely release of the SRP can enhance preparedness against predicted climate risks and hazards while proper application of it can reduce climate-related risks and enhance food production, security, and revenue generation, thereby improving the economy of the nation. The advisory also predicts periods of dry spells after planting. When the SRP is effectively applied, it reduces risk of crop failure and losses, increase agricultural productivity and food security thereby help in adapting and mitigating the impacts of climate change in Nigeria.

The aim of this chapter is therefore to show the evidence of climate change in Nigeria and to evaluate the actual performance of the NiMet SRP model between 2012 and 2017 in predicting the growing season over Nigeria. It intends to assess its skill by using different skill metrics to determine the extent to which the model is reliable as a climate smart agricultural tool over Nigeria. The result is also expected to build the confidence of the users of the NiMet SRP and advisory particularly in farming of Maize and Millet grown in almost every part of the country. Such valuable information will no doubt enable early planning of the farming season, timely land preparation, and planting. It will enhance correct choice of crop varieties, proper management of water resources, and planning in advance for irrigation needs (Omotosho et al. 2000).

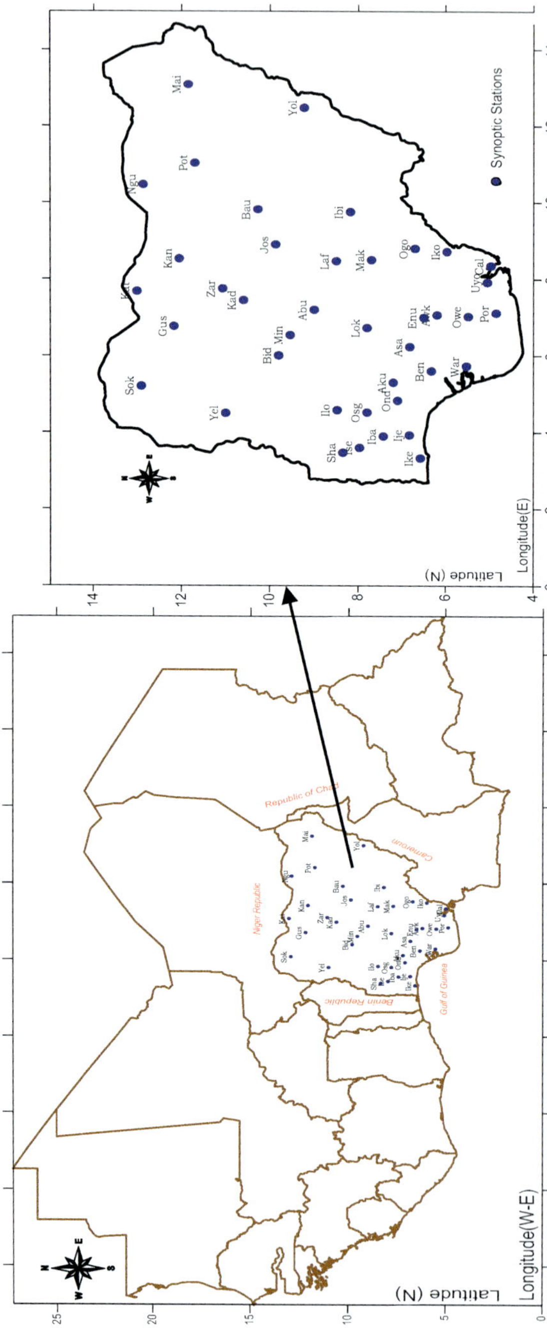

Fig. 1 Map of West Africa (left) showing study area and locations (blue dots) of NiMet synoptic stations (right) from where data used for the study was collected

Study Area Description

Nigeria is located between latitude 4–14°N and longitude 3–15°E. It is bounded by Niger republic to the north, Chad to the northeast, Cameroun to the east, Benin Republic to the west, and the Gulf of Guinea to the south (Fig. 1). The country occupies a land mass of about 923,768 km^2 (Nigeria Embassy 2018) with a population of about 180 million. Rainfall pattern in Nigeria is variable from one location to another and controlled by the movement of the intertropical discontinuity (ITD) which is the meeting zone of the maritime tropical (mT) winds from the Atlantic Ocean and the and the dry Continental Tropical (cT) air mass from the Sahara desert (Eludoyin and Adelekan 2013). The movement and position of the ITD during the year determines the season per time. In addition to these two major air masses, equatorial easterlies also contribute in modulating the rainfall pattern over Nigeria (Odekunle et al. 2008). Two seasons: wet and dry characterize Nigeria. Generally, rainfall occurs at a distance of about 200 km south of the surface location of the ITD. Figure 2 shows a typical northward latitudinal movement of the ITD from its lowest position of about 4°N in January to its highest position of about 22–23°N in August when it starts retreating back to its lowest position. The northward movement is usually accompanied by the flow of the moist mT air mass from the Atlantic Ocean. This air mass predominates and favors rainfall during the wet season. The southward retreat to the initial lowest position is accompanied with the flow of dry cT air mass from the Sahara desert during the dry season. This explains the reason why rainfall cessation in the North is earlier than that of the South, and onset of rain earlier in the South than North. The rapid southward retreat of the ITD to its lowest position signals the end of the rainy season in October over the North, while the harmattan season (dry, dusty, and cold period) starts November to March every year. The severity of the dry season is higher and longer over the North due to its proximity to the Sahara desert compared the South where rainfall could last for a period of 8–10 months (March–December) in a season. Consequently, annual mean rainfall is higher in the South (1200–3000 mm) than the North (400–1100 mm). The pattern of rainfall in the North is unimodal with a peak in August while the South experiences a bimodal pattern with two peaks in June and

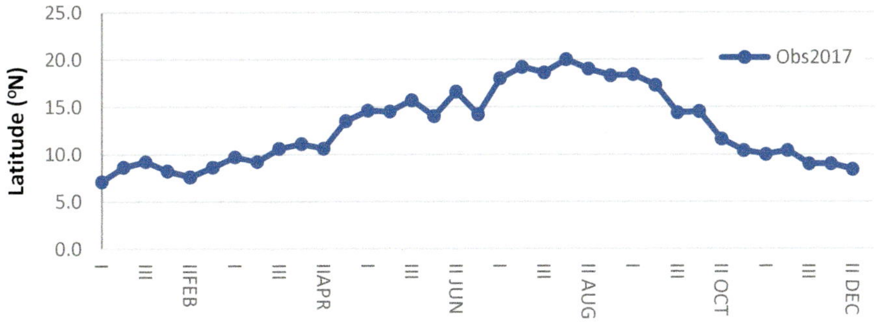

Fig. 2 Mean decadal positions of the ITD in 2017 over Nigeria

September (Ismaila et al. 2010), such rainfall amount is adequate for farming in Nigeria (Verheye 2010; Ismaila et al. 2010). Mean air temperature is between 25.0 °C and 30.0 °C during wet, 20.0 °C and 30.0 °C during the dry (harmattan) season (Ayanlade et al. 2013), while average monthly maximum temperature fluctuate between 28 °C and 40 °C (Terdoo and Adekola 2014). The vegetation of Nigeria is 70% Savanna grass land in the North (Sahel Sudan, Guinea,) while the South are the tropical rainforest (Eludoyin et al. 2014).

Literature Review

Climate change affects different regions, ecosystems, and sectors of the economy in many ways, depending not only on the sensitivity of those systems to climate change but also on their ability to adapt to risks and changing conditions. One adaptation tool that finds wide application now is the practice of climate smart agriculture. Climate smart agriculture is an emerging concept and practice that seeks to adapt agricultural production to climate change and weather variability, while maintaining agricultural productivity, biodiversity, and the ecosystem that sustains food security, livelihoods, and economic development. It seeks to enhance productivity, water conservation, livelihoods, biodiversity, resilience to climate stress, and environmental quality (FAO 2010, 2013; Neufeldt et al. 2011). The concept reflects an ambition to improve the integration of agriculture development and climate responsiveness and aims to achieve food security and broader development goals under a changing climate and increasing food demand. It also refers to improvement of practices and developing capacity to withstand extreme events of weather and climate such as floods, droughts, changes in onset, and cessation dates, etc. and reduce greenhouse gases (Terdoo and Adekola 2014; Arslan et al. 2015). The overall expected outcome of climate smart agriculture is to improve farmer's livelihood and incomes, promote food security and sustainable agricultural development by ensuring that agricultural production systems are monitored to respond to the challenges of climate change and variability (Olawale et al. 2016). Resilience of agriculture to climatic changes and variability could boost agricultural production and broadly contribute to sustainable development (Nwanze and Fan 2016).

Several attempts have been made by climate scientists to design or develop such tools in compliance with climate smart agricultural practices that are based on well informed and science based decisions and advisories aimed at mitigating or adapting to the effects of climate change (Olawale et al. 2016). Studies have shown that different methods have been used at different times by different scientists for seasonal prediction of rainfall onset, cessation, length of season, and seasonal rainfall amount in different parts of the world. Some of these methods are: prediction of seasonal rainfall amount over the USA using dynamic and empirical approaches. This method used NMME-2 dynamic model combined with an empirical method of weighted multiensemble mean to improve the skill of predicting rainfall from 30% correlation coefficient to about 70–80% (Wanders and Woods 2016). A study conducted by Dunning et al. (2016) determined the onset and cessation of rainfall over Africa by using six different satellite measured precipitation data sets such as Global Precipitation Climatology Project (GPCP) and Climate Hazard Group Infrared Precipitation (CHIRPS). Climatological averages, anomalies, and cumulative average anomalies of daily rainfall for the period 1997–2014 were extracted from the datasets and used to calculate the onset and cessation dates over the continent. The result showed that onset dates over the coastal areas of Nigeria were earlier (March 1st–April 1st) compared to the extreme northern parts (June 1st to July 1st). In Kenya, empirical method involving the use of surface synoptic and upper air data was used to design a model which used anomalies of equivalent potential temperature and saturated equivalent potential temperatures to predict rainfall onset, cessation, and amount of rainfall (Kerandi and Omotosho 2008). The results showed that the model performed better in predicting rainfall onset than cessation. The predicted cessation dates showed less agreement with observed, than the onset while prediction of rainfall amount showed correlation of 54–64%. Similarly, Omotosho et al. (2000) employed the use of equivalent potential temperature (Theta E) to determine rainfall onset, cessation, and amount for the growing season in Kano. His result showed that the use of surface data alone has limitations particularly in the prediction of rainfall amount. Odekunle (2004) used cumulative percentage mean rainfall, and daily rainfall probability to determine the onset, cessation, rainfall amount, and length of growing season. His result showed that crop germination can only be sustained from 2nd dekad of March, end of May and 1st dekad of July over Ikeja (south-western coast), Kaduna (central part), and Kano (north), respectively. Odekunle et al. (2005) and Matthew et al. (2017) also reported other methods such as rainfall evapotranspiration used by Benoit and Thornthwaite, and by Kowal and Knabe. These studies revealed best skills in prediction of onset over the Sahel and Savanna zone with biases of 7–15 days, and poor skill over the forest zone of Nigeria where biases were 19–32 days. Another method involving the use of upper wind data was employed by Beer et al. and Omotosho in the early 1990s. Ati et al. (2002) also compared different methods and identified the strength and weaknesses of each method. He proposed that an agrometeorological institution be constituted and empowered by the government. Such institution should be vested with the responsibility of predicting onset and cessation dates, interact with farmers, and provide advisories that will be disseminated to farmers. This will enable them take informed decisions about sowing and choice of crop varieties. Consequently, NiMet has been mandated by the government to carry out this task and has successfully met the expectations of most farmers who have used the advisory and forecast from the NiMet Seasonal Rainfall Prediction model (SRP) in the last 7 years. The SRP is considered successful due to the fact that many farmers and stakeholders now have access to the forecast through collaboration of NiMet with farmers and over 25 state governments to downscale the seasonal forecast specific for each of these states and local governments. Positive reports have also been shared by the users on the efficacy of this model in predicting seasonal rainfall onset and cessation dates, length of season, and rainfall amount.

Methodology

Materials and Method

In this chapter, the evidence of climate change have been constructed from collected information on temperature, rainfall, and other weather variables from 44 measuring synoptic stations across Nigeria. The analysis was compared with long-term average 1981–2010 using standardized anomalies to facilitate the comparison of observations at these locations across the country. This is necessary as an anomaly of 1 °C or 1 mm may be insignificant in one location and highly significant in another. A standardized anomaly of say, 1 °C or 1 mm has the same relative significance at both locations. The Nigerian Meteorological Agency (NiMet) seasonal rainfall prediction (SRP) model was accessed and used to produce a reforecast of seasonal rainfall onset, cessation, length of season, and total annual rainfall amount for the period 2012–2017. The model used input data such as historical climatic data of at least 30 years for which data was available and updated up to 2016 for the reforecast and 2017 for evaluation. The climatic data includes daily rainfall amount, maximum and minimum temperature, and solar radiation all provided by NiMet. Other input data include soil information that best represents each location, historical and latest record of El-Nino Southern Oscillation (ENSO) phases. ENSO is an oscillating pattern of warming and cooling of temperature of the eastern and central Pacific Ocean of the tropics and has strong influence on the weather, climate, and rainfall pattern in different parts of the world (NOAA National Oceanic and Atmospheric Administration USA). The statistical model also incorporates farming practices specific for maize and millet crop which are staple food crops produced in almost every part of the country particularly in the Savannah region. The NiMet SRP model is a Decision Support System for Agro-technology Transfer (DSAAT) model comprising of three basic components of data (weather, soil, and crop) and another component called CPT INSTAT (Climate Prediction Tool – Istituto Nazionale Di Statistica (Italian Statistics Institute)) for predicting dry spells as shown by the schematic diagram (Fig. 3). The model uses all these information provided and written as a program to calculate dates of rainfall onset, cessation, length of the season, and rainfall amount based on water balance of the soil for each location. For evaluation purposes, the observed onset date for a particular station and year is calculated by the model from observed daily rainfall values provided as input data. Onset date is established when the daily rainfall has accumulated to at least 30 mm from the start of the rainy season and the rainfall is consistent without dry spells for up to 7 days. It also considers when the amount of water in the root zone of the crop attains 50% available soil moisture level in the root zone of the crop, sufficient to sustain germination, and growth (NiMet 2013, 2014). Cessation is reached when the available soil moisture at the root zone of the crop is calculated from the water balance to be about 10%. Onset and cessation is therefore not the first and last daily rainfall, respectively but the time when these conditions are satisfied. The length of season is calculated from the difference between the onset and cessation dates while the total rainfall amount is the cumulative daily rainfall total for the year.

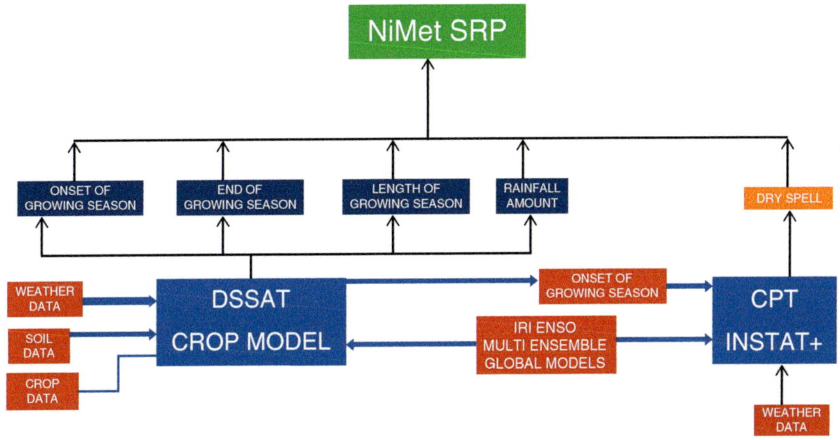

Fig. 3 Schematic of the NiMet SRP model used to predict onset, cessation, and length of season, rainfall amount and dry spells

The forecast on the other hand is calculated using the daily rainfall and temperature climatic data for at least 30 years as well as the ENSO phase and soil data. The average value is calculated, plus or minus the margin of error to arrive at the predicted mean parameter and the range (lower and upper limits) of the forecast values of onset, length of season, or cessation. The total predicted seasonal rainfall is calculated from the mean total values for all the El-Nino years plus or minus margin of error. In this chapter, the mean forecast values are added to margin of error (ME) at 95% confidence level to arrive at the range of values that were used to compare with the observed values. The margin of error for any parameter was however calculated by dividing the standard deviation (σ) of the parameter by the square root of the total number of ENSO years (N).

The standard deviation is given by:

$$\sigma = \sqrt{\frac{1}{N}\sum_{i=1}^{N}(x_i - \mu)^2}$$

where μ is the mean value of the parameters' number of ENSO years (N) and χ_i is the individual parameter's seasonal mean value for each station.

The margin of error (ME) $= 1.96 * \sigma/\sqrt{N}$

where σ is the standard deviation, N is the total ENSO years, and 1.96 is the z value at 95% confidence level. Mean predicted value of any given parameter + ME therefore gives the lower and upper limits at 95% confidence level. Observed values which fell within the range (lower and upper limits) of the predicted values for each station where considered accurate forecast while values which fell outside the range were considered poor forecast. The overall performance of the model (Fig. 11) was determined by dividing the total number of stations for which forecast was correct by the total number of stations for which the forecast failed and then expressed as a percentage.

Once the model run is initiated, it is able to generate, evaluate, and update rainfall and temperature forecast. The results are displayed for each of the stations in tabular format upon completion of the run. The choice of a reforecast rather than the actual forecast for the period is to provide uniform conditions of forecast for the period under review as against varying conditions used in the actual forecast from year to year which may not provide a level playing ground for assessing the true performance of the model. For instance, total of 38 synoptic stations used in 2012 increased to about 42 by 2017. Soil data improved and increased from 6 soil types used in 2012 to over 40 different soil types representing all the stations in 2017. Also, during the actual forecast, predicted ENSO phase is used as against observed ENSO phase used in the reforecast. Data used for this chapter is climatic data for a period of at least 30 years for which data was available, was collected from 41 synoptic stations located at different parts of the country.

Results and Discussion

Evidence of Climate Change in Nigeria

Rising global temperatures have been accompanied by changes in weather and climate. Small changes in the average temperature of the planet can translate to large and potentially dangerous shifts in climate and weather. Many places have witnessed changes in temperature and rainfall, resulting in more floods, droughts, or intense rain, as well as more frequent and severe heat waves (Zoellick 2009). The result of the analysis of temperature and rainfall across the country over the period of 1981–2017 shows some of the evidences of the changing climate; these are presented in the form of standardized anomalies.

Temperature

Temperatures vary from night to day and between seasonal extremes in the Northern and Southern parts of Nigeria. This means that some parts of the country are quite cold while other parts are hot. In agreement with the observed global trend, the concept of a global average temperature is convenient for detecting and tracking changes in Earth's energy budget. Figure 4a, b show that there is an increasing trend in annual mean maximum and minimum temperature anomalies over Nigeria from 1981 to 2017. There has been a persistent increase in the maximum temperature anomalies for the past five (5) consecutive years (2013–2017), with the year 2016 and 2017 depicting the highest standardized anomalies of 1.9 and 1.6, respectively. This is an indication that the year 2016 is the hottest year, followed by year 2017. The trend also showed that maximum temperatures are increasing at the rate of 0.06 °C. Similarly, Fig. 4b shows that the minimum temperatures have also been on

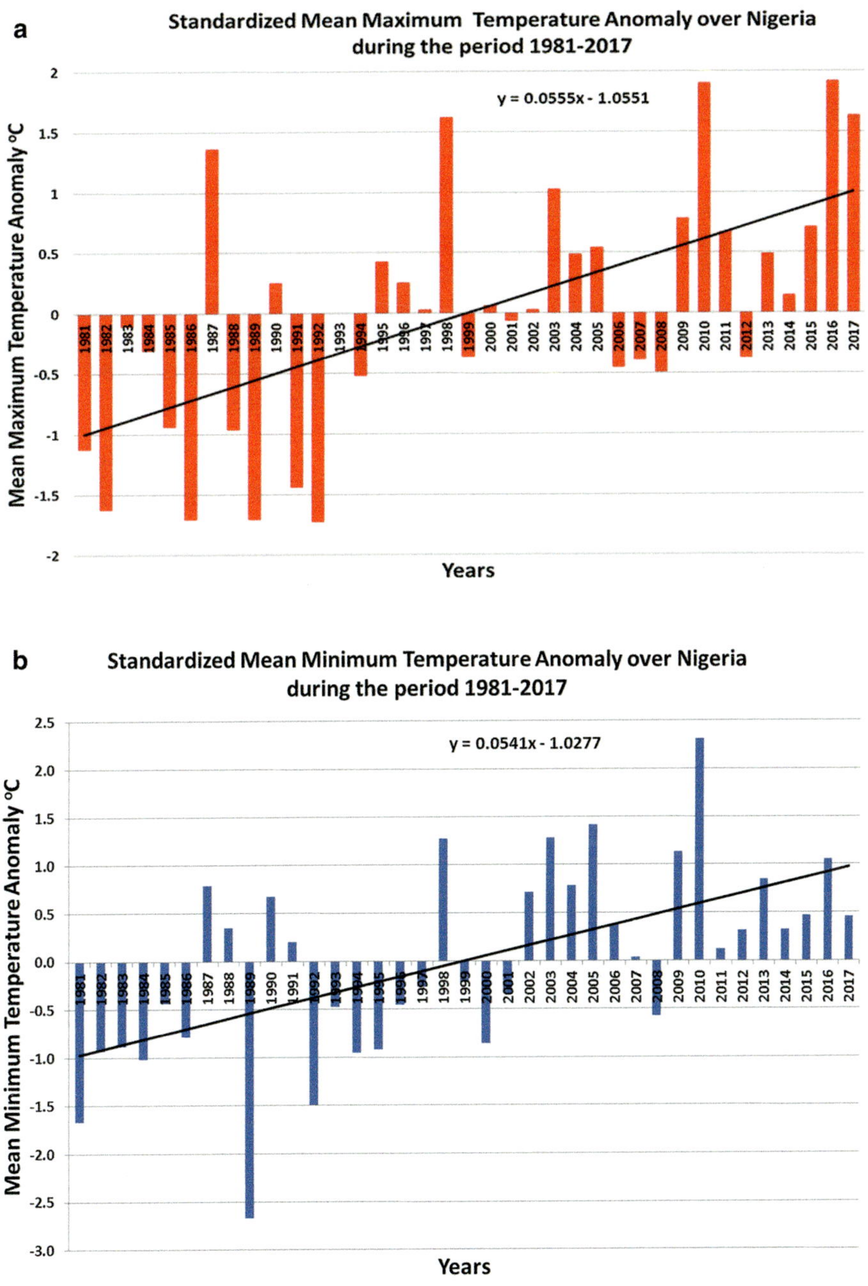

Fig. 4 (**a** and **b**) Mean maximum and minimum temperature anomaly over Nigeria

the increase for the past nine (9) consecutive years (2009–2017) and also increasing at the rate of 0.05 °C. These are clear indications of a warming climate and signals that average yearly temperature anomalies are on the increase with time.

Similar analyses done at the regional level for the northern and southern parts of Nigeria are presented in Figs. 5a, b and 6a, b. They show how much the northern and southern parts of Nigeria have warmed or cooled when compared with the long-term average (1981–2017). There are signals of warming both in the northern and southern parts of Nigeria. Though, the increase in maximum and minimum temperatures is more noticeable in the North than the South as revealed in the higher anomalies. The maximum temperature over the northern part appeared to be increasing at the rate of 0.06 °C per decade, (Fig. 5a), while that of the South is increasing by 0.04 °C per decade (Fig. 5b). Similarly, the minimum temperature of the northern part of the country is changing at the rate of 0.06 °C per decade (Fig. 6a), while that of the southern part is changing at the rate of 0.04 °C per decade (Fig. 6b). Generally, analysis of temperature over the country depict a rapid warming in the past 5 years, and all showed the last 2 years (2016 and 2017) as the warmest.

Rainfall

Effective drought and flood prediction can be enhanced by a clear understanding of the drivers of drought and flood. The long-term (1981–2017) mean annual rainfall of Nigeria is 1470.5 mm with a standard deviation of 133.2 mm and a coefficient of variation of 61.6%. The highest mean rainfall amount of 1755.0 mm over Nigeria for the period of 1981–2017 was recorded in the year 2012, while the lowest amount of 1080.3 mm was recorded in 1983. The regional rainfall distribution indicates that an average of 970.7 mm rainfall occurs in the northern part of the country with a standard deviation of 98.0 mm and 54.6% coefficient of variation during the period 1981–2017. The highest average amount of 1190.5 mm over the country was in 2016 and lowest average amount of 730.2 mm in the year 1983. Similarly, the southern part of the country had mean total rainfall of 1970.4 mm, with a standard deviation of 193.3 mm. The highest mean of total rainfall amount over the country was 2346.1 mm recorded in 2012 with the lowest mean amount of 1430.4 in 1983 for the period under review. However, mean annual total rainfall up to 4000 mm and above in 2012, 2013, 2014 was observed to be exceptionally higher than climatological mean in some coastal stations.

Figure 7 depicts the standardized anomalies over Nigeria, indicating both positive and negative anomalies and trend of the anomalies within the period 1981–2017. It clearly showed positive anomalies from 2006 to 2017 with about 10 years of the period having a standardized anomaly of 0.5 and above. These obviously correspond to wet/flood years recorded in Nigeria. It further revealed a changing pattern of rainfall with the period 1981–1994 indicating negative anomalies, except 1988 and 1991 which corresponds to dry/drought years observed in Nigeria. The rising trend of rainfall is in agreement with temperature increase. Increased in temperature results to increase in evapotranspiration and condensation thereby, producing more intense rains which results to flooding and other rainfall impacts. The analysis further showed that about 61.6% in the standardized anomalies is accounted by the year to year variation in rainfall over Nigeria. This confirms the increasing trend of rainfall in recent times in Nigeria due to climate change.

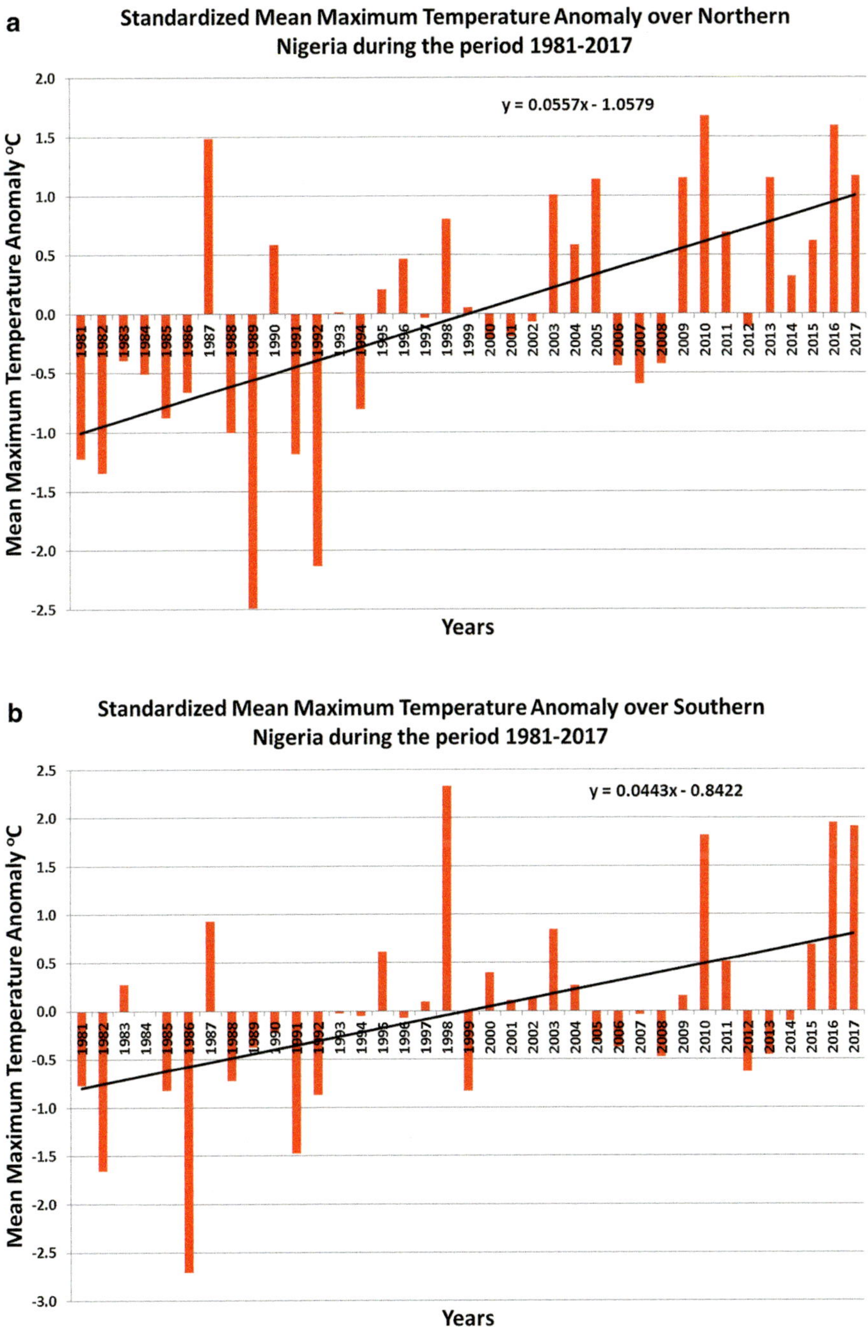

Fig. 5 (**a** and **b**) Maximum temperature anomaly over Northern and Southern Nigeria

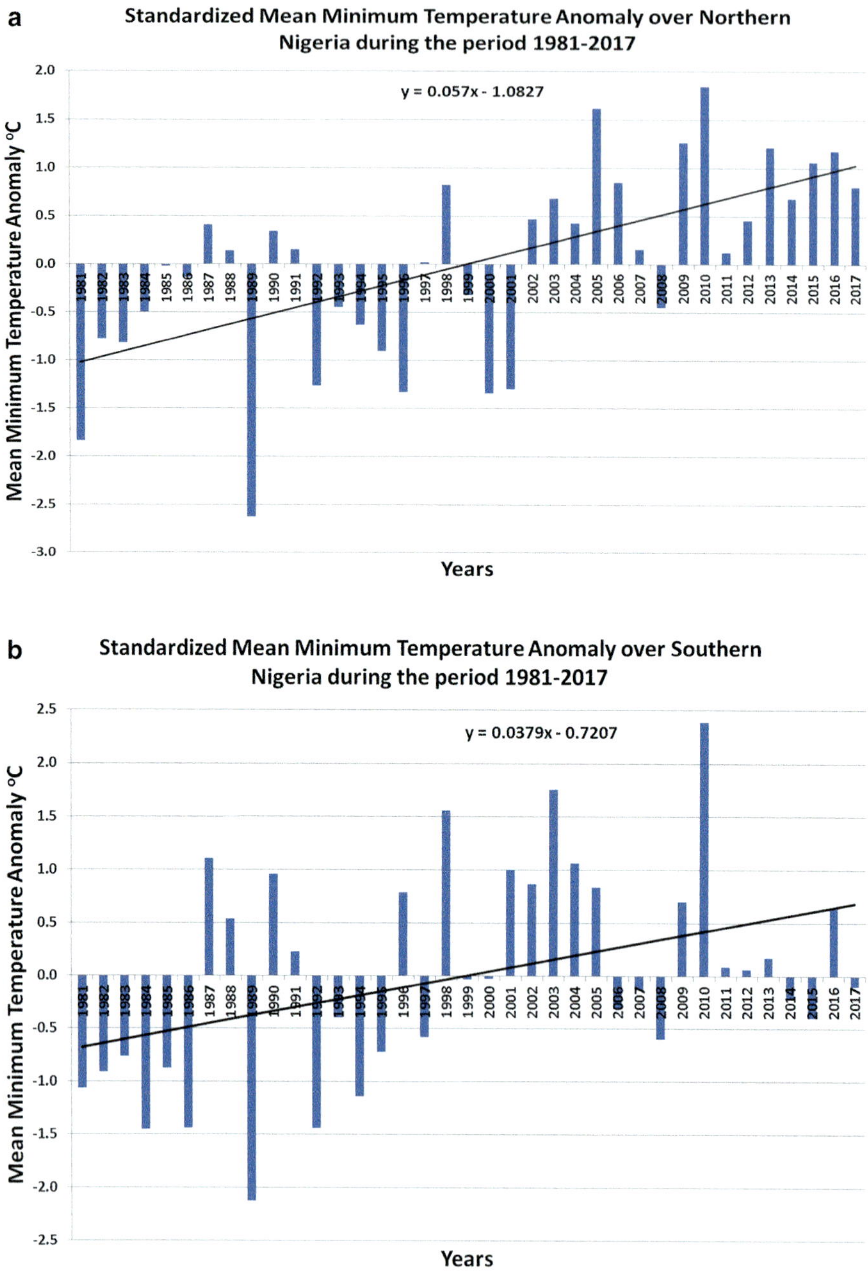

Fig. 6 (**a** and **b**) Minimum temperature anomaly over Northern and Southern Nigeria

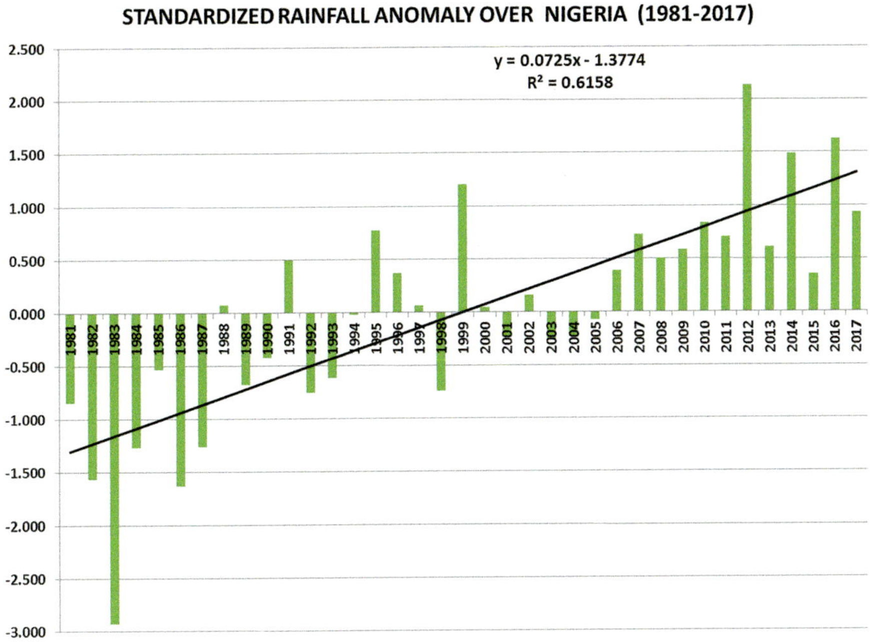

Fig. 7 Standardized rainfall anomaly over Northern Nigeria

Contrary to the few negative rainfall anomalies observed in the recent 12 years in the north, Fig. 6b showed that the most recent 12 years (2006–2017) in the South depict positive anomalies, and mostly above standardized value of 0.5. This is clear evidence that rainfall are on the increase (positive anomaly) in the south. From the coefficient of variation depicted in Fig. 8b, it is clear that the year to year variation in rainfall in the South contributes 49.9% to the upward trend in the rainfall pattern.

Figure 8a show the standardized rainfall anomaly and trend over the northern part of Nigeria, between 1981 and 2017. It is observed that recent years depict positive anomalies with years 2010, 2012, 2014, 2016, and 2017 depicting positive standardized anomalies of above 0.5. Only year 2008, 2011, and 2013 showed negative anomalies in the last 12 years. Year to year variations contribute 54.6% to the standardized anomalies of rainfall in the North.

Comparison of the Spread/Pattern of Predicted and Observed Rainfall Parameters

The analysis and results of evaluation of rainfall onset, cessation, length of season, and annual rainfall amount predicted using the NiMet SRP model from 2012 to 2017 are presented. The predicted results are compared with observed and discussed below.

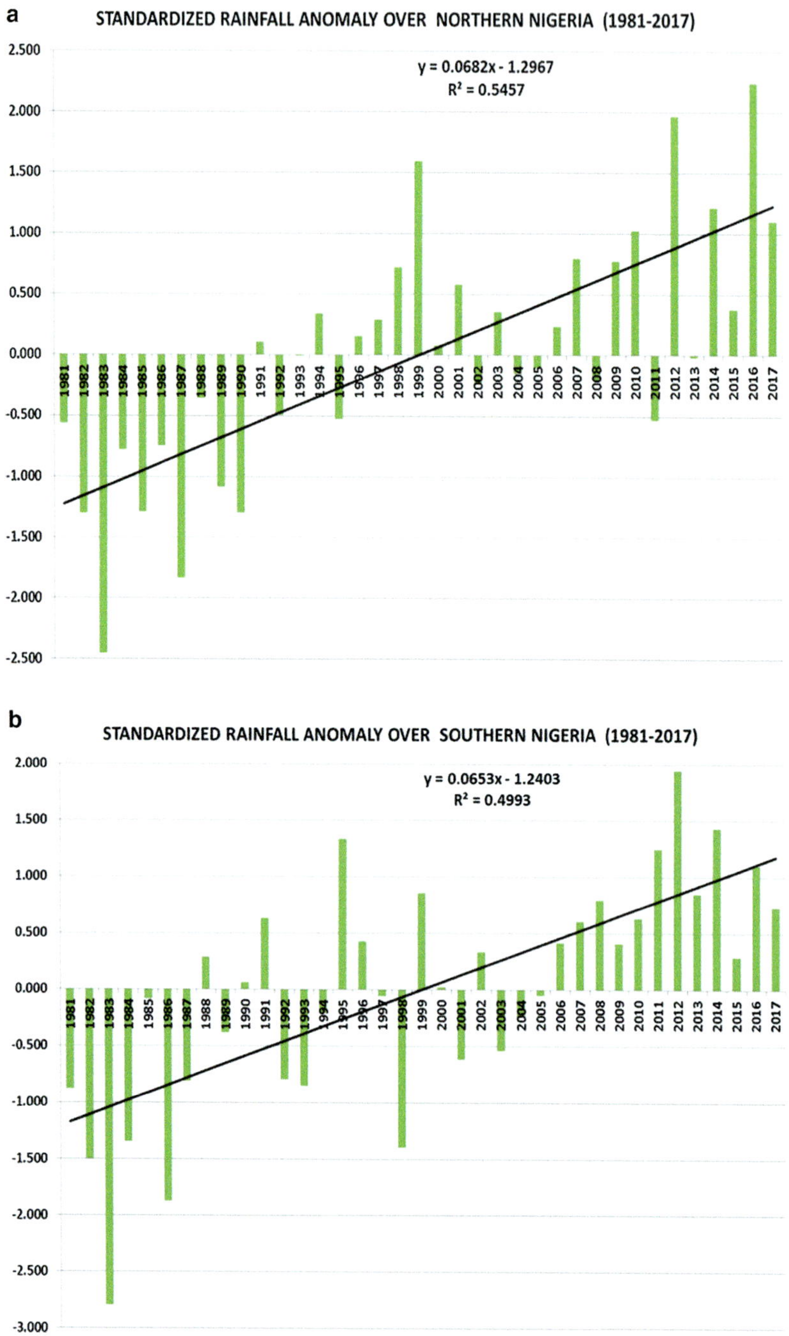

Fig. 8 (**a** and **b**) Standardized rainfall anomaly over Northern and Southern Nigeria

The predicted and observed rainfall onset, cessation, length of season, and rainfall amount (Fig. 9a–d) all show good agreement in the spread or distribution over Nigeria. Although the box plot does not reveal the individual variability from one station or location to another within the country, it is able to effectively reveal and compare between the predicted and observed distribution of these rainfall parameters and the possible mean and extremes of the dates and values within the country.

Rainfall Onset Dates in Nigeria

The plot (Fig. 9a) shows that rainfall onset is generally observed to start as early as the 60th day of the year (March 1st) in the most southern parts (Coast and tropical rainforest zone) of the country such as Warri, Port Harcourt, Lagos, and Uyo and as late as the 180th day (29th June) in the most northerly parts (Sahel) during the period from 2012 to 2017. In all the years, about 50% of the 41 stations observed mean onset dates between day 70 (10th March) and day 130 (10th May) shown by all the red box plots (Fig. 9). Less than 25% of the stations (mainly in the south) observed onset dates in the 1st and 2nd week of March. Less than 25% of the stations (northern stations) also observe mean onset dates between 130th and 176th day of the year (10th May–25th June) during the period.

A comparison of the predicted onset dates (blue box plots) with the observed (red box plots) arranged in pairs (Fig. 9a) shows that all the predicted onset dates for the most southerly states fell within the range of the observed during the period, thus implying an excellent forecast. The predicted onset dates for the most northerly states where however observed to be delayed by 4–6 days in 2013, 2015, 2016, and 2017 and more than 10 days in 2014. These marginal differences were accommodated in the margin of error allowable and still showed good forecast skills. In 2012, mean onset date was observed to be few days earlier than predicted in the country, and this further increased the confidence of the farmers to plant on the predicted date and reduced risk of losses after planting.

Variability in the onset dates is the difference between the 75th percentile (third quartile) and the 25th percentile (first quartile) defined by the box length. It can also be seen from Fig. 9a that the observed onset dates all have higher variability than the predicted onset dates across the country. Also, the variability in the predicted onset dates is contained within the variability of the observed in all the years except 2017. This also underscores the usefulness and reliability of the onset dates predicted by the model.

Rainfall Cessation Dates in Nigeria

Rainfall cessation in the country could be as early as 269th day (26th September) in the extreme North and as late as 365th day (December 31st) over coastal areas such as Warri, Port Harcourt, Lagos, and Uyo. The mean date for cessation of rainfall over the most northerly stations is 278th day (5th October), and 362nd day (28th December) over most southern station during the period 2012–2017 (Fig. 9b). A comparison of the mean predicted and observed cessation dates over the country

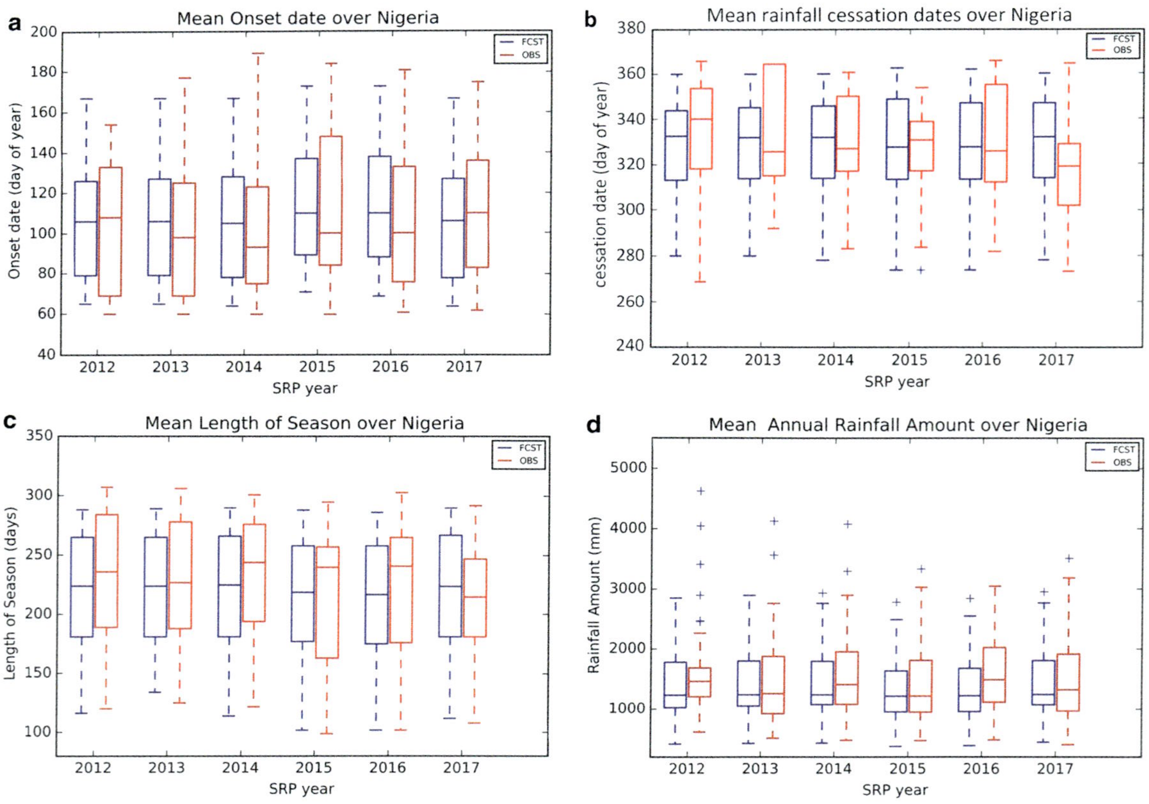

Fig. 9 Box plot comparing predicted (blue) and observed (red) rainfall (**a**) onset, (**b**) cessation, (**c**) length of season, and (**d**) total annual rainfall amount

showed great agreement with marginal differences of 1–9 days over the most southerly stations, and 5–12 days over the most northerly stations during the period. The predicted rainfall cessation date over the coastal area is observed to be delayed (prolonged cessation) during the period except in 2015 when it was earlier than predicted by few days. The extended period of cessation observed for most of the period is an advantage to farmers especially those who plant twice in a farming season. In the northernmost stations, predicted cessation dates were observed to be delayed (prolonged cessation) except in 2012 when earlier than predicted end of season was observed. The marginal differences observed are mostly accommodated within the margin of error and as such contributes positively to the skill of the performance of the SRP model.

Length of the Growing Season Over Nigeria

The northernmost stations such as Katsina, Sokoto, and Maiduguri observed length of season in the range of 100–125 days (average of 113 days) while the southernmost stations such as Warri, Uyo, Benin, and Port Harcourt observed length of season of 292–307 day (average of 301 days) between 2012 and 2017 (Fig. 9c). Fifty percentage of the 41 stations however have lengths ranging from 170 to 280 days while about 10% of the stations have lengths in the range of 100–120 days and 300 days and above, respectively in the extreme northern and southern parts of the country. A comparison of the predicted and observed length of season for the period showed very strong agreement with very little differences of 1–9 days over the extreme North and 2–17 days over the extreme South. These biases are however observed to be as a result of a prolonged season than predicted in most cases over these places. The observed extension of the length of season than predicted by the SRP model is also considered an advantage to the farmers and a strength in the model by ensuring that the predicted length of season is not over exaggerated to pose a serious risk and threat to users of the forecast.

Mean Annual Total Rainfall Over Nigeria

The lowest total annual rainfall was observed to range from 404 to 613 mm (mean 495 mm) over extreme northern stations such as Katsina, Sokoto, Nguru, and maiduguri, while the highest values ranged from 3046 to 4627 mm(mean 3785 mm) over places such as Uyo, Calabar, and Ikeja during the period (Fig. 9d). The highest value of 4627 mm was however recorded in 2012 over Uyo. Annual total rainfall amounts above 4000 mm were consistently recorded in Uyo in 2012, 2013, and 2014. The rainfall observed during the period showed unprecedented values above 3000 mm identified as outliers (blue crosses) in 2012, 2013, 2014, and 2017 (Fig. 9d red boxes). A comparison between the predicted and observed annual total rainfall also showed similar distribution and good level of agreement. However, large biases above 1000 mm were observed over coastal stations while the northernmost stations were observed to have the lowest biases of 10–300 mm. The observed large biases between the predicted and observed rainfall which is due to the uncommon rainfall identified as outliers in 2012, 2013, 2014, 2015, and 2017 for the southernmost stations no doubt have a negative effect on the skill of the NiMet

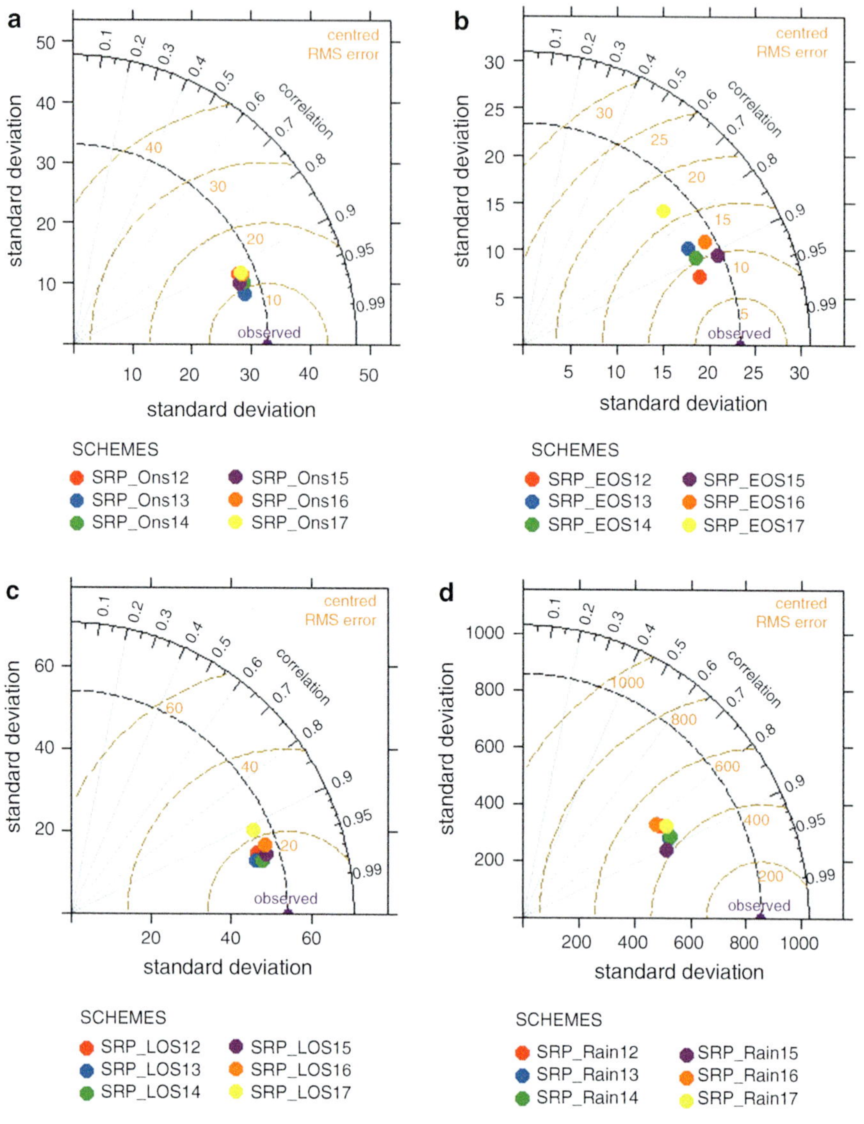

Fig. 10 Taylors diagram comparing standard deviation, correlation, and centered root mean square error between predicted and observed rainfall (**a**) onset, (**b**) end of season, (**c**) length of season, and (**d**) total annual rainfall amount

SRP model in predicting rainfall amount. This has made predictions largely variant when compared to previous predictions. However, the biases are mostly due to observed higher rainfall than predicted particularly in the South, and as such an advantage for farmers in most places in the North that are not threatened by higher rainfall than predicted. This situation therefore is a strong indication of the effect of climate change in recent time.

Evaluation of the Skill of the NiMet SRP Model Forecast: Standard Deviation, Centered Root Mean Square Error (RMSE), and Correlation Between Forecast and Observed

The forecasts Fig. 10a–d shows lower values of standard deviation than the observed shown by the dotted arc from origin. This implies a wider range of variability in the observed parameters than the predicted. The predicted dates of rainfall onset across the country between 2012 and 2017 show average standard deviation of about 30 days which is less than the observed standard deviation of about 33 days. On the other hand, predicted dates of rainfall cessation across the country show average standard deviation of about 22–24 days which is slightly less than the observed standard deviation of about 25 days. The predicted dates of length of season show average standard deviation of 52–53 days which is less than but very close to the observed standard deviation of about 55 days across the 41 stations in the country.

Predicted dates of annual rainfall amount show average standard deviation of about 600 mm which is less than the observed standard deviation of about 830 mm across the 41 stations in the country. The centered root mean square error values shown by the extent of departure of the brown arc lines from the observation point along the x-axis are lowest (about 12 days) for predicted onset dates (Fig. 10a), 10 days for end of season, 18 days for length of season and about 420 mm prediction of rainfall amount. The result of the RMSE and correlation coefficient (r) are in agreement with that obtained from model statistics of R-studio tabulated in details in Table 1.

Evaluation: Other Skill Metrics

Apart from standard deviation, centered root mean square error and correlation shown on the Taylor's diagram, other metrics according to CERC 2016 have also been used to determine the skill and evaluate the performance of the NiMet SRP model and they are: the mean bias (MB), mean gross error (MGE), normalized mean bias (NMB), normalized mean gross error (NMGE), and coefficient of efficiency (COE) which is based on Legates and McCabe (1999, 2013). It has values which range from −1 to +1 with values approaching +1 representing better model performance. The Index of Agreement (IOA) based on Willmott et al. (2012) is also used. This index also has a range between −1 and +1 with values approaching +1 representing better model performance (CERC 2016). FAC2 is Fraction of predictions within a factor of two while n represents the total number of station pairs used. RMSE is the root mean square error and r is the correlation coefficient. The forecast schemes refer to the model's predicted parameter and year. The results are summarized in Table 1 for onset, cessation, and length of season and rainfall amount.

Performance of the NiMet SRP

The general performance of the model is expressed as the skill in percentage and refers to the percentage of stations where the observed values fell within the range of the predicted values taking into consideration the margin of error allowable for each station (Fig. 11). The result shows that the highest skill and performance (above

Table 1 Summary of skill metrics used in evaluating performance of NiMet SRP model

Forecast schemes	n	FAC2	MB	MGE	NMB	NMGE	RMSE	r	COE	IOA
Onset date										
SRP_Onset_2012	41	1	1.2	9.3	0.0114	0.0887	12.5	0.92	0.7	0.83
SRP_Onset_2013	41	1	2.7	9.9	0.0256	0.0956	11.7	0.96	0.7	0.83
SRP_Onset_2014	41	1	5.2	9.1	0.0519	0.0905	12.4	0.94	0.7	0.84
SRP_Onset_2015	41	1	−0.8	12.7	−0.0066	0.1110	15.3	0.94	0.6	0.82
SRP_Onset_2016	41	1	6.9	11.3	0.0646	0.1063	14.2	0.93	0.6	0.80
SRP_Onset_2017	41	1	−3.9	9.4	−0.0350	0.0857	12.4	0.92	0.6	0.81
End of season										
SRP_Cessation_2012	41	1	−3.70732	7.2	−0.0111	0.02163	9.2	0.93	0.6	0.81
SRP_Cessation_2013	41	1	−3.26829	10.6	−0.0098	0.03173	13.3	0.87	0.5	0.77
SRP_Cessation_2014	41	1	−1.19512	6.6	−0.0036	0.019934	9.2	0.90	0.6	0.81
SRP_Cessation_2015	41	1	1.487805	7.5	0.0045	0.022914	9.6	0.91	0.5	0.75
SRP_Cessation_2016	41	1	−1.04878	8.5	−0.00319	0.025717	11.9	0.87	0.6	0.79
SRP_Cessation_2017	41	1	11.70732	13.8	0.0367	0.043292	19.3	0.73	0.2	0.58

Length of season										
SRP_Length_2012	41	1	−5.2682	13.2	−0.0229	0.0574	17.2	0.95	0.7	0.86
SRP_Length_2013	41	1	−4.9512	13.9	−0.02153	0.0604	18.5	0.96	0.7	0.86
SRP_Length_2014	41	1	−8.6585	12.9	−0.0371	0.0551	15.9	0.97	0.7	0.86
SRP_Length_2015	41	1	1.2439	12.6	0.0058	0.0590	16.3	0.96	0.7	0.87
SRP_Length_2016	41	1	−8.8292	15.4	−0.03949	0.0687	19.7	0.95	0.7	0.84
SRP_Length_2017	41	1	15.073	19.0	0.0718	0.0906	25.0	0.91	0.5	0.76
Rainfall amount										
SRP_Rain_amt_2012	41	0.96	−260	313	−0.15749	0.1898	550.4	0.83	0.5	0.73
SRP_Rain_amt_2013	41	1	−90.4	252	−0.06061	0.1687	417.4	0.88	0.6	0.80
SRP_Rain_amt_2014	41	0.98	−195	264	−0.12148	0.1651	416.2	0.87	0.5	0.80
SRP_Rain_amt_2015	41	1	−87	235	−0.06058	0.1648	314.0	0.91	0.6	0.80
SRP_Rain_amt_2016	41	0.93	−248	287	−0.15529	0.1796	432.0	0.82	0.4	0.71
SRP_Rain_amt_2017	41	0.98	−107	256	−0.07038	0.1680	396.5	0.85	0.6	0.80

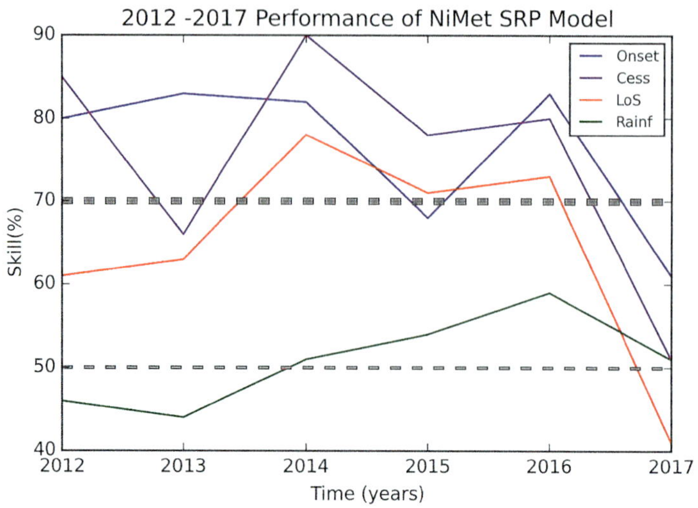

Fig. 11 Graph showing the general performance of the NiMet SRP model on predicting onset (blue line), cessation (purple), length of growing season (red), and rainfall amount (green) between 2012 and 2017

70%) of the NiMet SRP model is in predicting onset and cessation. This is followed by the length of season with an average skill of 60% while prediction of total annual rainfall amount has the lowest skill of about 50%. It can be seen that there are certain years when the skill reaches 80–90%, for instance, prediction of length of season in 2014 reached 90% skill while prediction of onset of rainfall scored between 80% and 85% in 2012, 2013, 2014, and 2016. The year 2017 recorded the lowest skills in predicting all the four parameters and this may be connected to the statistical characteristics of the model, which outweigh the dynamic characteristics and as such followed the pattern of climatology, and was unable to capture all the dynamism associated with changes in the weather and climate of 2017 bearing in mind that the year 2017 was adjudged the third hottest year in history and as such had a peculiar weather and climate due to increase in the effect of climate change in recent years.

Although the NiMet SRP model showed the lowest skill (51%) in predicting rainfall amount, this skill is far higher than skills displayed by most statistical and dynamic seasonal and subseasonal climate models used by most scientists and world class meteorological centers. For instance, the skill of US seasonal forecast for both precipitation amount and surface temperature shows correlation coefficient (r) less than 30% (Quan et al. 2006). The North American Multi-Ensemble Model phase 2 (NMME-2) also shows anomaly correlation of less than 40% for four (4) individual ensembles of the NMME model used in prediction of seasonal and subseasonal precipitation amount (Niko and Wood 2016). However, the mean correlation of 0.9 (90%) obtained from this work is in agreement with results obtained by Dunning et al. (2016) which also showed 90% correlation of ECMWF ERA predicted seasonal precipitation amount with the satellite observed CHIRPS precipitation amount.

Conclusion

This chapter has shown an increasing trend of maximum and minimum temperature in conformity with increasing global temperature trend. However, there is variability in trends of these parameters from year to year and from one location to another in Nigeria. Also an increasing trend of rainfall is observed in the country particularly in the last 10 years. Such variability and extreme weather and climate makes prediction a difficult task and leaves the people with great uncertainties of the safest or best period of practicing agriculture in a sustainable manner, hence the need for a climate smart approach to Agricultural practices in Nigeria. One of such approach is the use of the NiMet seasonal rainfall prediction as a climate smart tool for assessment of the growing season and guide to planning farming activities yearly. It is also considered an adaptation tool as information from the prediction will help minimize risk of losses and damage and as such help all stakeholders to cope with the variability and changes being experienced. The confidence and reliability of the NiMet SRP is evident in the results of assessment of the skill of this model such as the mean correlation of 90%, Coefficient of Efficiency of about 65% for onset, cessation and length of season, and Index of agreement of about 80% demonstrated by the model in predicting these very important rainfall parameters which suggests a good performance of the model. These results proved to be better than results reported earlier by most authors. The biases from the use of the SRP model are also found to be lower than very high onset date biases of 19–32 days for the southern part of the country by Matthew et al. 2017. The length of season predicted by the NiMet SRP model generally showed negative biases in most cases which suggest that the predicted length is mostly lower than observed and falls within the range of the observed at a time when the crops will be at maturity or approaching maturity, thus reducing risk of losses. The predicted length of season is always adequate for growing of cereal crops such as millet which can be grown to maturity between 65 and 90 days, wheat 95–175 days depending on variety used (Verheye 2010), and Maize 90–120 days. The results also prove that the model is not just reliable but a dependable tool in minimizing risk of losses in agricultural production especially in production of cereals in Nigeria and in reducing risk of dangers and losses associated with adverse weather. It is therefore a reliable climate smart tool helpful in maximizing food production, ensuring food security and adaptation to climate change. Application of NiMet SRP is made accessible to stakeholders, such as Farmer's associations, public and private risk managers, policy and decision makers, industry and investors in various aspect of economy that depend on rainfall. The SRP is therefore an invaluable tool which is capable of reducing the exposure of farmers to short-term risks, while also strengthening their resilience by building their capacity to adapt and prosper in the face of shocks and longer-term stresses, thereby ameliorating the unpleasant consequences of extreme weather and climate events in Nigeria.

References

Arslan A, McCarthy N, Lipper L, Asfaw S, Cathaneo A, Kokwe M (2015) Climate smart agriculture? Assessing the adaptation implications in Zambia. J Agric Econ. Paper presented at the international conference of agricultural economics, Milan, 8–14 August. https://doi.org/10.1111/1477-9552.1210

Ati OF, Stigter CJ, Oladipo EO (2002) A comparison of methods to determine the onset of the growing season in Northern Nigeria. Int J Climatol 22:731–742

Ayanlade A, Adeoye NO, Babatimehin O (2013) Intra-annual climate variability and malaria transmission in Nigeria. Bull Geogr Socio-Econ Ser 21(21):7–19

Buba AD (2004) Climate change and water problems in Chad Republic. J Arid Environ 3(2):24–27

CERC (2016) Cambridge Environmental Research Consultants (CERC) Ltd. Model evaluation toolkit user guide. http://www.cerc.co.uk/environmentalsoftware/assets/data/doc_userguides/Model_Evaluation_Toolkit_User_Guide.pdf

Dunning CM, Black ECL, Allan RP (2016) The onset and cessation of seasonal rainfall over Africa. J Geophys Res Atmos 121:11405–11424

Eludoyin OM, Adelekan IO (2013) The physiologic climate of Nigeria. Int J Biometeorol 57:241–264

Eludoyin OM et al (2014) Air temperature, relative humidity, climate regionalization and thermal comfort of Nigeria. Int J Climatol 34:2000–2018. https://rmets.onlinelibrary. wiley.com/doi/epdf/10.1002/joc.3817

Embassy of Nigeria (2018) The Embassy of Nigeria in United States of America: History of Nigeria. http://www.nigeriaembassyusa.org/index.php?page=test-sub. Accessed in 2018

FAO (2010) "Climate-smart" agriculture: policies, practices, and financing for food security, adaptation and mitigation. Food and Agriculture Organization of the United Nations, Rome. http://www.fao.org/docrep/013/i1881e/i1881e00.htm

FAO (2013) Climate-smart agriculture: sourcebook. Food and agriculture organization of the United Nations, Rome. http://www.fao.org/docrep/018/i3325e/i3325e.pdf

Farauta BK, Egbale CL, Idrissa YC, Agu VC (2011) Climate change and adaptation measures in northern Nigeria: Empherical situation and policy implications (No. 62). Nairobi, Kenya: African Technology Policy Studies Network. https://atpsnet.org/wp-content/uploads/2017/05/wps62.pdf. Accessed in 2018

IPCC (2015) Climate change 2014: synthesis report. In: Core Writing Team, Pachauri RK, Meyer LA (eds) Contribution of working groups I, II and III to the fifth assessment report of the intergovernmental panel on climate change. IPCC, Geneva, 151 pp

Ismaila U, Gana AS, Tswanya NM, Dogara D (2010) Cereals production in Nigeria: constrainsts and opportunities for betterment. Afr J Agric Res 5(12):1341–1350

Lipper L et al (2014) Climate-smart agriculture for food security. Nat Clim Chang 4:1068–1072

Matthew OJ, Imasogie OG, Ayoola MA, Abiye OE, Sunmonu LA (2017) Assessment of prediction schemes for estimating rainfall onset over different climatic zones in West Africa. J Geogr Environ Earth Sci Int 9(1):1–15

Mugalavai EM, Kikporir EC, Raes D, Rao MS (2008) Analysis of rainfall onset, cessation and length of growing season for western Kenya. J Agric For Meteorol 148:1123–1135

Legates DR, McCabe GJ (1999) Evaluating the use of "goodness-of-fit" measures in hydrologic and hydroclimatic model validation. Water Resour Res 35(1):233–241

Legates DR, McCabe GJ (2013) Short communication. A refined index of model performance: a rejoinder. Int J Climatol 33:1053–1056

Neufeldt H, Kristjanson P, Thorlakson T, Gassner A, Norton-Griffiths M, Place F, Langford K (2011) Making climate-smart agriculture work for the poor. Policy brief, vol 12. World Agroforestry Centre (ICRAF), Nairobi

NiMet (2013) Nigerian meteorological agency: seasonal rainfall prediction brochure. http://nimet.gov.ng/sites/default/files/publications/2013-seasonal-rainfall-prediction.pdf. Accessed Feb 2018

NiMet (2014) Nigerian meteorological agency: seasonal rainfall prediction brochure. http://www.nimet.gov.ng/sites/default/files/publications/SRP%20BROCHURE%20FINAL.pdf. Accessed Feb 2018

Nwanze KF, Fan S (2016) Climate change and agriculture: strengthening the role of small holders. Global food policy report. International Food Policy Research Institute, Washington, DC, pp 13–21

Odekunle TO (2004) Rainfall and length of growing season in Nigeria. Int J Climatol 24:467–479

Odekunle TO, Balogun EE, Ogunkoya OO (2005) On the prediction of rainfall onset and retreat dates in Nigeria. Theor Appl Climatol 81:101–112

Odekunle TO, Oniarah A, Aremu SO (2008) Towards a wetter Sudano-Sahelian ecological zone in twenty-first century Nigeria. J Roy Meteorol Soc Weather 63(3):66–70

Odjugo PAO (2007) The impact of climate change on water resources; global and regional analysis. Indones J Geogr 39:23–41

Olawale EO, Isaac KT, Labode P (2016) Differential impacts of rainfall and irrigation on agricultural production in Nigeria: any lessons for climate-smart agriculture? Elsevier B.V. Agric Water Manag 178(2016):30–36

Omotosho JA, Kerandi NM (2008) Seasonal rainfall prediction in Kenya using empirical methods. J Kenya Meteorol Soc 2(2):114–124

Omotosho BJ, Balogun AA, Ogunjobi K (2000) Predicting monthly and seasonal rainfall, onset and cessation of the rainy season in West Africa using only surface data. Int J Climatol 20:865–880

Quan X, Hoerling M, Whitaker J, Bates G, Xu T (2006) Diagnosing sources of U.S. seasonal forecast skill. J Clim 19:3279–3293. https://doi.org/10.1175/JCLI3789.1

Terdoo F, Adekola O (2014) Assessing the role of climate-smart agriculture in combating climate change, desertification and improving rural livelihood in Northern Nigeria. Afr J Agric Res 9(15):1180–1191

Verheye W (2010) Growth and production of maize: traditional low-input cultivation. In: Verheye W (ed) Land use, land cover and soil sciences. Encyclopaedia of Life Support Systems (EOLSS), UNESCO-EOLSS Publishers, Oxford. http://www.eolss.net

Wanders N, Wood EF (2016) Improved sub-seasonal meteorological forecast skill using weighted multi-model ensemble simulations. Environ Res Lett 11(9):4007

Willmott CJ et al (2012) Short communication: a refined index of model performance. Int J Climatol 32:2088–2094. https://rmets.onlinelibrary.wiley.com/doi/epdf/10.1002/joc.2419

Zoellick RB (2009) A climate smart future. The Nation Newspapers. Vintage Press Limited, Lagos, p 18

Educating Farmers in Rural Areas on Climate Change Adaptation for Sustainability in Nigeria

133

Benjamin Anabaraonye, Joachim Chukwuma Okafor, and James Hope

Contents

Introduction 2772
Definition of Terms 2774
  Adaptation 2774
  Climate Change 2775
  Climate Finance 2775
  Climate Resilience 2775
  Education 2775
  Farmers 2775
  Vulnerability 2776
Climate Change Education for Sustainable Development 2776
  Understanding Climate Change 2776
The Impacts of Climate Change in Nigeria 2777
  The Impacts of Climate Change on Agriculture in Nigeria: A Glance at Literature 2778
  Challenges Facing Farmers in Rural Areas Toward Climate Change Adaptation in Nigeria 2779
  The Necessity of Educating Farmers Toward Tackling Climate Change in Nigeria 2779
  Climate Change Education (CCE) in the Overall Context of Education for Sustainable Development (ESD) 2780
The Need for Educational Personnel to Educate Farmers in Rural Areas on Climate Change Adaptation in Nigeria 2781
  Practical Tools for Educating Farmers Toward Climate Change Adaptation 2783

B. Anabaraonye (✉)
Benjy Poetry and Music Global Concepts, Awka, Anambra State, Nigeria
e-mail: benjaminshines@gmail.com

J. C. Okafor
Department of Political Science, University of Nigeria, Nsukka, Enugu State, Nigeria
e-mail: Chukwuma.okafor@unn.edu.ng

J. Hope
Ahmadu Bello University, Zaria, Kaduna State, Nigeria
e-mail: jamhope@gmail.com

W. Leal Filho (ed.), *Handbook of Climate Change Resilience*,
https://doi.org/10.1007/978-3-319-93336-8_184

Conclusion and Policy Recommendations 2786
Further Recommendations 2787
References 2787

Abstract

Climate change is one of the global issues which we must of necessity tackle with alacrity in order to prevent global warming too unbearable for the survival of mankind. While the industrialized countries of the world, the major contributors to climate change, have the capacity to respond to the impact of climate change, most developing countries like Nigeria do not have sufficient adaptive capacity to climate change and global warming. They, therefore, need assistance from developed countries for climate change adaptation. Climate change education is therefore vital in every community in order to prepare us for climate change impacts and to learn how to adapt effectively. This chapter closely examines the impacts of climate change in Nigeria and most especially in the agricultural sector. It also explores the need to educate farmers in rural areas on climate change adaptation and the tools which can be creatively used for climate change education for farmers in rural areas in Nigeria to enable them to adapt effectively for sustainability.

Keywords

Adaptation · Climate change · Education · Farmers · Sustainability

Introduction

The challenging impacts of global warming, which leads to climate change, pose great threats not only to the socioeconomic environment which human beings live but adversely have exposed human lives to multifaceted vulnerability, namely, acid rainfall, drought, flooding, resource-related conflict, hunger, poverty, unemployment, loss of livelihood, and many others (DeSombre 2002; Ayoade 2003; Ekundayo 2008; Oladiran 2010; Adelekan 2011; Okafor 2011; Olufemi 2018). While chronicling the vulnerability of Nigerian society to the dangers of climate change, Olufemi (2018) argues that in 2017 alone, Nigeria witnessed series of climate-related disasters, ranging from the increased health risk, declining agricultural productivity, biodiversity loss, drying lakes, famine, conflicts or social unrest, poverty, worsening food insecurity situation, heat stress, declining soil capacity for agricultural production, increased natural disaster, and extreme weather events, among others. Thus the most peculiar effect of climate change in Nigeria is declining agricultural productivity arising from the irregular rain and unpredictable farming calendar.

There is a general consensus among environmental scholars and scientists that the world is experiencing a rapid global climate change but separately differs on its nature, dimension, and effects (Umar and Ozohu 2015; Tsojon 2017). However, much of its occurrence is attributed to anthropogenic activities, bringing into the limelight the influence of human activities on the nature and the environment over the years (Ladan 2014). It should be noted that there are numerous causes of climate change (Bast 2010). However, two causes stand out – the natural and human causes. The natural causes of climate change emanate from variations in earth's orbit, variation in ocean circulation, variation in albedo of the continents, as well as variation in solar radiation. The human causes are, however, results of deforestation, air pollution, and poor agricultural practices such as bush burning, excess and wrong application of inorganic fertilizer, burning of fossil fuels, urbanization, industrialization, and inefficient transport system, among others (Olufemi 2018).

Specifically in Nigeria, the contending challenges imposed by the climate change crisis are here with us and has multi-sectoral effects on Nigerian Society (Ayinde et al. 2011; Enete 2014; Iheoma 2015). Challenging this has become, while at the "one planet" Summit in Paris, France, on December 12, 2017, President Muhammadu Buhari submitted that Nigeria could not implement her Nationally Determined Contributions without adequate financial, technical, and capacity building support from developed countries. The implication is that high level of poverty, corruption and leadership crisis, unemployment, insecurity and insurgency, economic recession with other problems confronting Nigeria, and the preponderance of dearth of resources and technical and trained manpower have become a daunting challenge facing Nigeria in its quest toward reducing the impacts of climate change through various mitigation and adaptation strategies. The impacts and effects of climate change in Nigeria are challenging, enormous, and multi-sectoral. The sector most affected with far-reaching impacts is agricultural sector (Enete and Amusa 2010; Apata 2011; Ladan 2014; Olufemi 2018). Thus, the adverse impacts of climate change such as temperature rise, erratic rainfall, sand storms, desertification, low agricultural yields, drying up of water bodies like Lake Chad, gully erosion, and constant flooding are daily reality in Nigeria. Bassey (2017) and Olufemi (2018) have in consensus agreed that Nigeria is under the siege of climate change which has affected agriculture and separately demonstrated the extent of the country-specific effects. Flooding and drought affects soil quality arising from climate change impacts which endangers soil nutrients. These can cause desertification and soil degradation, change of rainfall pattern affects agriculture especially, the herding activities, access to land, water, dry land contribute to migration and displacement of population, increased temperature affects cereals like maize, millet, guinea corn. Statistics has shown that in 2012 alone, 6 million Nigerians were displaced and 300 deaths recorded due to effects of climate change. More than 100,000 persons were displaced by flooding in Benue alone in 2017 with states like Lagos, Borno, and Anambra also affected (Bassey 2017). The adverse impacts of climate change in Nigeria include the declining agricultural yield; the continued shrinking of Lake Chad which has remained the major source of water supply not only to Nigeria but to Cameroon, Chad, and Niger which has significant influence on agriculture; rise of disease and sickness; poverty and malnutrition among rural farmers; declining capacity of soil nutrient arising from massive flooding; farmers and herdsmen clashes in Benue, Taraba, Nasarawa, Ogun, Delta, and Enugu states; and outbreak of natural disaster and extreme weather events, among others (Olufemi 2018).

In conclusion, the impacts of climate change have posed the greatest challenge to the socioeconomic environment cum human lives not only in Nigeria, but around the world. Thus, with its excruciating effects on household livelihoods in Nigeria, many farmers are yet to be adequately informed or educated on the dangers of climate change (Pereira 2017). This is the central message of this chapter. In this chapter, we shall be exploring the practical tools that can be deployed in educating farmers in rural areas on climate change adaptation for sustainability in Nigeria. The rest of the chapter includes:

- The definition of terms
- Climate change education for sustainable development
- Understanding climate change
- The impacts of climate change in Nigeria
- The impacts of climate change on agriculture in Nigeria: a glance at literature
- Challenges facing farmers in the rural areas toward climate change adaptation in Nigeria
- The necessity of educating farmers toward tackling climate change in Nigeria
- Climate change education (CCE) in the overall contest of education for sustainable development
- The need for educational personnel to educate farmers in rural areas on climate change adaptation in Nigeria
- Practical tools for educating farmers toward climate change adaptation in Nigeria
- Case study of Poem "Plant a Tree" Review by Barrister James Hope and advanced by Benjy Poetry and Music Global Concepts for climate change education, adaptation, and sustainability in Nigeria
- Conclusion/policy recommendations

Definition of Terms

Adaptation

The Intergovernmental Panel on Climate Change (IPCC) defines adaptation as the "adjustment in natural or human systems to a new or changing environment. Adaptation to climate change refers to adjustment in natural or human systems in response to actual or expected climatic stimuli or their effects, which moderates harm or exploits beneficial opportunities. Various types of adaptation can be distinguished, including anticipatory and reactive adaptation, private and public adaptation, and autonomous and planned adaptation" (IPCC 2001).

Climate Change

The United Nations Framework Convention on Climate Change (UNFCCC) defines climate change as a change of climate which is attributed directly or indirectly to human activity that alters the composition of the global atmosphere and which is in addition to natural climate variability observed over comparable time periods (UNFCCC 1992). In addition climate change is identical to global warming.

Climate Finance

Climate finance refers to local, national, or transnational financing, which may be drawn from public, private, and alternative sources of financing. Climate finance is critical to addressing climate change because large-scale investments are required to significantly reduce emissions, notably in sectors that emit large quantities of greenhouse gases. Climate finance is equally important for adaptation, for which significant financial resources will be similarly required to allow countries to adapt to the adverse effects and reduce the impacts of climate change (UNEP 2015).

Climate Resilience

This can be generally defined as the capacity for a socio-ecological system to absorb stresses and maintain function in the face of external stresses imposed upon it by climate change and adapt, reorganize, and evolve into more desirable configurations that improve the sustainability of the system, leaving it better prepared for future climate change impacts.

With the rising awareness of climate change impacts by both national and international bodies, building climate resilience has become a major goal for these institutions (Wikipedia 2018).

Education

Education means the impartation of knowledge, skills, and ideas on individuals, people, or organizations for the purpose of equipping them to confront the challenges that life presents. This education could be informal or formal.

Farmers

These are people who work the land (till and cultivate land for planting) and rear and raise livestock either for consumption or commercial purposes.

Vulnerability

This is the degree to which a system is susceptible to or unable to cope with adverse effects of climate change, including climate variability and extremes (IPCC 2007). Vulnerability is a function of the character, magnitude and rate of climate change, and the degree to which a system is exposed, along with its sensitivity and adaptive capacity. It increases as the magnitude of climate change or sensitivity increases and decreases as adaptive capacity increases (OECD 2009).

Climate Change Education for Sustainable Development

Education is an essential element of the global response to climate change. It helps young people understand and address the impact of global warming, encourages changes in their attitudes and behaviors, and helps them adapt to climate change-related trends (UNESCO 2010).

Through its Climate Change Education for Sustainable Development program, UNESCO aims to make climate change education a more central and visible part of the international response to climate change. The program aims to help people understand the impact of global warming today and increase "climate literacy" among young people. It does this by strengthening the capacity of its member states to provide quality climate change education; by encouraging innovative teaching approaches to integrate climate change education in schools; and by raising awareness about climate change as well as enhancing nonformal education programs through media, networking, and partnerships. For more information on UNESCO's work on climate change, please visit UNESCO's website http://unesdoc.unesco.org/.

Understanding Climate Change

Scientists around the world have looked at data from satellites. They've studied weather patterns and measured changes in crop yields and soil composition. And what they've found is that our planet is, on average, getting warmer. Temperatures have been rising since the early twentieth century and most notably since the late 1970s. This warming reflects long-term changes to the Earth's climate. But it's important to understand that when we talk about climate, we are not referring to weather. They mean two different things. Weather can change from season to season, day to day, and even hour to hour. Sometimes when you least expect it! Weather reflects short-term conditions in the atmosphere. Climate, on the other hand, refers to the average temperatures and precipitation rates over an extended period of time. When these rates change over an extended period of time, it can result in profound impacts on our planet, impacts like rising sea levels; more extreme weather events, like droughts and floods; melting glaciers; shifts in ecosystems; as well as many others.

In the past, changes in our climate resulted from natural causes, such as differences in the sun's activity and volcanic eruptions.

However, scientists cannot explain the warming we experienced in the past 100 years without including greenhouse gases emitted by human activity. So what are these greenhouse gases? When scientists talk about "greenhouse gases," they are referring to several chemical compounds found in the Earth's atmosphere which would trap heat that would normally be released out to space. Some of these compounds occur in nature, such as carbon dioxide and methane. However, others are exclusively human-made and are found in sources like hair spray, paint fumes and aerosol sprays.

These greenhouse gases play a crucial role in warming the Earth's surface and making it habitable. However, too much human-generated greenhouse gas emissions upset the planet's natural balance, leading to an increase in warming. Now that we understand the meaning of climate change, let's take a look at how it actually works. The ugly trend is that greenhouse gas concentration in our atmosphere will continue to increase, continuing to warm the Earth – unless all nations reduce the billions of tons of greenhouse gas we produce each year. The good news is that the world is finally waking up to this global challenge, and there is much we can do to reduce our reliance on fossil fuels (Nakagawa 2015). Climate change brings about changes in precipitation levels, a rise in sea level, and more frequent and intense weather events. And it can threaten our access to such basic needs as food, water, health, and shelter. Let's look more closely at some of these impacts. Climate change menace is a problem ravaging our world. To the farmers, imparting or educating them with the knowledge, skill, and ideas of climate change-resilient strategies such as the mitigation and adaptation is vital to survival of agriculture in Nigeria.

The Impacts of Climate Change in Nigeria

Of the many problems generated by our economic and industrial progress, climate change is one that dominates media headlines in recent memory. The current warming trend of the Earth's atmosphere that started with the industrial revolution has given scientists, politicians, and individuals around the world cause for great concern. Dramatic reductions in seasonal sea ice at the Earth's poles, rising sea level, shifts in precipitation patterns leading to extended droughts or flooding, and more frequent extreme weather events are just some of the effects of global warming and changes to the global climate. The human toll in lost lives and livelihoods is mounting, particularly in Africa and the Small Island Developing States (SIDS). No country will remain unaffected, thus making it clear that climate change mitigation and adaptation are necessary to address one of the greatest challenges faced by the world today.

Climate change is a complex global problem because it is intertwined with many other issues, such as economic development and poverty reduction. Developing countries are the least responsible for climate change, yet the most at risk from its effects. Eradicating poverty and improving living conditions via the attainment of the Sustainable Development Goals (SDGs) must remain a priority. The challenge here is to accomplish the SDGs while reducing dependence on carbon, promoting climate resilience, and ensuring balanced economic development. The United Nations Secretary-General Ban Ki-Moon has proclaimed climate change to be the defining challenge of our time. It is a global problem, and yet each of us has the power to make a difference. Even small changes in our lifestyle and behavior can help reduce greenhouse gas emissions while ensuring a good quality of life.

The Impacts of Climate Change on Agriculture in Nigeria: A Glance at Literature

The greatest challenge facing agricultural development in developing countries, Nigeria inclusive, is the preponderance of climate change (Awojuola 2016), and its effects have been identified (Beyioku 2016). Nigeria's agriculture like any other developing country has been exposed to the dangers of climate change because her agricultural productions are systematically tied to rainfall. This rainfall has become irregular and unpredictable over a decade now leading to poor yield across the country (Olufemi 2018). Agricultural sector with its diverse dimensions stand out as one of the crucial sector of the economy that has inbuilt capacity when harnessed to employ millions of the people. Some scholars have described it as the engine room for socioeconomic development (Ayinde et al. 2011). However, these laudable characteristics of agriculture (farming) of having the capacity to promote speedy socioeconomic development are at risk of being eroded or fast being eroded by the debilitating impacts of climate change (Odjugo 2010; Nwajiuba 2012, 2014).

Bassey (2017) argues that though climate change has significant impacts on different sectors of the Nigerian society, the most affected is the agricultural sector which poses greater threat against the government and private sector efforts in achieving the Sustainable Development Goals number 2 which targets zero hunger in Nigeria by the year 2030. The constant flooding and drought experience in different part of Nigeria erode soil nutrients and reduce the level of agricultural yield. This development which climate change has brought is frustrating government efforts of encouraging local production of numerous food items which the country imports oversea. This has also led to astronomical increase in the prices of food products in recent times.

Adelekan (2009), Beyioku (2016), and Olufemi (2018) have separately agreed that the effects of climate change are here with us. Nigeria has suffered multiplicity of climatic forces which includes the increasing incidence of drought in the Northern Nigeria, erosion in the South East, continued cases of flooding in the North Central and South West, and the rising sea levels in the South – South of Nigeria is fast endangering the prospects of agriculture in Nigeria. Some of the impacts of climate change on agriculture in Nigeria highlighted are the declining agricultural productivity, the irregular rain and unpredictable farming calendar, biodiversity loss, drying lakes, famine, social unrest (farmers and herdsmen clash), poverty, worsening food insecurity situation, heat stress, declining soil capacity for agricultural production, increased natural disaster, and extreme weather events, among others.

Challenges Facing Farmers in Rural Areas Toward Climate Change Adaptation in Nigeria

The following are the peculiar challenges confronting farmers in Nigeria toward climate change adaptation.

1. **Lack of finance**. This makes the farmers unable to purchase climate-resilient seedlings and other necessary equipment which affects their crop yield.
2. **Lack of technical and capacity building support**. Many of the farmers in rural areas have not been informed and trained to adapt effectively to climate change. They have also not been sufficiently equipped with right tools and skills for climate change adaptation.
3. **Poverty**. Poverty is a deadly sting which affects the farmers and makes them unable to maximize their productivity in other to ensure sustainability. Arising from declining of productivity, they find it difficult to send their children to school, pay hospital bills, get the right educational materials to enhance their knowledge on climate change, etc.
4. **Insecurity**. Many farmers are distracted and disrupted from their farm work due to insecurity arising from farmers/herdsmen conflict, inter-/intra-racial or ethnic conflict, religious conflicts, etc. Constant clashes between farmers and herdsmen have brought about recently insecurity and lack of productivity in some states like Benue, Taraba, Delta, Enugu, etc.
5. **Corruption and lack of legislative oversight**. In Nigeria, policies meant to increase farmer's productivity are not taken seriously. Funds ear-marked for farmers' training, education, or sensitizations are sometimes hijacked by the policy implementers.

The Necessity of Educating Farmers Toward Tackling Climate Change in Nigeria

Environmental scholars and scientists have established a greater relationship between humans' increased levels of carbon dioxide emission into the environment and the intensity of global warming which lead to climate change (Sheykhi 2001; Ahmad et al. 2005; Gans and Jost 2005). Other scholars have argued that given the ongoing trends in global usage of energy, increased demand for development, and global population growth, the world should expect corresponding increase and effects of climate change (Ehrlich and Holdren 1971; Rozelle et al. 1997; Holden and Sankhayan 1998; Harte 2007; Zaman et al. 2011; Ray and Ray 2011; Gowland-Mwangi 2012; Maier 2015). Therefore, increasing temperature, increasing climate variability, increased rainfall and snowfall, changes in earth systems, and severe weather events have serious damaging effects on human health (Ruth 2017).

The level of climate change education among farmers, especially in developing countries like Nigeria in relation to adaptation, has been described as being inadequate (OECD 2012; Ashby and Pachico 2012; Pereira 2017). The inadequacy of awareness has prompted the question of the necessity of educating the farmers more toward tackling the menace of climate change. Why is farmer's education on climate change in Nigeria very critical and important for the survival of agriculture in Nigeria? The following necessities have informed the needed education.

i. The increasing low level of agricultural produce, with increasing high level of hunger and poverty among the rural farmers whose livelihood and survival depends on farm produce. Income from the sales of these farm produce is used for the payment of school fees of their children, hospital bills, etc.
ii. To increase their productivity while reducing the dangers of climate change. Knowledge, ideas, and skills gained through Climate change education will therefore increase their productivity while reducing the dangers of climate change.

The necessity of educating farmers in rural areas in Nigeria enables them to adopt sustainable land management, climate-resilient agriculture, and efficient and effective water management which enhances their productivity. It enables the farmers to know the areas that are vulnerable to climate change. With the knowledge of their vulnerabilities, it exposes them to have alternative source of water outside rainfall which has become irregular and unpredictable over the years.

It exposes farmers to be aware that indiscriminate cutting down of trees exposes the soil to erosion and loss of nutrients thereof. Through environmental education, it enables the farmers to be conscious of their environment and the nature. Global warming which leads to climate change is anthropogenic (caused by excessive abuse of our environment). The education of farmers enables them to moderate their excessive behavior on their environment.

Climate Change Education (CCE) in the Overall Context of Education for Sustainable Development (ESD)

This program uses innovative educational approaches to help a broad audience (with particular focus on youth including farmers) understand, address, mitigate, and adapt to the impacts of climate change. It encourages changes in attitudes and behaviors needed to put our world on a more sustainable development path and build a new generation of climate change-aware citizens.

Often, people think that problems such as climate change are to be solved by the government or scientists. However, we need to be aware that each of us is involved as part of the problem and part of the solution. Climate Change Education for Sustainable Development (CCESD), therefore, has a central role to play in helping the general public and especially the next generations understand and relate to the issues, make lifestyle changes to reduce greenhouse gas emissions, and adapt to the changing local conditions. While CCESD at all levels and in both formal and nonformal settings is needed, instilling climate change awareness and understanding at a young age is ultimately the best way to change behaviors and attitudes. To ensure relevance and uptake, CCESD should be oriented according to the local context and prioritize passing traditional knowledge and practices to learners. To this end, UNESCO is uniquely positioned to leverage its expertise in education, natural and social sciences, culture, and communication in order to lead by example and raise the profile of CCESD on the international agenda through the Climate Change Initiative. The objectives of UNESCO on CCESD are as following:

1. Strengthen the capacity of member states to provide quality climate change education for sustainable development at primary and secondary school level through improved education policy, analysis, research and planning, teachers' education and training of education planners, and training on curriculum review/ reform.
2. Encourage and enhance innovative teaching approach to integrate quality climate change education for sustainable development in school through interdisciplinary practices, science education, technical and vocational education and training (TVET), and education on disaster risk reduction (which is very important for rural farmers).
3. Raise awareness about climate change and enhance nonformal education program through media, networking, and partnerships. CCESD-related curricula focus on the development of knowledge, skills, and competencies needed and contribute to the mitigation and adaptation to climate change. Further, it should introduce basic scientific concepts, theories, and projections of climate change. Themes such as sustainable consumption, disaster preparedness, environmental protection, recycling, water, desertification, and renewable energies should be discussed, taking into account their relevance in the specific national and local context (UNESCO 2010).

The Need for Educational Personnel to Educate Farmers in Rural Areas on Climate Change Adaptation in Nigeria

There is great need in Nigeria today for educational personnel to be well trained on how to educate farmers in rural areas on climate change adaptation. Promoting climate change adaptation through education requires that climate change educators' attitudes, motivation, and skills contribute fully to the effort. This requires them to understand the implications of such training and retraining to recognize the need for it and to have the competence and confidence to introduce it to their target audience which are farmers in rural areas.

Climate change is a global phenomenon and is evident in Nigeria. The negative impacts of climate change such as temperature rise, erratic rainfall, sand storms, desertification, low agricultural yield, drying up of water bodies, and flooding are real in the desert-prone 11 frontline states of Nigeria. Environmental degradation and attendant desertification are major threats to the livelihoods of the inhabitants of the frontline states in Nigeria. This leads to increasing population pressure, intensive agricultural land use, overgrazing, bush burning, and extraction of fuel wood and other biotic resources. Women and children are particularly the most vulnerable to the impacts of climate change. Government should improve their financial input toward the finding of solutions to the climate change issues in Nigeria. Inadequate funds hamper progress in achieving Nigeria's objectives on climate change. The Nigerian government and all the stakeholders involved in the global phenomenon need to increase public awareness, promote research, and establish a commission or an agency that will handle issues related to global warming and climate change. The federal, state, and local government, international agencies, and other development partners are required to fund climate change projects in Nigeria for sustainable development. The technology and use of the renewable sources of energy like the solar, wind, and hydro energy in homes and companies across the country should be greatly encouraged. Laws should be enacted to discourage deforestation and encourage afforestation and reforestation. Government, companies, institutions, and even individuals can initiate activities which will lead to the planting of trees in strategic places to help green our environment and release more oxygen into the atmosphere Charles (2012). Here are the benefits of tree planting:

1. Trees absorb odors and pollutant gases (nitrogen oxides, ammonia, sulfur dioxide, and ozone) and filter particulates out of the air by trapping them on their leaves and bark.
2. Trees provide oxygen.
3. Trees conserve energy.
4. Trees save water. Shades from trees slow water evaporation from thirsty lawns. As trees transpire, they increase atmospheric moisture.
5. Trees help to prevent soil erosion. On hillsides or stream slopes, trees slow runoff and hold soil in place (Charles 2012; Treepeople 2017).

Furthermore, individuals and institutions can help prevent global warming by reducing the demand for fossil fuels and by using energy more wisely. Let us do our part to reduce waste by choosing reusable products instead of disposables. Climate change is now affecting every country on every continent. It is disrupting national economies and affecting lives, costing people, communities, and countries dearly today and even more tomorrow. Affordable, scalable solutions are now available to enable countries to leapfrog to cleaner, more resilient economies. The pace of change is quickening as more people are turning to renewable energy and a range of other measures that will reduce emissions and increase adaptation efforts. But climate change is a global challenge that does not respect national borders. Emissions anywhere affect people everywhere. It is an issue that requires solutions that need to be coordinated at the international level, and it requires international cooperation to help developing countries move toward a low-carbon economy. To address climate change, countries adopted the Paris Agreement at the COP21 in Paris on 12 December 2015. The Agreement entered into force shortly thereafter, on 4 November 2016. In the agreement, all countries agreed to work to limit global temperature rise to well below 2 °C and, given the grave risks, to strive for 1.5 °C. You can learn more about the agreement here: https://www.un.org/sustainabledevelopment/blog/2015/12/the-paris-agreement-faqs/.

Implementation of the Paris Agreement is essential for the achievement of the Sustainable Development Goals, which provides a roadmap for climate actions that will reduce emissions and build climate resilience. Progress is needed on many fronts: development and transfer of green technologies, greenhouse gas emission reductions, and the establishment of effective government policies. It is of equal importance to provide education and training and promote public awareness to the targeted audience. There is urgent need to create an informed global citizenry, a knowledgeable workforce, and enlighten government officials on their roles in climate change adaptation.

Practical Tools for Educating Farmers Toward Climate Change Adaptation

The following tools could be deployed in educating farmers toward climate change adaptation in Nigeria.

- **Engagement of village town crier**. Engaging village town criers through the collaboration of various community authorities in our local government could be one of the fastest avenues of reaching the rural farmers. This should not be done any market day dear to the villagers as this may not appeal to them. This should be done when they are around.
- **Nongovernmental organizations (NGOs)**. Nongovernmental organizations (NGOs) that are climate change bias can be used as valuable tools in reaching the farmers in rural areas through their respective union in local governments and states in Nigeria. These NGOs can conduct seminars, symposiums, and workshops on climate change adaptation for farmers in rural areas in Nigeria.
- **Religious institutions**. These institutions act as a crucial platform for reaching the farmers in rural areas in their churches, mosques, and other traditional religious centers. This is because majority of Nigerians are religious including the farmers. This avenue will also be used to educate rural farmers whose belief systems about the consequences of climate change are as a result of anger from the gods. They are made to realize that humans beings themselves (farmers also) play vital roles on the impacts of climate change.
- **Use of radio and television jingles**. This is a valuable tool of using indigenous language in reaching the rural farmers through radio and television programs, jingles, adverts, etc. Furthermore, climate change poems can be used as well to educate farmers in rural areas on special radio and television programs.
- **Distribution of fliers**: Fliers which contain relevant information on climate change adaptation and resilience can be printed and distributed to farmers in rural areas. This will help to create proper awareness among the farmers and equip them with the skills needed for climate change adaptation. Furthermore, indigenous language can be used to communicate with farmers who are illiterate.
- **Educational blogs**: This is a tool which can be used to inform, enlighten, and educate farmers in rural areas in Nigeria especially the Internet-literate ones about climate change adaptation. These blogs which are meant to be highly interactive allow the farmers to also contribute their ideas, suggestions, and feedback to the

climate change educators and bloggers for sustainability. Through the educational blogs, farmers can connect with appropriate climate change NGOs based in Nigeria like Climate Change Mitigation and Adaptation Initiative (CCMAI), African Clean Up Initiative (ACI), etc. This will enable them to access partnerships, support, and funding for sustainability. The Project Green Initiative is a good example of an organization which is very passionate about educating the communities which include the farmers in Nigeria about climate change adaptation. The Project Green Initiative creatively uses the tool of poetry and educational blog which focuses on climate change to educate the community including farmers in rural areas on climate change adaptation in Nigeria. To find out more about the Project Green Initiative based in Nigeria, visit our educational blog: www.projectgreeninitiative.wordpress.com.

Below is one of our climate change adaptation poems which can be used to educate farmers in rural areas:

Plant a Tree
Trees provide us with oxygen.
Trees help keep the environment clean.
Trees help to purify the air;
Increasing moisture as they transpire.

Trees help to prevent water pollution.
Trees help to prevent soil erosion.
Trees conserve energy.
It is so beautiful to plant a tree.

Trees absorb odors and pollutant gases.
Trees provide shade for the masses.
Trees cool the streets and the cities.
Trees create economic opportunities.

Trees are like the lungs of the planet.
Preserver and greener of the earth.
Trees matter beautifully and truly.
Be dutiful to plant a tree.

Trees provide a canopy.
A shade for the hungry.
A shade from the raging heat.
A beautiful place for retreat.

Trees provide food.
Trees provide wood.
Trees combat climate change.
When properly managed.

Trees slow runoff and hold soil in place.
As beauty and bliss, they embrace.
Trees help us appreciate nature.
Plant a tree for the future.

Case Study: Poem "Plant a Tree" Review by Barrister James Hope
Benjamin Anabaraonye's poem "Plant a Tree" has ignited the global call to everyone to plant a tree. His call to plant a tree is in line with the mandate of Sustainable Development Goals (SDGs). The impact of man on the environment has been seen as a primary cause of climate change (global warming). In the poem, greater emphasis was made on the importance of having trees in our homes and societies. I can't agree less with Richard Barbe Baker when he said that "It is not merely that the world is bettered by saving, replacing, and multiplying trees. It is that an aim of this kind becomes an impulse towards developing a mood and an outlook which will increasingly make it natural to think for the future, for other people, for generations yet unborn. Planting a tree is a symbol of a looking-forward kind of action; looking forward, yet not too distantly."

Each nation has that responsibility and obligation to ensure that the environment is protected and sustained against further damages and destruction. He made this call by using the following poetic technique:

Theme
The central theme of the poem is centered on planting a tree for the betterment of the world and curbing the menace of climate change.

Tone
The tone of the writer is approving, hopeful, and demanding.

The writers' mood is inviting, candid, urging, emotional, and very passionate about the subject matter.

Imagery
Imageries were also used, such as "Trees help to purify the air" and "Trees provide shade for the masses."

Style
The poem contains seven stanzas with each having an end rhyme of AABB. The poet exhibited a style of capitalizing each beginning line.

Techniques
The use of simile "Trees are like lungs of the planet"
The use of so many personifications "Preserver and greener of the earth"
The use of repetition "trees"
The use of hyperbole "lungs of the planet"
The use of conjunction "Trees absorb odours and pollutant gasses"

For the call to combat climate change in the international world, I aligned myself with this stanza:

"Trees provide food
Trees provide wood
Trees combat climate change
When properly managed"

This poem is a way of encouraging the participation, cooperation, and understanding of the population toward combating climate change problems. The education of our youths and children about the values of sustainability is our most important ally for the huge cultural change we seek in the world today. Thank you for adding a voice to this.

James Hope Esq
Associate lawyer and Partner at Reformers League,
Abuja, Nigeria (Hope 2018)

Conclusion and Policy Recommendations

From the findings of the study, it is recommended that there is a need to educate the farmers on the benefits of joining cooperatives, thereby pooling their resources together to help in procurement of farm inputs as well as credit facilities that can enable them to adapt more effectively to climate change. However, much work is still needed in the aspect of the level of value attached to their farm produce both on the side of farmers and governmental agencies. It is necessary for the government and other stakeholders in the area to develop and implement policies that specifically address the individual needs of the different states rather than general policies. Risk management is a key feature of adaptation. Current approach dealing with climate risks has to be strengthened. They can range from risk reduction to risk sharing – e.g., in disaster management or water resources management. Climate projections will have to be taken into account and new methods and procedures are required. Adaptation may require complex governance processes. New stakeholders have to be involved, as climate change adaptation will require action by people who may have not explicitly considered climate change in their past decisions. Communication among the different selected communities will have to improve strategies in various areas which need to be efficiently coordinated and managed. Capacity development is key in such a complex and dynamic environment. Increasing the robustness of livelihoods or agricultural systems, for example, is an important contribution to adaptation in itself. In this context, specific techniques like efficient irrigation or watershed management already contribute greatly to the adaptive capacity of regions. Building with an understanding of key climate change risks, such techniques can be improved, targeted, or transferred to regions where they have not been previously employed.

Further Recommendations

1. Climate change NGOs should maximize the leverage of churches, mosques, and other social gathering in the villages to disseminate the knowledge/message of climate change adaptation strategies because a greater number of these farmers are from our rural areas.
2. Innovative loan scheme can be given to farmers by the climate change NGOs to our rural farmers to alleviate the risk and dangers associated with climate change. As these farmers access the loans, such loans can be used to purchase relevant books and audio/video teachings on climate change adaptation to better enlighten them.
3. Furthermore, poetry should be used as a valuable and effective tool to help in climate change education and adaptation for sustainability in Nigeria. Poetry has a unique way of communicating the message of climate change adaptation to farmers in rural areas for sustainability.
4. Our youths should be encouraged toward creativity and excellence in acquiring relevant artistic skills in poetry for the purpose of climate change education for sustainable development. It is good to know that climate education and adaptation has the capacity to provide employment for our teeming unemployed youths in Nigeria thus helping to boost our nation's economy. These youths especially the unemployed can be well trained and engaged as climate change educators to help farmers in rural areas to adapt effectively. There is, therefore, an urgent need for climate change educators which can be employed by the government to help educate farmers in rural areas on climate resilience and adaptation for sustainability of our economy in Nigeria.
5. Climate change adaptation summits should be held monthly or annually with the help of the climate change NGOs and government agencies in specific regions in rural areas in Nigeria to ensure that the farmers are well educated, equipped, and able to network with one another so as to adapt effectively.

References

Adelekan IO (2009) Vulnerability of poor urban coastal communities to climate change in Nigeria. Fifth urban research symposium, Lagos

Adelekan IO (2011) Climate change, weather extremes and society. The Seventeenth Faculty Lecture of Faculty of the Social Sciences, University of Ibadan. May 25

Ahmad MH, Azhar U, Wasti SA, Inam Z (2005) Interaction between population and environmental degradation. Pak Dev Rev 44(4):1135–1150

Anabaraonye B (2018) Plant a tree. https://projectgreeninitiative.wordpress.com/2018/02/19/plant-a-tree/

Apata TG (2011) Effects of global climate change on Nigerian agriculture: an empirical analysis. Central Bank Niger J Appl Stat 2(1):31–50

Ashby, J & Pachico, D. (2012) Climate change: from concepts to action. A guide for development practitioners. Baltimore Maryland: Catholic Relief Services. https://www.crs.org/sites/default/files/tools-research/climate-change-from-concepts-to-action.pdf

Awojuola E (2016) Climate change, major challenge to agriculture development. Department of Climate change, Federal Ministry of Environment, Abuja

Ayinde OE, Muchie M, Olatunji GB (2011) Effect of climate change on agricultural productivity in Nigeria: a co-integration model approach. J Hum Ecol 35(3):189–194

Ayoade JO (2003) Climate change: a synopsis of its nature, causes, effects and management. Vantage Publishers Limited, Ibadan

Bassey N (2017) Climate change impacts on Land and Food. Centre for Health of Mother Earth Foundation-HOMEF

Bast JL (2010) Seven theories of climate change. Heartland Institute Publication, Chicago

Beyioku J (2016) Climate change in Nigeria: a brief review of causes, effects and solution. Federal Ministry of Information & Culture, Abuja

Charles, P. (2012) To what extent could planting trees help solve climate change. Grantham Research Institute on Climate Change and the Environment, London School of Economics and Political Science. https://www.theguardian.com/environment/2012/nov/29/planting-trees-climate-change

DeSombre ER (2002) Global environment and world politics. Continuum Books Limited, London

Ehrlich PR, Holdren JP (1971) Impact of population growth. Science, New Series 171(3977):1212–1217

Ekundayo O (2008) Nature's gifts: essence, oils and plants. An Inaugural Lecture delivered at the University of Ibadan on Thursday, 27th November

Enete IC (2014) Impacts of climate change on agricultural production in Enugu State. Niger J Earth Sci Clim Change 5(9):234. https://doi.org/10.4172/2157-7617.1000234

Enete AA, Amusa TA (2010) Challenges of agricultural adaptation to climate change in Nigeria: a synthesis from the literature. Being a paper presented at an international conference on "Enhancing agricultural adaptation to climate change" held on the 27th of July 2010 at Enugu

Gans O, Jost F (2005) Decomposing the impact of population growth on environmental deterioration. Discussion Paper Series, No. 422

Gowland-Mwangi J (2012) Agricultural extension strategies for effective mitigation against the effects of climate change. Being the keynote paper of proceedings at the 17th annual national conference (AESON), 11–14 March

Harte J (2007) Human population as a dynamic factor in environmental degradation. Popul Environ 28:223–236. https://doi.org/10.1007/s11111-007-0048-3

Holden ST, Sankhayan PL (1998) Population pressure, agricultural change and environmental degradation in the Western Himalayan region of India. Forum Dev Stud 2:271–300

Hope J (2018) Poem "Plant a Tree" Review. https://projectgreeninitiative.wordpress.com/2018/02/20/poem-plant-a-tree-review-by-james-hope/

Iheoma CG (2015) Impact of climate change on agricultural production and sustainability in Nigeria. Asian J Agric Ext Econ Sociol 4(1):29–41

IPCC (2001) Climate change 2001: the scientific basis. Contribution of working group I to the third assessment report of the intergovernmental panel on climate change. Published by the press syndicate of the University of Cambridge, The Pitt Building, Trumpington Street, Cambridge, United Kingdom. First published 2001. https://www.ipcc.ch/ipccreports/tar/wg1/pdf/WGI_TAR_full_report.pdf

IPCC (2007) IPCC Fourth Assessment Report: Climate Change 2007 https://www.ipcc.ch/publications_and_data/ar4/wg2/en/frontmattersg.html

Ladan SI (2014) An appraisal of climate change and agriculture in Nigeria. J Geogr Reg Plann 7(9):176–184

Maier D. (2015). The great myth: why population growth does not necessarily cause environmental degradation and poverty. Being a Master;s thesis on Environment and Development, London School of Economics and Political Science. https://pdfs.semanticscholar.org/8caf/6c572acb2874189ec7097194d9fda76eb4b7.pdf

Nakagawa M (2015) The science of climate change. https://yali.state.gov/courses/climate

Nwajiuba C (2012) Does agriculture have a future in South East? Inaugural lecture no. 5. Imo State University Owerri

Nwajiuba C (2014) Nigeria's agriculture and food security challenges. http://nairametrics.com/wpcontent/uploads/2011/10/Nigerias-Agricultural-and-Food-Security-Challenges.pdf, https://ng.boell.org/sites/default/files/uploads/2013/10/agriculture_-_green_deal_nigeria_study.pdf

Odjugo PA (2010) General overview of climate change impacts in Nigeria. J Hum Ecol 29(1):47–55

OECD (2012) Farmers Behaviours, agricultural management and climate change. OECD Publishing, Paris

OECD (2009) OECD Work on Climate Change. https://www.oecd.org/env/cc/Work-on-Climate-Change-2013-14_web.pdf

Okafor JC (2011) Oil and the politics of greenhouse reduction: implication for developing countries. Unpublished Master's Dissertation, Department of Political Science, Faculty of the Social Sciences, University of Ibadan, Ibadan

Oladiran EO (2010) Atmosphere dynamics and regional contributions to the Earth's energy balance and global warming. An inaugural lecture delivered at the University of Ibadan on Thursday, 26th August

Olufemi M (2018) Climate change and effects on present-day Nigeria. www.tribuneonline.ng

Pereira L (2017) Climate change impacts on agriculture across Africa. https://doi.org/10.1093/acrefore/9780199389414.013.292

Ray IA, Ray S (2011) Impact of population growth on environmental degradation: case of India. J Econ Sustain Dev 2(8):72–77

Rozelle S, Huang J, Zhang L (1997) Poverty, population and environmental degradation in China. Food Policy 22(3):229–251

Ruth D (2017) Impact of climate change on health in Nigeria. www.Thisday.org

Sheykhi MT (2001) Environmental consequences of population growth: a socio-demographic study of Asia with special reference to Iran. Hum Ecol Special Issue 10(8):103–109

Treepeople (2017) Top 22 benefits of trees. https://www.treepeople.org/resources/tree-benefits

Tsojon JD (2017) Impact of climate change on agricultural production by farmers in Taraba state. Niger Int J Entrep Dev Educ Sci Res 4(1):178–190

Umar OJ, Ozohu OA (2015) Impact of climate change on agricultural production and food supply in Africa. Paper presented at the international conference on latest trends in food, biological & ecological sciences, Dubai. 11–12 Oct

UN (2015) Paris Agreement on climate change-the United Nations. www.un.org/sustainabledevelopment/blog/2015/12/the_paris-agreement-faqs

UNEP (2015) Climate finance. https://www.unenvironment.org/explore-topics/climate-change/what-we-do/climate-finance

UNESCO (2010) Climate change education for sustainable development. http://unesdoc.unesco.org/images/0019/001901/190101E.pdf

UNFCCC (1992) United Nations framework convention on climate change. https://unfccc.int/resource/docs/convkp/conveng.pdf

Wikipedia (2018) https://en.wikipedia.org/wiki/Climate_resilience

Zaman K, Khan H, Khan MM, Saleem Z, Nawaz M (2011) The impact of population on environmental degradation in South Asia: application of seemingly unrelated regression equation model. Environ Econ 2(2):80–88. http://nidm.gov.in/PDF/. Adaptation to Climate Change with a Focus on Rural Areas and India

Water Harvesting Technology for Enhancing Food Security Livelihood: The Case of Northern Katsina State, Nigeria

134

S. Jari, M. Muntaka, and R. T. Nabinta

Contents

Introduction 2792
Justification of the Research 2794
Methodology 2795
Description of the Study Area 2795
Methods 2796
Results 2797
Discussions 2799
Conclusion 2800
Recommendations 2800
References 2801

Abstract

Rising population density, high rainfall variability, and depletion of natural resources, particularly of soils and water, continue to accelerate desertification and low crop productivity in the Sahel zone of Katsina State (12°15′N 7°30′E), Nigeria. This study was conducted at 10 communities across Dutsin-Ma, Safana, Kurfi, and Batsari local government areas of Katsina State during the rainy season of 2015. The objectives of the study were to ascertain the effects of *Zaï* pit (planting pit) density/ha and addition of organic matter in enhancing sorghum yield, food security, livelihood, and reclaiming of degraded land. Ten farmers

S. Jari (✉)
Department of Crop Production and Protection, Federal University Dutsin-Ma Katsina State, Katsina, Nigeria
e-mail: sjari@fudutsinma.edu.ng

M. Muntaka · R. T. Nabinta
Department of Agricultural Extension and Rural Sociology, Federal University Dutsin-Ma, Katsina State, Nigeria
e-mail: mmamman1@fudutsinma.edu.ng; runabinta@fudutsinma.edu.ng

W. Leal Filho (ed.), *Handbook of Climate Change Resilience*,
https://doi.org/10.1007/978-3-319-93336-8_187

(one each from the communities) were purposively selected from the study area and given induction course on *Zaï* pit technology, and the skills acquired were applied in their farms. Treatment consisting of *Zaï* pit (15,000/ha) and addition of organic matter significantly increases all the yield parameters and yield of sorghum in both mother (researcher managed) and daughter (farmer managed) trials. Increasing the number of *Zaï* pit above 15,000/ha reduces the catchment area for runoff and total rainfall harvested which consequently reduces yield. All the 10 farmers that participated in the project recorded a significant increase of over 200% in sorghum yield when they adopted *Zaï* pit and with the addition of organic matter. The type of organic matter used does not have significant effect on sorghum yield. For sustainable assurance against crop failure due to drought and enhancing crop yield, food security, income, and livelihood in the study area, it is recommended that farmers should adopt *Zaï* pit technology (15,000/ha) in combination with 10 t of organic matter.

Keywords
Zaï pit · Organic matter · Degraded land · Food security · Livelihood

Introduction

The Sahel is characterized by wide climate variations with succession of dry and wet years and irregular rainfall between 300 and 600 mm. Temperatures in West Africa increased by an average of 0.5–0.7 °C between 1951 and 2001, with an increase of around 0.7 °C in the minimum temperature and 0.35 °C in the maximum temperature (Sarr and N'Djafa Ouaga 2009). This climate variability stands as the major constraint to assuring food security in the region (Kandji et al. 2006). Two major factors characterizing agriculture in the Sahel are (i) erratic climatic conditions with frequent periods of water shortages and (ii) the presence of large areas of inherently low fertility and crust-prone soils. This picture is worsened further by slash and burn practices, continuous cultivation with low inputs, and overgrazing and trampling by cattle. These have resulted in severe human-induced land degradation in the Sahel (Roose et al. 1994). In Burkina Faso and Niger, farmers have developed their own soil and water conservation techniques (*Zaï* technology) to reduce the ongoing processes of soil erosion and land degradation. Most of these methods are multifunctional and consist of rainwater harvesting and soil water conservation technologies. These technologies reduce soil erosion and at the same time maintain the soil organic matter status as well as the physical properties of the soil. When surface runoff is collected and stored in reservoirs, it can be used for growing crops during periods of low or no rainfall.

Zaï pits are indigenous practices used to rejuvenate degraded soils in Burkina Faso by breaking up the soil crust to improve water infiltration and adding manure to improve soil fertility. The practice consists of a grid of dug holes roughly 15 × 15 cm, which are then filled with manure to improve soil fertility. *Zaï* pits can increase soil water holding capacity by up to five times while collecting up to 25% of the runoff in the immediate area surrounding the hole. Sawadogo et al. (2001) reported that this technique reduces both runoff and soil erosion and reduces risks linked to erratic and declining rainfall, stores water from rainfall for longer periods as a buffer against midseason dry spells, and helps increase the efficiency of water and nutrient use by plants, thereby increasing productivity.

Crop yields in the Sudan-Sahel zone of northern Katsina State, Nigeria, are strongly and positively influenced by rainfall and evaporation as major climatic parameters affecting crop yields. This is so because soil moisture, which is often regarded as the single most important parameter in determining crop yields in semiarid environments, can be coarsely derived from a relationship between rainfall and evaporation. However, according to Enete and Amusan (2010), the main cause of poor yields is attributed to drought, poor soils, pest, and diseases. Food and water scarcity has negative economic impacts and exacerbates poverty, especially in rural communities that are located in arid and semiarid areas of sub-Saharan Africa. Water scarcity is often related to high intensity and short duration of rainfall, with large spatial and temporal variability, rather than to cumulative annual and seasonal rainfall (Falkenmark and Rockström 2004; Rockström and Falkenmark 2015; IWMI 2015; Nicol et al. 2015). Such irregular patterns result in high risk of drought and intra-seasonal dry spells. This in turn leads to unpredictable and depressed crop yields, perennial food shortages, and disruptive conflicts over use and access to existing water supplies (Ngigi 2003). The situation is being aggravated by the ensuing climate change and variability (Pachauri et al. 2014). Climate change models have projected a decrease in rainfall in sub-Saharan Africa, and empirical literature has already shown similar trends (Nyagumbo et al. 2009). Therefore, meeting current and future food demands would require upgrading of agriculture by adopting climate- and water-smart technologies such as rainwater harvesting and management (Rockström and Falkenmark 2015). Smallholder farmers in semiarid areas of sub-Saharan Africa often experience total crop failure once every 10 years and drastically reduced yields two to four times within similar period (EC 2012). Agricultural droughts, which occur during the crop-growing season, often have dire consequences on food and economic security of poor rural households (Malesu et al. 2012).

The potential of rainfed farming can be unlocked significantly in regions subjected to dry spells through rainwater harvesting and storage for small-scale irrigation during the dry season (Awulachew et al. 2005; Mutabazi et al. 2005; Mati et al. 2006; Mati 2007; Malesu et al. 2012).

The application of the Zaï technique can increase production by about 500% if well executed. Sawadogo et al. (2001) explained that pits have been used to diversify plant biomass in Burkina Faso, and the practice has helped improve soil fertility and crop yield in the area. *Zaï* pits are especially relevant to areas receiving 300–800 mm annual rainfall (Roose et al. 1999). Higher rainfall amounts could cause waterlogging of the pits. *Zaï* pit practice has the potential to increase cereal yields by a factor 10, and yields increase further with the application of organic amendments. *Zaï* pits allow collecting 25% of a runoff coming from five times its area (Malesu et al. 2012).

According to GIZ (2012), *Zaï* pit technology is an example of a successful intervention in rainfed agriculture in West Africa, which is transforming the livelihoods of many resource-poor smallholder farmers. Innovative and indigenous ways to achieve improved crop yields through integrated land and water management such as rainwater harvesting and soil water conservation have been successfully tested and, in some cases, adopted in West Africa. Integrated land and water management could play a key role in enhancing crop yield and livelihood and building resilience to climate change. Reports from GIZ (2012) indicated that appropriate technologies have been developed and are effective for increasing crop production in dry areas, but in view of severe land degradation in the Sahelian zone of West Africa, integrated rainwater harvesting and soil water conservation (RWHSWC) can also enhance the effectiveness of these technologies further. They can also be used to rehabilitate degraded land, retain moisture, and re-establish vegetation cover to improve crop growth in order to alleviate poverty and enhance food security. The objectives of this research are:

- To demonstrate that *Zaï* pit technology is a veritable tool for enhancing crop productivity and livelihood of small-scale farmers in the Sahel drought-prone areas of Katsina state, Nigeria.
- To establish the most optimum number of *Zaï* pits per hectare and the type of organic matter to use for maximum yield and productivity of sorghum.
- To stimulate farmer interest and subsequently mass adoption of *Zaï* technology as a viable tool for rehabilitation of degraded land, enhancing crop yield and livelihood, building resilience of small-scale farmers to climate change.

This chapter highlights the findings of our research on the effect of *Zaï* pit density/ha and addition of organic matter in some communities located in the Sudan-Sahel zone of Katsina State, Nigeria, and how it can contribute to enhancing crop yield, food security, livelihood, and rehabilitation of the degraded lands. Farmer participatory extension approach was employed in order to ensure mass adoption of the technology and upscaling of it to areas with similar agroecological condition in the region.

Justification of the Research

- To address the challenges of increasing food and nutrition insecurity and increasing poverty in the face of a rapidly changing climate and a degrading natural capital base, the region needs to move toward an integrated climate-smart agriculture approach to managing competing demands for land, water, and other natural resources.
- Climate remains a major factor in crop productivity, and as climate change continues to strain agricultural production, it is important for farmers to find

ways to use rainwater more effectively. In sub-Saharan Africa, a region where only 15–30% of rainfall is being productively used by crops, techniques that enhance soil fertility and water use are especially important for food security.
- The use of *Zaï* pit for rainwater harvesting as assurance against drought for small-scale farmers has not been exploited sufficiently in this area.
- To generate data and information to tackle the knowledge gap on the *Zaï* technology as a veritable tool for soil water conservation for enhancing crop productivity, food security, and rural livelihood in the study area.
- Limited knowledge and skills about *Zaï* technology which is linked to the absence of robust extension services in the area of study. There is a major regional need for capacity development and training in soil water conservation across national institutions, service providers (agric extension personnel), and farmers' organizations.

Methodology

Description of the Study Area

The study was carried out at Federal University Dutsin-Ma Research Farm Badole and 10 communities located in three local government areas of Katsina State, Nigeria, namely, Fadumawa, Kaware, and Tamawa, in Kurfi local government area; Akata, Shirgi, and Yandaka in Batsari local government area; and Barebari, Kukar Samu, and Runka in Safana local government area. These local government areas are all located in the Sudan-Sahel ecological zone of Katsina State, Nigeria.

Katsina State covers three agroecological zones: the Sahel, the Sudan, and the Northern Guinea Savanna zones (Ogungbile et al. 1999; Maiangwa et al. 2007). The rainfall pattern in the State is unimodal and ranges between 350 and 500 mm in the Sahel, between 600 and 850 mm in the Sudan Savanna, and between 900 and 1000 mm in the Northern Guinea Savanna (Ogungbile et al. 1999). The onset of rains, which marks the beginning of the growing season, starts in May in the Northern Guinea and June in the Sudan agroecological zones (Elemo et al. 1990). Concomitantly, the duration of the growing season spreads from a range of 140 to 200 days in the Northern Guinea, 95–140 days in the Sudan, to 68–102 days in the Sahel (Elemo et al. 1990). Consequently, crops of longer growing cycles dominate the Northern Guinea, while those requiring shorter growing cycles dominate the Sudan and Sahel zones (Elemo et al. 1990).

The soils in the sites of the study belong mostly to the ferruginous tropical soil group, and depending on hydrology, parent material, and age, soil types such as alfisols, entisols, inceptisols, regosols, and ferrisols are found (Elemo et al. 1990). Characteristically, most of the soils in the sites are generally low in organic matter and phosphorus levels; also, as a result of the dominant kaolinitic type of clay (due to the type of parent material), the cation exchange capacity is low, resulting in reduced buffering capacity (Elemo et al. 1990). In general terms, therefore, the soils are poor in fertility status and structure and are readily susceptible to degradation where the protective cover of vegetation is weakened or removed (Odunze 1999; Harrison 1990).

The major crops grown in the area under study are sorghum, millet, groundnut, cowpea, soybean, and sesame. Pepper, onion, and tomatoes are also grown during the dry season in the low-lying areas, the (*Fadamas*). Though sole-cropped fields of crops occur, intercropping is the dominant practice. There is, however, always one crop dominant on any mixed-crop field (Ogungbile et al. 1999). The relative importance of crops also varies according to location (Ogungbile et al. 1999). Livestock production is also an integral part of the farming system as both crops and animals are sources of food and cash income for farmers (Ogungbile et al. 1999). The livestock kept include cattle, goats, sheep, donkeys, horses, camels, and poultry.

Methods

A mother trial (researcher managed) was conducted at the University Farm Badole and the daughter trials on farmer field (farmer managed) in 10 locations in Katsina State, Nigeria, during the rainy season of 2015. The experiments were designed to assess the affectivity of *Zaï* pit (planting pits) density and addition of organic matter in enhancing crop yield and ameliorating crusted soils. The treatments consisted of $P_1 = 10{,}000$, $P_2 = 15{,}000$, and $P_3 = 20{,}000$ *Zaï* pits per hectare, respectively; the amendments consisted of F_1, no amendment no *Zaï* pit (control); F_2 and F_3, 10 t/ha of goat manure and cow dung, respectively; and F_4, *Zaï* pit only. Test crop used was (SAMSORG 01) short-duration sorghum variety. The 12 treatments ($1 \times 3 \times 4 = 12$) were factorized and randomized in a completely randomized block design (CRBD) and replicated 3 times. The plot size was 2.5 m× 3 m = 7.5 m^2, and 50 cm was allowed between plots within a block and 1 meter between replications. Two middle rows (1.5×3 m = 4.5 m^2) in each plot serve as net plot, and all plants from there were collected and analyzed for yield. The result was subjected to statistical analysis for correctness.

The mother trial at the University Farm was laid out together with the 10 collaborating farmers, one each from 10 communities that are participating in the project. Using the experience gained from their participation in the layout of mother trial, farmers were assisted to set up a similar trial (daughter trial) on their farms together with five farmers from their locality with at least one of the participants being a woman farmer.

The research team from the University visited each of the 10 sites (daughter trials) three times during the cropping season, i.e., at establishment, intensive vegetative growth, and harvesting/maturity stages. Farmers were taught how to collect crop samples for yield analysis. The samples collected from each farmer were further subjected to analysis at the University laboratory.

Table 1 Effect of *Zaï* pit density/ha and type of organic matter on yield parameters and yield of sorghum at FUDMA. University farm Badole, 2015

Treatment	No of head/plot	No of seeds/head	1000swt (G)_	Yield (Kg/ha)
Zaï pit				
P1	14.42b	1928.33	25.83	997.30b
P2	18.83a	1941.25	25.02	1246.60a
P3	21.17a	1950.00	24.91	1404.90a
Se±	0.857	55.787	0.696	79.646
Fertilizer				
F1	14.6c	1611.11d	23.76b	755.00
F2	1956ab	2277.22a	28.83a	1751.30a
F3	21.44a	2046.67b	24.61b	1426.20b
F4	17.00bc	1824.44c	23.80b	968.50c
Se±	0.989	64.418	0.804	91.967
Interaction				
P*F	NS	*	NS	NS

Means followed by unlike letters are significantly different
NS not significant, *significant @ 5%, P

Table 2 Interaction between treatments on sorghum yield (kg/ha) at FUDMA University Farm Badole (2015)

Treatment		Zaï pit	
Fertilizer	P1	P2	P3
F1	1616.70d	1530.00d	1686.70 cd
F2	2400a	2038.30abc	2393.30a
F3	1996.70bc	2106.70ab	2036.30abc
F4	1700.00 cd	2090.00ab	1686.30 cd
Se±		111.575	

Results

Yield analysis of the mother trial at University Farm Badole is presented in Tables 1 and 2, and the highest grain yield of sorghum (2393.3 kg/ha) was recorded with the treatment having 15,000 *Zaï* pit density/ha and 10 t of goat manure/ha; this is statistically different compared with the control. The works of many authors (Hassan 1996; Mathieu et al. 1996; Oweis et al. 1999; and Biazin et al. 2012) also reported an increase in yield of sorghum with increase in number of *Zaï* pit per hectare up to 15,000 pits per ha. Additional grain of 1638 kg/ha was achieved when compared with control.

The highest sorghum yields from the farmer-managed plots were 1570, 1557, and 1528 kg/ha recorded at Kukar Samu, Barebari, and Badole, respectively, under treatment combinations with 15,000 *Zaï* pits/ha and 10 t/ha goat manure. All the 10 communities obtained additional 1000 kg/ha of sorghum grains compared to the

Table 3 Effect of *Zaï* pit density and organic matter on yield of sorghum (kg/ha) at 10 locations (farmer managed) in the Sahel zone of Katsina State, Nigeria, 2015

Location	*Zaï* pit	F_1	F_2	F_3	F_4
Atake	P_1	450	968	1154	780
	P_2	600	954	1050	930
	P_3	560	1010	1300	827
Badole	P_1	352	972	1154	705
	P_2	502	1528	1291	908
	P_3	671	1403	1390	853
Barebari	P_1	510	1180	1100	790
	P_2	554	1066	1557	810
	P_3	800	1115	1040	850
Fadumawa	P_1	300	690	1080	740
	P_2	555	1085	1050	800
	P_3	690	1011	1200	900
Kaware	P_1	700	1250	1033	870
	P_2	800	1011	1175	800
	P_3	920	1150	1290	950
Kukar-Samu	P_1	560	1050	1070	900
	P_2	560	1570	1190	970
	P_3	890	1280	1300	980
Runka	P_1	308	1050	1005	900
	P_2	490	1280	1285	1040
	P_3	519	1250	1180	1080
Shirgi	P_1	580	1050	1040	750
	P_2	760	1396	1300	1050
	P_3	610	1080	1014	1110
Tamawa	P_1	590	1085	1010	870
	P_2	760	1290	1180	1000
	P_3	790	1210	1200	1090
Yandaka	P_1	290	1100	1080	980
	P_2	430	1380	1290	1010
	P_3	490	1190	1200	1009

*F_1 = No *Zaï* pit no organic matter, F2 = *Zaï* pit + organic matter (goat manure) 10 t/ha, F3 = *Zaï* pit + organic matter (cow dung) 10 t/ha, F4 = *Zaï* pit only

control. Similar results were reported in the works of Abdou Hassan (1996) in Niger Republic.

Application of *Zaï* pits alone (without addition of organic matter) at the mother trials (researcher managed) and at 10 locations (farmer managed) as presented in Tables 1, 2, and 3, respectively, indicated a significant increase in sorghum yield compared with the control.

Progressive increase in sorghum yield was observed as the number of *Zaï* pit/ha increases from 10,000 to 15,000 pits/ha, but subsequent increase to 20,000 pit/ha does not lead to increase in yield. Working on *Zaï* technology in Niger, Abdou

(2014) and Hassan (1996) reported a decrease in millet yield as the number of *Zaï* pit per ha increases up to 20,000 pit/ha.

The control treatment recorded the lowest sorghum yield from farmer-managed plots (Table 3) 290, 300, and 308 kg/ha at Yandaka, Fadumawa, and Runka communities, respectively. This may be attributed to severity of degradation at the experimental sites and drought during the growing period. Similar trend was reported by Araya et al. 2012, in Ethiopia.

Sorghum yield and all the yield parameters at the mother trial (researcher managed) were significantly affected by the number of *Zaï* pit/ha and addition of organic matter, but a contrasting result was reported by Hassan (1996) in Mali.

The sources of organic matter have no significant effect on yield and yield parameters of sorghum at the mother trials (researcher-managed plot). Both cow dung and goat manure positively impacted on yield and yield parameters of sorghum.

Discussions

Increasing the density of *Zaï* pits from 10,000 to 15,000 pits per hectare increases sorghum yield by increasing the total amount of rainwater harvested which in turn gives better assurance against drought and creates a favorable condition for growth and yield of sorghum. Higher yield obtained from this treatment could be attributed to higher number of heads per plot, higher number of seeds per head, and heavier 1000 seed weight. This observation could be due to improved soil productivity mainly through biophysical and biological processes induced by *Zaï* pits. Indeed, by breaking the soil crust, pit digging facilitates more water infiltration; also, runoff water is harvested due to the earthen bund formed down the slope of the pits; all these ensures stable moisture supply which in turn positively impacted on yield and yield parameters. *Zaï* pits and addition of organic matter (cow dung or goat manure) significantly improve all yield parameters of sorghum and improved all the biophysical and chemical properties of the degraded land. Addition of organic matter enhances cation exchange capacity of the soil and absorption of nutrients. Increasing the number of *Zaï* pit up to 20,000/ha reduces the quantity of runoff collected in each pit as the surface area is reduced with an increase in the number of pits.

Zaï pits alone gave an appreciable yield of sorghum, but to obtain maximum yield, organic matter should be added. *Zaï* pit and addition of organic matter can make an important contribution to food security in Sahel region of Katsina State, Nigeria, by assuring that some harvest can be reaped even in a year of very low rainfall. Similar conclusion was reached by Sawadogo et al. (2001) *in* Burkina Faso and Hassan (1996) in Niger.

This significant increase in yield that is observed when *Zaï* pits are combined with the addition of organic matter may be because the application of organic inputs not only enhances soil nutrient availability but also improves crop nutrient uptake from soil reserves. The applied organic matter attracts termites which develop nesting and foraging activities. This results in improved soil structure as they open up large and numerous macropores throughout the entire soil profile. The soil structure enhancement following soil fauna activity leads to increased water infiltration and drainage, lower runoff, and reduced soil resistance to root penetration. This improvement of water status in the soil and the increased decomposition and nutrient release result into a beneficial impact of the *Zaï* technology on crop performance under semiarid conditions. Results from our experiments both at the mother trials and farmer-managed plots in the 10 locations confirmed the above assertion. Several studies (Adewumi et al. 2002; Barry et al. 2008, Ackermann et al. 2014) in the Sahel region reported that applying compost with mineral fertilizers allowed substantial gain in sorghum grain yields from the *Zaï* pit.

The larger the planting pits and the bigger the spacing, the more water can be harvested from the uncultivated micro-catchment; however economically more viable yield is obtained when the density is at 15,000 *Zaï* pits/ha.

Conclusion

The results from our experiments both at the mother (researcher managed) and farmer-managed trials demonstrated that *Zaï* technology and addition of organic matter can enhance sorghum yields significantly, and since sorghum is the main staple food of the population as well as a cash crop in the area, increasing the productivity of this crop would enhance food security, income, and livelihood of small-scale farmers.

With the recurrent drought due to erratic rainfall, declining soil fertility, and severe land degradation in the area under study, *Zaï* pit technology and addition of organic matter could be a viable option for farmers to rehabilitate their degraded land; assure against drought; enhance soil fertility, crop yield, income, food security, and livelihood; and allow them to better adapt to climate variability and improve their resilience to climate change.

Recommendations

Based on the results of our experiment both at the researcher-managed and farmer-managed plots, we recommend the adoption of 15,000 *Zaï* pit/ha and addition of 10 t/ha organic matter as the best practice for optimum sorghum yield and reclamation of degraded land in the study area.

Zaï pit and application of organic matter offer an integrated and systemic response to the combined challenges of food security, adaptation and adoption of smallholder farming practices, and natural resource conservation; therefore, policy responses to climate change should support out-scaling of this technology to smallholder farmers in the area under study.

Zaï pit technology has proved to be an efficient and effective tool for rainwater harvesting and assurance against drought for enhancing crop yield of small-scale farmers in drought-prone areas of Katsina State, Nigeria; therefore these techniques should be upscaled in the affected areas.

Zaï pit and addition of organic matter are appropriate measures and practices that optimize soil and water conservation at farm level, and they may underpin sustainable smallholder productivity across ecologies and therefore should be upscaled to cover wider areas of the Sahel region of Nigeria.

More practical research and innovation is urgently needed to identify barriers (e.g., labor) to widespread adoption of those technologies, propose ways of alleviating them, and develop appropriate approaches to guide investments that will assure improved resilience, productivity, and sustainability in smallholder systems in the Sahel region of Nigeria.

Acknowledgments The authors wish to appreciate the funding support provided by the Tertiary Institutions Trust Fund (TETFUND) and authorities of FUDMA for this study.

References

Abdou M (2014) Practices, techniques and technologies for restoring degraded landscapes in the Sahel. African forest forum, working paper series, vol (2)3, 42p

Ackermann K, Nill D, Schöning A, Trux A, van den Akker E, Wegner M, Gerhartsreiter T (2014) Water and soil conservation practices in the Sahel: an analysis of their potential to increase resilience of rural livelihoods. Planet@Risk 2(1):14–20 Special issue on desertification

Adewumi JK, Kolawole A (2002) Water Harvesting for Dry Land Farming. Zaria Geographer 15(1):1–11

Araya T, Cornelis WM, Nyssen J, Govaerts B, Raes D, Saire KD Deckers J (2012). Building resilience against drought and soil erosion: impact of field water conservation in the drought prone Vertisol areas of northern Ethiopia. In: Agro Environ 2012, Wageningen

Awulachew SB, Merrey DJ, Kamara AB, Van Koppen B, Penning de Vries F, Boelee E, Makombe G (2005) Experiences and opportunities for promoting small-scale/microirrigation and rainwater harvesting for food security in Ethiopia. Working paper 98, v. 86p, IWMI, Colombo

Barry, Boubacar, Olaleye Adesola O, Zougmore R, Fatondji D (2008) Rainwater harvesting technologies in the Sahelian zone of West Africa and the potential for outscaling. IWMI working paper 126, International Water Management Institute (IWMI), Colombo, 32p

Biazin B, Sterk G, Temesgen M, Abdulkedir A, Stroosnijder L (2012) Rainwater harvesting and management in rainfed agricultural systems in sub-Saharan African – a review. Phys Chem Earth 47–48:139–151

Elemo, K.A., V. Kumar, J.O. Olukosi and A.O. Ogungbile, (1990). Review of research work on mixed cropping in the Nigerian Savanna. Samaru miscellaneous paper no. 127, Institute for Agricultural Research, Ahmadu Bello University, Zaria, Nigeria, p 125

Enete AA, Amusan TA (2010) Challenges of agricultural adaptation to climate change in Nigeria: a synthesis from the literature. Field actions science reports vol 4, 2010

European Commission (2012). Building resilience in the Sahel http://ec.europa.eu/echo/news/2012/sahel_conference_2012_en.htm Accessed 28 Nov 2012

Falkenmark M, Rockström J (2004) Balancing water for man and nature: the new approach to ecohydrology. EarthScan, London

GIZ (2012) Good practices in soil and water conservation. A contribution to adaptation and farmers' resilience towards climate change in the Sahel. p 60, Deutsche Gesellschaft fur international Zusammenarbelt (GIZ) GmbH, www.GIZ.de/expertise/download/GIZ2012-EN-soil-water-conservation

Harrison P (1990) Sustainable growth in African agriculture. The long-term perspective study of sub-Saharan Africa: economic and sectoral policy issues. The World Bank, Washington, DC

Hassan A (1996) Improved traditional planting pits in the Tahoua department, Niger. In: Reij C (ed) Sustaining the soil indigenous soil and water conservation in Africa. Earthcan, London, pp 56–61

International Water Management Institute (IWMI) (2015) Improving water management in Myanmar's dry zone for food security, livelihoods and health. International Water Management Institute (IWMI), Colombo 52p

Kandji ST, Verchot L, Mackensen J (2006) Climate change and variability in the Sahel region: impacts and adaptation strategies in the agricultural sector. UNEP, ICRAF, Nairobi

Maiangwa MG, Ogungbile AO, Olukosi JO, Atala TK (2007) Land degradation: theory and evidence from the north-west zone of Nigeria. J Appl Sci 7:785–795

Malesu M, Black J, Alex O, Cheregony K, Nyabenge M (2012) Rain water harvesting inventory of Kenya. The World Agroforestry Centre, Nairobi

Mati BM (2007) 100 ways to manage water for small holder agriculture in East and Southern Africa. SWMnet working paper 13. Improved management in eastern and southern Africa (IMAWESA)

Mati B, De Bock T, Malesu M, Khaka E, Oduor A, Nyabenge M, Oduor V (2006) Mapping the potentials for rainwater harvesting technologies in Africa: a GIS overview on development domains for the continent and ten selected countries. Technical manual no. 6. World Agroforestry Centre (ICRAF), Nairobi

Mutabazi K, Hatibu N, Senkondo E, Mbilinyi B, Tumbo D (2005) Economics of rainwater harvesting for crop enterprises in semi-arid areas: the case of Makanya watershed in Pangani River basin, Tanzania. Soil-Water Management Research Group, Nairobi

Ngigi SN (2003) Rainwater harvesting for improved food security: promising technologies in the greater horn of Africa. Greater Horn of Africa Rainwater Partnership (GHARP), Kenya Rainwater Association (KRA), Nairobi, 266p

Nicol A, Langan S, Victor M, Gonsalves J (eds) (2015) Water-smart agriculture in East Africa. Colombo. International Water Management Institute (IWMI). CGIAR Research Program on Water, Land and Ecosystems (WLE)/Global Water Initiative East Africa (GWI EA). 352p, Kampala, Uganda. https://doi.org/10.5337/2015.203

Nyagumbo I, Mvumi BM, Mutsamba E (2009) CA in Zimbabwe: socioeconomic and biophysical studies (Conference paper). In: Regional conference on sustainable land management, Windhoek, Namibia, pp 7–11, September 2009

Odunze AC (1999) Conservation management for sustainable crop and livestock production. Paper presented at the catholic resource center. Kaduna, Nigeria, Dec 1

Ogungbile AO, Tabo R, van Duivenbooden N (1999) Multi-scale characterization of production systems to prioritize research and development in the Sudan Savanna Zone of Nigeria. In: Information bulletin No. 56, international crop research institute for the semi-arid tropics (ICRISAT), Patancheru Andhna Pradesh, India

Ouedraogo M, Kabore V (1996) The Zai: A traditional technique foe rehabilitation of degraded land in the yatenga burkina faso. In: Reij C (ed), A Book "Sustaining The Soil" indigenous soil and water conservation in Africa. Earthcan, London, pp 80–84

Oweis T, Hachum A Kijne J (1999) Water harvesting and supplemental irrigation for improved water use efficiency. SWIM paper No. 7. IWMI, Colombo, p 41

Pachauri RK, Allen MR, Barros VR, Broome J, Cramer W, Christ R, Church JA, Clarke L, Dahe Q, Dasgupta P, Dubash NK (2014) Climate change 2014: synthesis report. Contribution of working groups I, II and III to the fifth assessment report of the intergovernmental panel on climate change

Rockström J, Falkenmark M (2015) Increase water harvesting in Africa. Nature 519:283–285

Roose E, Kabor ÈV. Guenat C (1994) The *Zaï* practice: a West African traditional rehabilitation system for semi-arid degraded lands. A case study in Burkina Faso. Paper presented at the seminar on the Rehabilitation of Degraded Land, Tunisia. November 1994

Roose E, KaborÈ V, Guenat C (1999) *Zaï* practice: a west African traditional rehabilitation system for semi-arid degraded lands: a case study in Burkina Faso. Arid Soil Res Rehabilitat 13:343–355

Sarr B, N'djafa Ouaga H (2009) Les tendances actuelles et futures du climat en Afrique sahélienne: une base pour agir. Centre Régional AGRHYMET/CILSS, Niamey

Sawadogo H, Hien F, Sohoro A, Kambou F (2001) Pits for trees: how farmers in semi-arid Burkina Faso increase and diversify plant biomass. In: Reij C, Waters-Bayer A (eds) Farmer innovation in Africa: a source of inspiration for agricultural development. Earthscan Publications Ltd, London

Index

A
Accelerated sea level rise (ASLR), 31, 38
Acid-sand soil quality, *see* Tropical acid-sand soils
Action plan, 1419, 1434, 1438, 1440
ADA Consortium, 1688
Adansonia digitata, 1206
Adaptation, 430, 432–433, 434, 436, 441–442, 1102, 1104, 1108, 1649, 1656, 1788, 1793, 2060, 2061, 2063, 2487, 2730
  action, 1195–1196
  climate change in potato farming, 2436
  climate change through, 775
  community-based climate change, 1422
  decision-making, 2488
  decisions, 2655, 2660, 2663, 2665, 2679
  definition, 1662, 2774
  engaging stakeholders, 432
  ex-post, prevalence of, 433
  and local communities, 540, 544, 552
  measures, rice farmers' response and constraints (*see* Rice farmers)
  mitigation and, 1418, 1445
  negative impacts of climate change, 2417
  planning, 1523–1524
  policies, climate justice, 827
  research on, 1790
  rural farmers' adaptation decision, in Niger Delta Region (*see* Rural farmers, in Niger Delta Region)
  scientific knowledge of, 1675
  strategic pathways, 1449
  strategies, 1420
  for sustainability in Nigeria, 2774
  traditional options, 778
  understanding stage of, 432
Adaptation Plan, 983, 993
Adaptation strategies, 848, 850, 1640
  of farmers of African indigenous vegetable, 851–862
Adaptation strategies, climate change
  maize based farmers, 679
  materials and method, 677–678
Adaptation to climate change, 826–827
  and agricultural practices, 2632–2635
  definition, 2627
  French climate and energy policies, 1582–1584
  Greater Grenoble Area, 1592
  intermunicipalities, 1588–1591
  Konso community, small farming households of (*see* Konso community, southern Ethiopia)
  measures, 2628
  medium-and long-term adaptation, 2629
  multifactorial difficulties, 1584–1588
  need for, 2628
  short term adaptations, 2629
  soil and land management, 2632
  water management, 2633–2634
  Young's notion of institutional fits and misfits, 1580–1582
Adivasi Munda Community
  Aila effects and climate stressors, 61
  autoethnography, 55
  intersectional dimensions of livelihood resilience, 68
  in Kalinchi, 55
  livelihood approaches, 63–68
  socio-environmental identities, 56–60
Adoption, 1104, 1108, 1109, 1612, 1616
Advisors, 2372
Advisory dissemination, 2372

W. Leal Filho (ed.), *Handbook of Climate Change Resilience*,
https://doi.org/10.1007/978-3-319-93336-8

Africa, 2625
African Green Revolution (*see* African Climate Resilient Green Revolution)
agricultural practices & climate adaptation measures in Nigeria, Sierra Leone and Liberia, 2635
CCIDPs in (*see* Climate change-induced internally displaced persons (CCIDPs), in Africa)
climate change adaptation (*see* Adaptation to climate change)
climate change impacts on agriculture, 2630
Lablab purpureus (L.) Sweet, cultivar selection and management strategies (*see Lablab purpureus* (L.) Sweet, cultivar selection and management strategies)
National innovation systems and agriculture innovation system, 2639–2641
use of GE crops in agriculture and as climate change adaptation measures in, 2634–2639
African Climate Resilient Green Revolution, 1845, 1846–1848, 1851, 1853, 1854
EBA, 1848
microfinance, remittances and drought resilience, 1849–1851
PHL and drought resilience, 1847
African coastal zones
Niger Delta (*see* Niger Delta region)
vulnerability of, 31–35
African Economic Research Consortium (AFJRC), 1789
African highland initiative, 2732
African indigenous vegetables (AIVs), 850
African labor force, 1103
Africa's climate challenge, 1861, 1862
Agbabu in Ondo state, Nigeria
age distribution of respondents, 450, 452
bitumen exploration on flora and fauna biodiversity, 455–457
degradation of the ecosystem, 455–456
gender and level of education, distribution of respondents, 450, 452
livelihood diversification of respondents, 453–454
location, 450
map, 451
marital status of respondents, 453, 454
occupation, distribution of respondents, 453
Agricultural intensification, 2402
Agricultural Promotion Policy (APP), 1029
Agricultural sustainability, 677
Agricultural Transformation Agenda (ATA), 1029
Agriculture, 476, 786, 789, 792, 795, 1143, 1148, 2328, 2329
climate change on, 766
climate resilience in, 1504
climate sensitive, 1508
climate variability and change, 1502
commercial, 1517
and deforestation, 478
de-reservation, 479
in Egypt, 766
expansion, 479
extent of dependence on, 1509
and forest-related activities, 483
and GDP, 1503
intensification of, 487
livestock breeding, 1385
in Nigeria, 487, 1270
and production, 767
productivity, 1504
rain fed, 1509
resilience and climate adaptation plan, 1522–1524
resource base and, 1513
AgriSeed in Tanzania, 2238
Agro-advisory system
feedback, sensitive and responsive, 2381
importance of, 2372
WOTR, 2370
Agro-dealers
capacity building of, 2237
identify and capacitate networks of, 2232
network of, 2241
SME seed companies and, 2233
strengthening, 2227
Agro dealers, 1133, 1134
Agro–ecological zones (AEZs), 2253
vs. agro-ecosystems, 2339
Chilmari, 2355–2357
climate resilient agricultural practices in, 2348–2351
Ganges tidal floodplain, 2352
Khaliajuri, 2357–2359
potato yield across, 2434
in Rangamati Sadar, 2354–2355
soil moisture determination, 2422
soil temperature, 2426
Tanore, 2353–2355
temperature belts, 2419
Agro-ecology, 2276
Agroecology map, Mpakeni, 1721
Agroecosystems, carbon sequestration and, 396

Agro-forestry, 2731
Airborne infections and climate change
  air quality, 1217–1218
  bacteria, fungi/viruses, 1219
  droplet infection, 1219
  fungal airborne diseases, 1220
  health, climate change implication on, 1218–1219
  Nigeria, climate change trend in, 1220
Air quality, 1217
Akwa Ibom State
  age distribution, 1989
  climate change awareness, 1992
  data collection, 1987
  demographic characteristic of farmers, 1989
  descriptive statistics, 1988
  educational level, 1991
  focused group discussion (FGD), 1993
  gender distribution, 1989
  marital status, 1990
  multistage sampling technique, 1987
  personal interview, 1994–1997
  population, 1986
  structured questionnaire, 1987
  study area, 1985–1986
  weather variation, 1992
Albizzia lebbeck, 2292–2293
Allermöhe eco-settlement, 2090
Alternate wet and dry method (AWD), 2358
Alternative livelihood, 487, 495
Analysis of variance (ANOVA), 13, 1087
Andhra Pradesh Farmer-Managed Groundwater Systems Project (APFAMGS), 252
Annual rainfall (AR), 2552, 2553, 2555, 2560–2562
Aquaculture technologies, in Malawi
  cage culture, 1163–1166
  extension services, role of, 1166–1167
  IAAS, 1166
  water harvest, in fish pond production, 1166
Arable farmers, demographic and psychographic elements of, 2537
  See Also Climate change, in Southwestern Nigeria
Arid and semi-arid climatic conditions, 1629, 1643
Arid and semi-arid lands (ASALs), 980, 986, 994, 1685, 1686
  ADA Consortium, 1688
  drought challenges in, 1687
  Kenya, 2402
Arnstein's ladder of citizen participation, 2312
Asian Green Revolution, 1847
Asia Pacific region, 1404
Asphalt, 2518, 2521, 2525, 2527
  concrete production, 2524
  Minitab paired t-test, 2526, 2527
  quality test, 2525
Assimilation model, 595
Assimilative capacity, 595
Atlantic meriodinal mode (AMM), 2616
Atlantic multi-decadal oscillation (AMO), 2616
Atlas of Risks of La Paz Municipality, 1567
Attribute importance, 1201
Australia
  Clean Energy Target, 1405
  low-carbon technologies, 1408
  national climate policy, 1404–1407
  National Energy Guarantee (NEG), 1405
  Talanoa dialogue process, 1408
Automated content management system (ACMS), 2368
Automatic Local Flood Warning System (ALFWS), 1926
Awareness, 2392

B
Bacteria
  inoculants and carbon sequestration, 388–389
  necromass, 389
Bambara groundnut, 1239
  as climate-resilient and food security crop, 1248
  commercial orientation of farmers' welfare, 1250–1252
  determinants, land share, 1249
  econometric model, 1243
  indigenous crop, 1250
Bangladesh, 53, 2338
  agro-ecological zones, 2340 (*see also* Agro-ecological zones)
  climate resilient agricultural practices in, 2351
  impact of climate change on agriculture, 2343–2347
  policies for climate resilient agriculture in, 2359
  study area, resilient practices, 2339–2341
Bangladesh Climate Change Trust (BCCT), 353
Barraging of river, 2589
Barriers of isolation, insurance and climate change, *see* Insurance and climate change

Behavioral change, 248, 249, 251
Benefit-cost ratio (BCR) analysis, 1064, 1078–1080
Benefits, 2176, 2178, 2179, 2183, 2190, 2192
Benin
  indigenous fruit tree (*see* Indigenous fruit tree)
  woody perennial parklands, 502
Best-worst scaling (BWS), 1198
Biodiversity, 77–79, 449, 455–457, 1592
  conceptual framework, 1379–1380
  implications for, 1379
  in West Africa, ecosystems and, 1380–1381
  West Africa's regional, 1378
Biodiversity farming system, in Uganda
  adaptation and adoption problems, 1104–1105
  climate change mitigation, 1104
  crops grown in, 1109
  food production, 1106
  high value crops, 1108
  impact of, 1109
  implementation strategies, 1103–1104
  objectives, 1105
  perennial crops, 1106
  social awareness and acceptability of, 1106
Biodiversity loss, 1385
Biological carbon mitigation (BCM), 9
Biological rehabilitation interventions, 2446–2457
Biomass energy, 1099
Bioremediation
  in control and polluted soil, 323
  efficient in, 325
Biotechnology, 2331
Bitumen, 449, 454, 455–457
Bivariate Linear regression analysis (BLRA), 2537
Bond, 1865
Bonds market players, 1873
Boswellia paprifera, 2452
Breusch and Pagan test, 668
Bronfenbrenner's ecological system theory, 1938
Bt cotton, 2635
Building Nigeria's Response to Climate Change (BNRCC) community based adaptation project
  Billeri community, 2009
  Bursali community, 2009
  contribution towards poverty reduction, 2018–2019
  Dashe and Kwaikong communities, 2009
  Daudu community, 2009
  diverse adaptation activities and impact, 2010–2012
  Esuk Idebe and Akwa Esuk Enyama communities, 2010
  Falgore community, 2009
  gender equality, 2016–2018
  Gorori community, 2009
  Iko Esai and Agoi Ibami communities, 2010
  implementation attributes, 2019–2021
  methodology and approach, 2006–2008
  objectives, 2006
  performance and sustainability, 2014–2016
  Sansan, 2008
  Tosha community, 2008
  Wula Ekumpuo and Bebi communities, 2010
Building resilience, to climate change, 245–266
Building sector, of New York State, 433
  building professionals, 432, 441, 442
  climate change information, 436–439
  engaging stakeholders, adaptation process, 432
  experience and perception, of climate change impacts, 439–440
  ex-post adaptation, 433
  qualitative method, 434–435
  quantitative method, 435–436
  understanding stage, of adaptation process, 432
Buno, 52
Bureau of Agriculture and Rural Development (BoARD), 405

C
Cage culture, 1163
  extension services, lack of, 1165
  guidelines, lack of, 1165
  high cost of feed, 1165
  information, lack of, 1163
  large scale, 1163
  medium and small-scale, 1163
  quality fingerings, lack of, 1165
  theft, 1166
Caging of pond, 2586
California irrigation system, 1060
Canopy surface, 1591
Capabilities, 31, 33, 37, 46
Capacity
  adaptive, 1293
  building, 2233, 2234, 2237, 2239
  index, 1303
Carbon credit, 559, 565

Carbon dioxide, 2276
emissions, 592
Carbon Disclosure Project (CDP), 1897
Carbon sequestration, 4, 10, 16, 23, 406, 408, 411, 413–414, 423, 424
and agroecosystems, 396
and cycle in terrestrial ecosystems, 383
growth-promoting bacteria, 397
microbial inoculants in, 391–392
in soil, 384
strategy, 397
Carbon taxes, 1400
Cash for work program, 2136
Cattle herders adaptation strategies, 741
CCS technology, 1401
Cell-fusion technologies, 2332
Central Kenya, 2253
climate change impact on coffee production, 2253
coffee production, 2253
dairy sector development, 2254
innovations in dairy sector, 2265
institutional innovations in coffee sector, 2262
Central Tanzania, 2701, 2703, 2718, 2719
CERES (Crop-Environment-Resource-Synthesis), 900
Cessation, 2747, 2750, 2757, 2759–2761
of rains, 2602–2608
Chiseling, 2568
Choice-based conjoint analysis, 1200–1202
Chow test, 596
Citizen participation, 2311
City Adaptation Programmes of Action (CAPAs), 1968
Civil conflict, 1028
Clean Development Mechanism (CDM), 1363, 1465
Clean Energy Act 2011, 1405
Clemborg, 2089
Climate adaptation, 2176, 2178
in post-war garden cities, 2183–2184
in urban city blocks, 2188–2189
Climate Bonds Standard (CBS), 1867
Climate change (CC), 74–75, 90, 270, 430, 434, 436, 438–440, 443, 1140, 1142, 1145, 1148, 1151–1153, 1344, 1599–1601, 1911, 1919, 1920, 1925, 1927, 1930, 2288, 2516, 2519, 2520, 2527, 2552–2553, 2626
adaptation, 1147, 1151
adaptation of hosting communities, 642–643
adverse effects of, 430, 1911
agriculture, food security and industrialization, 2275
aquaculture technologies, fish production in Malawi (*see* Aquaculture technologies, in Malawi)
buildings, impacts on, 438
in Cameroon, 573
carbon sequestration and, 388
cattle herders adaptation strategies to, 741
causes of, 579, 1397–1399
challenges, 433, 1400–1404
characteristics, 636
and climate variability in PRB, 282
crop varieties, mitigate effects of, 2277
definition, 4, 1910, 2274, 2775
and drought as wicked problems, 1812–1818
economics of, 2147
effect of honey on egg production, 719
effects in Africa, 1861
effects of, 2516, 2517
fight against, 383, 385
on food processing, storage and food quality, 2275–2276
and food security, 1021, 1027–1028
food security (*see* Food security)
gender differentiated impacts, coping strategies (*see* Gender differentiated impacts, coping strategies)
in global water grabbing, 279
health impacts of, 572
impact assessment on health, 573
impact of, 656, 1265, 1397–1398, 1788
implications on, 431
information for farmers with respect to food production and, 2282
and insurance sector, 2148–2150
knowledge of participants on, 578
Konso community, small farming households of (*see* Konso community, southern Ethiopia)
lack of resilience, 643
legislative and institutional arrangements, 982–984
and livestock production, 732–735
and migration, 638–642
mitigating against, 386
mitigation and adaptation, 1391
on natural resource, 1143
in Nigeria, 2772
online survey participants, 440
participants' perceptions toward, 648–650

Climate change (CC) (*cont.*)
perception of, 2532
perceptions toward, 654
policy documents, 996
resilience, 2517–2518, 2527
resilient to, 1348
risks, 2627
role of, 433
severity, 2533
Sorghum crop yield, eastern Dryland of Sudan (*see* Dryland farming (sorghum), eastern Sudan)
source of information on, 579
SRP (*see* Seasonal rainfall prediction (SRP))
vulnerability index, 1862
vulnerability to, 2627
Climate change adaptation, 963, 975
concept, 1538, 1539, 1541, 1543, 1544, 1547–1549, 1552
evaluation, 1535, 1536, 1539, 1541, 1546, 1547, 1549, 1553
mainstreaming, 145
in Mali, 1729, 1732, 1734, 1738
monitoring, 1535–1537, 1541
process, 1533, 1535, 1539, 1541–1543, 1547, 1552, 1553
quality management system, 1533, 1537, 1539
reporting, 1533, 1535
strategy, 1532, 1533, 1536, 1539, 1541, 1543, 1547, 1549, 1551, 1552
and sustainable development, 139–140
water resources planning, 140–141
in West Africa, 1958–1978
Climate change adaptation and resilience (CCAR), 1535, 1536
Climate change adaptation/disaster risk reduction (CCA/DRR), 701
Climate change and variability (CCCV), 136
Climate change communication, Nigeria
credibility of sources, 2319
focus group discussion and in-depth interview sessions, farmers, 2314
knowledge, attitudes and recommended practices, 2317–2319
multi stage sampling procedure, 2313
national development plans, 2310
one-way model of communication, 2314–2316
Osun, Oyo, and Ondo states, 2312–2314
participation model, 2311–2312
raising awareness, warning and informing farmers, climate change, 2316–2317
stakeholder participation, 2311
timeliness and complexity of messages, 2320
Climate change education, 1969
Climate Change Expenditure Tagging (CCET), 2136
Climate change governance, 1533, 1534, 1553
Climate change-induced internally displaced persons (CCIDPs), in Africa
community engagement, 2030
disaster preparedness, 2029–2030
documentation, 2030
early warning, 2029
humanitarian assistance, 2030
regional protection of, 2028–2029
remediation, 2030
Climate change, in Southwestern Nigeria
adaptation/coping strategies, 2544–2546
annual rainfall data analysis (1962–2012), 2542
annual temperature data analysis (1962–2012), 2543
data collection and methodology, 2536–2537
individual perception, 2540
night and daytime temperature perception, 2540–2541
rainfall change/variability, 2533–2535
rural farmers awareness, 2538
seasonal rainfall totals, 2540
sources of information, 2539
study area description, 2535–2536
temperature change and variability, 2535
Climate change maladaptation, 1848
Climate change policy, 1598
Climate change projection, 907, 911
Climate change psychology, 2205, 2208, 2213
Climate change reduction practices, 1944
estimated marginal means, 1948
Scheffe's pair wise comparison, 1947
Climate change resilience, 2081, 2094
Climate Change Strategy and Action Plan (CCSAP), 353
Climate finance, 2775
Climate fluctuation (variability), 1937, 1940
Climate information, 919, 920, 922, 931, 2487
Climate justice
adaptation policies, 827–838
debate, 824
sustainable development and, 839–840
Climate, Land Use, Energy and Water strategies (CLEWs) framework, 181
Climate migrants, 639, 656

Climate Moisture Index (CMI), 186
Climate proofing and domestication, 2328
Climate resilience, 33, 36, 38, 41, 44, 46, 1239, 1240
in agriculture, 1504
definition, 2775
Climate resilient and nutritious crops (CRENUT) project, 1148, 1149, 1151
Climate Resilient Green Economy (CRGE), 798
Climate resilient retrofitting, 2179–2192
Climate smart agricultural (CSA) technologies
analytical model, 1650–1653
data sources and collection methods, 1650
descriptive statistics, 1653–1654
regression results, 1654–1657
sampling technique, 1650
Climate smart agriculture (CSA), 2228, 2745, 2748
in Africa, 1614–1615
agronomic and information systems, 1121
business case for, 1120–1121
data from online survey, 2729
dat from household survey, 2729
definition, 2722
importance, 1612–1613
information and insurance, 1124
innovations, 2734
interventions, 2724
literature review, 2728
policies at global level, 1613
practices, 1118, 2734
within SADC, 1615–1616
in Zimbabwe, 1616
Climate smart crop production systems, 1195
Climate variability, 1766
regional, 305
soils quality evaluation, 309
sustainability, 305
UNICAL T & R farm, 311–313
Climate vulnerability and capacity analysis (CVCA) toolkit, 2704
Climate Vulnerability Index (CVI), 157
Climatic variability effect, North-central Nigeria, 1271
annual millet yield data collection, 1272
annual rainfall and millet yield maps, 1273, 1276, 1277, 1279, 1281
annual temperature and millet yield maps, 1275, 1282, 1287
geo-spatial analysis, 1273
geo-spatial variation map generation, 1273–1287
rainfall and temperature data collection, 1272
study area description, 1271–1272
Climatic zones/desertification, 1385
Climigration, 230, 233
and well-being, 233–237
Cloud seeding, 2137
Coastal and Marine Area Regulations, 44
Coastal erosion, and rising sea levels, 1385
Coastal resilience, 467–468
artisanal fisher folks, adaptation strategies, 468–469
measures, 468
CoDriVE-Visual Integrator (CDVI), 262
Coffee
Central Kenya, 2253–2254
institutional innovations in coffee sector, 2262–2264
Cognitive barriers, 1889
Combat desertification, 1391
Commercialization, bambara groundnut, 1250
Communicating Risk of sea-level rise and Engaging stakeholDers in framing community-based Adaptation strategies (COREDAR), 100–102
Communities, SLR, *see* Sea-level rise and floods, Jamaica Bay
Community-based adaptation (CBA), 2006, 2007
See also Building Nigeria's Response to Climate Change (BNRCC) community based adaptation project
Community based natural resource management (CBNRM), 252
Community-centric flood risk reduction approach, 129
Community livelihoods, REDD+ pilot project, *see* REDD+ pilot project, in Tanzania
Community resilience, 1446
definition, 1359
in Djibouti, 1364
financing of climate change adaptation projects, 1361–1364
in NYS and Hudson Valley, 1445–1446
in Yemen, 1367–1370
Comprehensive African Agricultural Development Programme (CAADP), 1345, 1614
Conflicts, 1020, 1022, 1024, 1027–1029, 1226, 1228, 1233
climate change and environmental, 1230–1231
herders and crop farmers, 1232

Conservation agriculture intervention, 2456–2458
Conservation tillage (CT), 2557, 2560, 2562, 2565, 2568, 2569
Constraints, and rice farmers' response, *see* Rice farmers, in Ivo LGA of Ebonyi State
Construction, 2393, 2395
Consumerism, 1601–1602
Contagious caprine pleuropneumonia, 1266
Contour farming, 2731
Conventional tillage, 2556, 2558, 2560, 2567
Cook Islands
  background and field research method, 2199
  climateproofing adaptation strategies, 2213–2221
  historic actual risk events, 2209
  maritime economy contribution, 2199
  risk perceptions and actual risks, 2206
  stakeholder demographic profile, 2205
  tourism operators, 2212
Coping strategies, gender differentiated impacts, *see* Gender differentiated impacts, coping strategies
Corporate bonds, 1865
Correlation matrix analysis, 597
Cost, 2175–2176, 2178, 2179, 2183, 2190, 2192
County integrated development planning (CIDP) process, 983, 989, 994, 1683, 1701
Coupled ocean-atmospheric system, 1608
Cover crops, 1094
Creditor guarantors, 1873
Crop, 787, 790, 793, 794
  cultivar input files, 901
  failures, 792, 795, 937
  farming, 549
  farming management, 2633
  insurance, 2137
  intensification technologies, 2403
  production in Nigeria, 789
  raiding, 549
Crop irrigation, households
  irrigation farming, challenges in, 1071
  mode of access of irrigation water, 1072
  non-practice of irrigation, 1070
  source of irrigation water, 1071
  type of irrigation systems, 1073
  water management association, functionality of, 1072
  water management association, membership to, 1072
Cross breeding, 2732–2735
Cross River State
  age distribution, 1989
  climate change awareness, 1992
  data collection, 1987
  demographic characteristic of farmers, 1989
  descriptive statistics, 1988
  educational level, 1991
  focused group discussion (FGD), 1993–1994
  gender distribution, 1989
  marital status, 1990
  multistage sampling technique, 1987
  personal interview, 1994
  population, 1986
  structured questionnaire, 1987
  study area, 1985
  weather variation, 1992
CSA technologies, southern Tanzania, *see* Climate smart agricultural (CSA) technologies
Cyclone Aila, 53, 61–62
Czech Republic
  community activities for disaster management, 129
  data analysis, risk perception of floods, 122–126
  floods and flood management in, 119–120
  research methodology, 122
  risk perception, 120–121, 126–127

D

Dairy
  institutional innovations in dairy sector, 2265–2266
  technological innovations in dairy sector, 2265
Data availability, 1148
Debt management office (DMO), 1874
Decision-making, 1150–1151
Decision Support System for Agro-technology Transfer (DSAAT) model, 2750
Deforestation, 476
  definition, 477
  drivers in Southwest Nigeria, 481
  and forest degradation, 477, 480
  land-use change and, 478
  in Nigeria, 479
Degradation
  definition, 448
  ecosystem, 448, 449, 455–456
  environmental, 448

Degraded land, 2794, 2799, 2800
Descriptive statistics, 2498
Developing countries (DCs), 30
Development of agriculture (DA), 2127
Development Plan of Baja California Sur (PED) 2015-2021, 1565
Diameter at breast height (DBH), 10, 20
Didahara, 2061, 2062
  climate change adaptation governance efforts in, 2064–2065
  climate change adaptation governance system of, 2065–2070
  climate change adaptation in, 2063–2064
  climate change in, 2061
Diets, 1139, 1140, 1153
Dijon, 1588
Disaster management, 119
  laws and policies in LDCs, 1334
  legislation, 1318–1336
  water and, 1340
Disaster resilience, 87, 331
  aspects of, 332
Disaster risk reduction (DRR), 2133
Displacement, 1023
Diurnal temperature, 10, 13, 21
Diversification process, 54
Diversity, 2327
Djibouti, 1364–1367
Doha, 1483
  education city, 1484
  Lusail city, 1484
  Msheireb downtown city, 1484
Domestic legislation, 1318, 1341
Double vulnerability, 1012
Dread risk, 120
Drip irrigation, 937, 938, 946, 950, 953, 955, 956, 1058
Drivers, 478
  of deforestation, 479
  degradation, 479
  ecological change, 480
  land-use change, 481
  in Southwest Nigeria, 481
Droplet infection, 1219
Drought, 246, 248, 251, 253, 254, 265, 1053, 1662, 1663, 1665, 1667, 1668, 1672, 1766, 1769, 2043
  avoidance, 2330
  escaping, 2330
  and flooding, 1385
  rankings of livestock deaths from, 1749
  resilience in Northern Kenya, 1744
  tolerance, 2330
Drought affected Upazila, 2344
Drought resilience, 937, 955, 1092–1094
  microfinance and remittances, 1849
  and PHL, 1847
Drought-tolerant maize (DTM) varieties, 2227, 2231, 2232
  awareness levels, 2237
  benefits of, 2238, 2240
  CIMMYT approach, 2233–2234
  scaling of, 2235
Dryland areas, 937, 955
Dryland farming (sorghum), eastern Sudan
  annual rainfall, variation in, 2561
  conventional tillage, 2556
  experimental design and methods, 2560
  grain yields, 2562–2563
  measurements, calculations and statistical data analysis, 2560–2561
  semi-mechanized farming, 2554
  site description, 2559–2560
  soil and water management, 2556–2559
  soil water contents at sowing and harvest, 2565
  sorghum cultivation and production, in Gadarif State, 2555–2556
  traditional farming, 2554
  water use, 2564–2565
  water use efficiency, 2564
Dry savannahs, *see* Nigeria Savannahs, forecast decisions in
Dry spells
  categories, 2613
  definition, 2613
  occurence, 2602–2609, 2614
DSSAT 4.6, 908
DTMASS Project Partners 2017, 2244

E

Early warning system (EWS), 1923
  catastrophic flooding, 1926
  components of, 1925
  primary objective of, 1925
  public awareness and education, 1926
  UCAR report, 1924
Eastern Africa, climate change impact, 1419
Eastern Nigeria agro-environment, 2275
EBA, *see* Ecosystems based adaptation (EBA)
Ebrié Lagoon
  climate variability and local water availability, 364–370
  coastal waters and community-based conservation strategies, 374–375

Ebrié Lagoon (*cont.*)
household water availability and waste (-water) disposal, 373–374
location, 361
map of, 363
meteorological data, 362–363
regional climate change impacts, 370–373
survey, 363–364
EcoCrop, 1175
Economic Community of West African States (ECOWAS), 360, 1961
Economic Community of West African States Agricultural Policy (ECOWAP), 1614
Economic development, and climate change, 1641–1643
Economic diversification, 1630, 1643
Economic evaluation, 1057
Economic Recovery and Growth Plan (ERGP), 36
Ecopedagogy, 1803, 1811, 1818–1821
Ecosystem, 448, 449, 455–457, 2632
diversity, 1380
SLR (*see* Sea-level rise and floods, Jamaica Bay)
Ecosystem-based adaptation (EBA), 1846, 1848–1849
advantages of, 1927
biodiversity, 1927
biological rehabilitation interventions, 2446–2457
challenges of, 2470
conservation agriculture intervention, 2456–2458
cost savings, 1927
definition, 701–702
impact of interventions, 2444–2475
interventions, 2443–2444, 2447
IWM interventions, 2464, 2467
land degradation, 2444, 2464
land rehabilitation, 2441, 2476
landscape scale, 1929
measures by gender, 706, 710–711
mobilizing public support, 2469–2475
NAP process, 1929
NDC, 1928
resilience in Africa, 1349
social, environmental and economic criteria, 1929
soil and water conservation interventions, 2460, 2462
soil fertility improvement interventions, 2464, 2465
in Tigray regional state (*see* Tigray regional state, in Ethiopia)
types, 2444
typology, 2445
water harvesting interventions, 2460, 2461
Ecosystem based adaptations (EBA), water-energy nexus
ADA Consortium, 1688–1689
government strategic planning, 1697–1698
impacts of, 1683–1685
improvement of waterpoints, 1699–1700
policies for adaptation, 1686–1687
resilience, in arid regions, 1685–1686
water and land management institutions, 1698–1699
Ecosystem carbon stock (ECS), 410–413
Ecosystem services (ES), 516, 518, 520, 521, 523
derivation of, 528
Fisher's exact test for, 526
forest cover change on, 526–528
forest degradation and provision of, 530
hydroelectric power production, 280
income and provision of, 282
for livelihood, 524
PRB affect availability of, 287
provision of, 286
water related, 270
water use conflicts, 270
Ecotourism, 539, 551, 552
farming, 551
households' human assets, improvement in, 551
integration of, 550
on protected areas, 541
social assets, 551
sustainable livelihoods perspective, 540–541
Eco-village Adaptation to Climate Change in Central Tanzania (ECO-ACT), 2701
Education, 2775
Effectiveness, 2014
Electricity storage, 1402
El Niño in Philippines
on agriculture and fisheries, 2127
in aquatic resources, 2127
commission of elections, 2139
damaged areas, 2127
department of agriculture report, 2127
El Niño program, 2133
farmers affected in provinces, 2128
farmers affected in regions, 2127–2128
food scarcity, 2130
global crisis, 2127

government systems, 2131–2133
institutional responses and gaps, 2127
poverty, 2130
regional consultations, 2126
sea surface temperature, 2126
El-Nino Southern Oscillation (ENSO), 2618, 2750, 2752
El Niño Task Force (ENTF), 2133
Emissions trading schemes, 1400
Empirical climate change adaptation studies, 1197–1198
Empirical model, 1246
End Drought Emergencies (EDE), 1683
Energy efficiency, 430, 437, 442
Energy systems analysis, 181, 185
Enonkishu conservancy, 80–81
Environment, 2687, 2688
challenge in global context, 587–589
and human-health capital development, 589–590
Environmental educators, 638, 647, 655
Environmental flow regulation, 198
Environmental footprint, 1143
Environmental impact assessment (EIA), 41
Environmental justice, 1607
Environmental law, 1360
Environmental Management and Coordination Act, 985
Environmental migration
climate change and herdsmen, in Nigeria, 1228
definition, 1227
Environmental performance index (EPI), 592
Environmental refugees in Sahel, 1388–1389
Environmental stress & endemic poverty in West Africa, 1389
Environmental stresses, 2327–2328
Environmental vulnerability, 751
Epizootic ulcerative syndrome (EUS), 2578
Equity, 251, 260
Estimating environmental flow (EFRs), 188
Ethiopian smallholders' crop income, 663
climate variability, 663
crop income and rainfall, 666
data analysis, 665–666
data collection methods, 664
Ethiopian Rural Household Surveys, 663
model diagnostic test, 667–668
Ethiopia's 2015-16 drought
Agricultural Sample Survey data, 873–874
annual and Meher growing season metrics, 876
data verification, 886–888
description, 872–873
intra-zonal effects of drought, 883–886
model variables, 877
regression methods, 888–892
remotely sensed data, 874–875
statistical review of data, 877–882
Ethno-botanical survey, indigenous fruit trees, 504–505
European Investment Bank (EIB), 1868, 2389
European Union's Emissions (EU) Trading System (ETS), 1400
Evaluation, 1533, 1535, 1536, 1539, 1541, 1546, 1547, 1549, 1553
Evapotranspiration (ET), 2560
Ewaso Ng'iro North Catchment (ENNCA), 1685, 1688, 1691, 1700
ASAL areas, drought challenges in, 1687–1688
domestic water and energy use, in arid and semi-arid counties, 1693
irrigation water and energy use, in Isiolo County and semi-arid counties, 1694–1695
livestock water and energy use, in arid and semi-arid counties, 1693–1694
Exclosures, in Tigray (northern Ethiopia)
aboveground biomass, species richness and diversity, 408–410
analytical framework, data collection and analysis, 406–408
baseline studies and carbon sequestration, 413–414
carbon stock restoration and social attributing variables, 415, 417–420
ECS, 410–413
soil condition, 414–416
Exposure, 1293, 1298
factors, 217
index, 1300
Extension services, 1128
Extreme events, 1573
Extreme rainfall events, 2174
amount of rainfall, 2175
duration, 2174
frequency of occurrence/return period, 2175
Extreme weather events (EWEs), 1218

F

Faidherbia albida, 1206
Farmer-responsive extension services, 2366
Farmers, 1226, 1228, 1230–1233, 2499, 2773, 2775

Farmers (*cont.*)
attitude towards climate change, 1665
perceptions of climate change, 1663
perceptions of climatic trends, 1667–1668
psychological impact of climate change, 1668
reliability of, SCF, 2506
rural farmers, in Niger Delta Region (*see* Rural farmers, in Niger Delta Region)
seasonal climate forecasts (*see* Seasonal climate forecasts (SCF))
Farmers and sustainable agricultural development, 2309
Farmers' knowledge of climate change, 2252, 2257
Farm plot level, CSA technologies, *see* Climate smart agricultural (CSA) technologies
Farm yard manure (FYM) application, 2378
Federal Capital Territory (FCT), 751
Federal Executive Council, 1221
Federal Government of Nigeria (FGN), 1876
Female-led micro, small and medium enterprise (MSMEs), Kenya, 1004
adaptive capacity, 1006
business journey, 1005
challenges, adaptation of, 1011–1014
contributions to family and social welfare, 1003
exposure to climate risk, 1007–1008
policy documents, 1004
ways of responding to climate risk, 1008–1011
Fertilization strategies, 920, 922, 931
Fertilizer application, 2378
Fit, 2494, 2495
Five factor model, 1941
Fixed effect model, 668
Flat high-rise post-war garden city, 2179–2180
climate adaptation in, 2183–2184
climate resilient designs, 2181
cost-benefits of rainwater, 2182–2183
permeable paving, 2182
street, retention in, 2181
swale, retention in, 2181
Flat urban city block, 2184
climate adaptation in, 2188–2189
climate resilient designs, 2186
cost-benefits of rainwater, 2187–2188
gully, retention in, 2186
storage in street foundation, permeable paving, 2187
street, retention in, 2186
Flood(s), 2507
affected Upazila, 2344
disaster, 470
Flooding, 1910, 1912, 1916, 1917, 1919, 1920, 1926, 1928, 1930
Flood-related property insurance, 119
Focus group discussion (FGD), 450, 453, 456, 560, 799, 800, 806
Food Crisis Prevention Network (RPCA), 2049
Food insecurity, 1099
Food prices, 1150
Food scarcity, 2130
Food security, 918, 1141, 1144, 1148, 1162, 1166, 1167, 1344, 1717–1722, 2288, 2308, 2327, 2332, 2403, 2723, 2740, 2792, 2794, 2795, 2799, 2800
adaptation and mitigation measures in climate change, 791–792
challenges in Nigeria, 792–794
definition, 788, 2044, 2274
and desertification, 1353
impacts of climate change, 789–790
indicators in countries of Sahel, 2045
and malnutrition, 1145
Nigeria food system, resilience in, 794
recommendations for National Research Institutes, 2283
recommendations to government and policy makers, 2285
research institutes to achieve security and industrialization, 2276–2282
and nutrition, 1239, 1252
stability, 1350–1351
and sustainability, 1142
Food security, in Nigeria, 1022
funding, credit, training and poor agricultural practices, problems of, 1025
policy inconsistency, 1023–1025
Food security, in North-Eastern Nigeria
climate change and conflict, 1027–1028
and ineffective agricultural policy, 1028–1029
Food security status
correlation analysis, 1768–1770
cross sectional survey design, 1759–1761
economic activities, 1765
farmer's perception of, 1765–1767
household information, 1762–1764
location, Tharaka-Nithi, 1759–1760
maize prices trend, 1767–1768
Food shortage coping strategies, at Kikombo and Idifu Wards, 2714
Food trade, 1140, 1143

Forced migration, 2024
Forecast information, 2655, 2657, 2661, 2663, 2667, 2669, 2670, 2672, 2678, 2680
Forecasting strategies, 611
Forest cover change, 530
  causes of, 526
  and ecosystem services, 520
  effects of, 518
  impacts of, 526–528
  Katimok, 535
Forest degradation, 476
  causes of, 477
  in Nigeria, 479, 480
Forest ecosystem, 517, 523
  degradation, 524
  significance, 516
  survival, 523
Fossil-fuel subsidies, 1401
Fourier, Joseph, 1216
Fuel, 2098, 2100, 2103–2115, 2118, 2119
Fungus
  inoculants and carbon sequestration, 389
  metabolites (Glomalin) on Carbon Sequestration, 390
  necromass, 389
Furrows, 2731
Future city planning, 1495

G

Gender
  and climate change adaptation, 1967
  ecosystem-based adaptation (EbA), 706
  participatory vulnerability assessment, 708–710
  vulnerability assessment, 705
Gender analysis frameworks (GAF), 2101, 2119
Gender differentiated impacts, coping strategies
  coping *vs.* adaptation to climate change, 2100
  gender analysis frameworks and tools, 2101
  methodology, 2102–2103
  Moser framework, 2101–2102
  PGN and SGN, 2116–2118
  water and fuel needs, at household level, 2103–2115
Generalized linear models, 505
General Law of Civil Protection (LGPC), 1566
General Law of Ecological Balance and Environmental Protection (LGEEPA), 1566
General Law on Climate Change, 1566
Genetically modified organism (GMO), 2627
Genetic diversity, 1380, 2327
Genetic manipulation, 2328
Geographical Information System (GIS), 1271, 1273
GIS based sea-level rise
  analysis, 101
  census block boundary preparation, 101
  data source, 101
  flood scenario extent preparation, 101
Global Action Programmes (GAP), 2692
Global alliance for resilience
  aims, 2049
  effects on food and nutrition security, Niger, 2052–2053
  implementation agenda, 2050–2051
Global circulation models (GCMs), 546
Global climate model (GCM), 1175
Global environmental change (GEC), 1910
Global warming, 676, 679
Governance Dynamic Framework (GDF), 250–251
Government bonds, 1865
Grain yields, 1092
Gravity irrigation systems, 1060
Grazing for Growth Program (G4G), 81–82
Grazing pattern of cattle, 742
Greek communities' resilience, 653, 656
Green bond
  definition, 1865
  market, 1872
  pricing, 1870, 1871
  principles, 1866
Green Bond Principles (GBP), 1867
Green cities, 1968
Green economy, 1391
Green growth, 1968
Greenhouse gases, 680, 1036, 1216, 2040
  emissions, 786, 789, 1140, 1145, 1148, 2691
Green revolution, 1804
Grenoble, 1582, 1592
Gresivaudan Valley Local Authority, 1589
Grewia asiatica, 2294–2296
Gross domestic product (GDP), 2552
Groundwater, 247
Groundwater resources, impacts of climate change on, 247–248
Groundwater Survey and Development Agency (GSDA), 247
Growth, 588–591
Growth and Transformation Plan (GTP), 814

Guinea Savannah ecological zone of Nigeria
cattle herders adaptation strategies, 741–743
demographic characteristics of respondents, 735
drought, 736–739
monthly rainfall and temperature data, 733
questionnaire and in-depth interviews, 733
study area, 732
temperature trend and events, 739–740
Gulf Cooperation Council (GCC), 1479, 1483, 1488
Gulf Organization for Research and Development (GORD), 1488

H

Harvesting, 2379
Hausman test, 668
Health and disease, 1265
Health-environment-economy model, 590
Heat related symptoms (HRS)
among households, 214
self-reported, 211
types of, 212
Heat stress, 718
exposure, 225
occupational, 209
vulnerability to, 209, 215
Heavy metals
bioremediate, 325
remediation of, 320
spent engine oil polluted soil, 320
Heckman probit selection model, 2661
Heifer Exchange Program, 1747
expected impacts of, 1747–1749
funds for purchasing heifers, 1750
limitations and challenges, 1754
methodology of, 1747
objectives, 1749
verification of effectiveness of, 1750–1753
Herdsmen, in Nigeria, 1228
climate change and migration, 1230
farmer-herdsmen crises, 1232
herdsmen-farmer clashes, 1231–1232
Hibiscus sabdariffa, 1208
High value crops, 1108, 1110, 1112
HIMA project
on cultural capital, 567–568
on financial capital, 565
on human capital, 563–564
on natural capital, 565–566
on physical capital, 562–563
on political capital, 567
on social capital, 561–562
Honey
effect on egg production, 719–724
uses, 719
Household food situation, in Kikombo and Idifu wards, 2713
Households' vulnerability to climate variability, in Nigeria
agricultural productivity, 760–761
correlation analysis, 753, 758
data limitation, 755
descriptive analysis, 753
HVI, 753–754, 760
multiple regression analysis, 754–755
PCA, 753, 759
socio-economic characteristics, 755–758
type and sources of data, 752
Household Vulnerability Index (HVI), 754, 760, 761
Household welfare, 1253
Hudson river, climate change risks to, 1444–1445
Human capital, 589–590
Human Development Index (HDI), 753, 1053
Human-health capital development, 589–590
Human welfare, 2329
Hunger, 786, 787, 792, 793
Hydric deficit, 955, 956
Hydrology, 186
Hydropower, 178, 180, 182

I

Ibadan metropolis, in Nigeria, 1913
annual rainfall and precipitation in, 1917
climate of, 1915
dam construction and river defense, 1921
EWS, 1923–1927
flooded buildings, analysis of, 1922
flood plains, buildings constructed on, 1923
flood vulnerability of, 1918
location, 1913
Ogunpa River system, 1913
perrenial flood occurneces, 1916
physical expansion and land coverage, 1913
rainfall trend and periodicity, 1920
river bed, buildings constructed on, 1924
schematic representation, 1913
total rainfall in, 1921
urban hydrology, 1920
vulnerability, 1928
Icelandic Financial Services Association (SFF), 1895

IGAD Drought Disaster Resilience Sustainability Initiative (IDDRSI), 1686, 1701
Income generating activities (IGAs), 561
Indian Human Development Survey, 2115
Indian Thar region, *see* Thar desert region of India
Indigenous fruit tree
  diversity in ecological zones, 507
  income from, 509–511
  index of knowledge and uses calculated for, 506
  socio-cultural importance of (*see* Socio-cultural importance of indigenous fruit trees)
  statistical analysis, 505–507
  study species, 502
Indigenous innovation, in production, 817–818
Indigenous knowledge (IK), 799, 968
Indigenous technological knowledge (ITK), 2579
Infection, 1219
Innovation, small holder system, 1714
  and age cohort, farmers, 1719–1720
  and education, 1720–1722
  environmental management, 1713
  gender and plant cultivation, 1722–1723
  household related social capital and uptake of, 1718
  and information sources, 1720–1722
  ordered logistic regression model, 1719
  and social capital, 1717–1719
  and space, 1723–1724
Innovation platforms (IPs), 1730, 1738, 1739
In-situ water harvesting (ISWH), 2557, 2559, 2560, 2562, 2565, 2567, 2569
Institute for Sustainable Cities (ISC), 108, 110
Institutional Analysis and Development (IAD) Framework, 249–250
Institutional capacity, and multilevel articulation, 1565
Institutional factors, 1579, 1589
Institutions, 36, 37, 40, 44, 1612, 1613, 2063, 2064, 2066, 2067, 2069
  and adoption of CSA practices, 1621
Insurance and climate change, 1884–1887
  cognitive barriers, 1889
  communication isolation, 1895–1896
  competitive isolation, 1896
  customer protection gap, isolation of, 1899
  geographical isolation, 1891–1893
  institutional size, isolation of, 1897–1899
  insurance business, 1890
  knowledge isolation, 1894
  overcoming barriers of isolation, 1900–1902
  political barrier, 1887
  political isolation, 1893–1894
  structural barriers, 1890
  technological isolation, 1896–1897
Insurance spending, 2150
Insurers
  global and local knowledge, 2160–2161
  and North and south, 2158–2160
  and policy making, 2157–2158
  and public awareness, 2161–2162
Integrated aquaculture agriculture systems (IAAS), 1166
Integrated assessment models (IAMs), 2391
Integrated soil fertility management and Central Kenya
  descriptive data analysis, 2406
  Ndia West division, 2404
  partially balanced incomplete block design, 2406
  soil physical characteristics, 2404
Integrated urban planning, 2077, 2092
Integrated watershed management (IWM) interventions, 2464, 2467
Intellectual property rights (IPRs), 1360
Intended nationally determined contributions (INDCs), 1862, 1876
Intergovernmental Panel on Climate Change (IPCC), 460, 1398, 1648, 1884, 2024, 2552
Intermunicipalities, 1579, 1584–1591
Internal displacement
  climate change response measures, 2026
  climate-exacerbated disasters, 2027
  climate-related internal conflicts, 2027
Internally displaced persons (IDPs), 2025, 2028, 2029, 2031
International climate-related commitments, 1397, 1409
International community, 1358, 1360, 1368
International Finance Corporation (IFC), 1869
International Food Policy Research Institute (IFPRI), 792
International Maize and Wheat Improvement Centre (CIMMYT), 2226
Interplay, 2494
Intersectionality, 68–70
Intersectoriality, 1561–1562
Inter-tropical discontinuity (ITD), 2597, 2747
Irrigation advisory, 2378
Irrigation in evening, 938, 941

Irrigation systems, 1073
- California, 1060
- drip, 1058
- gravity, 1059–1060
- manual, 1059
- on potato, shallots and tomato, 1072–1073
- sprinkle, 1058–1059

Isiolo County, 987–989, 995, 1691
- energy parameters in, 1692
- household water sources in, 1696
- irrigation water and energy use in, 1694
- tourism water and energy use in, 1695
- water and energy consumption, 1698
- water and energy demands, 1690, 1691

Isolation barriers, insurance and climate change, *see* Insurance and climate change

Ivo LGA of Ebonyi State
- rice farmers' response and constraints (*see* Rice farmers, in Ivo LGA of Ebonyi State)
- topography of, 1790

Ivorian east coast, 364

J

Jal Sevaks (water volunteers), 256–257

Jamaica Bay, New York City
- SLR and floods (*see* Sea-level rise and floods, Jamaica Bay)
- study area, 98–99

Jeezrael Valley Project, 2088

Jeseník nad Odrou, 122

Jurisdictional fragmentation, 1581

K

Kampala Convention, CCIDPs, in Africa, *see* Climate change-induced internally displaced persons (CCIDPs), in Africa

Katimok Forest Reserve, 518, 519

Kenya
- agricultural production in, 847
- agricultural sector in, 847
- agro-climatic zones in, 849, 851
- economy, 847

Kenya Integrated Household Budget Survey (KIHBS), 1693

Kenya National Climate Change Response Strategy, 985

Kenya's climate change
- adaptation assessment, 2390, 2394–2397
- impact on infrastructure, 2387–2388
- methodology, assessment of vulnerability and adaptation, 2391
- vulnerability assessment, 2389–2390, 2392–2394

Kharif crop, 696, 697

Kimanjo, 1260

Kisan Credit Card (KCC), 1516

Knowledge-action gap
- definition, 2150
- global and local knowledge, 2154
- in high-income and low-income countries, 2152
- insurers' role in, 2156–2163
- scientific knowledge and policy making, 2151
- scientific knowledge, public awareness and actions, 2154–2156

Konso community, southern Ethiopia
- annual average rainfall distribution, 809, 810
- conversion of cropland to tree plots, 817
- data collection, 799, 800
- environmental factors, 814
- erosion of cultural practices, 815
- events of climate changes, 806–807
- hoe-farming areas, 801
- indigenous soil and water conservation practices, 815–817
- intensity of rainfall, 811–814
- landholdings and fragmentation, 803–804
- land use and cover, 803–804
- local perceptions on changes, 805–807
- metrological data, 801
- micro-basins, use of, 818
- modification of farming tools, 818
- monthly rainfall distribution, in *Hagaya* season, 811
- monthly rainfall distribution, in *Katana* season, 812
- ox-ploughing areas, 801
- range of rainfall variability, 809, 810
- reclaiming rocky outcrops, 818
- reduction of access to resources, 814
- soils, 802–803
- sources of livelihood, 802
- technologies and polices, 814
- trends of rainfall distribution, 810
- water-harvesting technologies, new design of, 818

Kostolac lignite basin, 1469–1471

Kruskal-Wallis test, 506

Kyoto protocol, 1400, 1404, 2152

L

Lablab purpureus (L.) Sweet, cultivar selection and management strategies
  challenge, 1085
  data analysis, 1087
  data sources and collection methods, 1087
  drought resilience, 1092
  grain yields, 1092
  historical and cultural significance, 1084
  insect damage, 1092
  relative maturity, 1088–1092
  study area, 1086
  study site selection and sampling methods, 1086–1087
  weather, 1088
Lagos State, 1939
  See also Mentoring and field study strategy, Nigeria
Laikipia, 1263
Laikipa Permaculture Centre, 1825
Lake Victoria outflow, climatic impact on, 195
Land-based livelihoods, 1717
Land degradation, 2288
Land grabbing
  in PRB, 286, 289
  status in Tanzania, 283
  water and, 287
Land use
  areas 2001-2015, 528
  changes in, 527
  ground-truthing, 522
  human-induced, 517
  land cover areas 1984-2001, 526
  types, 535
Land-use change, 476, 478
  and deforestation drivers, 480
  in Nigeria, 479–480
  in Southwest Nigeria, 481
Land use land cover change (LULC), 136
La Paz
  multi-threat qualitative assessment, 1564
  municipality of, 1559
  population of, 1559
  urban footprint of, 1559
  vulnerability and threats to, 1561–1562
Large-Scale Adivasi Multiple Societies (LAMPS), 1516
Least Developed Countries (LDCs)
  climate change adaptation measures in, 1340
  disaster management legislation in, 1318
  domestic water laws, policies and plans in, 1323
  laws, policies and plans in, 1319
  management of water resources in, 1339
  priorities of, 1336
Least significant difference (LSD), 1176, 1177
Legumes, 1085, 1095
Length of season, 2748, 2751, 2757, 2760, 2763, 2766, 2767
Leucaena leucocephala
  characteristics, 320
  metals in, 324
  physicochemical properties, 323
Livelihood, 538, 540, 541, 543, 546, 2794, 2795, 2800
  construction and adaptation strategies, 549–552
  definition, 540
  outcomes, 544
  vulnerability context and impacts on, 549
Livelihood resilience
  description, 52
  intersectional dimensions of, 68
Livestock, 75, 79
  Enonkishu, 82
  Maasai, 77
  management techniques, 83
  market trading in Marsabit (*see* Heifer Exchange Program)
  owners, 80
  pastoralists' mindset, 1744–1747
  quality, 81, 89
Local Adaptation Programmes of Action (LAPAs), 1968
Local authorities, 1584
Local Flood Warning Systems, 1927
Local innovation, 799, 817, 818
Local level strategies, 839
Logging, 477, 480, 490
Long-range Energy Alternative Planning (LEAP), 1684, 1693, 1700
Lord's Resistance Army (LRA), 1103
Lorien, 1260
Low-carbon technologies, 1408
Low cost irrigation systems, 2633
Lusail city, 1484

M

Maasai population, 1634
Maharashtra Groundwater (Development and Management) Act 2009, 261–263
Maize-based farmers, 678
  adaptation strategies, 679
  sustainability level, 679–681

Maize regression models, 890
Maize (Zea mays), climate change, 898
Maladaptation, 2627
Malawi, aquaculture technologies, fish production, *see* Aquaculture technologies, in Malawi
Mali
climate change risks for, 1729
rural population of, 1729
small-scale farming in, 1737
vegetables production (*see* Vegetables production, Koulikoro and Mopti regions of Mali)
Malnutrition, 2047–2048
Mangrove forests, REDD+ pilot project on livelihoods, *see* REDD+ pilot project, in Tanzania
Mangrove zone, 1381
Mann-Whitney test, 506
Manual irrigation systems, 1059
Manual Local Flood Warning System (MLFWS), 1926
Marginal mixed farming (MMF), 1759
Margin of error (ME), 2751
Market, 76, 80
Marshall stability, 2525, 2527
Maundu ni Meri, 1260, 1264
Mean monthly flow (MMF), 190
Medium term plans (MTPs), 983, 991, 1686
Meghna Estuarine islands, 331, 335, 339
Meningitis, 1218
Mentoring and field study strategy, Nigeria, 1939
climate change reduction practices of students, 1950
description, 1941–1942
gender and school location, 1942–1945
paticipant selection, 1945
validity and reliabilityof instruments, 1946
Microbial biomass carbon (MBC), 393
Microbial community composition, 394
Microfinance, 1846, 1848, 1850, 1854
Migration, 2332
Millennium Development Goals, 798
Millet (*Pennisetum glaucum* (L.), 918, 1270
Ministry of Disaster Management and Relief (MoDMR), 330
Mini yam set technology, 2279
Mitigation, 430, 431, 433, 437, 441, 982, 2385, 2729, 2730
and adaptation measures, rice farmers' response and constraints (*see* Rice farmers, in Ivo LGA of Ebonyi State)
definition, 1789
Mixed crop/livestock production systems, central Ethiopian highland, 1172
climate change, farmers' adaptation measures to, 1186–1188
climate change, farmers' perception of, 1184–1186
data analysis, 1175–1176
DIVA-GIS, suitability analysis, 1175
meteorological data, 1173
potential wheat production areas, 1183
rainfall, 1188–1189
rainfall variability and trend, 1176–1177
relative humidity, 1179
socio-economic survey, 1173
study area, 1172–1173
sunshine hours, 1179–1180
teff and wheat production conditions, 1181–1182
teff production potential area, 1186
temperature, 1189–1190
temperature variability and trend, 1177–1179
wind speed, 1180–1181
Mixed farming, 2633
Modeling, 922
Moderate resolution imaging spectroradiometer (MODIS) satellite, 870
Mognori community, Mole National Park
agriculture, 546
community entry protocol, 547
data analysis and interpretation, 548
historical climatic trend, 546
location, 544
poverty level, 546
qualitative method, 544
recruitment of participants and data collection, 547–548
Monitoring, 1535–1537, 1541, 1545, 1547, 1549, 1551, 1553
Monitoring, reporting and evaluation (MRE), 1535, 1536, 1553
Monitoring and evaluation (M&E), 1533, 1535, 1536, 1686
Moser framework, 2101–2102
Mpakeni village, 1716–1717
Msheireb downtown city, 1484
Mulching
gravimetric soil moisture, 2425
non-significant effects of, 2424
optimum rate of, 2418
potato, 2418
Multilevel articulation, institutional capacity and, 1565–1566

Multinomial logit (MNL), 1038, 1040, 1045
Multiple enterprise model, 77, 87
Multi-stage random sampling, 1791
Multistage sampling procedure, 2537
Multi-stakeholder analysis, 1567–1568
Multivariate probit (MVP) model, 1652, 1654, 1656

N

Nadungoru, 1260, 1262, 1264
Narok, 1004
  exposure to climate risk, 1008
  MSMEs (*see* Female-led micro, small and medium enterprise (MSMEs), Kenya)
  one-day stakeholder workshop, 1005
National Adaptation Planning (NAP) process, 1686, 1701, 1929
National Adaptation Programme of Action (NAPA), 1966
National Adaptation Strategy and Plan of Action on Climate Change for Nigeria (NASPA-CCN), 37, 38, 794
National Agricultural Extension Research and Liaison Services, Zaria (NAERLS), 2279
National Agricultural Investment Programmes (NAIPs), 2050
National Agricultural Policy, 2360
National Bank for Agriculture and Rural Development (NABARD), 2369
National Bureau of Statistics (NBS), 752
National Disaster Risk Reduction and Management (NDRRM) Act, 2131–2133
National Drought Management Authority (NDMA), 1691
National Economic Development Authority (NEDA), 2133
National Energy Guarantee, 1405
National Environmental (Coastal and Marine Area Protection) Regulations 2011, 44
National Environmental (Soil Erosion and Flood Control) Regulations 2011, 45
National Environmental Standards and Regulation Agency (NESREA), 36
National Environmental (Wetlands, River Bank and Lakeshores) Regulations 2009, 45
Nationally Appropriate Mitigation Actions (NAMAs), 1466
Nationally Determined Contribution (NDC), 31, 982, 985, 1402, 1928
National Meteorological Services Agency (NMSA), 1173, 1176, 1177
National Metrological Agency (NMA), 801
National Oceanic and Atmospheric Administration (NOAA) Coastal Ocean Climate Application program, 106
National Plan of Action for Food and Nutrition in Nigeria, 1023
National Research Institutes (NARIs), 2283–2284
National Special Program for Food Security (NSPFS), 1023
Natural hazards, 627
  distribution of, 624
  in Sandwip Island (*see* Sandwip Island, Bangladesh)
Natural resources, 80, 1142, 1143, 1230
NEWater project, 2090
New York City Panel on Climate Change (NPCC2), 102
Ngorongoro, Tanzania
  development goals and income status, 1637–1638
  development of, 1630
  district and protected areas, 1629
  existing land use, 1635
  human population changes, 1631
  human population projections, 1637
  income status, 1638–1640
  livestock changes in, 1631
  livestock units (LU), 1632
  location, 1628
  Ngorongoro Conservation Area (NCA), 1628
  population trend, 1633–1634
  wealth classification, 1632
Niger Delta region, 39
  challenges, 43
  climate risks and vulnerability in, 35
  institutional respondents, characteristics of, 40–41
  institutional strengths, 41–42
  institutional weaknesses, 42–43
  regulatory frameworks, 44–45
  semi-structured interviews, 40
Nigeria, 461, 1145–1146, 1984
  adaptive innovations for pond fish farming, 2584
  adverse climate conditions, 1217
  agriculture, 1939
  caging of pond, 2586
  climate change adaptation in, 2781–2784

Nigeria (*cont.*)
climate change and environmental conflicts, 1230
climate change awareness and adaptation measures (*see* Akwa Ibom State; Cross River State)
climate change education for sustainable development, 2692–2693
climate change, herdsmen and migration in, 1228
climate change-induced flooding, 2583
climate change trend in, 1220–1221
CO_2 emissions in, 7–10
coastal management in, 36–37
coastal States description, 462–463
crop production in, 789
data sources, processing and analysis from fish farmers, 2582
ecological implications of climate change, 5–7
effects of climate change on fisheries, 465–466
farmer-herdsmen crises, 1232
flooding in, 2580
households' vulnerability to climate variability (*see* Households' vulnerability to climate variability, in Nigeria)
impacts of climate change in, 2777–2781
Lagos, Ogun, Oyo states, 2581, 2582
lessons of, 2693–2695
location, size, and characteristic relief, 1939
netting of ponds, 2586
policy and programs, 37–38
pond fencing, 2585
ponds' embankment modification, 2588
pond slope manipulation, 2588
regulatory response, 38
resilience in the Nigeria food system, 794
river embankment modification, 2589–2590
rural farmers (*see* Rural farmers, in Niger Delta Region)
sustainable development goals, 2694–2695
tree species, in SFRS (*see* Tree species, in SFRS)
vulnerability to climate change, 2689–2692
Nigeria Climate Change Policy Response and Strategy, 1221
Nigeria Meteorological Agency (NIMET), 793
Nigerian Bitumen Company, 449
Nigerian environmental peculiarity, 591–594
Nigerian honey, 724
Nigerian Meteorological Agency (NIMET), 2493, 2605
Nigerian Meteorological Agency seasonal rainfall prediction (NiMet SRP) model, 2745, 2750, 2751, 2757, 2763, 2767
evaluation, 2763
performance, 2763–2766
Nigeria Savannahs, forecast decisions in
adaptation decision process, 2665–2669
agricultural risk management, 2655–2657
annual average temperature and changes, 2653
climate resilience, 2657–2658
climate resilient practices, 2677
Climate Smart Agriculture National Adaptation "Action" Plan, 2654
climate smart practices, 2673–2674
descriptive statistics, 2660
Heckman probit selection model, 2661
policy issues, 2678–2680
resilient decision making, 2670–2672
resilient practices, 2674–2676
selection model, 2676–2677
socioeconomic characteristics respondents, 2662–2665
study area, sampling technique and data collection, 2658–2660
weighted average index (WAI) analysis, 2661
Niger's PRP-AGIR, 2052
Nile Delta, 768, 769
No/low regret adaptation, 1525
Non-governmental organizations (NGOs), 108
Non-metric multidimensional scaling (NMDS), 395
Nordic Insurance Climate Change (NICC), 1898
Normalized difference vegetation index (NDVI), 870, 874, 876
North American Multi-Ensemble Model phase 2 (NMME-2), 2766
Northern Kenya, 1744, 1747
See also Heifer Exchange Program
No-till, 2557, 2560, 2567
Nutrition, 1145, 1149, 1151
Nvivo 9, 548

O

Ocean heating, 1602–1604
Ombrothermic diagram, 364
Onion cultivation, 2373, 2376, 2378, 2379

Onset, 2747, 2751, 2757, 2759, 2762, 2763, 2766, 2767
Ordered logistic regression model, 1719
Organic matter, 2792, 2794, 2801

P

Pacific maritime economies, 2199
Palmer drought severity index (PDSI) method, 734
PANA-resilience, 1195
Paris Agreement, 982, 985, 991, 1318, 1336, 1339, 1340, 1402, 1409
  water issues, 1336
Participatory rural appraisal (PRA), 333, 334, 336, 341
Participatory stakeholder analysis, 2263
Pastoralism, 1258, 1628, 1630, 1634, 2062
Pastoralists, 1744–1747, 2060, 2061, 2063
Pathways to resilience in semi arid economies (PRISE), 1006
Pearl millet cultivation, 2376
Pearson correlation technique, 597
People's Survival Fund (PSF), 2135
Perceptions, Konso community, 805–809
Perennial crops, 1099, 1102, 1106, 1108, 1110, 1112
Perimeter pond fencing, 2585
Permaculture
  approach, 1817
  curricular claims, 1806
  curricular philosophy, 1809
  definition, 1803
  design & water harvesting, 1822–1824
  design system, 1805
  design, zones, 1810
  in East Africa, 1821–1833
  and eco-literacy, 1821
  ecosystems in Africa, 1808–1812
  in Kenya and Uganda, 1803
  local practitioners, 1808
  potential of, 1803
  in Praxis in East Africa, 1827
  principles, 1809
  teaching, 1819
  values, 1804
Pest and diseases, 2277
Pesticide application, 2379
Philippine Atmospheric Geophysical and Astronomical Services Administration (PAG-ASA), 2135
Phoenix Seed Company in Mozambique, 2239–2240
Physicochemical analysis, tyre oil and standard diesel oil, 2525
Pilot programme for climate resilience (PPCR), 2389
Plan for Modernization of Agriculture (PMA) program, 1101
Plan for Sustainable Urban Mobility, 1567
Plant breeding, 2331
Plant vegetable industries, 2278
Plot level, CSA technologies, *see* Climate smart agricultural (CSA) technologies
Podhadí nad Dyjí, 122
Poisson regression model, 1717
Policy
  actors and networks, 1971
  advice, 1734
  agricultural, 1150
  at city-scale level in West Africa, 1965
  climate change adaptation, 1968–1970
  climate change adaptation into urban planning, 1969
  and food security, 1022–1025, 1029
  formulation and implementation, climate change adaptation, 1970
  urban climate change adaptation research and, 1966–1970
Political Constitution of the United Mexican States (CPEUM), 1566
Political leadership, 1581
Pomona water reclamation plant, 2091, 2092
Pond fencing, 2585
Population, 1139, 1142, 1145, 1218, 1220
Population growth, 1630, 1636
  along PRB, 288
  and climate change, 1629
  and de facto exclusion, 1641
  global and local, 279
  and resource scarcity, 271
Post-harvest losses (PHL), 1846–1849, 1853
Potato
  and climate change, 2417, 2418, 2436
  soil temperature, 2417, 2418, 2423, 2424, 2426
  tuber emergence, 2423
  yield, 2418, 2434
Potential evapotranspiration, 891
Poultry production, 722, 724
Poverty reduction, and climate change, 1640–1641
Poverty, water and, 151
Practical gender needs (PGNs), 2098, 2099, 2101, 2102, 2116–2118
  assessment, 2102

Practical gender needs (PGNs) (*cont.*)
health and nutrition, 2117
hygiene and sanitation, 2117
work load, 2117
Precautionary approach, 1485
Precipitation, 2043
Preparedness Determinants (PDs), 250
Principal component analysis (PCA), 753, 759
Private sector adaptation, 1006
Productive Safety Net Program (PSNP), 799, 807, 814
Productivity, 589
labour, 590
Products
agricultural, 504
plant, 510
species, 510
Protein energy malnutrition (PEM), 2275
Protoplasts, 2332
Public private partnership
choice of partners, 1122
MOA in, 1121
Public trust doctrine (PTD), 1598
Pullets, 719, 720
Pyrolytic bitumen, 2527
bitumen extraction and asphalt concrete production, 2524
gas-fired furnace, solar-powered temperature sensor, 2522
gas storage bag, non-condensable pyrolysis gas, 2523
Minitab paired t-test, 2526
non-biodegradability, 2520
production of, 2519
pyrolysis reactor, 2521
pyrolysis system set-up, 2522
quality test, on extracted bitumen, 2524
temperature, 2521
utilization of, 2520

Q
Qatar
assessment and comparative context, 1492–1494
climate vulnerability and urban sector in, 1479–1482
urban climate change policies (*see* Urban climate change policies, Qatar)
urban transformation and sustainable design, 1482–1485
Qatar Green Building Council (QGBC), 1487
Qatar National Development Strategy (QNDS), 1485
Qatar National Master Plan (QNMP), 1485
Qualitative data analysis (QDA), 548
Quality declared seed (QDS) production, 2701, 2702, 2704, 2706, 2717
Quality management system (QMS), 1533, 1535, 1537, 1539, 1541–1543, 1547–1549, 1551–1553

R
Rabi crop, 696, 697
Radan Mountain, 1471
Rainfall amount, 2748, 2751, 2754, 2757, 2760, 2763, 2766
Rainfall change/variability, in Southwestern Nigeria, 2534
Rainfall onset, 2533
Rainfall variability index, 2598
Rain-fed cropping, 1759
Rainforest zone, 1381
Rain water harvest, 1166
Random effect model, 668
Randomized complete block design (RCBD), 2560
Random sampling scheme, 505
Rangamati Sadar, 2344
Rapid Rural Appraisal (RRA), 1987
REDD+ pilot project, in Tanzania
climate, 559 (*see also* HIMA project)
research design, sampling design and data collection, 560
study area, 559
Reducing Emissions from Degradation and Deforestation (REDD), 414
Regional guidance framework, 2050
Relative humidity, 1179
Relative maturity, 1088, 1091
Relocation, form of, 230
Renewable energy initiatives, 1391
Reporting, 1533, 1535, 1536, 1553
Resettlement, 232
coastal communities, 237, 239
Residue mulching, 2558, 2568, 2569
Resilience, 9, 24, 799, 800, 804, 811, 815, 819, 981, 986, 995, 1143, 1149, 1259, 1265, 1533–1536, 1543, 1553, 1714, 1715, 1722, 1744, 1748, 1964, 1969, 1975, 1977, 2005, 2015, 2152, 2157, 2161, 2199, 2202, 2210, 2212, 2217, 2219, 2222, 2654, 2657
adaptation and, 1351

of African farming systems, 1346
of agricultural production systems, 1344
autonomous/ integration planning, 618
climate, 1447
to climate change, 1419
climate indicators, 615
community, 1446
components, 1571
control variables, 618
definition, 643
EBA in Africa, 1349
indicators, 615–616
inherent, 1572
and integration, 618
of livelihoods and production systems, 1345
modeling simulation, 614, 619–622
plan, 1445, 1446, 1448, 1449, 1452, 1454, 1456, 1458, 1459
as recovery, 612
rural *vs.* urban areas, 613
spatial analysis of, 1572
spatial and temporal scales, 618
statistical analysis, 614–615, 619–623
strategies, 611
as sustainability, 612
thinking, 1399
as transform, 612
unit of analysis, 616
waterfront, 1447
watershed, 1447
Resilient societies, 622
Resource conservation, 2518
Ricardian approach, 665
Rice farmers, in Ivo LGA of Ebonyi State
constraints, 1796
manifestations and effects of climate change, on rice production, 1793–1794
mitigation and adaptation measures, 1794–1795
socio-economic characteristics of, 1791–1792
Ridge-furrow (RF), 2557
Ridging, 2557, 2560
Roofing, type of, 220
Root mean square error (RMSE), 2763
Rumuruti, 1264
Rural community, 582
Rural farmers, in Niger Delta Region
age of, 1042
analytical technique of data, 1038
educational status of, 1043
empirical results, 1043
farming experience of, 1044
information on climate change, 1047
marginal effects, 1045
marital status of, 1043
MNL model, 1038, 1040
rate of adaptation, climate change, 1046
sampling and data collection technique, 1037
sex of, 1042
study area, 1037
VIF test, 1041
Rural health, infrastructure, 225
Rural livelihoods, 1141, 1143, 1145, 1148, 1150

S

SACCOS, 2734
Sadat, Anwar, 1098
Sahara zone, 1380
Sahel, 918, 932
ecosystem, 1195
Sahelian zone, 1380
Saline agriculture, 2290
Salinity, 2288, 2300
Sample farmers characteristics, 2374
Sandwip Island, Bangladesh, 335
cyclone, 341
erosion, 339
high priority intervention issues, 342–348
household income, 348
inhabitation and farm livelihood vulnerability, 342
land use change in, 349
methodological framework, 337
migration status in, 348
resilient households, strategies of, 345–347
salinity and salinity intrusion, 339–341
sampling frame of research, 338
sea level rise and flood risk, 339
Unions, 335, 338
vulnerability issues, 342, 343
Savanna Forestry Research Station (SFRS)
map of, 11
relative humidity, 10
tree species in (*see* Tree species, in SFRS)
Sea-level rise and floods, Jamaica Bay
adaptation to coastal flooding, 106–108
community resilience capacity, 108–110
COREDAR, 100
GIS, 101–102
predicted impact, 101–106
Seasonal climate forecasts (SCF)
agro-climate of Nigeria, 2495, 2496

Seasonal climate forecasts (SCF) (*cont.*)
agroecological zones, study area, 2497
characteristics, 2490–2491
climate information, 2503–2504
climatic and agricultural characteristics, 2497
communication and knowledge exchange, 2492–2493
conceptual framework, 2494–2495
correlation, 2504
data collection, 2497–2498
decisions and adaptation strategies, 2501
demographic characteristics, 2499, 2500
description, 2487
descriptive statistics, 2498
disseminate information, 2505
empirical literature, 2490
ethical considerations, 2498
farmers decision-making, 2504–2505
forecast dissemination, 2502
forecast information, 2499–2502
forecast timing, 2506
individuals decision-making, 2488, 2489
objective and questions, 2488
policy making, 2489
production, in Nigeria, 2493–2494
reliability of farmers, 2506–2507
scientist–stakeholder interactions, 2508
state-level access, 2502
usability, 2487, 2489, 2490, 2507–2508
users' perceptions, 2491–2492
Seasonal rainfall amounts, 2598
Seasonal rainfall prediction (SRP), 2493, 2745
length of season, Nigeria, 2761
mean annual total rainfall, Nigeria, 2761–2762
NiMet SRP model forecast, 2763–2766
northern Nigeria, standardised rainfall anomaly, 2757, 2758
rainfall cessation dates, in Nigeria, 2759
rainfall onset dates, in Nigeria, 2759
southern Nigeria, standardised rainfall anomaly, 2758
temperature, 2752–2756
Sectoral system of innovation (SSI) approach
changes in rainfall and temperature, 2255
innovation and adaptation processes, 2256
interdependencies in adaptation to climate change, 2256
mobilization of, 2254
structural components, 2254–2256
Seed advisories, 2376
Seed business development, 2227, 2233
impact of, 2234–2238
Seed fairs, 1128, 1129, 1131, 1133, 1135
Selian Agricultural Research Institute (SARI), 1086, 1089, 1092, 1094, 1095
Semi-arid conditions, 2700, 2702, 2705, 2716, 2719
of Central Tanzania, 2702
Semi-arid lands (SALs), 1006–1007
Semi-arid regions, of Maharashtra, India, 210
Semi-mechanized farming (SMF), 2554
Semi-structured interview method, 505
Senegal, 2047
Sensitivity
agro-ecological exposure, 1303
analysis, 1301
data on exposure, 1295
definition, 1292
Serbia
Kostolac lignite basin, 1469
mitigation and adaptation measures, climate change issues, 1468–1469
Radan Mountain, 1471–1472
spatial planning, climate change, 1465–1468
Sesbania sesban, 2296
Shardi, 59
Sheep and goats
challenges faced by producers, 1262
grazing shoats, 1261
stakeholders, 1260
translocation, 1261–1262
Shyamnagar Upazila, 2348–2353
Site-specific fertilization, 897, 898
Situation analysis, 2705
Skyros 2017, 646
Slums
of Delhi, 158
ever-increasing number of, 173
flooding in, 169
in India, 155
rapport with women, 157
Smallholder, 1054, 1055, 1059, 1069, 1080, 1295
livestock, 1345
Smallholder farmers, 701, 708, 709
adaptation to climate change, 971–974
coping and adaptation by, 965–968
food insecurity with, 963
IKS indicators used by, 969
vulnerability of, 964–965
Small-scale farming, 1737

Small scale irrigation systems, in Mali
  economic efficiency, 1056–1057
  economic viability, 1057
  technical efficiency, 1057–1058
Smart crop-livestock diversification actions, 1198
  attibutes, 1199
  changes in farmers' living conditions, 1208–1209
  choice experiment, 1198
  mixed logit estimation with income, 1204
  willingness-to-pay estimation, 1208
Snowball sampling, 40
Social cognitive theory, 249
Social housing insurance, 2160
Social studies, 1937
Social variables, 408, 425
Socio-cultural importance of indigenous fruit trees
  ethno-botanical survey, 504
  exploratory survey, 504
Socio economic characteristics, rural farmers, 1041–1044
Socio-economic resilience, 624
Soil aggregation, 390
Soil and water conservation (SWC), 799, 803, 806, 807, 815, 817
  interventions, 2460, 2462
Soil erosion, 1099, 1102
Soil fertility improvement interventions, 2464, 2465
Soil moisture, 2418
  determination, 2422
  gravimetric, 2424
Soil organic matter (SOM)
  chemistry and concentrations, 393
  formation, 391
  stability, 393
Soil profile sampling, 297
Soil quality evaluation, 309
Soil quality index (SQI), 298, 312
Soli sustainability, 305–309
Sorghum, 918
  biomass and water use efficiency, 951
  and climate change, 955
  grown in dry and dry-wet season, 939
  growth performance, 942
  height, 945
  leaf area index (LAI), 950
  production, 937
  root zone, 940
  seedling emergence, 945
  stem thickness, 949
  varieties, 919
  water use efficiency, 940
Sorghum crop yield, eastern dryland of Sudan, *see* Dryland farming (sorghum), eastern Sudan
Southern Agricultural Growth Corridor of Tanzania (SAGCOT), 1649
Southwest Nigeria, 480
  deforestation drivers in, 481
  land-use change, 481–495
Special Initiative for Rebuilding and Resiliency (SIRR), 104
Species diversity, 1380
Spent engine oil
  bioremediate heavy metals, 325
  deleterious effects of, 320
  discharged into environment, 320
  heavy metals, 320
  mineral elements in, 322
  pollution with, 322
  seedlings in polluted soil., 320
  soil contamination by, 320
  solute contamination, 322
Sprinkling irrigation systems, 1059
Stakeholder(s), 1147, 1890, 1896, 1900, 1902, 2149, 2155–2159, 2283, 2284
Standard Precipitation Evapotranspiration Index (SPEI) version 1.6, 362
State of calamity, 2128
State responsibility, 1404
Strategic environmental assessment (SEA), 41
Strategic gender needs (SGNs), 2101, 2102, 2116–2118
  assessment, 2102
  education, skill development and training, 2117
  education, 2108, 2111, 2113
  income, 2118
  income generation, 2109, 2111, 2114
  schooling of children, 2118
Strategic Program for Climate Resilience (SPCR), 1369
Strengths and weaknesses, 40–43
Sub-Saharan Africa (SSA), 1345, 1648, 1844, 1846, 1848, 1850, 1854
Sub-Saharan African (SSA) countries, 1055
Sudano-Sahelian region of Nigeria
  influence of onset, cessation and dry spells on rainfall characteristics, 2615–2618
  meteorological stations, 2596
  occupation, 2597
  onset, cessation and dry spell occurence, 2602

Sudano-Sahelian region of Nigeria (*cont.*)
population and funds, 2597
rainfall and length of rainy season in, 2597–2602
Sudano-Savannah zone, 1381
Sudan's First National Communication, 2552
Sundarbans Forest, 56
Super-farmer, 874
Supply chain adaptation, 2202, 2203
Susceptibility factors, 215
Sustainability, 265, 1145, 1149–1150, 1968, 1976, 2014
adaptation for, 2787
funding for, 2784
index, 678
legal frameworks, 824
in Nigeria, 2774
vulnerability and resource, 838
Sustainable agriculture, 1804
Sustainable development, 81, 677, 1143, 1146, 1398
climate change adaptation and, 139
contribution to, 136
water resources planning, 140
Sustainable Development Goals (SDGs), 361, 1845, 1847, 1850, 1853, 1878
Sustainable land use, 2732
Sustainable livelihoods approach (SLA), 1504, 1515, 1519, 1522, 1524
Sustainable rangeland management (SRM), 75, 81, 83, 88
Sustainable urbanization, 1485
Swiss Agency for Development and Cooperation (SDC), 2369
Syrian refugee crisis, 645
Systematic review study approach, 1198
System of rice intensification principle (SRI), 2732
System's thinking approach, 1606

T
Tackle deforestation, 1392
Talanoa dialogue process, 1408
Tanzania
income status, 1638
location, 1628
per capita per year, 1637
population pressure, 1637
scenarios for, 1639
Tanzania Coffee Research Institute (TACRI), 1086, 1088, 1091, 1092
Task Force on Climate-related Financial Disclosures (TCFD), 1897
Tax credits, 1878
Teff, in Ethiopia
production potential area, 1186
suitability area, 1183
Territoriality, 1561
Tharaka-Nithi
food security (*see* Food security status)
household survey, 1761
location, 1759
Solvin's sampling formula, 1760–1761
Thar desert region of India
air temperatures, 687
changes and shift in rainfall, 693–694
changes in air temperature, 693–696
climate change scenarios, 692
crop water requirements, 696, 697
drought scenarios, 690–691
evapotranspiration rates, 689
rainfall distribution, 686
relative humidity, 688
secular trends in annual rainfall, 692
solar radiation, 687
wind regime, 688
Theory of planned behavior (TPB), 249
Threshold, 589, 596, 598–601
Tigray (northern Ethiopia), exclosures in, *see* Exclosures, in Tigray (northern Ethiopia)
Tigray regional state, in Ethiopia
administrative units, 2442
description, 2441
EBA interventions (*see* Ecosystem-based adaptation (EBA))
predatory journals, 2443
soil types, 2442
study area, 2441–2442
Timber extraction, 478
Timer trade, 58
Tokenism, 2311
Tomato Ebola, 2277
Tonga
affected communities in, 232
migration due to climate change, 231
Total carbon conversion factor, 13
Tourism, climate change impacts and adaptation in, 541
Trade, 504
Tradeoffs, 1140, 1143
Trade Related-Aspects of Intellectual Property Rights (TRIPS Agreement), 1361
Traditional farming (TF), 2554
Traditional knowledge, 501, 505

Tree species, in SFRS
carbon contents variation, in wood and bark of tree species, 13–14, 16–18
data analysis, 13
interspecific carbon content variations, in wood and bark of tree species, 15
interspecific carbon content variations, sampled tree species, 18–20
leaf carbon content, diurnal temperature, 21–24
sample collection for biophysical measurement, 12
sampling procedure, 10–11
total carbon conversion factor, determination of, 13
tree height, 12
trees diameter and carbon content, 16
Trench construction, 2632
Trend analysis of temperature and rainfall, 2258–2259
Tropical acid-sand soils
climate data, 299
climate variability, 305–309
correlation analysis, 298
evaluation technique, 298
laboratory analysis, 297–298
physico-chemical properties, 300–305
profile sampling, 297
quality evaluation, 309–313
site description and sampling, 297
sustainability, 305
Tropical Africa
climate change-mediated floods impact on fish farming and food security, 2578
need for indigenous technological knowledge, 2579
vulnerability of agriculture to climate change in, 2576
Tropical Pesticides Research Institute (TPRI), 1086, 1088, 1090, 1092, 1094, 1095
Two-way communication, 2311
Tyre pyrolysis, 2520

U

Uganda
climate change, 180
cultivatable land in, 179
electricity generation in, 178
electricity generation mix in, 196
hydropower in, 183
resource systems in, 181
sectors in, 180
study area, catchments, 184
substantial role in, 179
water resources management in, 182
Underweight, 2049
UNICAL T & R farm, 300
United Nations Environment Programme (UNEP), 1911, 2039
United Nations Framework Convention on Climate Change (UNFCCC), 31, 35, 1216, 1683, 1886, 1966
UN Refugee Convention, 2025
Urban areas, 1958, 1960, 1962, 1964, 1965, 1968, 1969, 1971, 1975, 1977
Urban climate change adaptation, 1972–1974
Urban climate change policies, Qatar
awareness, education and research, 1491–1492
certification and building codes, 1487–1489
infrastructure and urban supply, 1489–1491
sustainable urban development planning, 1485–1487
Urban climate responses, 1493
Urban dimensions, of climate change adaptation in West Africa, 1962–1965
Urban ecosystems and communities, SLR, *see* Sea-level rise and floods, Jamaica Bay
Urbanization, 153
levels of, 171
trends of, 155
Urban resilience, 1975
Urban retrofitting, climate proofing
benefits, 2176
cost, 2175–2176
extreme rainfall events, 2174–2175
green benefits, 2189
high-rise post-war garden city (flat), 2179–2184
neighborhood typologies, 2176–2179
in sloping areas, 2189–2192
urban city block (flat), 2184–2189
Urban typologies, 2176–2179
Urgent Priority Interventions Issues (UPI), 352
Usability gap, 2487
US Fourth National Climate Assessment, 2024
Utility function, 1201

V

Validation
of DSSAT model, 905
procedures, models calibration and, 900
Variance inflation factors (VIF), 1653
test, 1041

Varimax rotated factor matrix, 1796
Vegetables production, Koulikoro and Mopti regions of Mali
crop irrigation, households, 1069–1072
data analysis, 1064–1065
extension demonstration site, farmers' visit of, 1077
farmers' extension providers, 1075
farmers' sources of agricultural information, 1077
frequency of extension providers, respondents, 1075
household heads' age, 1065–1066
household heads' level of education, 1066–1067
household heads' sex, 1065
household size, 1067
irrigation training programmes, farmers participation in, 1076
land size, 1068–1069
non-farming activities, income earning, 1067–1068
occupation of farmers, 1067
production theory, 1061–1062
research design, 1062–1064
residential house, households, 1068
road conditions, in region, 1077–1080
study area, 1062
training topics, farmers, 1076
Vietnam
adaptation measures by Government, 701
climate change in, 700
culture in, 700
ecosystem-based adaptation, 706
existing ecosystems in Sơn Thọ commune, 703–704
participatory study, 702
participatory vulnerability assessment, 703
Sơn Thọ commune, 702
Village conservation committees (VCCs), 560, 561, 563, 565, 567, 568
Village Water Management Team (VWMT), 256–258
Vulnerability, 538, 539, 541, 542, 1027, 1029, 1218, 1222, 1928, 2776
adaptive capacity, 778, 970
of agriculture sector, sensitivity and, 767
assessment, 701, 703–706, 1348
assessment approach, 1298
assessment codes, 1515
assessment status, 1519
attributes of concern, 542
to climate change, 967, 2627
codes of areas, 1515
concept of, 1292
definition, 542, 1299, 1502
degree of, 1295
dimensions, 1308
dynamics of, 1347
evaluation, 1514–1517
extent and degree of, 777
to food price changes, 774
greenhouse gas, 962
hazard, 542
to heat stress, 209, 215
index, 1299
index map, 1309
indicators, climate, 1297
Konso community, small farming households of (*see* Konso community, southern Ethiopia)
level of, 154
of livelihoods, 1344
magnitude of, 1310
pastoralists' and farmers, 1294
political factors, 1293
and poverty, 1346
quantification of, 170
of slum dwellers, 165
smallholder farmers, 964
social, 542
socioeconomic, 1295
sub-indices map, 1300, 1302, 1304
as a system, 542
temporal reference, 542
theoretical framework, for vulnerability assessment, 543–544
vulnerability context and impacts on livelihoods, 549
Vulnerable and Nigeria, 461, 464
Vulnerable farmers, in Central Tanzania, 2701
Vulnerable islanders resilience
adaptive capacity, 333–334
high priority intervention issues, 334–335
Sandwip Island, Bangladesh (*see* Sandwip Island, Bangladesh)
vulnerability assessment, 333

W

Walkley-Black method, 19
Walter's method, 2605
Ward Climate Change Planning Committees (WCCPCs), 1689
Waste, 2518, 2521, 2524, 2525
Water, 2098, 2100, 2103–2116, 2118, 2119

availability, 264–265, 361
agriculture and, 1335
conservation, 2093
governance, 252–253, 260, 264–265
law, 1318
legislation, 1339–1340
management, 1055, 1059, 1071, 1074
Paris Agreement, 1336
and poverty, 151
saving plan, 260
scarcity, 937
sector and rainfed agriculture, 246–247
security, 247, 254, 265
vector-borne diseases, 579
women, 151–152
Water budget, 251, 257
preparation and implementation of, 258–259
preparation of village, 259
Water-Energy-Food Nexus, 2077, 2084, 2093
Water Evaluation and Planning (WEAP), 1684, 1685, 1693, 1700
Water harvesting, 2407, 2409
farmers' rating on, 2408
interventions, 2460, 2461
plan, 259
on sorghum, 2410–2411
techniques, 2405
Water harvesting technology, in Katsina State
crop yields, 2793
justification of research, 2794
methods, 2795–2796
rainfed farming, 2793
soil erosion and land degradation, 2792
Zaï pit technology, 2792, 2799
Water reclamation with resource recovery, 2077, 2080, 2081, 2084, 2092, 2094
Water Resource Management Authority (WRMA), 1687, 1693, 1695
Water resources, 1140, 1142, 1152
depletion, 136–137
Water resources management
described, 135
levels, 143
as means to adaptation, 138–139
monitoring and tracking, 145–146
Watershed development (WSD), 253
Watershed Organization Trust (WOTR), 248, 251, 2367
Watershed restoration, 2632
Watershed Water Resource Committee (WWRC), 261
Water Stewardship Initiative (WSI), 248–265
Water use efficiency (WUE), 1056, 2557, 2564, 2567
Weather, 789, 794
Weather Indexed Insurance, 1121, 1124, 1130
Weather variability, 2657, 2663, 2673
and climate change, 847
Weeding, 2379
Weighted average index (WAI) analysis, 2661
Weighted matrix index (WMI), 334, 338
West Africa
climate change adaptation in, 1958
climate change effects in, 1386–1387
emergence of climate change in, 1384–1385
environmental stress & endemic poverty in, 1389–1390
policy at city-scale level in, 1965
socio-economic profile of, 1381–1384
urbanization, 1960–1961
West African Sahel, 2036
annual gross domestic product, 2037
climate change and variability, 2039–2043
food security, 2044–2048
fragility of ecosystem, 2038–2039
global alliance for resilience, 2048–2053
recommendations, climate change mitigation and adaptation, 2053–2054
socio-economic characteristics, 2037–2038
West African Science Service Centre for Climate Change and Adapted Land Use (WASCAL), 360
West Africa's territorial ecosystems, 1382
Wheat, in Ethiopia
production potential areas, 1184
suitability area, 1182
Willingness-to-pay
definition, 1202
estimation, 1208
Women, 2098, 2103, 2112, 2115, 2116, 2118, 2119
adaptive capacity among, 778
in ancient Egypt, 770, 771
income generation, 2109, 2111
MSMEs in Kenya (*see* Female-led micro, small and medium enterprise (MSMEs), Kenya)
perception of climate change, 777–778
in rural Egypt, 776–777
World Food Programme (WFP), 1103
World Meteorological Organization, 790

Y

Yabello *Woreda*, 2062
Yemen, 1367

Z

Zaï pit technology, 2792, 2794, 2797, 2799, 2800
Zanzibar, REDD+ pilot project on livelihoods, *see* REDD+ pilot project, in Tanzania
Zimbabwe
  agriculture and climate change in, 963–964
  CSA in, 1613
  CSA policies in, 1616–1617
  government of, 1621
  policy in, 1622
  rural, 1617
  rural economic transformation in, 1613
  special funds, 1624
Zimbabwe Super Seeds (ZSS), 2240–2241

Printed by Printforce, the Netherlands